APPLIED ELECTRONIC DEVICES and ANALOG ICs

APPLIED ELECTRONIC DEVICES and ANALOG ICs

J. Michael McMenamin

Delmar Publishers™

1945 - 1995 50 years

I(T)P™ An International Thomson Publishing Company™

Albany • Bonn • Boston • Cincinnati • Detroit • London • Madrid
Melbourne • Mexico City • New York • Paris • San Francisco
Singapore • Tokyo • Toronto • Washington

NOTICE TO THE READER

Publisher does not warrant or guarantee any of the products described herein or perform any independent analysis in connection with any of the product information contained herein. Publisher does not assume, and expressly disclaims, any obligation to obtain and include information other than that provided to it by the manufacturer.

The reader is expressly warned to consider and adopt all safety precautions that might be indicated by the activities herein and to avoid all potential hazards. By following the instructions contained herein, the reader willingly assumes all risks in connections with such instructions.

The publisher makes no representation or warranties of any kind, including but not limited to, the warranties of fitness for particular purpose or merchantability, nor are any such representations implied with respect to the material set forth herein, and the publisher takes no responsibility with respect to such material. The publisher shall not be liable for any special, consequential, or exemplary damages resulting, in whole or part, from the readers' use of, or reliance upon, this material.

Cover photo courtesy of Analog Devices

Delmar Staff
Publisher: Michael McDermott
Administrative Editor: Wendy Welch
Developmental Editor: Mary Clyne
Project Editor: Barbara Riedell
Production Coordinator: Andrew Crouth
Art/Design Coordinator: Lisa Bower
Assistant Editor: Jenna Daniels

Printed in the United States of America

For more information, contact:

Delmar Publishers
3 Columbia Circle , Box 15015
Albany, New York 12212-5015

International Thomson Publishing Europe
Berkshire House 168 - 173
High Holborn
London WC1V 7AA
England

Cover Design: Anne Pompeo

Thomas Nelson Australia
102 Dodds Street
South Melbourne, 3205
Victoria, Australia

Nelson Canada
1120 Birchmount Road
Scarborough, Ontario
Canada M1K 5G4

International Thomson Editores
Campos Eliseos 385, Piso 7
Col Polanco
11560 Mexico D F Mexico

International Thomson Publishing GmbH
Königswinterer Strasse 418
53227 Bonn
Germany

International Thomson Publishing Asia
221 Henderson Road
#05 - 10 Henderson Building
Singapore 0315

International Thomson Publishing - Japan
Hirakawacho Kyowa Building, 3F
2-2-1 Hirakawacho
Chiyoda-ku, Tokyo 102
Japan

1 2 3 4 5 6 7 8 9 10 XXX 01 00 99 98 97 96 95

Library of Congress Cataloging-in-Publication Data

McMenamin, J. Michael
Applied electronic devices and analog integrated circuit/ J. Michael McMenamin
p. cm.
Includes index.
ISBN: 0-8273-5416-9 (text)
1. Electronics 2. Electronic circuits 3. Linear integrated circuits I. Title
TK7816.M38 1995 94-59
621.381 —dc20 CIP

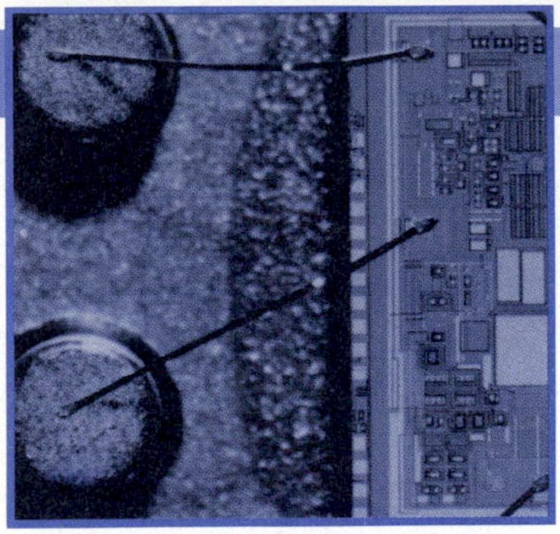

Contents

CHAPTER 3
TRANSISTOR AC AMPLIFIERS

CHAPTER 4
FIELD-EFFECT TRANSISTORS

CHAPTER 5
TRANSISTOR SWITCHING CIRCUITS

CHAPTER 6
INTRODUCTION TO OPERATIONAL AMPLIFIERS

CHAPTER 7
OP-AMP CIRCUITS

CHAPTER 15
TRANSDUCERS

CHAPTER 16
SIGNAL CONVERSION

CHAPTER 17 COMMUNICATION CIRCUITS

CHAPTER 18 GROUND LOOPS AND NOISE

CHAPTER 19
THERMAL CONSIDERATIONS

CHAPTER 20
CIRCUIT COMPONENTS

APPENDIX A
ELEMENTARY CALCULUS

INDEX

Preface

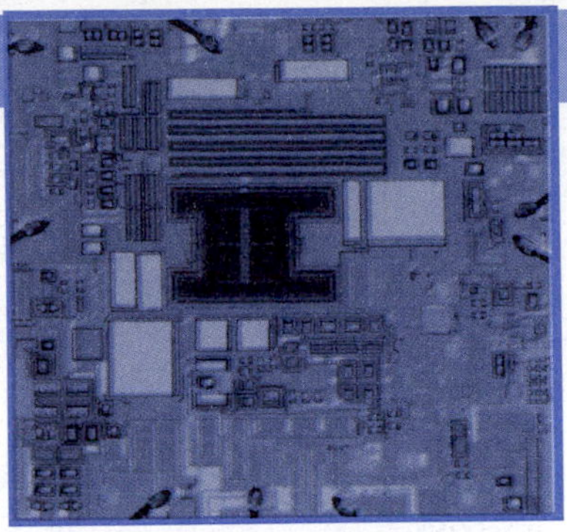

PEDAGOGIC APPROACH

This text is structured to present an integrated approach. By this I mean that a circuit is introduced; developed to a practical configuration (all components specified); analyzed for faults; modified in exercise problems; evaluated with a prelab using a SPICE program; and finally, built, tested, and reported in the laboratory. The objectives of this integrated approach are to

- Provide **reinforcement**
- Show **complete circuit designs**
- Enhance the students' **troubleshooting skills** using a familiar circuit
- Ensure that students perform **lab experiments on known circuits**

I believe the most important learning situation for the prospective technician is in the laboratory; the text should be directed at optimizing the lab experience. For this reason I have generated a laboratory experiments manual to complement the text. The labs are directed towards formalized reports at a more professional level. The emphasis is on forming conclusions that encourage the student to think and report on their knowledge of the experiment.

Supplemental programs for many of the experiments in the lab manual have been created in computer-aided circuit analysis programs. Instructors can choose among prelabs in PSpice, B^2Spice, and Electronics Workbench. The use of these versatile programs is strongly advised for dry running the laboratory experiments prior to the actual lab experience (prelab). In a classroom situation, the programs are very effective via a computer-driven overhead projector in demonstrating circuit operation and troubleshooting.

The level of this text is based on the assumption that the reader has completed DC and AC fundamentals along with college algebra. Calculus is used to derive some of the equations; however, it is not always necessary to understand the derivation in order to use the equation. Also, I feel there is a need to stimulate the reader's math interest by showing in the derivations the advantages of a higher level of math.

TOPICS

In selecting the topics, I have emphasized those the student needs to know rather than those that are nice to know. This is based on my own twenty-four years of practical work experience as a technician and electrical engineer, and also with much input from equipment and semiconductor manufacturers. The text presents topics that align with the needs of industry, and therefore places a much greater emphasis on integrated circuits. This emphasis has eliminated or reduced some traditional topics. The criteria I used for selecting the integrated circuit devices covered in this text were their popularity in current circuit designs, and with new devices, their predicted potential for use in future designs.

In the text the presentation of topics progresses from diodes and transistors to the most current linear integrated circuit devices. Starting with Chapter 12, the material is more specialized, and chapters are organized to satisfy the introductory needs of a particular electronics program. For example, as shown in the accompanying chart, a Communications Degree option could require Chapters 1 through 8, 10, 11 and then 16 through 18 and 20. An Industrial/Robotics program could require Chapters 1 through 13 and then 15, 16, 18 through 20.

Code	*Specialty*	*Suggested Chapters*																			
1	Communications	1	2	3	4	5	6	7	8	–	10	11	–	–	–	–	16	17	18	–	20
2	Indust/Robotics	1	2	3	4	5	6	7	8	9	10	11	12	13	–	15	16	–	18	19	20
3	Consumer Electronics	1	2	3	4	5	6	7	–	9	10	11	12	–	14	–	16	17	18	19	20
4	Automotive	1	2	3	4	5	6	–	–	9	10	11	–	–	–	15	16	–	18	19	20
5	Computer Repair	1	2	–	4	5	6	–	–	–	10	11	12	–	–	–	16	–	18	19	20
6	Elect Eng Technology	1	2	3	4	5	6	7	8	9	10	11	12	13	–	15	16	17	18	19	20

With a selection of program-appropriate chapters, it is anticipated that the material would be spread over two semesters or three quarters (three semesters for Electronics Engineering Technology).

It is also possible for instructors to integrate some of the specialty applications material of later chapters into the earlier chapters of this text. To help these instructors, the following system of icons has been developed to show where the applications material can be integrated:

Communications

Industrial/Robotics

Consumer Electronics

 Automotive

 Computer Repair

 Electrical Engineering Technology

For example, where you see the automotive icon, you can integrate material from chapters 15, 16, 18, 19, and 20.

Another use for the icons is to indicate which sections within a chapter should be covered for a given speciality.

FEATURES

- A new era text with less emphasis on discrete devices and more on the latest integrated circuits
- Integrated approach to presenting a circuit
- Circuit component values specified
- Stated objectives at the beginning of each chapter
- Summary of key points at the end of each chapter
- Definition of abbreviations given at the end of each chapter
- Answers to all problems at the end of each chapter
- Separate lab experiment manual closely correlated with the text
- Built-in computer-aided circuit analysis programs for prelabs and classroom instruction (in lab manual)
- Many lab circuits predrawn in Spice (in lab manual)

SUPPLEMENTS PACKAGE

In addition to the companion lab manual, a comprehensive supplements package has been developed to support this text. A complete solutions manual offers solutions to every text problem, as well as guidance for running the prelabs and lab projects smoothly. A computerized test bank contains approximately thirty questions per text chapter. A package of transparencies and masters is also available. Finally, Delmar offers an illustrated guide to using PSPICE.

ACKNOWLEDGMENTS

Most of this text was developed while I was teaching at Schoolcraft College in Michigan. This provided for student and faculty input into the text and lab material. For this I would like to thank the students and faculty at Schoolcraft College. Among the faculty members, I would like to extend special thanks to Professor Ronald McBride. I would also like to thank

Technical Dean Fernon Feenstra, Donald McVittie, and William Schlick for lab assistance, and Chuck Cossin, Darryl Nowacki, and Frank Wiltrakis of the Information Services department along with Sharon Szabo for their help in document preparation.

The constructive criticism of the following reviewers is greatly appreciated:

Thomas Bingham—St. Louis Community College

Paul Caravan

James Davis—Muskingum Area Technical College

Paul Goulart—RETS. Alabama

John Jellema—Eastern Michigan University

Clay Laster—San Antonio College

Gregory Maninakas—State University of New York

Lloyd Martin

Donald Montgomery—ITT Technical Institute

In addition, Delmar Publishers and the author gratefully acknowledge the contributions of the following reviewers:

Phillip Anderson

Richard G. Anthony

Dr. Harry M. Assenheim

Bob Bixler

Arnie Garcia

Huley G. Gill

Theodore Johnson

Dan Lookadoo

Robert E. Martin

Harbans Mathur

William Maxwell

Randall Reid

Lee Rosenthal

Hal Sappington

John Siena

Edward Wilson

A text of this magnitude requires considerable input from industry sources, and so I would like to thank the following application engineers:

Bob Marwin—Analog Devices

Tom Chrapkiewicz—Ford Motor Company, Philips

John Williams—Harris Semiconductor.

Henry Root—Hy-James Audio Engineers

Frank Schmid—Ingersoll Rand

Bill Vanting—Motorola Semiconductor

Fritz Wilson—Motorola Semiconductor

John Christenson—National Semiconductor

Gary Lachnette—National Semiconductor

Dennis Smith—National Semiconductor

Dave Johnston—Siemens

Christine Brennan—Giesting Associates

Finally I would like to give recognition to the following organizations for kindly giving me permission to include their technical information:

Analog Devices Corporation

Exar Corporation

Harris Semiconductor

Ingersoll Rand

"Copyright 1993 Hewlett-Packard Company. Reproduced with permission".

Maxim Integrated Products

"Copyright of Motorola, Inc. Used by Permission".

National Semiconductor Corporation

Philips Semiconductors

Sensym Inc

Turk Inc

J. Michael McMenamin

CHAPTER 1

Diodes

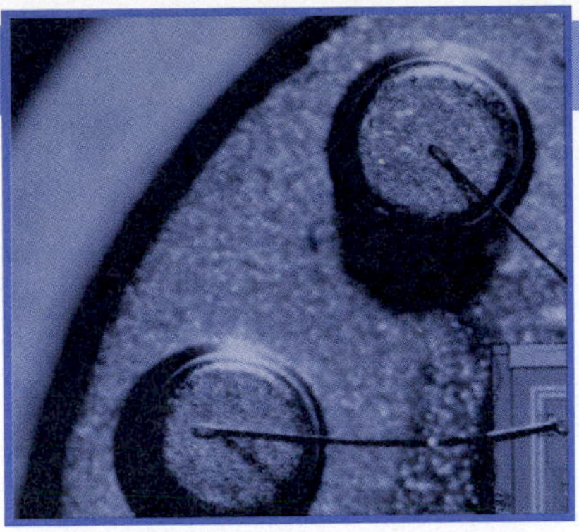

CHAPTER OBJECTIVES

After completing this chapter, you should be able to explain the following diode concepts:

- ❍ What is a semiconductor
- ❍ N-type and P-type semiconductors
- ❍ How a diode is constructed from semiconductors
- ❍ Why current only flows one way through a diode
- ❍ Forward voltage that has to be overcome before a diode conducts
- ❍ Reverse biasing a diode
- ❍ Diode leakage current
- ❍ Diode characteristic V–I curve
- ❍ Diode specifications
- ❍ Uses for a diode
- ❍ The zener diode
- ❍ Applications for a zener diode
- ❍ Photo diodes
- ❍ Applications for a photodiode
- ❍ Varicap diode
- ❍ Schottky diode
- ❍ Applications for a varicap diode
- ❍ Light-emitting diode
- ❍ Troubleshooting diode circuits

INTRODUCTION

In this chapter we will look at a device called a **diode**, which allows current to pass in only one direction. This characteristic is useful in controlling the current direction in a circuit. Diodes can also be specially manufactured to provide other functions such as providing a constant voltage (zener diode), detecting light (photodiode), generating light (light-emitting diode) or tuning a circuit (varactor diode).

SEMICONDUCTORS

Electricity is caused by the movement of free electrons. An electron is a tiny particle that has a negative charge and typically is found in a three-dimensional orbit around a central nucleus (Figure 1–1)—much like the planets rotate around the sun.

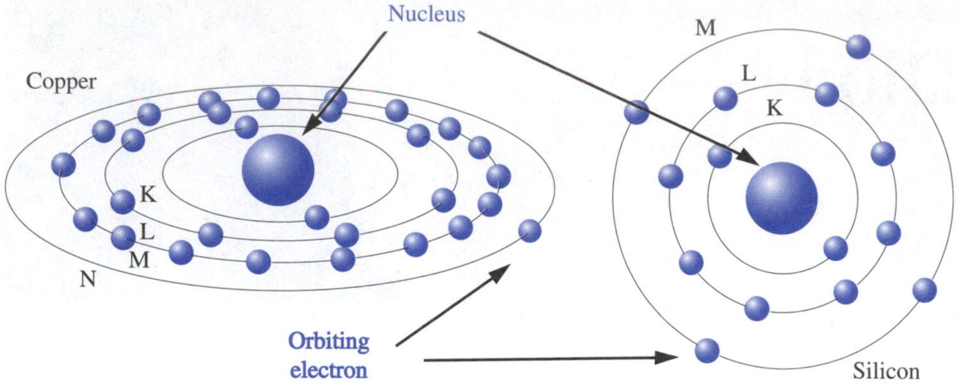

FIGURE 1–1 Electron orbits for the copper and silicon atoms

Notice that there are discrete orbits for the electrons and that there is a maximum number of electrons allowed in each orbit. The orbits are each identified by a letter, with *K* being the closest orbit to the nucleus, *L* the next closest, and so on. The maximum allowable number of electrons in the *K* orbit is two, in *L* is eight, in *M* is eighteen and in *N* is thirty-two. But the orbit furthest from the nucleus can have a maximum of eight electrons. For example, if the *N* orbit is the outermost, it will have a maximum of eight electrons rather than thirty-two. The inner orbits fill up with electrons; the outermost orbit, which is called the **valence band**, can have from one electron up to its maximum capacity of eight.

In our study of electron devices, we shall be concerned with the valence band electrons. These electrons, being furthest from the nucleus, can obtain sufficient energy from an external source (heat, light, etc.) to break free of their orbit and become free electrons in what is called the **conduction band**. The energy level required to generate a free electron increases with the number of electrons present in the valence band.

Referring back to Figure 1–1, notice that the copper atom has just one electron in its valence band, while the silicon atom has four. As expected, free electrons are easy to obtain in copper and more difficult to generate in silicon. In order for electricity to flow, free electrons (thermally freed—the hotter, the more free electrons) must exist in the material used to conduct the electricity. Materials that have many free electrons are called conductors—copper is a good example of an electrical conductor.

Materials like silicon (and germanium), with four valence electrons, form what is called a **covalent bond** between their atoms—which means that the electrons in the valence shell are shared between atoms, Figure 1–2.

Figure 1–2a illustrates covalent bonding—each atom shares its four electrons with adjacent atoms but at the same time effectively has eight electrons in its valence band. Covalent bonding prevents free electrons and causes silicon and germanium to be electrical insulators (nonconductors). However, both of these materials can be modified to become better conductors (semiconductors) of electricity by **doping** with another material (inserting a different atom) that releases free electrons or by doping with a material that creates an absence of electrons and therefore an attraction for electrons.

Since silicon is the most used semiconductor, let us look at the results of doping this material. If we dope silicon with arsenic atoms (which have five electrons in the valence band), the arsenic atoms form a covalent bond with the silicon atoms (Figure 1–2b), and we get a free electron for each atom of arsenic added. (Remember that eight are required for the bond.) So the arsenic creates (donates) an excess of electrons and is therefore called a donor material. Silicon doped this way is called an **N (negative)-type semiconductor**. The *N* refers to the free electrons, which have a negative charge.

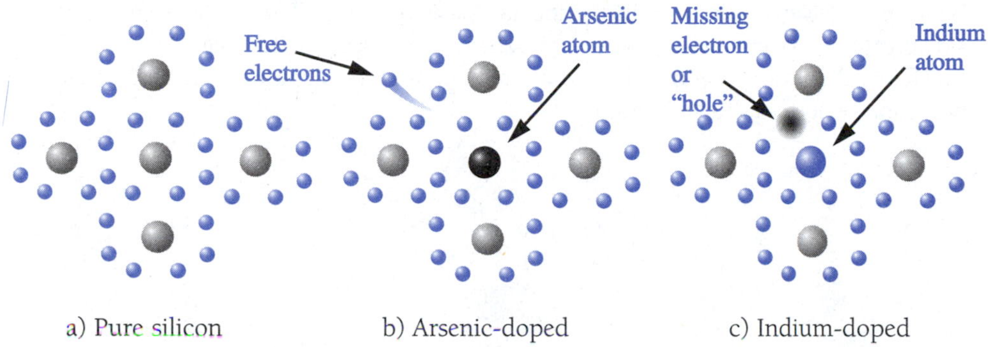

FIGURE 1–2 Covalent bonding

Adding indium atoms (which have three electrons in the valence band), to silicon (Figure 1–2c) creates a shortage of electrons (seven electrons in the shared valence band)—which results in an attraction for electrons (acceptor) to complete the covalent bond. Since electrons are attracted to this type of doped material, we call this a **P (positive)-type semiconductor**. It should be stated that doping does not change the total electrical charge on the material. It is still electrically neutral because the original undoped silicon and also the doping atoms are electrically neutral.

THE DIODE

A diode is an electronic device that allows current to flow in only one direction, as compared with a resistor, which allows current to flow either way. A resistor can be wired in a circuit without considering which end is connected to which point in the circuit; whereas a diode must be put in a certain way. In a circuit, a diode is marked to identify the anode (positive side) at one end and a cathode (negative side) on the other, Figure 1–3. Conventional current (I_{conv}) will flow from the anode to the cathode inside the diode—note that it will flow in the direction of the schematic symbol arrow; whereas electron current (I_{elect}) flows from cathode to anode.

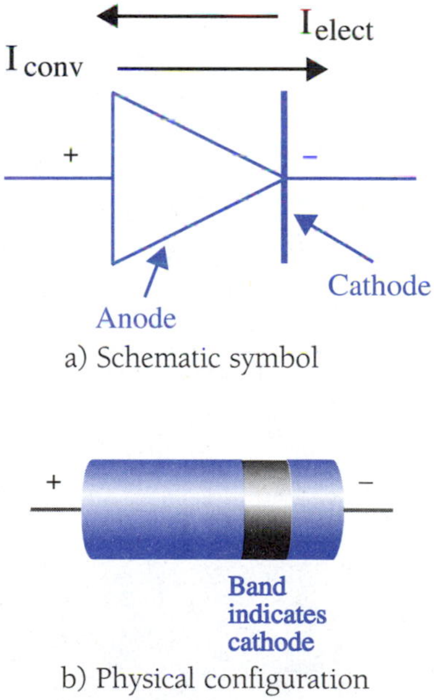

FIGURE 1–3 Current flow through a diode

DIODE CONSTRUCTION

We can make a diode by joining together a piece of N-type and a piece of P-type material. One method currently employed is to take a single piece of silicon material and dope one-half with an N-type donor material and the other half with a P-type acceptor material. Since the N-type side now has an excess of electrons and the P-type side a deficiency (hole), free electrons will be attracted across the junction from the N-type to the P-type material. As this happens, the N-side material becomes electrically positive because it has lost electrons, while the P-side material becomes electrically negative because it has gained electrons. The flow of electrons from the N side to the P side will continue until the P side is negative enough to prevent (repel) further electron flow. This point is reached when the P side is about 0.6V more negative than the N side (0.2V for germanium diodes).

The effect described occurs right at the boundary between the two materials; and the buildup of negative charge on the P side and positive charge (loss of electrons) on the N side creates a region between the two materials where there are no free electrons. This is called the **depletion region**. The diode junction appears like a 0.6V battery across the depletion region, as shown in Figure 1–4.

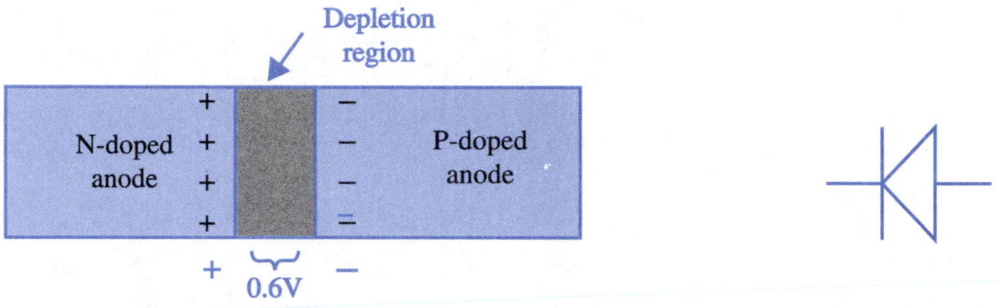

FIGURE 1–4 Diode junction

DIODE FORWARD BIASING

We can cause electrons to flow across the depletion region by connecting an external battery that will overcome the 0.6V junction voltage that acts as a barrier to current flow. The circuit is shown in Figure 1–5. When the external battery voltage V_{bb} is great enough to overcome the internal junction voltage (greater than 0.6V), electrons will flow across the depletion region to the P side. This is called **forward biasing** the diode; the diode appears like a closed switch when forward biased.

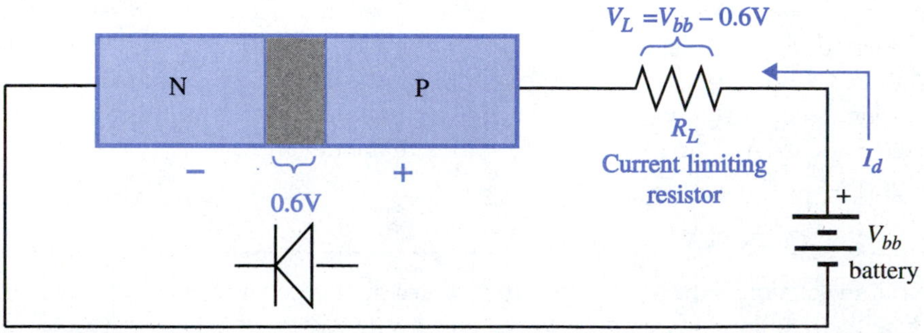

FIGURE 1–5 Biasing the P–N junction

Since power dissipation is the product of diode voltage and diode current, we must use a current limiting resistor (R_L); otherwise, the current will increase too much during forward biasing and damage the diode junction.

Once we overcome the 0.6V junction voltage, the remainder of the battery voltage is used to drive current through R_L. Notice that with forward biasing, electrons flow from the N-type to the P-type—this is called **majority carrier current flow**.

EXAMPLE

The battery voltage V_{bb} used in the diode circuit of Figure 1–5 is 10V. Find the value of the current limiting resistor R_L to limit the current through the diode to 2mA if the diode forward drop V_d is 0.6V.

SOLUTION

Writing loop equation in Kirchhoff law form:

$$V_{bb} - IR_L - V_d = 0$$

$$IR_L = V_{bb} - V_d$$

Solving for R_L:

$$R_L = \frac{V_{bb} - V_d}{I}$$

$$R_L = \frac{10V - 0.6V}{2mA}$$

$$R_L = 4.7K\Omega$$

EXAMPLE

Find the current flowing through the diode in Figure 1–5 if the battery voltage V_{bb} is 12V, the diode drop V_d is 0.6V, and $R_L = 3K\Omega$

SOLUTION

Writing Kirchhoff loop equation:

$$V_{bb} - IR_L - V_d = 0$$

$$IR_L = V_{bb} - V_d$$

Solving for I:

$$I = \frac{(V_{bb} - V_d)}{R_L}$$

$$I = \frac{(12V - 0.6V)}{3K\Omega}$$

$$I = 3.8mA$$

DIODE REVERSE BIASING

If we reverse the connections on the battery, the reverse bias voltage at the junction will increase—which causes the depletion region to widen. However, at a given temperature, some electrons attached to atoms in the depletion region can receive enough thermal energy to break their bonds and become free. These free electrons are repelled by the negative charge built up on the P side but are attracted to the electron-deficient N side of the depletion region.

When the electrons leave their atoms in the depletion region, they leave a **hole** in the atom, which causes the atom to become positively charged. This positive charge can attract one of the excess electrons on the P side. The net result is a small reverse flow of electrons from the P side to the N side (forward biasing causes electrons to flow from the N side to the P side), and this is called **minority current flow**. This current will increase as temperature is increased because greater thermal energy allows more electrons to escape from their atoms in the depletion region. At room temperature, typical reverse currents are $\approx$ 10nA (nA = 10^{-9}A) for commonly used diodes.

If we neglect the small reverse current, the diode appears like an open switch when reverse biased.

DIODE DEFINITION

The diode is a device, then, that easily allows electrons (called majority carrier current) to flow from the N side to the P side when the junction is forward biased. Or stating it another way, it will allow conventional current (I_f) to flow from the P side to the N side (in the direction of the diode symbol arrow) when forward biased. But when the junction is reverse biased, there is a small flow of electrons (called minority carrier current) from the P side to the N side—or conventional current (I_r) from N to P.

Figure 1–6 clarifies these statements for the **conventional current direction**.

Notice in Figure 1–6a that when the diode is forward biased, both forward and reverse current flow at the same time (free electrons are still attracted to the N side). But the forward current is much greater than the reverse current, so the net current is forward. With the diode reverse biased, there is only the reverse current, as shown in Figure 1–6b.

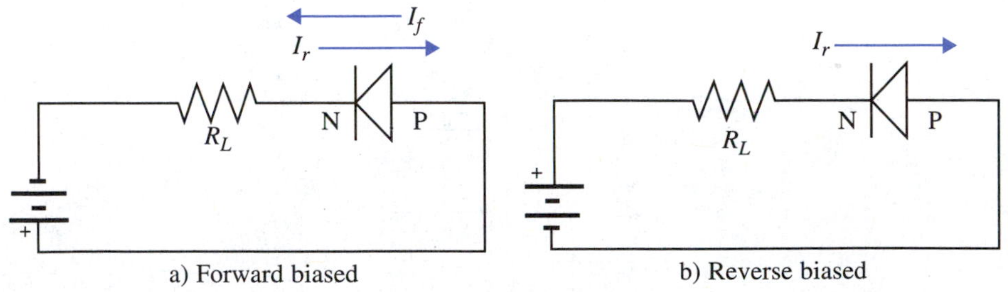

FIGURE 1–6 Current direction through a diode

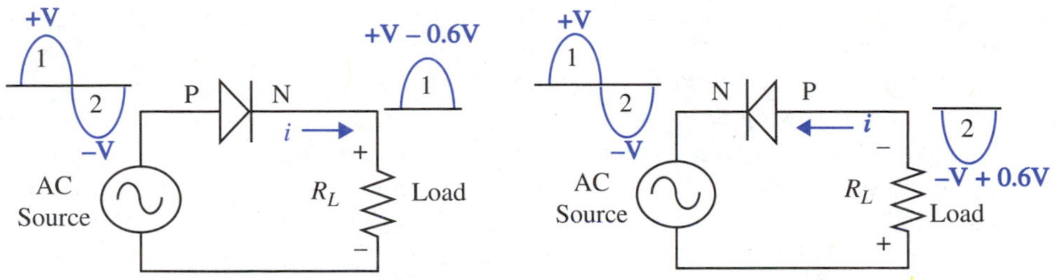

a) Positive alternation conducted b) Negative alternation conducted

FIGURE 1–7 Diode rectifying circuit

DIODE APPLICATIONS

What can we use a diode for in a circuit? To establish current in only one direction (neglecting the extremely small reverse current). One of the most basic uses for a diode is in a circuit that converts a current that is going positive and negative to a current that is going only in one direction (called **rectifying**). Such a circuit is shown in Figure 1–7. In this figure, the diode schematic symbol is given—the arrow is the P side (anode) of the junction and the straight line is the N side (cathode). Again, conventional current flows in the direction of the arrow.

The AC source provides a waveform that has both positive and negative voltage excursions. In Figure 1–7a, the positive portion is passed—actually 0.6V of the positive wave will be dropped across the diode in order to forward bias the junction.

Changing the diode around in Figure 1–7b allows only the negative part of the waveform to develop across the load. This type of circuit is used to convert the alternating current (AC) from the power line to the pulsating unidirectional current used in electroplating, e.g., chrome plating; and to convert AC to DC needed for DC power supplies (we will cover this concept in a later chapter).

A specific use for a diode is indicated in the alarm circuit of Figure 1–8. In the circuit, the negative (cathode) side of the diode is connected to a voltage divider that provides 17.4V

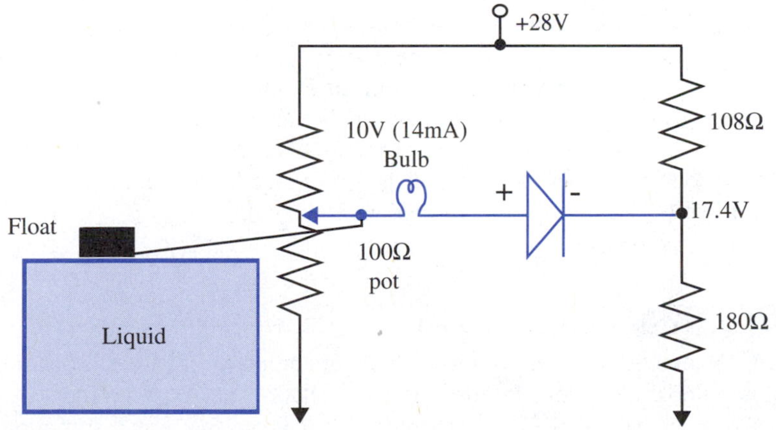

FIGURE 1–8 Liquid-level indicator

The positive side (anode) of the diode is connected to a potentiometer, the center tap of which is moved by the level of liquid in a tank. As the tank fills, the voltage on the center tap of the potentiometer rises. But there is no current through the diode and light bulb—because the diode is reversed biased. However if the liquid in the tank reaches the level where the voltage on the potentiometer is 18V with the 17.4V provided at the diode cathode, the diode will now forward bias. Further increase in liquid level will cause the increased voltage to appear across the light bulb because the drop across the diode stays at about 0.6V. Illumination of the bulb will serve warning that the tank level is approaching the full point (maximum bulb illumination).

EXAMPLE

Find the voltage across the bulb for the circuit of Figure 1–8 if the tank is 7/8 full.

SOLUTION

With the tank 7/8 full, the voltage at the center tap of the potentiometer V_p will be:

$$V_p = \frac{7}{8} \times 28\text{V} = 24.5\text{V}$$

If we subtract the 17.4V at the divider and the diode drop of 0.6V from this voltage, we have the voltage V_b across the bulb.

$$V_b = 24.5\text{V} - 17.4\text{V} - 0.6\text{V}$$

$$V_b = 6.5\text{V}$$

DIODE CHARACTERISTIC CURVES

We can show the way a diode reacts to an applied voltage by drawing a graph of voltage versus current, Figure 1–9. As we increase the forward bias (positive on anode) past 0.6V, the current increases greatly, and the diode acts like a low-resistance R_f. The more current we pass through the diode, the lower its effective resistance ($R_f = V_f / I_f$).

Reversing the power supply voltage causes the current to drop to the small reverse value. Making the voltage more and more negative, we reach a point where the electrons are forced to cross the depletion region from the P side to the N side, and a large reverse current flows. The voltage at which this occurs is called the **reverse breakdown voltage**, and if the current is not limited by an external resistor, the junction will overheat and be destroyed. Diode reverse breakdown voltages can range from about twenty volts to several thousand volts, depending on the diode construction.

DIODE EQUATION

To aid in the practical understanding of diode operation (and later, transistor operation), we will now look at the equation that gives the relationship between the forward voltage drop across a diode and the diode current. We will see that our assumption of a 0.6V drop across the diode is good at only one particular value of diode current.

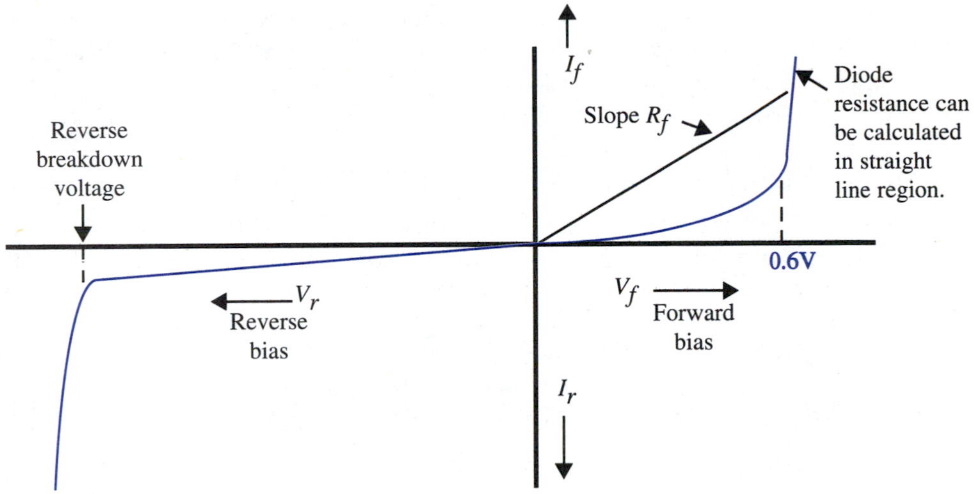

FIGURE 1–9 Diode voltage–current relationships

The forward current that flows through a diode can be expressed by the equation:

$$I_f = I_r\,[e^{(qV_f/nKT)} - 1] \qquad \textbf{(equation 1-1)}$$

Where:

- I_r is the reverse minority current.
- e is the natural constant 2.71828.
- q is the electrical charge of an electron $(1.6 \times 10^{-19}$ coulombs).
- V_f is the forward voltage across the diode.
- n is a diode construction factor ($n = 2$ is a good choice for a signal diode).
- K is Boltzmann's constant 1.38×10^{-23} joules/°Kelvin.
- T is the absolute temperature in degrees Kelvin (Celsius plus 273°).

EXAMPLE

Find the forward diode current I_f at 28°C if $V_f = 0.6$V, $I_r = 10$nA, and $n = 2$. Also compute with a forward voltage of 0.7V.

SOLUTION

The temperature 28°C is (28° + 273°) or 301°Kelvin. So the factor $KT/q = (1.38 \times 10^{-23} \times 301°\text{K} / 1.6 \times 10^{-19}) = 26$mV, and with the assumed value for I_r of 10nA and forward voltage of 0.6 volts, we can find I_f using equation 1-1;

$$I_f = I_r\,[e^{(qVf/nKT)} - 1]$$

$$I_f = 10^{-8}\ [e^{\,(0.6\text{V}/2\times 26\text{mv})} - 1]$$

$$I_f = 1.03\text{mA}$$

Increasing the voltage across the diode to 0.7V causes a forward current of 7.02mA. You can see from these results how much the current increases with a small increase in forward voltage—a seven-times increase in current with a 0.1-volt increase in the diode forward voltage.

DIODE VOLTAGE DROP VERSUS TEMPERATURE

The forward voltage drop across a diode reduces with increased temperature. This is not apparent from equation 1–1, which indicates rather that the voltage will increase with temperature. The reduction in diode voltage drop is caused by an increase in the reverse leakage current I_r which doubles for every 10°C increase in temperature. For a silicon diode, this causes the voltage to reduce 2.2mV for each degree centigrade increase in temperature. So at higher temperatures, we will see less than the expected voltage 0.6V drop across the diode; at lower temperatures, we will see a greater voltage drop.

EXAMPLE

If we assume the voltage drop is 0.6V at 25°C (room temp), what will be the drop at 125°C, which is a 100°C increase in temperature?

SOLUTION

$$V_f = 0.6\text{V} - (2.2\text{mV}/^\circ\text{C})(\text{temp change})$$

$$V_f = 0.6\text{V} - (2.2\text{mV}/^\circ\text{C})(100^\circ\text{C})$$

$$V_f = 0.6\text{V} - 0.22\text{V}$$

$$V_f = 0.38\text{V}$$

If we refer back to the tank filling alarm circuit of Figure 1–8, we realize that the bulb will start illuminating at a slightly lower liquid level at higher temperatures.

DIODE SPECIFICATION

If we plan to use a diode in a circuit, we need the required specifications for the device.

We might ask: How much current do we require the diode to pass in the forward direction? What is the maximum reverse voltage the diode will experience in the circuit?

The specifications for a common signal diode, the 1N4148, is given in Table 1–1. The 1N (one N) prefix means the device is a single junction device, i.e., diode.

From the specifications, we see that the maximum continuous forward current we can pass through the 1N4148 diode is 200mA, and the maximum reverse voltage (*PIV*) we can apply across the diode before possible breakdown is 75V.

The worst-case forward voltage drop across the diode is 1V at 10mA. Compare this with 0.72V, which can be computed with the ideal diode equation, and we see that an actual diode can have a larger voltage drop than the ideal device. In other words, its cur-

TABLE 1–1 1N4148 SPECIFICATIONS

MAXIMUM RATINGS		
PIV	Peak Inverse Voltage	75V
I_{dc}	DC Forward Current	200mA
T_A	Operating Temperature	–65° to +200°C
P_d	Power Dissipation	500mW
ELECTRICAL CHARACTERISTICS		
V_F	Forward Voltage Drop @ 10mA	1.0V
I_R	Reverse Current @ VR = 20V, 25°C	25nA
V_B	Min Breakdown Voltage with I_R = 5µA	75V
V_B	Min Breakdown Voltage with I_R = 100µA	100V
C	Capacitance	4pF

rent-versus-voltage curve is not as steep; meaning the 1N4148 has a higher forward resistance (see R_f in Figure 1–9). This forward resistance can be represented by a resistance (called **diode bulk resistance**) in series with the ideal diode, and this increases the voltage drop across the diode at a given current.

The maximum power dissipation listed is 500mW, and the maximum reverse leakage current at room temperature is 25nA with a reverse voltage of 20V.

Notice if we increase the reverse voltage to greater than the maximum of 75V, there is a large increase in reverse current—100µA at 100V. Reverse current also increases with temperature because of the generation of more free electrons.

EXAMPLE

A 1N4148 diode is connected in series with a 100KΩ resistor, and 100V is applied to the circuit with a polarity that reverse biases the diode. Determine the voltage drop V_d across the diode if the reverse leakage current is 100µA at 100V.

SOLUTION

If we initially assume all the 100V is across the reverse-biased diode, then the leakage current will be 100µA. This current, however, will cause a voltage drop V_r across the 100KΩ resistor:

$$V_r = 100\mu A \times 100K\Omega$$

$$V_r = 10V$$

The voltage drop across the diode V_d will be:

$$V_d = 100V - 10V$$

$$V_d = 90V$$

Since this voltage is 10 percent lower than 100V, we can expect the leakage current to be a little lower than 100µA, which means a little less voltage drop across the 100KΩ resistor.

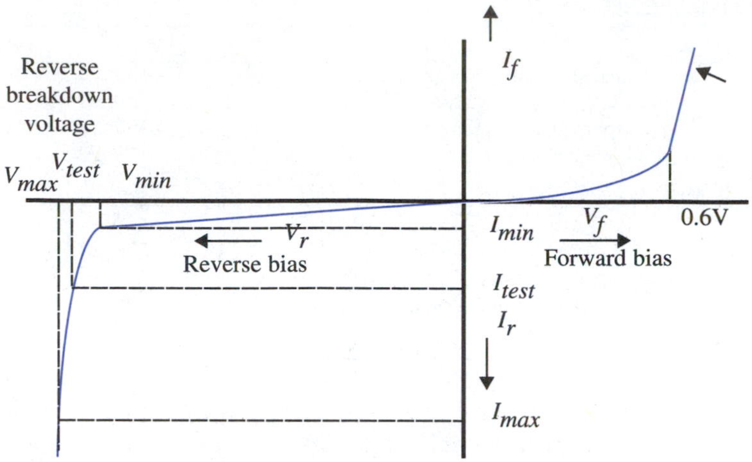

FIGURE 1–10 Zener diode characteristics

ZENER DIODE

A **zener** (pronounced *Z–ner*)-**type diode** is a device designed to have a specific and very constant breakdown voltage in the reverse direction, as shown in Figure 1–10. The diode is operated in the reverse mode to provide a very steady reference voltage. Notice that a small reverse current I_{min} has to flow (usually less than 1mA) before the constant breakdown voltage is achieved. The test values of breakdown voltage and reverse current are given in specifications.

Figure 1–11 shows how the zener diode is connected in a circuit. Also notice that two additional lines are added to the regular diode symbol.

In the figure, the diode is connected in reverse relative to the battery polarity, and we assume a diode reverse breakdown voltage (zener voltage) of 10V. If the variable supply is set for less than 10V (below the breakdown voltage), then this lower voltage appears across the diode. However, if the variable supply is raised above 10V, the diode breakdown occurs, and the 10V holds across the diode—even if the power supply voltage is further increased.

The resistor R_d limits the current so the product of diode current and zener voltage, the wattage rating of the diode, is not exceeded when operating at the breakdown voltage.

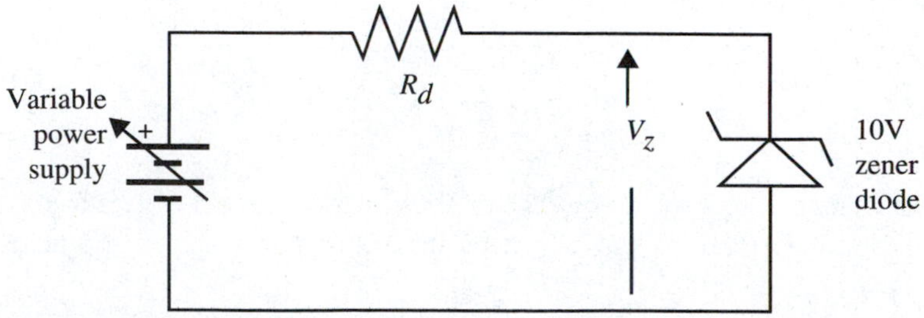

FIGURE 1–11 Basic zener diode circuit

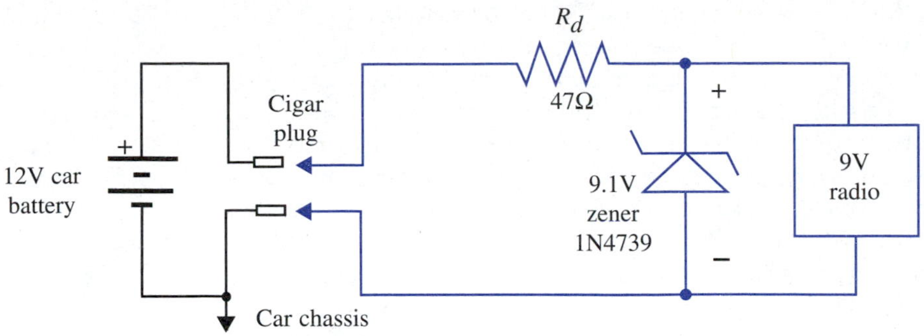

FIGURE 1–12 Transistor radio supply

We can see then that even though the supply voltage changes, the voltage across the diode is constant once breakdown is achieved. The zener diode, then, can be used as a reference voltage for power supply circuits (as explained later in the chapter on power supplies) or whenever we need a predictable voltage in a circuit.

EXAMPLE

We want to operate a 9V, 30mA, transistor radio from our car battery (via the cigar lighter plug). The circuit shown in Figure 1–12 could be used.

SOLUTION

In the circuit, a 9.1V zener is used to maintain the transistor radio voltage constant at the required value despite variations in the battery voltage of the car. The resistor in series limits the amount of current through the zener diode and drops the voltage difference between the supply voltage and the zener voltage.

EXAMPLE

Show the circuit for an automobile voltmeter that will read the battery voltage over a range from 10V to 18V using a 1mA meter.

SOLUTION

Figure 1–13 shows a possible circuit. This circuit makes maximum usage of the meter face by having a reading of 10V on the left side of the meter and 18V on the right side (expanded scale meter) for good readability.

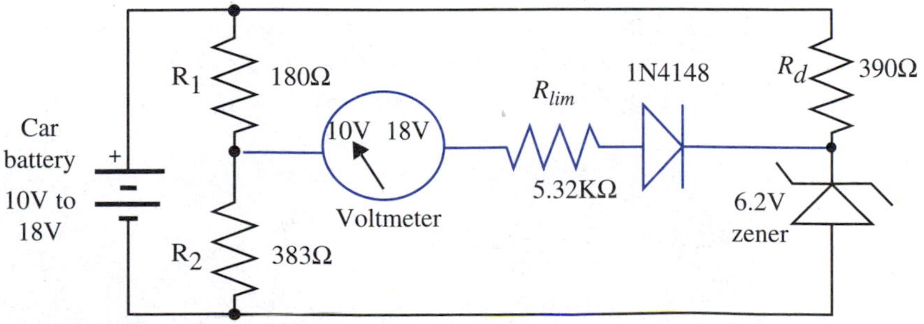

FIGURE 1–13 Automobile voltmeter circuit

The 6.2V zener diode is used for a voltage reference in the circuit because the power supply voltage (car battery) varies from 10V to 18V. R_d is selected to provide about 10mA through the zener diode when the battery voltage is 10V.

The R_1:R_2 divider provides 6.8V (across R_2) at the left of the meter when the battery voltage is 10V (no meter current). When the battery voltage is 18V, we can use Thevenin's theorem (which converts a complex circuit to simply one resistor and a voltage source) to replace the battery voltage and R_1 and R_2.

The Thevenin voltage is determined by opening the voltmeter portion of the circuit (leaves just the voltage divider R_1 and R_2) and determining the voltage across R_2:

$$V_{th} = \left(\frac{18V}{R_1 + R_2}\right)R_2$$

$$V_{th} = \left(\frac{18V}{180\Omega +}\right)383\Omega$$

$$V_{th} = 12.24V$$

The Thevenin resistance is determined by shorting the battery and determining the resistance seen looking back into the divider circuit. Now R_1 and R_2 are in parallel and

$$R_{th} = R_1 \parallel R_2$$

$$R_{th} = 122\Omega$$

When the battery voltage is 18V, we want the full-scale current of 1mA to flow through the meter circuit. This current flows through R_{th} causing a voltage drop V_d of

$$V_d = 1mA \times 122\Omega$$

$$V_d = 122mV$$

The voltage across R_2 is now:

$$V_{R2} = 12.24V - 0.122V$$

$$V_{R2} = 12.12V$$

This is an increase across R_2 of 5.32V (12.12V – 6.8V) from when the battery voltage was 10V, and this increase appears across R_{lim} in the meter branch of the circuit. This checks because it requires 5.32V across the 5.32KΩ resistor to provide 1mA.

Looking at it another way, when the battery voltage is 18V (full-scale reading):

$$R_{lim} = \frac{(12.12V - 0.6V - 6.2V)}{1mA}$$

$$R_{lim} = 5.32K\Omega$$

The diode in series with the meter prevents the meter from trying to read backwards when the battery reads below 10V (cold engine cranking). R_2 could be a rheostat to allow calibration of the meter circuit.

ZENER RESISTANCE

Zener diodes are made in various voltages, ranging from about 3V up to 200V. The voltage across a zener will change slightly if the current through it is changed (Figure 1–10). This is because of the internal resistance r_z (also called zener impedance) of the device.

We can show a zener as a constant voltage source with the zener resistance in series, Figure 1–14. The lower this zener resistance, the lower the voltage change across the zener with current, and the more constant our reference voltage. Zener diode data sheets provide values of r_z at a given test current for each specific type.

EXAMPLE

Find the actual voltage across the 10V zener diode in Figure 1–14.

SOLUTION

If we calculate the circuit current from the voltage drop (30V–10V) across the 1KΩ resistor, we have:

$$I = \frac{(30V - 10V)}{1K}$$

$$I = 20mA$$

This current flowing through the zener internal resistor of 5Ω will give us a voltage drop of 5 × 20ma or 100mv. The actual voltage, then, across the zener terminals will be

$$10V + 0.1V = 10.1V$$

Note that we used 10V across the zener for computing the current, but it is actually 10.1V. We had to do it this way because we did not know the actual zener voltage until

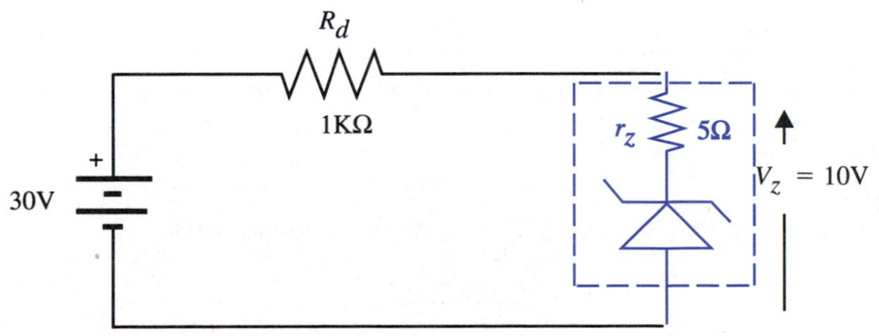

FIGURE 1–14 Effect of zener resistance

we determined the current. However, doing it this way created only a 1 percent (0.1V) error in the readings.

Zener voltages are usually specified at some test current value in order to take into account the internal resistance drop. This test current is achieved by selecting the supply voltage and the series voltage dropping resistor (R_d in Figure 1–14).

If we connect a zener diode backwards in a circuit, it will be a forward-biased diode, and the drop across the diode will be 0.6V.

EXAMPLE

What will be the current through R_d if the zener diode in Figure 1–14 is put in backwards?

SOLUTION

Now with just 0.6V across the zener, the voltage across R_d will be

$$30V - 0.6V = 29.4V$$

The current is

$$I = \frac{29.4V}{1K\Omega}$$

$$I = 29.4mA$$

If we were using a half-watt resistor for R_d, it could be damaged because the power dissipation in R_d with the zener in backwards is

$$P_d = 29.4mA \times 29.4V = 0.864W.$$

EFFECT OF LOAD CURRENT ON ZENER VOLTAGE

If we connect a load resistor R_L across the zener diode as shown in Figure 1–15, the current through the zener diode will be reduced by the amount of the load current. The reason for this is that the current through R_d is constant because the voltage across this resistor does not change when the load is connected. Therefore current required for the load is taken from the zener diode.

EXAMPLE

Find the current through the zener diode in Figure 1–15 if the load resistor is disconnected, and then with the load resistor connected.

SOLUTION

We first need to find the current I_d through the dropping resistor R_d:

$$I_d = \frac{(V_{bb} - V_z)}{R_d}$$

$$I_d = \frac{(25V - 15V)}{270\Omega}$$

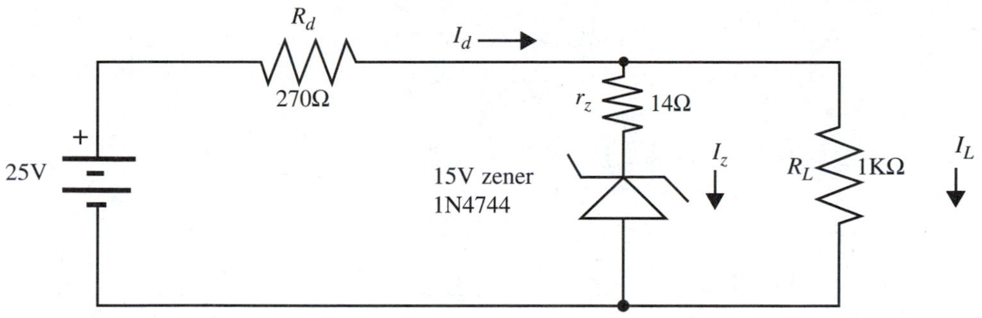

FIGURE 1–15 Zener diode with load resistor

$$I_d = 37\text{mA}$$

With the load resistor not connected, all this current must flow through the zener diode, so the "no load" zener current (I_{znl}) will be 37mA..

If we now connect the 1KΩ load resistor, the current I_L through this resistor, with the zener voltage of 15V across it, will be

$$I_L = \frac{V_Z}{R_L}$$

$$I_L = \frac{15\text{V}}{1\text{K}\Omega}$$

$$I_L = 15\text{mA}$$

The new zener current, with the load resistor connected (I_{zl}), will be the load resistor current subtracted from the current flowing through R_d, or

$$I_{zl} = 37\text{mA} - 15\text{mA}$$

$$I_{zl} = 22\text{mA}$$

So there is a change (decrease) of 15mA in the current through the zener when the load is connected.

EXAMPLE

Find the *change* in the zener diode voltage (ΔV_z) when the load resistor is disconnected in Figure 1–15.

SOLUTION

From the previous example, we found that the zener diode current increases by 15mA when the 1KΩ load resistor is disconnected. The change (increase) in zener voltage is caused by this increased current flowing through the 14Ω zener resistance r_z, or

$$\Delta V_z = I_L \times r_z$$

$$\Delta V_z = 15\text{mA} \times 14\Omega$$

$$\Delta V_z = 0.21\text{V}$$

This is a $\left(\frac{0.21\text{V}}{15\text{V}} \times 100\%\right)$ or 1.4% increase in the zener voltage.

ZENER POWER RATING

Zener diodes are power rated from small glass units of under a watt to large metal case units of over a hundred watts. The power dissipation of a zener is simply the zener voltage times the zener current.

EXAMPLE

What must be the minimum power rating for the 15V zener diode used in the circuit shown in Figure 1–15?

SOLUTION

The current through the zener (I_z) is the current through R_d minus the load current. Load current is

$$I_L = \frac{15\text{V}}{1\text{K}\Omega}$$

$$I_L = 15\text{mA}$$

Zener current is

$$I_z = 37\text{mA} - 15\text{mA}$$

$$I_z = 22\text{mA}$$

The power dissipation in the diode is

$$P_d = 15\text{V} \times 22\text{mA}$$

$$P_d = 330\text{mW}$$

However, the worst-case power dissipation in the zener diode occurs when the load resistor is disconnected and all of I_d (37mA) flows through the zener diode; so

$$P_{zmax} = 37\text{mA} \times 15\text{V} = 0.555\text{W}$$

To be conservative, we would 50 percent de-rate and use a 1W zener diode.

PHOTODIODES

A **photodiode** is a special type of diode that provides current proportional to the power level of light falling on the device.

When discussing the diode reverse current, it was stated that it resulted from electrons attached to atoms in the depletion region acquiring enough thermal energy to break away and become free electrons. If we expose the depletion region to light, the energy contained in the light will also cause free electrons. A photodiode is a diode that uses this phenomenon to detect light level. It has an optical window over the depletion region. Increasing the intensity of light (watts) falling on the diode depletion region, via the optical window, increases the reverse current.

The diode has to be operated with a reverse voltage across it when used as a photodetector; otherwise, the forward current would "swamp out" the reverse current since it has a much greater value. A circuit that will provide an output voltage related to light level is shown in Figure 1–16.

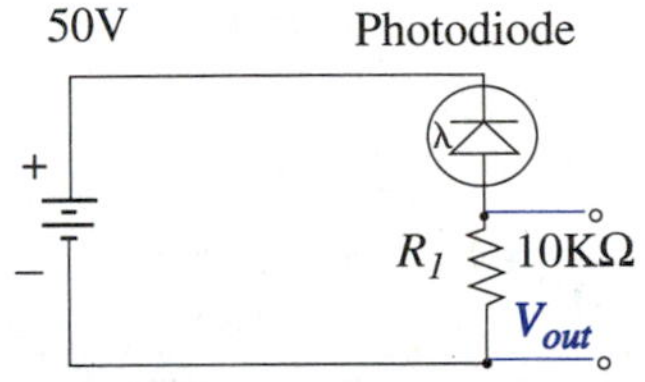

FIGURE 1–16
Photodiode circuit

The symbol for the photodiode contains the Greek letter λ (lambda) to indicate wavelength. The silicon diode photodetector responds to light wavelengths from 0.3 to 1.2 microns (millionths of a meter). This includes the eye's visible spectrum of 0.4–0.7 microns.

If the photodiode in Figure 1–16 is kept dark (covered), the voltage across the 10KΩ resistor will be low because a very small reverse current called **dark current** will flow. Increasing the light level will increase this reverse current and therefore the voltage across the 10KΩ resistor.

Sensitivity (S) of the typical photodiode to light is roughly 0.5μA/μW, which means if 1μW of light power (P_o) falls on the junction, the current will be 0.5μA.

EXAMPLE

If the light level falling on the total depletion region of the diode shown in Figure 1–16 is 100μW, find the voltage developed across the 10KΩ resistor.

SOLUTION

Using the given sensitivity of 0.5μA/μW, then the photocurrent *Ip* with 100μW falling on the junction will be

$$I_p = P_o \times S$$

$$I_p = 100\mu W \times 0.5\mu A/\mu W$$

$$I_p = 50\mu A$$

This current will cause a drop across the 10KΩ resistor of

$$V = 50\mu A \times 10k\Omega = 0.5V$$

By connecting a voltmeter across the 10KΩ resistor, we can effectively measure the light level. Increasing the value of R_1 will give a higher output voltage for a given light level.

Because of their small junction size and consequent small capacitance, photodiodes can respond very quickly to changes in light level. With a low value of series resistance used to reduce capacitance effects, response times of less than ten nanoseconds can be achieved. This makes the photodiode a good choice for high-speed fiber-optic data communications.

VARICAP (VARACTOR) DIODE

Varactor symbol

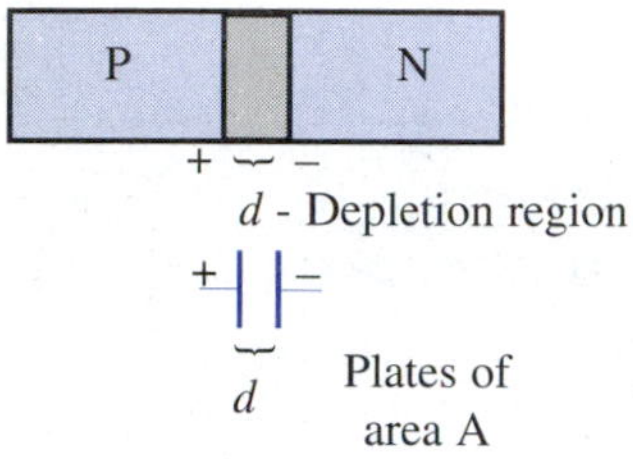

FIGURE 1–17
Diode viewed as a capacitor

We shall now look at a diode that acts like a capacitor and is called a **varicap** or **varactor diode**. Its great advantage is that the capacitance can be changed by changing the voltage across the device.

The depletion region of a conventional diode has very few free electrons and can be considered an insulator between the P doped side of the diode and the N doped side (see Figure 1–17).

Considering the P and N sides as conductors (or plates), the depletion region of width d is an insulator (dielectric) between two conductors—this is the physical requirement of a capacitor, which is a device that can store electrical charge. The equation for a capacitor is

$$C = \frac{KA}{d}$$

where:

C is the capacitance measured in farads

K is called the dielectric constant

A is the area of the conductors

d is the spacing between conductors

In a diode, d is the width of the depletion region. It was mentioned earlier that the depletion region widens if we increase the reverse voltage.

From the equation for capacitance, we see that the capacitance of a diode will reduce if we increase the reverse voltage which increases d. So we have a voltage variable capacitor—which is the varicap diode. However, the change in capacitance is not linearly related to the reverse voltage. The relationship to C is inversely proportional to the square root of the reverse voltage, or to the cube root—depending on the diode construction. With the square root, if the reverse voltage is increased four times, the capacitance will be halved. A cube root relationship means increasing the reverse voltage by a factor of eight will halve the capacitance.

The capacitance values of varicaps are in the picofarad range (5–100), and they can be used in a resonant circuit with an inductor for electronic tuning of radios and televisions. Setting the reverse voltage across a varicap to the required resonant value will tune in a particular station; changing the voltage will tune another station. This enables changing a station without using switches. This concept is used for remote control of televisions and VCRs.

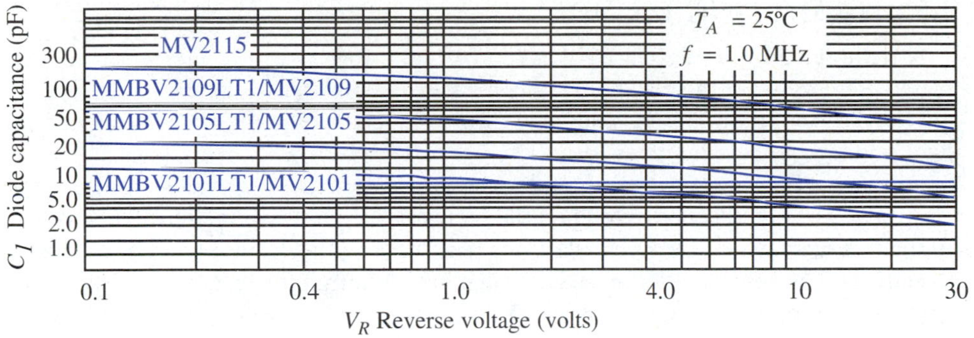

FIGURE 1–18 Varicap characteristics

A typical varicap is the MV2101, which has a nominal capacitance (±10%) of 6.8pF with a reverse voltage of 4V. The graph of capacitance versus reverse voltage for this device is shown in Figure 1–18. A design weakness of varicaps is their small capacitance range.

EXAMPLE

Find the voltage required across the MV2101 varicap to cause a parallel resonance with a 12μH coil at 20MHZ.

SOLUTION

The required capacitance for a resonance is when the reactance of the capacitor is equal to the reactance of the inductance, or

$$2\pi f L = \frac{1}{2\pi f C}$$

Solving for C;

$$C = \frac{1}{(2\pi f)^2 L}$$

At 20MHZ

$$C = 5.28\text{pF}$$

From the graph of Figure 1–18 we read a required reverse voltage of about 8V.

SCHOTTKY DIODE (SILICON AND GALLIUM ARSENIDE)

A diode that is used in switching and high-frequency applications is the **Schottky diode**. In a switching diode configuration, the Schottky diode can switch in picoseconds (10^{-12} seconds). As a diode for RF applications (mixers, harmonic generation—see Chapter 17), the gallium-arsenide-type Schottky diodes can operate up to about 45 gigahertz (10^9 Hz).

The Schottky diode uses a metal–semiconductor barrier formed by depositing a metallic layer on the semiconductor layer. This makes a diode that has no minority carriers to limit the switching speed. The forward voltage drop for a silicon Schottky diode is lower than the conventional diode with a forward voltage drop of about 0.3V at 1mA—whereas the gallium-arsenide-type Shottky is closer to the conventional diode with a drop of 0.7V at 1mA.

Figure 1–19 indicates the basic construction and symbol for the Schottky diode.

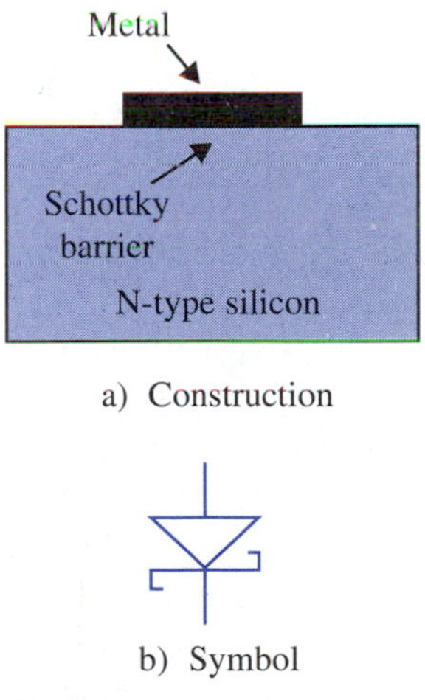

FIGURE 1–19 Schottky diode

LIGHT-EMITTING DIODES

A **light-emitting diode (LED)** is a special diode that gives off light when forward biased. Rather than being made from silicon, as are the previously discussed diodes, LEDs are formed from gallium semiconductors. These materials have the property of emitting light under the right conditions.

When an LED is forward biased, the electric field from the bias voltage causes some electrons to be excited enough to jump the energy gap from the valence band of the atom to the conduction band. As these electrons relax back to the valence band, they give up the energy they absorbed from the electric field in the form of light.

LEDs are formed from gallium arsenide phosphide (GaAsP), gallium phosphide (GaP), or gallium arsenide phosphide lens on a gallium phosphide LED (high-efficiency red—HER).

Table 1–2 shows the characteristics of diodes made from these materials, and Figure 1–20 indicates the light output at various wavelengths—along with the human eye response.

The intensity of the light from an LED increases with the forward current; with maximum currents ranging from 5mA to 100mA, depending on the size of the junction.

EXAMPLE

Calculate the resistor R_1 that has to be connected in series with a red LED in order to limit the current to 10mA when the circuit is connected to a 12V supply.

SOLUTION

From Table 1–2, the red LED has a voltage drop of 1.6V, which means 10.4V must be dropped across the limiting resistor R_1.
Using Ohm's law

$$R_1 = \frac{10.4\text{V}}{10\text{mA}}$$

$$R_1 = 1.04\text{K}\Omega \quad (\text{use } 1\text{K}\Omega)$$

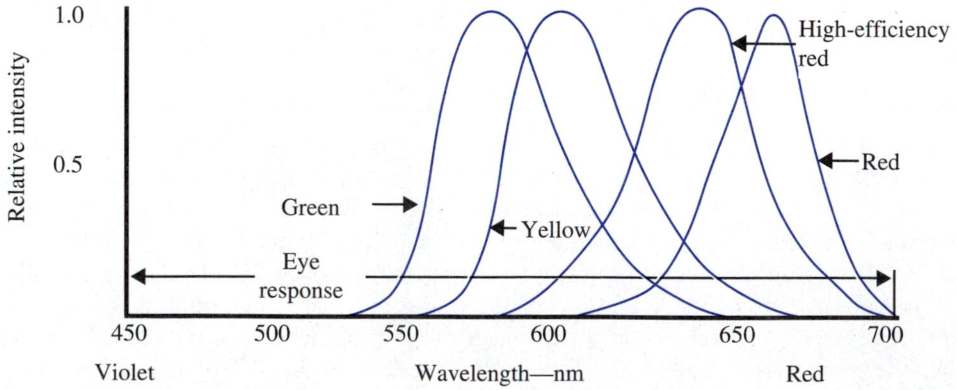

FIGURE 1–20 LED Intensity versus wavelength

TABLE 1–2 LED Colors and foreward voltages

MATERIAL	*COLOR*	*FORWARD VOLTAGE*
GaAsP4 on GaP	Hi Eff Red (HER)	2.0V
GaAsP4	Red	1.6V
GaAsP5	Yellow	1.8V
GaP	Green	2.0V

By connecting two junctions in parallel, but reversed P to N and N to P, we can have an LED that will illuminate with an AC voltage applied. One diode will light with the waveform positive, while the other will light when the waveform is negative. Sometimes LEDs of a different color are connected back to back in a package. By changing the polarity of the DC voltage, we can change the color, for example, from red to green.

The LED displays for numbers and letters used in some calculators, clocks, and other devices are made from LEDs arranged in seven segments, as shown in Figure 1–21.

In Figure 1–21a, the construction of an individual segment is indicated; in Figure 1–21b, the seven-segment arrangement is shown (with *S* illuminated) along with the standard letter segment identification.

Lastly in Figure 1–21c, the diode connections are given. Notice that all diodes are connected by a common anode connection (some displays are common cathode) that is connected to a positive supply. Each segment can be individually turned on by completing the circuit (turning on the switch) to the power supply common via a current limiting resistor (*R*). Switch closures to illuminate an *S* are shown.

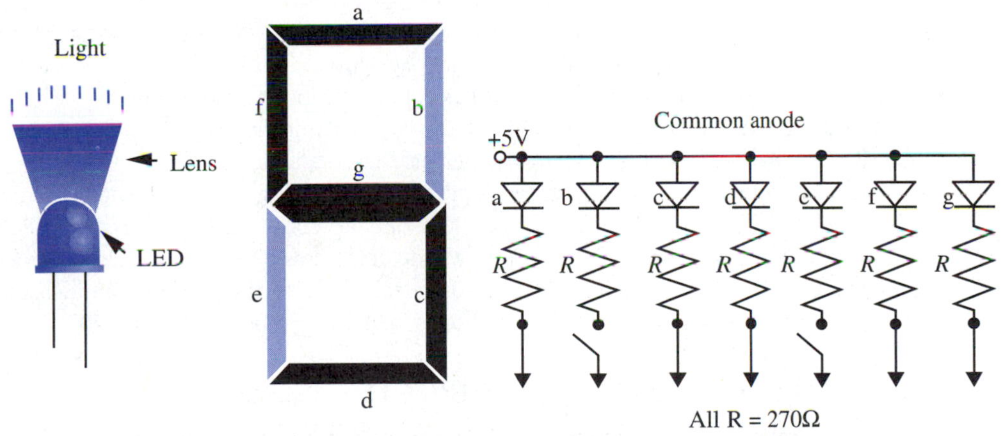

a) Segment side view **b) Segment ID** **c) Common anode connection**

FIGURE 1–21 Seven-segment display

THE VARISTOR (TRANSIENT SURGE SUPPRESSOR)

A **varistor** is a semi-conductor device that will safely conduct very heavy currents for a short period of time and reduce the damaging effects of high-voltage transients.

To understand how the varistor operates, we need to be aware that the varistor is connected across the voltage source (in parallel with the load) and forms a voltage divider

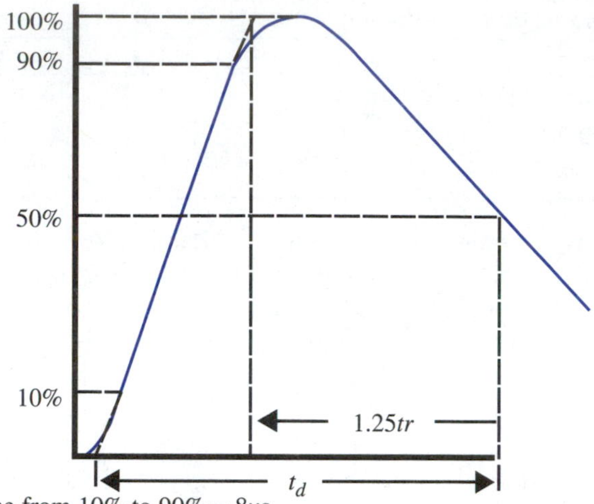

tr = time from 10% to 90% = 8μs
td = impulse duration = 20μs

a) Test pulse

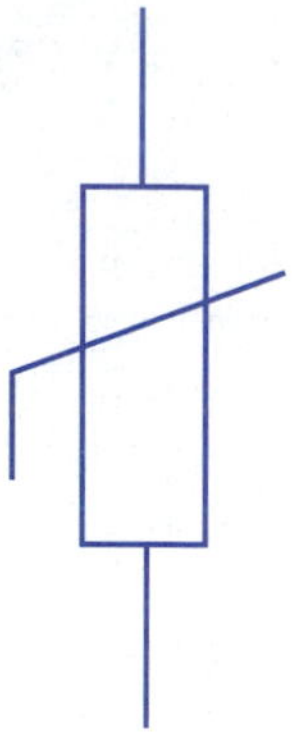

b) Symbol

FIGURE 1–22 Varistor

with the source resistance (or impedance). When a voltage spike occurs across the source, the varistor's resistance drops and conducts a heavy current. This heavy current causes a large voltage drop across the source resistance, which prevents the transient from developing a high voltage across the varistor and therefore the load.

The varistor consists mainly of zinc oxide grains that form semiconductor-like diodes at their boundaries. These boundaries prevent conduction at low voltages but become the source of nonlinear conduction at high voltages (the effective resistance drops). The fact that the diodes are distributed throughout the material means they are able to absorb more energy than, for example, a single zener diode.

An example of a varistor is the Harris Semiconductor V150LA1. This device is formed in a 7mm disc (like a disc capacitor) and can operate continuously across a 120VAC power line. It can conduct up to 1200A for a time defined as 8/20μs, which is a standard varistor test pulse with 8μs rise time and 20μs decay time to the 50% amplitude point. The test pulse is shown along with the varistor schematic symbol in Figure 1–22.

TROUBLESHOOTING DIODE CIRCUITS

With electronic circuits containing diodes, problems can occur if the diode fails to operate. The diode can fail if too much current is passed through it or if too much reverse voltage is applied across the diode. The failure modes for the diode can be open circuited (no conduction in either direction through the diode) or shorted (diode conducts both ways).

Checking a Diode in Circuit

If the diode is normally conducting in the circuit, as shown in Figure 1–23, we can determine whether it is good by checking to see if the forward voltage drop is approximately 0.6V (silicon diode). If we are checking an LED, we should measure between 1.6V and 2V in the forward direction.

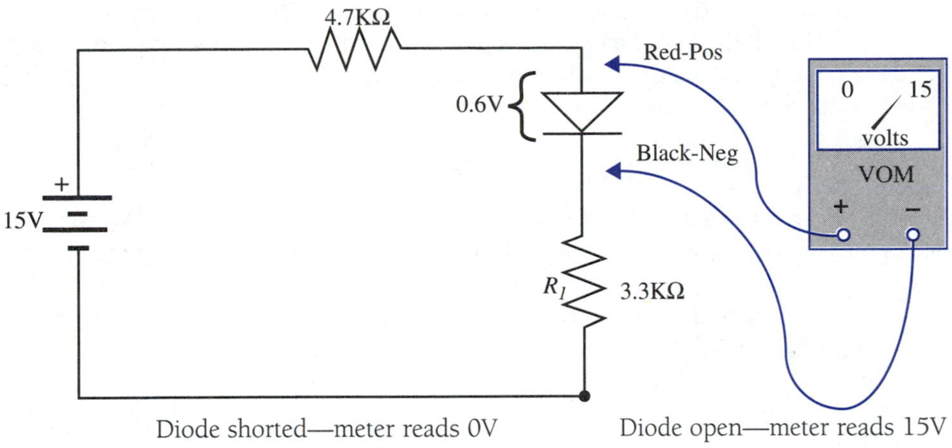

FIGURE 1–23 Checking a diode in a circuit

Testing a Diode with a VOM

With the diode out of the circuit, we can check it to see whether it is good by measuring the forward and reverse resistance with a volt–ohm meter (VOM), as shown in Figure 1–24.

The forward resistance we measure with the VOM for a good diode is a function of the resistance scale setting—remember, the more current we pass through a diode, the lower the diode resistance becomes (see Figure 1–8).

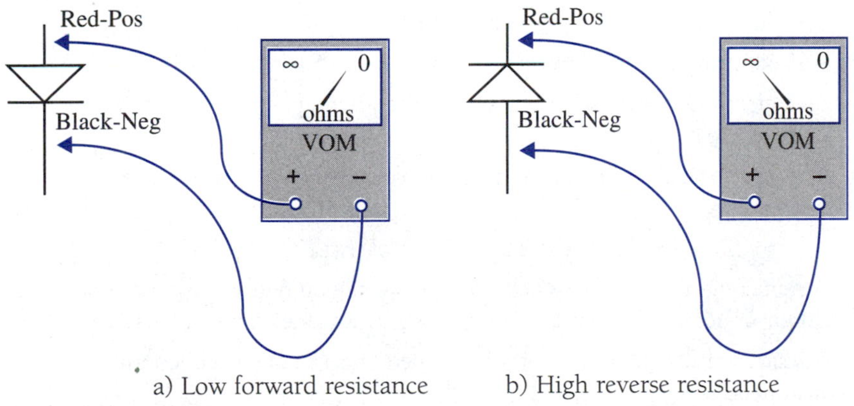

FIGURE 1–24 Checking a diode out of circuit

For example, if we use a Simpson 260 VOM on the R × 1 setting, the forward resistance for a 1N4148 diode will measure about 15Ω. On the R × 100 setting (lower current passes through the diode) it will read around 700Ω.

The reverse resistance measured with the VOM should be infinite for a good diode.

Testing a Diode with a DMM

Take care when using a digital multimeter (DMM) to check a diode. Verify that the voltage supplied by the meter is sufficient to forward bias the diode when checking forward resistance. Most DMMs have a diode check position that displays the forward voltage drop across the diode. With the leads to the diode reversed, the DMM displays the meter battery voltage (about 1.5V).

Checking an LED

All of the diodes we have studied can be checked out of circuit with the VOM or DMM. But because of its high forward voltage drop, we must check the LED on a meter that provides greater than 2V on the ohmmeter setting. If in doubt about the meter, check the voltage across its ohmmeter terminals with a second VOM (or DMM) set to the volts scale.

Determining the Reverse Breakdown Voltage for a Diode

If we wish to check the reverse breakdown voltage for a diode without destroying the device, we need to put a high resistance (usually 1MΩ) in series with the power supply. We connect the power supply to reverse bias the diode and connect a DMM across the diode. As we increase the power supply voltage, we note where the diode (DMM) voltage shows very little further increase. This voltage is the diode reverse breakdown value.

SUMMARY

- Electrons are in orbit around the nucleus of an atom, like planets around the sun.
- The outer orbit of the electrons is called the valence band.
- A maximum of eight electrons can occupy the valence band.
- Free electrons are those that have acquired enough energy to break free of the valence shell.
- The fewer the electrons in the valence band, the better the electrical conduction.
- Electron flow (current) is the movement of these free electrons.
- A semiconductor is a poor conductor, like silicon that has been doped with donor or acceptor atoms to provide a surplus or deficiency of electrons.
- A diode consists of a junction of P-doped and N-doped semiconductor materials.
- Current can flow only one way through a conventional diode.

- Conventional current flow through a diode is from the P side (anode) to the N side (cathode).
- The cathode can be indicated by a colored band on the body of the diode.
- The forward drop across a conducting diode is approximately 0.6V for a silicon device and 0.2V for a germanium device.
- We can find the current through a conducting silicon diode by subtracting 0.6V from the supply voltage and dividing the result by the limiting resistor.
- A diode is used whenever we wish to allow current to flow in only one direction in the circuit.
- The diode curve shows that as we increase beyond a forward voltage of 0.6V, the current increases rapidly.
- As we increase the reverse voltage across a diode, a breakdown voltage is reached where significant reverse current starts to flow.
- A diode can be destroyed by internal heating from allowing too much forward current to flow (*R* too small) or by applying too much reverse voltage and again allowing too much current to flow (*IrVr*).
- Specifications enable us to determine key diode parameters.
- The 1N prefix identifies a device as a diode.
- A zener diode is designed to break down in reverse at a specific voltage.
- Because of the internal resistance (zener impedance), the voltage across a zener diode can change slightly as the current through the zener changes.
- As with a regular diode, a resistor must always be connected in series with a zener diode to limit the current.
- A photodiode can be made because light falling on a reverse-biased junction increases the reverse leakage current.
- Because of their small size and consequent low interelectrode capacitance, photodiodes can respond very quickly to changes in light level.
- Since the depletion region of a diode is an insulator, a reversed-biased diode can be used as a voltage variable capacitor (varicap or varactor).
- A Schottky diode can operate at a much higher frequency than a conventional diode.
- A light-emitting diode is made from gallium-semiconductor material rather than from silicon.
- The LED is operated in the forward direction with a voltage drop of 1.6V to 2V.
- A seven-segment display is formed from seven individual LEDs arranged in a figure eight.
- We can tell if a diode is operating in circuit if 0.6V is measured across when it is forward conducting, or the full reverse voltage is measured when it is reverse biased.

- A good diode should read low resistance on an ohmmeter with the positive lead of the meter connected to the anode (+) of the diode and the negative lead connected to the cathode (–). Also, a high resistance should be measured if the leads are reversed.

EXERCISE PROBLEMS

1. Describe the doping of silicon and state the purpose.
2. Explain how a diode is constructed.
3. Describe why conventional current flows from the P side to the N side of a diode.
4. What is the value of the voltage that must be overcome before forward current flows in a silicon diode? A germanium diode?
5. If someone handed you a diode, how would you tell which end is the cathode?
6. Calculate the series resistor to limit the current through a 1N4148 diode to 10mA if the power supply voltage is 14V.
7. With a sketch, explain how a diode rectifier works.
8. Using the diode equation, calculate the current through a diode at 28°C if the forward voltage across the diode is 0.8V. Assume $n = 2$ and $I_r = 10\text{nA}$.
9. If the zener resistance is 2Ω, compute the voltage change across the 6.2V zener of Figure 1–A when the car battery voltage goes from 10V to 18V. Meter is 1mA full scale. ***Hint:* Consider meter current change.)**

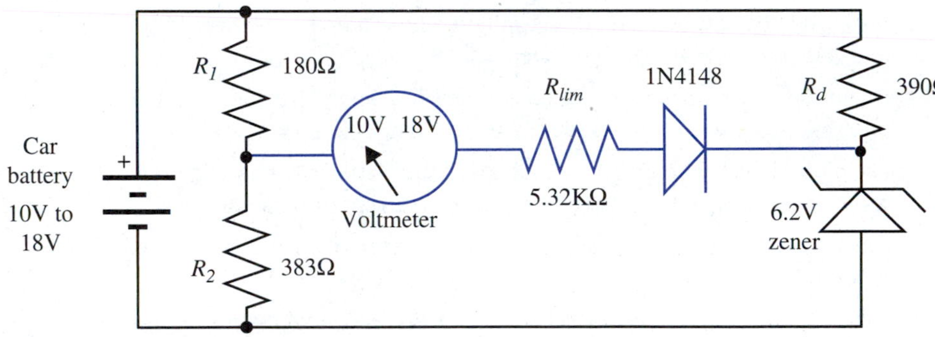

FIGURE 1–A (text figure 1–13)

10. The voltage across a zener is 5.6V with a current of 45mA, and the zener impedance is 5Ω. Calculate the new voltage across the zener if the current is changed to 20mA.
11. A 8.2V zener diode is to be operated from a power supply of 15V. Determine the value of the series resistor to be used to limit the zener power dissipation to 500mW.

12. Calculate the new voltage across the zener of problem 10, with 45mA of current, if a 560Ω resistor is placed in parallel with the zener. (*Hint:* The resistor will steal current from the zener.)
13. Calculate the power dissipation in the zener diode of Figure 1–B if the load resistor is disconnected.

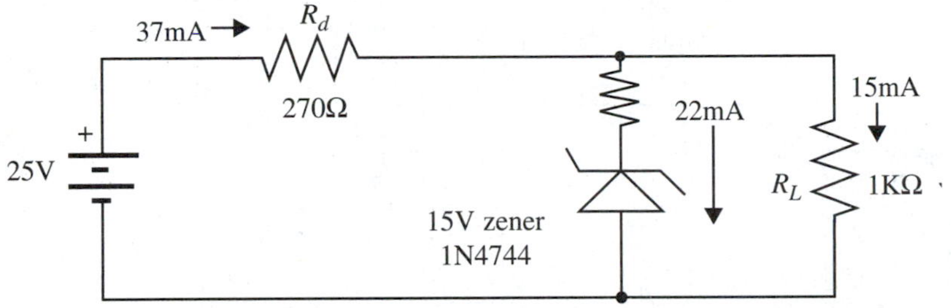

FIGURE 1–B (text figure 1–15)

14. Explain how a photodiode works.
15. If the sensitivity of a photodiode is 0.6μA/μW, find the photocurrent flowing with an incident light level on the photo junction of 50μW.
16. Find the voltage drop across the 10KΩ resistor in the photodiode circuit of Figure 1–C if 100μW of light falls on the diode junction and S = 0.5μA/μW.
17. Explain why you can change the capacitance of a varactor diode by varying the diode reverse voltage.
18. Determine the required reverse voltage across a MV2101 varicap to resonate with a 15μH inductor at 18MHZ.
19. Determine the resistor value in series with a yellow LED to limit the current to 20mA when operating from a 5V supply.
20. If the forward voltage drop for the LEDs in Figure 1–D is 1.6V, find the current through each diode with the switch closed and total power drawn from the 5V power supply with all segments illuminated.

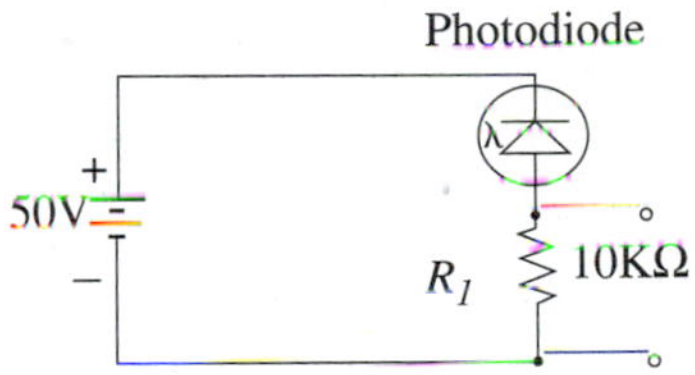

FIGURE 1–C
(text figure 1–16)

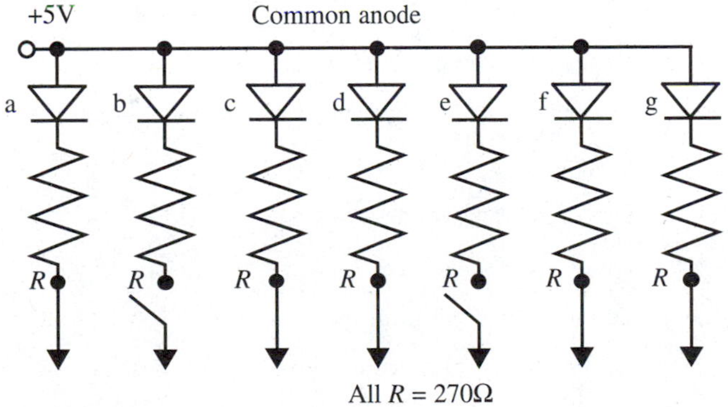

FIGURE 1–D (text figure 1–21c)

21. What happens to the light bulb in the circuit of Figure 1–E if the diode short-circuits?

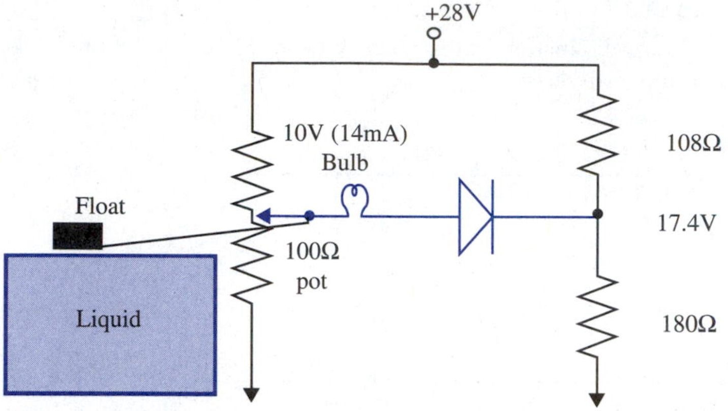

FIGURE 1–E (text figure 1–8)

22. Determine the voltage at the juncture of R_d and R_L in Figure 1–B if the zener diode opens.

23. State how you would check an LED with a meter. What type of meter would you use and why?

24. Describe how you would check a zener diode out of circuit.

ABBREVIATIONS

I_d = Zener dropping resistor current

I_f = Diode forward current

I_L = Load current

I_r = Diode reverse current

I_p = Photocurrent

I_z = Zener current

I_{zl} = Zener current with load resistor connected

I_{znl} = Zener current without load resistor connected

V_b = Battery voltage

V_f = Diode forward voltage drop

V_r = Diode reverse voltage

V_{znl} = Zener voltage without load

V_{zl} = Zener voltage with load

R_f = Diode equivalent forward resistance

R_L	= Load resistor
R_d	= Zener dropping resistor
r_z	= Zener resistance
P_o	= Light power
S	= Photodiode sensitivity

ANSWERS TO PROBLEMS

1. See text.
2. See text.
3. See text.
4. Si = 0.6V *Ge* = 0.2V
5. Band is on cathode end.
6. R_L = 1.34KΩ
7. See text.
8. I_f = 49mA
9. Change is 43mV.
10. With reduction of 25mA, V_z =5.475V
11. R_d = 112Ω
12. V_z =5.55V
13. P_z = 0.555W
14. See text.
15. I_p = 30μA
16. V_r = 0.5V
17. Depletion region width changes.
18. V_R = 8V
19. R = 160Ω
20. I_f = 12.6mA; P_t = 441mW
21. Bulb burns out when the liquid level drops.
22. V = 19.7V
23. See text.
24. Check front to back resistance like a regular diode.

CHAPTER 2

Introduction to Bipolar Transistors

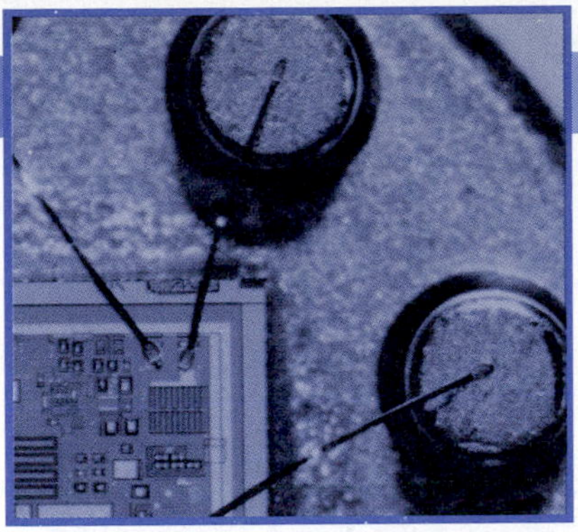

CHAPTER OBJECTIVES

After completing this chapter, you should be able to describe the following bipolar transistor concepts:

- Semiconductor construction of a transistor
- NPN and PNP transistor configurations
- Purpose of base, emitter, and collector
- Current gain of a transistor
- Voltage gain using a transistor
- Transistor biasing configurations
- Temperature effects on transistor biasing
- Determination of the internal base–emitter resistance
- Transistor circuit load line
- Troubleshooting a basic transistor circuit
- Transistor mechanical configurations

INTRODUCTION

A **bipolar transistor** is a device that amplifies (increases) current. This means that a very small current from a microphone can be used to control a larger current capable of powering a loudspeaker. In this chapter we will first look at how a transistor is able to amplify current, how we establish the DC (bias) currents necessary to allow the transistor to amplify, and then we shall look at a practical transistor biasing circuit.

TRANSISTOR CONSTRUCTION

You will recall that a diode was constructed by doping a piece of silicon with elements to create an N region and a P region. If we carry this process a step further by adding another N or P region, we can create a device called a (bipolar) transistor. Figure 2–1 shows **NPN transistor construction**.

There are three connections to a transistor: emitter, base, and collector. Since we have two junctions between the N, P, and N regions, we essentially have two diodes, as shown in Figure 2–1b. The center P region, attached to the base, is lightly doped and very narrow so the two "diodes" interact with one another.

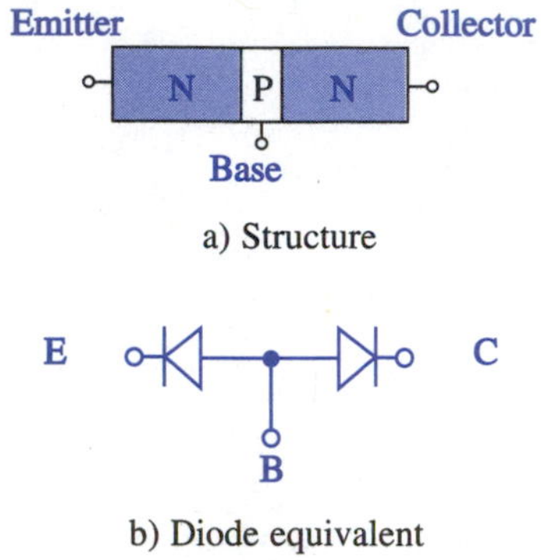

FIGURE 2–1
NPN transistor

TRANSISTOR OPERATION

To enable a transistor to amplify, the base-emitter diode junction must be forward biased (brought into conduction) and the collector-base diode junction reverse biased, as shown in Figure 2–2.

If we consider the junctions separately, the base supply battery V_{bb} provides a positive voltage at the base P region and a negative voltage at the emitter N region. This means that conventional current flows from the base to the emitter, while electrons flow from emitter to base. The collector supply battery V_{cc} is at a higher potential than V_{bb} and provides a positive voltage at the N region collector. This junction therefore is reverse biased and acts like a reverse-biased diode, so only a small reverse leakage current flows (nano-amps).

Now let us consider the interaction between the two junctions. With the base-emitter junction forward biased, many electrons are emitted from the emitter into the narrow and lightly doped base region. Being lightly doped means there are few atoms in the base region that are short of electrons, and therefore the capturing of electrons is very small.

The emitted electrons build up in the base region and can move two ways—sideways through the narrow base region to the positive terminal of V_{bb} or across the collector-base junction to the positive terminal of V_{cc}. Since the collector battery terminal has a much more positive voltage than the base terminal, and the path to the collector is shorter, the majority of the electrons jump over to the collector (it collects the electrons). The result is a small current I_b through the base terminal and a much higher current I_c through the

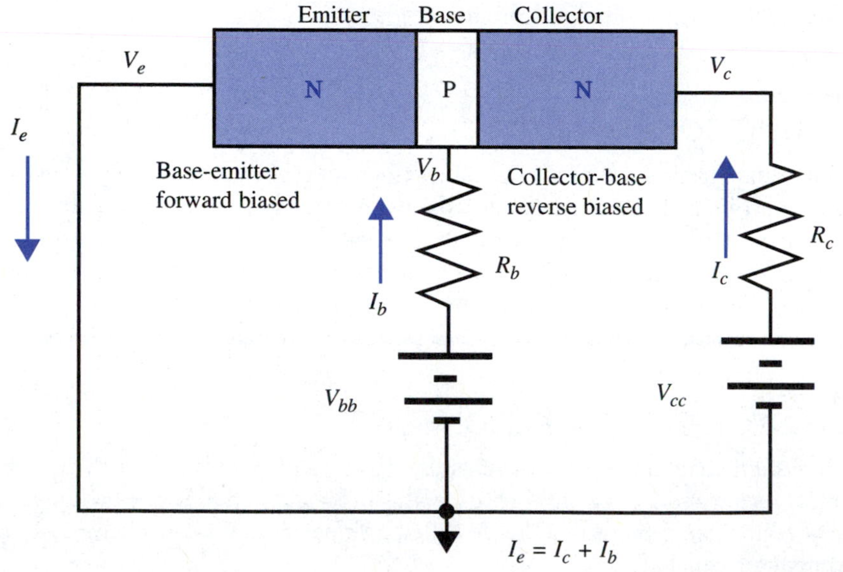

FIGURE 2–2 NPN transistor biasing

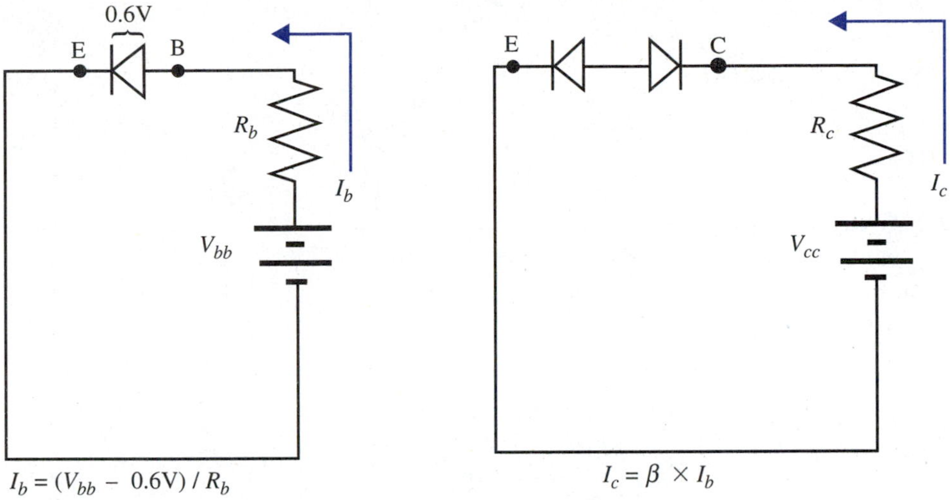

a) Base-emitter b) Collector-emitter

FIGURE 2–3 Base-emitter and collector-emitter equivalent circuits

collector terminal. These two currents are related by a factor called beta (β), which is the ratio of the two currents.

$$\beta = \frac{I_c}{I_b} \qquad \text{(equation 2–1)}$$

β indicates the ratio of the collector current to the base current, and this is specified on transistor data sheets (also called h_{fe}). β, being a ratio of currents, has no units.

We can look at this ratio another way and say that the amount of collector current is

$$I_c = (\beta)(I_b) \qquad \text{(equation 2–2)}$$

This relationship shows that the collector current is the base current amplified (increased) by the factor of β.

Let us now break out from Figure 2–2 the base–emitter loop circuit consisting of the base supply battery V_{bb} in series with the resistor R_b and the base–emitter diode. This is shown in Figure 2–3a.

The collector–emitter loop circuit is shown in Figure 2–3b, and current flows from the collector supply battery V_{cc}, through the resistor R_c and the reverse-biased collector-to-base diode, and finally through the forward-biased base-emitter diode. We can't find the collector current by using Ohm's law as we can with the base current, because we don't know the collector-to-emitter reverse voltage. But we can use the relationship

$$I_c = \beta I_b$$

EXAMPLE

If the value of I_b is 20μA and β is 150, what is the collector current I_c?

SOLUTION

Using equation 2–2

$$I_c = \beta \times I_b$$

$$I_c = 150 \times 20\mu A$$

$$I_c = 3mA$$

EXAMPLE

What base current will provide a collector current of 20mA if β is 200?

SOLUTION

Solving equation 2–2 for I_b:

$$I_b = \frac{I_c}{\beta}$$

$$I_b = \frac{20mA}{200}$$

$$I_b = 100\mu A$$

Looking back to Figure 2–2, we can write a Kirchhoff's current loop equation by looking at the direction of the arrows on the diagram.

$$I_e = I_c + I_b \qquad \textbf{(equation 2–3)}$$

Since

$$I_b = \frac{I_c}{\beta}$$

the equation now becomes

$$I_e = I_c\left(1 + \frac{1}{\beta}\right) \qquad \textbf{(equation 2–4)}$$

Another way of expressing the relationship between I_e and I_c, rather than equation 2–4, is to use the term *alpha* (α), where:

$$\alpha = \frac{I_c}{I_e} \qquad \textbf{(equation 2–5)}$$

Since the emitter current is the sum of the base and collector currents, α is always less than unity (1).

EXAMPLE

If the β of a transistor is 200 and the base current 30μA, find the collector current, emitter current, and α.

SOLUTION

First using equation 2–2

$$I_c = \beta \times I_b$$

$$I_c = 200 \times 30\mu A$$

$$I_c = 6mA$$

Now using equation 2–3

$$I_e = I_c + I_b$$

$$I_e = 6mA + 30\mu A$$

$$I_e = 6.03mA$$

Note: Since β is usually 100 or greater, we can say for practical purposes that

$$I_e \cong I_c$$

From equation 2–5

$$\alpha = \frac{I_c}{I_e}$$

$$\alpha = \frac{6mA}{6.03mA}$$

$$\alpha = 0.995$$

The things to remember from our look at the construction of the transistor are that, in operation, the base–emitter junction is forward biased (0.6V drop) and the base–collector junction reverse-biased. But most importantly, because of the β factor, the collector current is controlled by the base current, and since β can range from about 50 to about 300, (for small signal transistors) a small base current controls (generates) a larger collector current.

If we change the base current by a small amount (by changing V_{bb}), this change is multiplied (amplified) by the collector current ($\beta \times I_b$)—that is, the change in the base current is multiplied by β!

If we had built our transistor with two P regions and one N region, we would have had a **PNP transistor**. The current flows will be reversed for the PNP, but we get the same β relationship between I_c and I_b.

AN ACTUAL TRANSISTOR CIRCUIT

Let us look further at transistor action by studying an actual NPN transistor circuit. The collector, base, and emitter are identified by *C*, *B*, and *E* respectively in Figure 2–4; voltages at these terminals are measured with respect to power supply common (ground).

Since it is always represented by an arrow, the emitter can always be identified on a schematic. If the emitter arrow points out, as in Figure 2–4, then it is an NPN transistor. Arrow pointing inwards signifies a PNP transistor. Like the diode, the emitter arrow points in the direction of conventional current.

The resistor in the base circuit is R_b while the resistor between the collector and the power supply voltage is R_c.

Just as *R* is the designator for a resistor, *Q* is the identifier for a transistor (Q_1 in Figure 2–4)

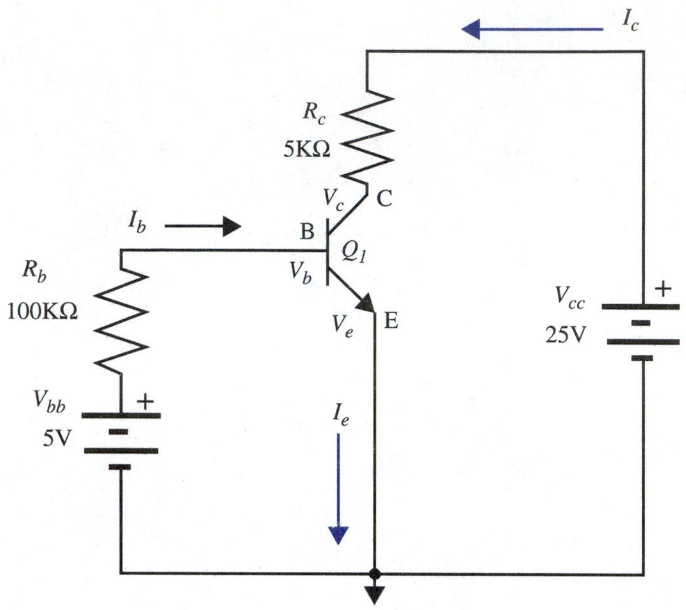

FIGURE 2–4 NPN transistor circuit

EXAMPLE

Find the base current and collector current for the circuit of Figure 2–4. Also if the base battery voltage V_{bb} is decreased by 0.1V, find the change in the base and collector currents and the change in the collector voltage.

SOLUTION

We can determine the base current by writing a base circuit loop equation:

$$V_{bb} = I_b R_b + V_{be}$$

Solving for I_b:

$$I_b = \frac{(V_{bb} - V_{be})}{R_b}$$

$$I_b = \frac{(5V - 0.6V)}{100K\Omega}$$

$$I_b = 44\mu A$$

Then

$$I_c = \beta \times I_b$$

$$I_c = 100 \times 44\mu A$$

$$I_c = 4.4mA$$

Now if we drop the base battery V_{bb} by 0.1V to 4.9V, the new base current is

$$I_b = \frac{4.9V - 0.6V}{100k\Omega}$$

$$I_b = 43\mu A$$

Base current change is

$$(44\mu A - 43\mu A) = 1\mu A$$

The new collector current is

$$I_c = \beta \times I_b$$

$$I_c = 100 \times 43\mu A$$

$$I_c = 4.3mA$$

Collector current change is then

$$4.4mA - 4.3mA = 0.1mA$$

The change in the collector voltage (ΔV_c) is the current change (ΔI_c) times R_c or

$$\Delta I_c R_c = \Delta V_c$$

$$\Delta V_c = 0.1mA \times 5K\Omega$$

$$\Delta V_c = 0.5V$$

The change of 0.1V in the base circuit resulted in a 0.5V change in the collector circuit. This is considered a voltage gain of 5. So current gain can be translated into voltage gain! Note that the decrease in voltage in the base circuit caused an increase in the collector voltage—this is called inverting voltage gain and is indicated by a negative sign (–).

Note: The voltage at the collector, or the value of the collector resistor, has no effect on the collector current, provided the collector-to-emitter voltage stays greater than about 2V to allow transistor action (collection) to occur.

VOLTAGE AMPLIFICATION WITH A TRANSISTOR

Transistors are more often used to obtain voltage gain rather than current gain. We shall use the abbreviation A_v for voltage gain (amplification of voltage).

Let us start our consideration of transistor gain by looking at the configuration we used in Figure 2–4. This circuit is called the *common emitter circuit* (so called because there is no input signal or output signal on this terminal, i.e., the emitter is AC grounded).

In Figure 2–5a, the signal voltage v_s to be amplified is connected to the base lead, and both junctions are biased (base–emitter forward, collector–base reverse) by the single power supply voltage V_{cc}. Notice, that V_{cc} is used to designate the supply voltage, V_c the collector voltage, V_b the base voltage, and V_e the emitter voltage—all measured to ground, i.e., power supply common.

Base current I_b is determined by V_{cc}, R_b, and the base-emitter drop (0.6V). The collector current is found from I_b times the β taken from data sheets for the particular transistor. We shall find we are interested in the DC voltage at the collector (V_c) because this determines how much AC voltage swing we can develop at the collector when amplifying AC signals.

EXAMPLE

Find the collector voltage V_c for the circuit of Figure 2–5a with the assumed β of 100.

SOLUTION:

First we need to find the base current I_b. Neglecting the small voltage drop across r_e, which we shall discuss later, and writing the base circuit loop equation (Figure 2–5b)

$$V_{cc} = I_b R_b + V_{be}$$

Solving for I_b:

$$I_b = \frac{V_{cc} - V_{be}}{R_b}$$

$$I_b = \frac{25\text{V} - 0.6\text{V}}{390\text{K}\Omega}$$

$$I_b = 62.6\mu\text{A}$$

But from equation 2–2

$$I_c = \beta \times I_b$$

$$I_c = 100 \times 62.6\mu\text{A}$$

$$I_c = 6.26\text{mA}$$

This current will cause a voltage drop V_{R_c} across R_c of

$$V_{R_c} = I_c \times R_c$$

$$V_{R_c} = 6.26\text{mA} \times 2\text{K}\Omega$$

$$V_{R_c} = 12.5\text{V}$$

The collector voltage V_c is equal to the supply voltage V_{cc} minus the drop across R_c (V_{R_c}), or

$$V_c = V_{cc} - V_{R_c}$$

$$V_c = 25\text{V} - 12.5\text{V}$$

$$V_c = 12.5\text{V}$$

So the collector voltage for this circuit is at one-half of the supply voltage. We shall see that a collector voltage equal to one-half of the supply voltage allows an AC

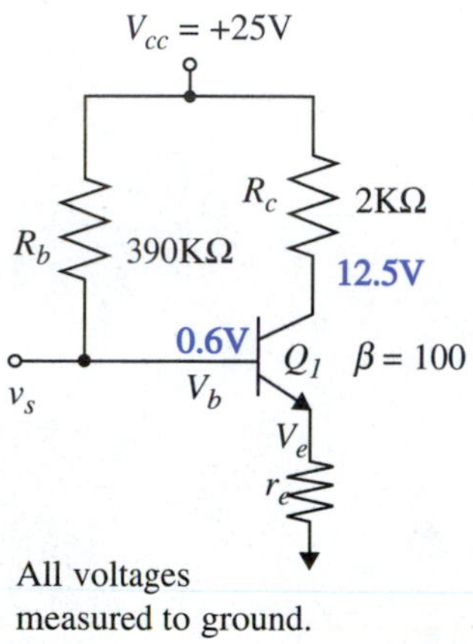

a) Actual circuit

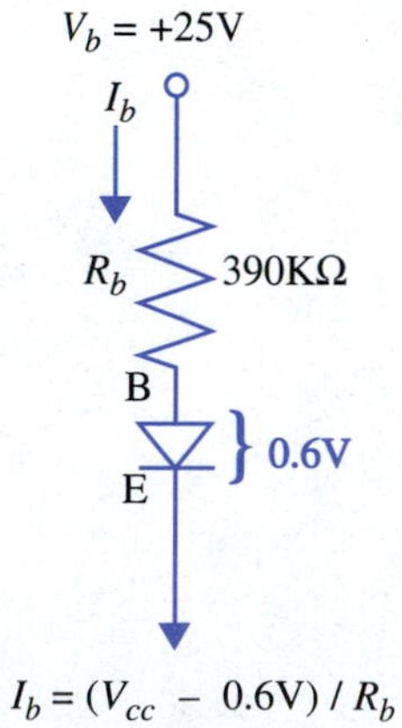

b) Base equivalent

FIGURE 2–5
Common emitter circuit

signal to swing equally positive and negative around this quiescent (Q) collector DC voltage level.

Before we can consider how the signal v_s is amplified, we need to define the additional resistor r_e in the emitter circuit of Figure 2–5. This resistor is the internal ohmic resistance of the emitter region and can be determined from the diode equation as follows:

$$I_f = I_r\,[e^{(qV_f/KT)} - 1] = I_r e^{(qV_f/KT)}$$

since

$$e^{(qV_f/KT)} >> 1$$

Taking the calculus derivative (see note below):

$$dI_f = \frac{I_r\, q}{KT}\, e^{(qV_f/KT)}\, dV_f.$$

dI_f and dV_f are used to represent small changes in the forward voltage and current. The diode resistance r_d is dV_f/dI_f, which is the reciprocal of the slope of the diode curve (see Figure 1–7).

$$r_d = \frac{dV_f}{dI_f} = \frac{KT}{qI_r\, e^{(qV_f/KT)}}$$

Substituting

$$I_f = I_r\, e^{(qV_f/KT)}$$

$$r_d = \frac{KT}{qI_f}$$

and since at 28°C,

$$KT/q = 26\text{mV (25mV at 20°C)}$$

$$r_d = \frac{26\text{mV}}{I_f}$$

Now relating this to the transistor

$$r_d = r_e \text{ and } I_f = I_e$$

So

$$r_e = \frac{26\text{mV}}{I_e} \text{ at 28°C)}$$ **(equation 2–6)**

(*Note:* It is not necessary to understand the derivation of equation 2–6. It is provided for those who prefer mathematical derivations of equations.)

EXAMPLE

Find the internal ohmic resistance r_e for the transistor used in the circuit of Figure 2–6.

SOLUTION

First we need to find the emitter current in order to use equation 2–6. We do this by first finding the base and then the collector current.

Base current

$$I_b = \frac{V_{cc} - V_{be}}{R_b}$$

$$I_b = \frac{25V - 0.6V}{390K\Omega}$$

$$I_b = 62.6\mu A$$

and

$$I_c = \beta \times I_b$$

$$I_c = 100 \times 62.6\mu A$$

$$I_c = 6.26mA$$

and since β is 100,

$$I_e \approx I_c.$$

$$I_e = 6.26mA$$

$$r_e = \frac{26mV}{6.26mA}$$

$$r_e = 4.15\Omega$$

Note: The larger the value of the emitter current, the smaller the value of r_e.

The resistor r_e in Figure 2–6 will have a DC voltage drop across it of 6.26mA × 4.15Ω = 26mV, which is small compared to the 0.6V diode drop from base to emitter. We shall see that this resistor does have an effect, however, when we later look at the transistor voltage gain equations.

EFFECT OF AN AC SIGNAL AT THE BASE

Let us assume that a positive going input signal voltage (v_s in Figure 2–6) is applied between the base and circuit ground. The base will rise up more positive than the 0.6VDC base-emitter bias voltage by this amount v_s. But the base-emitter diode drop is fixed at 0.6V, so where does the increased voltage v_s go? It must appear across the internal r_e, as is shown in Figure 2–6.

Now if we assume that the input signal v_s is +10mV, the peak voltage on the base will be 0.6V + 10mV = 610mV; while the peak voltage across the internal emitter resistance r_e will be 26mV + 10mV = 36mv. This 10mv increase in voltage across the internal emitter resistance r_e will cause 10mV/4.15Ω or 2.41mA additional emitter current to flow. Since $I_e \approx I_c$, this additional current (ΔI_c) will also appear in the collector circuit and increase the voltage drop (ΔV_{Rc}) across the collector 2KΩ resistor by

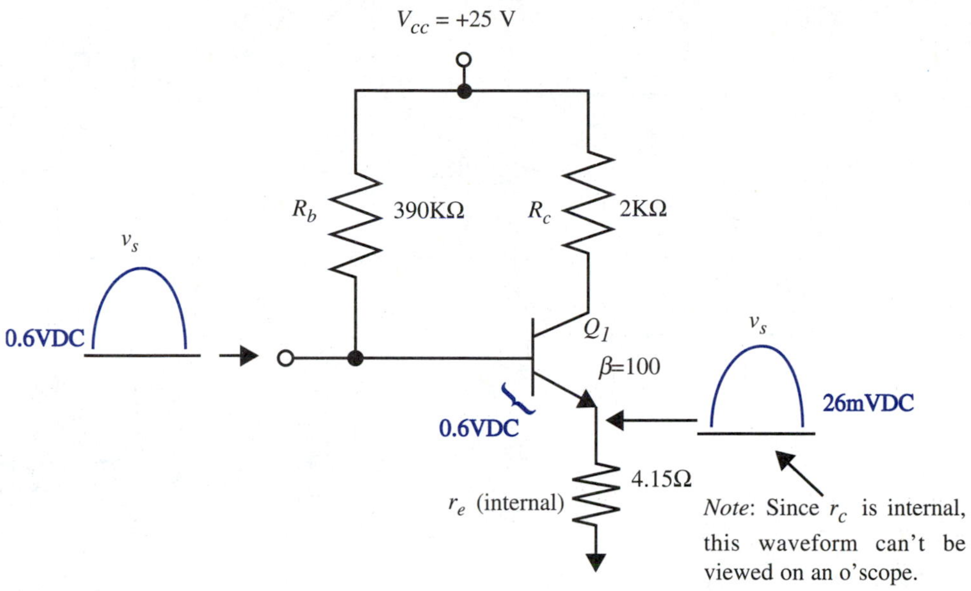

FIGURE 2–6 Common emitter with signal applied

$$\Delta V_{Rc} = \Delta I_c \times R_c$$

$$\Delta V_{Rc} = 2.41\text{mA} \times 2\text{K}\Omega$$

$$\Delta V_{Rc} = 4.82\text{V}$$

We can start to get a feel from this result about the voltage gain (A_v) possibilities of a transistor, because our increase of 10mV at the base became a 4.82V change at the collector for a voltage gain A_v of

$$A_v = \frac{\Delta V_{Rc}}{v_s}$$

$$A_v = \frac{4.82\text{V}}{10\text{mV}}$$

$$A_v = 482$$

So what is the maximum voltage change can we get at the collector of a transistor? Well, if we drop the base input voltage by 26mV to the point where there is no longer any voltage across r_e then the emitter current will go to zero and the transistor will turn off. With no current flowing through R_c, the collector voltage (V_c) will be the same as the supply voltage V_{cc}.

Now, if we increase the input signal voltage v_s at the base, the transistor base current and therefore the collector current will increase, and more voltage drop will occur across the collector resistor. A point is reached where almost all the supply voltage (V_{cc}) is dropped across the collector resistor and very little (0.2V) is across the transistor collector-to-emitter terminals—the transistor is then considered in **saturation**. If the col-

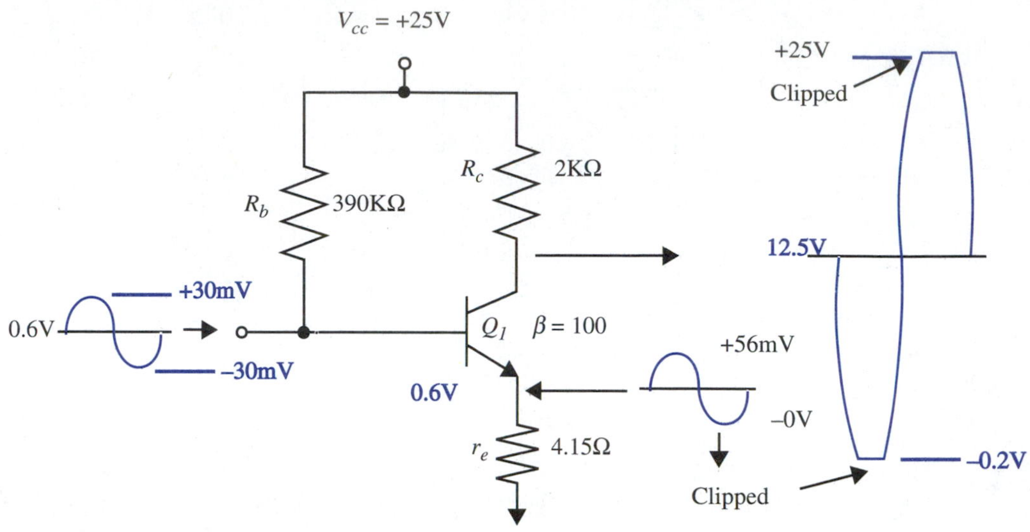

FIGURE 2–7 Overdriving the base circuit

lector voltage drops too low (less than about 1–2V), it becomes a less-efficient collector of electrons, and the current gain (β) drops. So when amplifying input voltages, we should allow a maximum voltage swing on the collector from the supply voltage (V_{cc}) when the transistor is turning off, to about 2V above the emitter voltage with the transistor turned on.

If our input voltage to the base is a sine wave, and we allow the amplified collector voltage to swing too far, it will cause clipping of the sine wave output. We can demonstrate this by putting into the base a sine wave voltage v_s that swings 30mV positive and 30mV negative from the 0.6V bias value. (See Figure 2–7.)

On the positive part of the input waveform, the peak current change (ΔI_e) through the emitter resistor is

$$\Delta I_e = \frac{v_s}{r_e}$$

$$\Delta I_e = \frac{30\text{mV}}{4.15\Omega}$$

$$\Delta I_e = 7.23\text{mA}$$

This gives a peak voltage change (ΔV_{R_c}) across the collector resistor of:

$$\Delta V_{R_c} = 7.23\text{mA} \times 2\text{K}\Omega$$

$$\Delta V_{R_c} = 14.46\text{V}$$

But before we put in the sine wave at the base, the collector DC voltage was 12.5V—so we can't drop the collector by more than this voltage. If we do, the result is that the bottom of the sine wave on the collector is clipped off at about 0.2V above ground (saturation level).

Consider now the negative portion of the input sine wave. (Remember, without the sine wave input, the DC voltage across the emitter resistance r_e was 26mV.) Now, if the input waveform drops 30mV, the transistor turns off because there is no longer any voltage across the resistance r_e. Therefore there is no I_e or I_c. The result is clipping at the top of the output waveform.

So we have learned from the material just covered that with too much base input signal (>±26mV), the output signal is amplified but distorted. (*Note:* We cannot measure the true voltage gain of a transistor amplifier if the output waveform is distorted.)

$$\text{Voltage gain} = A_V = \frac{\text{Undistorted output}}{\text{Undistorted input}} \qquad \textbf{(equation 2–7)}$$

Also we have learned that for maximum output voltage swing at the collector, the collector Q (quiescent) bias voltage V_c should be about one-half of the supply voltage.

TRANSISTOR BIASING CIRCUITS

Choosing the resistor values for desired values of DC voltages at collector, base, and emitter is called **biasing** the transistor. Biasing enables us to achieve maximum output voltage swing without distortion, or it provides a specific DC voltage at the collector as required by a second circuit that is connected to the collector.

We saw in our analysis of Figure 2–7 that having the collector DC voltage at one-half of the supply voltage allows the desired equal AC voltage swings in the both the positive and negative directions. In this circuit, the base current was established by the value of the power supply voltage (V_{cc}) and the base resistor (R_b). The collector current was β times the base current. Finally the collector voltage was found by multiplying the collector current by R_c and subtracting this value from the power supply voltage. In this analysis, we assumed, however, a β value of 100—but β can vary from about 50 to 300 for the same type of signal transistors.

EXAMPLE

Compute the collector voltage for the circuit of Figure 2–7 if β is 50 and also if β is 300.

SOLUTION

$$V_c = V_{cc} - I_c R_c$$

but

$$I_c = \beta \times I_b$$

so

$$V_c = V_{cc} - (\beta \times I_b \times R_c)$$

Our previous calculations determined that

$$I_b = 62.6\mu\text{A}$$

If we now compute the collector voltage with a β of 50,

$$V_c = 25V - (50 \times 62.6\mu A \times 2K\Omega)$$

$$V_c = 18.74V$$

which is quite a bit higher than the desired 12.5V.

Assuming a β of 300,

$$V_c = 25V - (300 \times 62.6\mu A \times 2K\Omega)$$

$$V_c = -12.56V$$

This negative voltage indicates that the drop across the collector resistor is too great and the transistor is in saturation!

Both of these collector bias voltages would cause severe clipping if we had an input signal at the base that caused a large voltage swing on the collector.

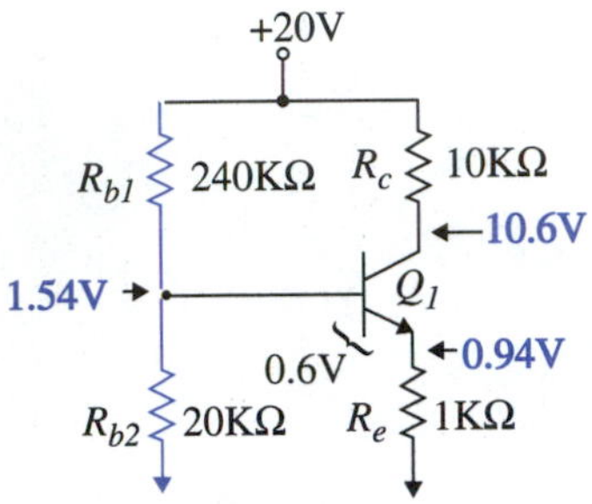

FIGURE 2–8

β independent (voltage divider) biasing

We need, then, a circuit that provides a DC voltage at the collector but is less sensitive to transistor β variations and temperature effects. A circuit that meets this requirement is shown in Figure 2–8. This is sometimes called **voltage divider biasing** or **universal biasing**. This circuit has an added resistor in the base circuit (R_{b2}) and also an added emitter resistor (R_e). The reason the biasing of this circuit is less sensitive to β changes is that the base voltage divider sets up a fixed voltage at the base (V_b) which in turn provides a fixed voltage V_e across R_e of

$$V_b - 0.6V$$

This voltage V_e establishes the value of the emitter current, which is V_e / R_e, and therefore the collector current since

$$I_e \approx I_c$$

This collector current sets the collector bias voltage V_c since

$$V_c = V_{cc} - I_c R_c.$$

EXAMPLE

Find the voltage at the collector for the circuit in Figure 2–8.

SOLUTION

We shall first assume with this circuit that the base current is negligible and compute the voltage V_b at the base:

$$V_b = \frac{V_{cc} \times R_{b2}}{(R_{b1} + R_{b2})}$$

$$V_b = \frac{20V \times 20K\Omega}{240K\Omega + 20K\Omega}$$

$$V_b = 1.54V$$

Subtracting the 0.6V drop across the base–emitter diode junction, we are left with an emitter voltage of

$$V_e = 1.54V - 0.6V$$

$$V_e = 0.94V$$

Now the emitter current I_e is

$$I_e = \frac{V_e}{R_e}$$

$$I_e = \frac{0.94V}{1K\Omega}$$

$$I_e = 0.94mA$$

Assuming again that with high β, $I_c \approx I_e$, then 0.94mA is also flowing in the collector circuit. This causes a 9.4V drop across the 10KΩ R_c resistor; so the collector voltage is

$$V_c = V_{cc} - (I_c \times R_c)$$

$$V_c = 20V - 9.4V$$

$$V_c = 10.6V$$

This voltage is roughly one-half of V_{CC}.

Notice that in the computations in the example, we did not use a value of β. So our collector voltage is independent of β! But we did ignore the base current when we computed V_b. Let us now see how much our results change if we include the base current in our calculations.

In order to consider the effect of the base current on the voltage at the base, we need to consider the input circuitry, as shown in Figure 2–9. The actual circuit is shown in Figure 2–9a.

We can move the emitter resistor to the base circuit by multiplying it by β ($R_t = \beta R_e$). The reason for doing this is to have the same current I_b through all components in the equivalent circuit, which makes it easier to compute the voltage drops.

$$R_t = (\beta)(R_e) \qquad \textbf{(equation 2–8)}$$

The equivalent circuit shown in Figure 2–9b has the base–emitter diode in series with the equivalent resistor R_t.

$$V_b = 0.6V + V_e$$

$$V_e = (I_b)\ (R_t).$$

$$R_t = \beta R_e$$

$$V_e = I_b\ (\beta R_e)$$

which, if $I_e \approx I_c$, is the same as

$$V_e = I_e R_e$$

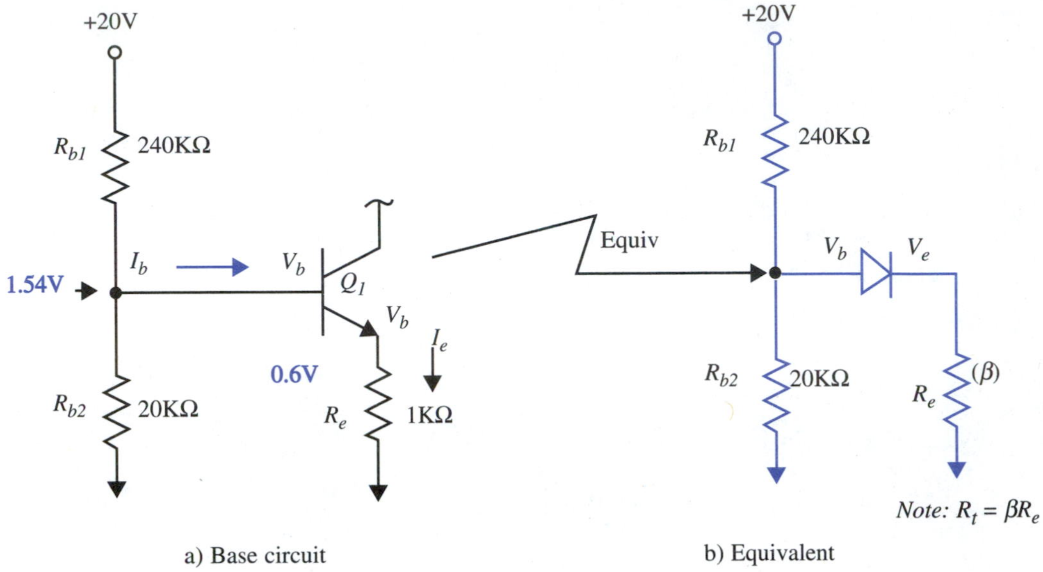

FIGURE 2–9 Base input circuitry

What this means is with the emitter resistor translated to the base circuit, we get the same voltage drop (V_e) with I_b flowing through a resistor βR_e as we did with I_e flowing through R_e.

To determine the actual base voltage V_b, we can make a Thevenin's equivalent of R_{b1} and R_{b2}, as shown in Figure 2–10. The Thevenin's voltage V_{th} is simply the voltage at the midpoint of the R_{b1} and R_{b2} divider—while the Thevenin's resistance R_{th} is determined by shorting out the voltage source and finding the parallel resistance of R_{b1} and R_{b2}.

EXAMPLE

Find the base current I_b and base voltage V_b for the circuit in Figure 2–9 with β values of 50, 100, and 300.

SOLUTION

We first need to find the Thevenin's equivalent of R_{b1} and R_{b2} (see Figure 2–10):

$$R_{th} = R_{b1} \parallel R_{b2} \qquad V_{th} = \frac{V_{cc} \times R_{b2}}{R_{b1} + R_{b2}}$$

$$R_{th} = 240\text{K}\Omega \parallel 20\text{K}\Omega \qquad V_{th} = \frac{20\text{V} \times 20\text{K}\Omega}{20\text{K}\Omega + 240\text{K}\Omega}$$

$$R_{th} = 18.5\text{K}\Omega \qquad V_{th} = 1.54\text{V}$$

Writing the loop equation for the equivalent base circuit from Figure 2–10:

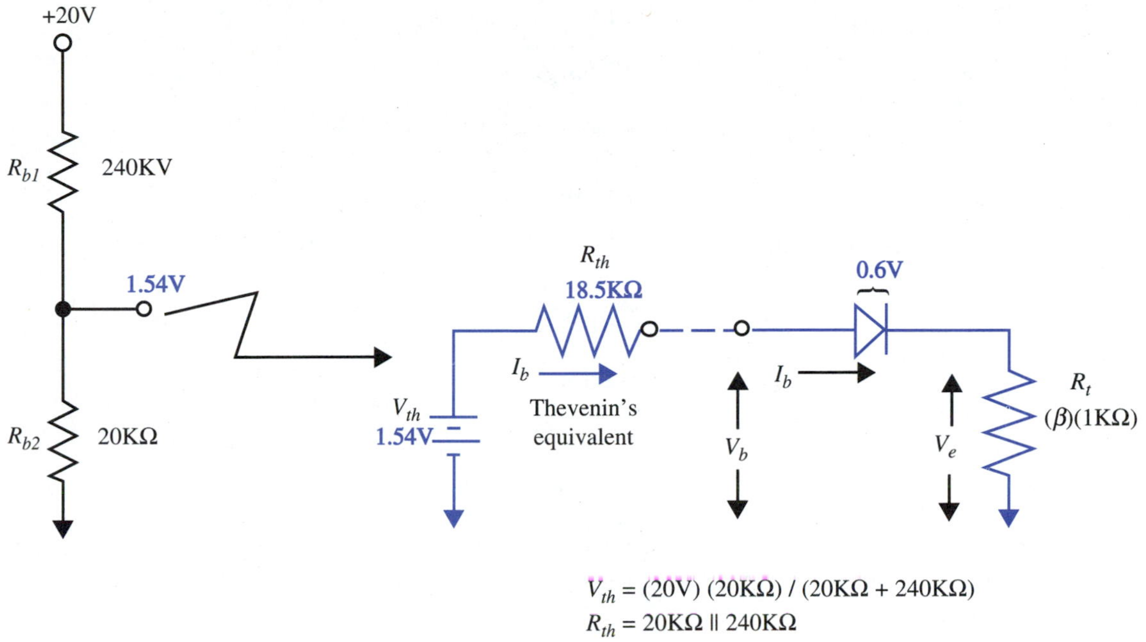

FIGURE 2–10 Base equivalent circuit

$$V_{th} = (I_b \times 18.5\text{K}\Omega) + 0.6\text{V} + (I_b \times \beta \times 1\text{K}\Omega)$$

Solving for I_b:

$$I_b = \frac{V_{th} - 0.6\text{V}}{18.5\text{K}\Omega + (\beta)(1\text{K}\Omega)}$$

With β = 50

$$I_b = \frac{1.54\text{V} - 0.6\text{V}}{18.5\text{K}\Omega + 50\text{K}\Omega}$$

$$I_b = 13.7\mu\text{A}$$

Base V_b is:

$$V_b = V_e + 0.6\text{V}$$

but

$$V_e = (I_b)(R_t) = (I_b)(\beta)(1\text{K}\Omega)$$

$$V_b = (I_b)(\beta)(1\text{K}\Omega) + 0.6\text{V}$$

$$V_b = (13.7\mu\text{A})(50)(1\text{K}\Omega) + 0.6\text{V}$$

$$\mathbf{V_b = 1.29V}$$

With β = 100

$$I_b = \frac{1.54\text{V} - 0.6\text{V}}{18.5\text{K}\Omega + 100\text{K}\Omega}$$

$$\mathbf{I_b = 7.93\mu A}$$

$$V_b = (7.93\mu\text{A})(100)(1\text{K}\Omega) + 0.6\text{V}$$

$$\mathbf{V_b = 1.39V}$$

With β = 300

$$I_b = \frac{1.54\text{V} - 0.6\text{V}}{18.5\text{K}\Omega + 300\text{K}\Omega}$$

$$\mathbf{I_b = 2.95\mu A}$$

$$V_b = (2.95\mu\text{A})(300)(1\text{K}\Omega) + 0.6\text{V}$$

$$\mathbf{V_b = 1.49V}$$

From these results we conclude that a higher β results in a lower base current and therefore a higher base voltage because the reflected resistance $(\beta)(R_e)$ is higher.

EXAMPLE

Find the collector voltage with β values of 50, 100, and 300 for the circuit in Figure 2–8.

SOLUTION

$$V_c = V_{cc} - (R_c \times I_c)$$

But

$$I_c = \beta \times I_b \qquad V_c = V_{cc} - (R_c \times \beta \times I_b)$$

Using the values of I_b determined for the various β values in the previous example:

For **β = 50**

$$V_c = 20\text{V} - (10\text{K}\Omega \times 50 \times 13.7\mu\text{A})$$

$$\mathbf{V_c = 13.2V}$$

For **β = 100**

$$V_c = 20\text{V} - (10\text{K}\Omega \times 100 \times 7.93\mu\text{A})$$

$$\mathbf{V_c = 12.1V}$$

For $\beta = 300$

$$V_c = 20\text{ V} - (10\text{K}\Omega \times 300 \times 2.95\mu\text{A})$$

$$\mathbf{V_c = 11.2V}$$

So a higher β results in a lower collector voltage because the collector current is larger.

We can summarize the results of the previous two examples in Table 2–1. The values of the emitter voltages were determined by subtracting 0.6V from the base voltages.

TABLE 2–1 **β Effect on Voltage Divider Bias Voltages**

β	50	100	300
V_c	13.2V	12.1V	11.2V
V_c	0.69V	0.79V	0.89V
V_c	1.29V	1.39V	1.49V

From Table 2–1 we conclude that even with a drastic change in β, we see a small change in our collector voltage operating point. This confirms our original claim that the circuit in Figure 2–8 was much less sensitive to β variations than the circuit shown in Figure 2–6.

PNP TRANSISTOR BIASING

We can recall that the PNP transistor had junctions that are of the opposite type from those of the NPN transistor—the base being of N material and the collector and emitter being of P material. So we would expect that the bias voltages would be of reversed polarity from those used with the NPN transistor.

Figure 2–11 shows how a PNP transistor is connected with the same circuit component values as the NPN transistor in Figure 2–8.

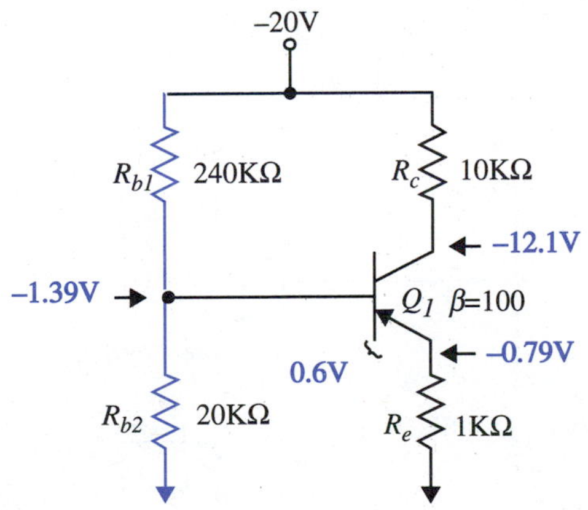

FIGURE 2–11 PNP transistor biasing circuit—negative supply

Notice that the only difference is that a negative 20V supply is used rather than a positive supply. The calculated base voltage with a β of 100 will be –1.39V rather than the +1.39V calculated previously with the NPN transistor. The emitter voltage is 0.6V more positive than the base or –0.79V, and the collector is –12.1V.

Suppose we required that both NPN and PNP transistors be used in a circuit and only a +20V supply is available; then the PNP transistor would be connected as shown in Figure 2–12.

Notice that the transistor is inverted in this circuit. The emitter being closest to the positive supply and the bottom end of the collector resistor is tied to the power supply common.

EXAMPLE

Find the voltages at base-emitter and collector for the circuit in Figure 2–12 with an assumed β of 100.

SOLUTION

Let us first generate a Thevenin's equivalent of R_{b1} and R_{b2} as shown in Figure 2–13.

Writing the voltage loop equation from the equivalent circuit

$$V_{cc} = (I_b)(R_t) + 0.6V + (I_b)(R_{th}) + V_{th}$$

Solving for I_b:

$$I_b = \frac{V_{cc} - 0.6V - V_{th}}{R_t + R_{th}}$$

Since

$$R_{th} = 18.5K\Omega$$

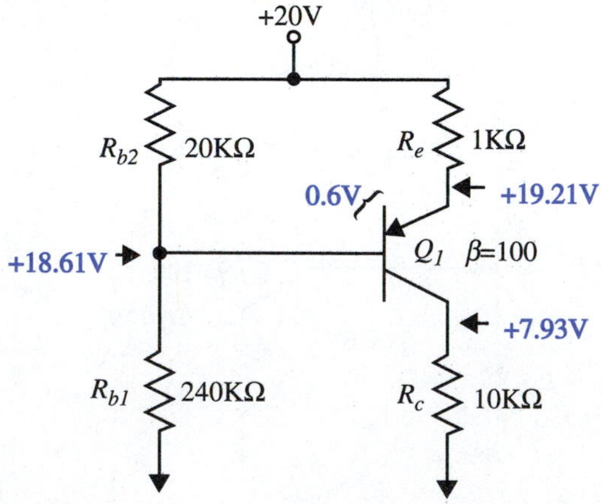

FIGURE 2–12 PNP transistor biasing circuit—positive supply

and

$$R_t = (\beta)(1K\Omega)$$

$$I_b = \frac{20V - 0.6V - 18.46V}{(100)(1K\Omega) + 18.5K\Omega}$$

$$I_b = 7.93\ \mu A$$

But

$$V_e = V_{cc} - (I_b)(R_t)$$

$$V_e = 20V - (7.93\mu A)(100)(1K\Omega)$$

$$V_e = 19.21V$$

For a PNP

$$V_b = V_e - 0.6V$$

$$V_b = 18.61V$$

$$V_b = (I_c)(R_c)$$

but

$$I_c = (\beta)(I_b)$$

$$V_c = (100)(7.93\mu A)(10K\Omega)$$

$$V_c = 7.93V$$

These values of V_e, V_b, and V_c are shown on the schematic of Figure 2–12.

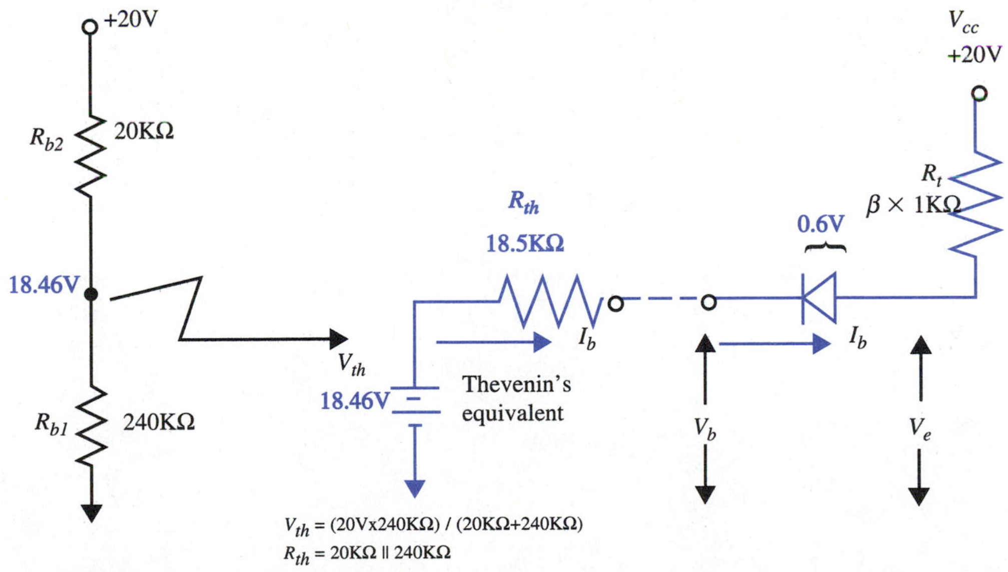

FIGURE 2–13 PNP Base equivalent circuit

COLLECTOR FEEDBACK BIASING

Another form of transistor biasing is the collector feedback type, shown in Figure 2–14. As you can see from the figure, this circuit has the advantage of using two fewer resistors than the fixed bias circuit discussed previously. However, the feedback bias method is not as stable with β changes as the β independent biasing method discussed previously.

Notice, in the figure, that the top end of the feedback resistor is connected to the collector rather than to the supply. The reason for this is that if the collector voltage drops, the following sequence occurs:

- Base current decreases.
- Collector current decreases.
- Collector voltage rises because less drop across R_C.

So the drop in collector voltage is compensated for by a resultant rise in collector voltage.

We can determine the sensitivity of the collector voltage to changes in β by developing an equation for V_c in terms of β.

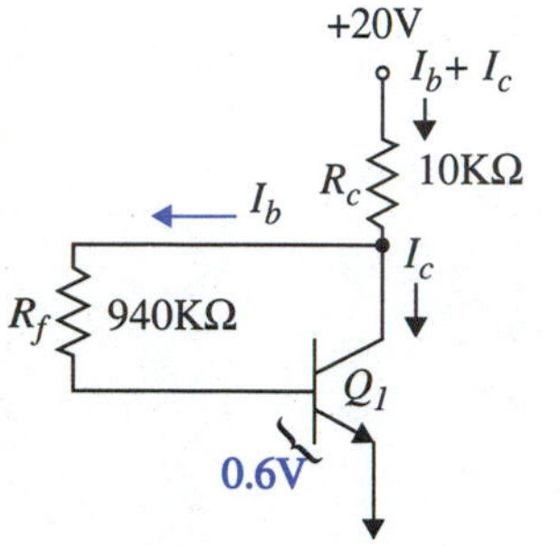

FIGURE 2–14

Collector feedback biasing

EXAMPLE

Compute the change in the collector voltage if we use β values of 50 and 300 for the transistor in the collector feedback bias circuit in Figure 2–14 and compare the results with the β independent biasing of Figure 2–8. Assume a midpoint collector voltage V_c of +10V (for maximum output voltage swing) when a transistor with a typical β of 100 is used.

SOLUTION

We shall determine R_f at the nominal β of 100.
From Figure 2–14

$$V_c = V_{CC} - (I_b + I_c)R_c$$

But

$$I_c >> I_b$$

so

$$V_c = V_{CC} - I_cR_c$$

Solving for I_c

$$I_c = \left(\frac{V_{CC} - V_c}{R_C}\right)$$

$$I_c = \frac{(20V - 10V)}{10K\Omega}$$

$$I_b = 1mA$$

And

$$I_b = \frac{I_c}{\beta}$$

$$I_b = 10\mu A$$

To find R_f we divide the voltage across this resistor ($V_c - 0.6V$) by the base current

$$R_f = \frac{(V_b - 0.6V)}{I_b} \qquad \textbf{(equation 2–9)}$$

$$R_f = \frac{(10V - 0.6V)}{10\mu A}$$

$$R_f = 940K\Omega$$

Now with a β of 50 we write the loop equation for the voltage drops from the power supply through the base circuit to ground remembering that

$$I_b = \frac{I_c}{\beta} = \frac{I_c}{50}$$

$$20V = (10K\Omega \times I_c) + (940K\Omega \times \frac{I_c}{50}) + 0.6V$$

Solving for I_c, with β of 50

$$I_c = \frac{20V - 0.6V}{10K\Omega + \frac{940K\Omega}{50}}$$

$$I_c = 0.674mA$$

Collector voltage

$$V_c = V_{cc} - (R_c \times I_c)$$

$$V_c = 20V - (10K\Omega \times 0.674mA)$$

$$V_c = 13.26V$$

Solving for I_c, with β of 300;

$$20V = (10K\Omega \times I_c) + (940K\Omega \times \frac{I_c}{300}) + 0.6\ V$$

$$I_c = 1.48mA$$

Collector voltage

$$V_c = 20V - (10K\Omega \times 1.48mA)$$

$$V_c = 5.2V$$

One can see from comparing these results of this example with Table 2–1 (on page 53) that collector feedback biasing doesn't provide the same bias stability afforded by the circuit of Figure 2–8. With a β of 50, the collector was 13.26V, which is +3.2V away from our desired value of V_c = 10V; with a β of 300 it is 5.2V or –4.8V away from the 10V value. The trade–off then is more resistors and better bias stability or fewer resistors (2) and greater change in the collector bias voltage if the transistor is changed.

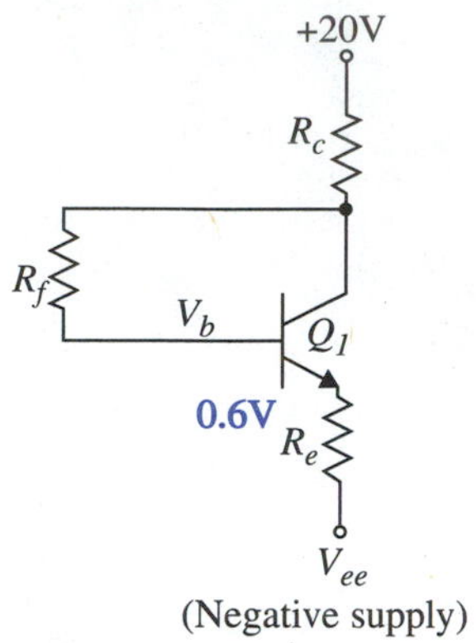

FIGURE 2–15
Improved collector feedback biasing

Adding a resistor R_e from emitter to ground in the circuit in Figure 2–14 increases the stability of the collector feedback circuit because it makes it closer to the β independent configuration.

Another method to stabilize the collector voltage is to connect the bottom end of this resistor R_e to a negative supply V_{ee} (positive supply with a PNP) rather than to ground, as shown in Figure 2–15. This allows us to use a larger value for R_e, and the emitter current (and therefore the collector current and collector voltage) becomes more a function of the value of R_e and the negative supply rather than the β of the transistor, as indicated in the following equation:

$$I_e = \frac{V_b - 0.6\text{V} - V_{ee}}{R_e}$$

With

$$V_{ee} >> V_b \text{ or } 0.6\text{V}$$

$$I_e \approx \frac{-V_{ee}}{R_e}$$

Since $I_e \approx I_c$ and β doesn't appear in the equation, the collector voltage is independent of β.

TEMPERATURE EFFECTS ON BIAS VOLTAGES

We saw earlier in Chapter 1 that the voltage drop across the silicon diode junction reduces 2.2mV for every one degree centigrade temperature increase. If we assume a 50°C increase in temperature from 25°C (77°F) to 75°C (167°F), the base–emitter diode drop is reduced from 0.6 V to

$$(0.6\text{V} - 2.2\text{mV} / °\text{C} \times 50°\text{C}) = 0.49\text{V}$$

If we recalculate the collector voltage for the β independent bias circuit (Figures 2–8 and 2–10) for this new base–emitter voltage with a β of 50, we get

$$I_b = \frac{V_{th} - V_{be}}{R_{th} + R_t}$$

$$I_b = \frac{1.54\text{V} - 0.49\text{V}}{18.5\text{K}\Omega + 50\text{K}\Omega}$$

$$I_b = 15.33\mu\text{A}$$

$$I_c = \beta \times I_b = 50 \times I_b = 0.766\text{mA}$$

And

$$V_c = V_{cc} - I_c R_c$$

$$V_c = 20\text{V} - (0.766\text{mA} \times 10\text{K}\Omega)$$

$$\boldsymbol{\beta = 50}$$

$$\mathbf{V_c = 12.34V} \text{ (was 13.2V @ 25°C)}$$

With a β of 300 we calculate

$$\beta = 300$$

$$\mathbf{V_c = 10.11V} \text{ (was 11.2V @ 25°C)}$$

From these results we can see that with the fixed bias (β independent) method, the collector voltage is relatively stable with temperature. Let us now look at the collector feedback bias changes with the same temperature increase.

Writing the loop equation from Figure 2–14

$$V_{cc} = R_c I_c + R_f \left(\frac{I_c}{\beta}\right) + V_{be}$$

First with a β of 50:

$$20\text{ V} = 10\text{K}\Omega \times I_c + 940\text{K}\Omega \left(\frac{I_c}{50}\right) + 0.49\text{V}$$

Simplifying

$$19.57\text{V} = 28.8\text{K}\Omega \times I_c$$

$$I_c = 0.677\text{mA}$$

and

$$V_c = 20\text{V} - 0.677\text{mA} \times 10\text{K}\Omega$$

$$\beta = 50$$

$$\mathbf{V_c = 13.2V} \text{ (was 13.26V @ 25°C)}$$

With β of 300 we calculate:

$$\beta = 300$$

$$\mathbf{V_c = 4.98V} \text{ (was 5.07V @ 25°C)}$$

These results show that the collector feedback bias, although less stable with β changes, is more stable with voltage 5.14V than the fixed bias configuration.

TRANSISTOR LOAD LINE

One way to help in the understanding of transistor operation is by use of a graphical aid that indicates how the collector voltage will change as a function of base current for given values of R_c and V_{CC}. Such an aid is a transistor load line, which is shown drawn on hypothetical transistor characteristics in Figure 2–16b. The end points for the load line are determined by first assuming the transistor is turned off ($I_c = 0$), which makes the collector voltage equal to V_{CC}, which is 20V; then it is assumed the transistor is full on ($I_c = V_{CC}/R_c$) which makes the collector voltage approximately equal to zero and the collector current 20V/10KΩ or 2mA.

We can look at R_c, the resistance in series with the collector, and the transistor effective resistance R_T (i.e., V_{ce}/I_c) as forming a voltage divider (V_{CC} is always divided between them—in other words, $V_{CC} = I_c R_c + I_c R_T$). The transistor then acts like a resistor—the value

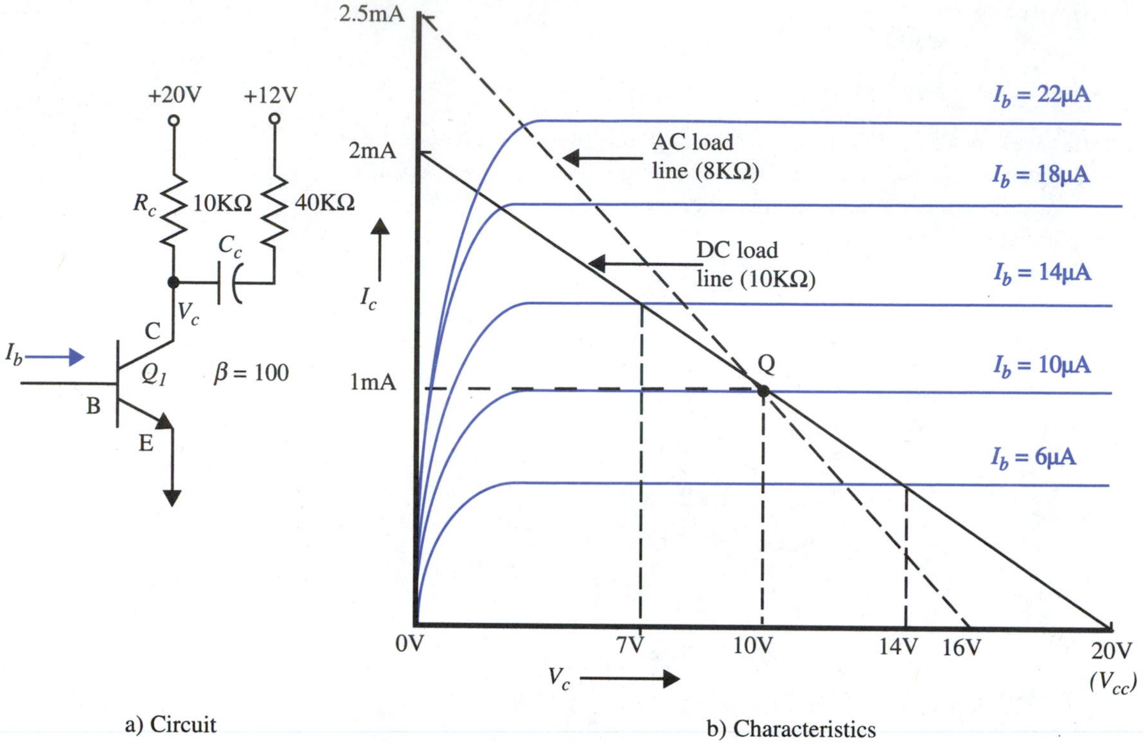

FIGURE 2–16 Transistor load line

of which varies with the base current. As the base current increases, the transistor resistance drops, which puts more voltage across R_c.

The *Q* (quiescent) point on the load line indicates the collector voltage V_c at the bias point set up by the base bias current (assumed to be 10μA). By dropping a vertical line at the *Q* point in the figure, we can see that the collector voltage is at 10V, which is one-half of V_{CC} and allows the equal positive and negative voltage swings on the collector required for high-level signals.

As the base current changes from the *Q* point bias value, we can determine the new collector voltage by drawing a vertical line from the intercept of the new base current and the load line.

EXAMPLE

Find the collector voltage for the circuit in Figure 2–16a if the base current is 14μA and also if the base current is 6μA.

SOLUTION

From the intercepts of the load line and these base currents in Figure 2–16b, we draw a vertical line to the V_c axis and read the collector voltage values.

For

$$I_b = 14\mu A$$

we read

$$V_c = 7V$$

With

$$I_b = 6\mu A$$

we read

$$V_c = 14V$$

So the lower the base current, the higher the effective resistance of the transistor, and the higher the value of V_c.

A load line can also be used with an AC load; this is not the same as a DC load line because of the different effect of capacitors and inductors on an AC signal. An AC load line is drawn by using the DC Q point as one point and the maximum current determined from V_{cc}/R_{eq} (R_{eq} is the equivalent collector resistance in the AC circuit) as the second point.

EXAMPLE

Sketch the AC load line in Figure 2–16b if a coupling capacitor connects a 40KΩ load resistor (R_L) in parallel with R_c, as shown in the circuit in Figure 2–16a.

SOLUTION

The AC equivalent resistance R_{eq} seen at the collector is

$$R_{eq} = R_c \parallel R_L$$

$$R_{eq} = \frac{(10K\Omega \times 40K\Omega)}{(10K\Omega + 40K\Omega)}$$

$$R_{eq} = 8K\Omega$$

The maximum current is

$$I_{max} = \frac{20V}{8K\Omega}$$

$$I_{max} = 2.5mA$$

This current occurs when V_c is equal to zero, and so this point is put on the vertical current axis. We can now draw the AC load line through this point and the DC Q point, as shown in Figure 2–16b.

Notice the intercept with the voltage axis is not 20V as with the DC load line but is at 16V. The reason for this is that just when the transistor turns off ($I_c = 0$), the 20V is divided between R_c and R_L, and the voltage V_o across R_L is

$$V_0 = \frac{20V \times 40K\Omega}{10K\Omega + 40K\Omega}$$

$$V_0 = 16V$$

In other words, we have lost 4V of the AC output across R_c.

The load line concept was much used for vacuum tube circuits design because characteristics were provided for each tube and were very precise (at least for a new tube) because of the repeatability with mechanical construction of vacuum tubes. Transistors are rather sloppy devices with β variations in new devices of as much as 6 to 1. It is rare, then, for a manufacturer of bipolar transistors to provide I_b, I_c, and V_c characteristic curves so the load line technique is not used in circuit design but rather as an aid to understanding the R_c or load resistor/transistor voltage divider concept. However, we will use load lines when we study field effect transistors in a later chapter.

TROUBLESHOOTING TRANSISTOR BIAS CIRCUITS

The key thing to remember when troubleshooting a transistor bias circuit is that if the transistor is conducting, there is always an approximate 0.6V from base to emitter. The base is 0.6V more positive than the emitter for an NPN transistor, and the emitter is always 0.6V more positive than the base for a PNP transistor.

Out-of-Circuit Testing

With the transistor out of the circuit, we can check it with a VOM or DMM by considering the transistor as two diodes connected back-to-back, as shown in Figure 2–17.

When checking an NPN transistor, connecting the positive lead of the VOM (DMM) to the base and the negative lead to the emitter should give a low resistance reading. The same result should occur with the negative lead moved to the collector.

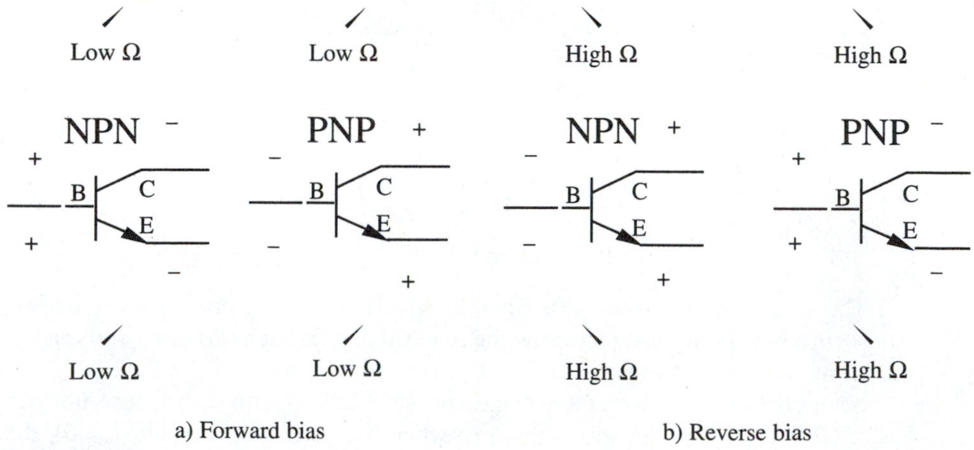

FIGURE 2–17 Checking a transistor out of circuit

If we now connect the negative lead to the base, the VOM should read very high resistance with the positive lead connected to either the emitter or collector. For the PNP transistor, the polarity senses are reversed for the same resistance readings.

Identifying Transistor Leads

Sometimes we have no information on a transistor as to whether it is NPN or PNP or which lead is base, emitter, or collector. We can determine the base connection by using the ohmmeter check indicated in Figure 2–17 since the base connection will read low resistance to both the emitter and collector. We can also determine whether we have an NPN or a PNP. If the low resistance is obtained with the positive lead on the base, we know we have an NPN; if with the negative lead on the base, we have a PNP.

We have found the base connection, but how can we determine which lead is the emitter and which the collector? A way to determine the difference is to take advantage of the fact that the reverse breakdown voltage from emitter to base is lower than the breakdown voltage from collector to base. The emitter–base break down is typically from 4V to 10V. But the collector breakdown is the maximum specified collector voltage and ranges from 20V up to about 300V for some transistors.

If the emitter–base breakdown is less than 9V, we can use a VOM that uses a 9V battery on the highest resistance range (10KΩ range on a Simpson 260 VOM). Using this setting, the emitter-base junction will break down and read a lower resistance than the collector-base junction. This does not damage the transistor because the current, and therefore the power level, are limited by the high internal resistance of the VOM.

If the VOM internal battery voltage is not enough to break down the emitter–base junction, we need to use a separate power supply and a 1mΩ resistor, as illustrated in Figure 2–18. Again, the resistor limits the current and prevents transistor damage. Since we are using a high-value resistor, it is recommended that a high-input-resistance (10mΩ)

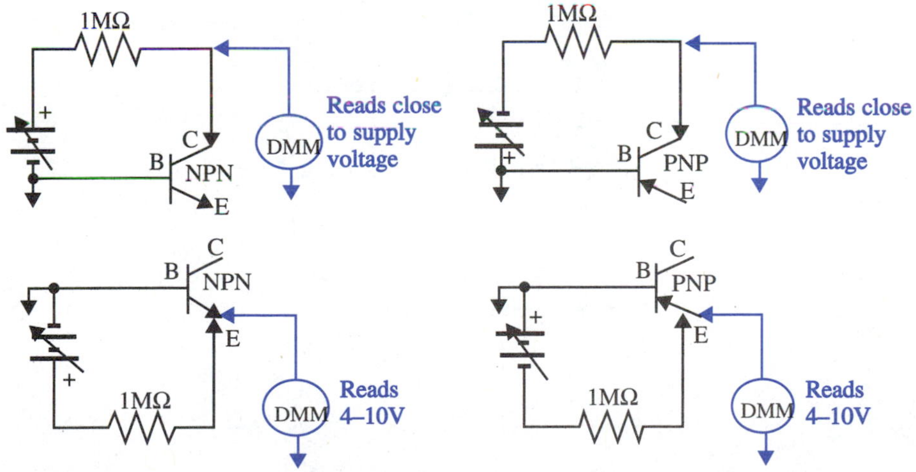

Note: Some VOMs will provide enough voltage on the high-resistance scale to break down the emitter-base junction without the use of an external power source.

FIGURE 2–18 Determining emitter and collector connections

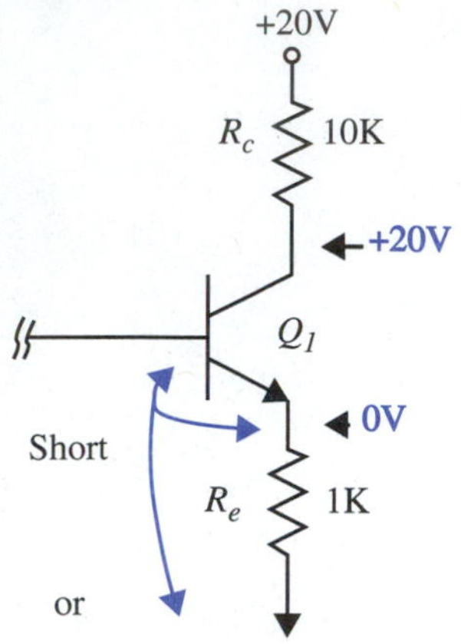

a) Turning transistor off

With transistor saturated, V_e and V_c are calculated by assuming R_c and R_e form a voltage divider.

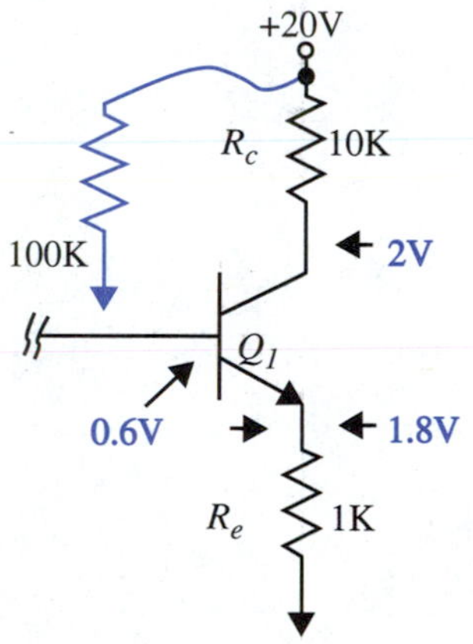

b) Turning transistor on

FIGURE 2–19
Checking a transistor in circuit

DMM be used to make the voltage readings. In this test, the junction showing the lowest voltage (breakdown) will be the emitter–base.

In-Circuit Testing

To check whether we have a good transistor in a circuit, we need to determine whether we can change the collector voltage by changing the base voltage.

We can connect a short (jumper) between the base and the emitter to stop transistor action (see Figure 2–19a). The transistor will now turn off, and in the case of an NPN transistor, the collector will be at the supply voltage.

If our initial measurement indicated the transistor is already turned off because of bias resistor problems, then we need to determine whether the transistor is good by turning it on. This we can do by connecting a resistor with a value of approximately ten times the collector resistor between the base and V_{CC}. This should cause the transistor to go into saturation with about 0.2V between the collector and the emitter. Figure 2–19b shows the procedures.

When checking a PNP transistor, we can still turn it off by connecting a short between base and emitter, but if we are using a positive supply (see Figure 2–11), then the collector will go to ground potential. To turn the transistor on, we connect a resistor of ten times the collector resistor from the base to ground. The collector voltage should rise to 0.2V less than the emitter voltage.

TROUBLESHOOTING OTHER CIRCUIT COMPONENTS

If the transistor checks good in a defective circuit, the problem could be an open resistor. We shall analyze the effect of an open resistor using the NPN circuit in Figure 2–8, which is repeated in Figure 2–20 with the DC voltages shown for a β of 100.

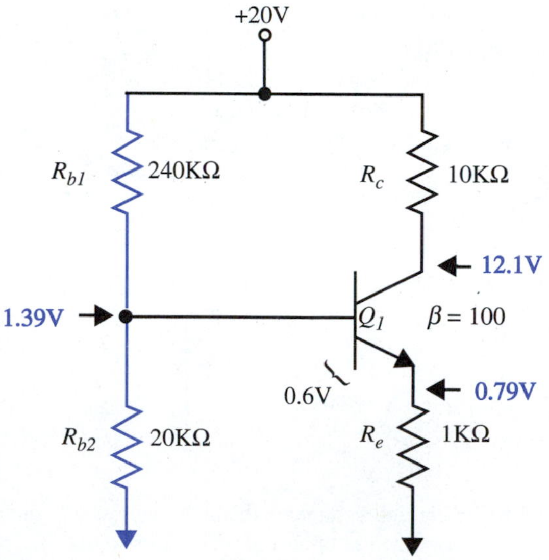

FIGURE 2–20 Troubleshooting Circuit Example

We shall assume a failure of one component at a time and also that resistors always fail open-circuited, not short-circuited.

R_{b1} Open

With R_{b1} open (which means no current through R_{b2}), the base of Q_1 will drop to 0V, and the transistor will turn off, causing $V_e = V_b = 0V$ and $V_c = V_{cc}$ or +20V for our example.

R_{b2} Open

If R_{b2} opens, the current that previously passed through this resistor will now enter the base of Q_1 and, being much larger than the normal base current, it will cause the transistor to go into saturation. With Q_1 saturated, we will have only about 0.2V (or less) across this transistor, and so we can consider R_c and R_e as a simple voltage divider for calculating the collector and emitter voltages. The emitter voltage V_e will be

$$V_e = \frac{20V \times 1K\Omega}{10K\Omega + 1K\Omega}$$

$$V_e = 1.82V$$

and

$$V_c = V_e + 0.2V$$

$$V_c = 2.02V$$

also

$$V_b = V_e + 0.6V$$

$$V_b = 2.42V$$

With these voltage conditions, no AC signal will appear at the collector unless the input signal is large enough to take the transistor out of saturation—then a distorted signal will appear.

R_e Open

If R_e opens, the transistor will turn off, the voltage at the collector will rise to 20V, and no AC signal will be present at the collector. The DC voltage at the base (with no base current) will be (20V x 20KΩ)/260KΩ) or 1.54V. The emitter DC voltage will depend on the resistance provided to ground by the measuring instrument and will range from close to 1.2V down to about 1V.

R_c Open

Finally, if R_c opens, Q_1 will turn off, and with no transistor action, the base-emitter diode in series with the emitter resistor will appear in parallel with R_{b2}. Figure 2–21 shows this circuit. The voltage seen at the base is found by making a Thevenin's equivalent of R_{b1} and R_{b2} and then connecting the diode/emitter resistor circuit. From the figure, we see that

$$V_b = 0.648V$$

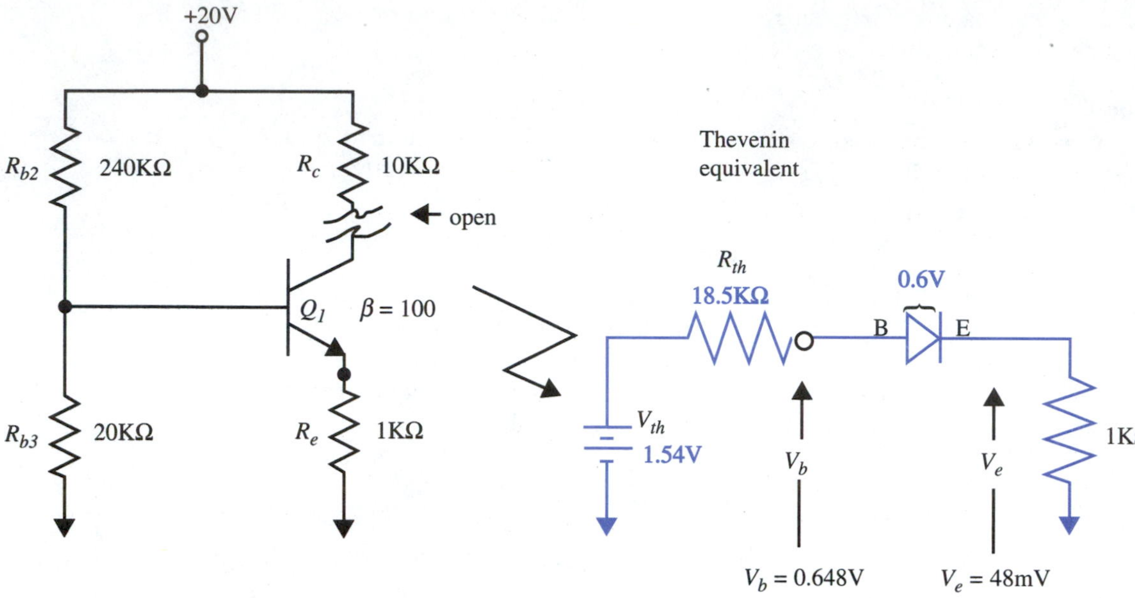

FIGURE 2–21 Transistor circuit open with collector resistor

and

$$V_e = 48\text{mV}$$

TRANSISTOR MECHANICAL CONFIGURATIONS

Transistors come in various mechanical packages, depending on the power rating, heat sinking, and hermetic sealing. The circuit connections affect the shape of the package, and thermal considerations could require a metal case or mounting tab to ensure maximum heat flow out of the device (see Chapter 19). Sealing is required to prevent moisture from contaminating the semiconductor material and changing its characteristics or causing corrosion of the circuit connections.

The most used transistor packages are shown in Figure 2–22 along with their type designation (old designations in parens).

- TO–226AA (TO–92)—This is the cheapest and most common transistor package. Its disadvantages are low power rating (about 500 mW) and lack of standardization of the lead connections.
- TO–206AA (TO–18)—About the same size as the TO–226AA (TO–92) but mounted in a metal case for better sealing and electrical shielding. The collector is connected to the case for optimum heat flow. When viewing the lead side

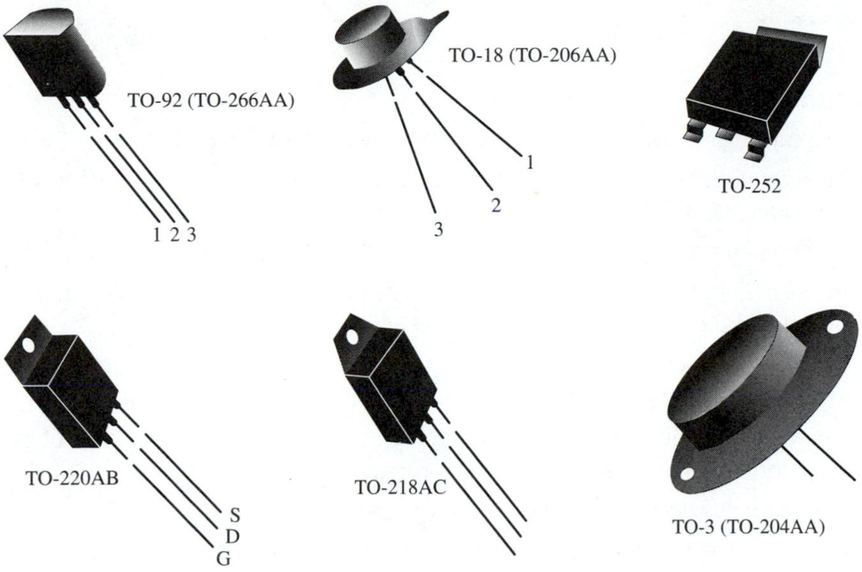

FIGURE 2–22 Transistor packages

(bottom) of the transistor, the emitter is closest to the tab, and going in a clockwise direction, the base is next, followed by the collector.

- TO–205AD (TO–5)—This package is metal and about 1.5 times the size of the TO–206AA (TO–18). Power rating is about 1W, and the lead identification is the same as the TO-18.
- TO–252—The bent leads identify this plastic package as a surface mount device.
- TO–218AC—A plastic package mounted to a metal tab is used in this configuration. A hole in the tab allows mounting the transistor to a heat sink, and the tab is electrically connected to the collector to allow heat flow away from the collector–base junction. If the heat sink has to be at a different voltage than the collector, a mica insulator is mounted between the tab and the heat sink. Power rating is about 150W.
- TO–220AB—This configuration is similar to the TO–218AC but is smaller and has a power rating around 40W.
- TO–204AE (TO–3)—The largest package with a steel or aluminum case. It has two leads, one for the emitter and the other for the base. The collector is connected to the case—again for better heat sinking. Connection is made to the collector by a terminal lug that is attached to a case mounting screw. Power rating is over 250W.

SUMMARY

- A transistor is a current gain device.
- The transistor is created by NPN or PNP doping of a semiconductor.
- The center region of the three-layer semiconductor is called the base, while one end is the emitter and the other the collector.
- Base-to-emitter forms one diode junction and base-to-collector another.
- In operation as an amplifier, a transistor has the base–emitter junction forward biased and the base–collector junction reverse biased.
- With an NPN transistor, electrons are "emitted" from the emitter into the base region, and the high positive voltage of the collector "collects" them.
- A few of the emitted electrons flow through the base lead, and the ratio of the collector current to base current is a constant called β ($\beta = I_c/I_b$).
- The ratio of collector current to emitter current is called α.
- A transistor is an amplifier because a small base current change causes a large collector current change.
- The voltage across a forward-biased silicon base-emitter junction is about 0.6V.
- The voltage across the reverse-biased base-collector junction cannot be directly determined using Ohm's law.
- Ohmic resistance of the emitter region is called r_e and is determined at room temperature by dividing 26mV by the DC emitter current.
- Increasing the voltage across the base–emitter of a transistor increases the collector current and therefore the voltage drop across a resistor in the collector circuit. This is the basis for transistor voltage gain.
- A transistor is "biased" to bring the collector to the correct operating point for the desired voltage swing on the collector.
- Ideal biasing is independent of the β variation between transistors.
- A PNP transistor can use the same biasing circuit as an NPN if we just use a power supply of the opposite polarity.
- We can troubleshoot a transistor in-circuit by forcing it "off" by shorting base to emitter, or "on" by connecting a resistor (ten times R_c) from the power supply to the base.
- Out-of-circuit, we can check a transistor with an ohmmeter by treating it as back-to-back diodes with the common lead at the base.
- We can determine the emitter from the collector on an unmarked transistor by testing for the lower breakdown voltage at the base-emitter junction.

EXERCISE PROBLEMS

1. With reference to Figure 2–A, explain how transistor action occurs.

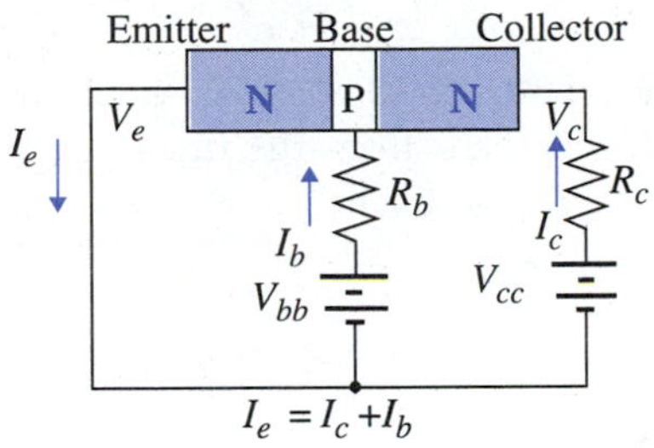

FIGURE 2–A (text figure 2–2)

2. Why does the collector of an NPN transistor collect more of the emitted electrons than the base?
3. What is the ratio of electrons flowing to the collector divided by electrons flowing to the base called?
4. What is meant by the current gain of a transistor, and why is it useful?
5. Describe how a transistor can provide voltage gain.
6. What is meant by the *alpha* (α) of a transistor?
7. The base current flowing into a transistor is 10µA. Find the current flowing into collector if the transistor β is 80.
8. With reference to Figure 2–B, find the voltage change at the collector if V_{bb} is increased from 5V to 5.2V.

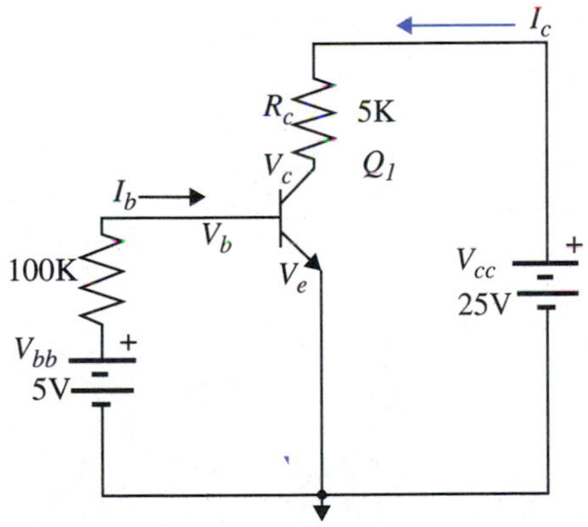

FIGURE 2–B (text figure 2–4)

9. What is r_e, and how can you determine its value?
10. What is a "common emitter" circuit?

11. How low can the voltage from collector to emitter be and maximum collection by the collector still occur?

12. If in the circuit in Figure 2–C V_s is –10mV, what would be the change in the voltage across the collector resistor from the quiescent value. Compare the change with the result with V_s of +10mV.

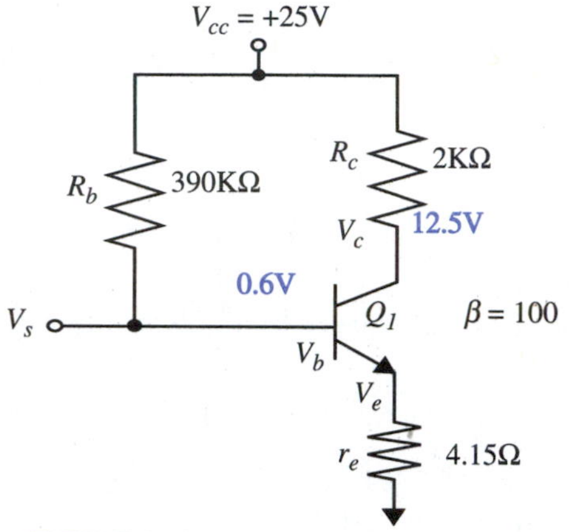

FIGURE 2–C (text figure 2–5)

13. What is the maximum voltage swing (positive to negative) that can occur at the base of the circuit in Figure 2–C without clipping the output signal?

14. Explain why the β independent biasing circuit of Figure 2–D allows for changing transistors without causing a large change in the collector voltage (*Hint*: Consider the voltage across R_e.)

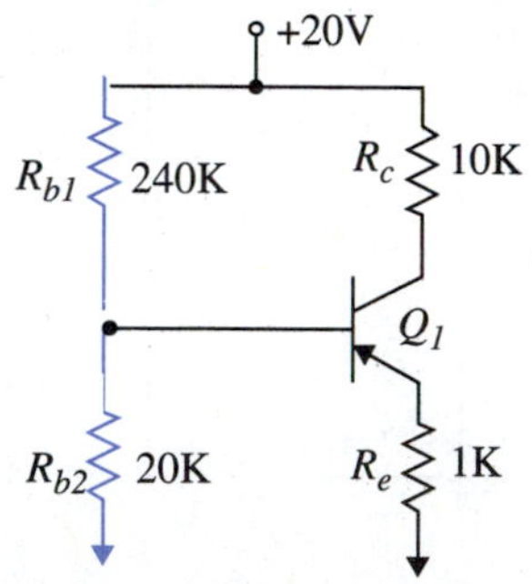

FIGURE 2–D (text figure 2–8)

15. Determine the voltage at the base, emitter, and collector of the circuit of Figure 2–D, considering the effect of base current, if the transistor β is 150.

16. Determine the voltage at the base, emitter, and collector of the circuit of Figure 2–E, considering the effect of base current, if the transistor β is 200.

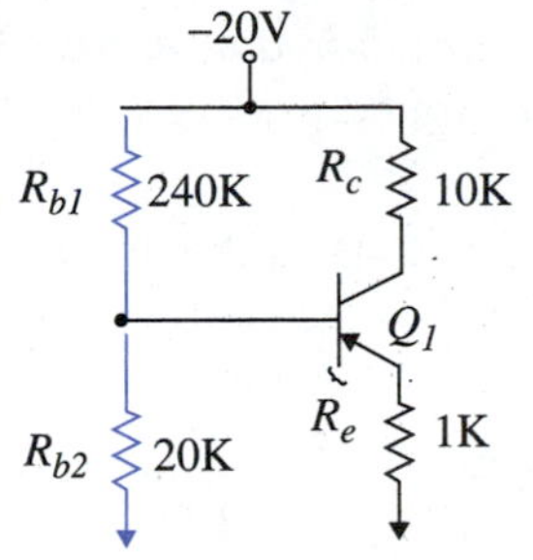

FIGURE 2–E (text figure 2–11)

17. Determine the voltage at the base, emitter, and collector of the circuit of Figure 2–F, considering the effect of base current, if the transistor β is 150.

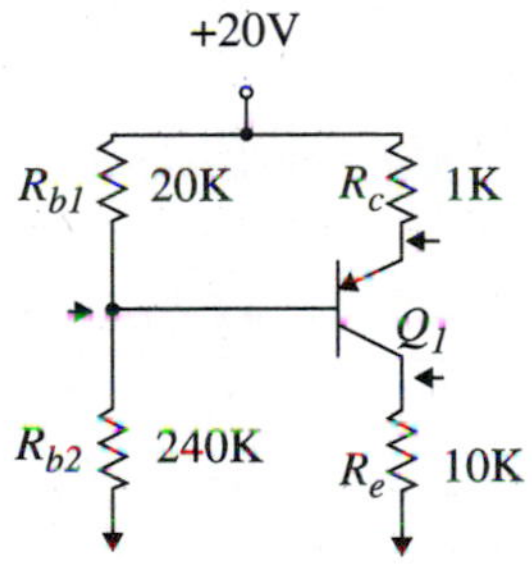

FIGURE 2–F (text figure 2–12)

18. Determine the collector voltage V_c for the circuit in Figure 2–G if β is 150.

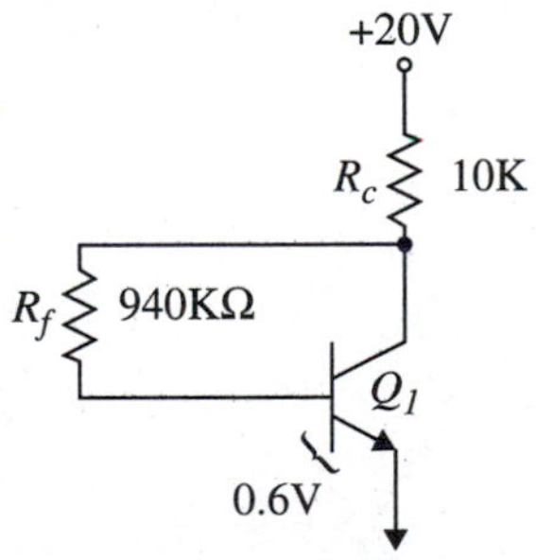

FIGURE 2–G (text figure 2–14)

19. You are troubleshooting an NPN transistor circuit and you determine the transistor is conducting, because you measure a voltage drop across the collector

resistor R_C. If the base voltage measured to ground is +1.88V, what should the emitter voltage to ground read?

20. You are troubleshooting a PNP transistor circuit and you determine the transistor is conducting, because you measure a voltage drop across the collector resistor R_C. If the base voltage measured to ground is +18.8V, what should the emitter voltage to ground read?

21. You are troubleshooting the circuit of Figure 2–D, and you measure the following voltages to ground V_c = 20V, V_e = 0V, and V_b = 1.56V. Which circuit component is defective?

22. You are troubleshooting the circuit of Figure 2–D, and you measure the following voltages to ground: V_c = 20V, V_e = 0.048V, and V_b = 0.648V. Which circuit component is defective?

23. You are troubleshooting the circuit of Figure 2–F, and you measure the following voltages to ground: V_c = 18V, V_e = 18.2V, and V_b = 17.6V What is the problem?

24. You are troubleshooting the circuit of Figure 2–E, and you measure the following voltages to ground: V_c = –20V, V_e = –20V, and V_b = –20V. What is the problem?

25. The maximum collector current flowing through a transistor in a circuit is 3A, and the voltage across the transistor from collector to base is 15V. Which transistor package would be used in this application to handle the power dissipation?

26. When would a TO–206AA be used rather than the cheaper TO–226AA package?

ABBREVIATIONS

I_b = Base current

I_e = Emitter current

I_c = Collector current

α = Ratio $\frac{I_c}{I_e}$

β = Current gain $\frac{I_c}{I_b}$; also called h_{fe}

V_{cc} = Power supply voltage

V_c = Collector voltage

V_b = Base voltage

V_e = Emitter voltage

V_{be} = Base to emitter voltage

V_{Rc} = Drop across collector resistor

R_b = External base resistor

R_e = External emitter resistor

R_c = External collector (load) resistor

r_c = Internal emitter resistor

R_t = Base input resistance (βR_e or βr_e)

Q point = Collector DC bias voltage

ANSWERS TO PROBLEMS

1. See text.
2. Higher positive potential.
3. β
4. See text.
5. See text.
6. Ratio of I_c over I_e
7. $I_c = 0.8\text{mA}$
8. $\Delta V = 1\text{V}$
9. Internal emitter resistance = $26\text{mV}/I_e$
10. No signal is applied or taken from the emitter.
11. About 2V
12. $\Delta V_c = +4.82\text{V}$
13. $V_s = 52\text{m}V_{p\text{–}p}$
14. Fixed drop across R_e
15. $V_b = 1.437\text{V}$, $V_e = 0.837\text{V}$, $V_c = 11.63\text{V}$
16. $V_b = -1.46\text{V}$, $V_e = -0.86\text{V}$, $V_c = -11.4\text{V}$
17. $V_b = 18.56\text{V}$, $V_e = 19.16\text{V}$, $V_c = 8.37\text{V}$
18. $V_c = 8.12\text{V}$
19. $V_e = 1.28\text{V}$
20. $V_e = +19.4\text{V}$
21. Transistor, open base–emitter junction
22. Q_1 open collector–base junction
23. R_{b2} open-circuited
24. Open ground connection
25. TO–218AC
26. Better sealing/shield

CHAPTER 3

Transistor AC Amplifiers

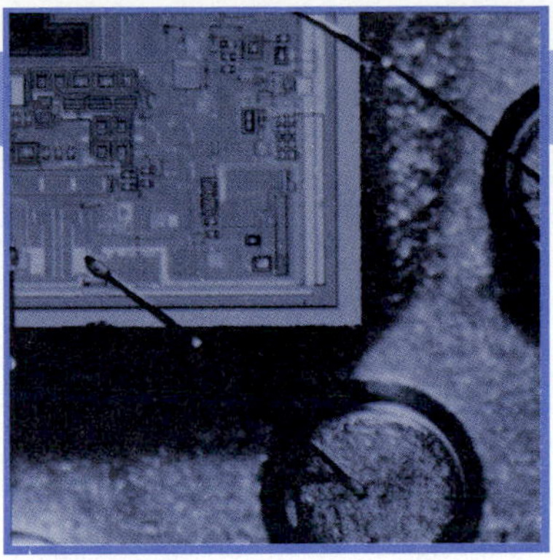

CHAPTER OBJECTIVES

After completing this chapter, you should be able to describe the following concepts pertaining to transistor amplifier circuits:

- Common emitter amplifier configuration
- Common collector (voltage follower) amplifier configuration
- Darlington configuration
- Common base amplifier configuration
- Amplifier gain equations
- Amplifier input and output resistance
- Purpose and selection of a coupling capacitor
- Purpose and selection of a bypass capacitor
- Use of a PNP transistor
- Two-stage voltage amplifier
- Transistor amplifiers with transformer and inductive loads
- Push–pull amplifiers
- Complementary symmetry amplifiers
- Class A, B, C, and D amplifiers
- Troubleshooting AC amplifiers

INTRODUCTION

In the previous chapter, we saw how a transistor is biased to have a desired DC voltage at the output of the circuit in order to obtain maximum collector voltage swing. In this chapter, we shall see that by varying this output voltage with time in response to a small input voltage, we achieve AC voltage gain.

We can consider the AC signals as being superimposed on the DC. The AC will cause the DC bias currents to increase or decrease, depending on the polarity of the input AC signal.

We shall study the three connections for transistor amplifiers: the common emitter, common collector (emitter follower), and the common base.

COMMON EMITTER AC AMPLIFIER

As mentioned previously, the common emitter circuit is so named because there is no input or output signal across the emitter resistor. Figure 3–1 is the same as the common emitter circuit we studied in the previous chapter but with the addition of an input AC coupling capacitor (C_1) which blocks DC, and a bypass capacitor (C_2) which effectively shorts R_e for AC signals.

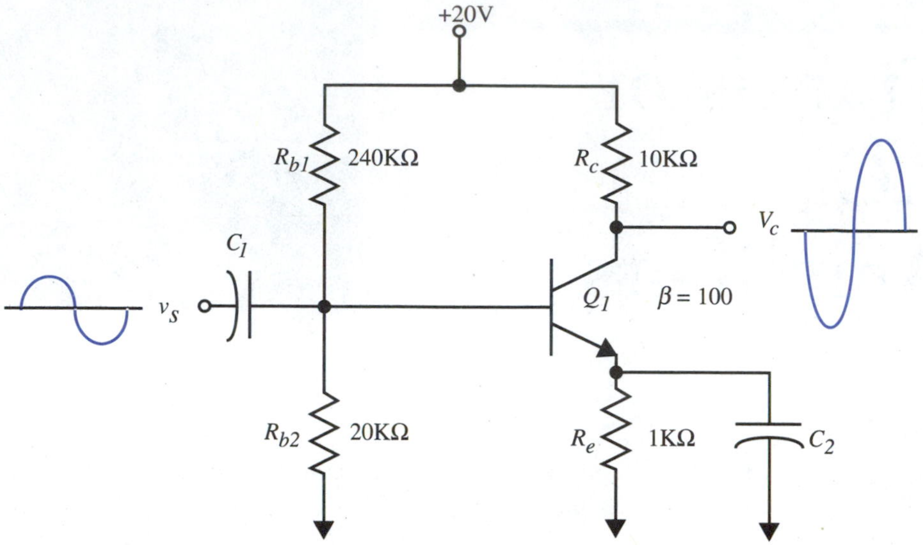

FIGURE 3–1 AC common emitter amplifier

We have seen the effect of changes in the DC voltage on the base causing an amplified change on the collector. As expected, a time-varying (AC) signal will also be amplified by this circuit (see Figure 3–1).

The purpose of the input coupling capacitor C_1 is to prevent the internal DC resistance of the AC signal source from effecting (loading down) the DC bias voltage on the base. The value of this capacitor is picked to have a low reactance compared to the transistor circuit input resistance at the frequency of interest, otherwise, too much AC voltage drop will occur across C_1 and less will be available at the amplifier base.

COUPLING CAPACITOR DETERMINATION

To determine the size of the coupling capacitor, we need to compute the circuit input resistance. We must first recognize that the top end of R_{b1} is at AC ground potential since a DC power supply has very low AC resistance to ground (typically less than one ohm if we keep the power supply leads short). So we can assume that the top end of R_{b1} and the bottom end of R_{b2} are connected together, i.e., they are in parallel. Also in parallel with these two resistors is the transistor input resistance R_t, which again is βR_e (neglecting r_e which is usually much smaller than R_e).

Assume first that C_2 is not connected, and the complete equivalent AC input resistance is shown in Figure 3–2.

From the figure, we can see that the input circuitry is simplified to a single resistor (R_{in}). A design rule of thumb is to use a coupling capacitor C_1 with a reactance at the lowest input frequency equal to one tenth of R_{in}.

EXAMPLE

Assuming that the lowest input frequency is 100Hz, compute the value for the coupling capacitor C_1 in Figure 3–1.

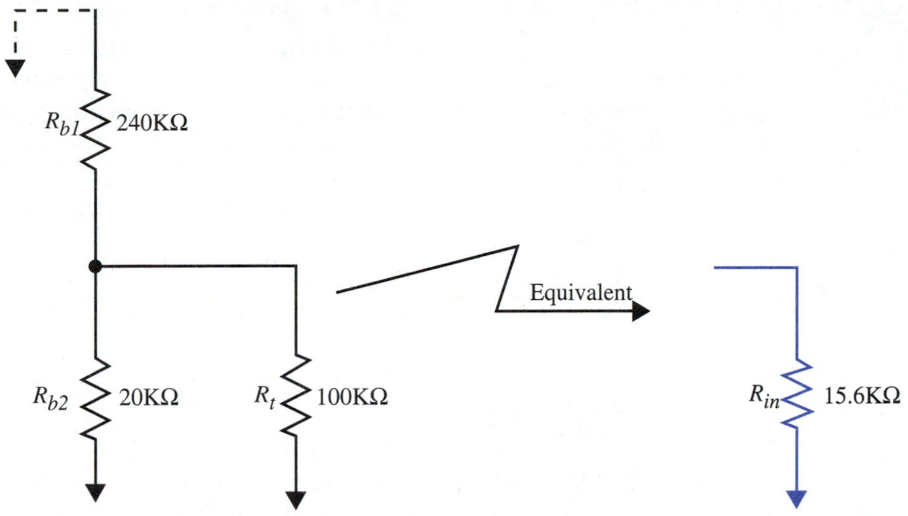

FIGURE 3–2 Transistor AC input circuitry

SOLUTION

From the capacitor reactance (X_c) formula

$$X_c = \frac{1}{2\pi fC}$$

and

$$C_1 = \frac{1}{2\pi fX_c}$$

But we assumed

$$X_c = \frac{R_{in}}{10}$$

$$X_c = \frac{15.6\text{K}\Omega}{10}$$

$$X_c = 1.56\text{K}\Omega$$

So

$$C_1 = \frac{1}{2\pi \times 100 \times 1.56\text{K}\Omega}$$

$$C_1 = 1\mu\text{F}$$

We shall see the purpose of the C_2 bypass capacitor (which is in parallel with R_e) if we first compute the circuit voltage gain without C_2 connected (see Figure 3–3a) and then with it connected.

(We shall use lowercase letters for AC voltages and currents.)

COMMON EMITTER CIRCUIT VOLTAGE GAIN

From our earlier look at transistor gain in the previous chapter, we found that the input signal voltage also appears across the emitter resistance r_e (the DC bias overcomes the 0.6V base-emitter diode drop). Now, we have an additional resistor R_e, and the input voltage v_s is across the combination $R_e + r_e$ and is equal to

$$v_s = i_e(R_e + r_e)$$

where i_e is the AC signal current flowing in the emitter circuit.

The **voltage gain** A_v for the circuit (see Figure 3–3a) is defined as the output signal seen at the collector (v_o) divided by the input signal applied to the base (v_s), or

$$A_v = -\frac{v_o}{v_s}$$

The common emitter circuit voltage gain is negative because the output signal at the collector v_o is 180° out of phase with the input base signal v_s. In other words, if the input signal on the base goes in the positive direction, it increases the collector current, which causes the collector voltage to go in the negative direction.

Remember the V_{cc} power supply is at AC ground potential, so v_o is the voltage developed across R_c due to the signal current i_c through this resistor, and is the AC voltage measured at the collector; so

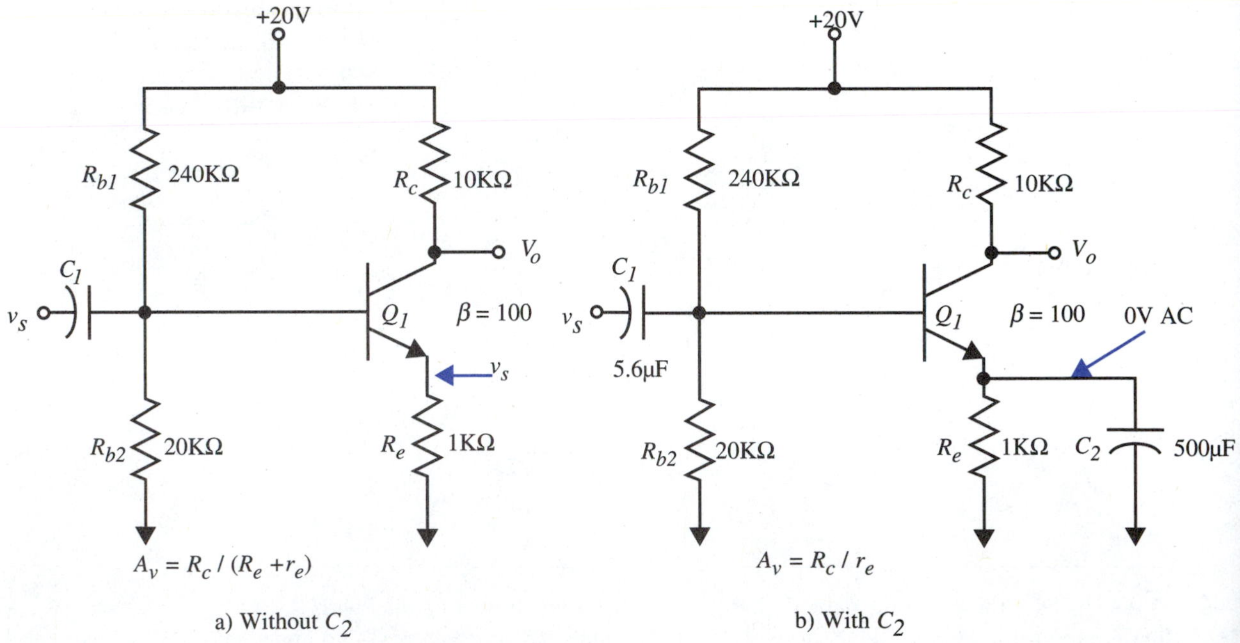

FIGURE 3–3 Computing circuit gain

$$v_o = v_c$$

Since $i_c \approx i_e$

$$v_o = i_e R_c$$

From before

$$v_s = i_e(R_e + r_e)$$

and

$$A_v = \frac{-v_{out}}{v_s}$$

$$A_v = \frac{-i_e R_c}{i_e(R_e + r_e)}$$

$$A_v = \frac{-R_c}{R_e + r_e} \text{ (without } C_2\text{)}$$ **(equation 3–1)**

Remember

$$r_e = \frac{26\text{mV}}{I_e}$$

where I_e is the DC emitter current.

EXAMPLE

Find the voltage gain A_v for the circuit in Figure 3–3a, assuming a β of 100.

SOLUTION

From our analysis of the biasing of this circuit in the previous chapter with a β of 100, we found that $I_e = 0.79$ mA; therefore

$$r_e = \frac{26\text{mV}}{0.79\text{mA}}$$

$$r_e = 33\Omega$$

Since in this example

$$r_e << R_e \text{ (with } R_e = 1\text{K}\Omega)$$

$$A_v = -\frac{R_c}{R_e}$$

$$A_v = -\frac{10\text{K}\Omega}{1\text{K}\Omega}$$

$$A_v = -10$$

EFFECT OF C_2 IN THE CIRCUIT

Now if we connect C_2 as a **bypass capacitor** in parallel with R_e (as shown Figure 3–3 b) and make the reactance of this capacitor much smaller than r_e (Design rule of thumb: Use an actual reactance value one-tenth of r_e at the lowest operating frequency, i.e., $X_{c2} \leq r_e/10$, then R_e is effectively shorted for the AC signal, and the top end of R_e is at AC ground or 0V AC)—and with C_2 in the circuit, then all the input AC signal appears across r_e, and the gain now is simply

$$A_v = -\frac{R_c}{r_e} \qquad \textbf{(equation 3–2)}$$

We would get the same gain as given by equation 3–2 without R_e in the circuit (emitter at DC ground), but remember we added R_e for DC bias stability!

EXAMPLE

Find the required value of C_2 for the amplifier circuit in Figure 3–3b and the circuit gain (A_v) if we desire to amplify signals down to 100Hz.

SOLUTION

From our previous computation

$$r_e = 33\Omega$$

Then, from our rule

$$X_{c2} = \frac{r_e}{10}$$

$$X_{c2} = \frac{33\Omega}{10}$$

$$X_{c2} = 3.3\Omega$$

$$C_2 = \frac{1}{2\pi f X_{c2}}$$

$$C_2 = \frac{1}{2\pi \times 100\text{Hz} \times 3.3\Omega}$$

$$C_2 = 482\mu\text{F (use } 500\mu\text{F)}$$

Using equation 3–2

$$A_v = -\frac{R_c}{r_e}$$

$$A_v = -\frac{10\text{K}\Omega}{33\Omega}$$

$$A_v = -303$$

This is a considerable increase over the gain of 10 without the capacitor C_2 bypassing R_e; but we will find that high-level waveforms are more distorted when C_2 is used because r_e changes as the input signal changes the emitter current:

$$r_e = 26\text{mV}/I_e$$

The trade-off, then, is gain versus distortion. (We shall cover this further when we study operational amplifiers in a later chapter.)

So the addition of the capacitor C_2 has increased the gain, but it also has the negative effect of increasing distortion for high-level signals and it also causes a reduction of the transistor input resistance R_t seen at the base because with R_e bypassed

$$R_t = \beta r_e \text{ (with } C_2\text{)} \qquad \textbf{(equation 3–3)}$$

rather than

$$R_t = \beta(r_e + R_e)$$

which we got without C_2.

EXAMPLE

Compute the transistor input resistance R_t for the circuit in Figure 3–3a and 3–3b. Assume $\beta = 100$.

SOLUTION

For Figure 3–3a without C_2

$$R_t = \beta(R_e + r_e)$$

but

$$r_e << R_e$$

$$R_t = 100 \times 1\text{K}\Omega$$

$$R_t = 100\text{K}\Omega$$

For Figure 3–3b with C_2

$$R_t = \beta r_e$$

$$R_t = 100 \times 33\Omega$$

$$R_t = 3.3\text{K}\Omega$$

which is much less than the 100KΩ we get without the capacitor C_2.

The effect of this lower transistor input resistance is that the coupling capacitor (C_1) has to be increased when C_2 is added to the circuit; otherwise the low frequency response of the amplifier will be affected.

EXAMPLE

Find the equivalent input resistance R_{in} with C_2 in the circuit (Figure 3–3b) and determine the required value of C_1 at 100Hz.

SOLUTION

The equivalent R_{in} is now

$$R_{in} = 240\text{K}\Omega \parallel 20\text{K}\Omega \parallel 3.3\text{K}\Omega$$

$$R_{in} = 2.8\text{K}\Omega$$

Using our rule of thumb that

$$X_c = \frac{R_{in}}{10}$$

$$X_c = 280$$

and

$$C_1 = \frac{1}{2\pi \times 100 \times 280\Omega}$$

$$C_1 = 5.6\mu\text{F}$$

Without C_2, we had a higher transistor input resistance and only required $C_1 = 1\mu\text{F}$ We can see the complete circuit with values defined in Figure 3–3b.

If

$$\beta \times r_e$$

is much smaller than

$$R_{b1} \parallel R_{b2}$$

the circuit input resistance

$$R_{in} \approx \beta \times r_e$$

and

$$X_{C1} = \beta r_e/10$$

Since

$$X_{C2} = r_e/10$$

then it follows that C_1 and C_2 are related by the approximation

$$C_2 = \beta \times C_1$$

In our previous examples at 100Hz, with C_1 determined to be 5.6µF, then C_2 would be 568µF, using this approximation, rather than the 482µF we computed.

EMITTER FOLLOWER

The second most common transistor circuit configuration, after the common emitter, is the emitter follower (common collector), shown in Figure 3–4. This circuit has a maximum voltage gain of only 1 (unity)—but the current gain of the transistor enables us to provide more current into a required load resistor (R_L). The input resistance (R_{in}) for the follower is high, while the output resistance (R_o) is low. This allows for power matching a load to the driving circuit.

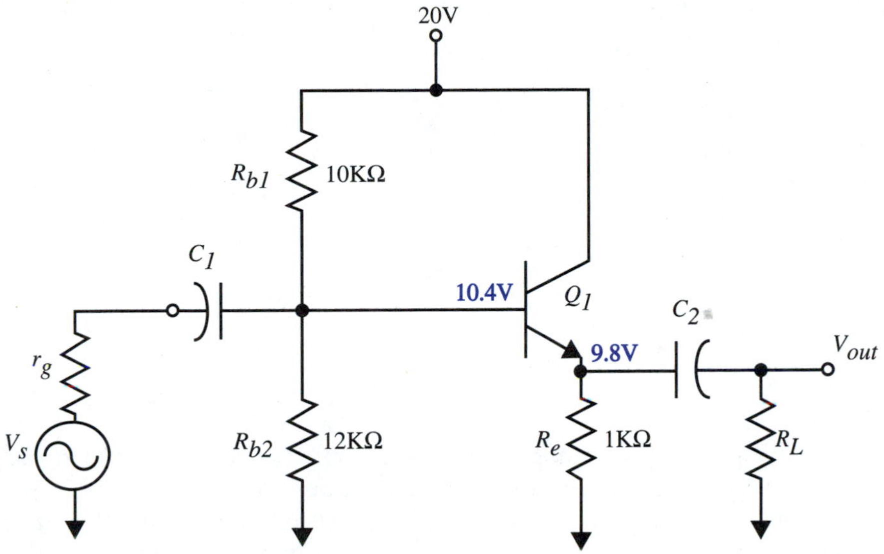

FIGURE 3–4 Emitter follower

Let us find the DC bias voltages for this configuration. We first can find the base voltage by determining the Thevenin's equivalent circuit of R_{b1}, R_{b2}, and the transistor input circuit, as shown in Figure 3–5.

If we assume

$$\beta = 100$$

and use Ohm's law, we get a base current of

$$I_b = \frac{V_{th} - V_{be}}{R_{th} + R_t}$$

Remembering

$$R_t = \beta R_e = 100 \times 1\text{K}\Omega$$

$$I_b = \frac{10.91\text{V} - 0.6\text{V}}{5.45\text{K}\Omega + 100\text{K}\Omega}$$

$$I_b = 98\mu\text{A}$$

The emitter voltage

$$V_e = I_b \times 100\text{K}\Omega \text{ (or } I_e \times 1\text{K}\Omega)$$

$$V_e = 9.8\text{V}$$

With the base emitter drop of 0.6V, we have a base voltage V_b of

$$V_b = (V_e + V_{be})$$

$$V_b = (9.8\text{V} + 0.6\text{V})$$

$$V_b = 10.4\text{V}$$

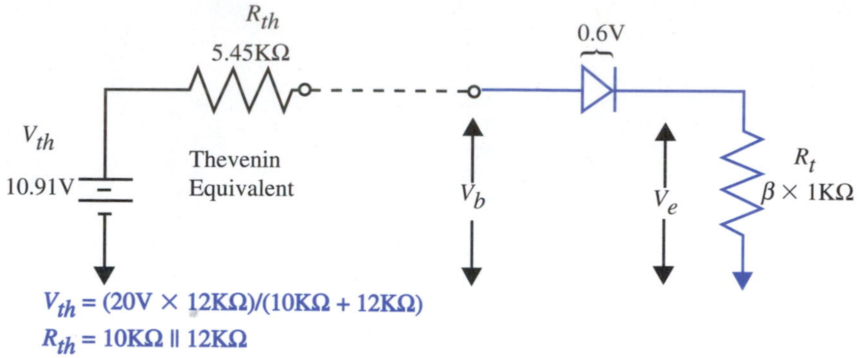

FIGURE 3–5 Follower base equivalent circuit

An input voltage swing of about ±9V can be accommodated with this emitter voltage. On the positive 9V swing, the emitter rises up from the bias voltage of 9.8V to 18.8V, i.e., (9 + 9.8)V; with a V_{cc} of 20V, this gives 1.2V between collector and emitter. The negative 9V swing drops the emitter voltage from 9.8V down to (9.8V – 9V) or 0.8 volts. Both swings will be accomplished without clipping the signal.

FOLLOWER INPUT RESISTANCE

As stated previously, the follower has a high input resistance. This feature, along with a low output resistance, makes the follower ideal for matching a low-resistance load to a high-resistance signal source.

The input resistance R_{in} for the follower can be determined with reference to Figure 3–4 and is the same as the input resistance to the common emitter configuration with the exception that R_L is now in parallel with R_e:

$$R_{in} = R_{b1} \,||\, R_{b2} \,||\, R_t \qquad \textbf{(equation 3–4)}$$

where

$$R_t = \beta(r_e + R_e \,||\, R_L)$$

The AC input resistance of the follower is typically higher than the common emitter configuration because R_e in the common emitter configuration is often AC bypassed (C_2) so

$$R_t = \beta r_e$$

EXAMPLE

Find the AC input resistance for the follower in Figure 3–4 if the load resistor R_L is 100Ω and β is 100.

SOLUTION

First

$$R_t = \beta\,(r_e + R_e \,||\, R_L)$$

In our previous analysis for the circuit in Figure 3–4, we determined that

$$V_e = 9.8\text{V}$$

Since this DC voltage is across the 1KΩ emitter resistor,

$$I_e = 9.8\text{V}/1\text{K}\Omega = 9.8\text{mA}$$

So

$$r_e = (26\text{mV}/9.8\text{mA}) = 2.65\Omega$$

and

$$R_t = 100(2.65 + 1\text{K}\Omega \parallel 100\Omega)$$

$$R_t = 9.36\text{K}\Omega$$

From equation 3–4:

$$R_{in} = R_{b1} \parallel R_{b2} \parallel R_t$$

$$R_{in} = 10\text{K}\Omega \parallel 12\text{K}\Omega \parallel 9.36\text{K}\Omega$$

$$R_{in} = 3.45\text{K}\Omega$$

FOLLOWER OUTPUT RESISTANCE

To get a feel for the current drive capability of a follower, we need to find the output resistance R_o seen by looking back into the transistor emitter.

The output resistance R_o of the follower is the base circuit resistors (including generator resistance) in parallel, transferred to the emitter circuit, and therefore divided by β, and added to r_e (see Figure 3–6)

$$R_o = r_e + \frac{(R_{b1} \parallel R_{b2} \parallel r_g)}{\beta} \qquad \textbf{(equation 3–5)}$$

The output resistance R_o is in parallel with R_e—but if R_o is much smaller (usually the case), we can neglect R_e. What is the purpose, then, of R_e in the circuit? It sets up the DC bias point and determines the emitter current (I_e) and, therefore, the value of r_e since r_e = 26mV/I_e

EXAMPLE

Find the output resistance R_o for the circuit in Figure 3–4 if r_g is 600Ω (neglect R_e).

SOLUTION

Looking back at the resistor values used in Figure 3–4, and using the value of 600Ω for r_g and 9.8mA for I_e, as previously determined, we have:

$$R_o = r_e + \frac{(R_{b1} \parallel R_{b2} \parallel r_g)}{\beta}$$

$$R_o = \left(\frac{26\text{mV}}{9.8\text{mA}}\right) + \frac{(10\text{K}\Omega \parallel 12\text{K}\Omega \parallel 600\Omega)}{100}$$

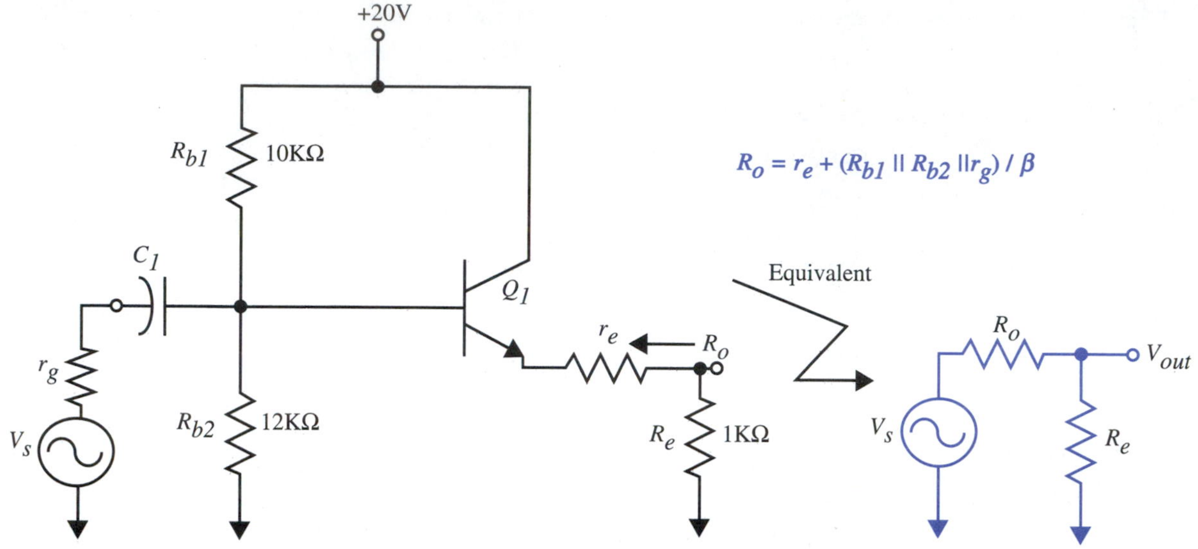

FIGURE 3–6 Follower equivalent output resistance

$$R_o = (2.65\Omega + 5.4\Omega)$$

$$R_o = 8.1\Omega$$

This is the resistance that is now effectively in series with the signal voltage v_s rather than the generator 600Ω without the follower! The actual resistance load on the 600Ω generator is R_{in}.

So the transistor follower has transformed the 600Ω generator resistance into a follower output resistance of 8.1Ω

Another use for the follower is power matching of the load resistor (R_L) to the generator. It will be recalled from basic DC circuit theory that maximum power transfer occurs when R_o is equal to R_L. If our load happens to be an 8Ω speaker, our previous circuit with its 8.1Ω output resistance will allow power matching.

If we wanted to match a 4Ω speaker, we could reduce r_e by increasing I_e (reduce R_e) or reduce the resistors in the base circuit to transfer a lower resistance to the emitter circuit.

DARLINGTON CONNECTION

The current drive, input resistance, and output resistance of a follower can be improved by connecting a follower in a Darlington connection, as shown in Figure 3–7.

What the Darlington connection does is to make the transistor configuration have an effective β of $\beta_1 \times \beta_2$. We can also use this connection with the common emitter and common base amplifiers.

The input resistance is increased because r_e and R_e, when transferred to the base circuit, are now multiplied by $\beta_1 \times \beta_2$. Similarly, the output resistance of a follower is

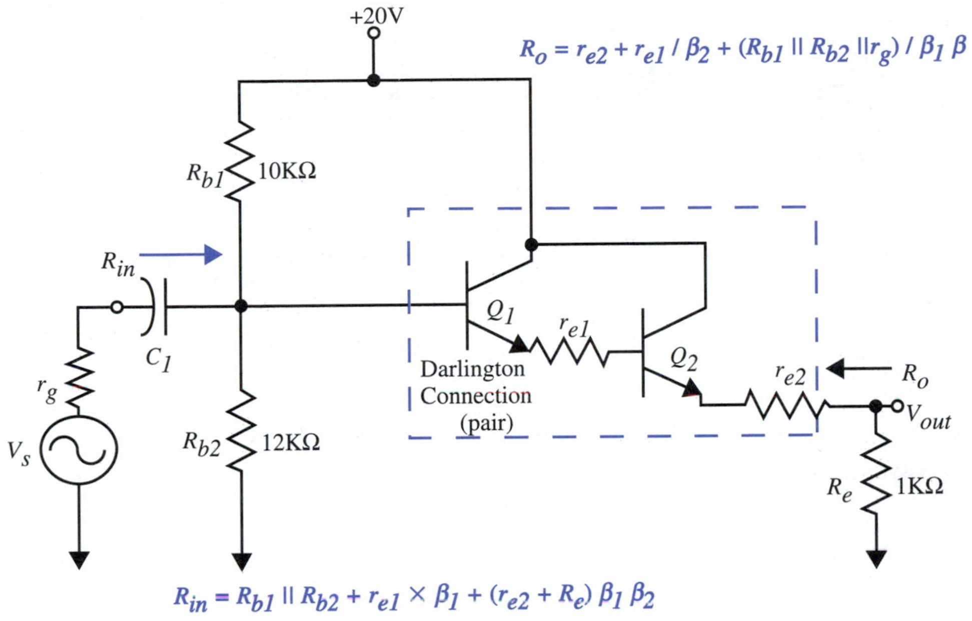

FIGURE 3–7 Darlington connection for a follower

reduced because the resistance in the base circuit is divided by $\beta_1 \times \beta_2$ when we transfer it over to the emitter side. The equations in Figure 3–7 reflect these new relationships.

EXAMPLE

If in the Darlington circuit in Figure 3–7, Q_1 has a β of 100, and Q_2 has a β of 50, what is the effective β for the circuit?

SOLUTION

Since β_1 is 100 and β_2 is 50, the circuit β is then

$$\beta_1 \times \beta_2 = 100 \times 50 = 5000$$

COMMON BASE CIRCUIT

The final transistor configuration we shall study is the common base circuit shown in Figure 3–8. We shall find that this circuit has non–inverting voltage gain (output in phase with input) and has a low input resistance.

Examining the DC biasing for this circuit, we shall see that it is the same as the common emitter circuit in Figure 3–1.

Let us analyze the AC operation of this circuit. The input voltage v_s appears across R_e and R_t in parallel (R_t the input resistance of common base is calculated the same as R_o the output resistance for the follower—but without the generator resistance). This causes an

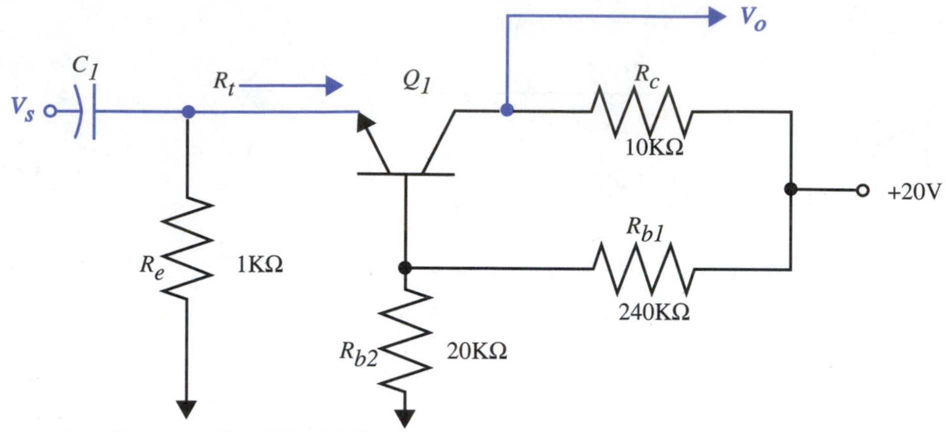

$R_t = R_o = r_e + (R_{b1} \parallel R_{b2}) / \beta$ $A_v = R_c / R_t$

FIGURE 3–8 Common base circuit

emitter signal current i_e of

$$i_e = \frac{v_s}{R_t}$$

so

$$v_s = i_e \times R_t$$

Again assuming

$$i_e \approx i_c$$

this current causes a voltage drop (v_o) across R_c of

$$v_o = (i_e)\,(R_c) \qquad \text{(with no phase shift)}$$

$$v_o = \frac{v_s}{R_t} \times R_c$$

In terms of common base circuit voltage gain A_v

$$A_v = \frac{V_o}{V_s}$$

$$A_v = \frac{i_e R_c}{i_e R_t}$$

$$A_v = \frac{R_c}{R_t} \qquad \textbf{(equation 3–6)}$$

EXAMPLE

Find the voltage gain A_v for the common base circuit in Figure 3–8 if β is 100.

SOLUTION

From equation 3–6

$$A_v = \frac{R_c}{R_t}$$

Since

$$R_t = R_o \text{ (for follower)}$$

we can use equation 3–5, but without the generator resistance:

$$R_t = r_e + \left(\frac{R_{b1} \parallel R_{b2}}{\beta}\right)$$

Our analysis of this circuit in the previous chapter gave an I_e of 0.79mA and an r_e of 33Ω with an assumed β of 100. Then,

$$R_t = 33\Omega + \frac{20K\Omega \times 240K\Omega}{(20K\Omega + 240K\Omega) \times 100}$$

$$R_t = 218\Omega$$

Voltage gain

$$A_v = \frac{R_c}{R_t}$$

Since

$$R_c = 10K\Omega$$

$$A_v = \frac{10K\Omega}{218\Omega}$$

$$A_v = 46$$

This gain is positive, which means no phase shift from input to output.

Notice that the input resistance seen by the source V_s is low, 218Ω in parallel with an R_e of 1KΩ which yields 179Ω. C_1 is selected to have a reactance of one tenth of this value (17.9Ω) at the lowest operating frequency.

Let us put a bypass capacitor C_2 from the base of the transistor in Figure 3–8 to ground and make its reactance

$$X_{C2} < \beta r_e/10.$$

Doing this effectively, AC shorts out both R_{b1} and R_{b2} because the power supply side of R_{b1} is at AC ground. This makes R_t equal to r_e, so the voltage gain equation is now simply

$$A_v = \frac{R_c}{r_e} \text{ (with base bypassed with } C_2\text{)} \qquad \textbf{(equation 3–7)}$$

EXAMPLE

Compute the required value of C_2 at 100Hz to bypass the base resistors and find the new voltage gain for the circuit of Figure 3–8 with the base AC bypassed.

SOLUTION

With

$$X_{C2} = \frac{\beta r_e}{10}$$

and

$$r_e = 33\Omega$$

from our previous analysis

$$X_{C2} = \frac{100 \times 33\Omega}{10} = 330\Omega$$

$$C_2 = \frac{1}{2\pi \times 100\text{Hz} \times 330\Omega}$$

$$C_2 = 4.82\mu\text{F (use } 5\mu\text{F)}$$

Using equation 3–6 to solve for A_v,

$$A_v = \frac{R_c}{r_e}$$

$$A_v = \frac{10\text{K}\Omega}{33\Omega}$$

$$A_v = 303$$

We use the common base circuit when we need to match a low-resistance signal source, i.e., antenna (typically less than 600Ω) to the input resistance of a transistor receiver circuit.

SUMMARY OF TRANSISTOR CIRCUIT CONFIGURATIONS

It is helpful now that we have studied all three transistor configurations to list key characteristics of each circuit.

CIRCUIT TYPE	*VOLTAGE GAIN*	*INPUT RESISTANCE*	*OUTPUT RESISTANCE*
Common Emitter	$-\frac{R_c}{R_e + r_e}$ or with R_e bypassed, $-\frac{R_c}{r_e}$	$R_e\beta \parallel R_{b1} \parallel R_{b2}$ or $r_e\beta \parallel R_{b1} \parallel R_{b2}$ (with R_e bypassed)	R_c
Emitter Follower	Approx. unity or $\frac{R_L}{R_L + r_e}$ (with $R_L << R_e$)	$(R_e\beta \parallel R_{b1} \parallel R_{b2})$	$R_o = r_e + \frac{R_i}{\beta}$ where $R_i = R_{b1} \parallel R_{b2} \parallel r_g$
Common Base	$\frac{R_c}{R_t}$ or $\frac{R_c}{r_e}$ (with R_{b2} bypassed)	$R_t = r_e + (R_{b1} \parallel R_{b2})\frac{1}{\beta}$	R_c

- Since R_t is usually less than R_e, the common base circuit has the highest voltage gain. However if we bypass R_e in the common emitter circuit, and we also bypass R_{b2} in the common base circuit, then the voltage gains for the common base and common emitter are the same.

- The common base has the lowest input resistance while the follower has the highest.
- The output resistance of the common base and emitter circuits is R_c, provided R_c is much less than the collector internal resistance (r_c) of about 1MΩ.

PNP TRANSISTOR AMPLIFIER CIRCUITS

Most transistor circuits use NPN transistors, but there are times when it is desirable to use the PNP types. Use of a PNP transistor can result in a simpler circuit, using fewer components. Figure 3–9 shows two common emitter amplifier circuit configurations using a PNP transistor.

We can see that we can use the same circuit as Figure 3–1a, but the power supply voltage is –20V rather than +20V. However, we get the same gain as with the NPN since A_v is still equal to R_c/R_e.

We can use a positive supply, as shown in Figure 3–9b. This circuit is preferred if we use both PNP and NPN transistors in the same circuit.

COMBINING PNP AND NPN TRANSISTORS

NPN and PNP transistors can be combined to provide a two-amplifier configuration without the use of coupling capacitors.

Consider the two circuits of Figure 3–10. The use of a PNP transistor for Q_2 in circuit (b) saves the two bias resistors required for Q_2 in circuit (a) plus the coupling capacitor

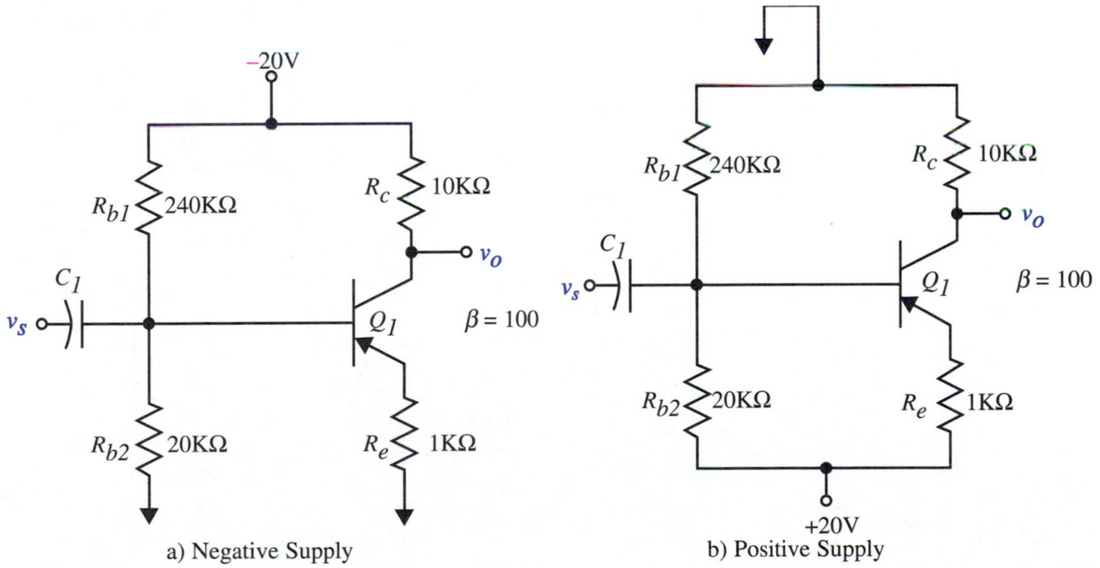

FIGURE 3–9 PNP common emitter circuit

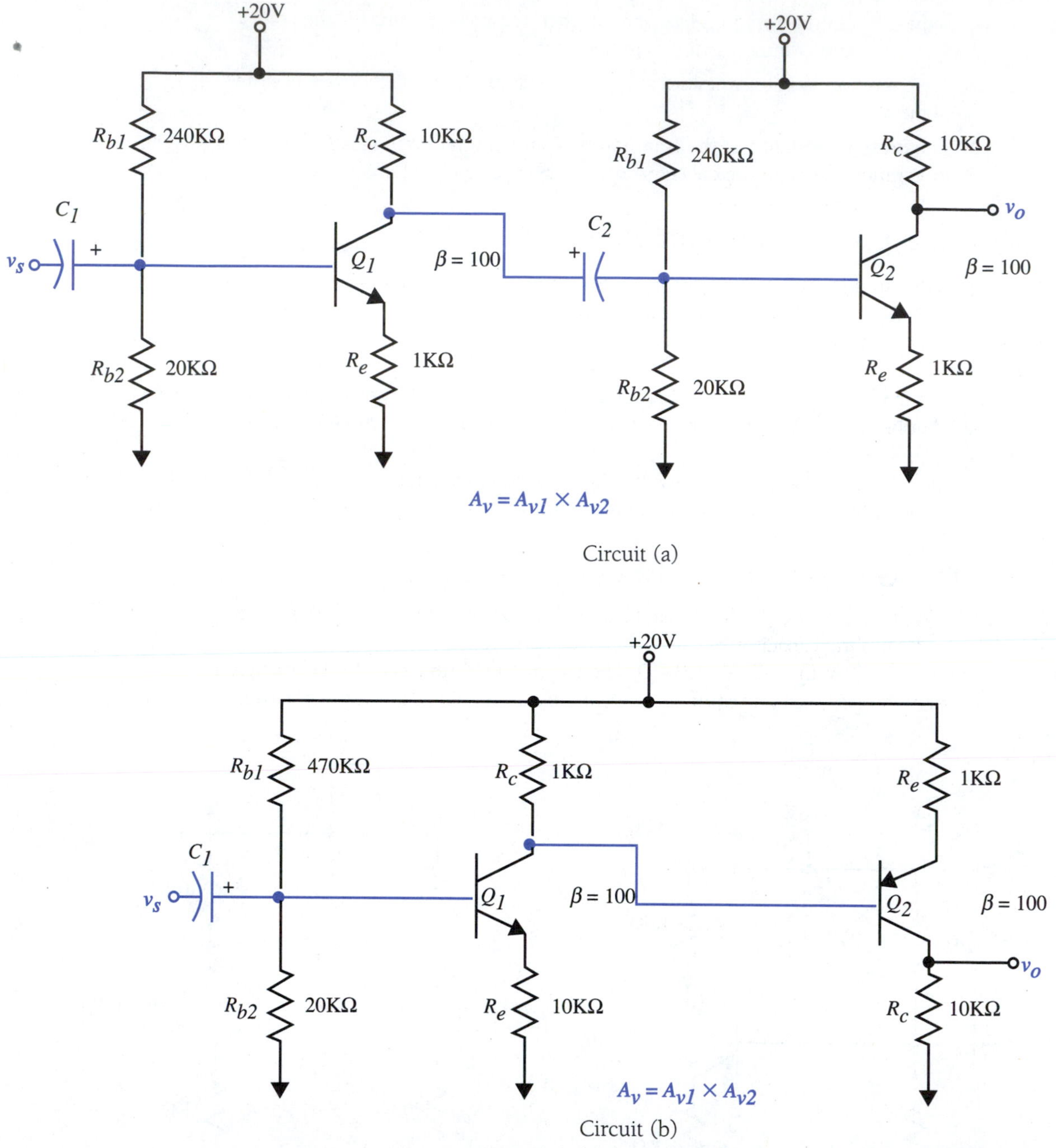

FIGURE 3–10 Two-stage amplifiers

C_2. In both circuits, each transistor stage is set up for a gain of 10. However, notice in circuit (a) that the 10KΩ output resistance (R_c) of Q_1 is loaded down by the 15.6KΩ input circuit R_{in} resistance (20KΩ || 240KΩ || 100 × 1KΩ) of the second stage Q_2.

EXAMPLE

Find the overall voltage gain for the circuit of Figure 3–10a.

SOLUTION

The overall gain of the circuit is the gain of the first stage times the gain of the second stage. The first stage has an effective resistor $R_{c(eff)}$ of 10KΩ in parallel with the 15.6KΩ input resistance of the second stage.

$$R_{c(eff)} = 10K\Omega \;||\; 15.6K\Omega$$

$$R_{c(eff)} = 6.1K\Omega$$

First stage gain is

$$A_{v1} = -\frac{R_{c(eff)}}{R_e}$$

$$A_{v1} = -\frac{6.1K\Omega}{1K\Omega}$$

$$A_{v1} = -6.1$$

Second stage gain is

$$A_{v2} = -\frac{10K\Omega}{1K\Omega}$$

$$A_{v2} = -10$$

So overall gain is

$$A_{v1} \times A_{v2} = +61$$

In the circuit of Figure 3–10b, the loading of the collector of the first stage is reduced because there are no biasing resistors (R_{b1} and R_{b2}) for the second stage. This allows a greater gain for the first stage.

We have increased R_{b1} in this circuit to change the bias point so that the base of Q_2 is at 18.18V (1.82V below the supply voltage) and the emitter is at 18.78V. This gives a 1.22V drop across the emitter resistor of Q_2 and, therefore, a 12.2V drop across the collector resistor ($I_c = I_e$). So now the v_{out} bias point is at +12.2V—which, again, is roughly one-half the supply voltage. (Use of standard value resistors in the Q_1 base circuit prevents a Q_2 collector voltage of +10V from being achieved.)

Looking back at Q_1 biasing, the change in R_{b1} puts Q_1 base voltage at 0.782V. With a base emitter drop of 0.6V, we have 0.182V across the 1KΩ emitter resistor and, therefore, a 1.82V drop across the Q_1 collector resistor. Subtracting this voltage from the 20V supply voltage gives us 18.18V—close to the desired 18.4V on the base of Q_2.

EXAMPLE

Compute the overall gain for the circuit of Figure 3–10b.

SOLUTION

Again, the overall gain is the product of the gain of the individual stages. For the first stage, the 0.181V drop across the 1KΩ emitter resistor means the emitter current I_e is

$$I_e = \frac{V_e}{R_e}$$

$$I_e = \frac{0.182\text{V}}{1\text{K}\Omega}$$

$$I_e = 0.182\text{mA}$$

Finding the internal resistance r_e,

$$r_e = \frac{26\text{mV}}{I_e}$$

$$r_e = \frac{26\text{mV}}{0.182\text{mA}}$$

$$r_e = 143\Omega$$

This is not negligible, compared to the 1KΩ emitter resistance, so we should use the gain equation 3–1 for computing the gain of the first stage. The effective collector resistance for the first stage $R_{c1(eff)}$ is,

$$R_{c1(eff)} = R_{c1} \parallel \beta R_{e2}$$

where R_{e2} is from the second stage.

$$R_{c1(eff)} = 10\text{K}\Omega \parallel \beta \times 1\text{K}\Omega$$

$$R_{c1(eff)} = 9.1\text{K}\Omega$$

Using equation 3–1,

$$A_{v1} = -\frac{R_{c1(eff)}}{(R_{e1} + r_{e1})}$$

$$A_{v1} = -\frac{9.1\text{K}\Omega}{(1\text{K}\Omega + 143\Omega)}$$

$$A_{v1} = -7.96$$

Second-stage gain is

$$A_{v2} = -\frac{10\text{K}\Omega}{1\text{K}\Omega}$$

$$A_{v2} = -10$$

Overall gain is

$$(A_{v1})(A_{v2}) = (-7.96) \times (-10) = +79.6$$

This is higher than the gain achieved with the circuit of Figure 3–10a because there is less loading of the first transistor collector resistor R_{c1}, and it therefore has a higher effective value ($A_{v1} = R_{c1(eff)}/R_{e1}$).

We can see from the comparison of the circuits in Figure 3–10 that in certain circuit applications, it is advantageous to use the PNP transistor. We could argue that an NPN transistor be used for the second stage, but this would require that the first-stage collector be biased to 1.6V. This means that the first stage would have to have much more collector current to provide the extra voltage drop across R_c.

A further advantage of the circuit of Figure 3–10b is that it can amplify DC signals. A disadvantage is that shifts in bias on the first stage are amplified by the second stage and upset the second-stage collector Q point.

PUSH–PULL OPERATION

The AC circuits described previously typically had the collector output biased to one-half of the supply voltage for maximum output voltage swing. A disadvantage to this approach is the circuit power efficiency (ratio of power put into the circuit to power taken out) is low. Because even without an AC signal present, the DC current and voltage drop across the transistor represent a power loss.

AMPLIFIER EFFICIENCY

So let us find the maximum efficiency of the transistor amplifier stage we have studied with the collector Q bias voltage point at half the supply voltage. Let the power supply voltage be V_{cc} and the load resistor R_c.

With the Q point at $V_{cc}/2$, the transistor is like a resistor of value R_c. So we have two resistors R_c in series between the supply and ground for a combined resistance of $2R_c$. Total power into the circuit is then

$$P_{in} = \frac{V_{cc}^2}{2R_c}$$

The maximum peak-to-peak sine-wave swing that can exist across the load resistor R_c is equal to the supply voltage V_{cc}. The RMS value of this p–p sine wave is

$$V_{rms} = \frac{V_{cc}}{2\sqrt{2}}$$

So power out is

$$P_{out} = \frac{V_{rms}^{2}}{R_c}$$

$$P_{out} = \frac{V_{cc}2}{8R_c}$$

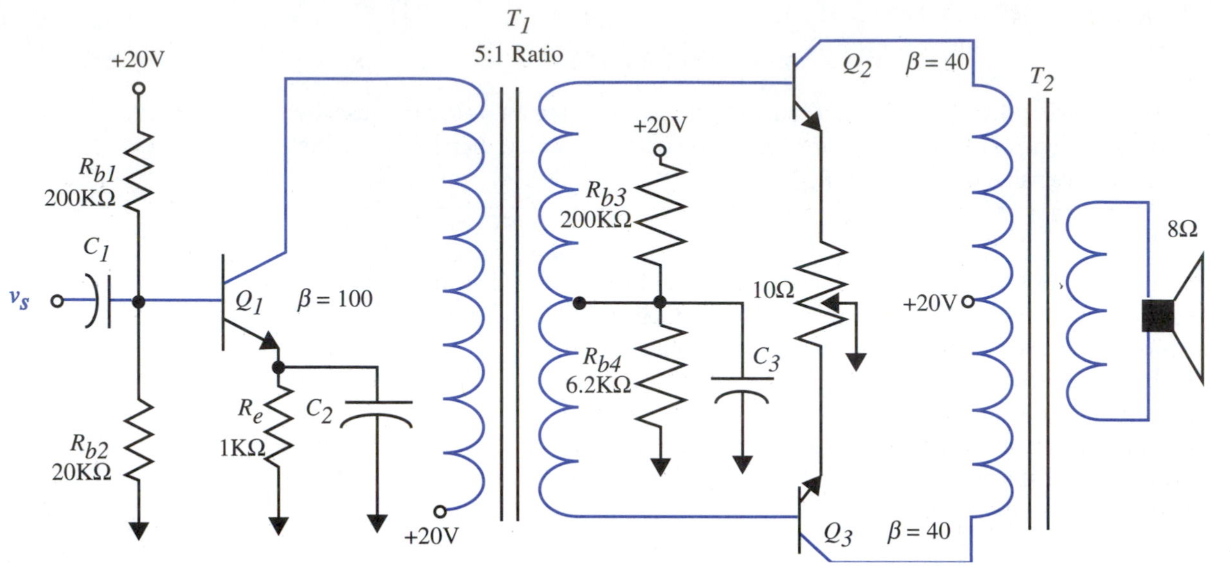

FIGURE 3–11 Push-pull output stage

Circuit efficiency is the ratio of useful power out over the DC power supplied by the power supply.

$$\%eff = 100 \times \left(\frac{P_{out}}{P_{in}}\right)$$

$$\%eff = 100 \times \frac{\left(\frac{V_{cc}^2}{8R_c}\right)}{\left(\frac{V_{cc}^2}{2R_c}\right)}$$

$$\%eff = 100 \times 1/4$$

$$\%eff = 25\% \text{ (maximum)}$$

Circuit efficiency becomes important in the design of the final power driver stage of an amplifier chain since this stage draws the most current from the power supply. The result is that the output stages are somewhat more complex to save power. Consider the push–pull output circuit shown in Figure 3–11.

The term **push–pull** means that Q_2 will turn on for one–half of the input sine–wave while Q_3 turns on for the other half. Biasing is set so that each transistor is just at the point of turn on, i.e., each transistor is just turned off.

With this circuit, the only time that current is drawn from the power supply by the output transistors is when an AC signal is present. This means a great improvement in circuit efficiency.

EFFICIENCY OF A PUSH–PULL CIRCUIT

The maximum efficiency for a push-pull circuit occurs when the peak of the output voltage across the load V_p equals the supply voltage V_{cc}.

The input power (P_{in}) to the circuit is the supply voltage times the average current. If we are dealing with a sine wave signal, the average value of a half sine wave is $2/\pi$, or 0.637, of the peak value, so

$$P_{in} = V_{cc} \times 0.637 I_p$$

$$P_{in} = V_{cc} \times \frac{0.637 V_p}{R_L}$$

The output power P_o is

$$P_o = \frac{\left(\frac{V_p}{\sqrt{2}}\right)^2}{R_L}$$

$$= \frac{V_p^2}{2R_L}$$

The efficiency η is with $V_p = V_{cc}$

$$\eta = P_o/P_{in}$$

$$\eta = \frac{\frac{V_p^2}{2R_L}}{V_p \times \frac{0.637 V_p}{R_L}}$$

$$\eta = 0.785 \text{ or } 78.5\%$$

The power dissipation P_d in each transistor during its conduction is the power in minus the power out:

$$P_d = P_{in} - P_o$$

$$P_d = V_{cc} \times \frac{0.637 V_p}{R_L} - \frac{V_p^2}{2R_L}$$

We can determine the output voltage amplitude for the maximum power dissipation in the transistors by calculus differentiation (dP_o/dV_p):

$$\frac{dP_o}{dV_p} = \frac{0.673 V_{cc}}{R_L} - \frac{2V_p}{2R_L}$$

Making dP_o/dV_p equal to 0 to find the maximum transistor power dissipation, and solving for V_p in terms of V_{cc}, we have:

$$V_p = 0.637 V_{cc} \qquad \textbf{(equation 3–8)}$$

Which shows that the maximum transistor dissipation doesn't occur with maximum output voltage!

BASIC OPERATION OF A PUSH–PULL CIRCUIT

In operation, if we assume a positive input signal on the base of Q_2, then this transistor will conduct and draw current through the upper half windings of the output transformer(T_2). With a positive signal on the base of Q_2, the base of Q_3 is negative because it is connected to the other end of transformer T_1. So this transistor is turned off. When the input signal changes polarity, Q_3 will be turned on and Q_2 turned off. Note that only half of T_2 is used at a time, so its effective turns ratio (N), when we compute the voltage across the transformer secondary, is cut in half.

$$v_s = \frac{v_p}{\left(\frac{N}{2}\right)} \qquad \textbf{(equation 3–9)}$$

OPERATION OF THE PUSH-PULL DRIVER CIRCUIT

Transistor Q_1 serves as a driver for transformer T_1. Since this amplifier stage draws current all the time, its efficiency will be lower than the push–pull output stage. However, being a lower-level stage, the power dissipation is relatively low.

Q_1 must pass both halves of the input sine wave, so it must be biased on. To prevent distortion, the DC biasing current through this transistor must be at least half of the maximum signal current anticipated through the primary of T_1. This may seem odd when we consider that, with the previous biasing used, the collector was always at half the supply voltage. However, when we use a transformer (or inductor) as a collector load, we must consider collector current bias rather than voltage.

Biasing of Q_1 is accomplished by providing 1.6V on the base, which puts 1V on the emitter. With the 1KΩ emitter resistor, this means that the emitter bias current, and therefore the collector current, is 1mA. If we assume the DC resistance of transformer T_1 primary is less than 100Ω, then the DC drop across this winding with this 1mA is negligible, and the full +20V supply voltage will be at the collector.

The actual load at the collector of the transistor Q_1 is the AC resistance reflected from the secondary of T_1. (This assumes a low primary DC resistance compared to the transferred secondary resistance; moreover the primary reactance is high.) The secondary resistance R_s is the approximately 5Ω emitter resistance R_e (balance pot at middle of range) of the output transistors multiplied by β (assume β of 40 for Q_2 and Q_3 power transistors):

$$R_s = \beta \times R_e/2$$

$$R_s = 40 \times 5\Omega$$

$$R_s = 200\Omega$$

Since C_3 AC bypasses R_{b3} and R_{b4} (top of R_{b3} is at AC ground), the total AC resistance seen at the secondary of T_1 is this 200Ω. If the step-down ratio of the transformer is 5:1, and recognizing that only one-half of T_1 secondary is being used at a time (either Q_2 or Q_3 is on), then the effective transformer turn's ratio is 10:1. So the AC load R_{c1} at the collector of Q_1 is

$$R_{c1} = (N)^2 \times R_s$$

$$R_{c1} = 10^2 \times 200\Omega$$

$$R_{c1} = 20K\Omega$$

If we apply a positive-going AC signal to the base of Q_1, the collector current will increase, and if the increase becomes 1mA, we shall have a 20V drop (20KΩ × 1mA) across the T_1 primary. Since our supply voltage is 20V, this will take Q_1 into saturation.

With a negative-going AC input signal, the current will decrease, and if it goes negative by 1mA, the net current through the transistor collector will be zero (DC bias current of 1mA minus 1mA signal current).

When the collector current increased by 1mA, we got a 20V negative swing at the collector. When the current decreased by 1mA we should expect a 20V positive swing at the collector. That means the collector will raise 20V over its original value or to +40V! This is twice the supply voltage—so where does the extra voltage come from? When the 1mA bias current was flowing through the inductance of the T_1 primary, it set up a magnetic field. This field collapses when the current in the primary is reduced, and this induces a voltage back into the winding and provides the extra 20V. So transistors with inductive collector loads can get an output voltage swing of up to twice V_{cc}.

EXAMPLE

Find the required voltage swing at the base of Q_1 in Figure 3–11 to provide a maximum collector swing of 20V if the effective collector resistance $R_{c1(eff)}$ is 20KΩ.

SOLUTION

The required input voltage v_s is the output voltage v_o divided by the gain (20V/A_v), where A_v is

$$A_v = \frac{R_{c1(eff)}}{r_e}$$

where

$$r_e = \frac{26\text{mV}}{1\text{mA}} \text{ or } 26\Omega$$

$$R_{c1(eff)} = 20\text{k}\Omega$$

$$A_v = \frac{20\text{K}\Omega}{26\Omega}$$

$$A_v = 769$$

So the required base drive voltage v_s is

$$v_s = \frac{20\text{V}}{769} = 26\text{mV}$$

which agrees with our analysis of Chapter 2 that determined a required signal of 26mV across r_e to cause transistor saturation.

OPERATION OF THE PUSH–PULL OUTPUT STAGE

Moving to the output stage consisting of Q_2, Q_3 and T_2, we can see that the 20V peak swing on the collector of Q_1 is stepped down by the 10:1 ratio to 2V peak at the base of

Q_2 or Q_3. Because these transistors are biased right at the point of turn on (0.6V), the full 2V will appear across the 5Ω (10Ω/2) emitter resistors. This means the peak collector currents of Q_2 and Q_3 will be 2V/5Ω or 400mA. With this peak current, we shall assume that 18V (20V supply minus 2V across the emitter resistor) is dropped across one-half of the transformer primary winding. This indicates our voltage gain is 18V/2V or 9V.

Solving for R_c in the gain equation

$$R_c = (A_v)(R_e)$$

our collector (Q_2, Q_3) AC resistance is

$$R_{c2} = R_{c3} = 9 \times 5\Omega = 45\Omega$$

For maximum power transfer, we use the transformer T_2 to match the 8Ω resistance of the speaker to this 45Ω collector resistance. Remembering that resistance is transformed from primary to secondary by the square of the turns ratio,

$$N^2 = \frac{R_{pri}}{R_{sec}}$$

$$N^2 = \frac{45}{8}$$

$$N^2 = 5.625$$

So

$$N = 2.37$$

This is half the overall turns ratio of 4.74 because, again, only one-half of the transformer primary is conducting current at a time.

With the effective turns ratio of one-half of 4.74, or 2.37, the 18V swing on the primary of T_2 is stepped down to 18V/2.37, or 7.59V, peak across the 8Ω speaker. If this is the peak of a sine wave (V_p), the RMS value is $V_p\sqrt{2}$ (same as 0.707 V_p), and the power at the speaker is

$$P_s = \frac{V_p{}^2}{2R}$$

$$P_s = \frac{(7.59\text{V})^2}{2 \times 8}$$

$$P_s = 3.6\text{W}$$

Each transistor, Q_2 and Q_3, contributes one-half of the output sine wave.

COMPLEMENTARY SYMMETRY

The push–pull configuration just described was very common with vacuum tube amplifiers, because the output transformer enabled matching the high output resistance of the vacuum tube to the low AC resistance of the speaker.

At this time, the push–pull circuit is restricted to applications like AC motor drive (servo) circuits where a step–up transformer is used at the output.

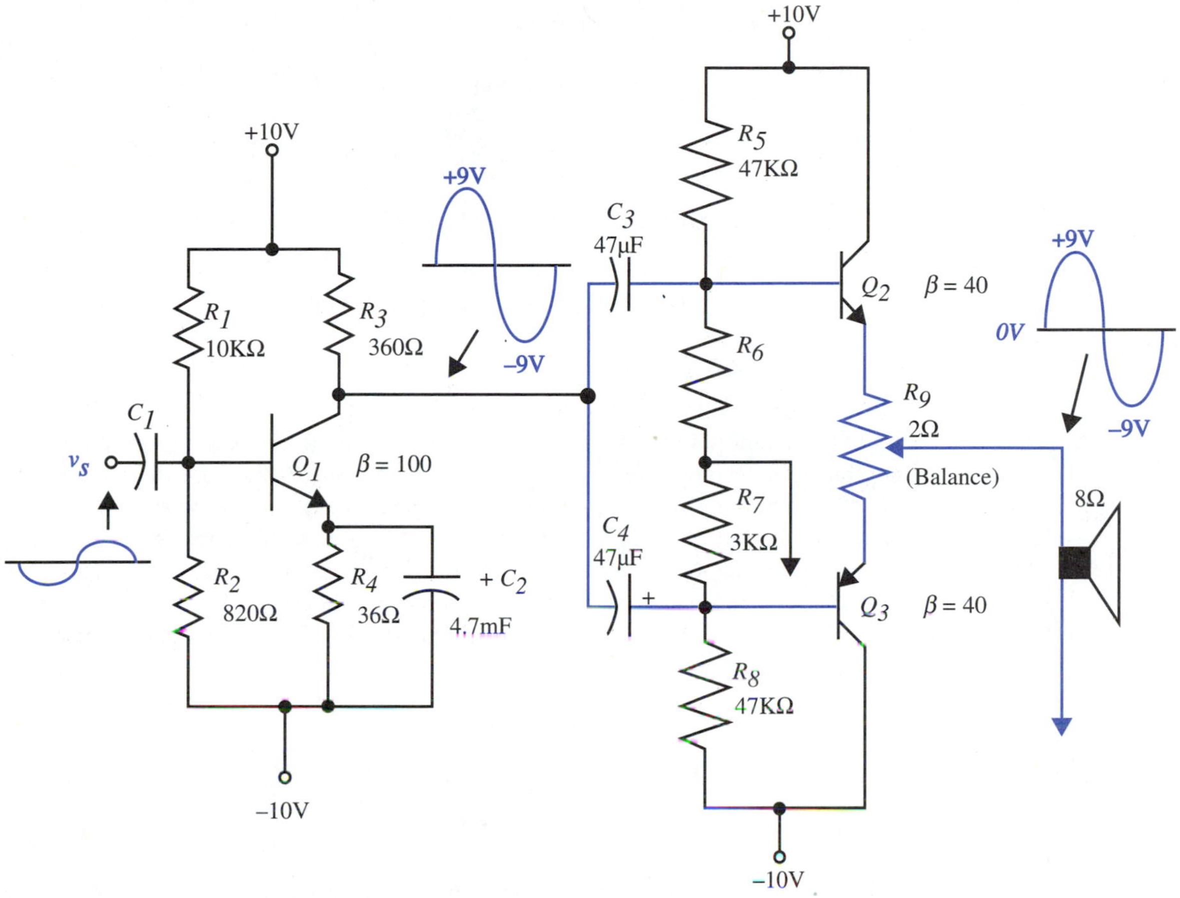

FIGURE 3–12 Complementary symmetry push-pull

Since we can get a very low output driving resistance from a transistor stage if we take our output from the emitter (emitter follower), we can eliminate the need for a transformer to drive low impedance loads. We show a push–pull output stage that doesn't use transformers in Fig 3–12—this stage is called a complementary symmetry (one transistor NPN is the complement of the other PNP) output stage.

Changing transistor Q_3 to a PNP type eliminates the need for an input transformer (T_1 in Figure 3–11) since Q_2 turns on with the positive alternation of the input sine wave, and Q_3 with the negative. As before, both output transistors are biased to just at the point of turn on.

COMPLEMENTARY SYMMETRY DRIVER STAGE

To determine the effective collector resistance of the Q_1 driver stage, we need to find the loading resistance R_t of the output stages. The input resistance R_t of the output follower stage is the 8Ω speaker resistance plus one-half of the 2Ω pot (R_9 wiper set at mid–point) times β.

$$R_t = R_e \times \beta$$

$$R_t = (8\Omega + \frac{2\Omega}{2})40$$

$$R_t = 360\Omega$$

This is in parallel with both sets of the 3KΩ and 47KΩ biasing resistors (selected to provide 0.6V at the bases).

$$R_{in} = 360\Omega \parallel 3K\Omega \parallel 47K\Omega \parallel 3K\Omega \parallel 47K\Omega$$

$$R_{in} = 287\Omega$$

From an AC standpoint, this resistance is in parallel with the collector resistor of Q_1, and the combination becomes the effective collector load resistor. We base the choice of the actual collector resistor (R_3) for Q_1 by trying to make it low enough to swamp out the effect of changes in the β of the output stages that will change R_{in} and therefore the gain of Q_1. If, however, we make R_3 too small, it will require that Q_1 conduct a heavy DC bias current that increases the power dissipation. The compromise choice was to use a resistor close to the value of R_{in} or the 360Ω shown in the schematic. The effective Q_1 collector resistance $R_{c(eff)}$ is R_{in} and R_3 in parallel.

$$R_{c(eff)} = R_{in} \parallel R_3$$

$$R_{c(eff)} = 287\Omega \parallel 360\Omega$$

$$R_{c(eff)} = 160\Omega$$

Looking at the Q_1 voltage amplifier circuit in greater detail, we see that the stage is like the common emitter stage used with the push–pull stage; but the resistor values have been reduced by a factor of roughly 25 to provide a low output resistance of 360Ω (R_3) from the collector of Q_1 in order to drive the low input resistance of the output stages.

If we DC bias the collector of Q_1 at about one-half the supply voltage, then there will be a 10V drop across R_3. The DC collector current I_{c1} will be

$$I_{c1} = \frac{\left(\frac{V_{CC}}{2}\right)}{R_3}$$

$$I_{c1} = \frac{\left(\frac{20V}{2}\right)}{360\Omega}$$

$$I_{c1} = 27.8mA$$

We select R_1 and R_2 to provide 1.6V on the base of Q_1, and with the 0.6V base emitter drop, there will be 1V on the emitter. Using a value of 36Ω for R_4 will achieve our required DC collector current of 27.8mA.

Since R_4 is AC bypassed, we need to determine the value of r_e to find the AC voltage gain of Q_1

$$r_e = \frac{26\text{mV}}{I_e}$$

$$r_e = \frac{26\text{mV}}{27.8\text{mA}}$$

$$r_e = 0.935\Omega$$

For Q_1 voltage gain

$$A_{v1} = \frac{R_{c(eff)}}{r_e}$$

$$A_{v1} = \frac{160\Omega}{0.935\Omega}$$

$$A_{v1} = 171$$

COMPLEMENTARY SYMMETRY OUTPUT STAGE

Looking now at the output stage, the maximum conservative voltage swing at the collector of Q_1 is about 18V, with the positive 9V turning on Q_2 and the negative 9V swing turning on Q_3. We have emitter follower action with both these transistors, which means that both the positive and negative cycles will appear across the speaker, giving an 18V peak-to-peak output signal. If we are using a sine wave, the maximum power delivered to the speaker is determined by converting the peak-to-peak to RMS.

$$v_{rms} = \frac{v_{p-p}}{2\sqrt{2}}$$

Speaker power is

$$P_s = \frac{(v_{rms})^2}{R_s}$$

$$P_s = \frac{(v_{p-p})^2}{8R_s}$$

$$P_s = \frac{(18\text{V})^2}{8 \times 8\Omega}$$

$$P_s = 5.1\text{W}$$

The voltage (v_s) required at the base of Q_1 in order to get this power out is

$$v_s = \frac{v_{p-p}}{A_{v1}}$$

$$v_s = \frac{18\text{V}}{171}$$

$$v_s = 0.105V_{p-p}$$

For both the push–pull and complementary circuits, the transistors Q_2 and Q_3 should be beta-matched, otherwise one-half of the output will be larger than the other half—

which means a distorted output signal. The potentiometer at the emitter of these transistors allows for some compensation for these differences.

The output resistance of the output stage is

$$R_o = (R_c \parallel R_{b1} \parallel R_{b2} \parallel R_{b1} \parallel R_{b2})/\beta + r_e + R_9/2$$

$$R_o = (360\Omega \parallel 3K\Omega \parallel 47K\Omega \parallel 3K\Omega \parallel 47K\Omega)/40 + r_e + 1\Omega$$

$$R_o = 7.17\Omega + r_e + 1\Omega$$

With our emitter current in the high current range, we can neglect r_e. So, R_o is very close to the desired output resistance of 8Ω which is required for maximum power transfer into the speaker.

SOLVING COMPLEMENTARY AMPLIFIER TEMPERATURE PROBLEMS

A problem can occur with the biasing of the output transistors for the push–pull and complementary symmetry configurations. As temperature increases, the voltage from base to emitter required to turn on the transistors decreases (2.2mV per degree C). So at higher temperatures, the transistors could turn on; which means DC current will flow through them. This can cause the transistors to heat up and further reduce the base emitter turn on voltage, giving a "thermal runaway" condition. The result can be burned out transistors and/or damage to the load caused by the DC current.

A solution to thermal runaway is to replace the R_{b2} resistor with a diode, as shown in Figure 3–13. The diode develops a forward drop of 0.6V at 25°C that is just at the point of transistor turn on. As temperature increases, however, the diode drop reduces, which keeps the transistor still biased at the point of turn on. Mounting the diode close to the output transistor ensures that both the diode and the transistor are at the same temperature.

CLASSIFICATION OF AMPLIFIERS

Operation of a transistor such that it turns on for half the input sine wave is called **Class B operation**. The push–pull circuits we have just studied are examples of this type of circuit. When the transistor stage is biased on so that the complete sine wave appears at the single transistor output (collector of Q_1 in the push–pull circuits), the mode is **Class A Operation**. Another mode of operation is **Class C operation** where the transistor is turned on for less than half the input sine wave (base–emitter reverse biased). The newest mode of operation is **Class D** which is used for high-efficiency audio amplifiers. With this mode, the AC waveform is converted to variable width pulses, where the width of the pulse is related to the sampled amplitude of the AC waveform.

Class C Amplifier

As can be expected, the output of a Class C stage doesn't resemble a sine wave—so why is it used? It is used in radio frequency circuits with a tank circuit (parallel resonant circuit) as the collector load. The current pulse, during transistor turn on, excites the tank,

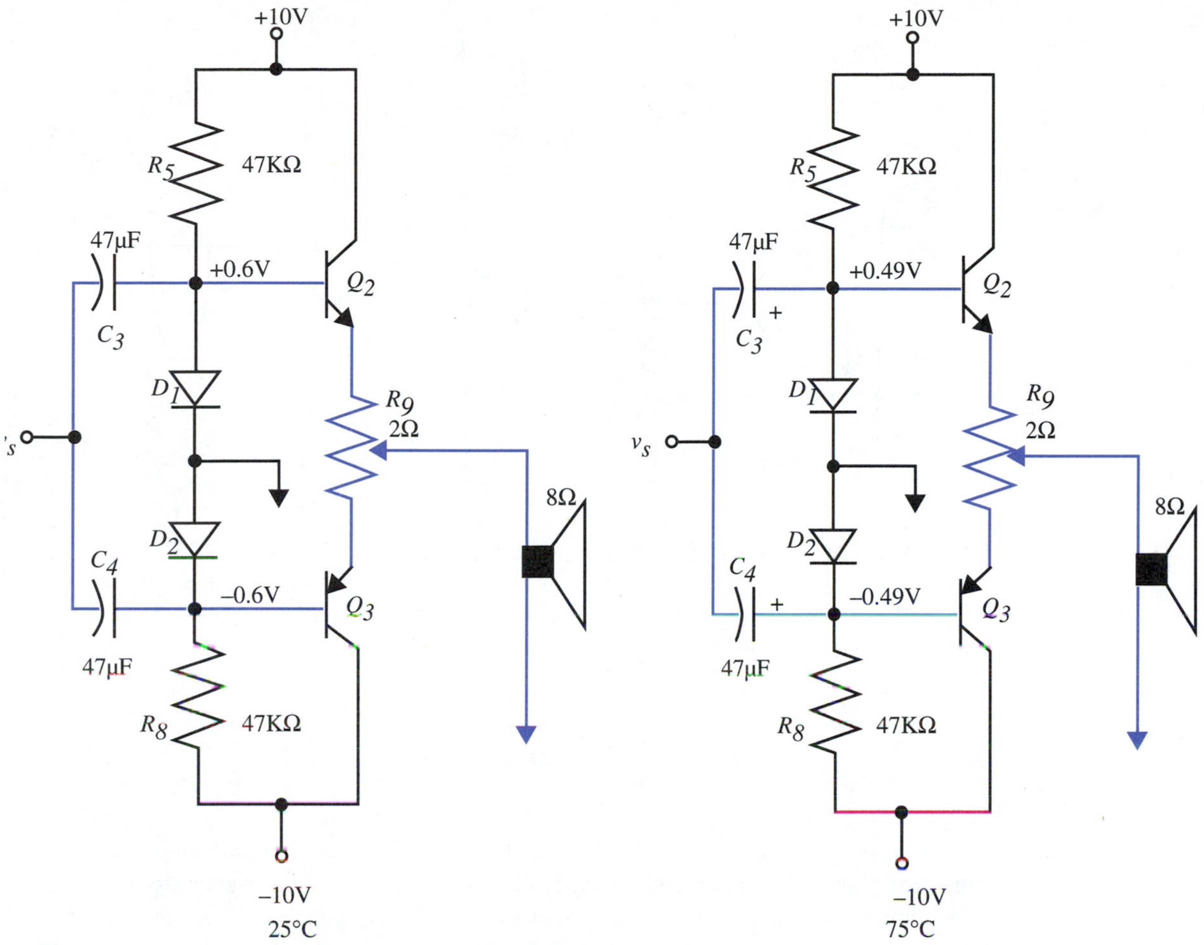

FIGURE 3–13 Complementary circuit with temperature compensation

and an output sine wave corresponding to the resonant frequency is generated. The wider the current pulse, the greater the amplitude of the output sine wave. Figure 3–14 shows a Class C stage.

Class C is an efficient output stage because the transistor operates closer to a switch-type operation. A switch has full voltage across it and no current when open—so there is no power dissipation; and when the switch closes, there is full current but no voltage across the switch—again no power dissipation.

In operation (see Figure 3–14), the capacitor (*C*) charges when the transistor is switched on but no current flows through the inductor (*L*) After the transistor is turned off, the capacitor discharges through the inductor and builds up a field around the inductor. When the field collapses, an induced voltage causes current flow back into the capacitor and charges the capacitor in the other direction. This oscillatory condition creates the sine wave at the collector. The resonant frequency for the tank circuit is determined by

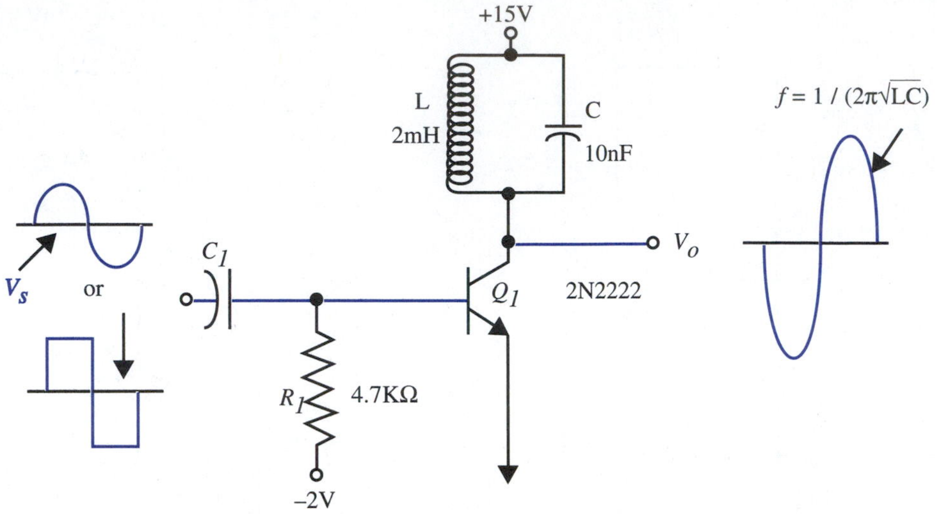

FIGURE 3–14 Class "C" output stage

$$f_o = 1/2\pi(LC)^{1/2}$$

and for the values given in Figure 3–14

$$f_o = \frac{1}{2\pi\sqrt{2 \times 10^{-3}\text{H} \times 10^{-8}\text{F}}}$$

$$f_o = 35.6\text{KHz}$$

If we turn on the transistor when the collector voltage swings to its lowest value, the transistor power dissipation will be least, and we will reinforce the oscillating waveform (like someone pushing a swing at the right time). This requires that the input frequency be at the same frequency as the tank circuit or a submultiple of the tank frequency (used for frequency multiplication).

Class D Amplifier

A Class D amplifier is also called a switching amplifier. Switching refers to the fact that the amplifier output is switched between two voltage levels. This accounts for its high efficiency. We can see how a Class D amplifier operates from the diagram and waveforms of Figure 3–15.

The input sine wave is converted to a varying pulse width signal. The most positive amplitude of the sine wave creates the widest pulse, and the most negative creates the narrowest. Pulse width, then, contains the amplitude information of the input signal, and the rate of change of the width of the output signal contains the frequency information.

We can consider the power amplifier as essentially a switch that is controlled by the width and polarity of the input signal. The width controls how long the switch is closed, and the polarity determines which supply is connected to the output.

Finally, the sharp transitions in the waveform are removed by a low pass filter (LC), and the amplified sine wave is applied to the speaker.

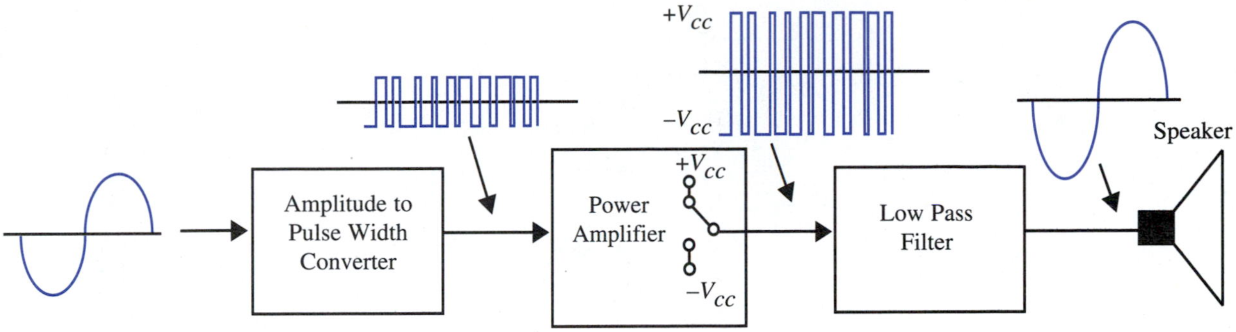

FIGURE 3–15 Class D amplifier

The following chart is a summary of the four amplifier classes.

CLASS	*LINEARITY*	*EFFICIENCY*
A	Best	Worst
B	Good	Good
C	Worst	High
D	Good	High

TRANSISTOR OSCILLATOR CIRCUITS

An **oscillator** is a circuit that self-generates a signal (typically a sine wave) without an input signal being applied. All that is required for the oscillator circuit to operate is a DC power supply. In order to oscillate, a circuit must satisfy the following conditions:

- A portion of the output signal must be fed back to the input.
- The portion fed back must reinforce the output (called positive feedback).
- The amount of fed-back signal must be sufficient to maintain the oscillation (overall gain greater than unity).
- A frequency-selective network is used to ensure that oscillation occurs just at one frequency.

A transistor oscillator circuit can be inbuilt using several different circuits. We shall look at three common types—Hartley, Colpitts, and Pierce.

HARTLEY OSCILLATOR

Figure 3–16 shows the circuit for a Hartley oscillator. Conditions for oscillation are satisfied by feeding back, in phase, a portion (10% to 25%) of the output signal developed across L_{1a} (*Note*: the +22V is at AC ground). The parallel resonant circuit (tank circuit) consisting of L_1 and C_1 ensures oscillation at the desired frequency in accordance with the familiar equation:

$$f_o = \frac{1}{2\pi\sqrt{L_1C_1}} \quad \text{(equation 3–10)}$$

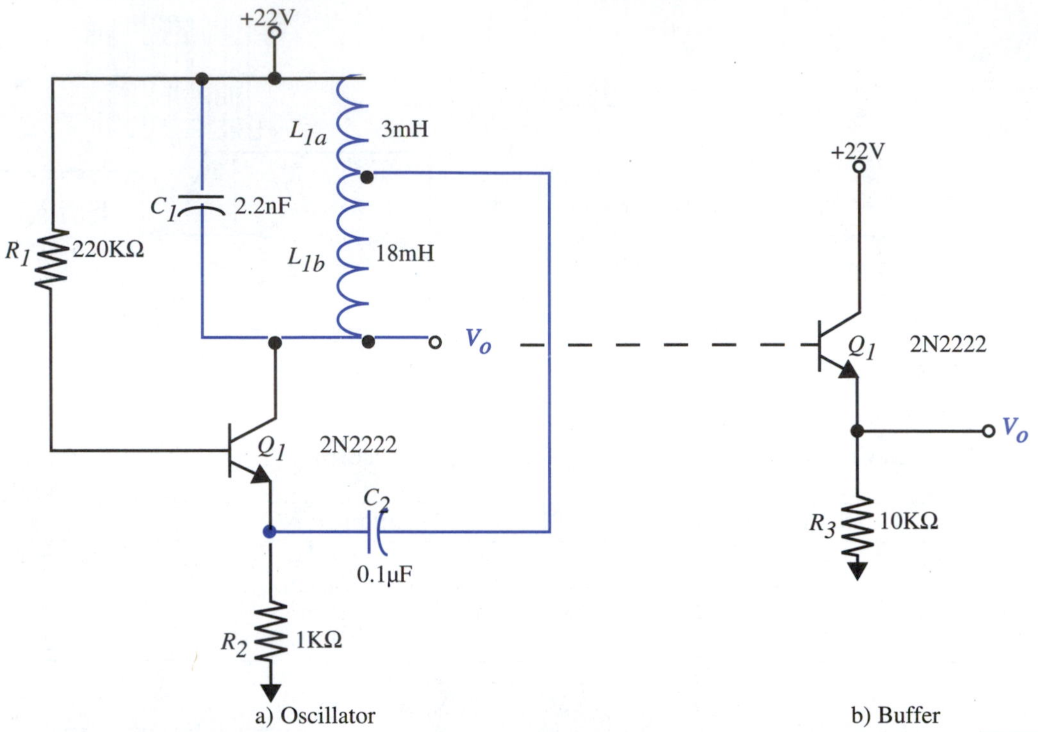

FIGURE 3–16 Transistor Hartley oscillator

Frequency of the oscillator can be fine-adjusted by having a small variable capacitor (variable up to about 100pF) connected in parallel with C_1. L_1 can be a single tapped inductor or two separate inductors.

The oscillator output is taken from the collector and fed into a high-input resistance buffer stage (usually a follower). The buffer is used to prevent loading the tank circuit or adding capacitance, which could shift the frequency.

EXAMPLE

Find the frequency of oscillation for the Hartley oscillator circuit of Figure 3–16 (L_1 = 3mH + 18mH)

SOLUTION

Using equation 3–10,

$$f_o = \frac{1}{2\pi\sqrt{L_1 C_1}}$$

$$f_o = \frac{1}{2\pi\sqrt{21\text{mH} \times 2.2\text{nF}}}$$

$$f_o = 23.4\text{KHz}$$

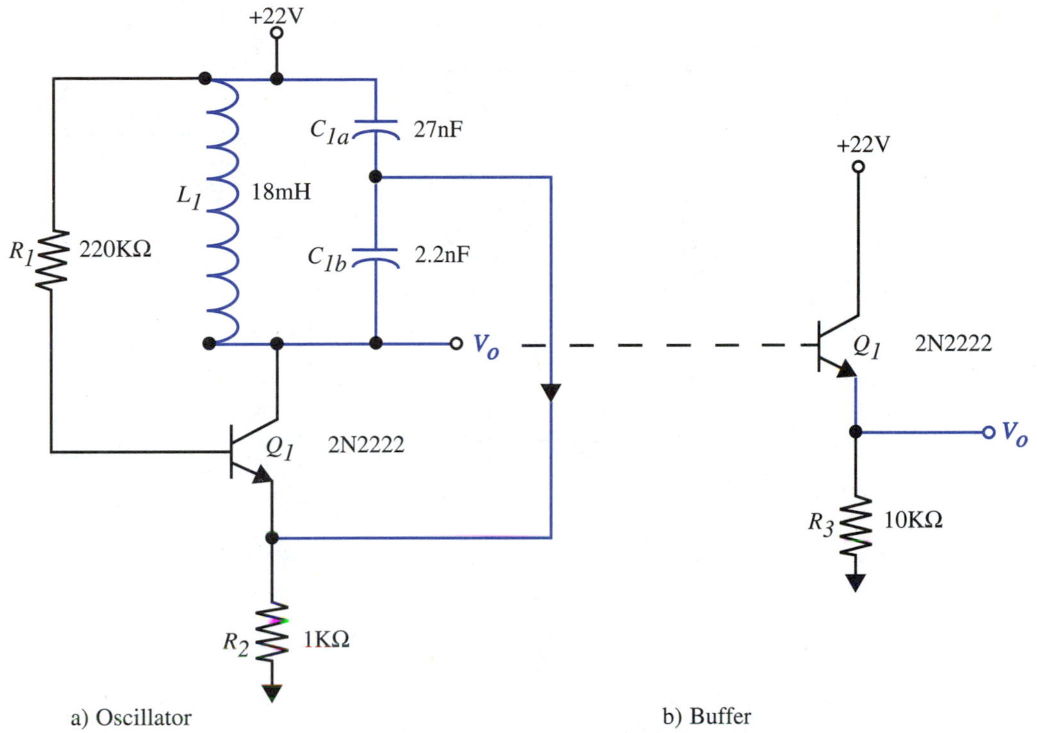

FIGURE 3–17 Colpitts oscillator

COLPITTS OSCILLATOR

The Colpitts oscillator circuit is similar to the Hartley oscillator except it has the advantage that the portion of the output fed back to the input is taken from a capacitive divider rather than an inductive divider.

Figure 3–17 shows the circuit for this configuration.

Equation 3–10 can also be used to determine the frequency of oscillation for the Colpitts oscillator.

EXAMPLE

Find the frequency of oscillation for the Colpitts oscillator circuit of Figure 3–17.

$$\left(C_1 = \frac{C_{1a} \times C_{1b}}{C_{1a} + C_{1b}}\right)$$

SOLUTION

First finding C_1,

$$C_1 = \frac{C_{1a} \times C_{1b}}{C_{1a} + C_{1b}}$$

$$C_1 = \frac{27\text{nF} \times 2.2\text{nF}}{27\text{nF} + 2.2\text{nF}}$$

$$C_1 = 2.03\text{nF}$$

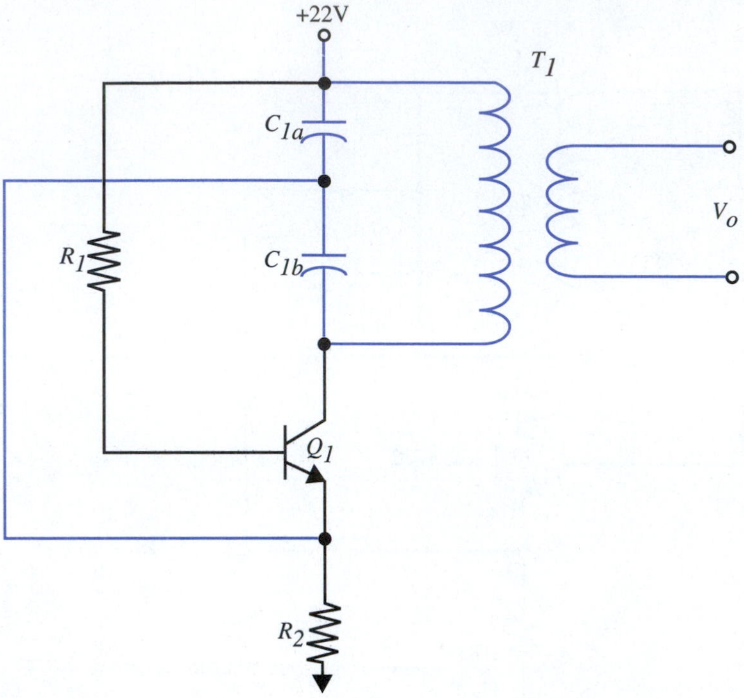

FIGURE 3–18 Oscillator transformer coupled unit

Now using equation 3–10

$$f_o = \frac{1}{2\pi \, L_1 C_1}$$

$$f_o = \frac{1}{2\pi \sqrt{18\text{mH} \times 2.03\text{nF}}}$$

$$f_o = 26.33\text{KHz}$$

The inductor of both the Hartley and the Colpitts oscillators can be the primary of a transformer, which is used to match the output impedance of the transistor to the load. This configuration is shown in Figure 3–18 using a Colpitts oscillator.

CRYSTAL CONTROLLED OSCILLATORS

A quartz crystal is used for precise control of the frequency of an oscillator. It exhibits the piezoelectric effect, which means the crystal mechanically deforms when a voltage is applied across its terminals, and similarly, it generates a voltage when it is mechanically deformed. The crystal mechanically vibrates at a precise frequency, determined by the size and cut of the crystal when excited by the voltage impressed across it. During this vibration, energy is absorbed by the crystal and then returned to the circuit—much like an electrical resonant circuit. The equivalent electrical circuit for a crystal is shown in Figure 3–19.

The crystal is described by the series branch of resistance, inductance, and capacitance—while the capacitor C_t in parallel represents the capacitance across the crystal faces and the contact capacitance. At low frequencies, the series branch is capacitive so the crystal appears like two capacitors in parallel. At the series resonant frequency, the series crystal branch impedance is simply R_C. This loss element is very low in a crystal (high Q), so C_t is effectively shorted. At frequencies above the series resonant frequency, the series branch is inductive. This inductive component parallel resonates with C_t at a frequency slightly higher than the series resonant frequency causing maximum voltage to be developed across the crystal.

An oscillator circuit that uses a crystal (Xtal) is the Colpitts oscillator shown in Figure 3–20. In this circuit, the crystal's inductive component beyond series resonance, resonates with the C_{1a}, C_{1b} combination. In other words, the crystal inductance replaces the L_1 inductor used with the Hartley and Colpitts oscillators.

The high Q and stability of the crystal mean that the oscillator frequency will be very precise. Crystal oscillators are used in radio station transmitters to keep the transmitted signal frequency within the Federal Communication Commission (FCC) band allocation.

TROUBLESHOOTING AC AMPLIFIERS

When troubleshooting an AC amplifier, we use the same basic techniques described for transistor bias circuits to determine whether the transistor itself is good. However, with the AC amplifier, our concern is with AC signal flow through the circuit, i.e., is the sig-

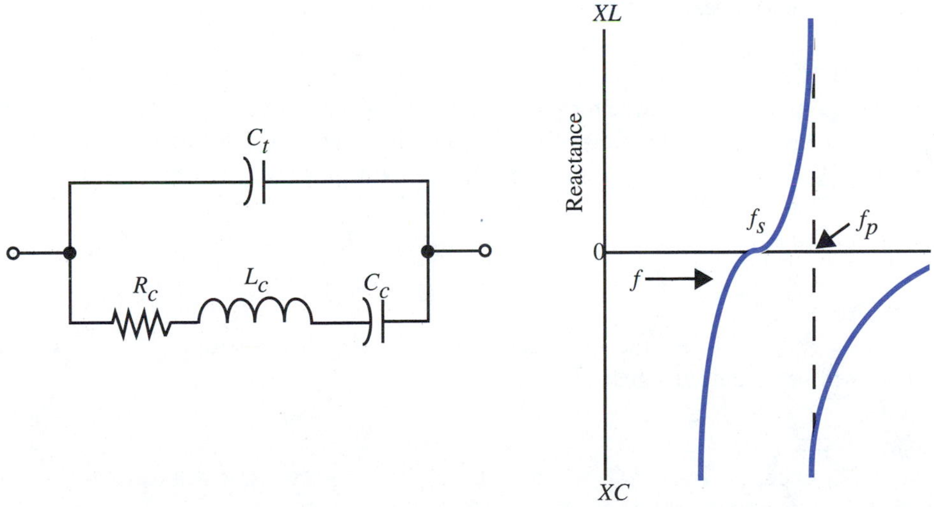

FIGURE 3–19 Crystal equivalent circuit and response

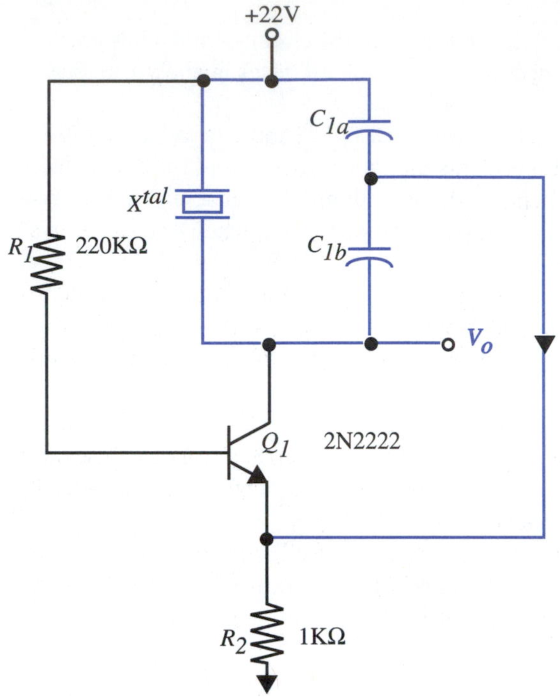

FIGURE 3–20 Crystal controlled oscillator

nal passed from one amplifier stage to the next? Is the stage gain correct? Is the waveform distorted?

Let us first look at problems that can occur with the single gain stage we studied in Figure 3–1 and repeat in Figure 3–21.

C_1 Shorted

Starting with the input capacitor C_1, if it is shorted, the 600Ω DC resistance of the generator will be in parallel with the 20KΩ bias resistor, and the base DC voltage will drop from 1.39 V to 48mV, which will turn off Q_1. We will then measure 0V on the emitter and +20V on the collector of Q_1. If the input voltage swing is sufficient to turn the transistor back on, then a clipped signal will appear at the collector of Q_1.

C_1 Open

With C_1 open circuited, the AC signal will not appear at the base of Q_1, and therefore there will be no output signal.

R_{b1} Open

With R_{b1} open, the base of Q_1 will drop to 0V, and the transistor will turn off (same DC conditions as with C_1 shorted).

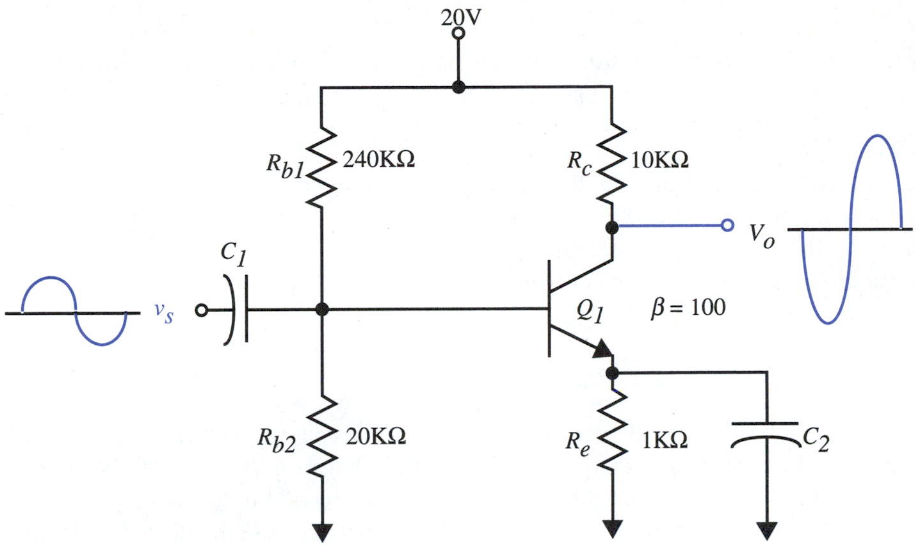

FIGURE 3–21 Trouble–shooting a single stage amplifier

R_{b2} Open

If R_{b2} opens up, all the current from R_{b2} will enter the base of Q_1 and cause it to go into saturation. With Q_1 saturated, we will have only about 0.2V across this transistor, and so we can consider R_c and R_e and a simple voltage divider for calculating the collector and emitter voltages. So the emitter voltage V_e will be

$$V_e = \frac{20V \times 1K\Omega}{10K\Omega + 1K\Omega}$$

$$V_e = 1.82V$$

and

$$V_c = V_e + 0.2V$$

$$V_c = 2.02V$$

also

$$V_b = V_e + 0.6V$$

$$V_b = 2.42V$$

With these voltage conditions, no AC signal will appear at the collector unless the input signal is large enough to take the transistor out of saturation—then a distorted signal will appear.

C_2 Open

If C_2 opens, the AC gain will drop to the DC gain, i.e., 10 for our circuit example.

C_2 Shorted

If C_2 shorts, the emitter of Q_1 will drop to ground potential, and the base will be 0.6V. The current now flowing into the base will be (20–0.6)V/240K or 81μA. Multiplying this current by the assumed β of 100 gives a collector current of 8.1mA. With this predicted current and a collector resistor of 10KΩ, we can see that the transistor will be in saturation.

R_e Open

If R_e opens, the transistor will turn off, the voltage at the collector will rise to 20V, and no AC signal will be present at the collector. The DC voltage on the base will be ((20V × 20KΩ)/260KΩ) or 1.54V. The emitter DC voltage will depend on the resistance provided to ground by the measuring instrument and will range from close to 1.2V down to about 1V.

R_c Open

Finally, if R_c opens, Q_1 will turn off, and with no transistor action, the base emitter diode in series with the emitter resistor will appear in parallel with R_{b2}. Figure 3–22 shows this circuit. The voltage seen at the base is found by making a Thevenin's equivalent of R_{b1} and R_{b2} and then connecting the diode/emitter resistor circuit. From the figure, we see that

$$V_b = 0.648\text{V}$$

and

$$V_e = 48\text{mV}$$

TROUBLESHOOTING A MULTISTAGE AMPLIFIER

If we are troubleshooting a multistage amplifier, as shown in Figure 3–10, we can examine each amplifier stage as we did previously, but first we should determine which stage is defective. We can do this by signal tracing. First, we determine that we have the correct signal input (wave shape, amplitude, and frequency). Then, since the reason we are troubleshooting is that an incorrect or no signal appears at the output, we look at the AC signal in the middle of the amplifier chain (for the two-stage amplifier the output of the first stage).

Now, if the signal is not present in the middle of the amplifier chain, we know the problem stage precedes the middle. If the signal is fine (based on a rough estimate of the total gain to this point), then the problem must be in a later stage.

We continue this process of looking at the middle stage of the suspected portion of our overall amplifier until we isolate the defective stage. Once we have found the problem cir-

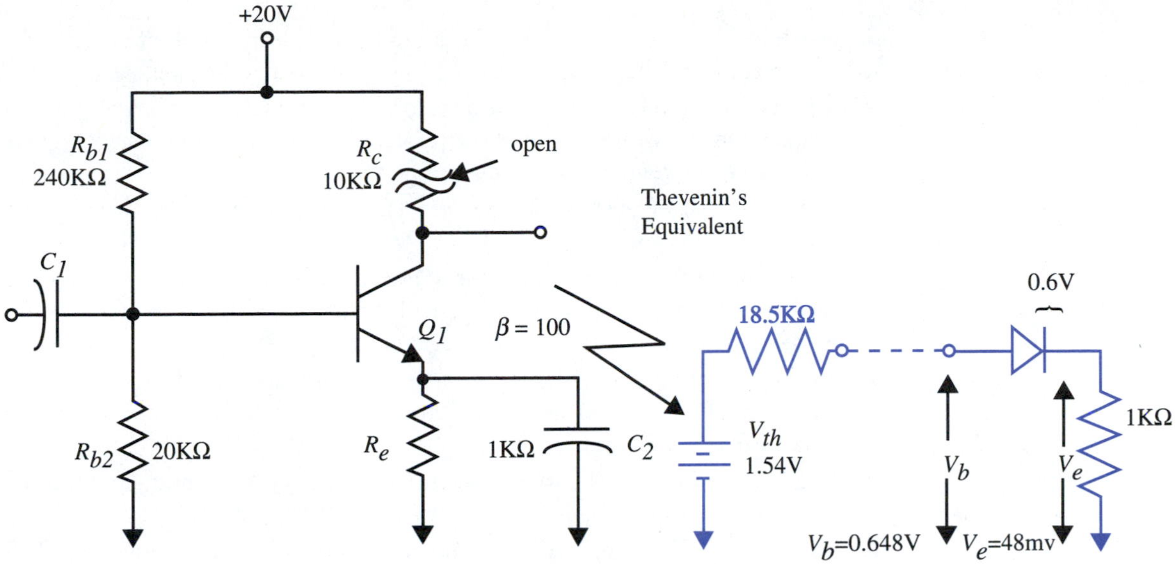

FIGURE 3–22 Transistor circuit with open collector resistor

cuit, then we can analyze it, using the symptoms determined previously for the signal stage amplifier to isolate the defective component.

SUMMARY

- An AC amplifier will not amplify DC signals.
- We can identify an AC amplifier by the presence of the coupling capacitor (C_1) between stages.
- Common emitter means no signal is applied or taken from the emitter.
- The gain of a common emitter amplifier is R_c/R_e without C_2 and R_c/r_e with C_2.
- The coupling capacitor C_1 prevents the previous stage from effecting the transistor biasing.
- C_2 AC bypasses the emitter resistor R_e, which is used for DC bias stabilization.
- If C_2 is used, the transistor input resistance (R_{in}) seen at the base is r_e times beta (β).
- The follower provides a voltage gain of unity but provides a current gain of β.
- A follower is used when we want to drive a low-impedance load.
- The output resistance of the follower is $r_e + (R_g \parallel R_{b1} \parallel R_{b2})/\beta$.
- The common base amplifier is used when we want to match a low-resistance source (antenna, etc.) to the transistor input resistance.

- PNP transistors are useful when we want to DC couple from one amplifier to the next to eliminate the need for the coupling capacitor and the base bias resistors.
- A push–pull amplifier uses two transistors that are biased just at the point of turn on. One transistor amplifies the positive portion of the input signal, while the other amplifies the negative portion.
- The advantage of push–pull is greater efficiency, since power is used only when a signal is present.
- To eliminate the need for transformers, the complementary symmetry circuit is used. This circuit uses a matched pair of transistors: an NPN for the positive swing and a PNP transistor for the negative swing of the signal.
- A problem with the push–pull type circuits is thermal drift, which can turn on the output transistors. This can be cured by using compensating diodes in place of the R_{b2} resistors.
- A class A amplifier has current flow through the transistor and amplifies the complete sine wave.
- A class B amplifier is biased just at the point of turn on and amplifies one-half of the input sine wave.
- A class C amplifier is biased below turn on (base–emitter junction reverse biased) and effectively acts like a switch.
- Class C amplifiers have a tuned collector load (L,C) to reconstruct a sine wave.
- When troubleshooting an AC amplifier, we need to be aware of the anticipated gain of the stage as well as the DC voltages at each point.
- Multistage amplifiers are checked by verifying the correct input, and if no signal appears at the output, then the signal is checked in the middle of the amplifier chain. This process of always checking in the middle of the suspected part of the circuit continues until the defective stage is found.

EXERCISE PROBLEMS

1. What is the purpose of the coupling capacitor C_1 in the AC amplifier circuit of Figure 3–A?

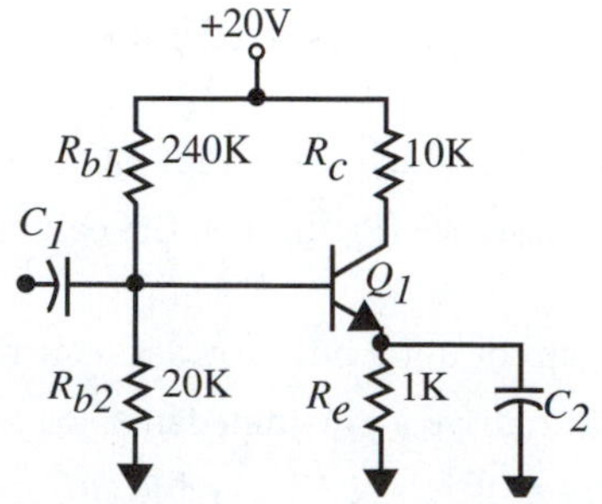

FIGURE 3–A (text figure 3–1)

2. How is the value of C_2 in Figure 3–1 determined?
3. Why are R_{b1} and R_{b2} in parallel from an AC standpoint?
4. What is the purpose of C_2 in the AC amplifier, and how is the relationship $C_2 = \beta C_1$ derived?
5. What happens if the value of C_2 in Figure 3–A is lowered?
6. Go through the steps to show that the AC gain of the common emitter amplifier, without C_2, is $-R_c/R_e$ if $R_e >> r_e$.
7. Why is the gain of the AC common emitter amplifier, with C_2, equal to R_c/r_e?
8. Calculate the required value of C_2, at 200Hz, for the circuit of Figure 3–A.
9. What is the purpose of the emitter follower circuit?
10. Find the base voltage for the circuit of Figure 3–B if R_{b1} is 15KΩ, *Rb*2 is 22KΩ, R_e is 2KΩ, β is 150, and V_{cc} is 24V.
11. Find the circuit AC input resistance for the follower of Figure 3–B with C_2 and R_L disconnected, and with this condition, also the required value of C_1 at 50Hz ($\beta = 100$).

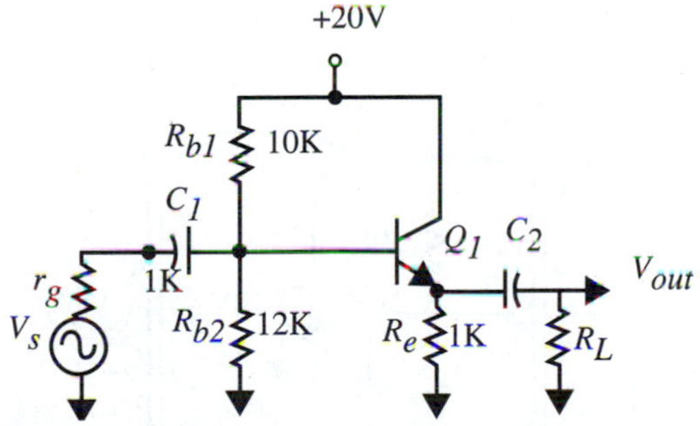

FIGURE 3–B (text figure 3–4)

12. Calculate the output resistance R_o for the circuit defined by problem 10 if the generator resistance is 5KΩ.
13. State the purpose of the common base circuit.
14. Calculate the input resistance R_t for a common base circuit if R_{b1} is 150KΩ, R_{b2} is 15KΩ, I_e is 2mA, and β is 100.
15. Calculate the gain for the circuit of problem 14 if R_c is 4.7KΩ
16. Calculate the gain of the circuit of problem 15 if the R_{b2} resistor is effectively AC shorted with a capacitor. (*Hint*: consider effect on R_t.)
17. Calculate the overall gain of the circuit of Figure 3–C if all resistors are decreased by a factor of 10.

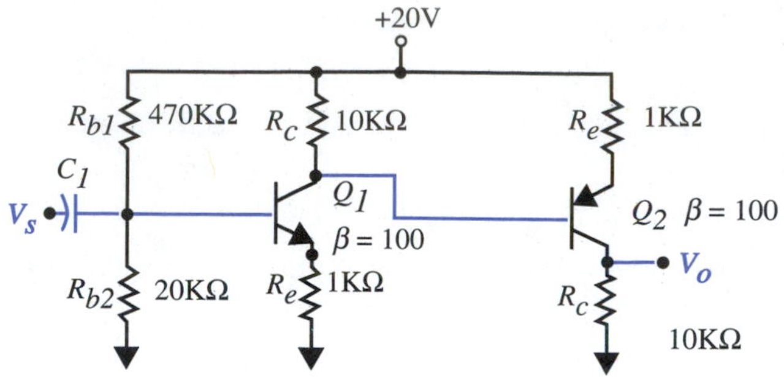

FIGURE 3–C (text figure 3–10)

18. Why is a push–pull circuit used?
19. Why is only one-half of the transformer turns ratio used when computing the secondary?
20. With reference to Figure 3–D, why must the DC bias current through the Q_1 collector be one-half the maximum expected transistor current? What is the DC voltage on the collector with no signal applied?

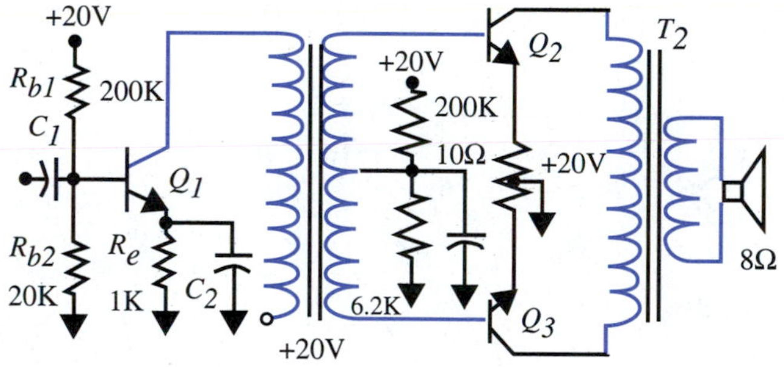

FIGURE 3–D (text figure 3–11)

21. What is the AC load on the collector of Q_1 in Figure 3–D if the T_1 step-down ratio is changed to 10:1?
22. Calculate the power level at the speaker of Figure 3–D if its impedance is 4Ω rather than 8Ω and the T_2 ratio is changed to match this new load.
23. State the advantages and disadvantages of the complementary-type over the regular push–pull circuit.
24. Explain why the Q_1 transistor for the complementary circuit (Figure 3–E) draws more current than the Q_1 transistor used in the regular push–pull circuit (Figure 3–D).

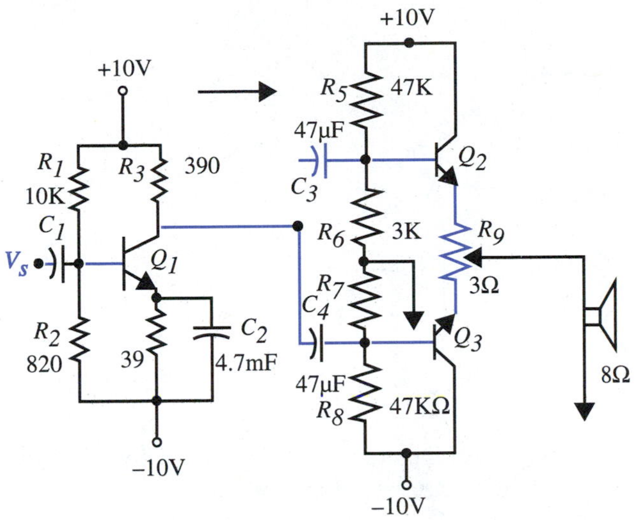

FIGURE 3–E (text figure 3–12)

25. Why are the push–pull output stages R_{b2} resistors sometimes replaced with diodes?
26. Describe Class A, B, and C types of amplifiers.
27. Why does a Class C amplifier require a tuned circuit at the output?
28. State the conditions required to be satisfied for an oscillator to operate.
29. Describe the difference between a Hartley and a Colpitts oscillator.
30. Why is a crystal used in an oscillator circuit?
31. Determine the DC voltages at base and emitter if C_1 in Figure 3–B shorts. Assume r_g is 1KΩ and β is 100.

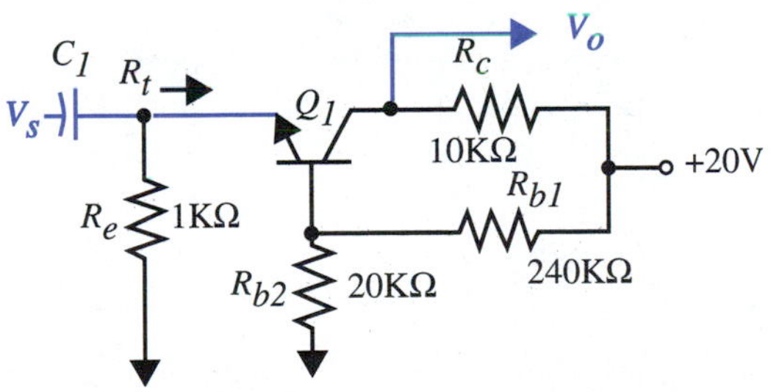

FIGURE 3–F (text figure 3–8)

32. Determine the DC voltages at collector, base, and emitter if R_e in Figure 3–F open circuits.
33. Determine the voltages at base, emitter, and collector if the 20KΩ resistor in Figure 3–G open circuits.

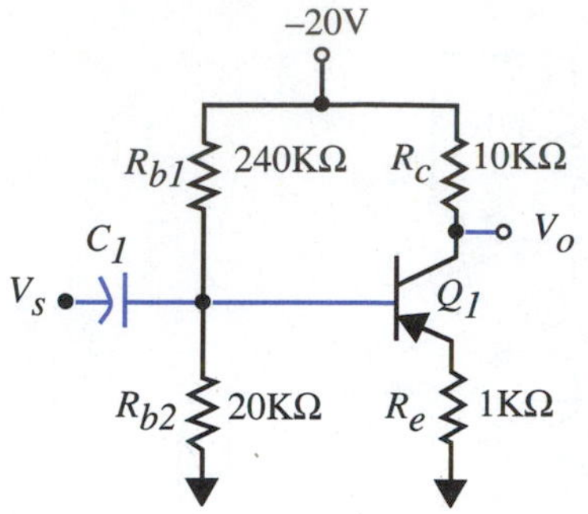

FIGURE 3–G (text figure 3–9)

34. Determine the voltages at base, emitter, and collector of both Q_1 and Q_2 in Figure 3–C if the collector resistor R_c of Q_1 open circuits.

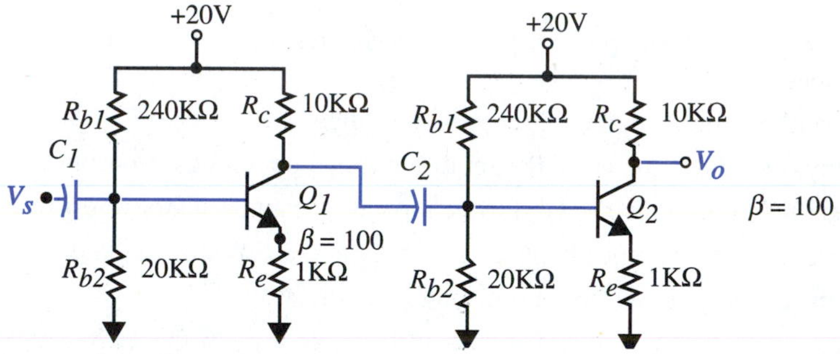

FIGURE 3–H (text figure 3–10)

35. Determine the DC voltages at base, emitter, and collector of both Q_1 and Q_2 in Figure 3–H if the coupling capacitor C_2 short circuits. (Hint: First consider Q_2 voltages.)

ABBREVIATIONS

R_{b1} = Upper base biasing resistor

R_{b2} = Lower base biasing resistor

R_e = External emitter resistor

R_c = External collector resistor

R_{in} = Base circuit input resistance

R_t = Transistor input resistance

R_o = Follower output resistance

r_g = Generator internal resistance

V_s = Generator output voltage

V_o = Circuit output voltage

A_v = Circuit gain (V_o/V_s)

ANSWERS TO PROBLEMS

1. To prevent DC loading if the base bias divider.
2. Set C_2 reactance equal to 1/10 of r_e at lowest operating frequency.
3. Top end of R_{b1} is at AC ground.
4. See text.
5. Gain is reduced.
6. See text.
7. C_2 AC shorts R_e.
8. $C_2 = 241\mu F$
9. Current gain with high input resistance and low output resistance.
10. V_b =13.9V; V_e = 13.37V
11. R_{in} =5.17KΩ; C_1= 6.16μF
12. R_o = 25.3Ω
13. Low input resistance matches low-resistance sources.
14. R_t = 149Ω
15. A_v = 31.5
16. A_v = 362
17. A_v = 79.5 (gain stays the same)
18. More efficient
19. Only half energized at a time.
20. To allow the ± current swing. V_c = +20V
21. R_c = 80KΩ
22. P_s = 3.6W
23. No transformers required.
24. Resistors are reduced to drive low resistance of next stage.
25. Temperature stability
26. See text.
27. Switch-type operation creates distortion which is filtered by tank circuit.
28. Frequency-sensitive, positive feedback must be present.
29. Hartley uses inductive divider for feedback; Colpitts uses capacitive.
30. Crystal is a precise frequency-selective component.
31. V_b = 1.69V; V_e = 1.09V
32. V_c = 20V; V_b = 1.54V; and V_e depends on the type meter used to measure voltage.
33. V_b = –2.42V; V_e = –1.82V; V_c = –2.02V.
34. V_{b1} = 0.781V;
    V_{e1} = 0.181V;
    V_{c1} = V_{b2} = 17.6V;
    V_{e2} = 18.2V;
    V_{c2} = 18V;
    Q_2 saturates.
35. V_{e2} = 1.8V; V_{b2} = 2.4V;
    V_{c2} = 2V; V_{c1} = 2.4V;
    V_{b1} = 1.39V; V_{e1} = 0.79V

CHAPTER 4

Field–Effect Transistors

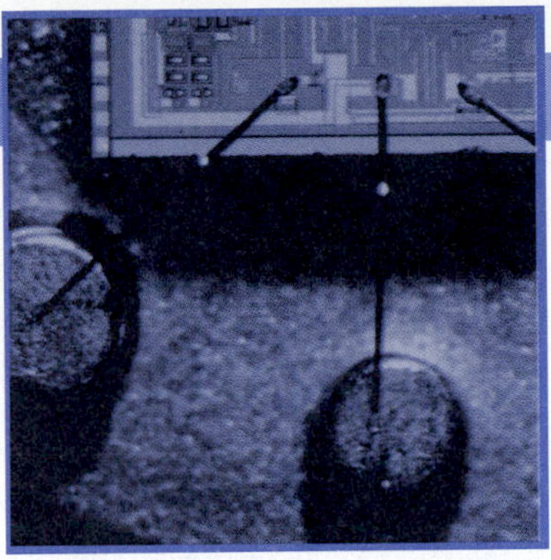

CHAPTER OBJECTIVES

After completing this chapter, you should be able to describe the following FET concepts:

- The differences between an FET and a bipolar transistor
- Construction of a junction (J) FET
- How to bias a JFET
- Operation of an FET amplifier
- The source follower configuration
- The common gate amplifier
- Similarity of an FET to a vacuum tube
- Construction of a metal oxide semiconductor (MOS) FET
- Biasing an MOSFET
- Enhancement and depletion modes of operation
- Power MOSFETS
- Care in the handling of MOSFETS
- Troubleshooting FETs

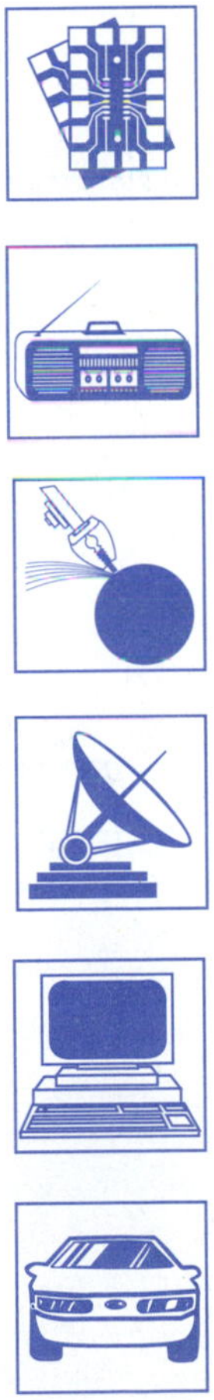

INTRODUCTION

The transistor we have been studying up to this point is called a bipolar type because it contains two (bi) diode junctions. Now we shall look at a transistor that has a single diode junction and is operated with the diode in the reverse-bias mode. This type of transistor is called a **junction field** (voltage) **effect transistor** or **JFET**.

Whereas the bipolar transistor is operated by causing current to flow into the base/emitter junction, the JFET is operated by a voltage (reverse biased) applied to its input terminal (gate).

What is the advantage of a JFET over a bipolar transistor? The fact that the JFET input is a reverse-bias diode means the current drain into the device is very small—so there is little loading of the signal source that is being amplified.

CONSTRUCTION OF A JFET

Junction FETs were the first type developed. The construction and circuit symbol for this type of FET (N channel) are given in Figure 4–1. (The gate arrow faces out for a P channel JFET.)

The N-doped channel can be considered a resistor between what is called the FET source (source of electrons) and the FET drain. Now if we apply a negative voltage (relative to the source) to the gate, the P–N junction between the FET gate and the channel will become reverse biased. We can recall from our study of the diode that a depletion region (insulator) is created at the reverse-biased junction. The greater the reverse bias, the wider the depletion region. Applying more reverse bias causes the depletion region to extend further and further into the N channel. Since we are narrowing the channel, the resistance of the channel increases. If we apply enough reverse voltage (called the cutoff voltage V_{co}), we can extend the depletion region across the whole channel, causing its resistance to go very high.

So we have a device where we can control the channel resistance by an applied voltage (field).

A complete JFET circuit is shown in Figure 4–2 .

If we consider the circuit as a voltage divider consisting of the drain resistor R_d and the FET channel resistance, then the more reverse bias we apply to the gate, the higher the channel resistance and therefore the higher the value of the output (drain) voltage. Conversely, reducing the bias causes the output voltage to drop because we are lowering the channel resistance.

Since we are considering the N channel as a resistor, a voltage drop will occur along the channel whenever current flows. If we tie the gate and source together so that there is no reverse bias and connect a positive voltage to the drain, then as we move along the channel away from the source terminal, the voltage becomes more positive because of the increased $I \times R$ drop. This makes the channel close to the P–N junction more positive than the P side or gate, and the diode junction becomes self-reverse-biased. (See Figure 4–3.)

This FET self-reverse-bias also creates the depletion region, which, again, raises the channel resistance. The more current in the channel, the greater the voltage drop and so the greater the self-bias. A current level can be reached where the depletion region becomes wide enough to prevent further current increase, and the current becomes constant. The drain-to-source voltage at which this current limiting occurs (current stops

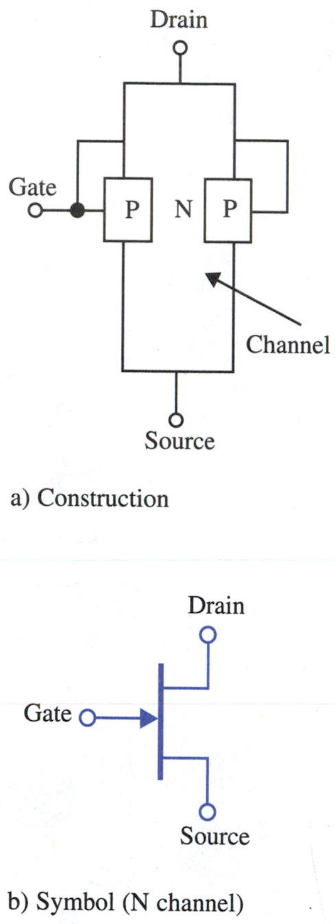

FIGURE 4–1 Junction field effect transistor

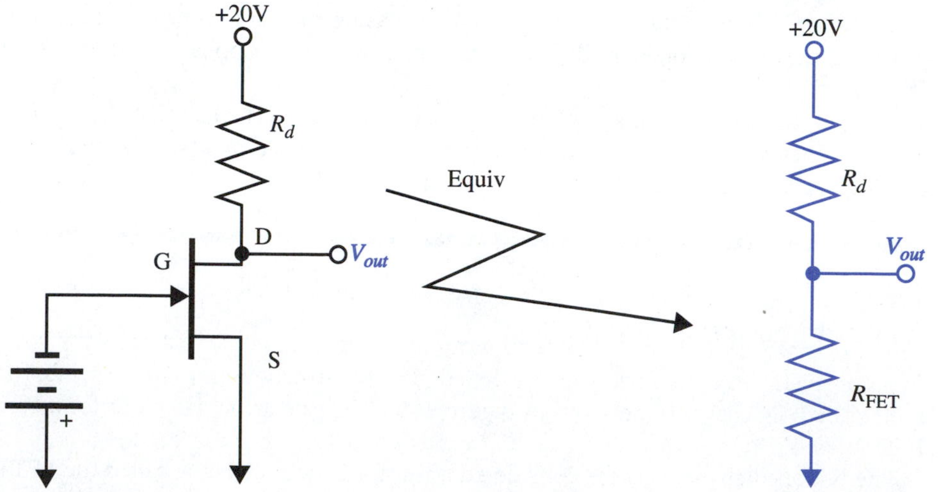

FIGURE 4–2 JFET circuit

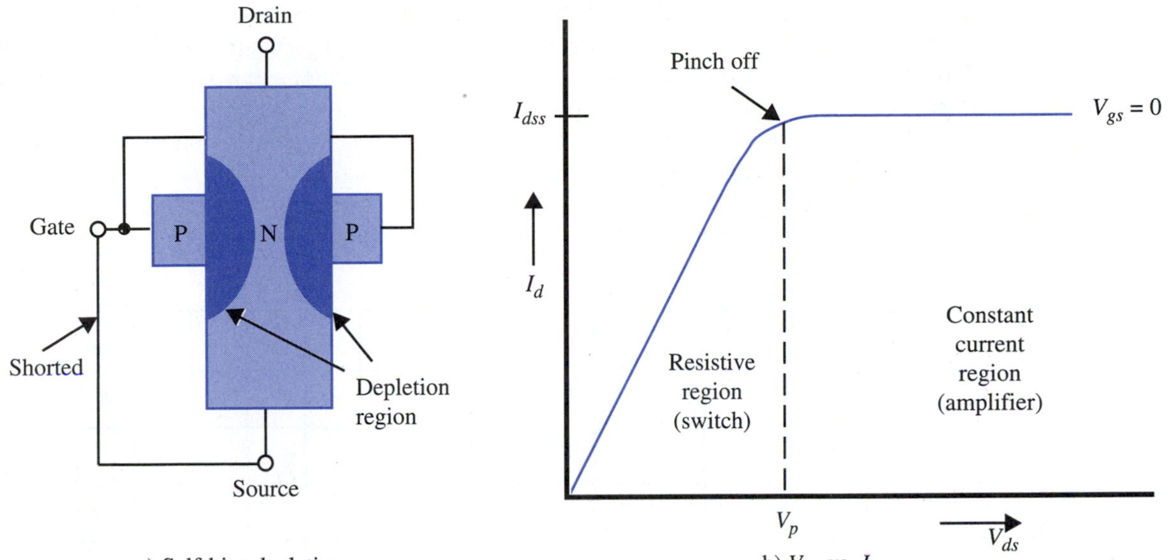

a) Self-bias depletion b) V_{ds} vs. I_d

FIGURE 4–3 JFET pinch off (With $V_{gs} = 0$)

increasing) is called the FET **current pinch-off voltage** V_p. The drain-source current at this limit point, with the gate-to-source shorted (*ss*), is called I_{dss}. We shall see that the JFET is typically operated in the constant current region (at a drain–source voltage greater than pinch off) when used as an amplifier and in the resistive region when used as a switch (at a drain–source voltage less than pinch off).

Since the gate input signal sees a reverse-biased diode, the input current to the gate is negligible. This, again, is the main difference between the FET and the conventional (bipolar) transistor. The conventional transistor requires base current to operate, whereas the FET works from the input voltage with essentially no current.

The term that is similar to the bipolar β that is used for a FET is **transconductance** (g_m or Y_{fs}) and is the ratio of the channel current change (ΔI_d) divided by the gate-to-source voltage change (ΔV_{gs}):

$$g_m = Y_{fs} = \frac{\Delta I_d}{\Delta V_{gs}} \quad \text{(Units } 1/\Omega \text{ or S)} \qquad \textbf{(equation 4–1)}$$

The transconductance relationship between input voltage and channel current, given above, is only valid beyond the pinch-off voltage, i.e., in the constant current region (see Figure 4–3b).

BIASING THE JUNCTION FET

Biasing a FET is not as easy as biasing a bipolar transistor, since the voltage from gate to source is not fixed like the 0.6V value from base to emitter. So we have to use the FET characteristic curves from the device data sheets in order to find the gate-to-source voltage V_{gs}.

Figure 4–4a shows, for the 2N5459 JFET, the relationship of the drain current I_d and the drain voltage V_d for a given gate-to-source voltage V_{gs}. Figure 4–4b indicates the relationship of the drain current to gate-to-source voltage at a fixed V_{ds} of 15V, which puts us in the constant current region.

Notice in these graphs that V_{gs} is negative. Remember, with subscripts we assume that the first letter shown (g) is more positive than the second letter (s). If we always put the positive lead from the DMM on the location indicated by the first subscript letter, we will read the correct polarity.

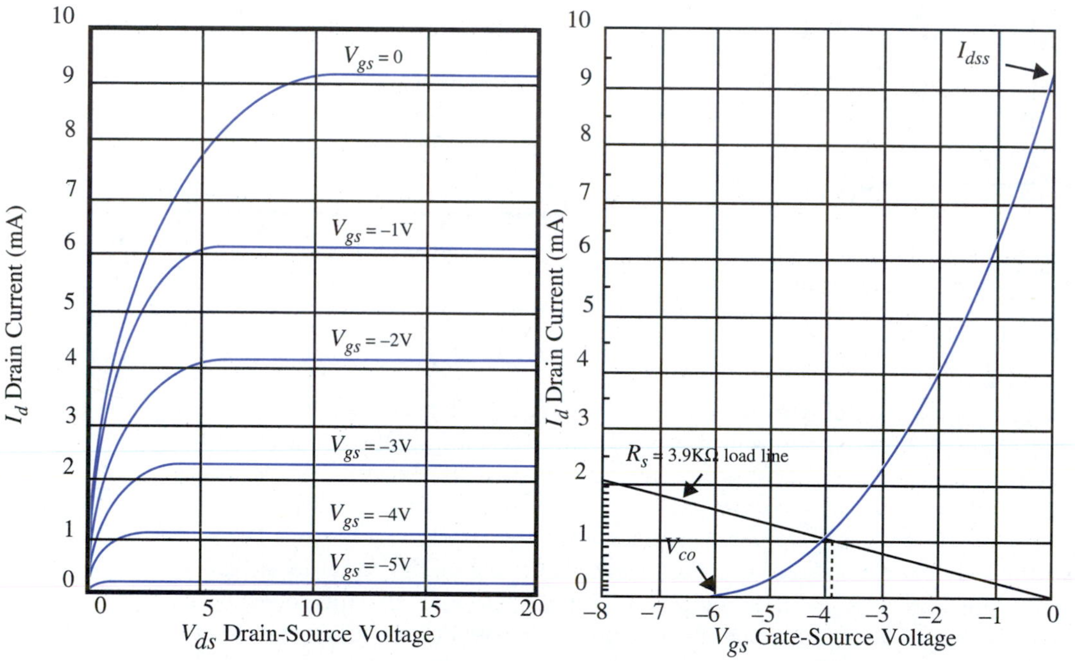

a) Drain characteristics

b) Transfer characteristics

FIGURE 4–4 2N5459 characteristics

The graph of Figure 4–4b has a parabolic form, which can be approximated by the following equation:

$$I_d = I_{dss}\left(1 - \frac{V_{gs}}{V_{co}}\right)^2 \qquad \textbf{(equation 4–2)}$$

where

I_d = Drain current
I_{dss} = Drain current with V_{gs} of 0V
V_{gs} = Gate-to-source voltage
V_{co} = Cut off voltage—the gate-source voltage required to reduce the drain current to zero (also called $V_{gs(off)}$).

In the constant current region of the 2N5459 curve of Figure 4–4a,

$$I_{dss} = 9.2\text{mA} \ (I_{dds} \text{ is } I_d \text{ with } V_{gs} \text{ of } 0\text{V})$$

From the 2N5459 curve of Figure 4–4b,

$$V_{co} = -6\text{V} \ \ (\text{for } I_d \text{ of } 0\text{V})$$

If we solve equation 4–2 for the gate-to-source bias voltage V_{gs},

$$V_{gs} = V_{co}(1 - \sqrt{\frac{I_d}{I_{dss}}}) \qquad \textbf{(equation 4–3)}$$

For biasing, we can either use the curves of Figure 4–4a or equation 4–2 or 4–3 to determine the required circuit bias voltage V_{gs}.

Before we can find V_{gs}, we need to determine the required value of I_d, which in turn is a function of the required drain voltage bias point, the drain resistor, and supply voltage V_{dd}.

We can first select the drain resistor R_d, and for maximum output voltage swing, set the drain voltage V_d at roughly one-half of V_{dd}. We then can determine the drain current I_d from

$$I_d = \frac{V_{dd} - V_d}{R_d}$$

We now enter this current I_d on the vertical axis of Figure 4–4b and run a horizontal line until we intercept the curve. At the intercept point, we drop a vertical line to the horizontal gate–source voltage axis and read the required gate-to-source voltage V_{gs}.

EXAMPLE

Determine the drain current, V_{gs} bias voltage, and value of source resistor for a JFET amplifier circuit. The JFET is a 2N5459, and a V_{dd} of +20V is available, the specified value of R_d is 8.2KΩ, and it is desired to have a drain voltage V_d of +11.8V.

SOLUTION

To set the drain at +11.8V, the value of I_d must be

$$I_d = \frac{V_{dd} - V_d}{R_d}$$

$$I_d = \frac{20\text{V} - 11.8\text{V}}{8.2\text{K}\Omega}$$

$$I_d = 1\text{mA}$$

Entering this current on the vertical axis of Figure 4–4b and moving across horizontally to the intercept point with the transfer characteristic curve, we drop a vertical line (shown dotted in Figure 4–4b) and read a V_{gs} of –3.9V on the horizontal axis. We could have solved for V_{gs} using equation 4–3 and with I_{dss} at 9.2mA and V_{co} at –6V,

$$V_{gs} = V_{co}\left(1 - \sqrt{\frac{I_d}{I_{dss}}}\right)$$

$$V_{gs} = -6V\left(1 - \sqrt{\frac{1mA}{9.2mA}}\right)$$

$$V_{gs} = -4.02V$$

We shall choose to use the –3.9V figure from the graph. To develop this –3.9V of gate bias, we can connect the gate resistor to a –3.9V supply, as shown in Figure 4–5a, or use a self-biasing resistor R_s, as shown in Figure 4–5b.

If we choose the self-bias approach, we must develop +3.9V across R_s, which in effect makes the source 3.9V more positive than the gate—or in other words, the gate is the desired 3.9V more negative than the source.

Since the source current is equal to the drain current (1mA), the required value of R_s is

$$R_s = \frac{3.9V}{1mA}$$

$$R_s = 3.9K\Omega$$

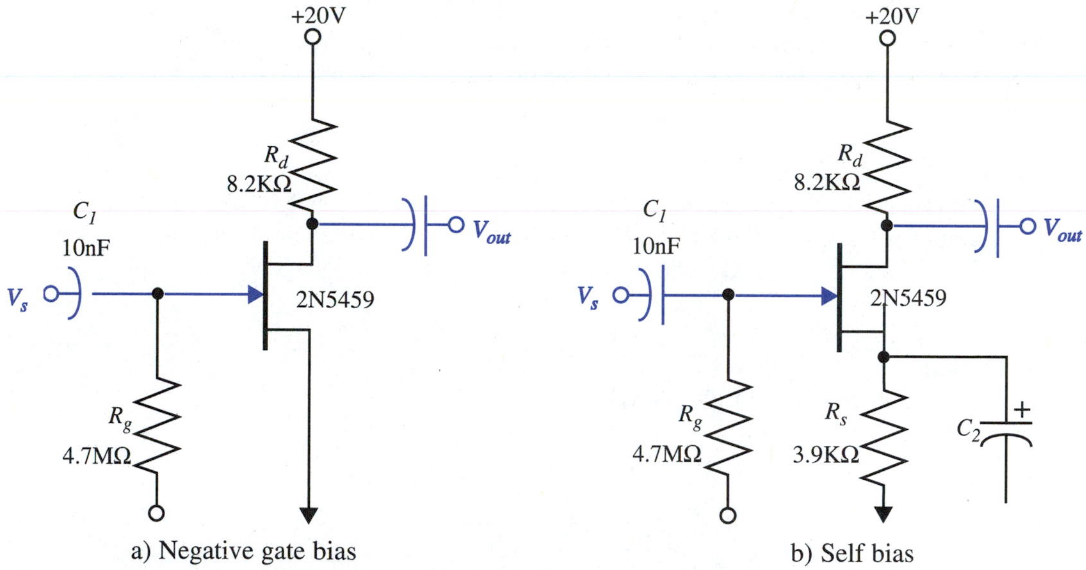

FIGURE 4–5 JFET AC amplifier

One problem with FET biasing is that the graph provided in Figure 4–4b is for a typical device. So if I_d changes at a given gate-to-source voltage when we change FETs, then the bias voltage at the drain will change.

Adding two additional curves to the graph of Figure 4–4b, shown in Figure 4–6, provides the range of variation that can occur in drain current versus gate-to-source voltage for a 2N5459 JFET.

The new curves show that the gate-to-source cutoff voltage V_{co} can range from –2V (curve C) to –8V (curve A), while the source current at a gate-to-source voltage of 0V (I_{dss}) ranges from 4mA for curve C to 16mA (projected off the graph) for curve A.

In order to determine the range of drain current that can occur with the variations indicated by the A and C curves, we need to draw a load-line graph for the source resistor (R_s). We can do this by picking a gate-to-source voltage (V_{gs}) and dividing by the value of R_s to find the drain current. We then put a point on the graph of Figure 4–4b corresponding to these values of V_{gs} and I_d. If we do this for two values of V_{gs} (and determine two values for I_d), we can generate the R_s load line by drawing a straight line through these two points.

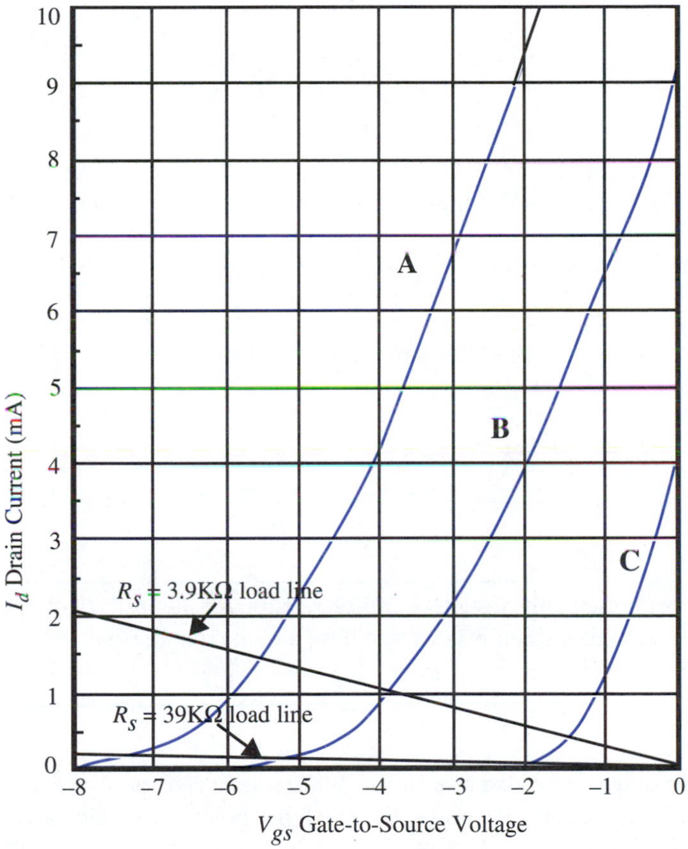

FIGURE 4–6 I_d Bias variations

EXAMPLE

Let us assume that we change FETs in the circuit of Figure 4–5b and the new FET has a cutoff voltage V_{co} of –8V (leftmost curve of Figure 4–6), and then we replace with a FET that has a cutoff voltage of –2V (rightmost curve of Figure 4–6). Find the new drain currents I_d and voltages V_d with each FET.

SOLUTION

In order to find the drain voltage for the new FETs, we must find the new drain current. We can do this by drawing a load line on the graph of Figure 4–6 for the 3.9KΩ source resistor. Two data points are needed to draw the line: We can use a V_{gs} of 0, which makes I_s equal to 0 (no voltage across R_s so therefore no current); the other point we can pick is a V_{gs} of –8V, which means that I_s is equal to 8V/3.9KΩ or 2.05mA. Drawing the line on the graph between these two points, as shown, we find that the intercept point of this line with the leftmost (A) curve occurs at a source (drain) current of 1.5mA (draw horizontal line from the intercept point) and a V_{gs} of –5.6V (draw vertical line from the intercept point). Using this drain current, the new drain voltage is

$$V_d = V_{dd} - (I_d \times R_d)$$

$$V_d = 20\text{ V} - (1.5\text{mA} \times 8.2\text{K}\Omega)$$

$$V_d = 7.7\text{V}$$

This is a change of 7.7 V – 11.8 V = –4.1 V (–35%) from the previous drain bias voltage we got by using the nominal drain current of 1mA from the 3.9KΩ load-line intercept with curve B.

The intercept of the rightmost curve (C) with the source resistor load line occurs at I_d of 0.4mA and V_{gs} of –1.5V, which provides a drain voltage of

$$V_d = 20\text{V} - (0.4\text{mA} \times 8.2\text{K}\Omega)$$

$$V_d = 16.72\text{V}$$

which is a change of 16.72V – 11.8V = +4.92V (+42%) from our nominal value.

EXAMPLE

A technician mistakenly installs a 39KΩ resistor for R_s instead of the required 3.9KΩ. Find the new drain bias voltage if we assume the FET has the characteristic of curve B of Figure 4–6.

SOLUTION

We need to draw a new load line for the 39KΩ source resistor. Again, one point we can use is I_s of 0mA and V_{gs} of 0V. For the other point, let us pick a value of –8V for V_{gs} and solve for I_s.

$$I_s = \frac{V_s}{R_s}$$

$$I_s = \frac{8\text{V}}{39\text{K}\Omega}$$

$$I_s = 0.205\text{mA}$$

Using these two points, we can draw the 39KΩ load line on the graph as shown in Figure 4–6. At the intercept of curve B with the new load line, we draw a vertical line to the V_{gs} axis and read a value of –5.5V. Since this voltage is developed across the 39KΩ source resistor, we can determine the drain (source):

$$I_d = I_s = \frac{5.5\text{V}}{39\text{K}\Omega}$$

$$I_d = 0.141\text{mA}$$

Now the new drain voltage will be

$$V_d = 20\text{V} - (0.141\text{mA} \times 8.2\text{K}\Omega)$$

$$V_d = 18.84\text{V}$$

A MORE PREDICTABLE BIASING CIRCUIT

A biasing circuit which is more independent of the FET used and therefore provides a more predictable bias is shown in Figure 4–7a.

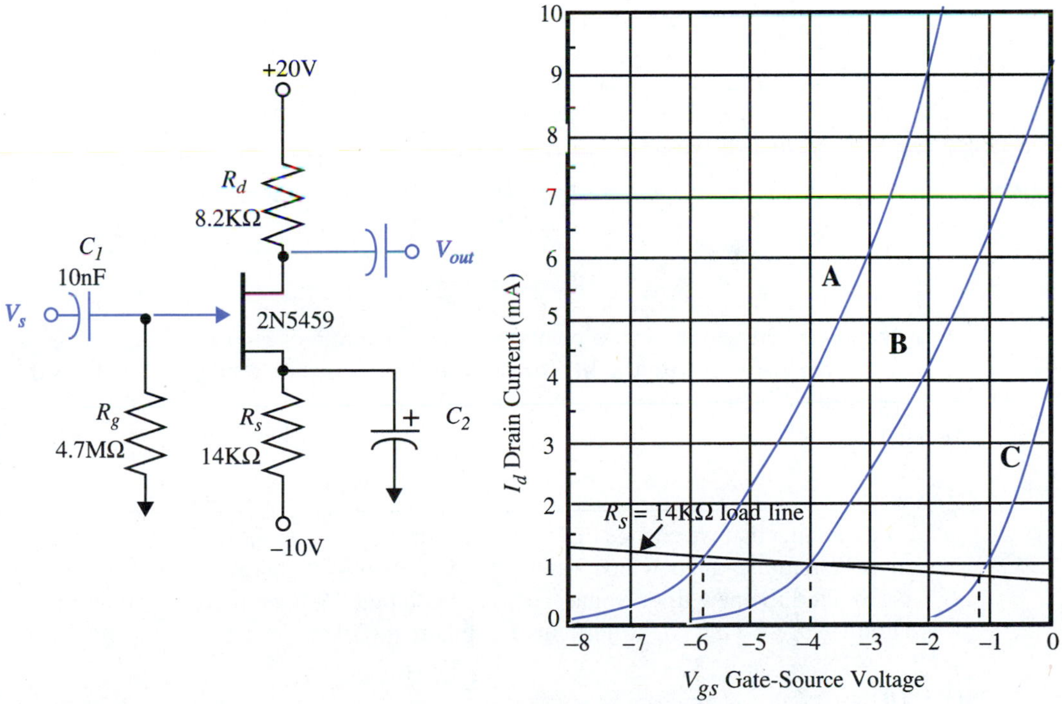

a) Circuit b) Bias load line

FIGURE 4–7 More stable FET biasing circuit

Notice that the change from the previous circuit of Figure 4–5b is that resistor R_s has been increased from 3.9KΩ to 14KΩ, and the bottom end of this resistor is taken to a –10V supply rather than to ground. The drain voltage is still at the nominal value of 11.8V, but we shall see that this arrangement will make the drain current more independent of the particular FET. This is apparent when you consider that if the bottom end of R_s is taken to a more negative voltage supply (V_{ss}), the bias voltage V_{gs} has less effect on the drain current since

$$I_s = \frac{V_{ss} + V_{gs}}{R_s}$$

but with

$$V_{ss} >> V_{gs}$$

$$I_s \approx \frac{V_{ss}}{R_s}$$

In order to find the drain voltage for the circuit of Figure 4–7a, we must again draw a load line on the graph of Figure 4–6. The new graph is shown in Figure 4–7b. The end points of the load line can be determined by first assuming that V_{gs} is equal to 0V, which means we have a 10V voltage drop across R_s, and this provides a source current of

$$I_d = \frac{10V}{14K\Omega}$$

$$I_d = 0.71mA$$

If we then assume that V_{gs} is equal to –8V (which means we have an 18V voltage drop across R_s), the source current is

$$I_d = \frac{18V}{14K\Omega}$$

$$I_d = 1.285mA$$

Drawing a line between these two points provides the graph of Figure 4–7b.

Notice that the load line with the 14KΩ resistor is more horizontal than the 3.9KΩ load line, which makes the intercepts with the A, B, and C curves close to the same drain current value. So with less change in I_d, there is less change in V_d when FETs are replaced.

EXAMPLE

Let us assume that we change FETs in the circuit of Figure 4–7a, and because of tolerance variations, the new FET has a cutoff voltage of –8V (leftmost curve of Figure 4–7b). Find the new drain voltages V_d. Also compute V_d if we use a FET that has a cutoff voltage of –2V (rightmost curve of Figure 4–7b).

SOLUTION

For the FET with a –8V cutoff, the drain voltage is determined by taking the intercept of the leftmost curve with the 14KΩ load line. Now we read the I_d value from the intercept point on the graph and get a value of 1.15mA, so the drain voltage is

$$V_d = V_{dd} - (I_d \times R_d)$$

$$V_d = 20V - (1.15mA \times 8.2K\Omega)$$

$$V_d = 10.57V$$

This is a change of

$$10.57V - 11.8V = -1.23V$$

(–10%) from our nominal drain voltage value of 11.8V.

The intercept with the –2V curve yields a drain current of 0.85mA, which provides a drain voltage of

$$V_d = 20V - (0.85mA \times 8.2K\Omega)$$

$$V_d = 13.03V$$

This is a change of

$$13.03V - 11.8V = 1.23V$$

(+10 %) from our nominal drain voltage value of 11.8V.

These results show that the source circuit of Figure 4–7a, with its higher-value source resistor, acts more like a constant current source and provides a more predictable drain voltage.

We conclude, then, from these examples, that the circuit of Figure 4–7a should be used if bias stability is important.

THE JFET AMPLIFIER

Now that we have biased our FET amplifier to the correct operating point, we can apply an AC signal. The gain of this signal is related to the transconductance (g_m) of the FET, which again is the change in drain current resulting from a change in gate-to-source voltage or $\Delta I_d/\Delta V_{gs}$.

Referring back to the characteristics of the 2N5459 in Figure 4–4a, we see that the V_{gs} curves are closer together where the voltages are more negative. So in this region, for a given change in V_{gs}, there is less change in I_d, and therefore the g_m is lower. Since the drain current is lower where the V_{gs} curves are closer together, we conclude that at lower drain currents, we will get a smaller value for g_m.

Since

$$g_m = \Delta I_d/\Delta V_{gs}$$

it is the slope of the drain current versus gate-to-source voltage curve. We can generate an equation for the slope at any point on a curve by performing the calculus operation of **differentiation** on the equation that represents the curve (equation 4–2).

Differentiating equation 4–2, we get

$$g_m = 2(I_{dss})\frac{(V_{gs} - V_{co})}{V_{co}^2} \qquad \textbf{(equation 4–4)}$$

This equation indicates that the lower the gate-to-source bias voltage (V_{gs}), the higher the value of g_m (remember that V_{gs} and V_{co} are both negative).

EXAMPLE

Find the nominal value of g_m for the 2N5459 at our drain bias current of 1mA, using equation 4–4.

SOLUTION

With the drain current of 1 mA, the value of V_{gs}, from the center (nominal) curve of Figure 4–7b is –4.2V. Also, from this graph V_{co} equal –6V; from Figure 4–4a, I_{dss} equals 9.2mA.

Entering these values into equation 4–4,

$$g_m = 2(I_{dss})\frac{(V_{gs} - V_{co})}{V_{co}^2}$$

$$g_m = 2(9.2\text{mA})\frac{(-4.2\text{V} + 6\text{V})}{(-6\text{V})^2}$$

$$g_m = 920\mu\text{S}$$

FET VOLTAGE GAIN EQUATION

Let us now develop the voltage gain equations for an FET amplifier. If we assume that v_s in Figure 4–7a is a sine wave input of 100mV peak, and we have a nominal g_m of 920µS, then the output current (i_d) at the drain (and the source) will be from equation 4–1.

$$i_d = v_s \times g_m = i_s \qquad (g_m = \frac{i_s}{v_s}) \qquad \textbf{(equation 4–5)}$$

$$i_d = 100\text{mV} \times 920\mu\text{S}$$

$$i_d = 0.092\text{mA peak}$$

The output voltage v_o will be this current times the drain resistor value.

$$v_o = i_d \times R_d \qquad \textbf{(equation 4–6)}$$

$$v_o = 0.092\text{mA} \times 8.2\text{K}\Omega$$

$$v_o = 0.754\text{V (peak)}$$

So our voltage gain (A_v) is the 0.754V output divided by the 100mV input for a gain of 7.54. If we substitute for i_d in equation 4–6 using the relationship from equation 4–5, we get

$$v_o = v_s \times g_m \times R_d \qquad \textbf{(equation 4–7)}$$

but

$$A_v = -\frac{v_o}{v_s}$$

(minus sign indicates 180° phase shift between the input signal and output signal). Therefore

$$A_v = -g_m \times R_d \qquad \textbf{(equation 4–8)}$$

If we let

$$1/g_m = r_s$$

then

$$A_v = -\frac{R_d}{r_s} \qquad \textbf{(equation 4–9)}$$

This equation is similar to the equation

$$A_v = -\frac{R_c}{r_e}$$

used for the bipolar transistor, and we can consider that r_s is a resistor in series with the source lead just as r_e is a resistor in series with the emitter lead.

A good assumption is that the g_m for a practical FET can vary ±50% from the nominal value, so the circuit gain will also vary by this amount when using equations 4–8 or 4–9.

EXAMPLE

Compute the expected voltage gain variations for the circuit of Figure 4–7a if the nominal value of g_m is 920μS.

SOLUTION

With a ±50% variation from the nominal value, the range of g_m is from 460μS to 1.38mS.

If g_m is equal to 460μS, and using equation 4–8,

$$A_v = -g_m \times R_d$$

$$A_v = -460\mu S \times 8.2K\Omega$$

$$A_v = -3.77$$

With g_m equal to 1.38mS,

$$A_v = -1.38mS \times 8.2K\Omega$$

$$A_v = -11.3$$

Remember the nominal gain was –7.54.

If we compare the gain of our FET circuit with that of a bipolar transistor, we see that the FET gain is much lower—in the tens rather than hundreds.

We can raise the gain of our circuit by increasing the drain resistor (R_d) as is indicated by equations 4–8 and 4–9. But if we stay with the same drain bias current, then the drain DC voltage will drop from the V_d value we established for maximum output voltage swing. This means the amplifier would be used for amplifying lower-level voltages, which would not require such a large drain voltage swing.

A circuit with an increased value of drain resistor is shown in Figure 4–8. This circuit also has the advantage of using a single power supply. The slightly higher drain resistor value means a little higher voltage gain; but the slightly lower source resistor gives a little lower bias stability if the FET is replaced.

The voltage divider in the gate circuit of Figure 4–8 raises the gate to +6V. The 10KΩ source resistor, with 1mA flowing through it, develops a 10V drop; putting the source at +10V. This gives a gate-to-source voltage of 6V – 10V = –4V (a little off from the –4.2V read on the graph of Figure 4–7b at I_d equal to 1mA). The 1mA through R_d gives a drain voltage V_d of

$$24V - 10V = +14V.$$

Included in the schematic in series with R_s is r_s which, remember, has a value of $1/g_m$. From our previous calculations, g_m ranged from 460μS to 1.38mS, which means r_s ranges from 2.17KΩ to 725Ω.

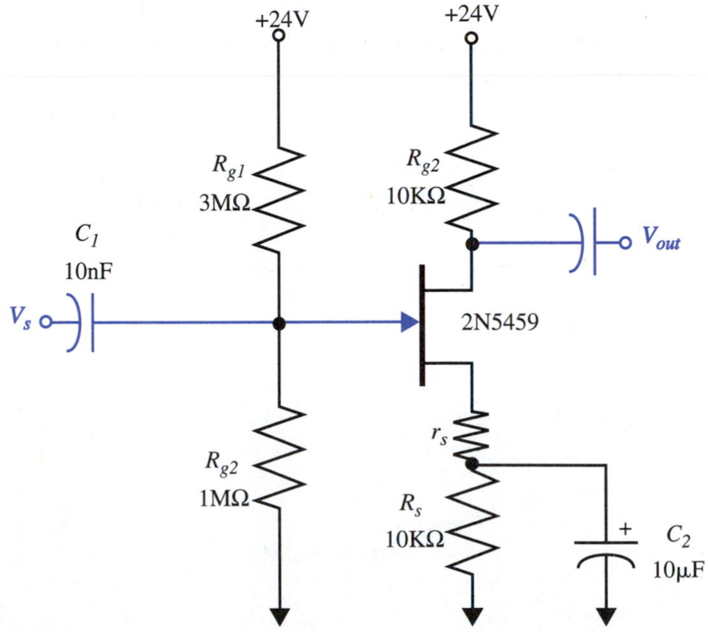

FIGURE 4–8 Optimized FET amplifier circuit

Again, capacitors used in our circuit were selected by making the reactance less than one-tenth of the associated resistor at an assumed lowest frequency of 250Hz. The loading resistor used to calculate the coupling capacitor (C_1) is 3MΩ in parallel 1MΩ (FET gate resistance is assumed very high). So

$$C_1 = \frac{1}{2\pi f\left(\frac{3\text{M}\Omega \,||\, 1\text{M}\Omega}{10}\right)}$$

$$C_1 = \frac{1}{2\pi \times 250\text{Hz} \times 75\text{K}\Omega}$$

$$C_1 = 8.49\text{nF (use 10nF or larger)}$$

For C_2, the reactance must be one-tenth of r_s so that R_s is effectively bypassed. Since r_s has a worst-case low value of 725Ω, X_{c2} must be less than 725Ω/10 or 72.5Ω.

$$C_2 = \frac{1}{2\pi \times 250 \times 72.5\Omega}$$

$$C_2 = 8.78\mu\text{F (use 10}\mu\text{F or larger).}$$

The big advantage of the FET over the bipolar transistor is the high input resistance at the gate. In our last example, our generator resistance could be as high as 75KΩ (one-tenth of 750KΩ) with little effect on the voltage gain. Compare this with the bipolar amplifier of the previous chapter.

Looking back at the 2N5459 characteristics of Figure 4–4a, we see that the spacing between the curves of gate voltage is not linear, i.e., the spacing from 0V to –1V is much greater than the spacing from –1V to –2V. What this means is that if we swing the input voltage too far, we get a distorted output signal. The negative portion of the output sine wave will be greater than the positive portion! This limits the output swing we can get from this amplifier and requires that we limit the application to small signals if FET distortion is to be kept low.

SOURCE FOLLOWER

Similar to the bipolar transistor emitter follower is the FET source follower. The input resistance of the source follower is high, and the output resistance low. We shall see that the gain is approximately equal to unity (1), just as it is with emitter follower. The source follower circuit we shall analyze is shown in Figure 4–9.

Again, biasing is set at a selected drain/source (I_d) current of 1mA with –4V bias. This is the case because the divider in the gate circuit raises the gate to +8V and the 12KΩ source resistor with 1mA flowing through it raises the source to +12V (half the supply voltage) and the remainder of the supply voltage (24V – 12V) is dropped across the FET drain–source channel.

Let us first derive the source follower voltage gain without R_L connected.

We can express the input voltage as

$$v_s = v_{gs} + (i_s \times R_s)$$

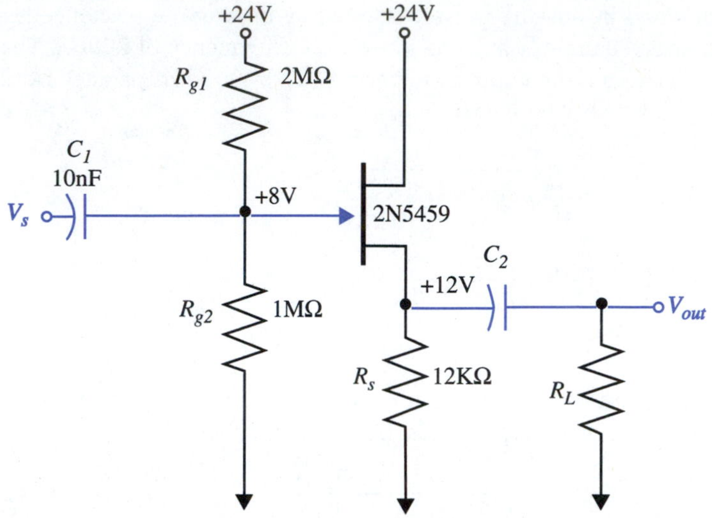

FIGURE 4–9 Source follower

but

$$v_{gs} = i_s \times r_s$$

so

$$v_s = i_s(r_s + R_s)$$

and

$$v_o = i_s \times R_s$$

Gain A_v is

$$A_v = \frac{v_o}{v_s}$$

$$A_v = \frac{(i_s \times R_s)}{i_s(r_s + R_s)}$$

Source follower

$$A_v = \frac{R_s}{(r_s + R_s)} \qquad \textbf{(equation 4–10)}$$

If

$$R_s >> r_s$$

then

$$A_v = \frac{R_s}{R_s} = 1 \qquad \textbf{(equation 4–11)}$$

EXAMPLE

For our circuit of Figure 4–9 with g_m of 940μS, what is the voltage gain without R_L connected?

SOLUTION

$$r_s = \frac{1}{g_m} = 1.06\text{K}\Omega$$

$$A_v = \frac{R_s}{(r_s + R_s)}$$

$$A_v = \frac{12\text{K}\Omega}{(1.06\text{K}\Omega + 12\text{K}\Omega)}$$

$$A_v = 0.919$$

If we use a higher value of R_s, the gain will be closer to unity. But connecting R_L puts it in parallel with R_s, and this lowers its effective resistance, and the gain is reduced.

The portion of the input signal voltage that doesn't appear at the output is the voltage dropped across the resistance r_s. This resistance appears from gate to source and therefore is the output resistance of the FET follower. It should be noted that it is higher than the output resistance of a bipolar follower ($1/g_m$) versus r_e). The circuit output resistance R_o is this resistance in parallel with R_s (R_s is not so much greater than r_s that it can be ignored).

FET output resistance

$$R_o = r_s \,||\, R_s \qquad \textbf{(equation 4–12)}$$

There is no transfer of resistance from the gate circuit to add to r_s, as with the emitter follower, because there is no current flowing into the gate terminal.

EXAMPLE

Determine the output resistance (R_o) for the circuit of Figure 4–9 with the load resistor disconnected if g_m is 1200μS. ($r_s = \frac{1}{1200\mu\text{S}}$)

SOLUTION

The output resistance is

$$R_o = r_s \,||\, R_s \quad \text{(Remember, } r_s \text{ is } 1/g_m\text{.)}$$

$$R_o = 833\Omega \,||\, 12\text{K}\Omega$$

$$R_o = 779\Omega$$

Compare this with the very low ohms we got with the bipolar follower.

COMMON GATE AMPLIFIER

The FET common gate amplifier is shown in Figure 4–10, and it should be noted that this configuration looks similar to the common base amplifier. Its characteristics are also similar: a low input resistance and a relatively high output impedance.

The input resistance of the common gate stage is the same as the output resistance of the follower or

$$r_s \parallel R_s$$

This low resistance makes the common gate useful for an input stage for radio receiver circuits to match the low resistance of the antenna. (A FET is preferred in this application because the single junction creates less noise than the two junctions of the bipolar transistor.)

We will now go through the steps to derive the gain equation for the common gate amplifier.

Since the gate is AC bypassed to ground, the input voltage is

$$v_s = v_{gs}$$

Because the right side of R_d is at AC ground, the output voltage is

$$v_o = (i_d)(R_d)$$

$$v_o = (v_{gs})(g_m)(R_d)$$

The gain A_v is

$$A_v = (v_{gs})(g_m)\left(\frac{R_d}{V_{gs}}\right)$$

$$A_v = g_m \times R_d$$

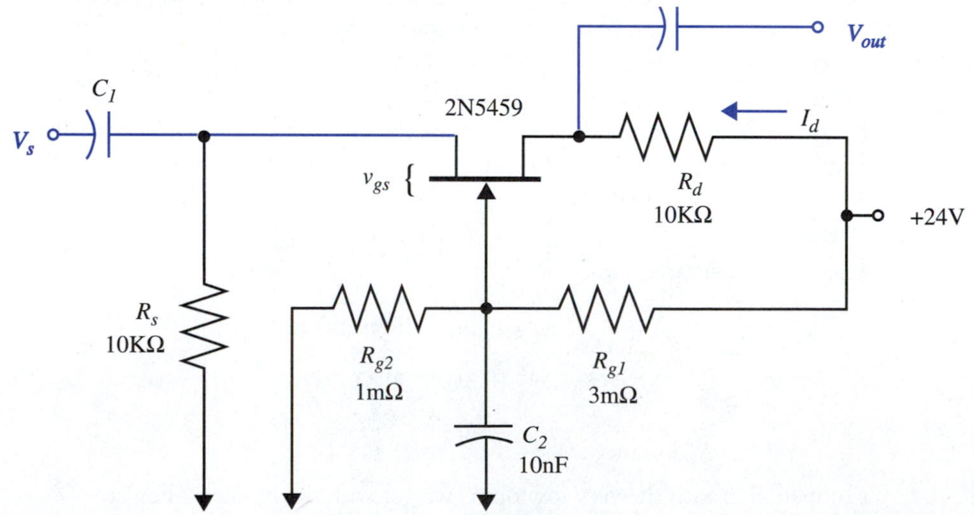

FIGURE 4–10 Common gate amplifier

which again is the same as common gate gain

$$A_v = \frac{R_d}{r_s} \qquad \textbf{(equation 4–13)}$$

For our circuit of Figure 4–10, the input resistance ranges from 2.17KΩ to 725Ω, with the g_m range for the 2N5459 of 460 to 1380μS at our drain current of 1mA. The gain, as with the common source amplifier, will range from 4.6 to 13.8. The gain can be stabilized, but reduced, by adding a resistor in series with r_s. However, this will also raise the input resistance.

EXAMPLE

Assume for the common base of Figure 4–10 that we have an FET with a g_m of 1000μS. Compute the voltage gain and the input resistance for the amplifier.

SOLUTION

The gain is

$$A_v = (g_m)(R_d)$$

$$A_v = (1000\mu S)(10K\Omega)$$

$$A_v = 10$$

Input resistance

$$R_{in} = r_s \parallel R_s$$

and

$$r_s = \frac{1}{g_m}$$

$$R_{in} = 1K\Omega \parallel 10K\Omega$$

$$R_{in} = 909\Omega$$

FET OSCILLATOR CIRCUITS

The high gate resistance of an FET makes it easy to apply feedback to this terminal (less loading), and so oscillators are easily constructed using an FET amplifier.

Figure 4–11 shows a commonly used FET configuration for a crystal oscillator. The crystal is operated in the series resonant mode, i.e., the signal fed back, and therefore sine-wave oscillation, occurs at the series resonant frequency of the crystal. This particular configuration is called a Pierce oscillator, and its obvious advantage is its simplicity.

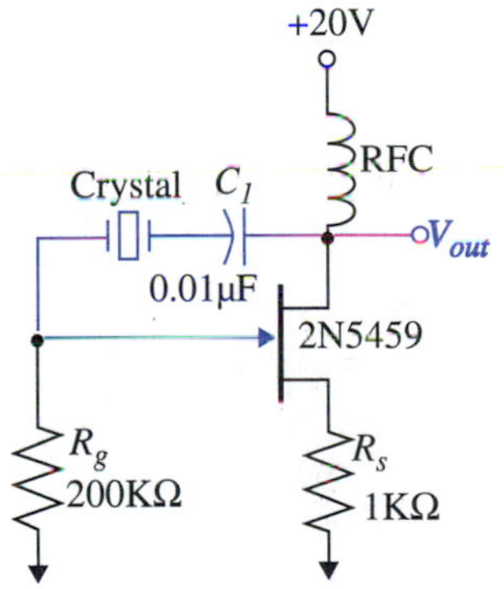

FIGURE 4–11

Pierce crystal oscillator

The capacitor C_1 prevents DC from appearing across the crystal, causing crystal strain. R_s is added to provide some negative feedback. This reduces the signal amplitude at the drain, but it results in a cleaner sine wave.

The inductor in the drain circuit labeled RFC is a Radio Frequency Choke. This coil isolates the drain from the power supply (which is a low AC impedance point) and allows an AC voltage to be developed at the drain. The reactance of RFC should be at least ten times the impedance of R_g in parallel with about 10pF of capacitance—computed at the frequency of resonance.

COMPARING THE FET WITH THE VACUUM TUBE

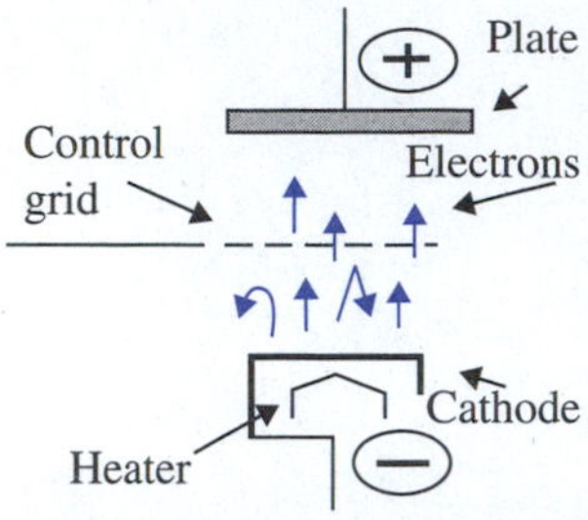

FIGURE 4–12
Triode vacuum tube

The original electronic amplifiers used vacuum tubes. The basic operation of the vacuum tube is very similar to that of the FET. Both devices are controlled by an input voltage that causes the output current to change. Figure 4–12 shows the basic construction of the triode (three-element) vacuum tube.

In operation, the cathode heater gives the electrons sufficient energy to be emitted from the cathode surface. The control grid potential is negative relative to the cathode, and this tends to drive some of the free electrons back to the cathode. However, the plate (anode) is at a very high positive potential—which attracts the electrons. By making the control grid negative enough, we can effectively prevent any electrons from reaching the plate. If we make the grid at the same potential as the cathode, most of the electrons will reach the plate. So the grid voltage controls current in the vacuum tube much like the gate voltage does in the FET.

As with the FET, the control effect of the grid in the vacuum tube is specified in terms of transconductance (g_m), which, again, is the ratio of the output current (i_p) change divided by the input grid voltage change (v_g).

$$g_m = \frac{\Delta i_p}{\Delta v_g}$$

The tube amplifier gain is

$$A_v = \frac{v_p}{v_g}$$

Since

$$v_p = i_p \times R_p \quad (R_p \text{ is the plate load resistor.})$$

then

$$A_v = g_m \times R_p$$

Vacuum tubes are still used in high-power radio transmitters; and of course, the CRT used in televisions, monitors, etc., is a vacuum tube.

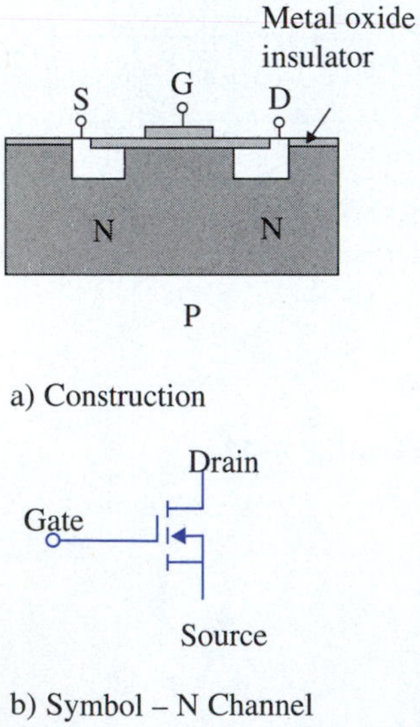

FIGURE 4–13 MOSFET construction

METAL OXIDE SEMICONDUCTOR FETS (MOSFETS)

An FET with a slightly different construction from the JFET is the **MOSFET**, shown in Figure 4–13. This FET configuration is also called an "E" type.

Notice that the N channel is in two parts, separated by P material; so the connection from source to drain is like two back-to-back diodes (NP and PN). This means that no current can flow from source to drain without a gate voltage. If we now put a positive voltage on the gate, electrons will be attracted from the P material and gather below the gate on the other side of the metal oxide (MOS) insulator. This will cause the P material to

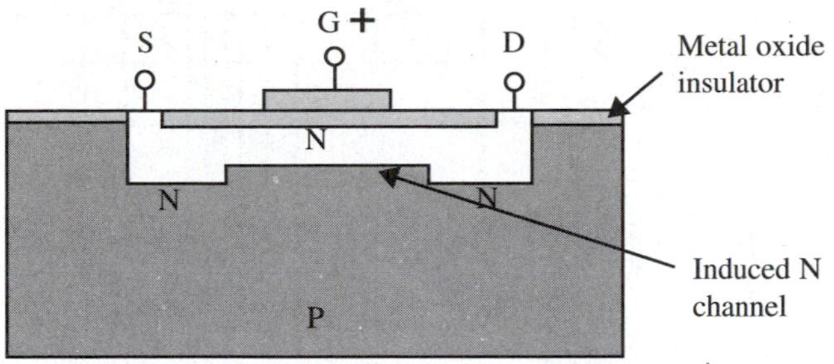

FIGURE 4–14 Positive gate bias on N-channel MOSFET

become *N* in this region. Now we have a continuous N-type path (channel) from source to drain and current can flow. (See Figure 4–14.)

The greater the positive voltage on the gate, the wider the channel and the more current between drain and source.

Since we don't have the JFET reverse-biased diode connection from gate to channel in a MOSFET, but actually a metal oxide insulator between gate and channel, the input resistance to a MOSFET is extremely high, and the gate appears as a capacitor.

The type of MOSFET shown in Figures 4–13 and 4–14 is called an **enhancement MOSFET**, because increasing the gate voltage enhances (increases) the current. This is not like the JFET we studied previously, where increasing the gate voltage (negatively) reduced the channel width and reduced the current.

BIASING THE MOSFET

Biasing of the enhancement MOSFET is somewhat different than the junction FET; because without any bias voltage, the enhancement type has no current flow—so we can't use self-biasing via the source resistor. Our biasing choices are voltage divider bias on the gate or a negative voltage on the source.

The characteristics for a 2N4351 N-channel enhancement MOSFET are shown in Figure 4–15. We can see from the curves of Figure 4–15a that if, for example, we pick a gate-to-source bias voltage (V_{gs}) of +2V, we get a drain current of about 1.8mA at the higher drain-to-source voltages.

In order to achieve this bias level, we can use one of the biasing schemes illustrated in Figure 4–16. Both circuits are set up to have a gate-to-source voltage of +2V. We can determine the drain current for this gate-to-source voltage by drawing the drain resistor (5.6KΩ) load line on the Figure 4–15a graph. Again, we need to select two points to draw the load line. If we set the drain current equal to zero, the drain voltage will be equal to the supply voltage of 20V (no voltage drop across R_d). The other point we can use is with the drain to source voltage equal to zero which means the drain current will be 20V/5.6KΩ or 3.57mA. The load line using these two points is shown in Figure 4–15a.

The intercept of the load line with the +2V V_{gs} curve gives us a drain current of 1.8mA. The drain voltage for the circuit of Figure 4–16a is

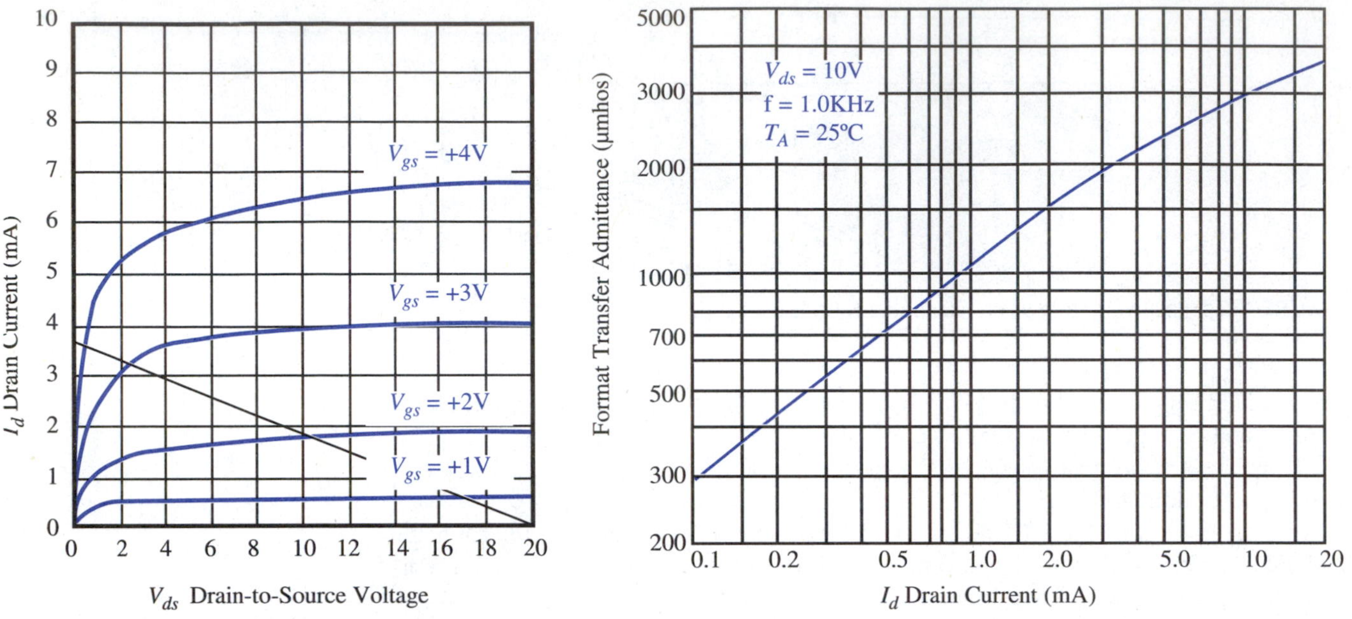

a) V_{gs} effect on I_d and V_d

b) g_m versus I_d

FIGURE 4–15 Characteristics for 2N4351 enhancement MOSFET

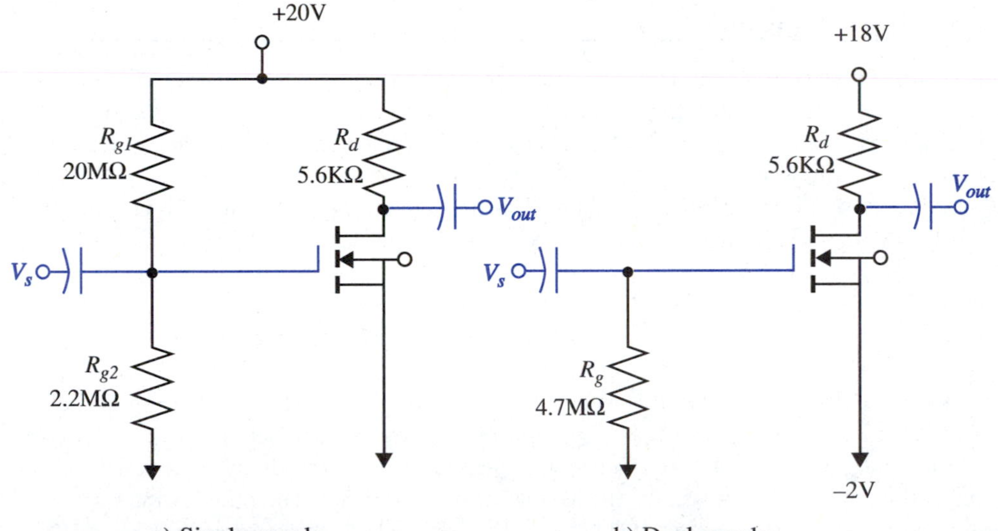

a) Single supply

b) Dual supply

FIGURE 4–16 Biasing the enhancement MOSFET

$$V_d = V_{dd} - (I_d)(R_d)$$

$$V_d = +20\text{V} - (1.8\text{mA})(5.6\text{K}\Omega)$$

$$V_d = +10\text{V}$$

For the circuit of Figure 4–16b,

$$V_d = +18\text{V} - (1.8\text{mA})(5.6\text{K}\Omega)$$

$$V_d = +8\text{V}$$

Both circuit drain voltages are at half of the total power supply voltage.

The voltage gain of the two circuits is, as before, $-g_m R_d$. From Figure 4–15b, we get a g_m of 1300µS at a drain current of 1.8mA, which gives a voltage gain value of

$$A_v = -(g_m)(R_d)$$

$$A_v = -1300\mu\text{S} \times 5.6\text{K}\Omega$$

$$A_v = -7.3$$

EXAMPLE

If the gate–source bias voltage is changed to +1V, what are the new values of I_d, R_d, g_m, and voltage gain if we still require a +10V bias voltage at the drain?

SOLUTION

From Figure 4–15a, the drain current with a gate-source bias of +1V is 0.5mA. To obtain the required 10V drop in the drain circuit requires a drain resistor R_d of

$$R_d = \frac{10\text{V}}{0.5\text{mA}}$$

$$R_d = 20\text{K}\Omega$$

Using the 0.5mA drain current, we read a g_m of 700µS from Figure 3–14b. We can now compute the circuit gain.

$$A_v = -g_m \times R_d$$

$$A_v = -700\mu\text{S} \times 20\text{K}\Omega$$

$$A_v = -14$$

So changing the gate-to-source bias voltage has almost doubled the voltage gain!

DEPLETION/ENHANCEMENT MOSFETS

There is another type of MOSFET in which making the gate voltage negative relative to the source reduces the drain current—this is called a **depletion FET** and is shown in Figure 4–17. In this device, a lightly doped N region is provided between an N region at the source and the drain—so current can flow with no voltage on the gate.

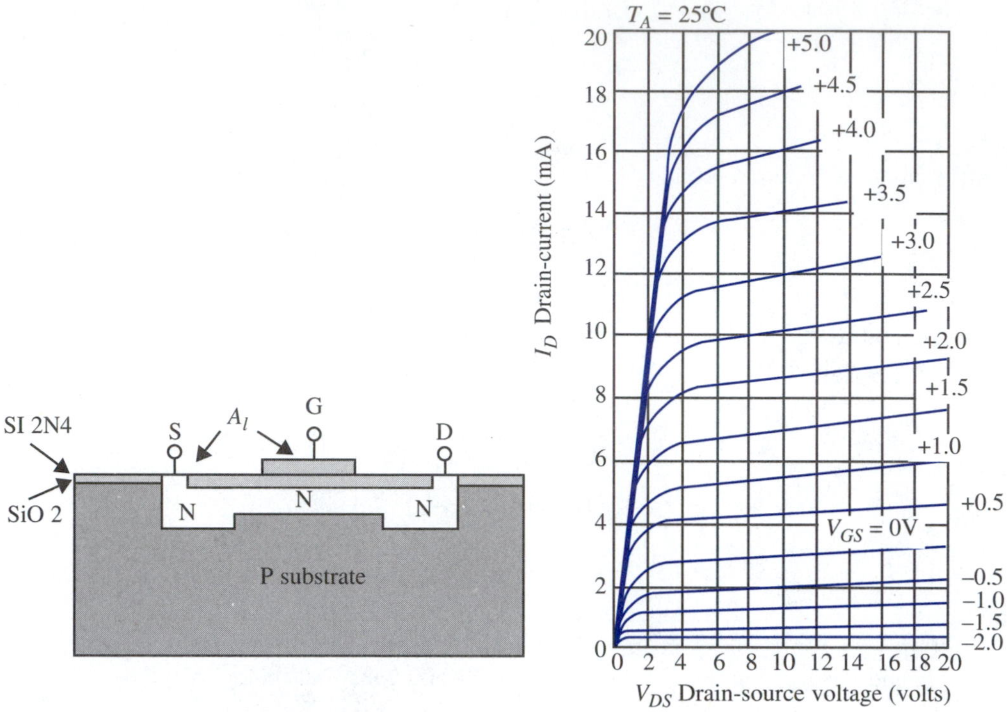

a) Construction b) V_{gs} effect on I_A

FIGURE 4–17 Depletion mode MOSFET (2N3797)

Now if we put a negative voltage on the gate, electrons will be driven (repelled) from the connecting *N* region, and it will narrow—which will reduce current flow. If the gate voltage is made negative enough (V_{co}), the connecting channel will disappear, and current will be cut off.

By putting a positive voltage on the gate of a depletion-type MOSFET, we can cause the channel to widen, and the current will increase. So depletion MOSFETS can also be operated in the enhancement mode. Figure 4–17b shows the characteristics for a 2N3797 depletion/enhancement MOSFET. Notice that with a drain-source voltage of 10V and 0V on the gate, about 3mA of drain current will flow.

To turn the FET off, the gate voltage is taken typically to −5V. In the enhancement mode, with +5V on the gate, 20mA of drain current will flow. The advantage of this device is that we can operate without a biasing voltage, which simplifies the circuit configuration.

Let us now look at the circuit shown in Figure 4–18. In this circuit, we are using a supply voltage of +18V and have picked a required drain voltage of +9V.

EXAMPLE

What is the voltage gain for the circuit of Figure 4–18?

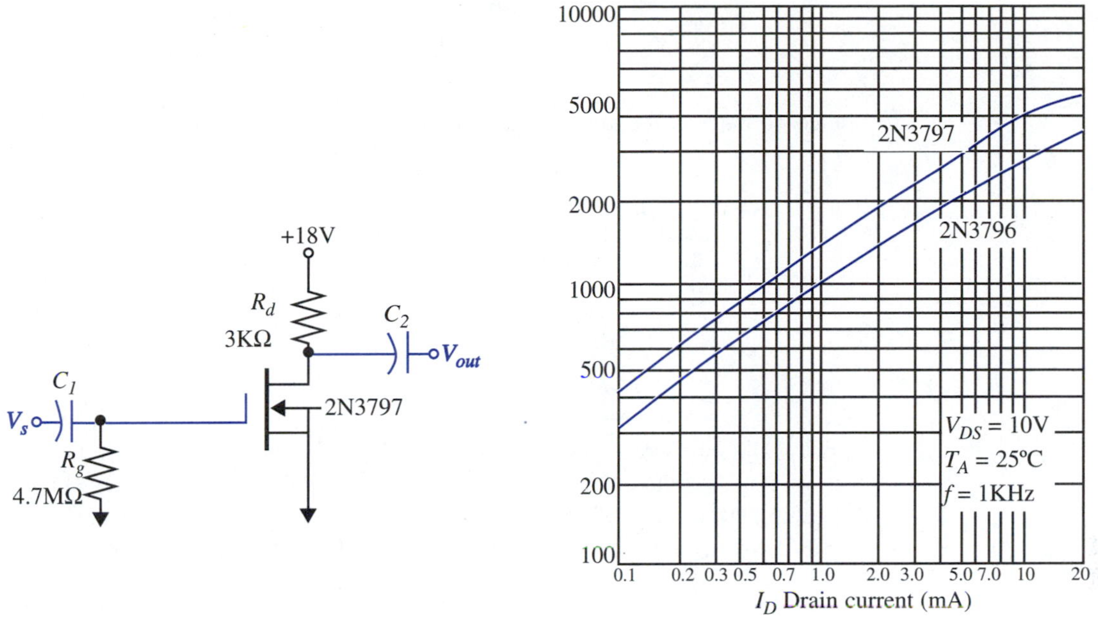

a) Circuit b) Transfer characteristic

FIGURE 4–18 Depletion/enhancement MOSFET circuit

SOLUTION

Since we have 0V bias on the gate, the drain current from the Figure 4–17b is 3mA (V_d is 9V), and the 3KΩ drain resistor is used to get the required 9V drop.
The voltage gain equation for our circuit is $-g_m R_d$.
From Figure 4–18b we read a g_m of 2300μS at 3mA. So our voltage gain is

$$A_v = -g_m \times R_d$$

$$A_v = -2300\mu S \times 3K\Omega$$

$$A_v = -6.9$$

EXAMPLE

What is the voltage gain and the new drain voltage for the circuit of Figure 4–18 if the drain resistor is changed to 5.6KΩ?

SOLUTION

We need to draw a load line on the graph of Figure 4–17b for the 5.6KΩ resistor. If we assume that all the 18V supply voltage is dropped across the 5.6KΩ resistor (drain–source voltage is 0), then the drain current would be

$$I_d = 18V/5.6K\Omega$$

$$I_d = 3.2mA$$

The coordinates of the first point are then

$$I_d = 3.2mA$$

$$V_{ds} = 0$$

The second point is

$$I_d = 0$$

$$V_{ds} = 18V$$

Drawing a line between these two points, and finding the intercept with the $V_{gs} = 0$ line, we read a drain current of about 2.7mA and a drain voltage of 3V. Entering this drain current into the graph of Figure 4–18b, and finding the intercept with the 2N3797 graph, we read a g_m of about 2100µS. The voltage gain is then

$$A_v = -g_m \times R_d$$

$$A_v = -2100\mu S \times 5.6K\Omega$$

$$A_v = -11.8$$

It is probably apparent at this point of our study of the MOSFET that the voltage gain is like the JFET and typically much less than that of the bipolar transistor. We can raise the value of R_d in an attempt to get higher voltage gains; but in order to maintain the same drain bias voltage point, we have to reduce the drain current by changing the gate bias voltage. The problem is that the g_m of an FET drops as the drain current is reduced—so our gain increase is not directly proportional to the increase in R_d.

The reduction in g_m with reduced drain current was illustrated in Figures 4–15b and 4–18b, which shows the forward transfer (input gate voltage to output drain current) characteristics.

From Figure 4–18b, we see that if we change our drain current from 3mA to 0.3mA, our transconductance (g_m) drops from 2300 to 800. If we still wanted to maintain the same +9V on the drain of our circuit with the 0.3mA drain current, we would increase R_d from 3KΩ to 30KΩ. Our voltage gain would now be

$$A_v = -g_m \times R_d$$

$$A_v = -800\mu S \times 30K\Omega$$

$$A_v = -24$$

We have increased our gain, but now from the graph of Figure 4–17b, we see that we have to provide a bias voltage of about –2V in order to have a drain current of 0.3mA.

This has taken away our advantage of the no gate bias requirement.

To repeat, the main advantage of FETs is their higher input impedance, lower noise, and as we shall see when we look at switching transistors, their potential lower "on" resistance.

HIGH-FREQUENCY GALLIUM ARSENIDE FETS

As mentioned in Chapter 1 when Schottky diodes were discussed, using gallium arsenide rather than silicon provides a semiconductor with no minority current. This allows for faster electron current flow, and therefore we can create an FET with much higher frequency response. Figure 4–19 indicates the somewhat complex construction of a GaAs FET.

The N-type channel required for depletion-mode operation is created by selective silicon ion implantation into an undoped GaAs substrate. This makes an FET with improved pulse response as well as radiation resistance and also contributes to device uniformity.

Harris Semiconductor provides the HMF–12100 GaAs FET with specified performance up to 18GHz. It is a depletion-mode N-channel FET with typical pinch–off voltage of –2.1V.

Disadvantages of the GaAs FET are higher cost and increased noise (rising with $1/f$) below 100MHz.

C–MOSFETS

All the specific FETs we have discussed have been N channel—but P-channel devices are also used. The basic difference between them is a reversal of the doped material. With the JFET, for example, the P-type has a P-doped channel, and the gate is connected to N-type material.

In the circuit, the P-channel FET will have a negative drain supply, and the gate will be driven positive relative to the source for cutoff.

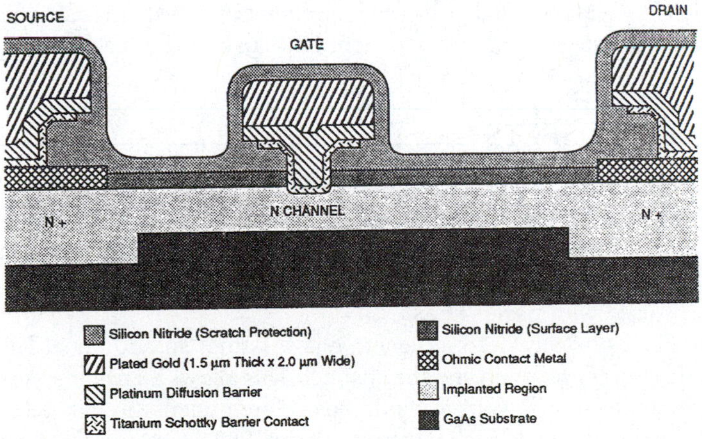

FIGURE 4–19 GaAs FET construction

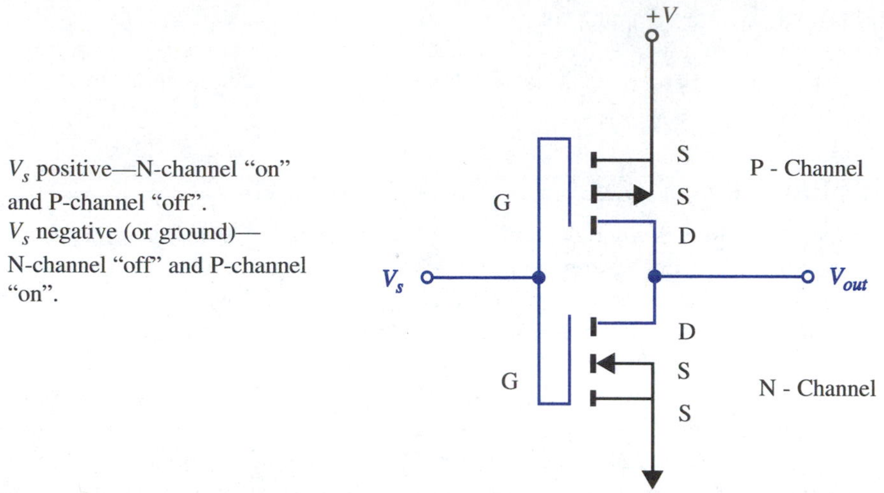

FIGURE 4–20 C–MOS connection (enhancement FETs)

The complementary connection of an N-channel and P-channel MOSFET is called **C–MOS** and is similar to the complementary bipolar transistor connection we studied previously. This C–MOS connection is shown in Figure 4–20 and consists of enhancement-type MOSFETS.

With the positive supply used, the upper MOSFET is a P-channel (arrow facing out) while the bottom MOSFET is an N-channel (arrow facing in).

Since we are dealing with enhancement-type MOSFETS, a positive voltage applied to the shared gate terminal will turn on the N channel and turn off the P channel, causing the output to go low (drop).

Taking the gate voltage negative will reverse the situation, with the upper MOSFET turned on and the lower off, and the output will go high.

An application for C–MOS is in digital switching circuits where it is a basic building block for logic gates. Its advantage is extremely low power drain and chip packaging density.

POWER-TYPE MOSFETS

The construction of the FETs we have studied results in a long channel, which makes the channel resistance rather high. This is because the source and drain connection are on the same side of the semiconductor chip. If the drain is moved to the other side, the resistance of the channel can be reduced and the current-handling capacity of the FET increased. This technique is used with power FETs.

Since power FETs typically need a higher voltage rating, an additional lightly doped region is placed between the drain and the channel. This allows a greater spreading of the depletion region under reverse voltage conditions before punch–through occurs.

Figure 4–21 shows the construction of two power FETs that use these concepts. In Figure 4–21a, we see the V–MOS structure, and in Figure 4–21b the more cost-effective

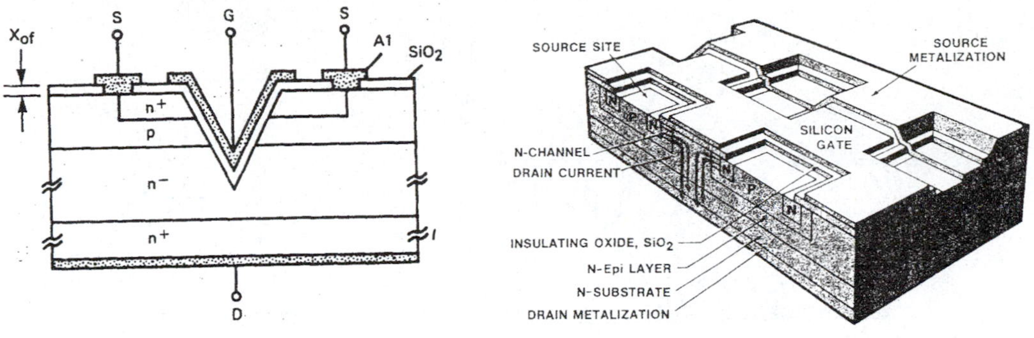

FIGURE 4–21 Power MOSFETS

Motorola T–MOS. These power MOSFETS can have "on" resistances of less than an ohm and can handle supply voltages up to 1000V. Some of the lower-voltage devices can pass currents close to 100 amps.

Since the power required to operate the gate is very low, and the "on" resistance is so low, these power-type MOSFETS are close to the ideal switch, i.e., they can go from a high "off" resistance to a low "on" resistance. They are useful for power control and can also be used for class C amplifiers in radio transmitters and class D audio amplifiers.

Also, Power C–MOS is being used as complementary output stages for linear amplifiers and shows advantages of being more linear and efficient than the bipolar devices.

CARE IN THE HANDLING OF MOSFETS

The insulated gate of a MOSFET, being a very high impedance, will allow transient voltages to develop on the gate. Since the insulating metal oxide layer is so thin, these voltages can cause breakdown from gate to channel. This requires extreme care when handling MOSFETS, particularly when the ambient air is dry—which allows static voltages to develop.

A shorting ring is provided with new MOSFETS to keep all three terminals at the same potential. Sometimes an internal zener diode is provided that gives limited protection for the gate-to-channel insulator.

The Motorola Corporation has provided a concept for a workstation to be used when handling MOSFETS; this is shown in Figure 4–22. Also in the figure is the human body equivalent as a transient voltage source, which reinforces the concern about handling these devices.

TROUBLESHOOTING FET AMPLIFIERS

First, let us look at how we can test an FET out-of-circuit.

If we have a JFET, we can check the gate-source connection with a DMM or VOM by remembering that it is a diode, and therefore an N-channel FET will read low resistance

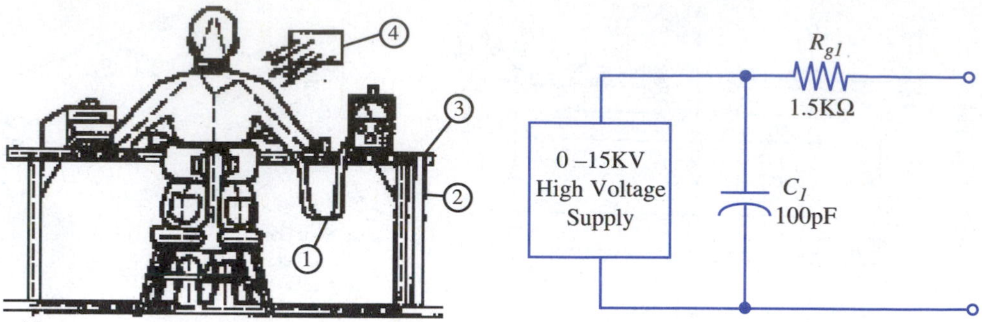

Notes:

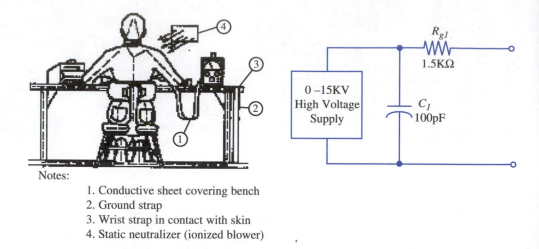

1. Conductive sheet covering bench
2. Ground strap
3. Wrist strap in contact with skin
4. Static neutralizer (ionized blower)

a) Work station b) Human body equivalent network

FIGURE 4–22 Care in the handling of MOSFETS

with the positive lead on the gate and the negative lead on the source. When we switch the leads, the resistance will read high.

With a P-channel JFET, the connections would be reversed for the same relative gate-source resistance readings. Since the channel is between the source-to-drain connection, we would expect to read the same resistance both ways. The resistance will be fairly low (value depends on type) with no reverse bias applied to the gate.

These techniques for checking JFETs are illustrated in Figure 4–23. MOSFETS are more difficult to test out-of-circuit than JFETs.

If we have a depletion-type MOSFET, then we should read resistance (same) both ways from source to drain and an open circuit from gate to source or gate to drain. However, with an enhancement type, there is no conducting path from source to drain until a bias voltage is applied to the gate so we will read high resistance between any two terminals.

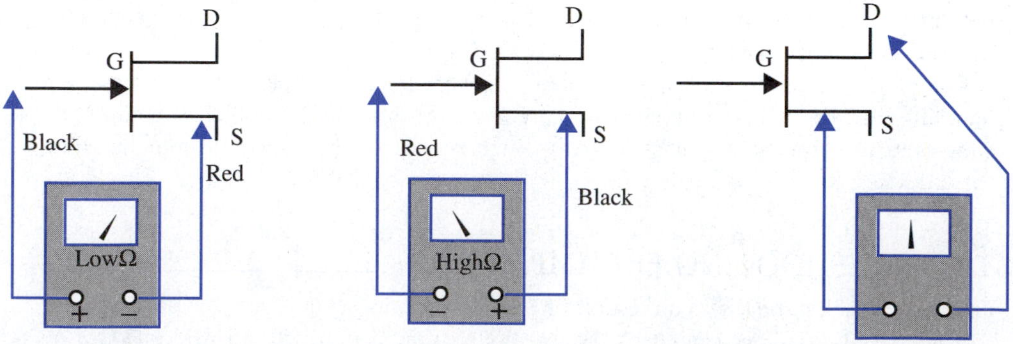

FIGURE 4–23 Checking a JFET

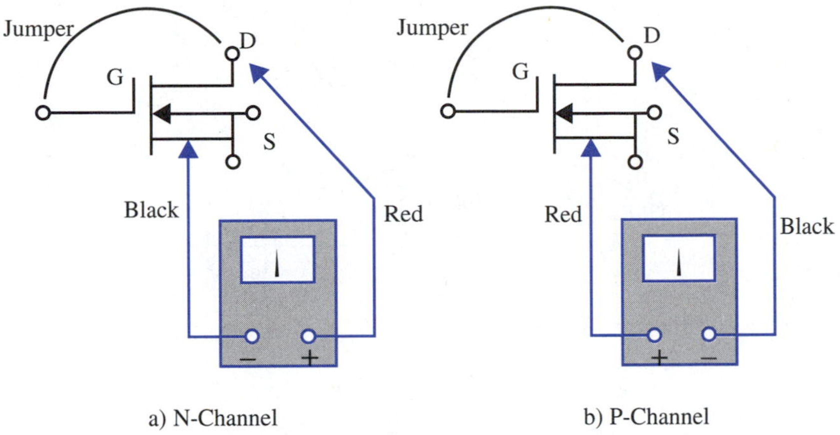

FIGURE 4–24 Checking an enhancement MOSFET

To lower the drain-to-source resistance and check for enhancement, we can use the VOM-supplied voltage, on the ohms scale, to bias the gate, as shown in Figure 4–24.

Since the gate is insulated, we should read high resistance from gate to source in both directions for all types of MOSFETS. In circuit, FETs are a little more difficult to troubleshoot than bipolar transistors.

With bipolar transistors, we can tell if the transistor should be turned on by looking to see whether there is 0.6V from base to emitter; but FETs are not this straightforward, since the gate-to-source voltage can vary with bias conditions.

We also need to know whether we have a JFET or MOSFET in the circuit, and if a MOSFET, whether it is an enhancement- or depletion/enhancement-type before we can interpret the gate voltage.

Let us first troubleshoot a JFET amplifier circuit and repeat the circuit of Figure 4–8 as our example. Figure 4–25 shows the circuit again.

Determining FET Bias Voltages

Starting with the gate circuit, we can estimate from the resistor values used that the gate will be at one-fourth of the supply voltage or about 6V. When we measure this voltage, however, we must take into account the loading from the DMM or 'scope since these devices have an input resistance of 10MΩ. This meter resistance, since we are measuring the voltage to ground, will be in parallel with the 1MΩ. The voltage we will actually read with instrument loading will be about 5.6V. Keeping in mind that even though we measured 5.6V at the gate, it is really at 6V. We can now measure the voltage at the source. This voltage should read more positive than the gate voltage. Also, the drain voltage should read more positive than the source voltage.

Open Gate Connection

A failure we could experience could be an open gate connection, which would cause the FET to fully turn on. In our circuit, this will cause both the source and drain to be close to one-half the supply voltage or 12V.

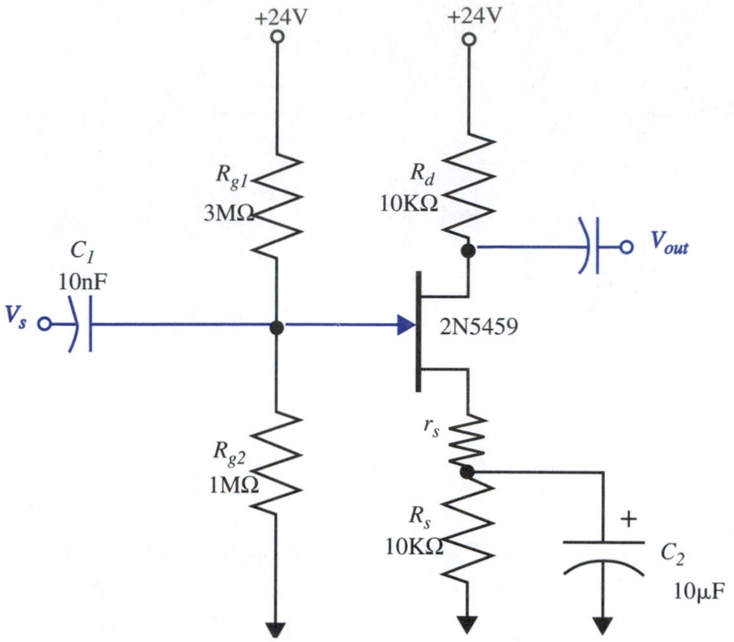

FIGURE 4–25 Troubleshooting a JFET amplifier circuit

Open Drain Connection

If the drain connection opened circuited, the gate–to–source diode would forward bias. Using a Thevenin's equivalent of the gate circuit, we can determine that we would read 0.67V on the gate and 70mV on the source; and of course, we would read the supply voltage on the drain.

Open Source Connection

With an open circuit of the source connection, we would see 0V at the source, 5.6V (6V) at the gate, and the supply voltage at the drain. If the gate-to-source diode is shorted, we would see roughly half the supply voltage on the gate, source, and drain terminals.

Open R_{g1}

If the R_{g1} resistor becomes open, the +6V bias voltage at the gate will go to zero volts. To determine the actual new voltages at source and drain, we need to draw a 10KΩ R_s load line on the transfer characteristic of Figure 4–4b. The points will be

$$V_{gs} = -8\text{V}$$

$$I_d = 0.8\text{mA}$$

and

$$V_{gs} = 0\text{V}$$

$$I_d = 0\text{mA}$$

The intercept of the load line with the transfer curve yields a V_{gs} of –4.5V and a drain current I_d of 0.5mA. From these values, we can compute a source voltage of +5V (was +10V) with the gate at 0V and a drain voltage of +24V – (0.5mA × 10KΩ) or +19V (was +14V). Even without making these computations, we can estimate that loss of gate bias will cause the source voltage to reduce (less voltage across R_s and therefore less source current) and the drain voltage to increase.

Open R_{g2}

With R_{g2} open, the gate-source diode will forward bias. Current will flow from the +24V supply through R_{g1}, the gate-source diode, and R_s to ground. Considering the relative values of R_{g1} and R_s, we can estimate that the source voltage will be tenths of a volt, the gate voltage 0.6V, and the drain voltage determined by divider action of the channel resistance and R_d.

Open R_s

The drain-source current path will be interrupted if R_s opens. The gate voltage will remain at +6V, but the source and drain will rise to +24V.

Open R_d

If R_d becomes open, the gate-source diode becomes forward biased. Most of the +24V supply voltage is dropped across the gate resistors, and the gate voltage will be about 0.6V. There will be just tenths of a volt at the source and drain.

Open C_2

With C_2 open, R_s is no longer AC bypassed. This means the AC voltage gain A_v is now

$$A_v = \frac{R_d}{r_s + R_s}$$

Shorted C_2

With C_2 shorted, R_s will also be shorted, and the source will be at 0V. The gate-source diode is forward biased, which means the gate is at +0.6V, and the drain voltage determined by divider action of the channel resistance and R_d.

TROUBLESHOOTING ENHANCEMENT-TYPE MOSFETS

For our troubleshooting analysis of the enhancement-type MOSFET, we shall use the circuit of Figure 4–26, which is a repeat of Figure 4–16a. The approximate 9:1 ratio of gate resistors means that one-tenth of the supply, or +2V, should appear at the gate.

Loading of the measuring instrument is more critical in this circuit because the resistor from gate to ground is 2.2MΩ rather than 1MΩ. We would expect to see 0V at the source and a drain voltage greater than the minimum +2V required for good amplifier action. Possible problems could be an open gate, open source, or open drain, all of which will cause the drain voltage to rise to the supply voltage.

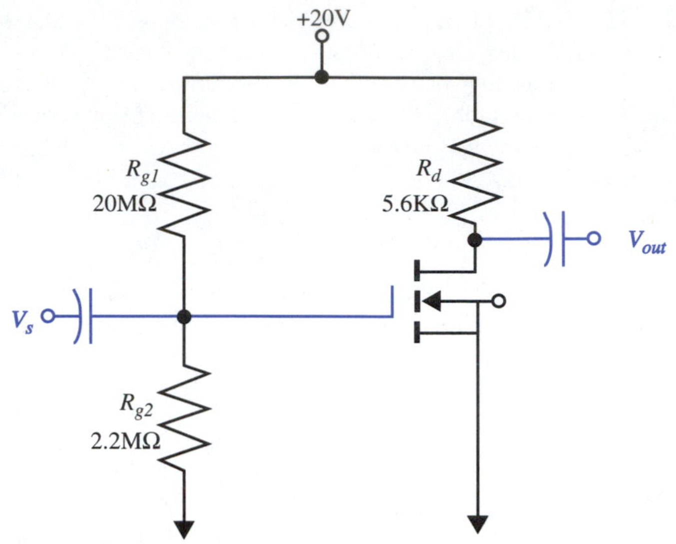

FIGURE 4–26 Troubleshooting an enhancement MOSFET

Shorted Gate

A shorted gate means the gate lead is connected to the P material beneath it, which will cause a diode to exist between the gate and the source. This diode will connect the gate to ground, so the gate voltage will drop to around 0.6V.

Open R_{g1}

If R_{g1} becomes an open circuit, there will be no forward bias on the gate, so the V_g would be 0V, and since the FET is turned off, V_d would be 20V.

Open R_{g2}

With R_{g2} open, the full supply voltage of 20V will be applied to the gate (V_{gs} is 20V). Examination of the characteristics of Figure 4–15a shows that the FET will be fully turned on with approximately 0V on the drain and a source current of 20V/5.6KΩ or 3.57mA.

Troubleshooting Depletion/Enhancement MOSFETS

For troubleshooting the depletion/enhancement MOSFET we shall repeat the circuit of Figure 4–18a in Figure 4–27.

MOSFET Bias Voltages

Since no bias voltage is used on the gate circuit, we would expect to see 0V on the gate and source. The expected drain voltage would be something in excess of +2V (we calculated +9V previously).

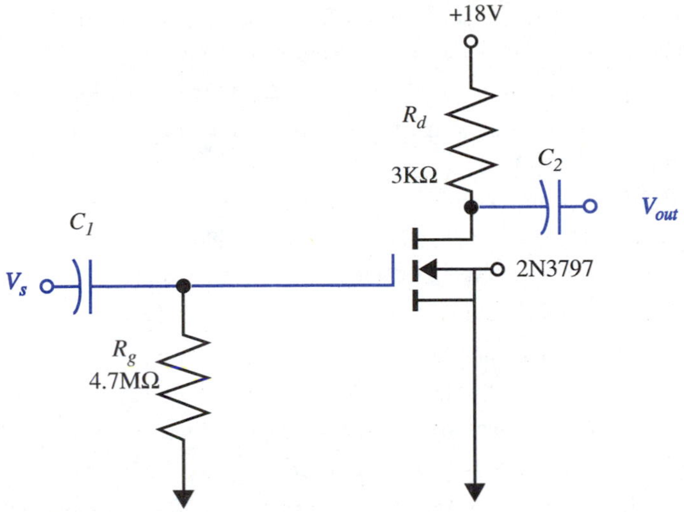

FIGURE 4–27 Troubleshooting a depletion MOSFET

MOSFET Open Circuit Analysis

If the gate connection open circuits, there would be no change in the bias voltages—but of course we could not amplify. With an open-circuited source or drain, the drain voltage would rise to the supply voltage. Because of the high resistance of the gate circuit, a shorted gate would again mean no change in the bias voltage at the drain.

SUMMARY

- A bipolar transistor is a current-operated device, but an FET is controlled by the voltage on the gate.
- The current drawn by the gate circuit is extremely small.
- A junction FET has a reverse-biased diode junction between gate and source and a channel between source and drain.
- Bipolar transistors come in NPN and PNP, whereas FETs are N channel or P channel.
- Applying a reverse voltage to the JFET gate causes the channel width to reduce, which reduces current flow.
- By adding a resistor in series with the source, we can self-bias a JFET.
- The larger the source resistor, the more stable the JFET bias.
- If the source resistor R_S is AC bypassed with a capacitor, the FET AC voltage gain is R_d/r_S.
- The input resistance to an FET source follower is high, and the output resistance is low and equal to $r_S \parallel R_S$.

- The common gate amplifier has a low input resistance, which is useful for matching the low resistance of a receiving antenna.
- JFETs generate less noise than bipolars because they only have one junction.
- Vacuum tubes are similar in operation to FETs because they are both controlled by an input voltage.
- A MOSFET has an insulated gate rather than the diode used in the JFET.
- An enhancement MOSFET has an insulated region between source and drain, which is made conductive by the appropriate gate voltage.
- Self-bias cannot be used with an enhancement MOSFET.
- A depletion/enhancement MOSFET has a lightly doped region between source and drain, which allows conduction with no bias voltage.
- FETs have gains in the tens, whereas bipolars have gains in the hundreds.
- The transconductance (g_m) of an FET varies with drain current.
- The high gate resistance of an FET simplifies the design of an oscillator.
- The advantage of a Pierce oscillator is its simplicity.
- Gallium arsenide FETs can operate at much higher frequencies than silicon types.
- The complementary connection of N-channel and P-channel MOSFETS is called C-MOS.
- C-MOS is used in digital switching circuits.
- Power FETs have the drain connection on the other side of the chip, which allows the channel to be shorter and therefore lower the source to drain resistance.
- Power C-MOS has the advantage of linearity and efficiency over bipolar.
- The high-resistance insulated gate of a MOSFET is subject to damage from transient voltages.
- Some MOSFETS have an internal transient protection diode attached to the gate, but this lowers the gate resistance.
- We can test a JFET out-of-circuit by remembering that the gate-to-channel connection is a diode.
- MOSFETS will read high resistance in both directions from gate to source.

EXERCISE PROBLEMS

1. Give the advantages and disadvantages of an FET over a bipolar transistor.
2. Sketch the basic parts of a JFET.
3. Describe the operation of a JFET.
4. What causes FET "pinch off"?
5. Explain FET internal self-bias.

6. Calculate the new drain voltage for the circuit of Figure 4–A if R_S is changed to 5.1KΩ and the FET has the characteristics shown in Figure 4–D.

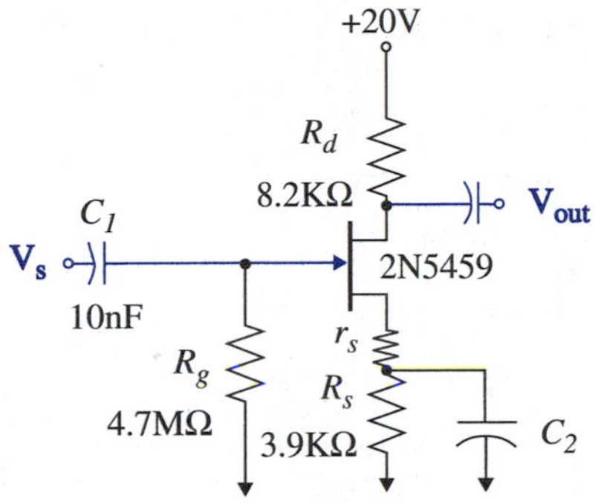

FIGURE 4–A (text figure 4–5b)

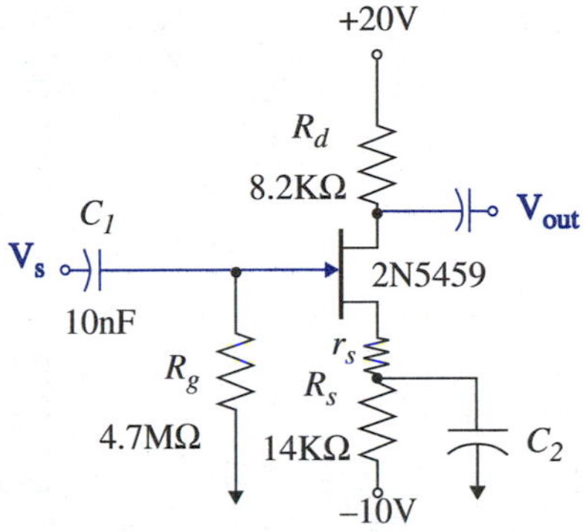

FIGURE 4–B (text figure 4–7a)

7. What is the voltage gain of a FET common source amplifier if g_m is 1200µS, R_d is 3.9KΩ, and R_s is AC bypassed?
8. Calculate the new voltage gain for the circuit of Figure 4–A if a value of R_s is used that changes I_d to 1.5mA and the FET has the characteristics of Figure 4–E. (Hint: Find V_{gs} and use equations 4–4 and 4–8.)
9. What is the drain voltage of the circuit in Figure 4–B if an 8.2KΩ resistor is accidentally used for R_S when the circuit is constructed, and the FET has the characteristics of the curve of Figure 4–E?

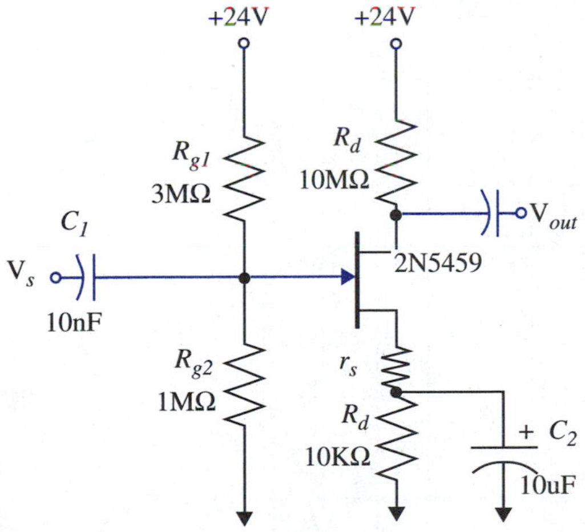

FIGURE 4–C (text figure 4–8)

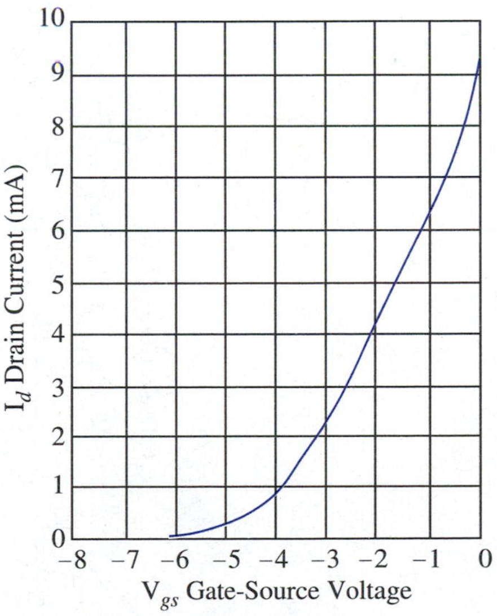

FIGURE 4–D (text figure 4–4b)

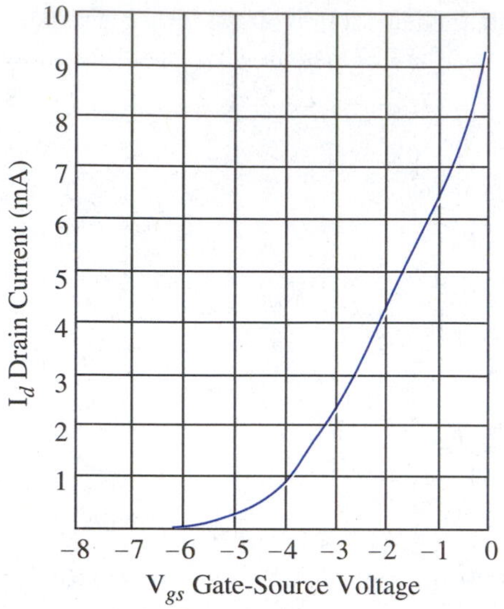

FIGURE 4–E (text figure 4–8)

10. Determine the g_m and voltage gain A_v for the circuit of Figure 4–B if the FET has the characteristics of the leftmost curve of Figure 4–F, with I_{dss} of 16 mA. (Hint: Use equation 4–4.)
11. Compute the range of voltage gains that can occur with the circuit of Figure 4–C if g_m has a nominal value of 920μS and can vary by ±50%.
12. Give the range of output resistance for a follower using a 2N5459 if g_m varies from 460μS to 1380μS. (Assume R_S>>r_s.)
13. State a use for the FET common gate amplifier.

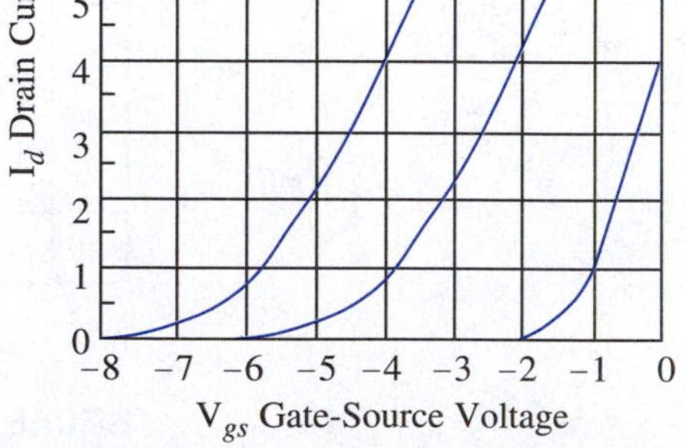

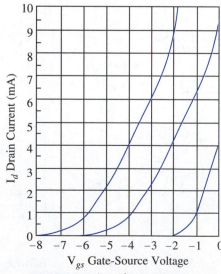

FIGURE 4–F (text figure 4–29)

14. What do the FET and vacuum tube have in common?
15. How does a MOSFET differ from a JFET?
16. What is the resistance between the source and drain of an enhancement MOSFET if there is 0V bias on the gate?
17. Calculate the DC bias voltages on the gate and the drain for the enhancement MOSFET circuit of Figure 4–G. (Hint: Use load line.)
18. Determine the voltage at the drain of the circuit of Figure 4–H if R_d is changed to 2.2KΩ.

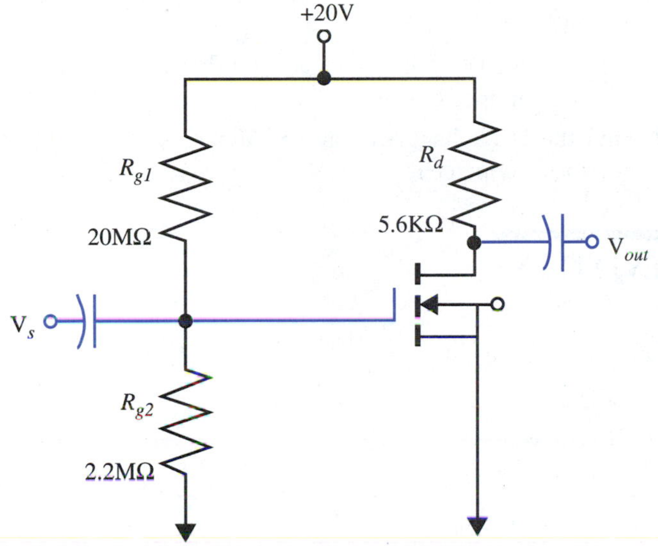

FIGURE 4–G (text figure 4–16)

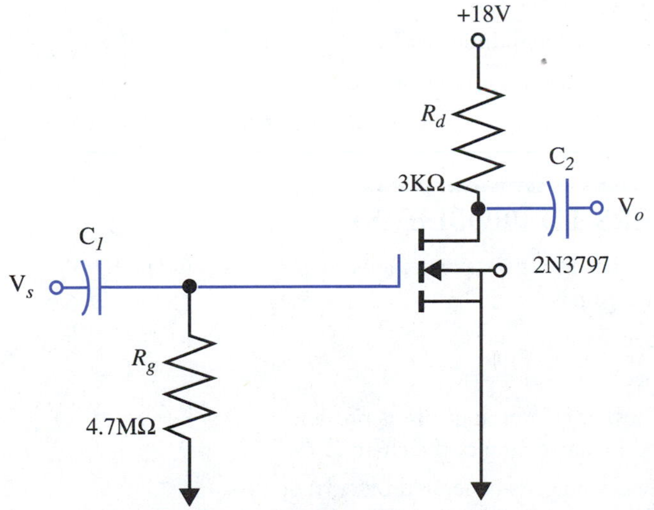

FIGURE 4–H (text figure 4–18)

19. Determine the transconductance (g_m) for a 2N3797 MOSFET at drain currents of 0.1mA, 1mA, and 10mA.
20. What is the difference between a regular MOSFET and a power MOSFET?
21. Why is an internal diode sometimes connected to the gate of a MOSFET?
22. State how you would check a JFET out-of-circuit.
23. State how you would check an enhancement MOSFET out-of-circuit.
24. Devise a test method for checking a depletion MOSFET out-of-circuit.
25. What is the voltage gain of the circuit of Figure 4–C if C_2 becomes an open circuit and g_m is 940µS?
26. What are the DC voltages at gate and drain for the circuit of Figure 4–G if R_d becomes an open circuit?
27. Find the DC voltages at gate and drain for the circuit of Figure 4–G if R_{g1} becomes an open circuit.

ABBREVIATIONS

gm = Transconductance = $\dfrac{\Delta I_d}{\Delta V gs}$

V_c = FET cutoff voltage

V_d = Drain voltage

V_g = Gate voltage

V_p = Pinch-off voltage

V_s = Source voltage

V_{gs} = Gate-to-source voltage

$V_{R_{s1}}$ = Voltage across R_{s1}

R_{s1} = External source resistor

R_d = External drain resistor

r_s = Internal source resistance = $1/g_m$

I_d = Drain current = I_s = Source current

ANSWERS TO PROBLEMS

1. Higher input resistance; lower voltage gain
2. See text.
3. See text.
4. Further current increase is prevented by narrowing of the channel.
5. Bias voltage is generated by current flow through R_s.
6. I_d = 0.8mA and V_d = +13.3V
7. A_v = –4.68
8. V_{gs} = –3.5 V; g_m = 1278µS; and A_v = –10.5
9. V_d = +6.7V
10. V_{gs} = –5.6V; g_m = 1.125mS; A_v = –9.23
11. $A_{v(min)}$ = –4.6; $A_{v(max)}$ = –13.8

12. R_o range 725Ω to 2.17KΩ
13. Low-resistance input stage for a receiver
14. Both are voltage controlled devices.
15. MOSFET has an insulated gate.
16. Very high
17. V_{gs} = 1.98V; V_d = 11V
18. V_d = 11.5V
19. g_m values 450μS, 1500μS, 3800μS
20. Power MOSFET has a shorter and lower-resistance channel.
21. To prevent static voltage from damaging the insulator
22. See text.
23. See text.
24. See text.
25. A_v = –0.904
26. V_g = 1.98V, V_d = 0V
27. V_g = 0V; V_d = 20V

CHAPTER 5

Transistor Switching Circuits

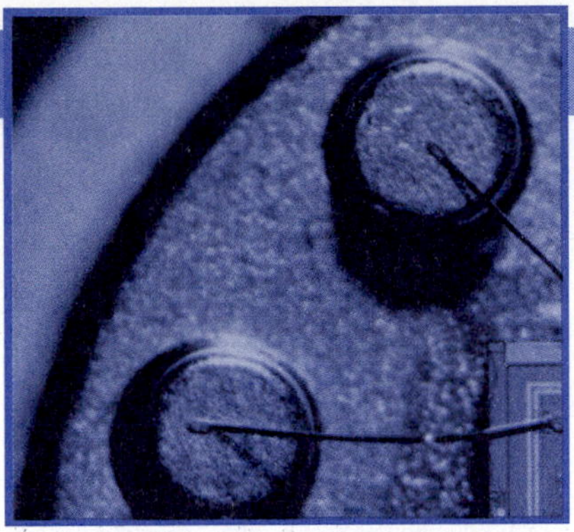

CHAPTER OBJECTIVES

After the completing this chapter, you should be able to explain the following switching transistor concepts:

- Why switching is used
- Effect of capacitance on switching time
- Effect of inductance on switching time
- How a bipolar transistor can be used as a switch
- Bipolar transistor switching characteristics
- Techniques for reducing switching time
- The reason "totem pole" circuits are used
- The FET as a switch
- Advantages of an FET over a bipolar transistor in switching applications
- Switching circuit measurement techniques
- How to troubleshoot a switching circuit

INTRODUCTION

The first question that might be asked when encountering switching transistors is "Why are they used and how do they differ from the transistors we have studied so far?" This chapter will answer these questions by first studying the need for electronic switching and then showing how both bipolar and FET transistors can be used for this application. Finally, actual switching circuits will be examined from an operational and troubleshooting viewpoint.

WHY ARE ELECTRONIC SWITCHING CIRCUITS USED?

We were first introduced to the switch when we studied DC circuits. It was a manually operated device used to prevent or allow current to flow.

It is not always convenient or possible to manually activate a switch. What if we wanted to turn on a backup transmitter on a satellite. Obviously, this cannot be a manual operation. Other times we need to switch "on" and "off" at a high rate of speed—for example, faster than a million times a second. Again, this cannot be a manual operation.

In its ideal form, there is no power dissipation in a switch. When it is in the "off" position, full supply voltage is across the switch, but there is no current, so power dissipation is zero. In the "on" position, there is current, but no voltage drop across the switch, so again the power dissipation is zero.

An electronic switching circuit gives both the capability of remote operation and high switching rates. The power dissipation in an actual electronic switching circuit is not zero, but it can be made quite low.

Before we look at switching devices, we need to consider the effect of capacitors and inductors on circuit response.

EFFECT OF CAPACITANCE ON SWITCHING TIME

Capacitance will exist when there is an insulating material between two conductors. This insulator can be air, the insulation on a wire, or even a diode junction. When there is capacitance in a circuit, be it an actual capacitor or stray capacitance, it has to be charged before the voltage can change (is switched) across the capacitor.

Consider the simple circuit of Figure 5–1.

When the switch is first closed, the voltage across the capacitor is zero (assuming there is no initial charge on the capacitor *C*). As current flows into the capacitor, a voltage develops across it that opposes the supply voltage. This causes the voltage across the resistor *R* to drop, and so the circuit current drops. The higher the voltage on the capacitor, the smaller the current, and therefore the slower the voltage buildup on the capacitor. This effect causes an exponential rise in voltage across the capacitor in accordance with the equation:

$$v_c = V_{cc}(1 - \epsilon^{-t/RC}) \qquad \textbf{(equation 5–1)}$$

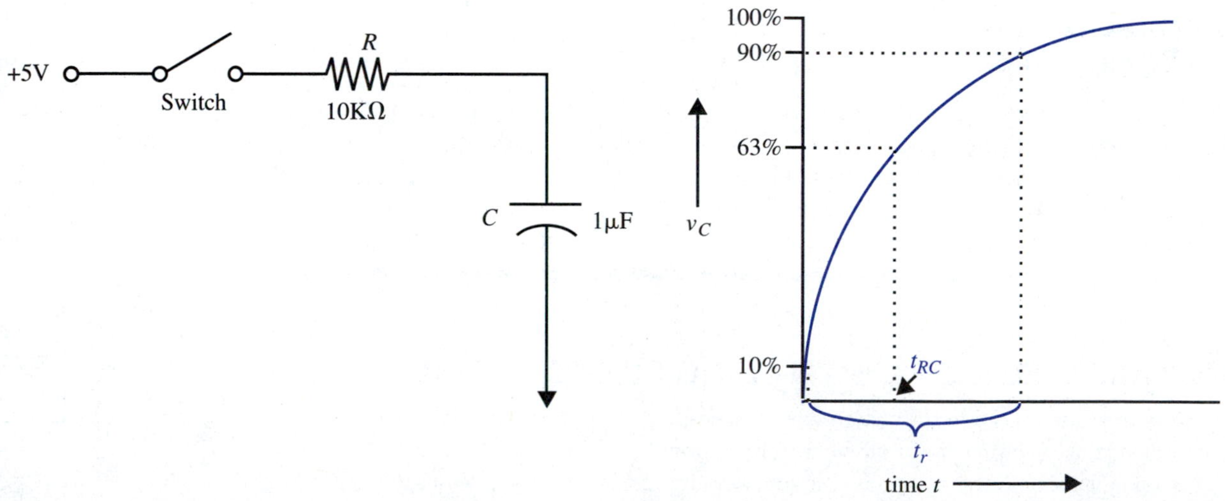

a) RC circuit

b) Time response

FIGURE 5–1 RC charging circuit

The graph in Figure 5–1b shows the time response of voltage in accordance with this equation. Key times on the graph are the time (t_r) for the voltage to increase from 10% to 90% of the maximum value—called the circuit **rise time**; and the time (τ) for the voltage to reach 63% of its maximum value—which is always equal to the product of $R \times C$ and is called the **time constant** for the RC circuit

By putting a value in equation 5–1 for v_c of $0.1V_{CC}$ and solving for the time to reach this voltage, and then the same with a $v_c = 0.9\ V_{CC}$, we can determine the time (rise time) to go from a voltage of 10% to 90%. Subtracting the time to reach 10% from the time to reach 90%, we end up with the following relationship for rise time:

$$t_r = 2.2RC \qquad \textbf{(equation 5–2)}$$

EXAMPLE

Calculate the rise time (t_r) and the time constant (τ) for the circuit of Figure 5–1.

SOLUTION

Using equation 5–2 and the values specified in the figure,

$$t_r = 2.2RC$$

$$t_r = 2.2 \times 10\text{K}\Omega \times 1\mu\text{F}$$

$$t_r = 22\text{ms}$$

For the time constant, we simply use the product of $R \times C$:

$$\tau = 10\text{K}\Omega \times 1\mu\text{F}$$

$$\tau = 10\text{ms}$$

We can use the time constant to approximate how long it will take for the voltage across the capacitor to reach a certain value. For example, if we set t at $3RC$ (three time constants) in equation 5–1, we find that the voltage across the capacitor is about 95% of the supply voltage. With t at $5RC$, the voltage is 99% of the supply voltage. It would take an infinite time for the capacitor to charge to the full supply voltage.

EXAMPLE

Calculate the actual time for the voltage across the capacitor in the circuit of Figure 5–1 to reach 99% of the +5V supply (4.95V).

SOLUTION

Using the relationship that to be within 1% of the supply voltage a time of $5RC$ must elapse,

$$t = 5RC$$

$$t = 5 \times 10\text{K}\Omega \times 1\mu\text{F}$$

$$t = 50\text{ms}$$

When a charged capacitor discharges through a resistor, the voltage across the capacitor drops also in an exponential manner. Figure 5–2 shows the circuit and the voltage waveform across the capacitor.

With the switch closed, the voltage on the capacitor decays according to the equation

$$v_C = V_{CC}\epsilon^{-t/RC} \qquad \textbf{(equation 5–3)}$$

The **fall time** (t_f) is the time it takes for the capacitor voltage to drop from 90% of its initial value to 10% of its initial value. Again this can be determined by using equation 5–2. A time (τ) equal to one time constant (RC) is the time it takes the capacitor voltage to discharge to 37% of its initial value.

EXAMPLE

Calculate the fall time (t_f) and the time constant (τ) for the circuit of Figure 5–2.

SOLUTION

Using equation 5–2 and the values specified in the figure,

$$t_r = 2.2RC$$

$$t_r = 2.2 \times 50\text{K}\Omega \times 1\mu\text{F}$$

$$t_r = 110\text{ms}$$

For the time constant, we simply use the product of $R \times C$:

$$\tau = 50\text{K}\Omega \times 1\mu\text{F}$$

$$\tau = 50\text{ms}$$

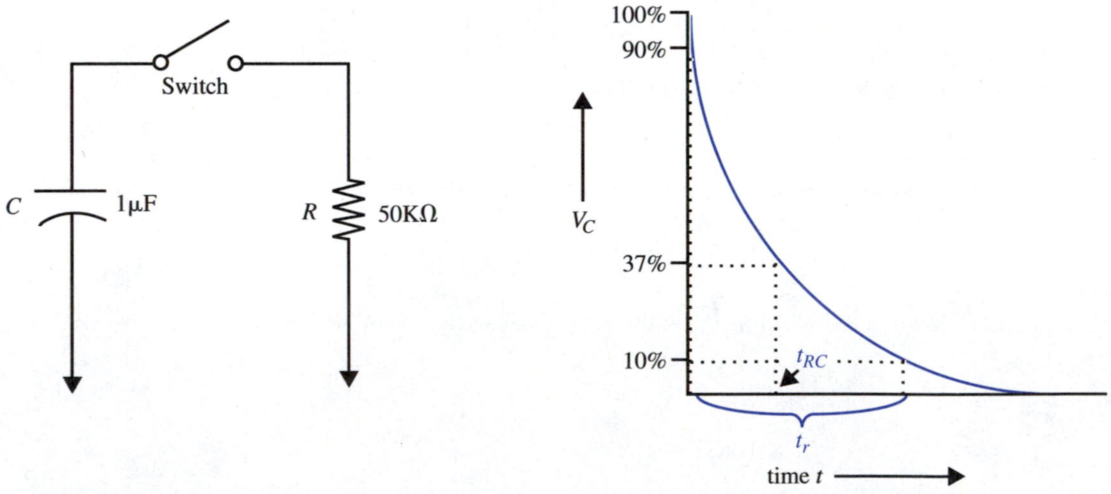

FIGURE 5–2 RC discharging circuit

EFFECT OF INDUCTANCE ON SWITCHING TIME

Because of its property of resisting current change by developing an induced "reverse voltage," inductance can slow circuit switching time. Inductance exists in any electrical conductor. It increases with the length of the wire and if the wire is coiled.

The circuit of Figure 5–3 demonstrates the buildup of current in a RL circuit after a voltage is applied to the circuit.

The current waveform, and therefore voltage across the resistor, follow an exponential just like the voltage buildup on a capacitor.

$$i = \frac{V_{cc}}{R}(1 - \epsilon^{-tR/L}) \qquad \textbf{(equation 5–4)}$$

the final or maximum current being V_{cc}/R.

Also:

$$v_R = V_{cc}(1 - \epsilon^{-tR/L}) \qquad \textbf{(equation 5–5)}$$

The time constant for the RL circuit is

$$L/R$$

and the rise time (t_r) from the 10% to 90% points is:

$$t_r = 2.2L/R \qquad \textbf{(equation 5–6)}$$

EXAMPLE

Calculate the rise time (t_r) and the time constant for the circuit of Figure 5–3.

SOLUTION

Using equation 5–6 and the values specified in the figure,

$$t_r = 2.2L/R$$

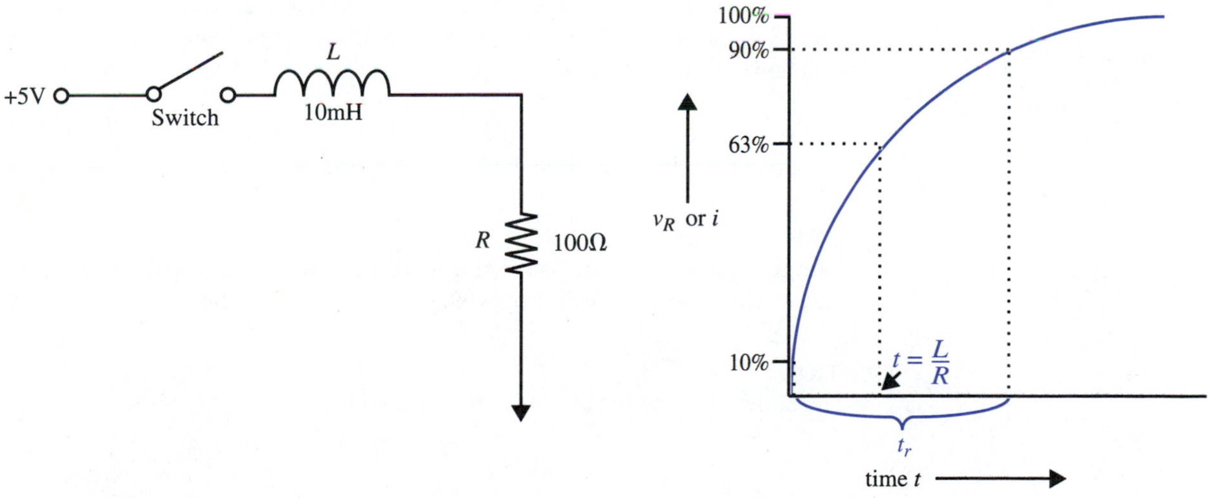

FIGURE 5–3 "RL" increasing-current circuit

$$t_r = 2.2 \times 10\text{mH}/100\Omega$$

$$t_r = 0.22\text{ms}$$

For the time constant, we simply use the quotient L/R:

$$\tau = 10\text{mH}/100$$

$$\tau = 0.1\text{ms}$$

After the switch in the inductive circuit of Figure 5–3 is closed for five time constants, the current is within 1% of its maximum or steady state value. If we now open the switch, the rapid change in current will cause a voltage v_L to develop across the inductor in accordance with the equation

$$v_L = Ldi/dt$$

What this equation indicates is that the faster the switch is opened, the greater the voltage across the inductor. The polarity of the induced voltage will be negative on the left (switch side) of the inductor in this circuit and positive on the right side.

If the energy stored in the inductor is not dissipated by passing the current generated by the induced voltage through a resistor, arcing will occur across and damage the switch. Figure 5–4 shows two methods to provide a current path and prevent switch damage.

With capacitive suppression, the capacitor across the switch provides a path for the transient current that flows through the power-dissipating resistor to ground, and back through the power supply to the capacitor, and through the capacitor to the left side of the inductor. This type of spark suppression is used in the ignition points system of older automobiles.

With diode suppression, a path is provided through the power-dissipating resistor to ground, and up through the diode to the left side of the inductor. Sometimes the suppression diode will be connected in parallel with the inductor. In this configuration, the diode must be able to absorb the inductor-stored energy without overheating and burning out.

If a "charged" inductor is connected across a resistor, the current decays in accordance with the following equation:

$$i = I\epsilon^{-Rt/L} \qquad \textbf{(equation 5–7)}$$

where I is the current flowing through the inductor just prior to connecting to the resistor.

EXAMPLE

Find the voltage drop across the resistor in the circuit of Figure 5–4b at 50μs after the switch is opened. Neglect the voltage drop across the diode.

SOLUTION

The current I at the time the switch is opened is

$$I = 5\text{V}/100\Omega$$

$$I = 50\text{mA}$$

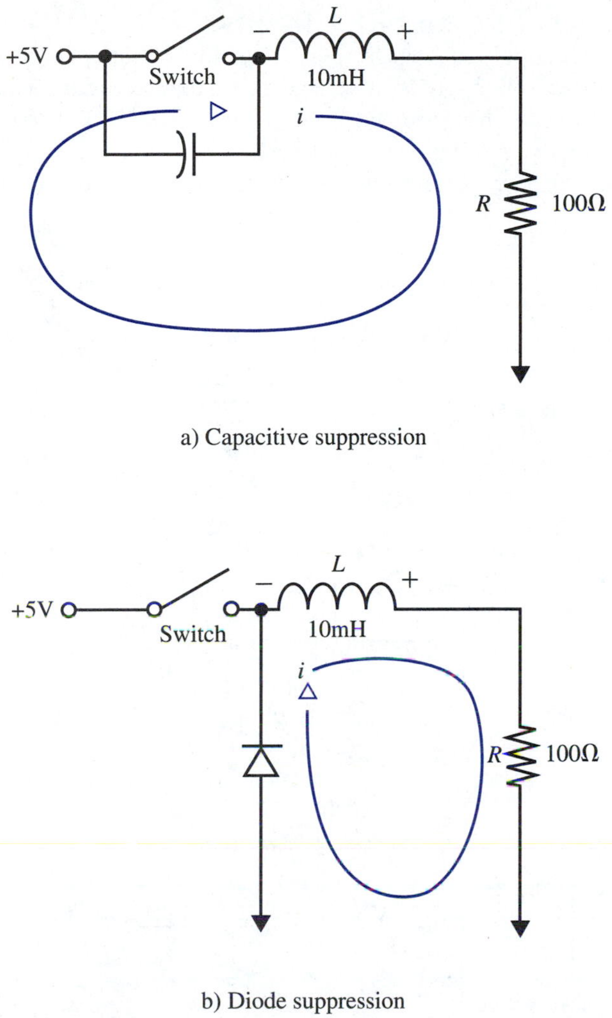

FIGURE 5–4 Arc-suppression techniques

The current at 50μs is, using equation 5–7:

$$i = I\epsilon^{-Rt/L}$$

$$i = 50\text{mA} \times \epsilon^{-100 \times 50\mu\text{s}/10\text{mH}}$$

$$i = 30.3\text{mA}$$

The voltage drop across the resistor is this current times the resistance value, or

$$v_R = 30.3\text{mA} \times 100\Omega$$

$$v_R = 3.03\text{V}$$

Pulse Response of RC and RL Circuits

If we close a switch supplying voltage to a circuit for a short period of time and then open it, we are effectively applying a voltage pulse to the circuit. The pulse response of RC and RL circuits we have studied is shown in Figure 5–5. The circuits are called integrators or differentiators, respectively, because the output waveshapes approximate those to be expected from performing these math operations on the input signal (see Appendix A).

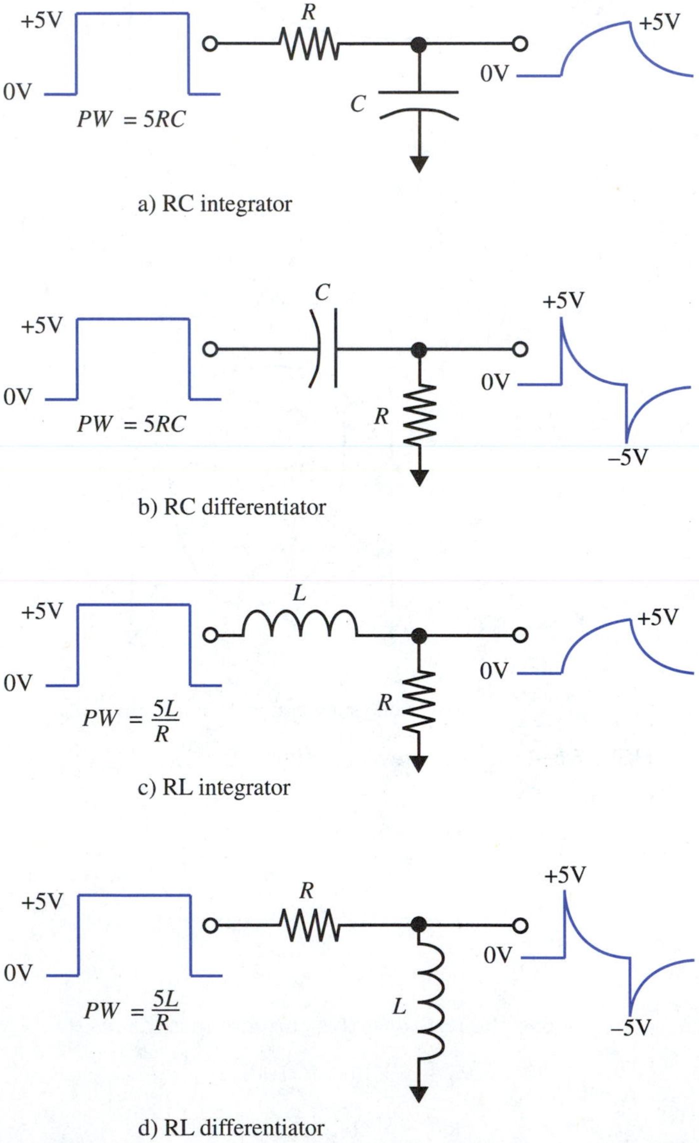

FIGURE 5–5 Pulse response of RC and RL circuits

Notice that the input pulse has a width of five time constants to allow the integrator output waveform peak voltage to equal the peak input value. If the input pulse is reduced, the output amplitude of the integrators will reduce—this is called **rise time limiting.**

The voltage swing (change) at the output of the differentiator is not a function of the input pulse width since no voltage drop can initially appear across the capacitor in Figure 5–5b or across the resistor in Figure 5–5d.

SWITCHING CIRCUIT WITH STRAY CAPACITANCE

Let us now consider an example of a switching circuit that has stray capacitance to ground across the output resistor.

EXAMPLE

Consider the circuit in Figure 5–6. If we close the switch, we would like to have 4V appear across R_2 immediately. But the stray capacitance (C_{st}) must be charged before this can happen. Again, the actual time to charge a capacitor to within 1% of its final voltage is

$$t = 5RC$$

Find the time for this circuit if

$$R = R_1 \| R_2$$

(Thevenin's equivalent of R_1 and R_2)

and

$$C = C_{st}$$

SOLUTION

The time required for the full voltage (within 1%) to appear across R_2 is t_c:

$$t_c = 5(R_1 \| R_2)C_{st}$$

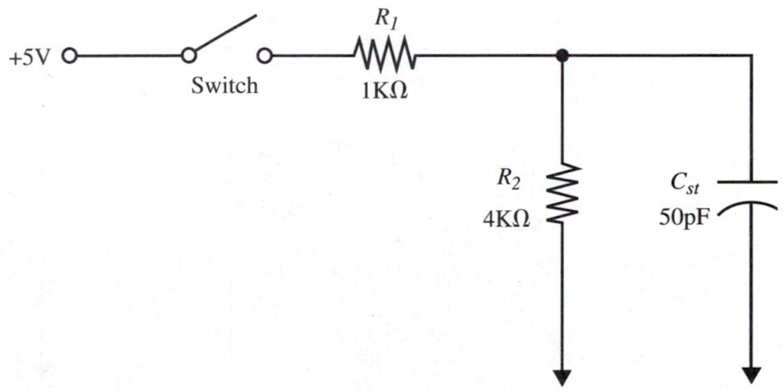

FIGURE 5–6 Simple switching circuit

$$t_c = 5 \times 800\Omega \times 50\text{pF}$$

$$t_c = 200\text{ns}$$

This may seem like a short time—until we consider that voltages in a computer are required to change in a few nanoseconds (10^{-9} seconds).

From this example, we conclude that the time required to switch (transfer) a voltage across a load is controlled by the amount of capacitance in the circuit and the effective source resistance. This becomes a key factor in the design of switching circuits.

How can we speed up the voltage transfer to a load? Let us see what happens when we add a "speedup" capacitor (C_1) across R_1 in the circuit of Figure 5–6. The new circuit is shown in Figure 5–7.

In order to understand the reason for the addition of C_1, we need to review the theory of charge on a capacitor. The equation for capacitor charge is

$$Q = CV \quad \text{(coulombs)} \qquad \textbf{(equation 5–8)}$$

Charge results from the flow of electrons and if two capacitors are connected in series, the charge on both capacitors will be the same. But if one capacitor has a smaller capacitance than the other, then we can see from equation 5–8 that more voltage has to be developed across this capacitor for the same charge.

In the circuit, we see that one-fifth of the 5V supply is dropped across R_1 (1V) and the remaining four-fifths across R_2 (4V). Now if we pick the correct value of C_1, one-fifth of the supply voltage can also be dropped across this capacitor and the rest across C_{st}. Since with the same charge, the size of the capacitor is inversely proportional to the voltage across it, the relationship between the two capacitors is

$$C_1 = 4\ C_{st}$$

$$C_{st} = 4 \times 50\text{pF}$$

$$C_{st} = 200\text{pF}$$

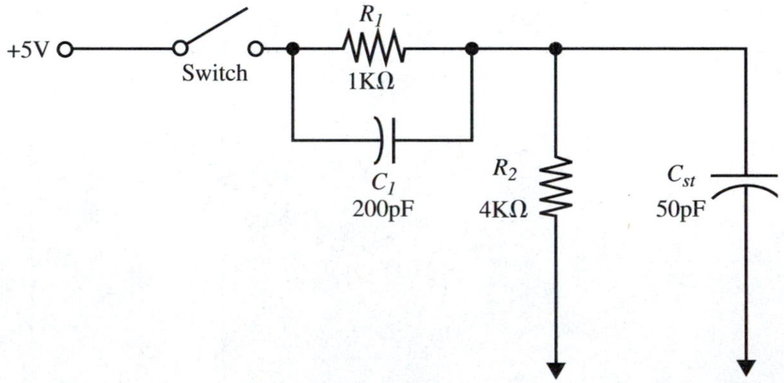

FIGURE 5–7 Use of a speedup capacitor

For the voltage across the capacitors to divide as we have stated, the current through R_1 must be the same as the current through R_2, and the charge on C_1 must be the same as the charge on C_{st}. These conditions will exist if the following relationship is maintained:

$$R_1 \times C_1 = R_2 \times C_{st} \qquad \textbf{(equation 5–9)}$$

Now the result of closing the switch with C_1 in the circuit will be that 4V will instantly appear across R_2 rather than delaying by the 200ns without C_1. This is why C_1 is called a speedup capacitor.

It is interesting to note that the relationship of equation 5–9 is used for the design of an oscilloscope probe—where R_1 is the probe resistance, C_1 the probe capacitance, R_2 the scope input resistance, and C_{st} the probe cable capacitance. This configuration makes the probe response independent of frequency.

THE BIPOLAR TRANSISTOR AS A SWITCH

A switching circuit that uses a bipolar transistor is shown in Figure 5–8. We shall look at this circuit first as a simple switch, without being concerned about speed.

Circuit operation occurs when base current is provided by the base supply voltage; the transistor will turn "on," and collector current will flow illuminating the bulb. Removing the base supply voltage will cause the base current to go to zero, and the transistor and therefore the bulb will turn "off."

EXAMPLE

Calculate the collector current for the circuit of Figure 5–8, assuming a β of 100 for Q_1.

SOLUTION

With +12V applied to the left of the 2.2KΩ base resistor, we have about 11.4V across this resistor—assuming 0.6V base-emitter drop. This gives a base current of 11.4V/2.2KΩ or roughly 5mA.

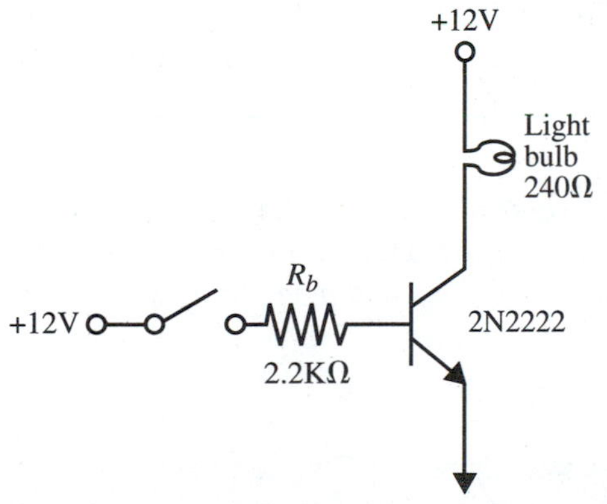

FIGURE 5–8 Bipolar transistor switching circuit

With the assumed β of the transistor of 100, then from our previous work, we would expect 500mA (5mA × 100) of collector current. But 500mA times the collector circuit bulb resistance of 240Ω gives a voltage drop (V_{bulb}) across the bulb of

$$V_{bulb} = 500\text{mA} \times 240\Omega$$

$$V_{bulb} = 120\text{V}$$

But how can this happen, since we only have a 12V supply in the collector circuit? Obviously, we will not have 500mA of collector current but 12V/240Ω or 50mA—assuming zero volts drop across the transistor.

What is happening is that the collector voltage drops as far as it can, but when it drops below about one volt, collection of electrons is drastically reduced, and β drops. The collector voltage will keep dropping until we have about 0.2V (called V_{sat}) from collector to emitter. This condition is called **transistor saturation**.

We can conclude, then, that just about all the voltage appears across the bulb and very little across the switch (transistor), which is what we wanted with our ideal switch to keep the switch power loss low.

The effective β for our circuit with the transistor in saturation is

$$\beta = \frac{50\text{mA}}{5\text{mA}}$$

$$\beta = 10$$

A good rule to follow in order to drive a signal (low-power) transistor into saturation—and therefore act like a switch—is to assume a saturated β of 10 and provide a base current of one-tenth of the required collector current.

POWER DISSIPATION IN A SWITCHED TRANSISTOR

Power dissipation in a transistor that is in saturation is the collector current times V_{sat} plus the base current times V_{be}. Expressed algebraically

$$P_d = (I_c \times V_{sat}) + (I_b \times V_{be})$$

EXAMPLE

An audible alarm requires 10V and 10mA (effective resistance 1KΩ) to provide the desired audio level. Show a transistor switch that can activate the alarm using a +5V base supply voltage, and compute the base current and the transistor power dissipation.

SOLUTION

The circuit is shown in Figure 5–9. To cause the transistor to go into saturation, we need—from our design rule—a base current of one-tenth the collector current, or 1mA. With the +5V base supply, the current I_b is

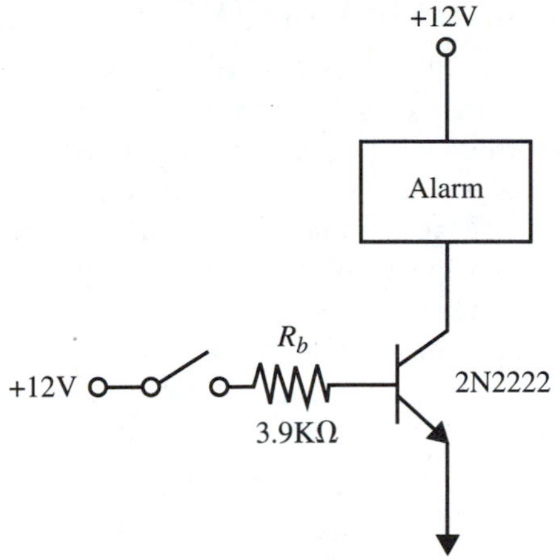

FIGURE 5–9 Alarm switching circuit

$$I_b = \frac{(5V - 0.6V)}{3.9K\Omega}$$

$$I_b = 1.128mA$$

This is a little greater than the 1mA we desired, but the collector current is limited by the effective 1KΩ resistance of the alarm to 10mA. The power dissipation in the transistor is the 0.2V collector-emitter drop times the 10mA (2mW) plus the base current of 1.128mA times the base-emitter drop of 0.6V (0.677mW). Total dissipation is

$$P_t = 2mW + 0.677mW$$

$$P_t = 2.677mW$$

This is low when compared to the maximum power dissipation for a 2N2222 at 25°C of 800mW.

We might wonder how the base voltage is switched on and off to activate the transistor switch. Typically the base voltage is supplied from some comparison circuit that provides an output in response to an event. For example, our audio alarm could be activated in response to a fire detector sensing smoke.

HIGH-SPEED TRANSISTOR SWITCH

When turning on lights and alarms, as in our previous examples, we are not concerned with the speed of switching, since the response time of our eyes and ears is in the millisecond range; but some circuits are required to change state very quickly, which requires a fast response. A factor that limits switching time is the circuit capacitance.

AN ACTUAL TRANSISTOR SWITCH

The time for a transistor to turn off or turn on is called the switching time. One of the reasons why switching time is important is that during this time, we are dissipating maximum power in the transistor—because we have current flow at the same time we have appreciable voltage across the transistor. Figure 5–10 shows the transistor switching waveforms. Notice that in the "transistor off" region, the collector voltage is maximum, but the current is zero—so transistor power dissipation is zero. Similarly, in the "transistor on" region, current is maximum but voltage is zero—again transistor power dissipation is zero. However, in the switching region, neither voltage nor current is zero. This means the power loss in the transistor, which is the product of voltage and current, is not zero during these times.

Switching power loss is only important when we are switching the transistor on and off at a fast rate.

EXAMPLE

Find the average power dissipation in a transistor if the power loss during a 1μs switching time is 1W, at a switching rate of 100Hz and also 100KHz.

SOLUTION

A switching rate of 100Hz gives a time between switching of 1/100Hz or 10ms. Average power P_a at 100Hz switching is

$$P_a = 1\text{W} \times 1\mu\text{s}/10\text{ms}$$

$$P_a = 0.1\text{mW}$$

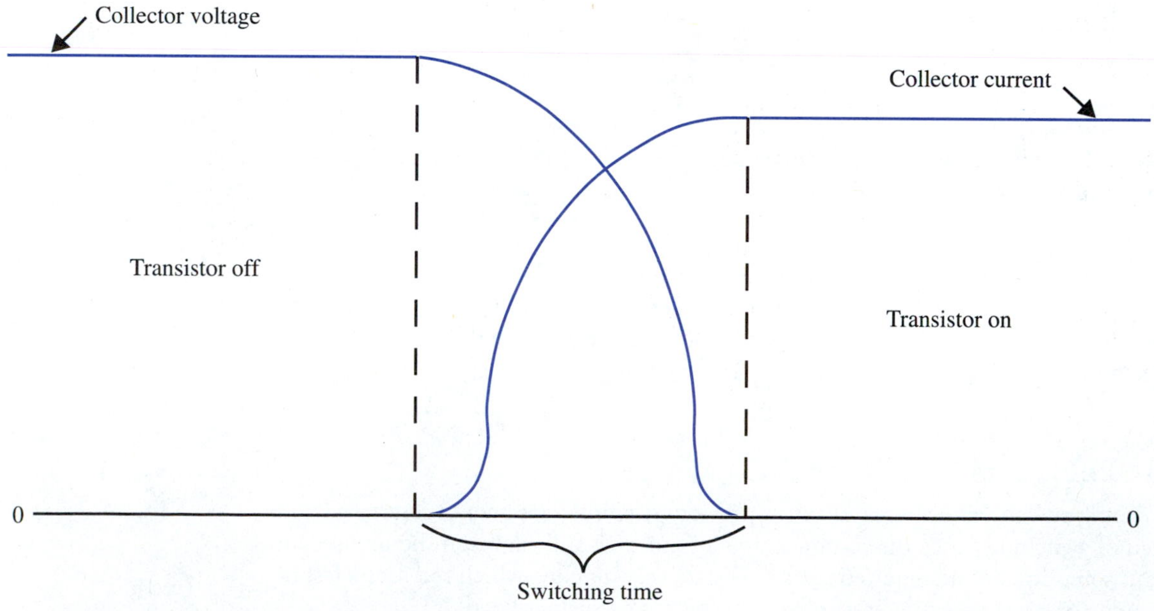

FIGURE 5–10 Transistor switching waveforms

With a switching rate of 100KHz, the time between switching is 1/100KHz or 10μs. Average power at 100KHz switching is

$$P_a = 1\text{W} \times 1\mu\text{s}/10\mu\text{s}$$

$$P_a = 0.1\text{W}$$

We can see from these results that the higher the switching rate (the more often we activate the switch), the higher the power dissipation in the transistor.

TRANSISTOR SWITCHING TIME

As was explained earlier, how fast a voltage can be transferred to a load is related to circuit capacitance, but with transistor switches, the response of the actual transistor must also be considered. In order to turn on a transistor, we must charge the internal capacitance from collector to base (C_{cb}) and capacitance from base to emitter (C_{be}), as shown in Figure 5–11a.

The delay to turn on (t_d in Figure 5–11b) is defined from the point when the switch is closed to when the collector voltage drops by 10%. This delay is due mainly to the time required to charge C_{be}.

Once the transistor starts to turn on, the collector-base capacitance C_{cb} must be allowed to discharge in order for the collector voltage to drop. This slows down the rate of fall of

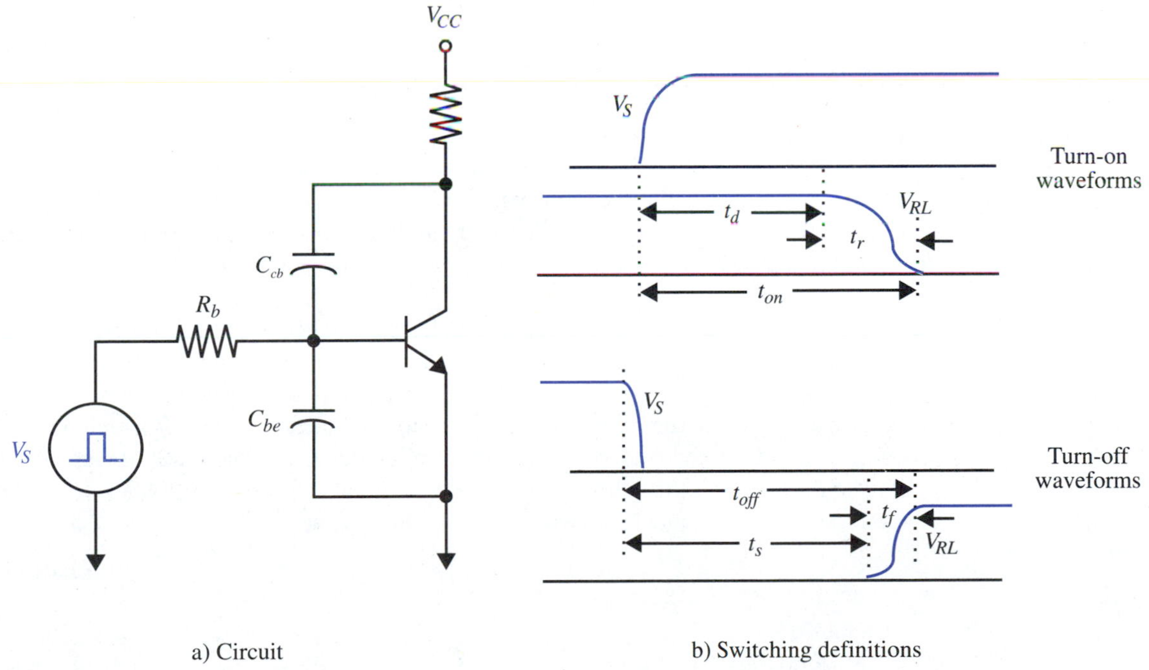

FIGURE 5–11 Transistor turn-on and turn-off waveforms

the collector voltage (takes time t_r). The total time, from the base switch closure to when the collector voltage has dropped by 90%, is called t_{on}. From Figure 5–7b

$$t_{on} = t_d + t_r \qquad \textbf{(equation 5–10)}$$

EXAMPLE

Determine the maximum turn-on time (t_{on}) for a 2N2222 transistor if the specification sheets list maximum values of t_d equaling 10ns and t_r equaling 25ns.

SOLUTION

From equation 5–10,

$$t_{on} = t_d + t_r$$

$$t_{on} = 10\text{ns} + 25\text{ns}$$

$$t_{on} = 35\text{ns}$$

t_r is basically the time the transistor is dissipating power during turn on.

What determines the values of t_d and t_r? The value of t_d is dependent on the time to charge the base-to-emitter capacitance, which in turn is a function of the base circuit *RC* time constant, i.e., the effective driving resistance (R_b) in the base circuit times the value of C_{be}. Since C_{be} is fixed by the transistor construction, we only have control over R_b. The smaller we can make R_b, the shorter the time t_d.

For t_r, we are concerned with the capacitance from collector to the base and its discharge path through the turning-on transistor. The harder we turn on the transistor (higher base current), the lower the effective transistor resistance—which means shorter t_r. Specification sheets always list values of t_d and t_r, which are optimized by using a test circuit that makes these times as short as possible.

If we reverse bias the base-emitter junction, the transistor switch will turn off. However, electron flow across the junctions will flow until the base region is clear of electrons. This creates a delay in turn off (t_s) called **storage delay** (electrons are trapped or stored in the base region), and it is measured from the time the base switch opens to when the collector rises by 10%.

At the completion of the storage delay, the capacitance of the collector-to-base junction now charges to the rising collector voltage. This charge time, from the 10% to 90% points, is called t_f (see Figure 5–7). But t_f is not only related to the collector-to-base capacitance, it is also a function of the value of the load resistor R_L. The reason for this is that with the transistor turned off, the only path for current to charge C_{cb} is via R_L. So the smaller we can make R_L, the faster the collector voltage will rise—which creates a smaller t_f. The sum of t_s and t_f is called the transistor switch turn-off time t_{off}, or

$$t_{off} = t_s + t_f \qquad \textbf{(equation 5–11)}$$

EXAMPLE

Find the turn-off time for a 2N2222 transistor if the specification sheets list maximum values of t_s equaling 225ns and t_f equaling 60ns (with R_L equaling 200Ω).

SOLUTION

Using equation 5–11,

$$t_{off} = t_s + t_f$$

$$t_{off} = 225\text{ns} + 60\text{ns}$$

$$t_{off} = 285\text{ns}$$

Comparing this with our previous example, we see that the turn-off time for the 2N2222 is about eight times longer than the turn-on time. Again, power dissipation occurs during the time t_f.

REDUCING SWITCHING TIME

How can we reduce the transistor switching times? Let us first look at a technique to reduce t_{on}. To reduce the delay-time (t_d) portion of t_{on}, we need to charge C_{be} quickly. As stated before, this requires a small value for R_b; but if we make R_b small, we have a large base current, which drives the transistor into heavy saturation. This increases the storage time when we try to turn the transistor off. What we can do is to select R_b such that we just saturate the transistor at the minimum β value, but also put a "speedup" capacitor across R_b, as shown in Figure 5–12. The charging current for this capacitor will flow into C_{be} and quickly charge this capacitor, which performs the desired function of reducing t_d.

To determine the required value for R_b, we can use the equation

$$R_b = (V_s - V_{be})\left(\frac{\beta}{I_c}\right) \qquad \textbf{(equation 5–12)}$$

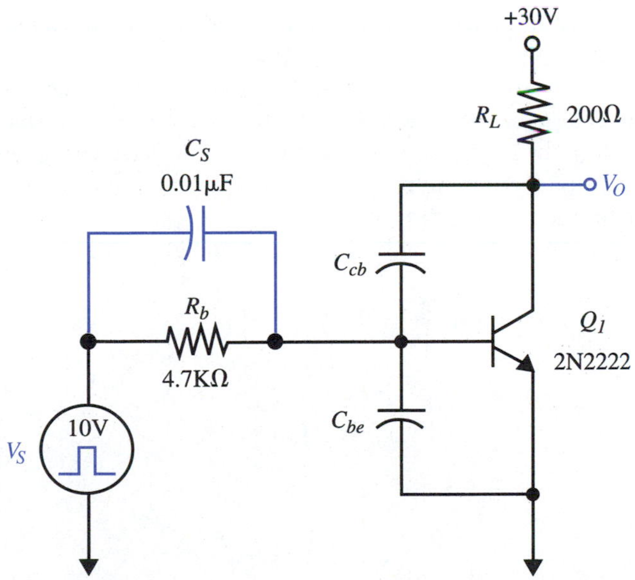

FIGURE 5–12 Circuit for faster switching

where V_s is the input voltage, V_{be} is 0.6V, β is the minimum beta of the transistor, and I_c is the current that just causes the output voltage to be zero ($I_c = V_{cc}/R_c$).

EXAMPLE

Find the required value of R_b for the circuit shown in Figure 5–12 if the minimum β for the transistor is 75.

SOLUTION

Since the supply is 30V, and R_c is 200Ω, I_c is 30V/200 or 150mA.
Using equation 5–12,

$$R_c = (10\text{V} - 0.6\text{V})\left(\frac{75}{150\text{mA}}\right)$$

$$R_c = 4.7\text{K}\Omega$$

This value will guarantee that the transistor saturates.

IMPROVING TRANSISTOR TURN-OFF TIME

When the input pulse V_s drops to zero, the base side of C_s will go negative by the amount of voltage developed across the capacitor during charging. This causes a reverse base current that helps remove the excess electrons from the base region. So the speedup capacitor not only speeds up the charging of C_{be} during turn-on but also helps to reduce the storage time (t_s).

Now, how do we determine the required value of the speedup capacitor C_s? A larger value reduces t_d and t_s by a greater amount, but if we are pulsing the transistor on at regular time intervals, C_s must be pretty well discharged before we pulse again, otherwise the voltage on the capacitor will buck (oppose) the turn-on pulse.

If we require that the capacitor be at least 90% discharged before the next turn-on pulse is applied, we can determine the value of C_s from the discharge time constant R_bC_s. It takes a time $2.3R_bC_s$ to drop the voltage across the capacitor by 90% (since $0.1 = \epsilon^{-t/rc}$ and solving for t). This must be the minimum time (t_m) between the trailing edge of one turn-on pulse and the leading edge of the next pulse.

If t_p is the pulse width and f_p the pulse repetition frequency, then t_m is

$$t_m = \left(\frac{1}{f_p} - t_p\right)$$

But

$$t_m = 2.3R_bC_s$$

Solving for C_s

$$C_s = \frac{\left(\frac{1}{f_p} - t_p\right)}{2.3R_b} \qquad \textbf{(equation 5–13)}$$

EXAMPLE

Find the value of C_s for a switching transistor circuit if R_b is 4.7KΩ, the pulse width is 1μs, and the pulse repetition frequency is 10KHz.

SOLUTION

Using equation 5–13,

$$C_s = \frac{\left(\frac{1}{10\text{KHz}} - 1\mu\text{s}\right)}{(2.3 \times 4.7\text{K}\Omega)}$$

$$C_s = 9.16\text{nF (use 10nF)}$$

TOTEM POLE CIRCUIT

To keep the value of the t_f portion of the turn-off time small, a low collector time constant is required. Since the collector capacitance is fixed, we need to make R_L as small as possible (200 Ω in our examples). But we are limited in how low we can make R_L, because a lower value means higher transistor currents. A way of effectively reducing R_L is to replace it with a transistor, as shown in the "totem pole" (one transistor is above the other) configuration of Figure 5–13.

The advantage of the totem pole configuration is that the collector capacitance is discharged via the low "on resistance" of a transistor—rather than through R_L.

With reference to Figure 5–13a, turning on Q_2 causes the collector-base capacitance C_{cb2} to discharge quickly through the small "on resistance" of this transistor. If we now turn off Q_2, its collector (the output voltage) will rise; and if at this time we turn on Q_1, C_{ce2} will quickly charge through the low "on resistance" of Q_1. We don't have to worry about the capacitance from collector to base of Q_1 in the output circuit because its collector voltage doesn't change; however, this capacitance is in parallel with the base capacitance C_{be}.

So, Q_1 has taken the place of R_L for the charging of C_{cb2} and, having a much lower "on resistance," gives us a much faster turn-off time.

Note that with the totem pole circuit, only one transistor can be turned on at a time, otherwise we will create a short across the power supply.

A simplified equivalent circuit for the charging and discharging of C_{cb2} is shown in Figure 5–13b. Because of its fast response time, the totem pole circuit is used as the output stage of digital logic circuits.

FET SWITCHING CIRCUITS

Using an FET as a switching element has some advantages over the bipolar transistor. The current and therefore the power required to "close" an FET switch is negligible, and for small currents, the "on resistance" of the FET is much lower. Also the "switch on" and "switch off" times are typically shorter than those of the bipolar transistor.

EXAMPLE

Compare the switching characteristics of a bipolar transistor and FET.

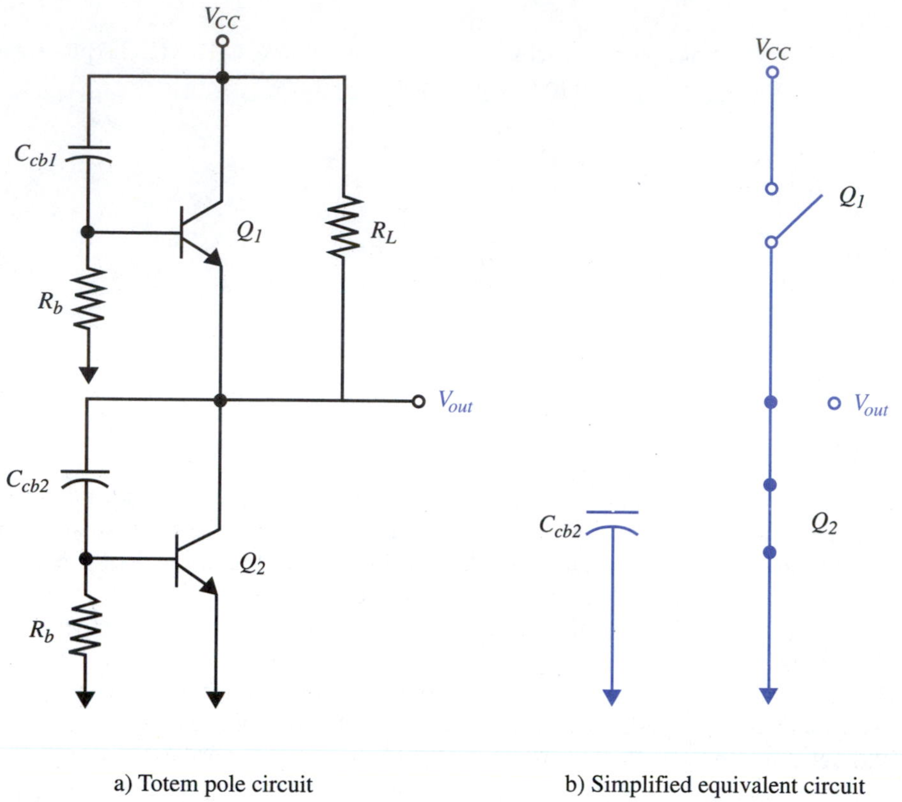

FIGURE 5–13 Totem pole switching configuration

SOLUTION

We shall compare a 2N2369 bipolar and a 2N4856 FET (both switching transistors) using the following data interpolated from manufacturers' data sheets.

Transistor	*Input R*	*Saturation Voltage*	t_{on}	t_{off}
2N2369 Bipolar	< 1KΩ	0.27V @ 1mA, 0.25V @ 10mA	35ns	30ns
2N4856 FET	> 1MΩ	0.04V @ 1mA, 0.4V @ 10mA	9ns	25ns

- As expected, the bipolar has a much lower input resistance.
- Saturation voltages across the transistors are lower for the FET at 1mA but higher at 10mA. This is because the FET responds more like a pure resistor—whereas the bipolar is very nonlinear and changes much less with current.
- Turn-on time for the FET is much faster than for the bipolar, and this is probably because all the input current to the FET is used to quickly charge the input capacitance. The input drive circuit to the bipolar must supply base current as well as capacitor charging current.

- We can expect the turn-off time for the FET to be faster since there is no minority current flow storage time as with the bipolar, and this is confirmed with the data.
- Remember, the faster the switching times, the lower the power dissipation in the switching device

---

Figure 5–14 shows a switching circuit, using an N-channel JFET. Reviewing the operation of the N-channel JFET, with zero bias (gate to source) the FET is fully conducting, and for the 2N4856 JFET, a bias voltage of −10V on the gate will ensure FET turn off (I_d is 0.25nA).

For the circuit of Figure 5–14, a positive input pulse of 10V, with fast rise and fall times, is used to switch on the FET by overcoming the −10V gate bias voltage. The current (I_o), when the FET is switched on, is determined from the power supply voltage (V_{dd}), the drain resistor (R_d) value, and, to a small degree, by the "on resistance" (r_{ds}) of the FET, or

$$I_o = \frac{V_{dd}}{(R_d + r_{ds})} \qquad \textbf{(equation 5–14)}$$

EXAMPLE

With reference to Figure 5–14, with the +10V supply voltage, R_d of 1KΩ, and a specified r_{ds} of 40 Ω, find the FET "on" current I_o.

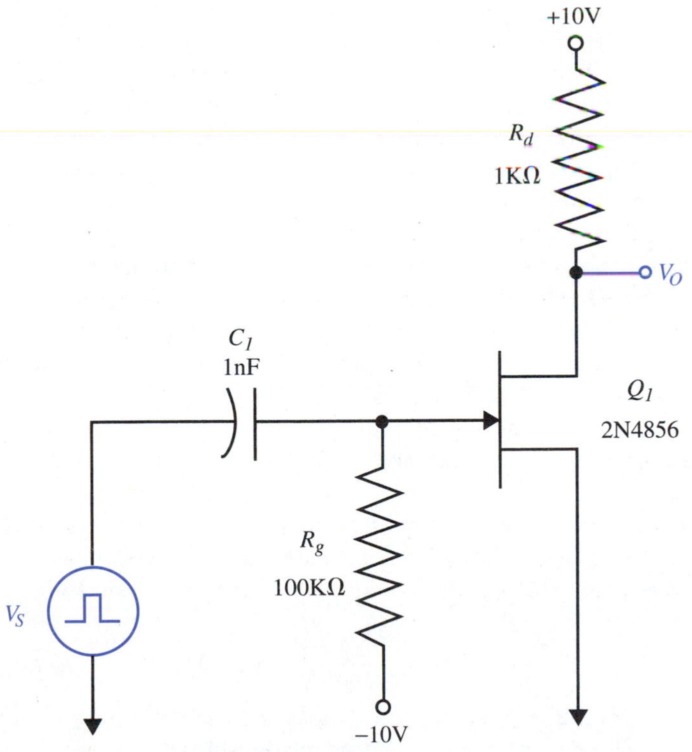

FIGURE 5–14 JFET switching circuit

SOLUTION

Using equation 5–14,

$$I_o = \frac{10V}{(1K\Omega + 40\Omega)}$$

$$I_o = 9.62mA$$

Without considering r_{ds}, we would have computed 10mA.

The FET switching times for the circuit of Figure 5–14 will not be as fast as the specification values of 9ns for turn on and 25ns for turn off, since the manufacturers' measurements were made with a driving source resistance of 50Ω and an R_d resistance of 460Ω. If our circuit is driven from 50Ω, and we take care not to add any stray capacitance, we probably can come close to the turn-on value of 9ns. However, our use of 1KΩ for R_d, rather than 460Ω, will more than double our expected turn-off time from 25ns to greater than 50ns. This is because we have increased the resistance that charges the drain capacitance.

Like bipolar transistors, FETs can be used in a totem pole configuration to reduce turn-off time.

SWITCHING CIRCUIT MEASUREMENT TECHNIQUES

The caution about keeping stray capacitance to a minimum in measuring FET turn-on time serves to introduce the topic of switching circuit measurements. We need to be aware of the special techniques required when measuring fast switching signals. Figure 5–15 shows how *not* to make high-speed switching circuit measurements.

Starting with the input circuit, the use of the low output resistance pulse generator is correct—but the use of a 92Ω coax rather than 50Ω coax to match the pulse generator output resistance of 50Ω results in pulse distortion. Also, the use of long leads at the end of the coax and no coax terminating resistor (50Ω for 50Ω coax) introduce inductance and changes the effective source resistance.

Moving to the output circuit, long power supply leads without a bypass capacitor add inductance in series with R_d, which will cause ringing on the output waveform (inductance resonates with the stray capacitance). Long component lead length also adds inductance. The use of a coax cable, without a probe, to measure the output pulse at the drain means we are adding about 30pF from the drain to ground for every foot of cable we use. Since the FET drain capacitance is 8pF, this will really increase the drain time constant. The FET manufacturer recommends using an oscilloscope probe with capacitance less than 2.5pF.

Finally, the rise time of the oscilloscope used to measure the switching waveforms should be at least three times faster than the rise time of the fastest waveform we wish to measure. This will provide a scope reading that is within 5% of the actual value. We determined this from the equation

$$t_{(measured)} = \sqrt{(t_p)^2 + (t_s)^2} \qquad \textbf{(equation 5–15)}$$

where t_p is the actual pulse rise time, and t_s the scope rise time. If we need to specify the bandwidth (*BW*) of the scope, then we use the relationship

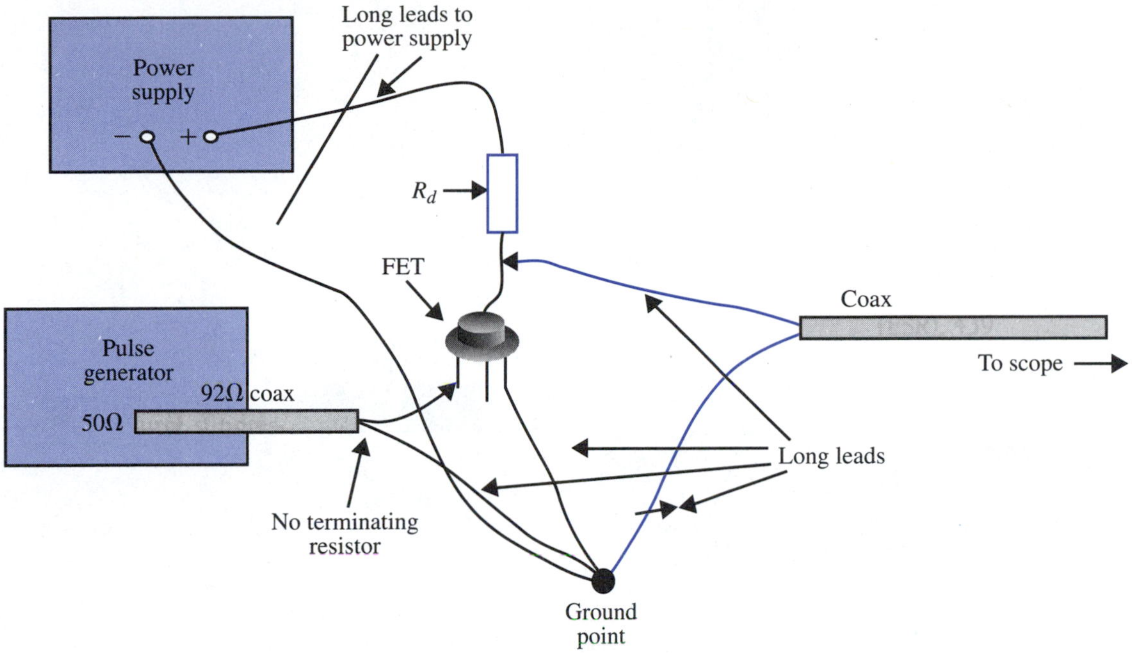

FIGURE 5–15 Incorrect switching measurement techniques

$$BW = \frac{0.35}{t_s} \qquad \textbf{(equation 5–16)}$$

(This equation is derived from the relationships t_s equals 2.2*RC* and *BW* equals $\frac{1}{2}\pi RC$.)

EXAMPLE

Determine the required scope rise time and bandwidth in order to measure a specified FET turn-on time of 9ns to within 5% of the actual value.

SOLUTION

The required scope rise time t_s is 9ns/3 or 3ns.
Required bandwidth is, from equation 5–16,

$$BW = \frac{0.35}{3\text{ns}}$$

$$BW = 117\text{MHz}$$

From the bandwidth result, we start to get a feel for the high frequencies associated with rise times in the nanosecond range, and why good pulse measurements are difficult to make.

Now, let us look at the correct measurement techniques, shown in Figure 5–16. All of the points covered heretofore are included in this figure, starting with the correct coax for the

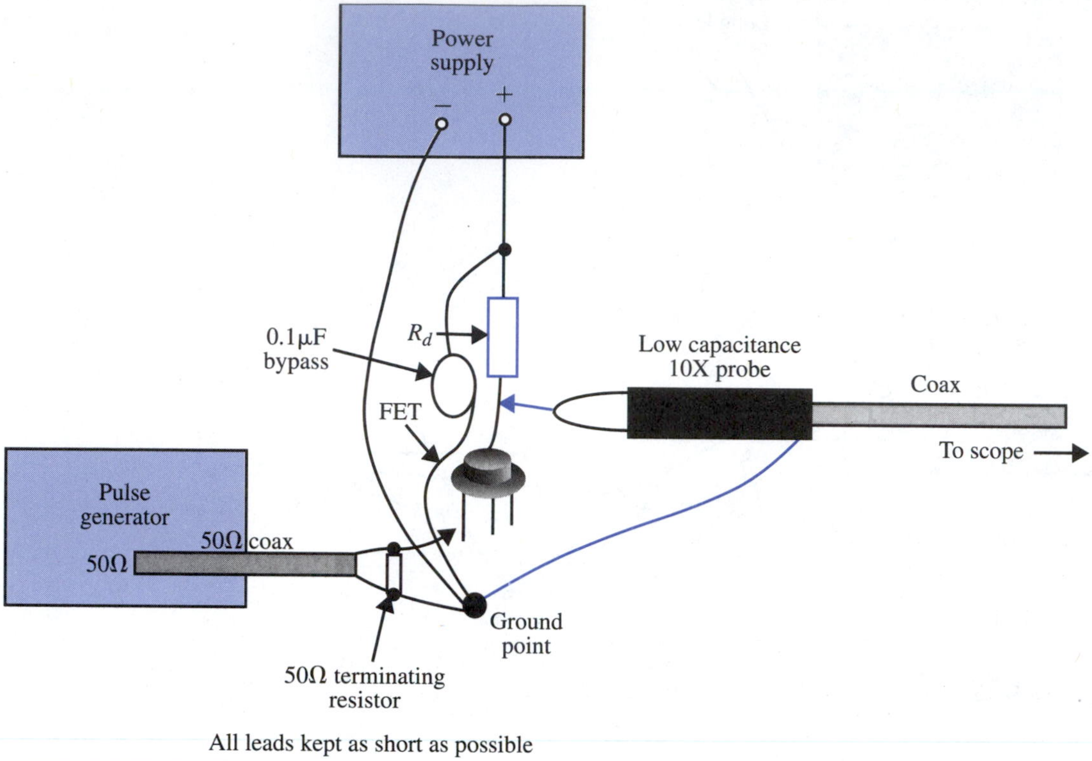

FIGURE 5–16 Correct switching measurement techniques

pulse generator along with the right termination for the coax. Component and test leads are kept short, and bypass capacitors are used. A low-capacitance scope probe is used, and the scope bandwidth is consistent with the measured rise time. If a low-capacitance probe is not available, then a voltage divider should be used to isolate the scope capacitance from the circuit.

We will expand on this voltage divider technique to reduce capacitive loading. The concept here is to isolate the loading capacitor, from the circuit being measured, with a resistor, but without the voltage divider affecting the rise time measurement. Figure 5–17 shows the divider connected to the FET drain.

For the probe capacitance not to affect the rise time seen at the probe tip, the time constant of $(R_1 \| R_2)C_p$ must be much shorter (one-tenth) than the drain time constant R_dC_d. Since R_1 is typically much larger than R_2 (also R_d), then the time constant at the probe tip is R_2C_p.

EXAMPLE

Find the value of divider resistors R_1 and R_2 to prevent capacitive loading by the probe if the probe's capacitance is 5pF, R_d is 470Ω, and C_d is 8pF.

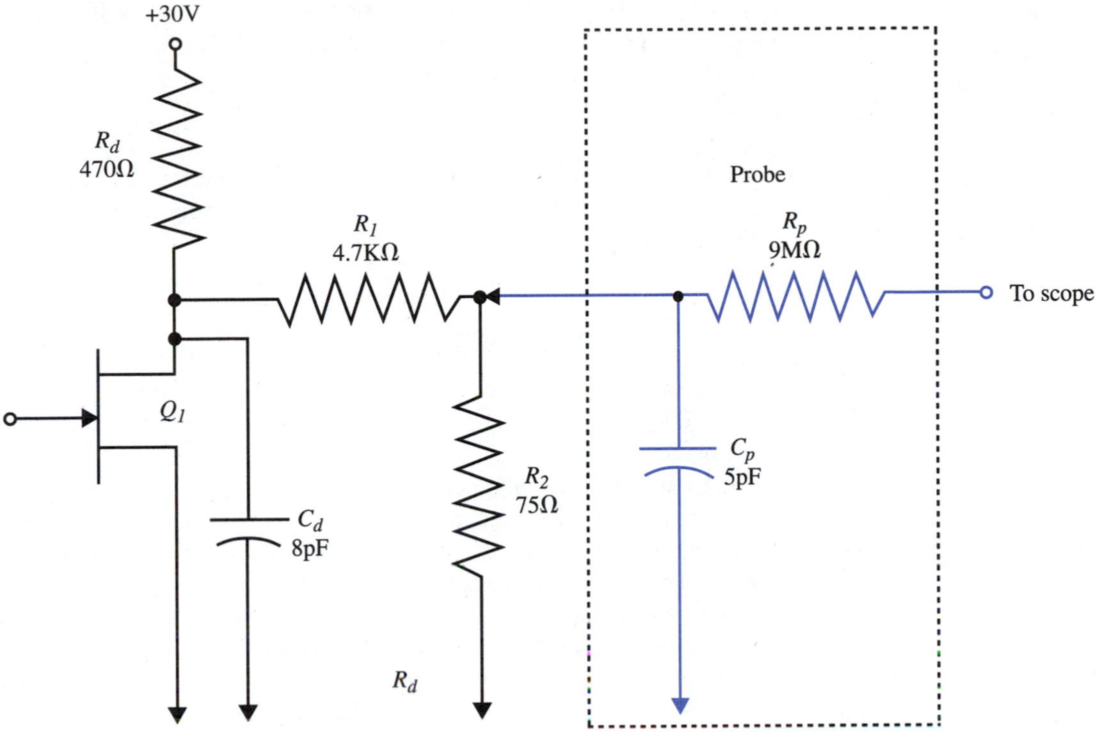

FIGURE 5–17 Capacitive loading reduction

SOLUTION

The drain time constant τ_d is

$$\tau_d = R_d \times C_d$$

$$\tau_d = 470\Omega \times 8\text{pF}$$

$$\tau_d = 3.76\text{ns}$$

The time constant at the probe tip (τ_p) should then be 3.76ns/10 or 0.376ns. Solving for R_2,

$$R_2 = \frac{\tau_p}{C_p}$$

$$R_2 = \frac{0.376\text{ns}}{5\text{pF}}$$

$$R_2 = 75\Omega$$

We can pick R_1 to be ten times larger than R_d to avoid resistive loading; then R_1 is 4.7KΩ. Notice that the signal we are measuring is now attenuated by the ratio of 4.7KΩ to 75Ω or 62.6 to 1. This will require a more sensitive scope setting. We must also verify that the scope bandwidth is sufficient for the measured pulse rise time.

The voltage divider technique, used to prevent adding capacitance to the circuit, should also be used when measuring across the frequency-determining capacitor of an oscillator circuit—but again keep the lead lengths as short as possible.

TROUBLESHOOTING SWITCHING CIRCUITS

Aside from the care in making switching measurements stressed above, troubleshooting both bipolar and FET switching circuits uses basically the same techniques as with amplifier versions of these devices.

We can check to determine whether the bipolar transistor in a switching circuit is good by applying a base current via a resistor connected to the power supply, to provide about one-tenth of the predicted collector current. This should cause the collector voltage to change state and drop to the saturation voltage (0.2V).

With the FET switching circuit, we again need to know whether we are dealing with a JFET- or MOSFET-type device. The schematic symbol should indicate the type of FET; however, the JFET will be depletion mode and the MOSFET enhancement mode; so if the circuit has a bias voltage present, we know we have a JFET circuit.

If we are checking a JFET circuit, then the switching circuit is "on" when the gate-to-source bias is zero. We can, therefore, check the FET by seeing whether we can turn it on by placing a short between gate and source. The MOSFET, being enhancement mode, is "off" with zero volts from gate to source.

SUMMARY

- An ideal switch has no power loss when it is open or closed.
- A capacitor or an inductor can slow up switching.
- A means must be provided to dissipate the energy stored in an inductor.
- An electronic switch gives both the capability of remote operation and high switch repetitions.
- Since it has no power loss when off and little power loss when on, a bipolar transistor can be used as an electronic switch.
- When used as a switch, a bipolar transistor is either off or in saturation.
- The voltage drop across a saturated bipolar transistor is about 0.2V.
- To ensure a bipolar transistor is in saturation, we provide a base current equal to one-tenth of the predicted collector current.
- Circuit capacitance limits high-speed switching time.
- A speedup capacitor allows us to quickly charge circuit capacitance.

- Most power loss occurs during transistor switching time, since neither voltage nor current is zero.
- The more often a transistor circuit is switched, the higher the transistor power dissipation.
- Turn-on time for a transistor consists of two parts: delay time t_d and rise time t_r.
- Delay time is the time it takes to charge the input capacitance to the turn-on threshold voltage.
- Rise time is the time it takes for the output capacitance to discharge by 90%.
- Turn-off time for a bipolar transistor consists of two parts: storage time t_s and fall time t_f.
- Storage time is the time it takes for electrons to clear the base region and is greater with heavier conduction.
- Fall time is the time it takes to charge the output capacitance to 90% of its final voltage.
- FETs are faster switching devices because they don't have storage delay.
- We can speed up transistor turn-on time by connecting a speedup capacitor across the source resistance.
- A totem pole circuit enables discharging the output capacitance through the low **on** resistance of a saturated transistor, resulting in a smaller t_f.
- An FET switch has the advantage of requiring less power to turn on and also has a lower **on** voltage drop for small currents.
- Since we are dealing with very high frequencies when we have fast switching speed, we must use special measurement techniques.

EXERCISE PROBLEMS

1. State the purpose of an electronic switch.
2. A series RC circuit consists of a 0.01μF capacitor and a 4.7KΩ resistor. Find the voltage across the capacitor 50μs after 10V is applied to the circuit.
3. A 1μF capacitor charged to 15V is connected across a 10KΩ resistor. What is the voltage across the resistor 5ms after the connection is made?
4. How long does it take for the voltage across a 22nF capacitor that is in series with a 100KΩ resistor to be within 1% of the applied voltage?
5. A series RL circuit consists of a 100mH inductor and a 3KΩ resistor. Find the voltage across the resistor 20μs after 20V is applied to the circuit.
6. A 330μH inductor conducting 1 amp is connected across a 1KΩ resistor. What is the voltage across the resistor 50ns after the connection is made?
7. What is the typical "on" voltage across a bipolar transistor switch?
8. If the "on" collector current for a transistor switch is 25mA, what is the required base current?

9. If the base current for a transistor switch is 0.5mA, what is the probable collector current?
10. How does capacitance affect switching time?

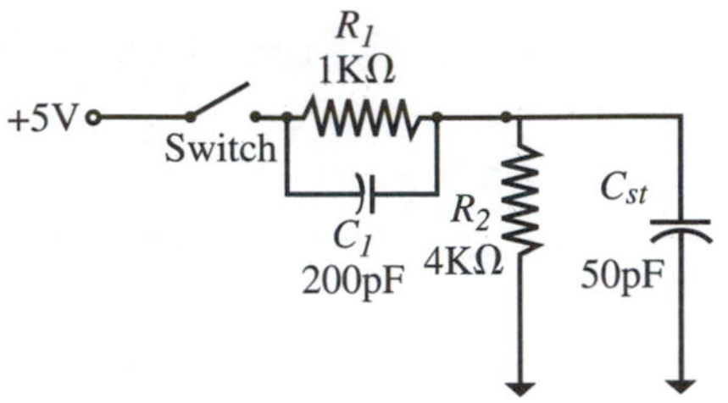

FIGURE 5–A (text Figure 5–4)

11. With reference to Figure 5–A, compute the required value of the speedup capacitor C_1 if R_1 is 3KΩ, R_2 is 10KΩ, and C_{st} is 30pF.
12. What is the significance of the transistor "switching time"?
13. Find the average power dissipation in a transistor if the power loss during a 200ns switching time is 200mW and the switching rate is 20KHz.
14. If the average power dissipation in a switching transistor is 10mW with a switching time of 100ns and a switching rate of 500KHz, find the power loss (P_s) during the switching time.
15. Define the following transistor switching terms: t_d, t_r, t_s, and t_f.
16. If specification sheets list a t_d of 15ns, t_r of 30ns, t_s of 100ns, and t_f of 50ns, what is the value of t_{on} and t_{off}?
17. How can we reduce the switching time for a transistor?
18. Find the value of R_b to maximize switching time if V_s is 5V, β_{min} is 50, and I_c is 30mA.
19. Find the value of the speedup capacitor C_s for a transistor switching circuit if R_b is 6.8KΩ, the input pulse width is 0.6μs, and the pulse repetition frequency is 50KHz.
20. Find the value of the speedup capacitor C_s for a transistor switching circuit if R_b is 3.3KΩ, the input pulse width is 0.2μs, and the pulse repetition frequency is 100KHz.
21. State why a totem pole circuit is used.
22. Give the advantages of using an FET in a switching circuit.
23. If the "on" resistance of an FET is 25Ω, R_d is 680Ω, and the supply is 25V, find the FET "on" current I_o.
24. The "on" current for an FET is 20mA, r_{ds} is 35Ω, and the supply is 10V. Find the value of R_d.

25. If the rise time of the switching signal we are measuring is 5ns, what is the bandwidth of frequencies associated with this rise time?
26. What is the problem with using long leads in a high-speed switching circuit?
27. Why shouldn't you use a coax cable test lead to view high-speed waveforms on the scope?
28. What rise time will you actually read on a 100MHz scope when measuring a waveform with a rise time of 6ns?
29. Determine the scope bandwidth required to measure a switching waveform with a rise time 5ns to within 5% of the actual value.
30. Find the value of divider resistors R_1 and R_2 to prevent capacitive loading by a scope probe when measuring the signal at an FET drain if the probe capacitance C_p is 3pF, the drain capacitance C_d is 10pF, and the drain resistance R_d is 680Ω.
31. List the methods for checking bipolar transistors, JFETs, and MOSFETs in switching circuits.

ABBREVIATIONS

BW = Bandwidth

C_{be} = Base-emitter capacitance

C_{cb} = Collector-base capacitance

C_{st} = Stray capacitance

f_p = Pulse repetition frequency

I_o = FET "on" current

P_a = Average power

P_d = Power dissipation

P_t = Total power

r_{ds} = FET "on" resistance

t_c = Time for $5RC$

t_D = Drain time constant

t_d = Transistor turn-on delay

t_f = Collector current fall time (collector voltage rise time)

t_m = Minimum time between pulses

t_o = Oscilloscope rise time

$t_{off} = t_s + t_f$

$t_{on} = t_r + t_d$

t_p = Pulse rise time

t_r = Rise time of collector current (fall time of collector voltage)

t_s = Base region storage delay

ANSWERS TO PROBLEMS

1. Allows faster switching and also remote switching
2. $v = 6.55V$
3. $v = 5.9V$
4. $t = 11ms$
5. $v = 9v$
6. $v = 859V$
7. $V_{sat} = 0.2V$
8. $I_b = 2.5mA$
9. $I_c = 5mA$
10. Slows it down
11. $C_1 = 100pF$
12. Faster switching time means less transistor power dissipation.
13. $P_a = 0.8mW$
14. $P_s = 200mW$
15. See text.
16. $t_{on} = 45ns$; $t_{off} = 150ns$
17. See text.
18. $R_b = 7.33K\Omega$
19. $C_s = 1.24nF$
20. $C_s = 1.29nF$
21. Faster rise time
22. Lower input power and smaller "on" resistance at low currents
23. $I_o = 35.5mA$
24. $R_d = 465\Omega$
25. $BW = 70MHz$
26. Added inductance
27. Cable capacitance
28. $t_r = 6.95ns$
29. $BW = 210MHz$
30. $R_2 = 227\Omega$; $R_1 = 6.8K\Omega$
31. See text.

CHAPTER 6

Introduction to Operational Amplifiers

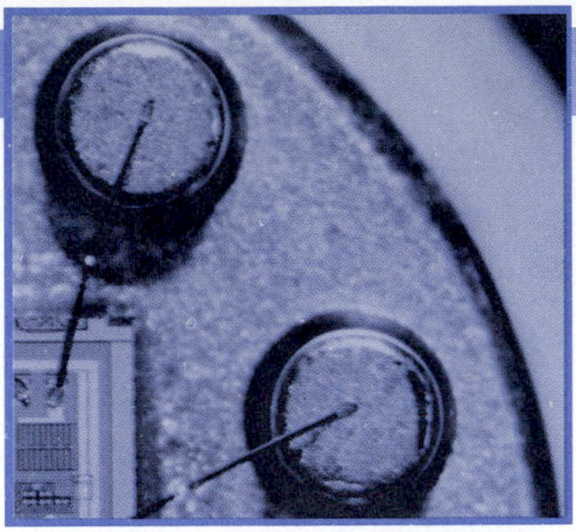

CHAPTER OBJECTIVES

After completing this chapter, you should understand the following operational amplifier concepts:

- How a difference amplifier amplifies the difference between two input signals
- The reason for operational amplifiers
- How an ideal operational amplifier operates
- Open loop operation
- The noninverting amplifier
- The inverting amplifier
- The op-amp follower
- Distortion reduction in a closed loop op-amp.
- Powering an op-amp
- Input and output resistance of an op-amp
- Operational amplifier follower
- Current boosting
- Operation with a single power supply
- How to troubleshoot the basic op-amp circuit

INTRODUCTION

In an earlier chapter, we studied transistor amplifier circuits that enable us to amplify (increase) weak voltages and currents up to usable levels.

The problem with transistor amplifiers is that biasing for the correct operating point is quite cumbersome. If we want stable amplifier gains, we have to restrict the gain of the transistor stage to about 10. Also, loading is a problem when connecting transistor stages together.

The original computers were analog rather than digital, and the math operations performed by the computers—such as multiplication, addition, subtraction, etc.—required amplifiers with very predictable and precise gains. To satisfy this requirement, an amplifier was developed consisting of several transistors connected in a unique configuration. Since the intent was to use it for math operations, it was called an **operational amplifier.** We shall see in this chapter how the operational amplifier is able to provide precise and stable gains. We shall also see how the actual gain value can be set and how the op-amp is powered.

DIFFERENTIAL AMPLIFIER

Before we look at the total operational amplifier, it is helpful to look at its input circuitry, which consists of two transistors connected in what is called a **differential amplifier configuration.** (See Figure 6–1.)

The term *differential amplifier* means that the difference between two input signals is amplified. In Figure 6–1, the difference between signals V_a and V_b is amplified and appears between the collectors of Q_1 and Q_2.

Let us assume that both bases are at ground potential (V_a and V_b are initially 0V) and also assume that β of Q_1 and Q_2 are equal—which is a basic requirement for a differential amplifier.

Initially with 0V on both bases, the common connection of the emitters will be 0.6V less positive than the base, or −0.6V (the negative voltage at the bottom end of R_e forward biases the base-emitter junctions); and the sum of the two emitter currents (I_e) through the shared emitter resistor R_e will be

$$I_e = \frac{-0.6\text{V} - (-12\text{V})}{1\text{K}\Omega}$$

$$I_e = \frac{11.4\text{ V}}{1\text{K}\Omega}$$

$$I_e = 11.4\text{mA}$$

The emitter conventional current flow will be towards the negative supply. This current is provided by Q_1 and Q_2.

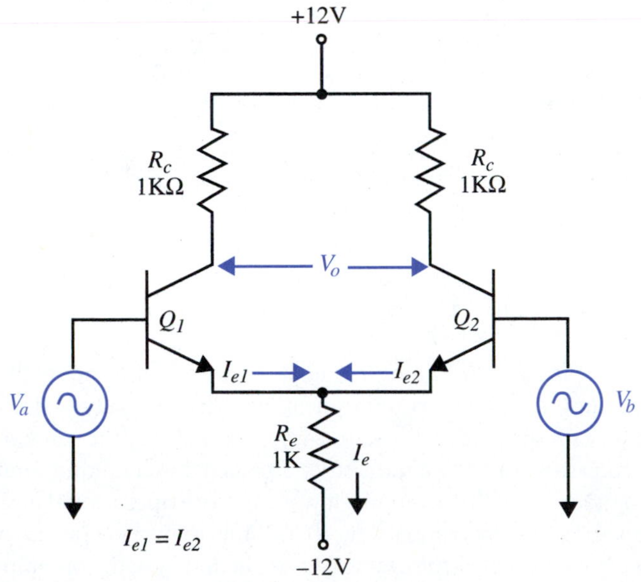

FIGURE 6–1 Differential amplifier

Since we have equal β for Q_1 and Q_2, the emitter currents will be equal, and therefore the collector current ($I_e \approx I_c$) for each transistor will be half this value or

$$I_{c1} = I_{c2} = 5.7\text{mA}$$

The voltage on each collector will be

$$V_c = +12\text{V} - (5.7\text{mA} \times 1\text{K}\Omega)$$

$$V_c = +6.3\text{V}$$

This means we shall see no voltage difference with a meter connected between the two collectors. Since the output signal from a differential amplifier is taken from the two collectors, the output signal V_o is zero volts.

If we now raise each base by 0.1V, the emitters will go to −0.5V, and the combined emitter current is increased to 11.5mA, giving individual collector currents of 5.75mA. The new collector voltages are 6.25V, and again we measure no voltage difference between them.

This demonstrates that with the same input voltages to a differential amplifier (no difference voltage), the output voltage between collectors is zero.

Now if we leave V_b at 0V and raise V_a to a positive voltage, the voltage at the emitters will rise up with V_a. This will reduce the Q_2 emitter-base bias voltage since the emitter rises and the Q_2 base is held at 0V. Reducing its base-emitter bias causes the Q_2 collector current to reduce, and its collector voltage to rise. At the same time, since its base voltage is raised, Q_1 collector current increases, which causes its collector voltage to drop. So, since it is drawing more current, the collector of Q_1 is at a lower voltage than that of Q_2. With different collector voltages on the two transistors, we now will see an output voltage between them which is not zero.

DIFFERENTIAL GAIN

Let us now calculate the actual voltage difference we see between the two collectors for a given difference in voltage between the two inputs. From this we will be able to compute the differential amplifier gain.

The circuits of Figure 6–2 will aid our derivation of the gain equation. Figure 6–2a differs from Figure 6–1 by the inclusion of the base-emitter resistances r_e.

Recall that the voltage gain of a common emitter stage is equal to the collector load resistor (R_c) divided by the total emitter resistance to ground ($r_e + R_e$). If R_e is bypassed with a capacitor, then the gain is simply R_c/r_e.

Considering the gain of Q_1 in Figure 6–2a, we see that in the emitter circuit, we have r_e in series with the parallel combination of R_e and the transferred common base circuit input resistance of Q_2. Since there is no biasing resistor in the base circuit of Q_2, the common base input resistance of Q_2 is simply r_e.

The equivalent circuit for determining the gain of Q_1 is shown in Figure 6–2b. Typically, the input resistance of Q_2 is much smaller than the resistance of R_e (which is in parallel), so we can ignore R_e, and the total Q_1 emitter resistance can be assumed to be the two r_e in series or $2r_e$. The Q_1 gain then is

$$A_{v1} = -\frac{R_c}{2r_e} \qquad \textbf{(equation 6–1a)}$$

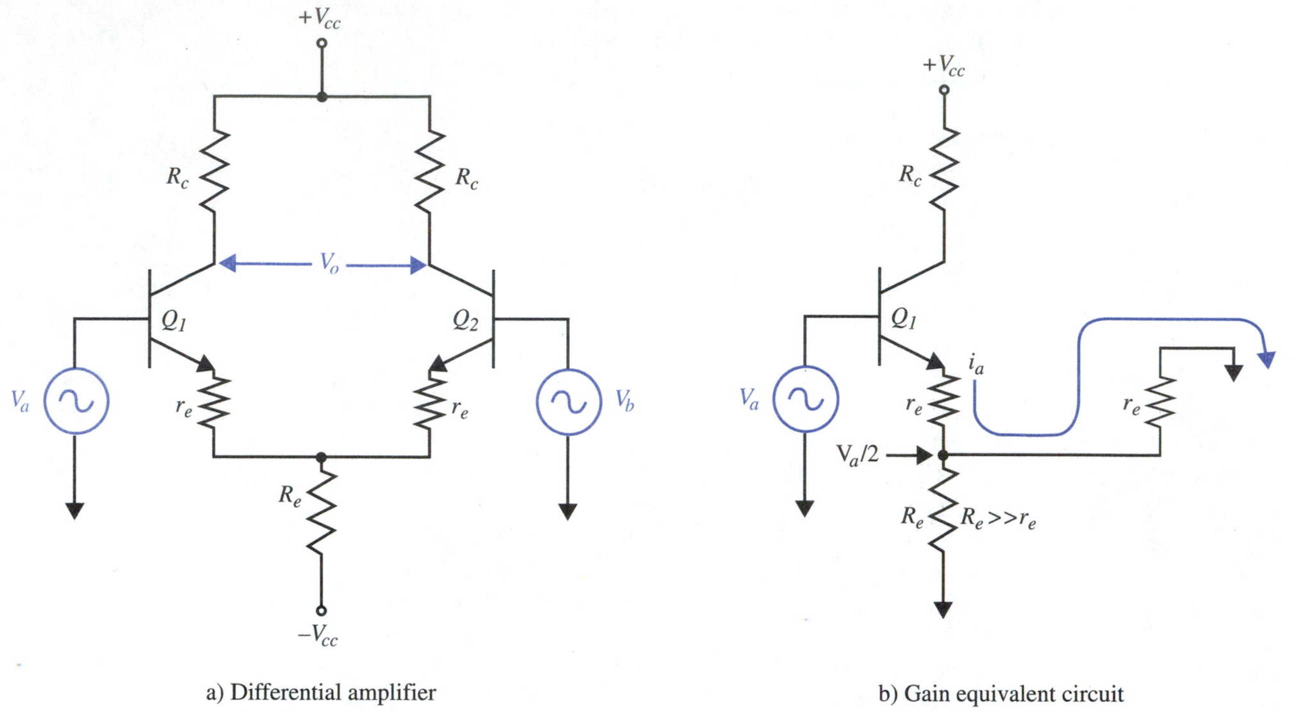

FIGURE 6–2 Circuits for differential gain derivation

The signal V_a applied to the Q_2 stage appears across both emitter resistances in series, i.e., r_e of Q_1 and the input resistance r_e of Q_2. So V_a is developed across $2r_e$ and applied to the common base configuration of Q_2, and therefore the voltage gain of Q_2 is

$$A_{v2} = \frac{R_c}{2r_e} \qquad \textbf{(equation 6–1b)}$$

Note that there is no phase shift between the input signal and the amplified signal on the collector of Q_2; but a 180° shift does occur to the amplified signal on the collector of Q_1. So if we measure the voltage between the two collectors, with one signal going positive while the other is going negative, we will see twice as much swing as we would from one collector to ground. The differential gain is then twice the gain of each stage, or

$$A_v = \frac{R_c}{r_e} \qquad \textbf{(equation 6–2)}$$

Considering that there are resistors in the base circuit for biasing, and a Thevenin resistance r_g from the generator (actually 1/2 r_g—see Instrumentation Amplifier Figure 7–11b in the next chapter), the gain equation must be modified to include R_b (parallel equivalent of biasing and generator resistors) divided by β as indicated in the common base circuit equation (see equation 3–5):

$$A_v = \frac{R_c}{r_e + \frac{R_b}{\beta}} \qquad \text{(equation 6–3)}$$

COMMON MODE REJECTION

Common mode (mode of operation) means that the same signal is applied to both differential inputs and **rejection** indicates how well the amplifier rejects this type of signal. As we shall see later, common mode rejection is important in rejecting noise pickup (EMI) on the input leads of an amplifier.

Let us now derive the equation for the common mode rejection of a differential amplifier. We shall refer again to Figure 6–2a, which is repeated in Figure 6–3a.

With the same signal V_{cm} applied to both inputs, the signal currents flowing in Q_1 and Q_2 are the same.

The emitter equivalent circuit differs from Figure 6–2b because at the point where the two emitters connect to R_e, one stage doesn't see the common base circuit loading of the other stage. The reason for this is that each input signal provides the same signal on its emitter. So, there is no signal current flow from one emitter to the other. In other words, each emitter sees no loading from the other emitter. This is the same concept that makes a resistor with the same voltage on both ends appear like an open circuit (no current). With

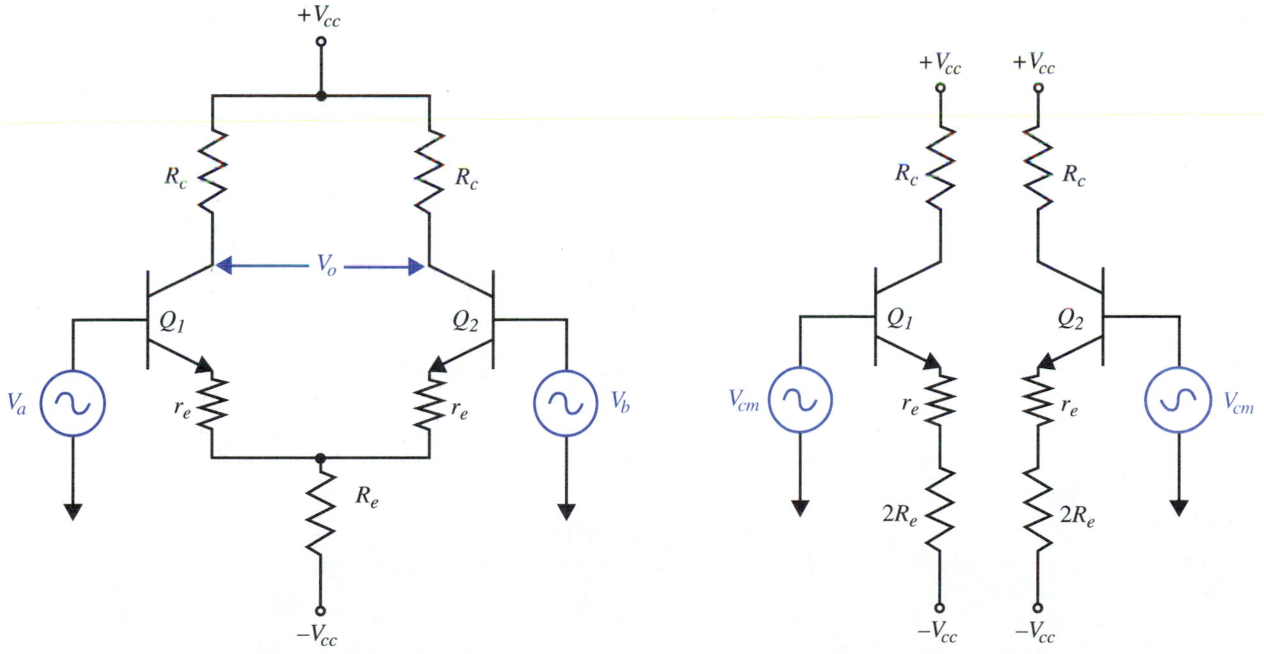

a) Differential amplifier b) Common mode equivalent

FIGURE 6–3 Derivation of common mode rejection

this situation, R_e is not now shunted out by the r_e of the common base input; so it becomes part of our equivalent circuit used to calculate the common mode gain.

Since the two signal currents through R_e are the same, this resistor can be divided into two resistors of twice the value of R_e—one for each transistor stage. The equivalent circuit is shown in Figure 6–3b.

The common mode voltage gain (A_{cm}) for each side from the figure is then

$$A_{cm} = \frac{R_c}{r_e + 2R_e} \qquad \textbf{(equation 6–4)}$$

This gain causes a voltage change on the collector of each transistor in response to the common mode (same) input voltage.

Since the gain on both sides is the same, we expect the same change on each collector, and the output voltage measured between them should be zero. However, slight differences in the two transistors could cause a small output voltage to appear. This effect can be reduced by keeping the common mode gain as low as possible. From equation 6–4, we see that one way of reducing A_{cm} is by raising R_e.

A simple way to raise R_e is to make it a constant current source—which has a high resistance. A practical method to obtain a constant current source is to use a transistor with a fixed base voltage and therefore a constant voltage across the emitter resistor. This makes the emitter current, and therefore the collector current, constant. Figure 6–4 illustrates the circuit.

EXAMPLE

Calculate the common mode gain of a differential amplifier circuit configuration of Figure 6–4 and compare it with the differential gain. Assume the effective resistance seen looking into the collector circuit (the R_e for our circuit) is typically 1MΩ.

SOLUTION

The voltage at the base of Q_3 is

$$V_{b3} = 10\text{V} - \left\{ \frac{[10\text{V} - (-10\text{V})]}{(24\text{K}\Omega + 2\text{K}\Omega)} \times 2\text{K}\Omega \right\}$$

$$V_{b3} = -8.46\text{V}$$

With −8.46V on the base of the constant current transistor Q_3, we then have −9.06V at the emitter and therefore (10V − 9.06V) or ≈1V across the 1KΩ emitter resistor. This means the emitter current is 1V/1KΩ or 1mA. For a high β transistor

$$I_c \approx I_e$$

and therefore the emitter current flowing in each differential transistor is one-half this value, i.e., 0.5mA. From this current, we can compute r_e:

$$r_e = \frac{26\text{mV}}{0.5\text{mA}}$$

$$r_e = 52\Omega$$

Using the value of 10KΩ for R_c from the schematic and substituting the values of R_c, r_e, and R_e (assumed 1MΩ) into equation 6–6,

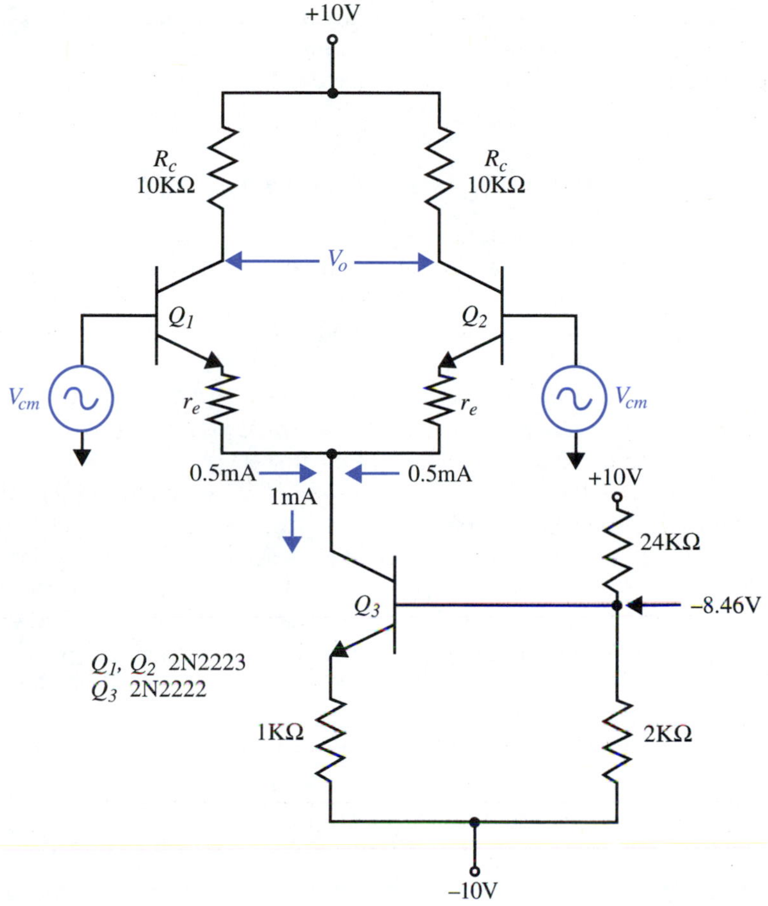

FIGURE 6–4 Differential amp with constant current source

$$A_{cm} = \frac{R_c}{r_e + 2R_e}$$

$$A_{cm} = \frac{10\text{K}\Omega}{52\Omega + 2\text{M}\Omega}$$

$$A_{cm} = 0.005$$

The differential amplifier gain is determined by using equation 6–2.

$$A_v = \frac{R_c}{r_e}$$

$$A_v = \frac{10\text{K}\Omega}{52\Omega}$$

$$A_v = 192$$

So comparing the two gains, we see that the differential gain is better than 38,400 times greater than the common mode gain. This gives some idea of how much a valid signal would be amplified versus an unwanted common mode noise signal.

COMPLETE OPERATIONAL AMPLIFIER CIRCUIT

The differential amplifier we just studied is the input stage of an operational amplifier. A simplified (without bias stabilization circuits) internal operational amplifier schematic is shown in Figure 6–5.

Notice in Figure 6–5 that the output of the input differential stage feeds into a second differential voltage amplifier stage (Q_3 and Q_4). However, the output from the second stage is taken from just one collector (Q_4). The reason for this is that the output from the complete operational amplifier is single ended, i.e., output voltage is referenced to ground rather to another lead as with the differential output of the first stage.

The output from the collector of Q_4 drives the base of a PNP amplifier (Q_5). This stage, in turn, drives a complementary-symmetry stage (ref Chap 3) consisting of Q_6 and Q_7. The two differential amplifier stages, along with the PNP stage can provide an overall voltage gain for the op-amp on the order of 100,000.

By using equal positive and negative power supplies (sometimes called power supply rails rather than supplies) for the amplifier, and setting the internal bias points, the output terminal is at 0V DC when both inputs have the same signal applied (no difference between inputs).

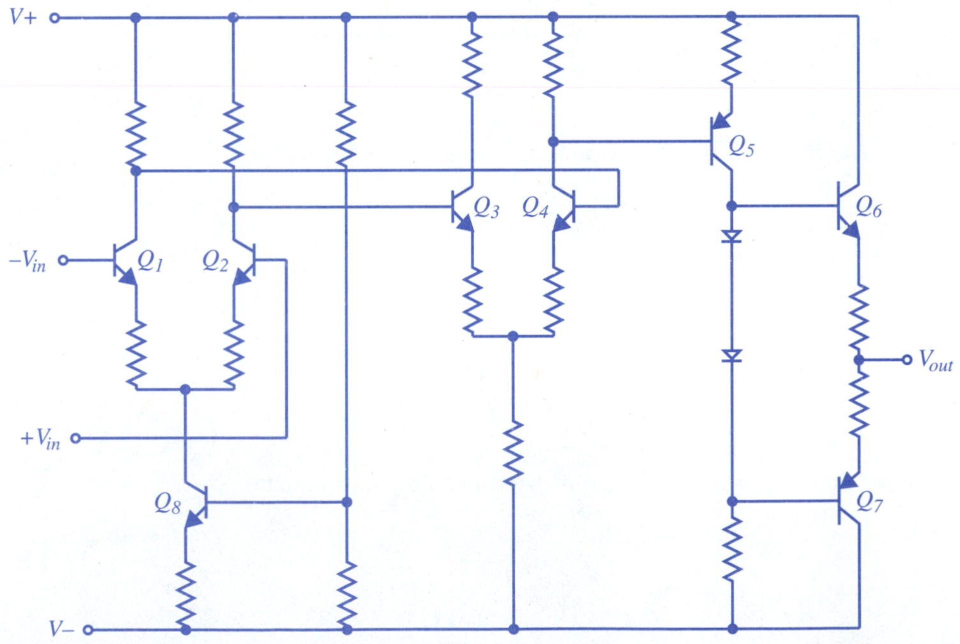

FIGURE 6–5 Complete operational amplifier schematic

OPEN LOOP OPERATION

When we apply the op-amp in circuits, we can configure it one of two ways: closed loop, where part of the op-amp output signal is fed back (connected) to the input; or open loop, with no feedback signal. Let us first study open loop operation. An op-amp in this configuration is shown in Figure 6–6.

Amplifiers are represented by a triangle in schematics. The two inputs are designated as + and −, with the positive sign indicating that the output is in phase (also called noninverting) with that input and the negative sign indicating a 180° phase shift (also called inverting) between that input and the output signal (see Figure 6–6 waveforms).

The op-amp maximum output voltage swing is one to two volts less than the positive supply (rail) and up to one to two volts greater (less negative) than the negative supply (rail). Actual op-amp output voltage (within the limits of the power supply voltages) is always the difference in voltage between the two inputs (V_{in}) multiplied by the open loop gain (A_{ol}) of the amplifier, or

$$V_o = (V_{in})(A_{ol}) \qquad \textbf{(equation 6–5)}$$

EXAMPLE

Op-amp has an open loop gain A_{ol} of 100,000, and the difference between the two inputs is 100μV. What is the output voltage?

SOLUTION

Multiplying the 100μV by the op-amp gain yields an output voltage V_o of

$$V_o = (V_{in})(A_{ol})$$

$$V_o = 100\mu V \times 100{,}000$$

$$V_o = 10V$$

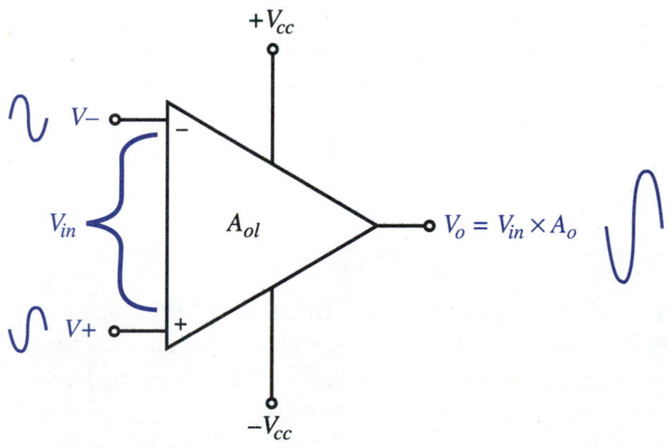

FIGURE 6–6 Open loop op-amp

EXAMPLE

The noninverting input (+) to an op-amp is +100.38mV, and the inverting input (−) is +100.4mV. What is the magnitude and polarity of the output voltage if the op-amp open loop gain A_{ol} is 200,000?

SOLUTION

The difference between the inputs is

$$V_{in} = 100.4\text{mV} - 100.38\text{mV}$$

$$V_{in} = 20\mu\text{V}$$

The output voltage is

$$V_o = (V_{in})(A_{ol})$$

$$V_o = 20\mu\text{V} \times 200{,}000$$

$$V_o = 4\text{V}$$

But what about polarity? Since the difference between the two inputs is 20μV, and inverting input is 20μV more positive than the noninverting input, we can consider that there is zero volts of signal on the noninverting input and +20μV on the inverting input. The reason we can do this is because the op-amp responds only to the difference between the two inputs. With +20μV on the inverting input, the output will be negative.

$$V_o = -4\text{V}$$

These examples show how little input signal is required to get a sizable output from an open loop op-amp. There are disadvantages to operating an op-amp open loop.

1) Since the open loop gain of the op-amp can vary considerably from device to device, the output voltage is not terribly predictable.
2) Small noise signals can be amplified by the high gain and disrupt the desired signal.
3) The range of signal frequencies that can be amplified is limited (more about this in Chapter 8).
4) Small DC input voltages will cause the output to saturate (go to one supply voltage).

OP-AMP CLOSED LOOP OPERATION

Closed loop operation means that part (sometimes all) of the output signal is fed back to the inverting input. This way of feeding the signal back is called positive feedback if the signal fed back is in phase with the input, or negative or degenerative feedback if the feedback signal is 180° out of phase with the input and therefore reduces (opposes) the input signal to the amplifier (see Figure 6–7). We shall find that positive feedback is used with oscillators (studied later) and negative feedback is used with amplifiers.

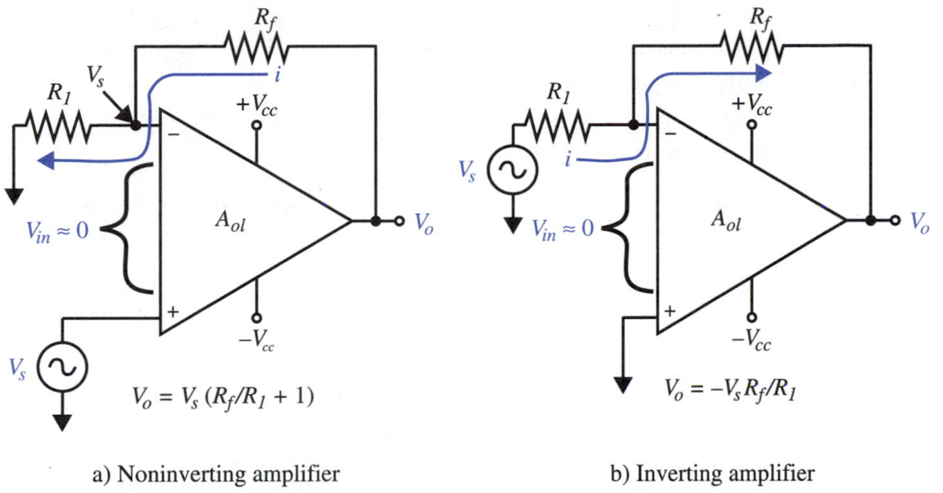

FIGURE 6–7 Ideal closed loop amplifiers

The advantages of amplifier closed loop operation using negative feedback are

1) Gain becomes independent of the op-amp.
2) Amplifier-created distortion is reduced.
3) Input resistance to the amplifier is increased.
4) Output resistance of the amplifier is reduced.
5) Frequency range of the amplifier is increased.

CLOSED LOOP GAIN

With a closed loop (negative feedback) amplifier, the input signal can be applied to either the inverting or noninverting input to the op-amp, as shown in Figure 6–7. The amplifier is named from which input is used. In Figure 6–7a, we have a noninverting amplifier (output in phase with input); in Figure 6–7b, an inverting amplifier (output 180° out of phase with input).

Notice in both amplifier configurations, a negative feedback path exists between the output and the inverting input provided by the dividing action of R_f and R_1. In Figure 6–7a, we see that the fed-back voltage is developed across R_1 and opposes the input voltage, and so V_{in} is reduced to a very small value. In Figure 6–7b, the fed-back voltage develops a current through R_1, which almost cancels out the signal current, and again V_{in} is reduced to a very small value. Let us first derive the closed loop gain for a noninverting amplifier.

GAIN OF NONINVERTING AMPLIFIER

We shall make the following assumptions for our derivation of gain:

1) The op-amp open loop gain is infinite, and therefore the voltage difference V_{in} between the signals on the op-amp inverting and noninverting inputs, required for any output voltage, is close to zero.
2) The op-amp input resistance is infinite, and therefore no current flows into either of the op-amp input terminals.

With the first assumption, we can say that the input voltage V_s applied to the noninverting input terminal of the op-amp must also appear on the inverting terminal (see Figure 6–7a).

Since the voltage V_s on the inverting terminal is developed across R_1, V_s can be expressed as

$$V_s = i \times R_1$$

The second assumption we made was that no current flows into the op-amp terminals, and so all the current (i) that flows through R_1 must also flow through R_f (see Figure 6–7a).

The output voltage from the op-amp V_o is the sum of the voltage drops across R_1 and R_f, so

$$V_o = i(R_f + R_1)$$

The closed loop voltage gain A_v is defined as the output voltage V_o divided by the input signal V_s, or

$$A_v = \frac{V_o}{V_s}$$

$$A_v = \frac{i(R_f + R_1)}{i(R_1)}$$

and

$$A_v = \frac{(R_f + R_1)}{R_1}$$

If we now separate terms, noninverting amp gain is

$$A_v = \frac{R_f}{R_1} + 1 \qquad \textbf{(equation 6–6)}$$

Equation 6–6 was derived with the assumption that the open loop gain of the op-amp was infinite, but in practice open loop gains can range from about 20,000 to over 1,000,000. The closed loop gain obtained from equation 6–6 is within 1% accuracy if the open loop gain is at least 100 times greater than the required closed loop gain.

EXAMPLE

Find the voltage at the output (V_o) of the noninverting amplifier shown in Figure 6–8 if the input voltage (V_s) is a sine wave of 20mV peak to peak. Also find the output voltage if R_f is halved.

SOLUTION

Since the given value of R_f is 200KΩ and that of R_1 is 2KΩ, the circuit gain using equation 6–6 is

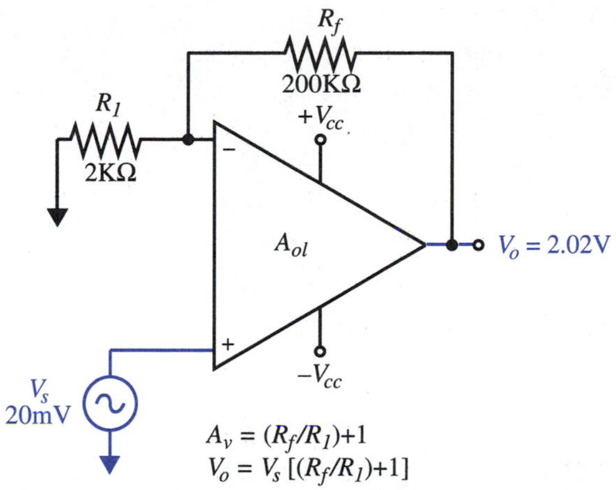

FIGURE 6–8 Noninverting amplifier gain example

$$A_v = \frac{R_f}{R_1} + 1$$

$$A_v = \frac{200\text{K}\Omega}{2\text{K}\Omega} + 1$$

$$A_v = 101$$

The output voltage

$$V_o = V_s \times A_v$$

$$V_o = 20\text{mV} \times 101$$

$$V_o = 2.02\text{V peak to peak}$$

What happens to the output voltage if we halve R_f? The new gain is

$$A_v = \frac{100\text{K}\Omega}{2\text{K}\Omega} + 1$$

$$A_v = 51$$

and

$$V_o = 20\text{mV} \times 51$$

$$V_o = 1.02\text{V peak to peak}$$

So halving R_f halves the gain!

EXAMPLE

The minimum open loop gain specified in the data sheets for a 741 op-amp is 20,000. What is the maximum closed gain we can use if we want the measured gain to be within 1% of the computed value using equation 6–6?

SOLUTION

Since the rule given is that the open loop gain be 100 times greater than the required closed loop gain, the maximum gain $A_{v(max)}$ we can use is

$$A_{v(max)} = \frac{A_{ol}}{100}$$

$$A_{v(max)} = \frac{20{,}000}{100}$$

$$A_{v(max)} = 200$$

GAIN OF AN INVERTING AMPLIFIER

We again make the assumption of infinite open loop gain and infinite input resistance in our derivation of the gain of the inverting amplifier. With reference to Figure 6–9, the assumption of infinite gain means that the difference between the two inputs to the op-amp is zero. Therefore, since the noninverting terminal is at ground potential (0V), the inverting input must also be at zero volts. We refer to the inverting terminal as being at **virtual ground**—it is at ground potential without being grounded.

With the inverting terminal at ground potential, the full input signal is developed across R_1. So

$$V_s = (i)(R_1)$$

Since the input resistance of the op-amp terminals is considered infinite, no current flows into these terminals, and so the current flowing through R_1 must also flow through R_f and

$$V_o = -i \times R_f$$

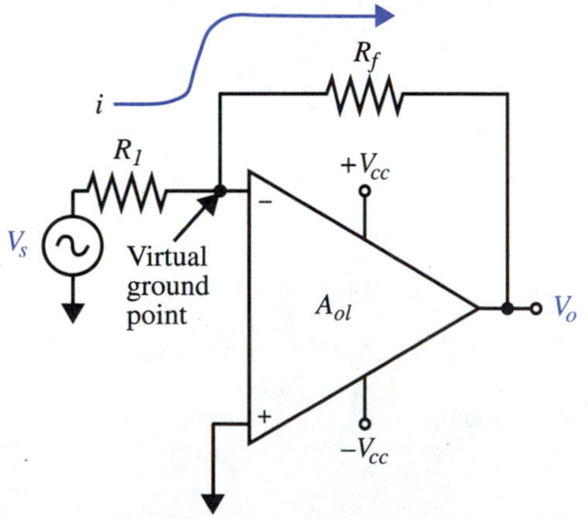

Closed loop gain $A_v = -R_f/R_1$

FIGURE 6–9 Gain of inverting amplifier

The negative sign indicates that current (conventional) is flowing via R_f towards the output terminal. The gain of the amplifier is

$$A_v = \frac{V_o}{V_s}$$

$$A_v = \frac{-iR_f}{iR_1}$$

(the "i" cancels)

$$A_v = -\frac{R_f}{R_1} \qquad \textbf{(equation 6–7)}$$

In this gain equation, the minus sign indicates that the input signal and output signal are 180° out of phase. This is why it is called an inverting amplifier.

EXAMPLE

The given value of R_f for an inverting amplifier is 270KΩ and R_1 is 3KΩ. Find the circuit gain.

SOLUTION

Using equation 6–7,

$$A_v = -\frac{R_f}{R_1}$$

$$A_v = -\frac{270\text{K}\Omega}{3\text{K}\Omega}$$

$$A_v = -90$$

POWERING AN OP-AMP

As with a transistor, the op-amp needs to have an external power source for operation. Most op-amps require two power supplies in order to operate: a positive supply and a negative supply. Using two supplies allows the output of the op-amp to be at a DC value of 0V when there is no input signal. This allows connection between different op-amp circuits without the use of coupling capacitors—which is required for amplification of DC signals.

Figure 6–10a shows the complete connections for a 741 op-amp used as an inverting amplifier set up for a voltage gain of 100. The op-amp pin numbers are identified on the schematic; and Figure 6–10b can be used to determine the actual pin on the op-amp chip. Note that pin numbers are identified from the top view of the chip starting with the dot in the plastic as number 1 and with the numbers increasing going in a counterclockwise (horseshoe) path.

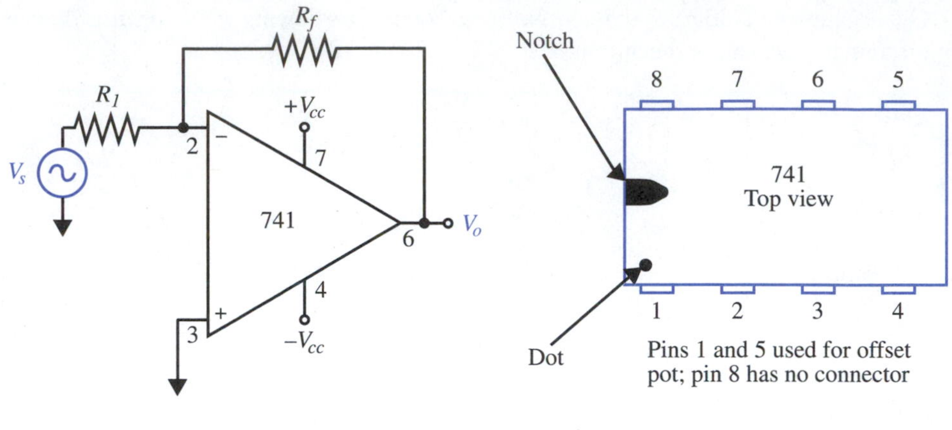

a) Pin identification b) Dip package

FIGURE 6–10 Connections for a 741 amplifier circuit

741 SPECIFICATIONS

The 741 op-amp is a low-cost device that has been used in many basic op-amp circuits. We shall use the 741 specifications as a base line for comparison with other op-amp devices.

Listed in the accompanying chart are some of the specifications for the 741C that we shall be using in this and later chapters.

Parameter	*741C Specification (25°C)*	
Open Loop Gain (A_{ol})	20,000 (min)	100,000 (typical)
Internal Input Resist (r_i)	0.3MΩ (min)	2MΩ (typical)
Internal Output Resist (r_o)		75Ω (typical)
Common Mode Rejection (CMR)	70db (min)	90db (typical)
Power Supply Rejection (PSR)	77db (min)	96db (typical)
Input Offset Voltage (V_{iv})	6mV (max)	2mV (typical)
Input Bias Current (I_b)	500nA (max)	80nA (typical)
Input Offset Current (I_{os})	200nA (max)	2nA (typical)
Slew Rate (SR)		0.5V/μs (typical)
Gain Bandwidth Product (GBP)		1 MHz (typical)

INPUT RESISTANCE OF AN OP-AMP

We are concerned with the input resistance of an amplifier because it can load down the stage that is driving the amplifier. If the input resistance of the amplifier is very large compared to the resistance of the driving circuit, then the full voltage of the driving stage appears at the amplifier input and overall circuit gain is maximized.

INPUT RESISTANCE OF A NONINVERTING AMPLIFIER

Let us now consider the input resistance of the noninverting amplifier. But first we shall define again the terms we shall be using:

r_i = amplifier open loop input resistance

R_{in} = amplifier closed loop resistance (what we are trying to find)

V_{in} = voltage between the two amplifier input terminals

V_s = input signal voltage

Figure 6–11 shows an internal resistance r_i inside the op-amp—which again has a typical value of 2MΩ for a 741 op-amp.

If we assume that the open loop gain of the op-amp is infinite, then the signal voltage V_s is the same at both input terminals of the op-amp and V_{in} is 0. With both ends of r_i at the same voltage, that means there will be no signal current (*i*) flowing into r_i, and the closed loop op-amp input resistance appears infinite!

We know that the gain of the practical op-amp is not infinite, and therefore the input resistance is also not infinite. So, we need to derive the equation for the actual input resistance R_{in} of a practical op-amp.

First we can develop an equation for the voltage across R_1 in terms of the input voltage V_{in} to the op-amp. The voltage V_1 across R_1 is the output voltage V_o reduced by the $R_f : R_1$ divider, or

$$V_1 = \frac{V_o R_1}{R_1 + R_f}$$

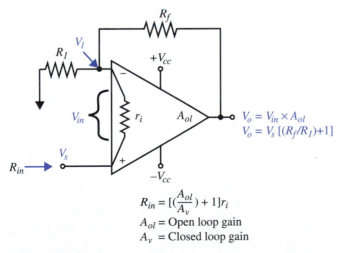

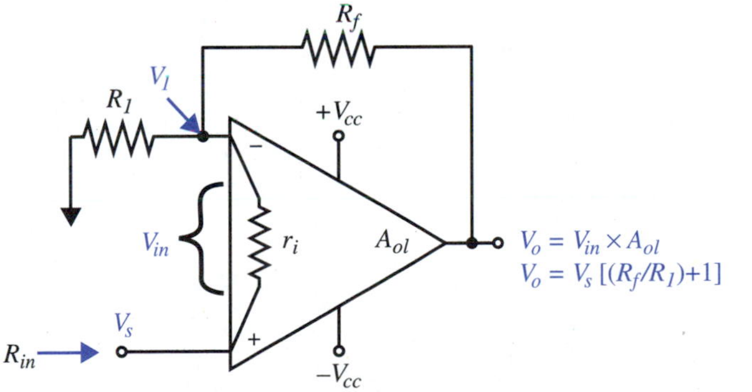

$$R_{in} = [(\frac{A_{ol}}{A_v}) + 1]r_i$$

A_{ol} = Open loop gain

A_v = Closed loop gain

FIGURE 6–11 Input resistance of a noninverting amplifier

Since

$$V_o = (A_{ol})(V_{in})$$

for all op-amp configurations,

$$V_1 = \frac{(A_{ol})(V_{in})(R_1)}{(R_1 + R_f)}$$

but

$$\frac{R_1}{(R_1 + R_f)} = \frac{1}{A_v}$$

So

$$V_1 = \frac{(A_{ol})(V_{in})}{A_v} \qquad \textbf{(equation 6–8)}$$

The equation for the noninverting op-amp circuit input resistance is

$$R_{in} = \frac{V_s}{i}$$

but since

$$V_s = V_1 + V_{in}$$

$$R_{in} = \frac{(V_1 + V_{in})}{i}$$

and since

$$i = \frac{V_{in}}{r_i}$$

$$R_{in} = \frac{(V_1 + V_{in})}{(V_{in}/r_i)}$$

Substituting equation 6–8 for V_1,

$$R_{in} = \frac{(A_{ol})(V_{in}/A_v) + V_{in},}{(V_{in}/r_i)}$$

Simplifying:

$$R_{in} = \left(\frac{A_{ol}}{A_v} + 1\right) r_i \qquad \textbf{(equation 6–9)}$$

Equation 6–9 indicates that the input resistance for the noninverting amplifier is the internal resistance r_i multiplied by the ratio of the open loop gain divided by the closed loop gain (this ratio is called the loop gain A_L) plus one. Since A_{ol} should be at least a hundred times greater than A_v, the internal resistance r_i is increased by at least this factor.

EXAMPLE

From the 741 op-amp specifications, r_i is typically 2MΩ, and the open loop gain A_{ol} is 100,000. Find the input resistance R_{in} if R_1 is 1KΩ and R_f is 99KΩ.

SOLUTION

Using equation 6–6,

$$A_v = \frac{R_f}{R_1 + 1}$$

$$A_v = \frac{99\text{K}\Omega}{1\text{K}\Omega + 1}$$

$$A_v = 100$$

Using equation 6–9,

$$R_{in} = \left(\frac{A_{ol}}{A_v} + 1\right) r_i$$

$$R_{in} = \left(\frac{100{,}000}{100} + 1\right) 2\text{M}\Omega$$

$$R_{in} = 2.002 \text{ gigaohms}$$

which is a very large resistance!

INPUT RESISTANCE OF AN INVERTING AMPLIFIER

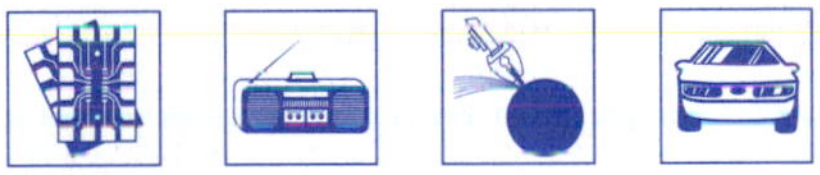

In the inverting amplifier configuration, the noninverting input to the op-amp is grounded, and the inverting input is at virtual-ground potential. So the input resistance R_{in} of the inverting amplifier is simply R_1, as shown in Figure 6–12.

$$R_{in} = R_1 \qquad \textbf{(equation 6–10)}$$

which can be a rather small resistance.

EXAMPLE

Find the input resistance for an inverting amplifier with R_1 of 2KΩ, R_f of 47KΩ, and A_{ol} of 200,000.

SOLUTION

We are only concerned with R_1 since the value of this resistor is the circuit input resistance, so

$$R_{in} = R_1 = 2\text{K}\Omega$$

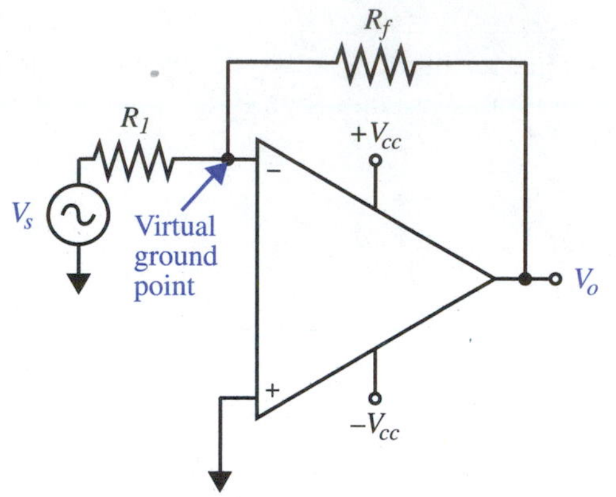

Because of virtual ground,
V_s sees R_1 as its load

FIGURE 6–12 Input resistance of the inverting amplifier

OUTPUT RESISTANCE OF AN OP-AMP

The output resistance of an amplifier causes the output voltage to drop as we draw more load current. So if we don't want the output voltage to change with loading, then the amplifier should have 0Ω output resistance. It is not possible to achieve a 0Ω value, but we shall see that with a closed loop op-amp, the output resistance is extremely small. We shall go through the derivation of the equation for the output resistance of an op-amp with reference to Figure 6–13.

The following terms are defined before we start the derivation:

R_o = closed loop output resistance (what we are trying to find)

r_o = open loop internal output resistance of the op-amp

A_{ol} = open loop gain of the op-amp

A_v = closed loop gain of the op-amp

V_{in} = voltage between the two op-amp input terminals

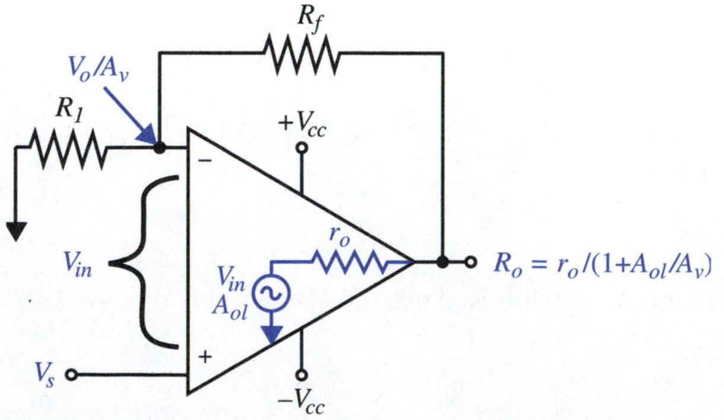

FIGURE 6–13 Output resistance of an op-amp

The closed loop output resistance R_o causes the output voltage to drop an amount ΔV (Δ—delta—indicates a small change in the value) when the current increases by an amount Δi, or

$$\Delta V_o = -(R_o)(\Delta i)$$

(The negative sign indicates voltage drops when current increases.) From Figure 6–13,

$$V_o = (A_{ol})(V_{in}) - ir_o$$

but

$$V_{in} = V_s - V_o/A_v$$

$$V_o = (A_{ol})(V_s) - \frac{(V_o)(A_{ol})}{A_v} - ir_o$$

Collecting like terms,

$$V_o\left(1 + \frac{A_{ol}}{A_v}\right) = (A_{ol})(V_s) - ir_o$$

We need to find the change in V_o when we change i, and this can be found by taking the derivative (calculus) of this equation. This operation causes $(A_{ol})(V_s)$ to go to zero, since this is not an i term. So

$$\left(1 + \frac{A_{ol}}{A_v}\right)\Delta V_o = -r_o\,\Delta i$$

But from above,

$$\Delta V_o = (-R_o)(\Delta i)$$

$$\Delta V_o - R_o\left(1 + \frac{A_{ol}}{A_v}\right)\Delta i = -r_o\,\Delta i$$

Solving for R_o,

$$R_o = \frac{r_o}{1 + (A_{ol}/A_v)} \qquad \textbf{(equation 6–11)}$$

But if

$$\frac{A_{ol}}{A_v} \gg 1$$

$$R_o = \frac{(r_o)(A_v)}{A_{ol}}$$

What this equation tells us is that the open loop output resistance of an op-amp is reduced by the factor A_{ol}/A_v when the op-amp is connected in any closed loop configuration.

EXAMPLE

A 741 op-amp has a specified open loop gain (A_{ol}) of 100,000 and an open loop output resistance of 75Ω. If the closed loop gain is 200, what is the closed loop output resistance?

SOLUTION

Using equation 6–11,

$$R_o = \frac{75\Omega}{\left(1 + \dfrac{100{,}000}{200}\right)}$$

$$R_o = 150 \text{ m}\Omega$$

A very small value when compared to the kilohms output resistance of a bipolar transistor collector or FET drain!

OP-AMP FOLLOWER

The high input resistance and low output resistance of an op-amp are used to advantage with an op-amp follower. This op-amp configuration is shown in Figure 6–14.

Since the voltage V_s applied to the noninverting input also appears at the inverting input, it therefore appears at the output.

So, the voltage gain of the follower is unity (1). Actually, the output is V_o/A_{ol} less than the input, but since this term is close to zero, our assumption of unity gain is valid.

EXAMPLE

Calculate the input and output resistance of an op-amp follower using a 741 with an open loop gain (A_{ol}) of 100,000, an open loop input resistance (r_i) of 2MΩ, and an open loop output resistance (r_o) of 75Ω.

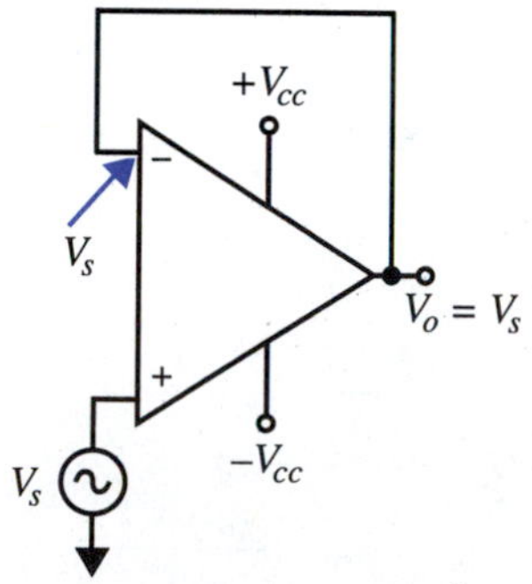

FIGURE 6–14 Op-amp follower

SOLUTION

Using equation 6–9 to find the input resistance,

$$R_{in} = \left(\frac{A_{ol}}{A_v} + 1\right) r_i$$

$$R_{in} = \left(\frac{100{,}000}{1} + 1\right) 2\text{M}\Omega$$

$$R_{in} = 200 \text{ giga-ohms}$$

Using equation 6–11 to find the output resistance,

$$R_o = \frac{r_o}{1 + (A_{ol}/A_v)}$$

$$R_o = \frac{75\Omega}{1 + (100{,}000/1)}$$

$$R_o = 0.75\text{m}\Omega$$

BOOTSTRAP CIRCUIT

If we use a capacitor to couple a signal into the noninverting input of an op-amp, we have to provide a resistor on the op-amp side of the capacitor to allow it to discharge. The need for the resistor will be clear if we remember that the input leads of the op-amp connect to the base of a transistor, which only will allow the capacitor current to flow one way. So the capacitor can charge but not discharge, and this will prevent an AC signal from being coupled into the op-amp.

This discharge resistor R_a is shown in the follower circuit of Figure 6–15a; but now our input resistance to the follower is not the gigaohms we computed previously but the kilohms value of R_a.

To restore the high input resistance of the follower, we can use the bootstrap ("lift by the bootstraps") circuit of Figure 6–15b.

The bootstrap raises the effective value of R_a by feeding back to the bottom end of R_a the same signal that is at the top end (output signal and input signal are the same). With 0V across R_a, there is no current flow through this resistor, and R_a appears like an open circuit for the input signal. However, the C_1 charging and discharging current can flow through the relatively low resistance of R_a and R_b.

Bootstrapping can be used with a noninverting amplifier (with gain) by dividing down the voltage fed back to the input using a resistor by setting R_1 equal to R_b and also putting a resistor equal to R_f in series with the feedback capacitor C_2.

DISTORTION REDUCTION BY THE CLOSED LOOP OP-AMP

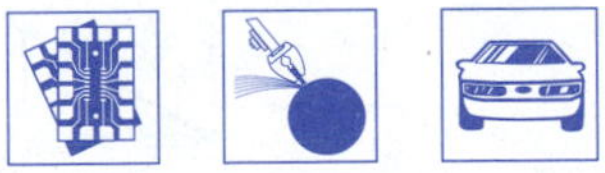

Distortion (difference between input signal and output signal waveshapes) occurring within an amplifier is reduced when we operate the amplifier in a closed loop configuration. We

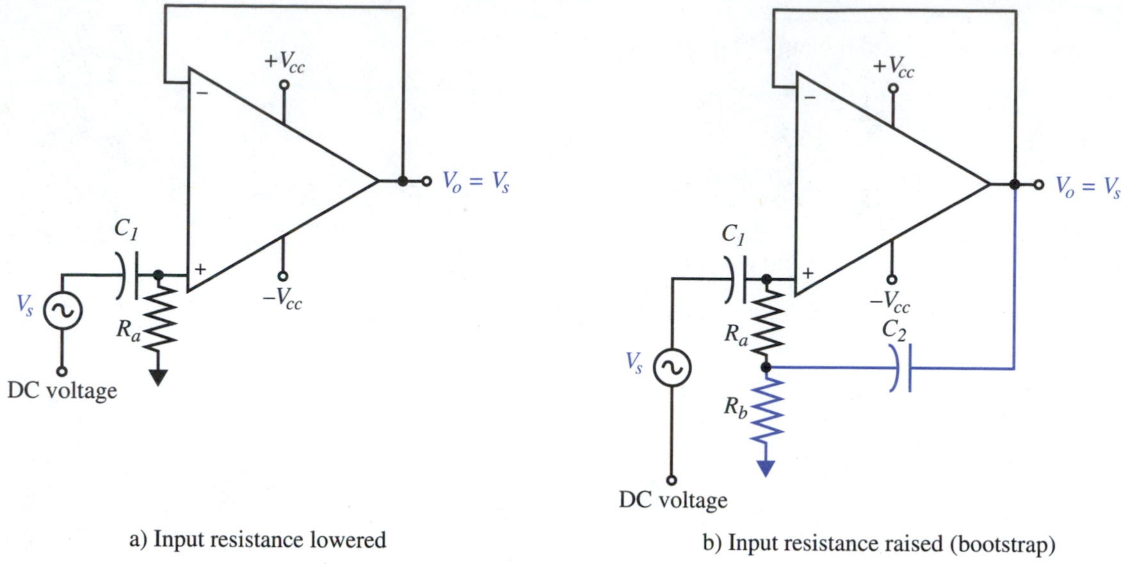

FIGURE 6–15 Bootstrapped follower

shall prove this by assuming that the distortion occurs in the high-level output amplifier of the op-amp and can be represented as a voltage V_d (a fraction of the input signal V_s). Figure 6–16 shows the voltages used in the derivation of the output distortion voltage.

The distortion voltage (V_{od}) at the output of the amplifier is equal to V_d minus the portion of V_{od} that is divided down by R_f and R_1 back to the inverting input and amplified by the open loop gain A_{ol}. The portion fed back to the inverting input by the divider R_f and R_1 is

$$V_i = V_{od}\frac{R_1}{(R_f + R_1)}$$

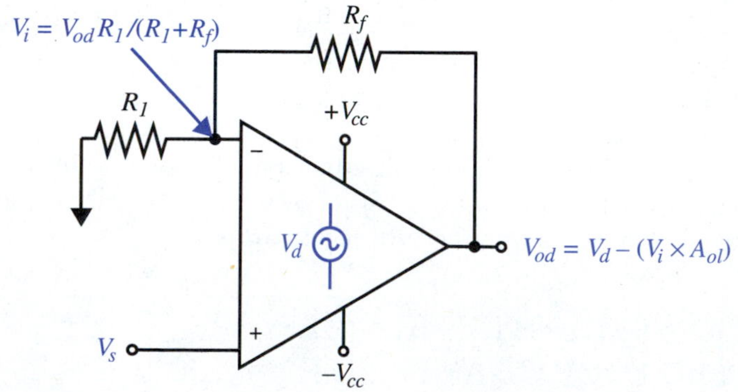

FIGURE 6–16 Distortion reduction in an amplifier

but

$$\frac{R_1}{(R_1 + R_f)} = 1/A_v$$

$$V_i = \frac{V_{od}}{A_v}$$

At the op-amp output,

$$V_{od} = V_d - V_i A_{ol}$$

Substituting for V_i,

$$V_{od} = V_d - V_{od}\frac{A_{ol}}{A_v}$$

Collecting V_{od} terms,

$$V_{od} = V_d\left(\frac{A_v}{A_{ol} + A_v}\right) \qquad \textbf{(equation 6–12)}$$

At the output, then, V_d is reduced by the factor

$$\frac{A_v}{A_{ol} + A_v}$$

If

$$A_{ol} \gg A_v$$

then the distortion, with the closed loop configuration, is reduced by the factor A_v/A_{ol}.

We can find the internal distortion voltage V_d if we know the open loop distortion factor (D) for the amplifier and the signal output voltage (V_{os}) by using the equation

$$V_d = (V_{os})(D) \qquad \textbf{(equation 6–13)}$$

EXAMPLE

An op-amp circuit has a closed loop gain A_v of 50 and generates 5% (D is 0.05) distortion in the op-amp. If the op-amp open loop gain is 100,000, and the input signal voltage is 0.1V, find the signal output voltage V_{os}, the distortion voltage V_{od}, and the percent distortion at the amplifier output.

SOLUTION

We can find the internal distortion voltage V_d from the output signal voltage V_{os}.

$$V_{os} = V_s A_v$$

$$V_{os} = 0.1\text{V} \times 50$$

$$V_{os} = 5\text{V}$$

Then from equation 6–13,

$$V_d = (V_{os})(D)$$
$$V_d = 5\text{V} \times 0.05$$
$$V_d = 0.25\text{V}$$

Now using equation 6–12,

$$V_{od} = \frac{(V_d)(A_v)}{A_{ol} + A_v}$$
$$V_{od} = \frac{0.25\text{V} \times 50}{100{,}000 + 50}$$
$$V_{od} = 125\mu\text{V}$$

From before, the output signal voltage V_{os} is 5V and this gives (125μV/5V) × 100 or 0.0025% distortion at the output compared to the specified 5% without feedback (open loop).

CURRENT BOOSTING

Considering the 741 op-amp as a typical device, it can provide up to 25mA into a short circuit (no output voltage swing). As the output voltage swing is increased, the shift in the internal operating point of the op-amp causes the available load current to decrease. For example, if its output voltage swing is to be within 1V of the supply voltage, the load current it is able to provide drops to 1.5mA. Within 2V of the supply voltage, a maximum load current of 6mA is available. Putting it another way, if the power supply voltages are ±12V, then with a 1.5mA load, our maximum output voltage swing is ±11V. A 6mA load current reduces our swing to ±10V. What this means is that if we can reduce the current drawn by the load from the op-amp, we can maintain a greater output voltage swing.

One way to reduce the current drawn from the op-amp is to connect a current-boosting transistor to the output, as shown in Figure 6–17. The current gain from base to emitter allows the circuit to supply a large load current without drawing much current from the op-amp.

Notice that the feedback connection for R_f is taken from the emitter of the transistor rather than from the output of the op-amp. What this does is to include the transistor as part of the op-amp, so we don't have to worry about the 0.6V drop from base to emitter (or its change with temperature), and we also maintain the low output resistance.

EXAMPLE

With reference to Figure 6–17, assume the β of the transistor is 50 and the load current through R_L is 50mA. Find the current drawn from the op-amp and the maximum output voltage swing if the power supply voltages are ±12V.

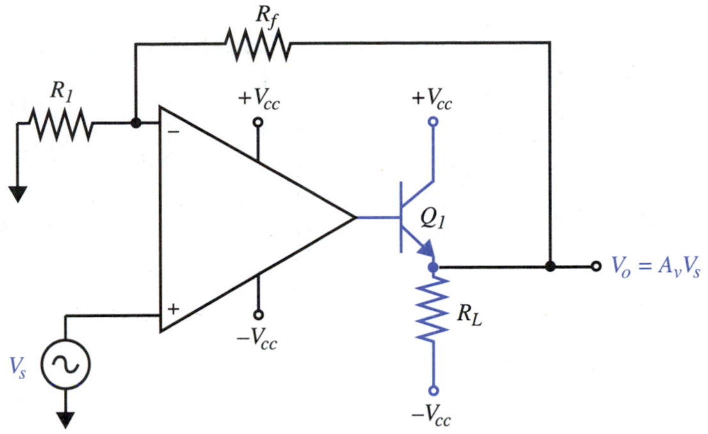

FIGURE 6–17 Op-amp current boosting

SOLUTION

We simply divide the load current of 50mA by the β of 50 to find a base current of 1mA, which is the required current from the op-amp. Since the current drawn from the op-amp is less than 1.5mA, we can expect our output voltage swing to be ± 11V.

With a short on the output of a 741 op-amp, the device will go into a current-limit mode at about 25mA. So even if we don't need a large voltage swing out of the op-amp, we are still limited by the amount of current the op-amp can supply, and therefore we could need a current-boosting transistor.

EXAMPLE

We need to supply a peak current of 500mA to a load with a voltage swing of ± 1V. The β of the current-boosting transistor is 40. Find the current drawn from the op-amp.

SOLUTION

Since the op-amp can supply a maximum of 25mA, we need to use a current-boosting transistor. Dividing the load current of 500mA by the β of 40 we obtain a base current of 12.5mA; so this is the current supplied by the op-amp.

OPERATION WITH A SINGLE POWER SUPPLY

In our earlier discussion of the op-amp, we assumed that both a positive and a negative power supply were available so that the output signal could swing equally positive and negative from a zero volt level. There are times, however, when only a single power source is available, e.g., the nominal +12V in an automobile.

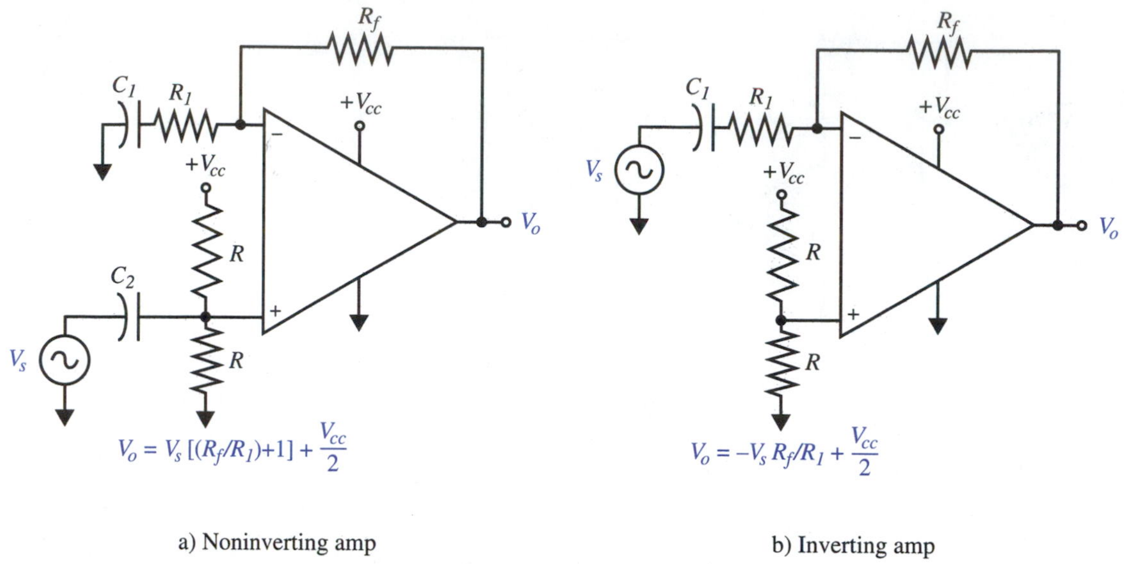

FIGURE 6–18 Single power supply operation

To operate a dual powered op-amp from a single power supply requires that the inputs and output be biased to one-half of the single power supply voltage. This is accomplished by using a voltage divider on the noninverting input to raise the voltage on this terminal to one-half of the power supply, as shown in Figure 6–18

From a DC standpoint, both circuits in Figure 6–18 act like followers since the capacitors C_1 on the inverting inputs isolate these inputs from DC ground and effectively take R_1 out of the circuit; so the DC output voltage is not divided down by R_f and R_1. This means the output DC voltage follows (is the same as) the voltage at the inverting input, i.e., one-half the supply voltage.

For AC signals, if we consider the capacitors to be short circuits, the first circuit has a gain of $(R_f/R_1) + 1$, and the second a gain of $-R_f/R_1$. The assumption that the capacitors are short circuits assumes that the reactance of C_1, in Figure 6–18b, at the lowest frequency used with the circuit, is less than one-tenth of the value of R_1, or

$$X_{c1} = \frac{R_1}{10}$$

Similarly for C_2 in Figure 6–18a,

$$X_{c2} = \frac{R/2}{10} \quad \text{(two } Rs \text{ in parallel)}$$

$$X_{c2} = \frac{R}{20}$$

Note that the input resistance seen by an AC source for the circuit of Figure 6–18a is not the extremely high resistance of a noninverting amplifier but *R*/2.

EXAMPLE

Find the voltage gain and the DC voltage at the output for the circuit of Figure 6–18a if R_1 is 2KΩ, R_f is 100KΩ, *R* is 200KΩ, and the power supply is +18V. Also determine the values of C_1 and C_2 if the lowest operating frequency for the circuit is 1KHz.

SOLUTION

Since we have a noninverting amplifier, the voltage gain is

$$A_v = \left(\frac{R_f}{R_1}\right) + 1$$

$$A_v = \left(\frac{100\text{K}\Omega}{2\text{K}\Omega}\right) + 1$$

$$A_v = 51$$

The two *R* resistors form a voltage divider that halves the 18V supply voltage, and therefore we have +9V at the noninverting input. Since the circuit is a follower for DC (C_1 isolates R_1 from ground), there will be +9V at the output. The load seen by C_2 is the two *R* resistors in parallel, so

$$X_{c2} = \frac{R/2}{10}$$

$$X_{c2} = \frac{200\text{K}\Omega/2}{10}$$

$$X_{c2} = 10\text{K}\Omega$$

The value of C_2 at 1KHz is then

$$C_2 = \frac{1}{2\pi f X_{c2}}$$

$$C_2 = \frac{1}{2\pi 10^3 \times 10^4}$$

$$C_2 = 15.9\text{nF}$$

C_1 reactance should be one-tenth of R_1, or

$$X_{c1} = \frac{R_1}{10}$$

$$X_{c1} = \frac{2\text{K}\Omega}{10}$$

$$X_{c1} = 200\Omega$$

At 1KHz,

$$C_1 = \frac{1}{2\pi 10^3 \times 200\Omega}$$

$$C_1 = 0.796\mu F$$

TROUBLESHOOTING BASIC OP-AMPS

Since the op-amp internal circuit is complex, we cannot check it out-of-circuit. The basic approach to in-circuit testing is to cause the output to change by forcing the inverting input—since this input is rarely grounded.

EXAMPLE

Determine how to check the op-amp in the configuration shown in Figure 6–19.

SOLUTION

A good approach is to force the inverting input by connecting it to either the positive or negative power supply. The choice of power supply depends on the present DC voltage polarity at the output.

If the op-amp output is positive, we connect the inverting input to the positive supply; if negative, we connect it to the negative supply. A change in the DC voltage at the output of the op-amp after the inverting input is forced indicates a good op-amp.

EXAMPLE

Using the method described for the op-amp in Figure 6–19, the op-amp checks good, but with the noninverting input grounded (0V), the output is at +11V. What is the problem?

SOLUTION

Either R_1 or R_f could be open circuited. If R_1 is open, the circuit configuration becomes a follower, and since we have 0V at the noninverting input, we would also see

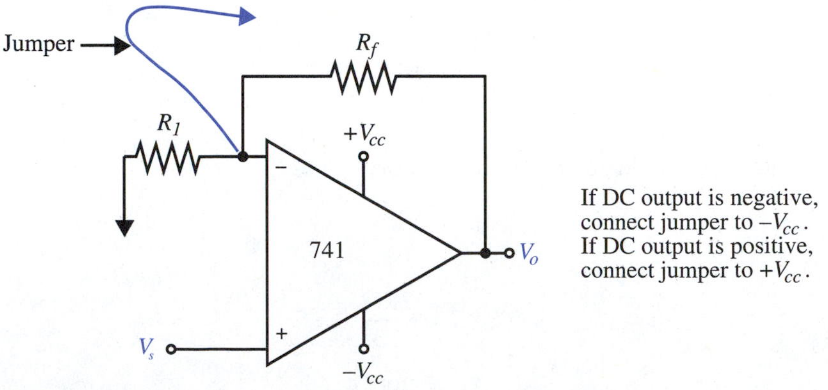

FIGURE 6–19 Checking an op-amp in a noninverting circuit

the 0V at the output. But we have +11V at the output, so R_1 is not open. With R_f open, the op-amp will be open loop, and any small difference in the bias of the input differential amplifier will cause the output to go to +11V or −11V. So the problem is an open R_f. This can be verified by temporarily connecting a resistor in parallel with R_f of the same ohmic value and noting whether the op-amp output goes to the required 0V.

SUMMARY

- The op-amp was originally designed to perform math operations in analog computers.
- A differential amplifier amplifies the difference between the two input signals.
- The two transistors in a differential amplifier share the same emitter resistor.
- The output voltage from a differential amplifier is taken between the two collectors.
- Common mode rejection is the rejection of the same signal applied to both inputs of a differential amplifier.
- Gain of common mode signals can be reduced by increasing the shared emitter resistor (constant current source).
- Open loop operation of an op-amp means there is no feedback of the output signal back to the input.
- The open loop gain of an op-amp is typically greater than 100,000.
- An op-amp has a differential input circuit.
- One input to the op-amp is inverted (−) at the output and the other (+) is not.
- The output signal from an op-amp is always the amplified difference of the signals applied to the two input terminals.
- The gain of a noninverting amplifier is $R_f/R_1 + 1$.
- The gain of an inverting amplifier is $-R_f/R_1$.
- A plus and a minus power supply are used with most op-amps to allow the output to be at 0V DC.
- Input resistance of a noninverting amplifier is $(1 + A_{ol}/A_v)r_i$.
- Input resistance of an inverting amplifier is R_1.
- Output resistance of all op-amp circuits is $r_o/(1 + A_{ol}/A_v)$.
- If the open loop gain is greater than a hundred times the closed loop gain, distortion occurring within the op-amp has negligible effect at the op-amp output.
- Current output from an op-amp can be increased by connecting a transistor follower at the output.
- An op-amp can be operated from a single power supply by biasing the noninverting input to one-half the available supply voltage.

EXERCISE PROBLEMS

1. If both inputs to the differential amplifier of Figure 6–A are at 0V, find the collector voltages if R_e is changed to 1.5KΩ.
2. Find the differential amplifier gain for the circuit shown in Figure 6–A.
3. Find the differential amplifier gain for the circuit shown in Figure 6–A if the power supply voltages are changed from ±12V to ±15V.
4. If V_a is 75mV and V_b is 50mV in the circuit of Figure 6–A, find the output voltage V_o.
5. What does "differential gain" mean?
6. What is common mode rejection?
7. Find the common mode voltage gain A_{cm} for the circuit of Figure 6–A.

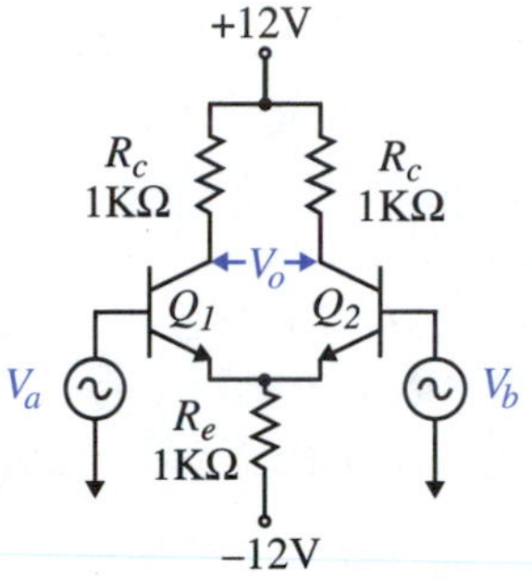

FIGURE 6–A (text figure 6–1)

8. Find the common mode voltage gain A_{cm} for the circuit of Figure 6–A if the supply voltages are changed to ±9V.
9. Determine the DC voltages at the bases, emitters, and collectors of Q_1, Q_2, and Q_3 in Figure 6–B with ±12V power supplies.

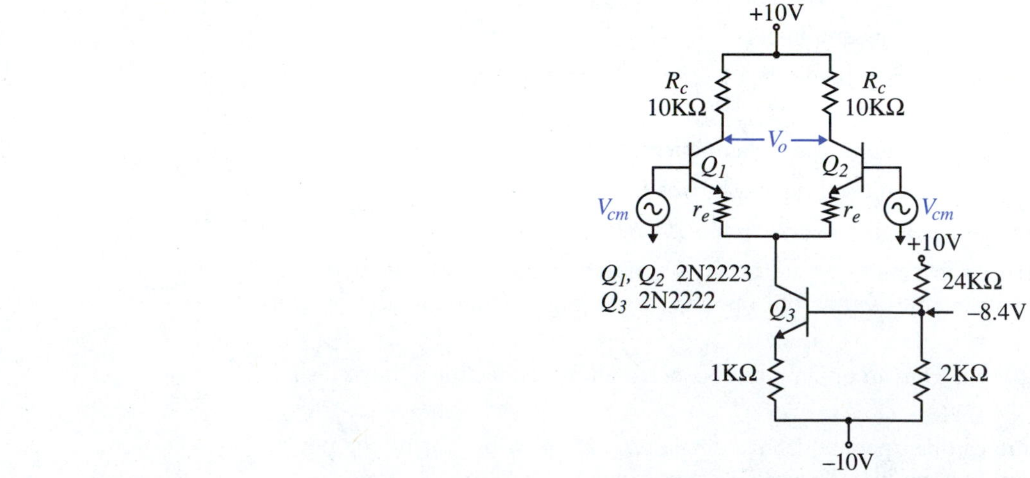

FIGURE 6–B (text figure 6–4)

10. Find the voltages on the collectors of the transistors of Figure 6–A if R_e becomes an open circuit.
11. Determine the common mode voltage gain for the circuit of Figure 6–B if the power supply voltages are changed to ±15V.
12. What is meant by "open loop operation of an op-amp"?
13. If the open loop gain of an op-amp is 200,000, find the output voltage V_o if there are +20μV on the inverting terminal and +50μV on the noninverting terminal.
14. If the open loop gain of an op-amp is 100,000, find the output voltage V_o if there are +30μV on the inverting terminal and −10μV on the noninverting terminal.
15. Give the two assumptions used to derive the closed loop gain of an op-amp.
16. Show the circuit for a noninverting amplifier with a gain of 15 if R_1 is 5KΩ.
17. Show the circuit for a noninverting amplifier with a gain of 60 if R_f is 150KΩ.
18. Find the output voltage from a noninverting amplifier if the input voltage is 20mV peak to peak, R_f is 120KΩ, and R_1 is 1KΩ.
19. Find the value of R_1 required if the value of R_f is 200KΩ and the desired output voltage is +5V with an input of +1V for a 741 amplifier.
20. Determine the gain of the circuit of Figure 6–C if R_1 becomes an open circuit.

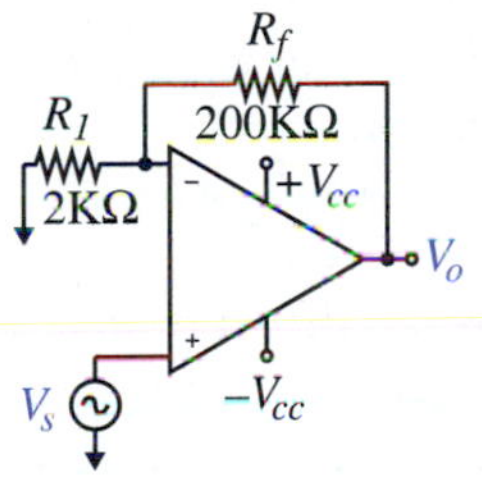

FIGURE 6–C (text figure 6–8)

21. Show the circuit and determine R_1 for a 741 inverting amplifier with a gain of 15 if R_f is 75KΩ.
22. Show the circuit and determine R_f for an inverting amplifier with a gain of 100 if R_1 is 6.8KΩ.
23. Find the DC output voltage from a 741 inverting amplifier if the input voltage is +68mV DC, R_f is 150KΩ, and R_1 is 3KΩ.
24. The desired output from an amplifier is +3V with an input of −0.01V. If R_1 is 2KΩ, draw the circuit and find R_f.
25. If the output voltage from an op-amp with an open loop gain of 100,000 is 2.42V, find the voltage V_{in} between the two input terminals.
26. Describe the output signal with a 100mV sine wave input if the R_f resistor used with a 741 inverting amplifier becomes an open circuit. Power supplies are ±12V.

27. Find the circuit input resistance R_{in} for an inverting amplifier with a gain of -100 and R_f of 200KΩ.
28. A noninverting amplifier has a closed loop gain of 5, an open loop gain of 100,000, and the op-amp internal input resistance r_i is 1MΩ. Find the circuit input resistance R_{in}.
29. A generator with an open circuit output voltage of 100mV and an internal resistance (r_g) of 1MΩ is driving a noninverting op-amp circuit. What is the actual voltage applied to the op-amp noninverting input if A_v is 5, A_{ol} is 100,000, and r_i is 2MΩ?
30. A generator with an open circuit output voltage of 100mV and an internal resistance of 5KΩ is driving an op-amp inverting amplifier circuit with R_1 of 5KΩ. What is the actual voltage applied to the input side of R_1?
31. If the closed loop gain for a non-inverting amplifier is 5, the open loop gain is 100,000, and the op-amp internal output resistance r_o is 75Ω, find the circuit output resistance R_o.
32. If the closed loop gain for an inverting amplifier is 15, the open loop gain is 200,000, and the op-amp internal output resistance r_o is 50Ω, find the circuit output resistance R_o.
33. An amplifier with an open loop gain A_{ol} of 100,000 is connected in a circuit with a closed loop gain A_v of 100. Find the distortion output voltage V_{od} if the distortion voltage V_d generated within the amplifier is 50mV.
34. An amplifier with an open loop gain A_{ol} of 200,000 is connected in a circuit with a closed loop gain A_v of 100. Find the distortion output voltage V_{od} and the percent output distortion if the percent distortion in the amplifier is 4% and the input signal V_s is 150mV.
35. We need to boost the current from a 741 op-amp used as a noninverting amplifier to 100 mA. Show the circuit assuming a required voltage gain of 30 and determine the value of R_f if R_1 is 10KΩ, and also find the minimum β of the boost transistor, if we wish to limit the op-amp current to 2mA.
36. Show the circuit to generate a +5V power supply capable of supplying 1 amp. A 5V zener diode (make nominal zener current 10mA), a 741 op-amp, and transistor and resistors are available. Op-amp is to be operated from a +12V supply. Specify the required β and power rating for the transistor if we want to limit the op-amp current to 10mA.
37. Show the circuit for a noninverting amplifier with a voltage gain of 50 using a single +20V supply. Let R_f be 150KΩ, R be 300KΩ, and for the selection of C_1 and C_2, assume the lowest operating frequency is 100Hz. (See Figure 6–D.)

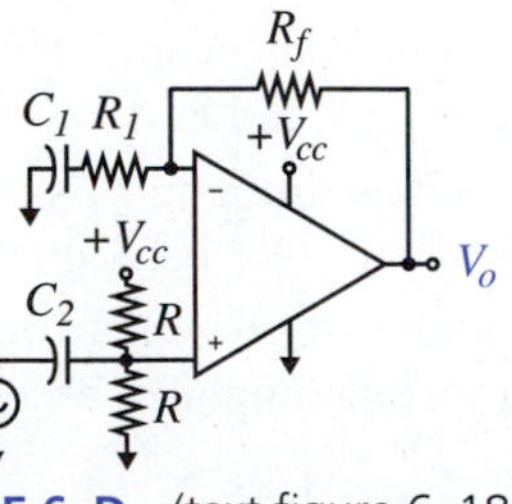

FIGURE 6–D (text figure 6–18a)

38. What would be the DC voltage at the output of the circuit of Figure 6–D if the resistor between the power supply and the noninverting input became an open circuit?

39. State how you would troubleshoot a noninverting amplifier with the input at 0VDC if the op-amp used is a 741 operating from ±12V supplies and you measured +11V at the op-amp output.

40. Identify the defective component for a noninverting circuit with an input voltage of +1V and a voltage gain of 5, if the op-amp used is a 741 operating from ±12V supplies and you measure +1V at the op-amp output.

ABBREVIATIONS

A_{ol} = Op-amp open loop gain
A_v = Op-amp closed loop gain
A_{cm} = Common mode signal gain
R_f = Op-amp feedback resistor
R_1 = Op-amp feedback divider resistor
R_i = Op-amp circuit input resistance
R_{in} = Internal op-amp input resistance
r_o = Internal op-amp output resistance
R_o = Op-amp circuit output resistance
V_{cm} = Common mode input signal
V_d = Internal op-amp distortion voltage
V_{od} = Op-amp circuit output distortion voltage
$V_o = V_{os}$ = Op-amp circuit output signal voltage

ANSWERS TO PROBLEMS

1. $V_c = 8.2V$
2. $A_v = 219$
3. $A_v = 277$
4. $V_o = 5.48V$
5. Amplification of the difference between the two inputs
6. Rejection of a signal that is the same on both inputs
7. $A_{cm} = 0.4989$
8. $A_{cm} = 0.4985$
9. $V_{b3} = -10.15V$; $V_{e3} = -10.75V$; $V_{c3} = V_{e12} = -0.6V$; $V_{b1} = V_{b2} = 0V$; $V_{c1} = V_{c2} = +5.75V$
10. $V_c = +12V$
11. $A_{cm} = 0.005$
12. No feedback from output to input
13. $V_o = +6V$
14. $V_o = -4V$
15. Infinite gain and input resistance
16. $R_f = 70K\Omega$
17. $R_1 = 2.54K\Omega$
18. $V_o = 2.42V_{p\text{-}p}$
19. $R_1 = 50K\Omega$
20. $A_v = +1$

21. $R_1 = 5K\Omega$
22. $R_f = 680K\Omega$
23. $V_o = -3.4V$
24. $R_f = 600K\Omega$
25. $V_{in} = 24.2\mu V$
26. $V_o = -11V$
27. $R_{in} = 2K\Omega$
28. $R_{in} = 20G\Omega$
29. $V_{in} = 99.998mV$
30. $V_s = 50mV$
31. $R_o = 3.75m\Omega$
32. $R_o = 3.75m\Omega$
33. $V_{od} = 49.95\mu V$
34. $V_{od} = 299\mu V$; %dist = 0.002%
35. $R_f = 290K\Omega$; $\beta = 50$
36. $\beta = 100$; $P_t = 7W$
37. $R_1 = 3.06K\Omega$; $C_1 = 5.2\mu F$; $C_2 = 106\mu F$
38. $V_o = 0V$ to $+1VDC$
39. Perform op-amp check, then check R_f.
40. Open R_1 resistor.

CHAPTER 7

Basic Op-Amp Circuits

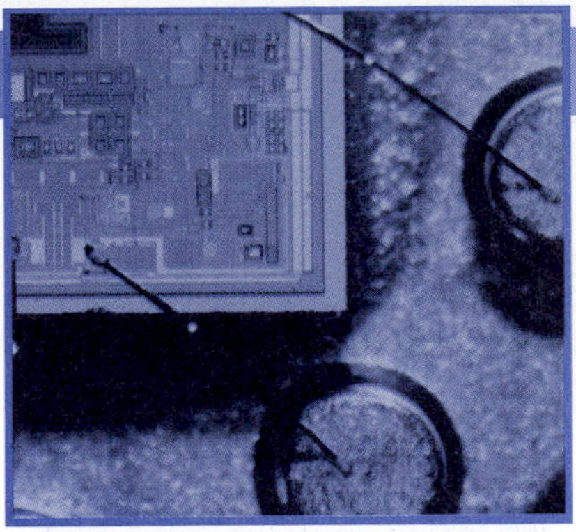

CHAPTER OBJECTIVES

After completing this chapter, you should be able to recognize and describe the operation of the following basic op-amp circuits:

- Summing amplifier
- Averaging amplifier
- Subtracting amplifier
- Integrator
- Differentiator
- Instrumentation amplifier
- Sine wave generator
- Square wave generator
- Pulse generator
- Diode circuits
- Filter circuits
- Current-to-voltage converter

INTRODUCTION

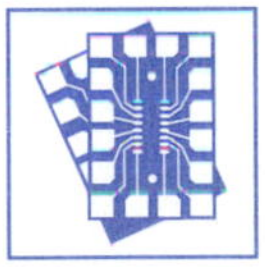

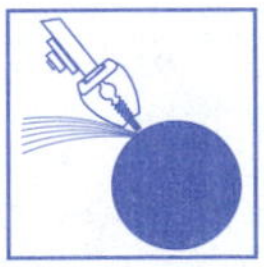

In this chapter, we shall look at some basic circuits that build on the concepts we learned in the previous chapter. These circuits will illustrate the ease of using the op-amp in practical applications.

We shall begin with circuits that demonstrate the math operations (which was the original use for the op-amp) and progress through instrumentation amplifiers, generators, diode and filter circuits.

Our coverage of op-amp circuits will be limited to those circuits that are best achieved with op-amps rather than other linear devices. This gets away from the "op-amp can do anything" approach. In those cases where another linear device can perform the same circuit function as illustrated with the op-amp, the linear device will be identified.

INVERTING ADDING (SUMMING) AMPLIFIER

An adding circuit will add together two or more input signals and provide the result at the output of the circuit.

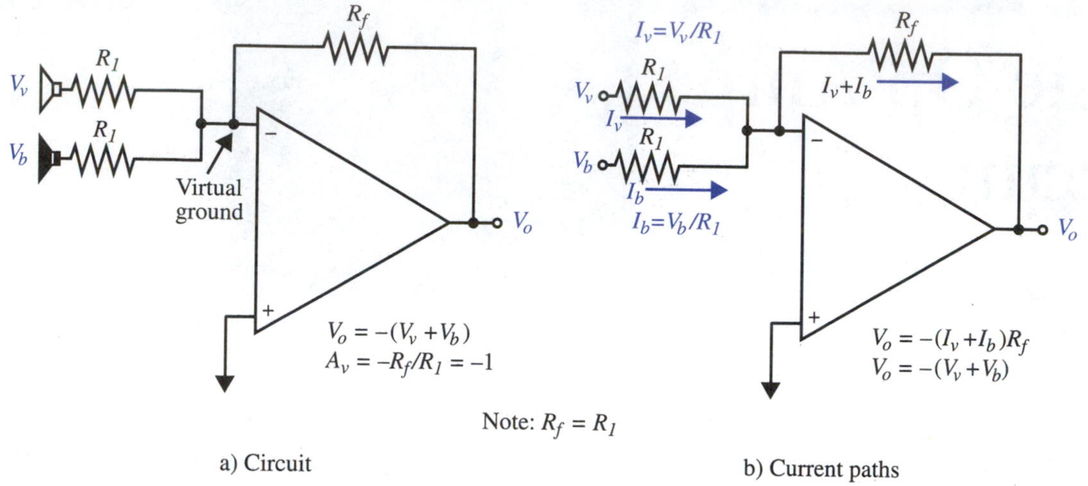

FIGURE 7–1 Inverting adding circuit

Audio mixing is an example of adding signals. We could have microphone outputs from various instruments that need to be combined to get the overall sound of the band; but for our example, let us restrict ourselves to a microphone for the vocalist and one for the band. Figure 7–1 shows how we can mix (add) the two signals together using an inverting amplifier.

The key to the op-amp inverting adding circuit is the virtual ground point at the inverting input. This prevents interaction between the two signals because the current from each microphone is only a function of the value of the associated R_1 resistor and is not affected by the signal level of the other microphone.

If we consider the vocal "mike" first, it generates an output voltage v_v that is amplified by the op-amp with a gain of $-R_f/R_1 = R$ (since $R_f = R_1$ gain is -1). Similarly, the band mike generates a signal v_b that is amplified with a gain of -1. At the output then, we have a signal V_o of

$$v_o = -1(v_v) - 1(v_b)$$

$$v_o = -(v_v + v_b)$$

We have, then, the negative sum of the two input signals at the output. The negative sign means a 180° phase shift that will have no effect on this output. A caution with adding circuits using op-amps is that the resultant output voltage should never be greater than about 2V less than the op-amp positive supply voltage or 2V greater than the negative supply voltage, otherwise the sum will be in error.

Another way to find the output of an inverting/adding circuit is to find the current generated from each mike's output signal by dividing the input voltage by the circuit input resistor (R_1), as shown in Figure 7–1b. Since there is no current flowing into the op-amp terminals, both currents must flow through the feedback resistor. With $R_f = R_1 = R$, the output voltage is then

$$v_o = -R(i_v + i_b)$$

$$v_o = -R\left(\frac{v_v}{R} + \frac{v_b}{R}\right)$$

$$v_o = -(v_v + v_b)$$

which is the negative sum as before!

What if the band is drowning out the vocalist? If we increase the input resistor on the band side, then the gain for the band signal will drop below unity (1), and the vocalist will be heard.

Keeping the input resistors fixed while increasing the feedback resistor will make the circuit gain greater than unity. We now have the amplified sum at the output.

EXAMPLE

The output from the vocal mike is 100mV RMS and from the band mike 200mV RMS. We would like both signals to be mixed (added), but we want each signal to be 200mV at the output of the adder.

SOLUTION

The circuit we can use is shown in Figure 7–2. The vocal input has an R_1 resistor of $R/2$, which gives a gain for this input of 2. The band gain is unity.

EXAMPLE

In a particular piece of equipment, we have a +5V supply and a −5V supply. We are not interested in whether they are both exactly at 5V, but we would like to know whether they are the same value.

SOLUTION

If the voltages are the same, and we feed the +5V into one input of an adding circuit and −5V into the other, the output will be

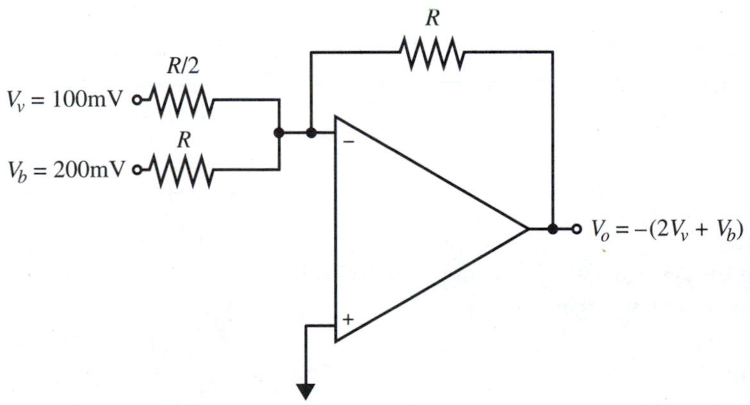

FIGURE 7–2 Signal mixing

$$V_o = -(+5\text{V} - 5\text{V})$$

$$V_o = 0\text{V}$$

If one is +5V and the other −4.9V, the output will be

$$V_o = -(+5\text{V} - 4.9\text{V})$$

$$V_o = -0.1\text{V}$$

So, if we see a DC voltage at the op-amp output, we know the power supply voltages are not the same. The polarity of the DC output tells us which supply voltage is larger—i.e., a positive output indicates that the negative supply is larger, and a negative output that the positive is larger.

AVERAGING CIRCUIT

We can easily convert an inverting/summing circuit into an inverting/averaging circuit by changing the value of the R_f resistor.

If we have a two-input summing circuit, we determined before that the output was with R_f being equal to R_1.

$$V_o = -\left(\frac{V_1}{R} + \frac{V_2}{R}\right)R$$

where V_1 and V_2 are the two input signals. Now if we make the feedback resistor R_f equal to $R/2$, then

$$V_o = -(I_1 + I_2)R$$

$$V_o = -\left(\frac{V_1}{R} + \frac{V_2}{R}\right)\frac{R}{2}$$

$$V_o = -\left(\frac{V_1 + V_2}{2}\right)$$

which is the average of the two inputs. If we have n number of inputs, then the feedback resistor is divided by n to find the average, or

$$V_o = -\frac{(V_1 + V_2 + \ldots\ldots\ldots\ldots\ldots V_n)}{n} \qquad \textbf{(equation 7–1)}$$

EXAMPLE

Find the average of the following three DC voltages: +5V, −3V, and +4V applied to a three-input averaging circuit.

SOLUTION

Using equation 7–1,

$$V_o = -\frac{(V_1 + V_2 + V_3)}{3}$$

$$V_o = -\frac{(+\ 5V - 3V + 4V)}{3}$$

$$V_o = -2V$$

But what if we want the noninverting average of the input signals. Then we need to use the noninverting averaging circuit, shown in Figure 7–3a.

The best way to analyze this circuit is to find the Norton's equivalent of the input circuit by making each input a Norton's generator and then combining them, as shown in Figure 7–3b. The Norton's equivalent current is determined by shorting the right-hand side of each resistor to ground and computing the resultant current. The Norton's equivalent resistor is determined by shorting out the voltage source and looking back into the circuit. The Norton's equivalent circuit is drawn by putting the current generator in parallel with the equivalent resistor.

The voltage drop across the equivalent resistor is the input voltage to the op-amp, and since we have a follower, it is also the op-amp output voltage.

EXAMPLE

Find the noninverting average of +3.00V, +6.00V and −4.00V using three 3.00KΩ input resistors.

SOLUTION

Using the circuit of Figure 7–3a, we first break the input up into three Norton's generators, as shown in Figure 7–4. Notice that the −4.00V input creates a reversed current generator. Combining the currents, we have

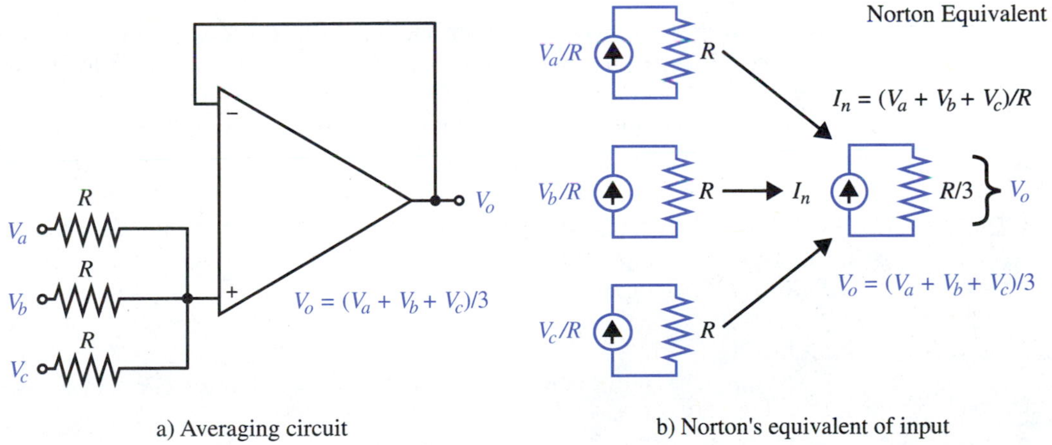

FIGURE 7–3 Noninverting averaging circuit

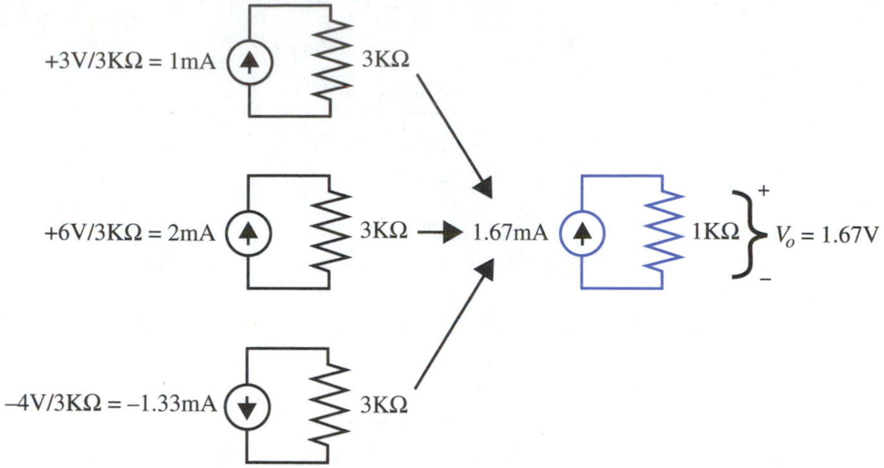

FIGURE 7–4 Norton's input equivalent

$$I_n = 1.00\text{mA} + 2.00\text{mA} - 1.33\text{mA}$$
$$I_n = 1.67\text{mA}$$

Multiplying this current by the equivalent resistance of 1KΩ, we get an input to the op-amp and therefore an average V_a of

$$V_a = 1.67\text{mA} \times 1\text{K}\Omega$$
$$V_a = 1.67\text{V}$$

NONINVERTING ADDING CIRCUIT

We saw earlier that, when summing audio signals, it made no difference if the output was inverted. However, if we are dealing with DC signals, then the polarity of the output usually is important. This brings up the need for a noninverting adding circuit.

We can create a noninverting adding circuit by simply adding gain to the noninverting averaging circuit. This can be done by using an R_f and R_1 resistor. The gain required is equal to the number of inputs—since we can convert from the average of a set of numbers to the sum by multiplying (gain) by the number of items in the set. For example, if we had four inputs, then the required gain is four, and R_f would be three times R_1 (noninverting gain $R_f/R_1 + 1$). (See Figure 7–5.)

EXAMPLE

Show the circuit to find the noninverting sum of four DC voltages.

SOLUTION

The circuit is shown in Figure 7–5 and the gain (4) is selected to increase the average voltage present at the noninverting input to the sum at the output.

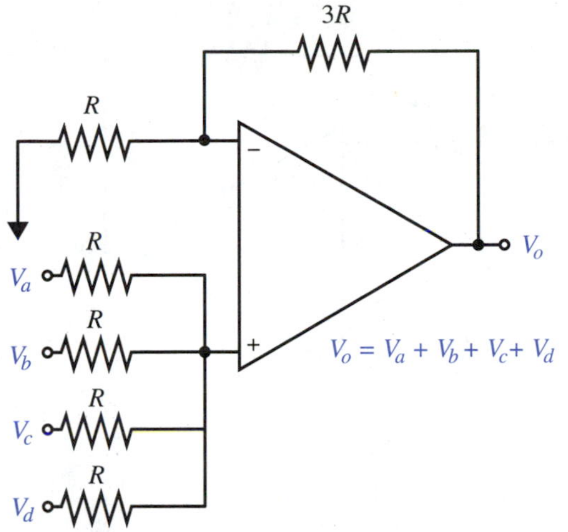

FIGURE 7–5 Noninverting adding circuit

$$\frac{(V_a + V_b + V_c + V_d)}{4} \times 4 = V_a + V_b + V_c + V_d$$

SUBTRACTING CIRCUIT (DIFFERENCE AMPLIFIER)

The subtracting circuit is used to find the difference between two signals. For example, if we have two pulses that are supposed to be identical in shape and time position, and we feed each one into an input to a subtracting circuit, we should get 0V at the circuit output. Any voltage at the output tells us how much the pulses differ.

The configuration of a subtracting circuit is shown in Figure 7–6. The way to analyze this circuit is to consider each input separately, i.e., with superposition. Considering first the input on the inverting side, if the input on the noninverting side V_a is shorted out, the output voltage V_{ob} will be

$$V_{ob} = -V_b \times \left(\frac{R}{R}\right)$$

$$V_{ob} = -V_b$$

Now if we short out V_b and consider the V_a input, we find that the input voltage V_a is divided down to one-half at the op-amp noninverting input by the two resistors on this side. This voltage $V_a/2$ will be amplified by the noninverting gain of the op-amp, or

$$V_{oa} = \left(\frac{R}{R} + 1\right)\frac{V_a}{2}$$

$$V_{oa} = V_a$$

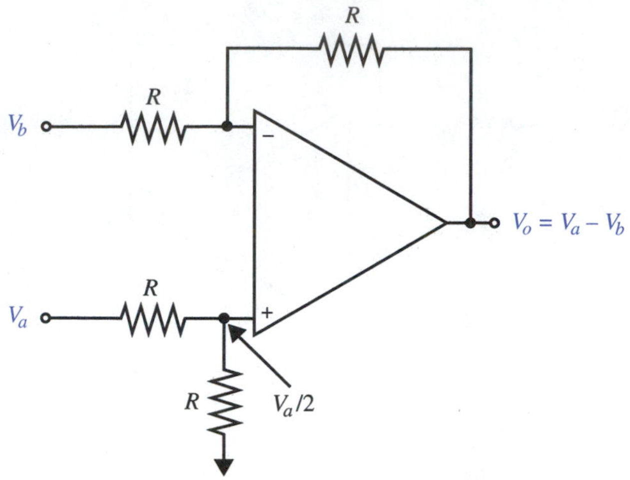

FIGURE 7–6 Subtracting circuit

The combined output is

$$V_o = V_{oa} + V_{ob}$$
$$V_o = V_a - V_b$$

which is the subtraction of the two inputs.

We can have a subtraction circuit with gain by making the op-amp feedback resistor R_f and the resistor from the noninverting circuit to ground equal to one another but larger than the two R_1 resistors.

EXAMPLE

Show a circuit that will amplify the difference between two signals by a factor of 5. Use 10KΩ for the input resistors.

SOLUTION

The circuit is shown in Figure 7–7. Since we require a gain of 5, R_f must be 5 times R_1, or 50KΩ. The V_a input on the noninverting side is divided down by the ratio $50K\Omega/(50K\Omega + 10K\Omega)$, so we have $V_a(5/6)$ at the noninverting input terminal. However, it is amplified by $(50K\Omega/10K\Omega) + 1$, or 6. At the output, then, we have

$$V_o = 5V_a - 5V_b$$
$$V_o = 5(V_a - V_b)$$

INTEGRATOR CIRCUIT

The word *integrate* means *bring together,* and, in calculus, integration is used to sum together many small rectangular areas in order to find the total area of a complex shape. (Appendix A contains a complete discussion of this topic.)

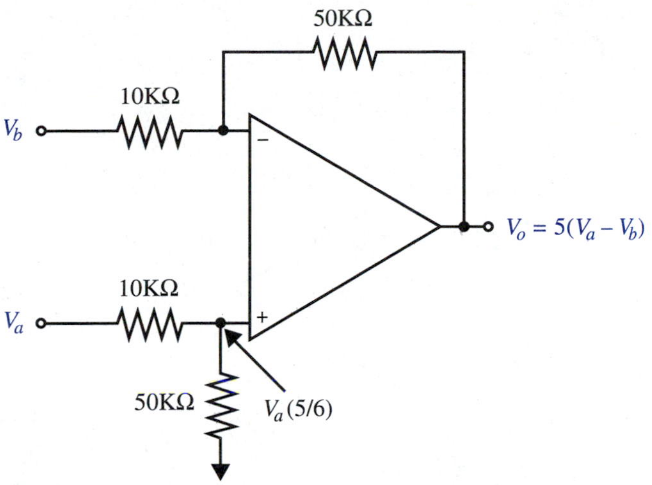

FIGURE 7–7 Subtracting circuit with gain

An integrator circuit provides an output voltage that is equal to the area (product of voltage amplitude and time) of the input waveform. If we feed a DC voltage V_s into the input of an integrator, the output will be a product of the input voltage, the elapsed time (t) since the DC input was connected, and a circuit constant K, or

$$V_o = (V_s)(t)(K) \qquad \textbf{(equation 7–2)}$$

If we assume that V_s and K are constant values, then the output voltage V_o is only a function of time. When V_s is first applied, t is 0 and therefore V_o is 0.

As time increases, the output voltage increases in direct proportion, which means V_o is a linear ramping waveform.

The op-amp integrator circuit is shown in Figure 7–8. Since both inputs are at ground potential, the current charging the capacitor from output to inverting input is

$$i_c = \frac{V_s}{R}$$

The voltage across a capacitor is

$$V_c = \frac{q}{C}$$

but

$$q = i_c \times t$$

$$V_c = i_c \times \frac{t}{C}$$

Substituting for i_c,

$$V_c = V_s \times \frac{t}{RC}$$

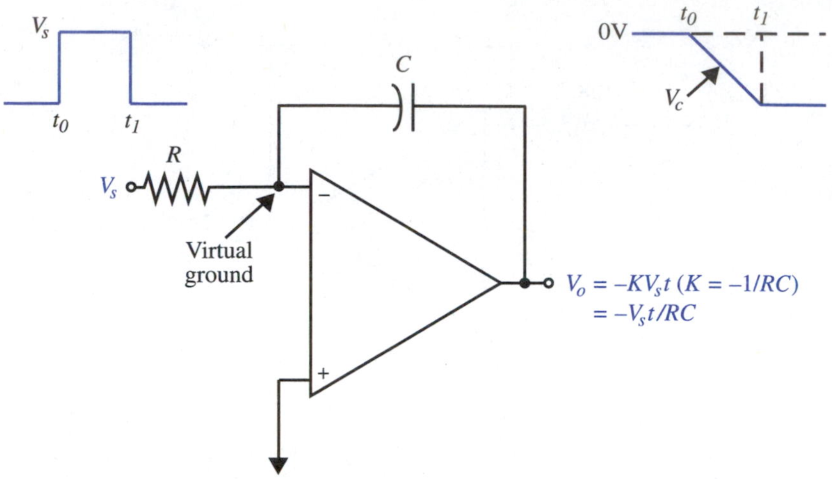

FIGURE 7–8 Op-amp integrator circuit

but

$$V_o = -V_c$$

So

$$V_o = -\frac{V_s t}{RC} \qquad \textbf{(equation 7–3)}$$

From our previous discussion, we can see that $-1/RC$ in equation 7–3 is the circuit constant *K*, and the equation satisfies the requirement for an integrator since the output is equal to the area (product of volts by time) of the input signal.

The integrator circuit is used whenever we want a very linear (straight line) ramp voltage. The reason the waveform is so linear is that the current charging the capacitor is not affected by the voltage developed across the capacitor as it is in a simple RC circuit. The current is only a function of the input voltage V_s and the value of *R* because, again, the right side of *R* is at virtual ground. If this resistor and the input voltage are both constant, then the charging current is also constant. We shall see the integrator used in a digital voltmeter in a later chapter.

In some applications, the input to the integrator V_s is not a constant voltage, and in order to find the predicted output waveform, we have to use integral tables or a calculator with integration capability. For example, using these tables, we find that if a sine wave is fed into an integrator circuit, the output is shifted 90° and becomes a cosine wave.

EXAMPLE

An integrator circuit has *R* of 100KΩ and *C* of 1μF and a DC input voltage of +40mV. Find the output voltage 100msec, and also 10sec, after the +40mV is applied.

SOLUTION

From equation 7–3 with t being 100ms,

$$V_o = -\frac{V_s t}{(R)(C)}$$

$$V_o = -\frac{(+40\text{mV})(100\text{ms})}{(100\text{K}\Omega)(1\mu\text{F})}$$

$$V_o = -40\text{mV}$$

With t being 10sec,

$$V_o = -\frac{V_s t}{(R)(C)}$$

$$V_o = -\frac{(+40\text{mV})(10\text{s})}{(100\text{K}\Omega)(1\mu\text{F})}$$

$$V_o = -4\text{V}$$

As with the summing circuits, we must not let the integrator output ramp to within 2V of the op-amp power supply voltages since it will no longer be linear.

DIFFERENTIATOR CIRCUIT

The math operation of differentiation (reference Appendix A) determines the slope ($\Delta y/\Delta x$) of a function (see Figure 7–9a). This can be useful for finding the maximum or minimum values in engineering problems since the slope is zero at these points. For example, the maximum height a projectile reaches above the earth can be determined from taking the **derivative** of the equation for the projectile trajectory and setting it (the slope) equal to zero. This works because the projectile is horizontal at its maximum height.

An op-amp differentiator circuit gives an output corresponding to the slope ($\Delta V_s/\Delta t$) of the input waveform. Figure 7–9b shows the differentiator circuit.

If the signal applied to the differentiator is a ramping voltage of slope $\Delta V_s/\Delta t$ (the Δ, again, means small change of the value), then the charging current will be

$$i_c = \frac{C\Delta V_s}{\Delta t}$$

Notice that the resistor and capacitor are switched in position from the integrator circuit.

If we suddenly apply a positive DC voltage from a zero-resistance source to the input of the differentiator circuit, the input capacitor will instantaneously charge, since the right side of the capacitor is at virtual ground. This surge of charging current will flow through R and cause a negative-going voltage pulse at the output. In theory, this spike will have infinite amplitude because it takes infinite current to charge a capacitor in zero time. Actually, because of resistance of the input voltage source and the limitations of the op-amp power supply, we will see a narrow pulse that will have an amplitude about 2V less than the power supply voltage.

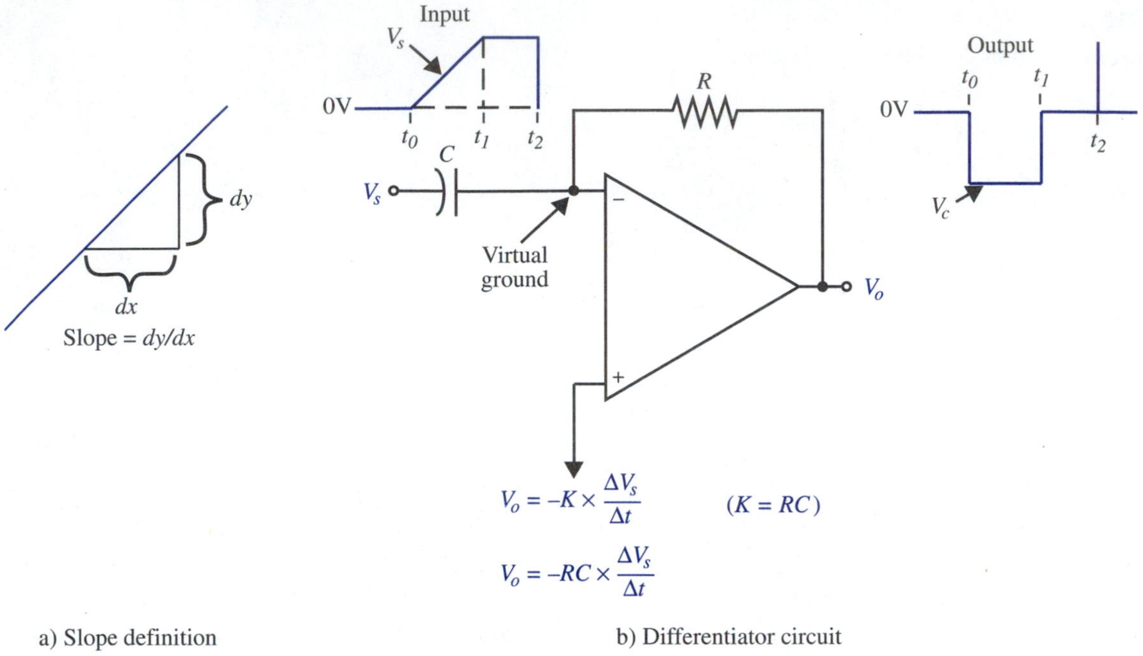

FIGURE 7–9 Op-amp differentiator

This current flowing through R will create an output voltage of

$$V_o = -i_c \times R$$

$$V_o = -RC\frac{\Delta V_s}{\Delta t} \qquad \textbf{(equation 7–4)}$$

This equation shows that an op-amp differentiator circuit provides an output that is proportional to the slope of the input. The proportionality constant K equals $-RC$ in equation 7–4.

Feeding a sine wave into a differentiator gives a cosine wave at the output. We can intuitively see that this is the case, since a cosine wave has its peak when the sine wave is going through zero—at its point of maximum slope.

EXAMPLE

A differentiator circuit has a resistance R of $1\text{M}\Omega$ and capacitance C of $0.1\mu\text{F}$. If the input signal has a slope of -6mV/ms, what is the output voltage V_o?

SOLUTION

Using equation 7–4,

$$V_o = -RC\,(\Delta V/\Delta t)$$

$$V_o = -(1\text{M}\Omega)(0.1\mu\text{F})\left(-\frac{6\text{mV}}{\text{ms}}\right)$$

$$V_o = +0.6\text{V}$$

So with a negative slope at the input, the output is a fixed positive DC level. If the input has no slope, the output signal is zero!

INSTRUMENTATION AMPLIFIER

As the name suggests, an instrumentation amplifier is used in measurement circuits. Specifically, it is used when we need to measure low-level signals in a noisy environment.

We can justify the need for an instrumentation amplifier by looking at the way a conventional amplifier responds to noise pickup on the input leads. Figure 7–10 shows a noninverting amplifier with noise-source pickup.

Magnetic current-noise pickup occurs when the noise field cuts an input wire and induces a voltage V_m. Electrostatic noise voltage V_e is generated in the wire because of a "capacitive voltage divider" action between the capacitance between the noise voltage source (C_s) and the wire and the capacitance from the wire to ground (C_w).

For this circuit, the output voltage will be

$$V_o = (V_s + V_m + V_e)\left(\frac{R_f}{R_1} + 1\right)$$

If we get high levels of noise pickup, it will be difficult to detect the wanted signal V_s at the output.

With an instrumentation amplifier, one side of the signal source is not connected to ground but instead is connected in a balanced configuration between two input leads to the amplifier, as shown in Figure 7–11. (Notice that this is basically the subtraction circuit of Figure 7–6, but all resistors are not of equal value because we want gain from this circuit.)

To show how effective this circuit is for noise reduction, we need to derive the gain equations for the noise signals and also for the wanted signal. In these derivations, we shall assume that the two R_1's are equal and that the two R_f's are also equal.

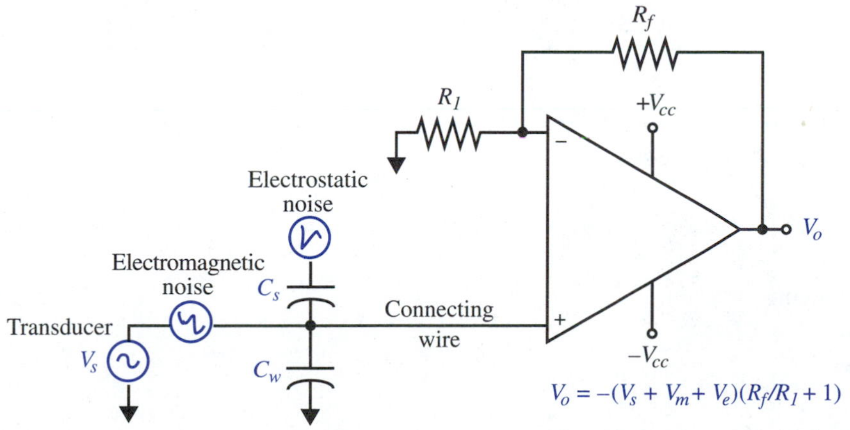

FIGURE 7–10 Noise coupling into a noninverting amplifier

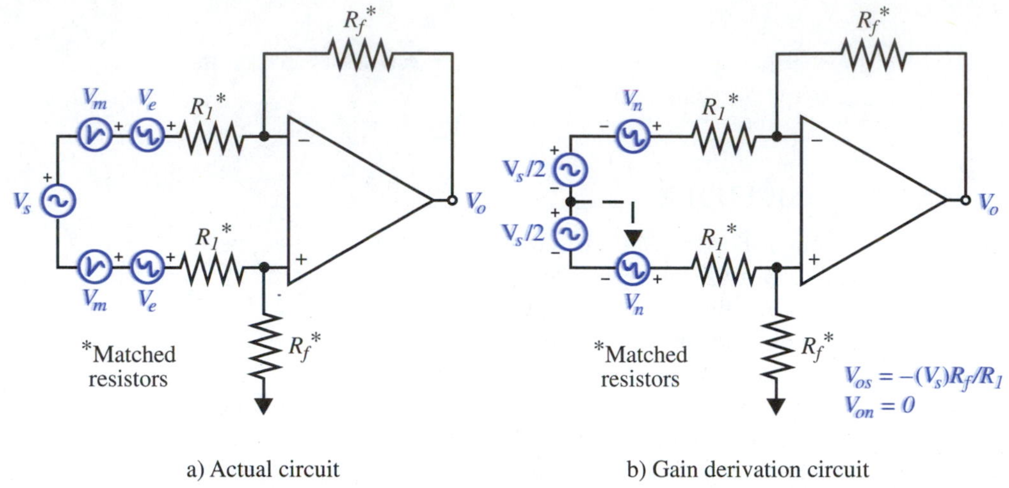

FIGURE 7–11 Basic instrumentation amplifier

Considering the noise signals first, we shall presume that the input wires are close (twisted) together so the same (common mode) noise signals are induced in each lead. Also we shall assume that the common mode gain (A_{cm}) of the op-amp is zero, and that the combined noise voltage is the square root of the sum of the squares, or

$$V_n = \sqrt{V_m^2 + V_e^2}$$

Using superposition, the noise signal output from the inverting side of the circuit is

$$V_{on-} = -(V_n)\frac{R_f}{R_1}$$

For the noninverting side, the combined noise signal is first divided down by the R_1 and R_f divider, so we have at the noninverting input

$$V_{n+} = (V_n)\left(\frac{R_f}{R_f + R_1}\right)$$

This signal is amplified by the noninverting amplifier gain $(R_f + R_1)/R_1$, and at the output of the op-amp we have

$$V_{on+} = (V_n)\left(\frac{R_f}{R_f + R_1}\right)\left(\frac{R_f + R_1}{R_1}\right)$$

Simplifying

$$V_{on+} = +(V_n)\frac{R_f}{R_1}$$

Combining the two outputs

$$V_{on} = (V_{on+}) + (V_{on-})$$

$$V_{on} = (V_n)\frac{R_f}{R_1} - (V_n)\frac{R_f}{R_1}$$

$$V_{on} = 0$$

So we see complete rejection of the noise signals!

Looking now at the wanted signal gain, we break the generator into two parts with output voltages of $V_s/2$, as shown in Figure 7–11b. Since we are balanced with respect to ground, the center point of the new generators is at ground potential. We can now use superposition to find the gain for each side.

For the inverting side with V_{s-} of $V_s/2$, the output signal is

$$V_{os-} = -(V_s) - \frac{R_f}{R_1}$$

$$V_{os-} = -(V_s)\frac{R_f}{2R_1}$$

The half of the generator signal on the noninverting side is negative with respect to ground, and it appears at the noninverting input to the op-amp divided down by the R_1 and R_f divider.

$$V_{s+} = -\left(\frac{R_f}{R_f + R_1}\right)\left(\frac{V_s}{2}\right)$$

At the output, this signal is multiplied by the noninverting gain.

$$V_{os+} = -\left(\frac{R_f + R_1}{R_1}\right)\left(\frac{R_f}{R_f + R_1}\right)\left(\frac{V_s}{2}\right)$$

Simplifying

$$V_{os+} = -\frac{V_s R_f}{2R_1}$$

Combining the two outputs

$$V_{os} = V_{os+} + V_{os-}$$

$$V_{os} = -\left(\frac{V_s R_f}{2R_1}\right) - \left(\frac{V_s R_f}{2R_1}\right)$$

$$V_{os} = -\frac{V_s R_f}{R_1} \qquad \textbf{(equation 7–5)}$$

We see that the gain of the differential amplifier for the wanted signal is simply equal to the gain of the conventional inverting amplifier!

It was assumed, when determining the gain of the noise signal, that the common mode gain was zero. From our earlier study of the differential amplifier, we know that this can't be true. Op-amp manufacturers specify the actual common gain for the op-amp, using the term "common mode rejection ratios" (CMRR). This can be expressed as a ratio or as common mode rejection (CMR) in decibels (decibels or db = 20 log CMRR). Ratios can range from about 1,000 (60db) to 1,000,000 (120db) for different types of op-amps. The higher ratios are for op-amps specifically designed for instrumentation amplifiers.

EXAMPLE

If the specified CMR for an op-amp is 80db, find the CMRR.

SOLUTION

$$\text{CMRR} = \text{Inv log}\,\frac{80}{20}$$

$$\text{CMRR} = 10{,}000$$

To compute the actual noise out of an instrumentation amplifier, we multiply the noise voltage by the inverting gain and divide by the common mode rejection ratio (CMRR).

$$V_{on} = -\left(\frac{V_n}{\text{CMRR}}\right)\left(\frac{R_f}{R_1}\right) \qquad \textbf{(equation 7–6)}$$

EXAMPLE

An instrumentation amplifier has R_1 values of 10KΩ and R_f values of 220KΩ, and the common mode rejection for the op-amp is 80db (10,000). If the signal level is 10mV and the total noise is 200mV, find the level of signal and noise at the output.

SOLUTION

We can determine the output signal V_{os} from equation 7–5.

$$V_{os} = -(10\text{mV})\left(\frac{220\text{K}\Omega}{10\text{K}\Omega}\right)$$

$$V_{os} = -220\text{mV}$$

For the output noise we use equation 7–6.

$$V_{on} = -(220\text{mV})\left(\frac{220\text{K}\Omega}{10\text{K}\Omega}\right)\left(\frac{1}{10{,}000}\right)$$

$$V_{on} = -0.44\text{mV}$$

Even though the noise at the input was twenty times greater than the signal, at the output the signal is almost five hundred times greater than the noise.

If our signal source is a high-resistance transducer, the instrumentation amplifier configuration of Figure 7–11 could cause signal loading. To get around this problem, nonin-

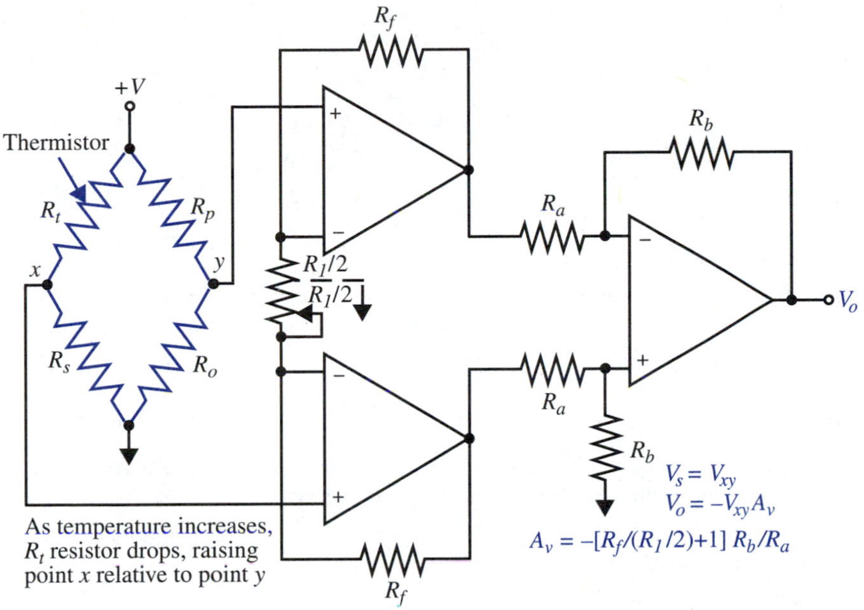

FIGURE 7–12 High-resistance instrumentation amplifier

verting amplifiers can be added to each input. Since the input resistance for this type amplifier is high (approximately $r_i \times A_{ol}/A_v$; with r_i the internal resistance of the op-amp, typically in the megohms), the loading for the input signal is negligible.

The circuit is shown in Figure 7–12. The signal source is a high-resistance bridge circuit that provides a balanced input signal. It is available with all the amplifiers contained within a single chip (Ref National Semiconductor LH0036).

Notice that the two input amplifiers share a common R_1 resistor. Because of the balanced circuit configuration, the midpoint of this resistor is at ground potential—so the effective resistance for gain computation for each amplifier is $R_1/2$. By making this common resistor a pot, we can achieve variable gain, and by causing the same gain change for both amplifiers, we can preserve the balanced configuration.

The overall gain equation for the circuit of Figure 7–12 is

$$A_v = \left(\frac{2R_f}{R_f} + 1\right)\left(\frac{-R_b}{R_a}\right) \qquad \textbf{(equation 7–7)}$$

which is the product of the gain for the individual stages!

EXAMPLE

A high-resistance instrumentation amplifier has the following resistor values: R_1 is 2KΩ, R_f is 20KΩ, R_a is 5KΩ, and R is 200KΩ. Find the overall circuit gain.

SOLUTION

Using equation 7–7, the gain is

$$A_v = \left(\frac{2R_f}{R_1} + 1\right)\left(\frac{-R_b}{R_a}\right)$$

$$A_v = \left(\frac{(2)(20\text{K}\Omega)}{2\text{K}\Omega} + 1\right)\left(\frac{-200\text{K}\Omega}{5\text{K}\Omega}\right)$$

$$A_v = -840$$

To find the common mode rejection of the high-resistance instrumentation amplifier, we are concerned with the CMRR of the two input amplifiers—since the common signals will be small going into the last amplifier.

EXAMPLE

Find the signal output and noise output from a high-resistance instrumentation amplifier with the following characteristics: R_1 is 1 KΩ, R_f is 20 KΩ, R_a is 10 KΩ, R_b is 100 KΩ, and CMR for the input amplifiers is 94db. The input signal V_s is 5mV, and the noise V_n is 100mV.

SOLUTION

The output signal is

$$V_{os} = V_s \times A_v$$

$$V_{os} = (5\text{mV})(A_v)$$

From equation 7–7,

$$A_v = \left(\frac{2R_f}{R_1} + 1\right)\left(\frac{-R_b}{R_a}\right)$$

$$A_v = \left(\frac{2 \times 20\text{K}\Omega}{1\text{K}\Omega} + 1\right)\left(\frac{-100\text{K}\Omega}{10\text{K}\Omega}\right)$$

$$A_v = -410$$

Then

$$V_{os} = -5\text{mV} \times 410$$

$$V_{os} = -2.05\text{V}$$

For the noise (refer to equation 7–6),

$$V_{on} = \frac{100\text{mV}}{\text{CMRR}} \times (A_v)$$

$$\text{CMR} = 94\text{db} = 20 \log \text{CMRR}$$

$$\text{CMRR} = \text{Inv} \log\left(\frac{94}{20}\right)$$

$$\text{CMRR} = 50{,}119$$

So

$$V_{on} = -(100\text{mV})\left(\frac{410}{50{,}119}\right)$$

$$V_{on} = -0.82\text{mV}$$

SINE WAVE GENERATOR

A sine wave generator is a source of sine waves. In other words, it is a circuit whose only input is the DC power supply voltage—but its output is a sine wave.

All generators (oscillators) must have positive feedback from the output of the circuit to the input, which means in our discussion, from the output of the op-amp to the noninverting input. There must be enough signal fed back to the noninverting input to maintain the oscillation—gain around the loop must be unity. Finally the signal fed back must not be phase shifted from the output.

A low-distortion sine wave generator for frequencies in the 1Hz to 1MHz range is the Wien (ween) bridge oscillator. A basic version of this circuit is shown in Figure 7–13a. The circuit consisting of R_a, C_a, R_b, and C_b forms a frequency-sensitive voltage divider, and at one frequency no phase shift occurs through this network. If we let R_a equal R_b and C_a equal C_b, the frequency at which no phase shift occurs—which is therefore the frequency of oscillation—is

$$f_o = \frac{1}{2\pi RC} \qquad \textbf{(equation 7–8)}$$

This equation is derived by solving for the voltage at the noninverting input to the op-amp in terms of V_o, the voltage divider formed by the impedances of the resistor/capacitor

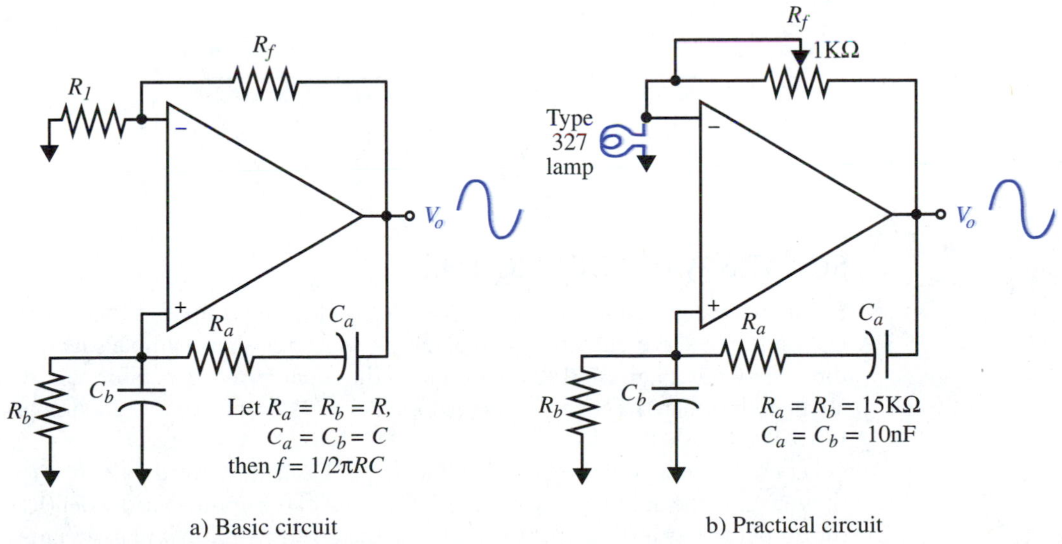

FIGURE 7–13 Wien bridge oscillator

combinations Z_a and Z_b and at the frequency where no phase shift occurs (reactive terms equal 0).

It is necessary to have a certain amount of negative feedback with a sine wave oscillator to keep the gain around the positive feedback path equal to unity, otherwise the output will be a clipped waveform (if we have too much negative feedback, there will be no oscillation).

The circuits for the negative and positive feedback form a bridge configuration, which is the basis for the name of this oscillator. At the frequency of oscillation, if we use equal values for the capacitors and the resistors, the voltage fed back to the noninverting input is one-third of the output voltage. This means that the voltage fed back to the inverting input should also be one-third, since there needs to be very little difference in voltage at the inputs to the op-amp in order to get an output because of the high internal gain (A_{ol}). Another way to look at this is that since the output is divided down by three at the noninverting input, we need a gain of three to get the right output—which requires that R_f be equal to $2R_1$.

The actual amount of negative feedback voltage can vary slightly because of component drift. So, in practical Wien bridge oscillator circuits, active compensation is used as in Figure 7–13b. Compensation is accomplished by the variation of the resistance of the lamp. If the output voltage is too high, the lamp filament gets warmer, and its resistance increases, which increases the amount of voltage feedback into the inverting terminal—which reduces the output voltage. If the output voltage drops, the bulb resistance lowers, reducing the negative feedback and increasing the output voltage. The bulb voltage is not sufficient for illumination.

EXAMPLE

Calculate the frequency of oscillation (f_o) for the Wien bridge circuit of Figure 7–13b.

SOLUTION

Using equation 7–8 and the values of the frequency determining resistors and capacitors from Figure 7–13b,

$$f_o = \frac{1}{2\pi RC}$$

$$f_o = \frac{1}{2\pi \times 15\text{K}\Omega \times 1 \times 10\text{nF}}$$

$$f_o = 1.06\text{KHz}$$

SQUARE WAVE GENERATOR

Square wave signals are useful, for example, as clocking signals in digital circuits and for test signals in audio circuits. Using op-amps, we can design a simple square wave generator as shown in Figure 7–14. The output of the square wave generator will swing between about 2V less than the supply voltages. If we use a 741 op-amp with 12V supplies, the swing is +10 volts to −10 volts.

Let us assume a point in time when the output has just changed to +10V. This will make the voltage at the noninverting input +1V because of the ten-to-one divider (R and $9R$) from the output to this terminal. The capacitor at the inverting input will now start to charge to-

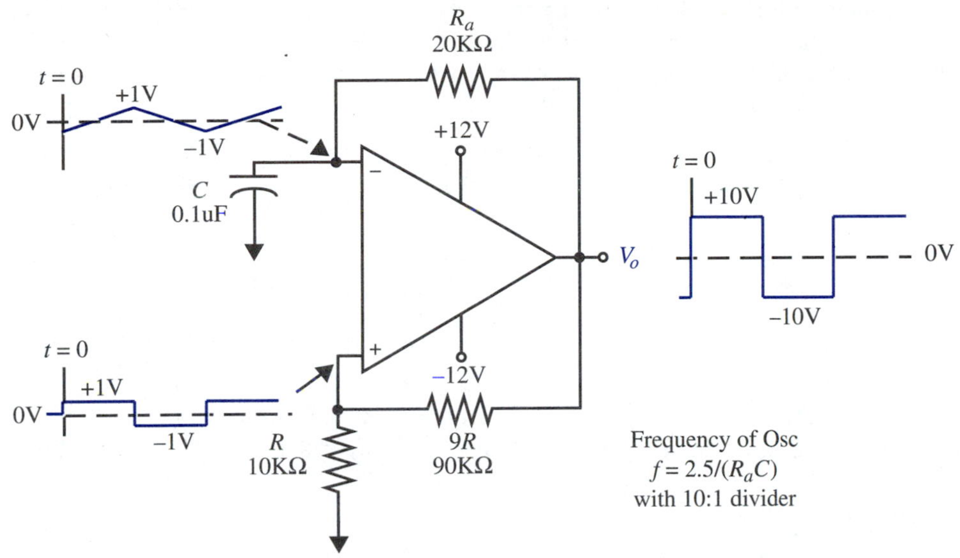

FIGURE 7–14 Square wave generator

wards the +10V output because of current through R_a. However, when the voltage buildup on this capacitor just exceeds +1V, the inverting input to the op-amp is now more positive than the noninverting side, and the output swings to −10V. Now the voltage on the noninverting input changes to −1V, and the capacitor now charges towards −10V. When the voltage on the capacitor goes a little more negative than −1V, the output swings to +10V, and the cycle repeats.

The waveforms at all op-amp terminals are shown in Figure 7–14. Let us derive the equation for the charging of the capacitor since this enables us to determine the frequency of oscillation. The voltage on the capacitor during charging is

$$V_c = 11\text{V}[1 - \varepsilon^{-(t/R_aC)}]$$

This equation assumes that initially there is −1V on C, and the capacitor is charging towards +10V (11V change). We need to determine the time for C to reach +1V from its initial value of −1V (a 2V change) so V_c is equal to 2V.

$$2\text{V} = 11\text{V}[1 - \varepsilon^{-(t/R_aC)}]$$

Solving for $\varepsilon^{-(t/R_aC)}$,

$$\varepsilon^{-(t/R_aC)} = \frac{9}{11}$$

To solve for t, we have to take the natural log of both sides.

$$-t/R_aC = \ln\left(\frac{9}{11}\right)$$

$$t = 0.2R_aC \qquad \textbf{(equation 7–9)}$$

This represents the time for generating one-half of the output square wave. The other half is generated when the capacitor discharges from +1V to −1V, and this of course takes the same amount of time. The period of the square wave, then, takes a time 2*t*, and so the square wave frequency (f_s) is

$$f_s = \frac{1}{2t}$$

$$f_s = \frac{1}{0.4R_aC}$$

$$f_s = \frac{2.5}{R_aC} \qquad \textbf{(equation 7–10)}$$

Remember, this equation only applies if a ten-to-one divider is used from the output to the noninverting input. Some square wave generators use a different divide-down ratio.

EXAMPLE

A square wave generator has the component values given in Figure 7–14, i.e., C is 0.1μF, R_a is 20KΩ, R_1 is 90KΩ, and R_2 is 10KΩ. If the power supply voltages are ±15V, find the frequency of the square wave and its amplitude.

SOLUTION

Using equation 7–10 to find the frequency

$$f_s = \frac{2.5}{R_aC}$$

$$f_s = \frac{2.5}{20\text{K}\Omega \times 0.1\ \mu\text{F}}$$

$$f_s = 1.25\text{KHz}$$

The output swings to about 2V less than the supply voltages or ±13V.

A square wave generator with excellent rise and fall times can be built using a 555 timer chip (covered in Chapter 9).

PULSE GENERATOR

Pulses are used to trigger an event at a fixed repetition rate. We can convert a square wave generator into a pulse generator by simply adding a resistor diode combination (R_b and D_2) in parallel with R_a and a diode (D_1) in series, as shown in Figure 7–15.

If, for example, R_b is one-tenth the value of R_a, then the charging time for C via R_b is also one-tenth. This means that the time the output signal is positive t_p is one-tenth of the time that it is negative t_d; so we have a positive pulse output.

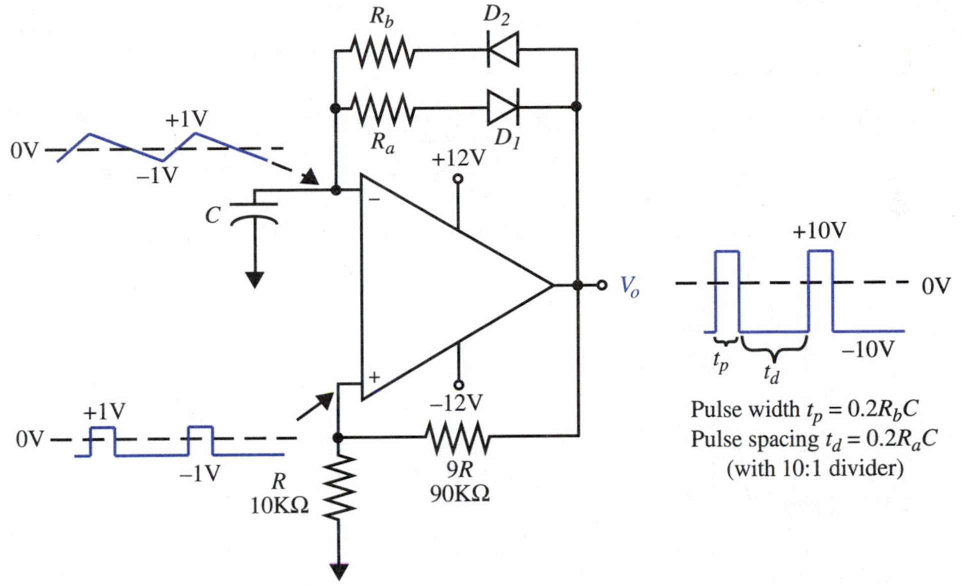

FIGURE 7–15 Positive pulse generator

Reversing the direction of D_1 and D_2 creates a negative pulse output. D_1 is added to the circuit to prevent R_a and R_b from being in parallel when D_2 is conducting. We can determine the charging time for C, and therefore the output pulse width (t_p) for the positive pulse generator, by using equation 7–9.

$$t_p = 0.2R_bC \qquad \textbf{(equation 7–11)}$$

Similarly, we can find the time between pulses (t_d) using the same equation.

$$t_d = 0.2R_aC \qquad \textbf{(equation 7–12)}$$

The pulse repetition frequency (f_p) is one (1) over the total time period ($t_p + t_d$), or

$$f_p = 1/(t_p + t_d)$$

$$f_p = \frac{1}{0.2C(R_a + R_b)} \qquad \textbf{(equation 7–13)}$$

EXAMPLE

A positive pulse generator is required with a 1ms pulse and a pulse repetition frequency (PRF) of 100Hz. If C is 1μF, determine the values of R_a and R_b.

SOLUTION

R_b can be determined by rearranging equation 7–11.

$$R_b = \frac{t_p}{0.2C}$$

$$R_b = \frac{1\text{ms}}{0.2 \times 1\ \mu\text{F}}$$

$$R_b = 5\text{K}\Omega$$

R_a can be determined by first finding t_d (remembering that the period is *1/f*) where

$$t_d = \frac{1}{f} - t_p$$

$$t_d = \frac{1}{100\text{Hz}} - 1\text{ms}$$

$$t_d = 9\text{ms}$$

Then using equation 7–12,

$$R_a = \frac{t_d}{0.2C}$$

$$R_a = \frac{9\text{ms}}{0.2 \times 1\ \mu\text{F}}$$

$$R_a = 45\text{K}\Omega$$

IDEAL DIODE CIRCUIT

We saw uses for diodes in the pulse generator circuit. Now we shall look at op-amp circuits where the key element is a diode. First we shall study a circuit shown in Figure 7–16 that overcomes the forward voltage drop across a diode and is therefore called an **ideal diode circuit.**

Since the closed loop op-amp always works to make the voltage at the two inputs the same, positive inputs will cause the op-amp output to raise to +0.6V plus the signal level to overcome the diode drop. At the circuit output, we will see only the positive portion of the input signal.

When the input signal goes negative, the diode is reverse biased, so we have no output. This circuit is useful for rectifying low-level AC signals, e.g., voltmeter circuits.

EXAMPLE

Determine the output waveform if the input to the ideal diode circuit of Figure 7–16a is a 10mV peak-to-peak sine wave.

SOLUTION

At the output, we shall see only the positive portion of the sine wave with a peak amplitude of 5mV.

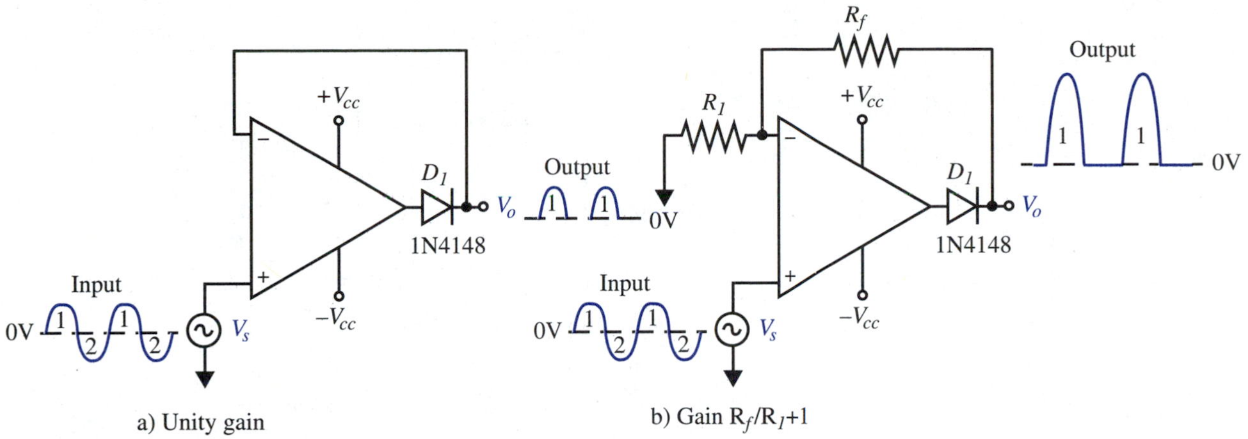

FIGURE 7–16 Ideal diode circuit

EXAMPLE

Determine the output waveform if the input to the ideal diode circuit of Figure 7–16b is a 10mV peak-to-peak sine wave and R_f is 27KΩ and R_1 is 1KΩ.

SOLUTION

At the output, we shall see the amplified positive portion of the input signal.

$$V_o = V_s(\text{peak})\left(\frac{R_f}{R_1} + 1\right)$$

$$V_o = (5\text{mV})\left(\frac{27\text{K}\Omega}{1\text{K}\Omega} + 1\right)$$

$$V_o = 140\text{mV}$$

INVERTING IDEAL DIODE CIRCUIT

Now let us look at an inverting amplifier version of the ideal diode circuit, as shown in Figure 7–17.

If the input signal is negative, diode D_2 is reverse biased, and the circuit acts like a conventional inverting amplifier with a gain $A_v = -R_f/R_1$ and D_1 is forward biased.

With a positive input signal, D_2 forward biases, and the op-amp output goes to negative to try to maintain 0V at the inverting input. Since D_1 is now reverse biased, no signal appears at the output.

EXAMPLE

Determine the output signal for an inverting ideal diode circuit if the input is a 10mV peak-to-peak sine wave, R_f is 33KΩ, and R_1 is 1KΩ.

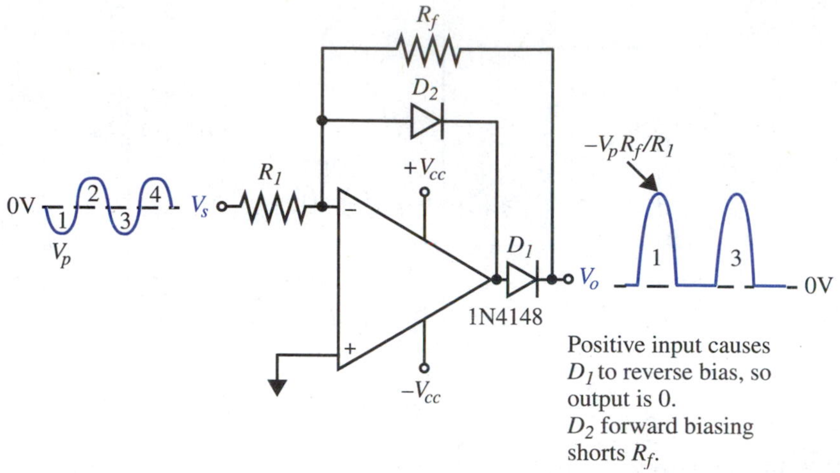

FIGURE 7–17 Inverting ideal diode circuit

SOLUTION

An output will only appear for the input negative 5mV peak, and this gets amplified by the circuit gain.

$$V_o = -(-5\text{mV})\left(\frac{33\text{K}\Omega}{1\text{K}\Omega}\right)$$

$$V_o = +165\text{mV}$$

PEAK DETECTOR CIRCUIT

If we add a capacitor to the output of a noninverting ideal diode circuit, it will charge to the peak output voltage from the circuit. Since the diode only allows current flow in one direction, the capacitor will hold the charge. A circuit of this type could be used to monitor the AC line for over-voltage conditions since the capacitor would store the maximum voltage encountered.

Figure 7–18 shows the basic peak detector circuit with additions to improve operation. A follower is added to monitor the capacitor voltage without discharging the capacitor, and a MOSFET (enhancement) switch is connected across the capacitor to allow for discharging the capacitor as required for circuit reset.

To maintain the charge on the capacitor for extended periods of time, the capacitor should be a low-leakage type, and the op-amp follower should be a high-input resistance, low bias current device (FET input).

ABSOLUTE-VALUE CIRCUIT

We can combine the inverting ideal diode circuit with a summing circuit and create a full-wave rectifying circuit or an absolute-value circuit (output is always one polarity—independent of the polarity of the input), as shown in Figure 7–19.

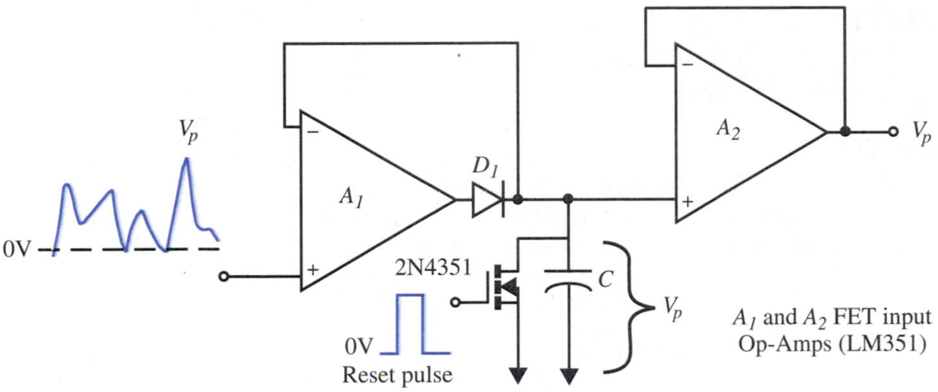

FIGURE 7–18 Peak detector circuit

The way to analyze this circuit is to separate the two inputs to the A_2 summing circuit. If we assume the circuit input is a sine wave with a peak V_p, then this signal is applied to the upper circuit to the summer (point **A**) and will provide an output from A_2 (point **C**) inverted and with unity gain.

The lower circuit will provide just the negative portion of the sine wave, inverted from the input, and amplified with a gain of unity (1) to the left of *R*/2 (point **B**). This signal will be amplified by minus two (−2) in the summing circuit.

At the output, we have the original sine wave from point **A** inverted (top waveform), and also twice the positive portion of the input from point **B** (middle waveform). Summing these signals together at the output, we get a positive full-wave rectified signal with a maximum value of V_p (bottom waveform). If we want a negative output, we simply reverse D_1 and D_2.

Making *R*/2 a potentiometer enables us to compensate for component tolerances and create equal amplitude alternations at the output.

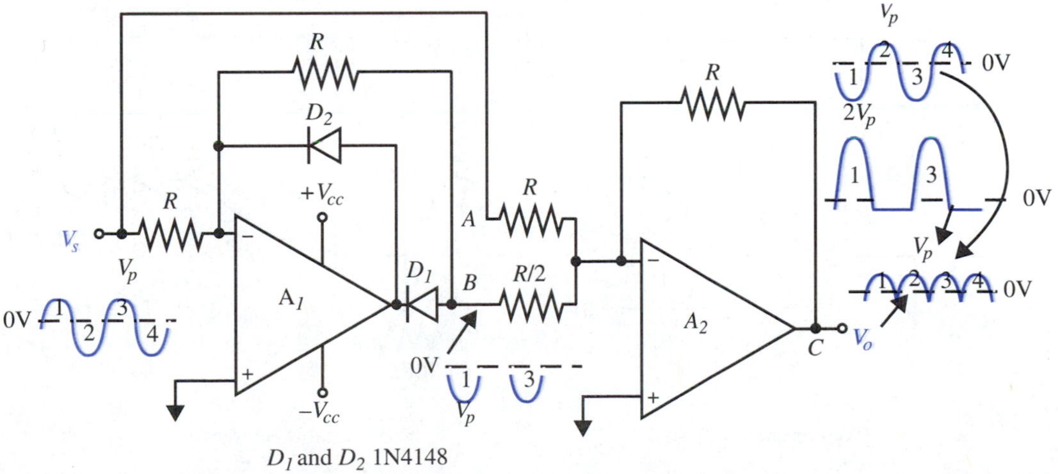

FIGURE 7–19 Absolute-value circuit

EXAMPLE

If the input signal to the absolute-value circuit of Figure 7–19 is a 100mV peak-to-peak sine wave, describe the signal at points **A, B, and C.**

SOLUTION

The signal at point **A** is the input. At point **B**, we have a negative half sine wave with a peak amplitude of 50mV. Finally at the output (point **C**), we have a full-wave positive rectified signal with peak amplitude 50mV.

DC RESTORER CIRCUIT

There are times when we need to convert a signal that is swinging positive and negative around ground (bipolar) to one that is all above ground or all below ground (unipolar). For example, we could have a square wave signal that is bipolar, but we would like to use it as a clock signal for digital TTL logic. This requires that the signal be only positive (0V to +5V). A circuit that can convert a bipolar signal to unipolar is a DC restorer. Figure 7–20 shows an op-amp version of this circuit. (DC restorers are also used in TV receivers to restore the average brightness level to a signal.)

Let us assume we have a sine wave of peak V_p that we wish to convert to a positive unipolar signal. If we assume the waveform originally goes negative, this portion couples through the capacitor *C*, and the inverting input moves slightly negative, causing the op-amp output to go positive. The diode D_1 now conducts, and the inverting input is forced back to the virtual ground potential (remember, the op-amp is always working in closed loop to maintain 0V between the two inputs).

With 0V on the inverting input, *C* quickly charges to the negative peak voltage ($-V_p$). This voltage developed across *C* stays across the capacitor and acts like a battery in series with the generator. So, the actual voltage applied to the inverting input, which appears at

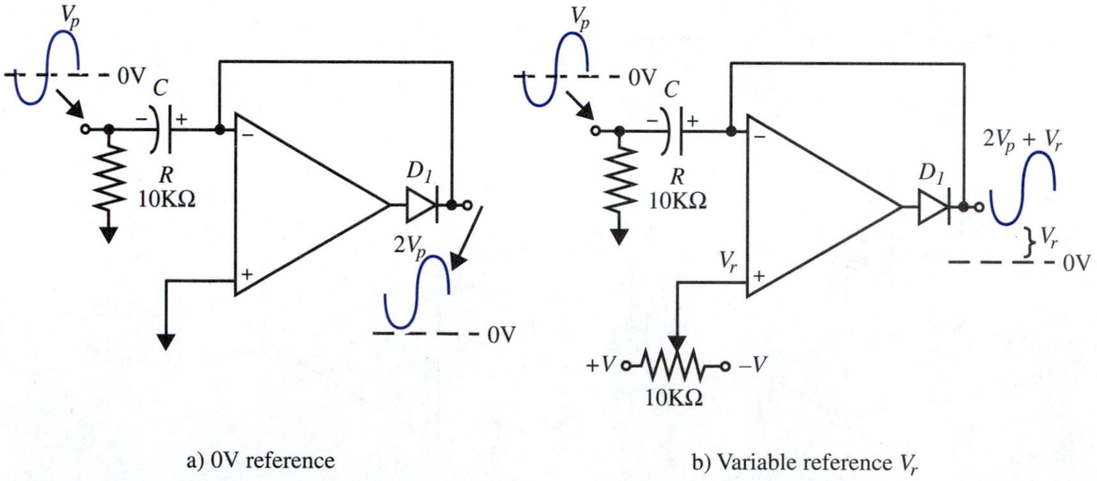

FIGURE 7–20 DC restorer circuit

the output, is the algebraic sum of the capacitor voltage (V_p) and the instantaneous generator voltage (v_s).

$$v_o = V_p + v_s \qquad \text{(equation 7–14)}$$

Let us now consider a square wave input. The range for v_s is from $-V_p$ to $+V_p$ for a change of $2V_p$. Putting these values, one at a time, into equation 7–14,

$$v_o \text{ ranges from 0V to } 2V_p$$

When the inverting input is going positive, the op-amp output goes negative, which reverse biases the diode D_1 effectively taking the op-amp out of the circuit. The only time the op-amp operates is if the input tries to go negative. The net result is a positive unipolar (one polarity) output. If we reverse D_1, the output will be negative unipolar.

Connecting the noninverting input of the circuit of Figure 7–20a to a reference voltage (V_r) as in Figure 7–20b will cause a DC shift in the output because the inverting input is now driven to V_r. The new equation for the output is

$$v_o = V_r + V_p + v_s \qquad \text{(equation 7–15)}$$

EXAMPLE

A sine wave of 3V peak is applied to the circuit of Figure 7–20a. Describe the output signal.

SOLUTION

The capacitor will charge to the 3V peak, so using equation 7–14 ($v_o = V_p + v_s$), the output sine wave signal will range from 0V to +6V.

EXAMPLE

A 10V peak-to-peak square wave is applied to the circuit of Figure 7–20b with the diode D_1 reversed and the reference voltage set at −3V. Describe the output signal.

SOLUTION

Using equation 7–15,

$$v_o = V_r + V_p + v_s$$

V_s ranges from +5V to −5V, so

$$V_o = -3V + 5V + V_s$$

$$V_o = -3V + 5V + 5V$$

$$V_o = +7V$$

or

$$V_o = -3V + 5V - 5V$$

$$V_o = -3V$$

Our output signal has a positive peak of +7V and a negative peak of −3V.

LIMITER CIRCUIT

Input signals to some circuits must not exceed a particular value, otherwise damage will occur. The limit could be a positive value, negative value, or both.

A circuit that limits both the positive and negative excursions of a signal is presented in Figure 7–21. For low-level signals, the input signal appears at the output amplified by $-R_f/R_1$. If the output signal exceeds the breakdown voltage of one zener and the forward drop of the other ($V_z + 0.6V$), then the effective R_f for the circuit will drop to a very low value, and the circuit gain will be essentially zero. This means the output signal will no longer increase as the input signal increases—so it is limited! With our particular circuit, this occurs on both the positive and negative swings of the output.

In the limiting region, the gain is

$$A_{vl} = \frac{r_z + r_d}{R_1} \qquad \textbf{(equation 7–16)}$$

where r_z is the zener impedance (typically 10Ω to 20Ω) and r_d the forward diode resistance ($26mV/I_o$). I_o is the current flowing through the diode combination and can be determined by dividing the input voltage at the point of limiting (V_{sl}) by R_1, or

$$I_o = \frac{V_{sl}}{R_1}$$

But at the limit point, the output voltage is $V_z + 0.6V$, so

$$V_{sl} = \frac{V_z + 0.6\text{V}}{A_v}$$

$$V_{sl} = \frac{(V_z + 0.6\text{V})R_1}{R_f}$$

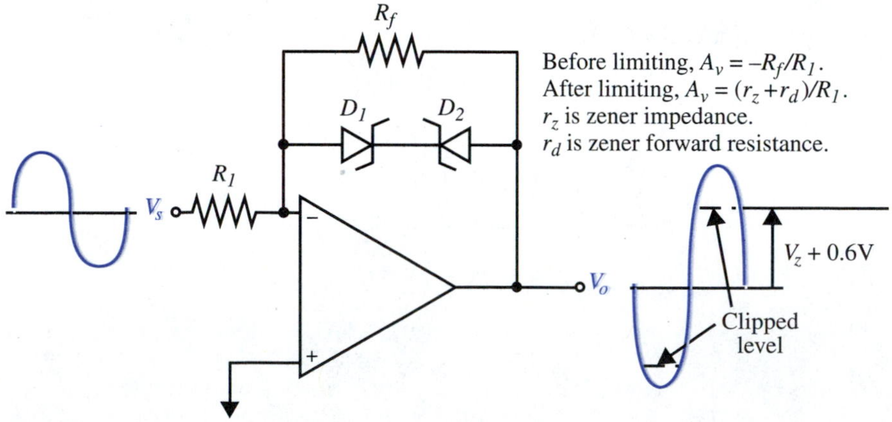

R_f
D_1
D_2
R_1
V_s
V_o
Before limiting, $A_v = -R_f/R_1$.
After limiting, $A_v = (r_z + r_d)/R_1$.
r_z is zener impedance.
r_d is zener forward resistance.
$V_z + 0.6$V
Clipped level

FIGURE 7–21 Double-ended limiter circuit

But

$$I_o = \frac{V_{sl}}{R_1}$$

And therefore

$$I_o = \frac{(V_z + 0.6V)}{R_f}$$

since

$$r_d = \frac{26mV}{I_o}$$

So

$$r_d = \frac{(26mV + R_f)}{(V_z + 0.6V)} \qquad \textbf{(equation 7–17)}$$

EXAMPLE

Find the gain before limiting and at the point where limiting starts for the circuit in Figure 7–21 if R_1 is 100Ω, R_f is 10KΩ, the zener voltage is 6.2V, and r_z is 15Ω.

SOLUTION

The gain before limiting is $-R_f/R_1$ or −100.

Using equation 7–17 to find r_d,

$$r_d = \frac{(26mV + 10K\Omega)}{(6.2V + 0.6V)}$$

$$r_d = 38\Omega$$

So from equation 7–16,

$$A_{vl} = \frac{(15\Omega + 38\Omega)}{100\Omega}$$

$$A_{vl} = 0.53$$

We can see that in the limiting region, the gain is reduced by a factor of about 200.

SIGNAL COMPRESSION CIRCUIT

Audio signals of voice and music have a tremendous variation between the peak and low levels. Another way of saying this is that they have a large **dynamic range.** If we wish to record or transmit this type of signal, we have to accommodate the peaks, even though they occur infrequently. This becomes inefficient, especially with transmitters, because equipment must be designed to handle the peak power. A solution is to compress the audio signal to reduce the audio dynamic range by passing the signal through an amplifier that has a lower gain for the higher-amplitude signals.

We can build a simple compression (log amplifier) circuit by replacing the R_f resistor in an inverting amplifier circuit with a diode. The relationship between current through and voltage across a diode follows a log characteristic in the low-current region of operation, i.e., as we increase the current through the diode, we see a smaller increase in the voltage across the diode. This is apparent from the diode characteristic curve in Figure 7–22a.

Figure 7–22b is a basic compression circuit using two diodes to handle both the positive and negative swings of the signal to be compressed. R_f is included in the circuit to provide an output when the diodes are cut off around the zero point of the input signal and to control the low-level signal gain before the diode resistance drops.

The logarithmic relationship between input voltage (or current) and output voltage for the higher-level signals can be determined from rearrangement of the basic diode equation.

$$V_d = V_o = -(26\text{mV})\ln\left(\frac{I_f}{I_s}\right) \qquad \textbf{(equation 7–18)}$$

where I_f is the forward current through the diode and is equal to the input voltage V_s divided by R_1. I_s is the reverse leakage current, and a rather good correlation with measured data is obtained using $I_s = 10\text{fA}$ (femtoamps [10^{-15}A]).

Substituting this value in equation 7–18,

$$V_o = -(26\text{mV})\ln\left(\frac{V_s}{R_1 \times 10\text{fA}}\right) \qquad \textbf{(equation 7–19)}$$

where V_s is the peak value of the input.

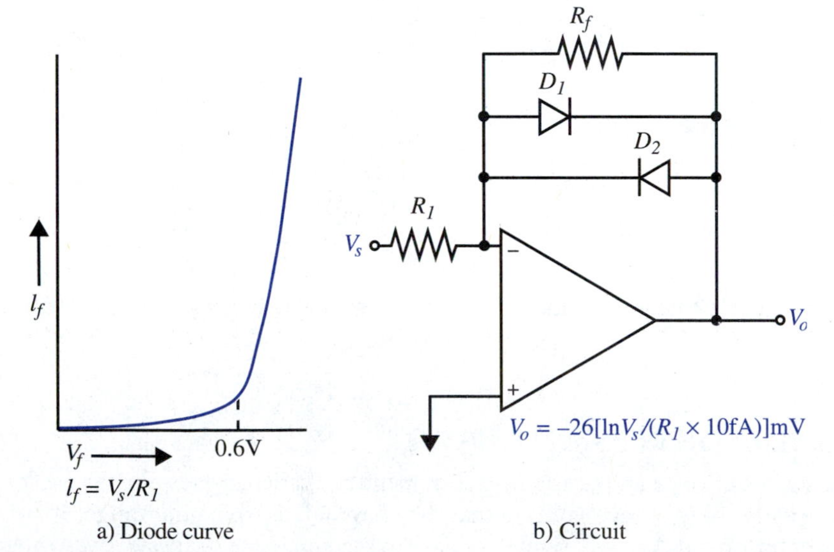

FIGURE 7–22 Signal compression circuit

EXAMPLE

A 10V peak to peak sine wave (5V peak) is applied to a signal compressor circuit with R_1 of 10KΩ. If we make the low level gain equal to unity, i.e., $R_f = R_1$, find the output peak signal. Compare the outputs if the input is reduced to 1V peak to peak (0.5V peak).

SOLUTION

Using equation 7–19,

$$V_o = -(26\text{mV}) \ln\left(\frac{V_S}{R_1 \times 10\text{fA}}\right)$$

$$V_o = -(26\text{mV}) \ln\left(\frac{5}{10\text{K}\Omega \times 10\text{fA}}\right)$$

$$V_o = -0.64\text{V}$$

With 0.5V peak input,

$$V_o = -(26\text{mV}) \ln\left(\frac{0.5}{10\text{K}\Omega \times 10\text{fA}}\right)$$

$$V_o = -0.58\text{V}$$

The two input signals have a ratio of 10:1, whereas the output signals ratio is 0.64/0.58 or 1.1:1 This shows compression of the higher-level signal.

A signal can be uncompressed (expanded) by passing through an expansion circuit. By moving the back-to-back diodes in Figure 7–22 to replace R_1, we can create such a circuit.

OP-AMP FILTER CIRCUITS

Filter circuits are used to pass wanted frequencies and to reject unwanted frequencies. This is important when we want to detect signals in a noisy environment. When we tune in a weak station on the radio, we are setting up a filter circuit that allows the desired signal to pass but rejects all others. It should be understood that noise within the pass range of the filter will not be rejected.

At radio frequencies, filters are usually resonant circuits using inductors and capacitors; but at lower frequencies, inductors are too large and expensive. So, low-frequency filters are usually restricted to resistors and capacitors. A problem with resistor/capacitor filters is the signal loss that occurs in the resistor portion of the filter.

Op-amp filters can provide gain to overcome the resistor losses of an *RC* filter as well as reduce loading of the signal source. Because of these advantages, the op-amp-type filter circuit has been much used over the last few years—but it is being replaced in many audio applications by the newer switched capacitor filter (SCF) chip. This device has the advantage of not requiring external capacitors. We shall study the op-amp-based filters in this section and the SCF in Chapter 14. Filter circuits are of four basic types:

1) Low Pass
2) High Pass
3) Band Pass
4) Notch

Ideal filters would completely pass wanted signals and completely reject unwanted ones. However, this would require a filter with a response with infinite slope at the transition point between wanted frequencies and unwanted ones. In reality, we have a sloping region where the unwanted signal is gradually reduced. The steepness of the slope is related to the number of filter circuits used. The more sections, the steeper the slope. We can illustrate this with the simple low-pass filter circuit shown in Figure 7–23.

In the figure, we see the amplitude response of the filter with a cutoff frequency f_c defined by $X_c = R$. At higher frequencies beyond the cutoff frequency, we see the sloping response that still allows unwanted signals to pass. We can determine the slope by realizing that the output signal in the slope region is reduced in half each time we double the frequency because X_c is reduced by half.

Doubling a frequency is called an **octave change**. If a signal is reduced to half, this is called a −6 decibel change (decibel or db = 20 log 1/2 = −6). So the slope for our single-section filter is described as having a slope of −6db/octave.

Following our low-pass filter with another filter creates a two-section filter with a slope twice as steep, or −12db/octave. A three-section filter will produce a slope of −18db/octave, and so on! The term **filter order** is sometimes used rather than *section* and refers to the highest order of exponent for the equation of the filter. So a two-section filter could be called a second-order filter. The higher the order, the better the rejection!

Another way of describing the slope of a filter is −20db/decade for a first order and −40db/decade for a second order. **Decade** means increasing the frequency by a factor of 10 (20db = 20 log 10), which causes the amplitude to drop to one-tenth.

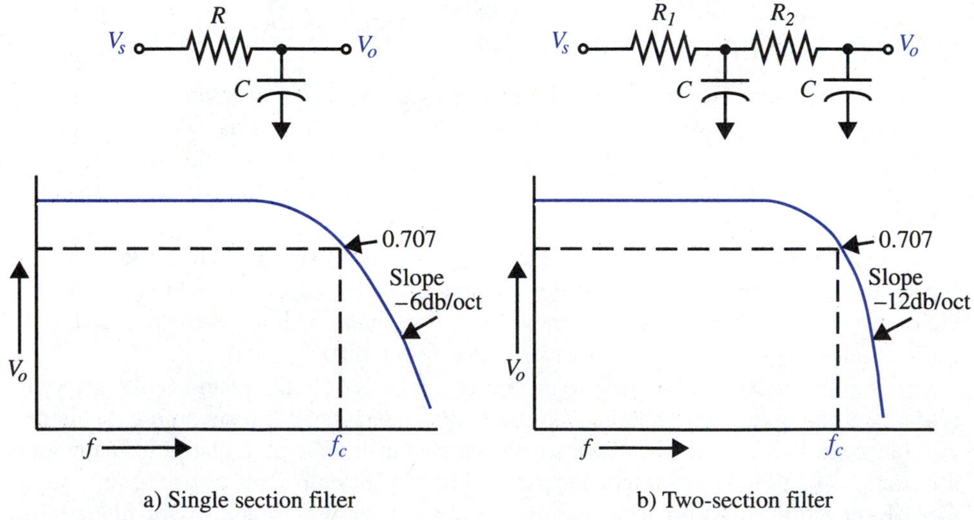

FIGURE 7–23 Filter basics

Filter circuits using amplifiers are called **active filters**, and these circuits can provide gain as well as filtering. Op-amps are a good choice for filter circuits because of their high input resistance and low output resistance. If a filter circuit output is connected to a follower, the cutoff frequency of the filter will not be changed by circuit loading. Similarly, connecting an op-amp output to the input of a filter will not affect the filter because the op-amp has a very low output resistance.

We shall look at various types of the most used second-order op-amp filter circuits, but because of their complexity, we will not derive the filter equations. Resistors and capacitors used in filter circuits should have a tolerance of 1 percent or less—which means that resistors should be metal film and the capacitors stable types like mica, NPO ceramic, polystyrene, etc. Where possible, if more than one capacitor is used, they should be made of equal value in order to simplify the equations and the parts selection.

Since we are considering relatively low-frequency circuits, we shall use what is called a **Butterworth damping factor**

$$\alpha = \sqrt{2}$$

where α is $1/Q$. This gives a response curve with the steepest slope around the cutoff frequency but with no overshoot.

LOW-PASS FILTERS

A second-order version of a low-pass filter is shown in Figure 7–24. It is called a multiple feedback filter (MF) type and is a rather common configuration. The filter variables are the cutoff frequency (f_c), gain (A_v), and damping factor (α) set for $\sqrt{2}$. The equations for these variables are given in Figure 7–24.

EXAMPLE

Determine the values of R_1, R_2, and R_3 for a second-order MF low-pass filter with a desired cutoff frequency f_c of 1KHz and a gain of 10. Use a 10nF capacitor for C_1 and α of $\sqrt{2}$.

SOLUTION

Using equations from Figure 7–24,

$$k = 2\pi f_c C_1$$

$$k = 2\pi \times 1\text{KHz} \times 10\text{nF}$$

$$k = 6.28 \times 10^{-6}$$

$$R_1 = \frac{\alpha}{2kA_v}$$

$$R_1 = \frac{\sqrt{2}}{(2 \times 6.28 \times 10^{-6} \times 10)}$$

$$R_1 = 1.13\text{K}\Omega$$

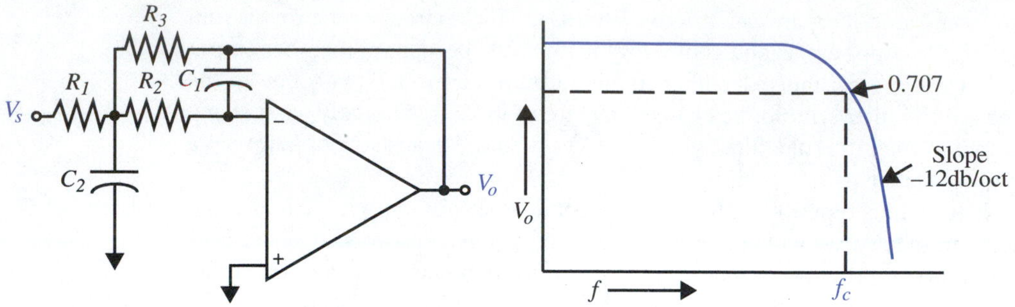

$k = 2\pi f_c C_1$

$$R_1 = \frac{\alpha}{2kA_v}$$

$$R_2 = \frac{\alpha}{2k(A_v + 1)}$$

$$R_3 = A_v R_1$$

$$C_2 = \frac{4(A_v + 1)C_1}{\alpha^2}$$

FIGURE 7–24 Second-order low-pass filter

$$R_2 = \frac{\alpha}{2k(A_v + 1)}$$

$$R_2 = \frac{\sqrt{2}}{(2 \times 6.28 \times 10^{-6})(10 + 1)}$$

$$R_2 = 1.02\text{K}\Omega$$

$$R_3 = A_v R_1$$

$$R_3 = 11.3\text{K}\Omega$$

$$C_2 = \frac{4(A_v + 1)C_1}{\alpha^2}$$

$$C_2 = \frac{4(10 + 1)10^{-8}}{2}$$

$$C_2 = 0.22\mu\text{F}$$

HIGH-PASS FILTER

A second-order MF type high-pass filter is shown in Figure 7–25. Again the variables are cutoff frequency (f_c), gain (A_v), and the damping factor (α).

EXAMPLE

Find the values of R_1, R_2, and C_3 for a second-order MF high-pass filter if the desired cutoff frequency f_c is 300Hz, the gain A_v is 2, and the damping factor $\sqrt{2}$. Let

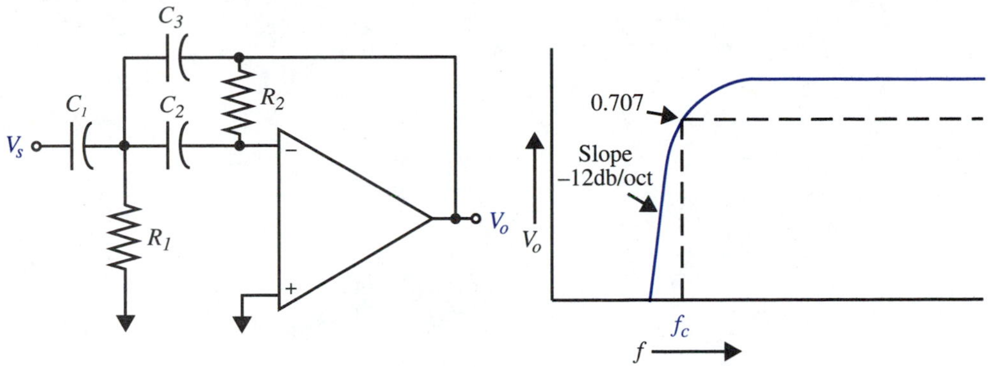

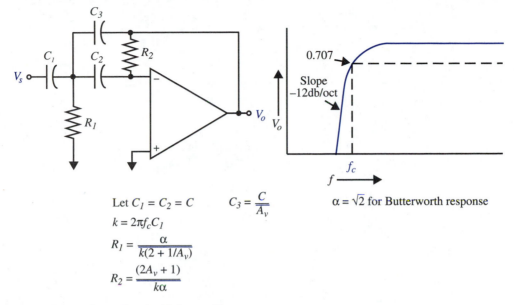

Let $C_1 = C_2 = C$ $\quad C_3 = \dfrac{C}{A_v}$ $\quad \alpha = \sqrt{2}$ for Butterworth response

$$k = 2\pi f_c C_1$$

$$R_1 = \frac{\alpha}{k(2 + 1/A_v)}$$

$$R_2 = \frac{(2A_v + 1)}{k\alpha}$$

FIGURE 7–25 Second-order high-pass filter

$$C_1 = C_2 = C = 0.1\mu\text{F}$$

SOLUTION

Solving for k first,

$$k = 2\pi f_c C_1$$

$$k = 2\pi \times 300 \times 10^{-7}$$

$$k = 1.885 \times 10^{-4}$$

Now, solving for R_1,

$$R_1 = \frac{\alpha}{k(2 + 1/A_v)}$$

$$R_1 = \frac{1.414}{(1.885 \times 10^{-4})(2 + 1/2)}$$

$$R_1 = 3\text{K}\Omega$$

Solving for R_2,

$$R_2 = \frac{(2A_v + 1)}{k\alpha}$$

$$R_2 = \frac{(4 + 1)}{(1.885 \times 10^{-4})(1.414)}$$

$$R_2 = 18.76\text{K}\Omega$$

Finally,

$$C_3 = \frac{C_1}{A_v}$$

$$C_3 = \frac{0.1\mu\text{F}}{2}$$

$$C_3 = 50\text{nF}$$

BAND-PASS FILTER

A band-pass filter passes a band of frequencies around a center frequency f_o. The bandwidth (BW) is defined as the difference between the lower cutoff frequency (f_1) and the upper cutoff frequency (f_2), or

$$\text{BW} = (f_2 - f_1) \qquad \textbf{(equation 7–20)}$$

The Q of a band-pass filter indicates the selectivity of the filter (the higher the Q, the more selective the filter) and is determined from the relationship

$$Q = \frac{f_o}{\text{BW}} = \frac{f_o}{(f_2 - f_1)} \qquad \textbf{(equation 7–21)}$$

Figure 7–26 shows the circuit for an MF band-pass filter.

EXAMPLE

Find the values for R_1, R_2, and R_3 for a band-pass filter if the desired center frequency (f_o) is 2KHz, the Q is 10, and the required gain 5. Assume that C is 0.01μF.

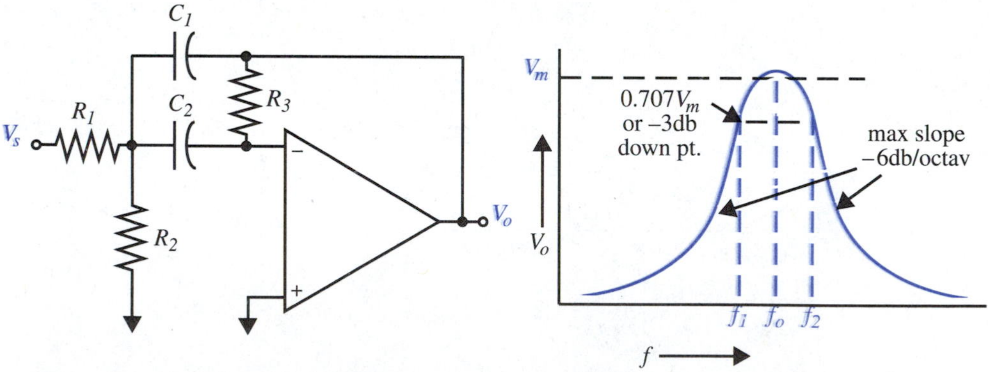

Let $C_1 = C_2 = C$, and $k = 2\pi f_o C$, also remembering that $Q_o \; \frac{f_o}{(f_2 - f_1)}$

$$R_1 = \frac{Q}{kA_v} \qquad R_2 = \frac{Q}{(2Q^2 - A_v)k} \qquad R_3 = \frac{2Q}{k}$$

FIGURE 7–26 Band-pass filter

SOLUTION

First solving for k,

$$k = 2\pi f_c C$$

$$k = 2\pi(2 \times 10^3)(10^{-8})$$

$$k = 1.26 \times 10^{-4}$$

Solving for R_1,

$$R_1 = \frac{Q}{kA_v}$$

$$R_1 = \frac{10}{(1.26 \times 10^{-4})5}$$

$$R_1 = 15.9\text{K}\Omega$$

Solving for R_2,

$$R_2 = \frac{Q}{(2Q^2 - A_v)k}$$

$$R_2 = \frac{10}{(200 - 5)1.26 \times 10^{-4}}$$

$$R_2 = 407\Omega$$

Finally

$$R_3 = \frac{2Q}{k}$$

$$R_3 = \frac{20}{1.26 \times 10^{-4}}$$

$$R_3 = 159\text{K}\Omega$$

NOTCH FILTER

The last filter circuit we will study is the notch filter. This circuit rejects a band of frequencies around a center frequency (f_o). The width of the band is defined by the lower and upper cutoff frequencies, much like the band-pass filter.

We can generate a notch filter by feeding the input signal and output signal from a unity gain band-pass filter (A_1) into a summing circuit (A_2), as shown in Figure 7–27. Since the output from the band-pass filter (v_B) is 180° out of phase with the input (v_s), this signal will subtract from the input when it passes through the summing circuit. We can set the gains for both signals entering the summing circuit to cause a cancellation at the center frequency f_o.

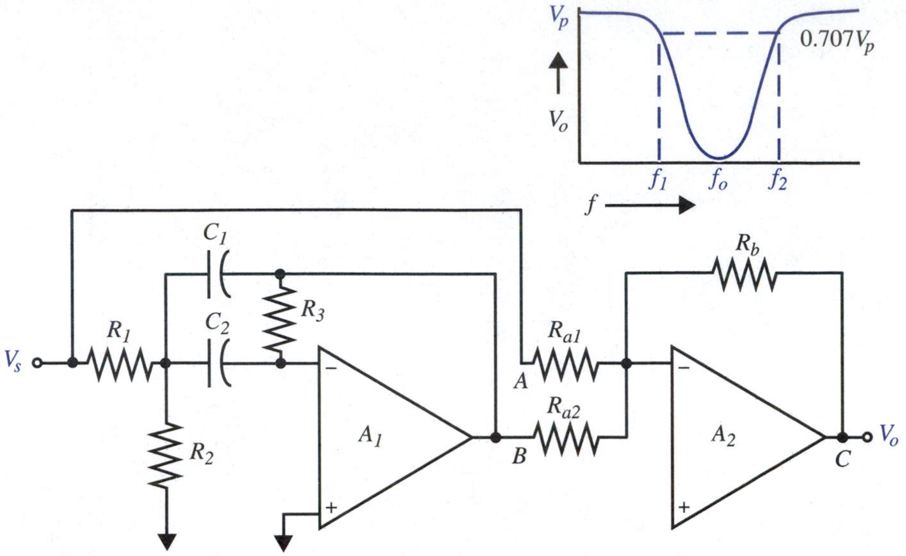

FIGURE 7–27 Notch filter circuit

For simplicity, we have made the band-pass filter gain equal to unity in Figure 7–27 so both input (R_a) resistors to the summing circuit can be the same.

EXAMPLE

We desire a notch filter with a center frequency of 500Hz, with a Q of 8 (band pass), and gain for out-of-notch frequencies of 10. Let

$$C = 0.1\mu\text{F}$$

and

$$R_{a1} = R_{a2} = 10\text{K}\Omega$$

working through the band pass equations with a band pass

$$A_v = 1$$

SOLUTION

Solving for k,

$$k = 2\pi f_o C$$

$$k = 2\pi \times 500 \times 10^{-7}$$

$$k = 3.14 \times 10^{-4}$$

Solving for R_1,

$$R_1 = \frac{Q}{kA_v}$$

$$R_1 = \frac{8}{(3.14 \times 10^{-4})(1)}$$

$$R_1 = 25.5\text{K}\Omega$$

Solving for R_2,

$$R_2 = \frac{Q}{(2Q^2 - A_v)k}$$

$$R_2 = \frac{8}{(128 - 1)(3.14 \times 10^{-4})}$$

$$R_2 = 201\Omega$$

Finally

$$R_3 = \frac{2Q}{k}$$

$$R_3 = \frac{16}{(3.14 \times 10^{-4})}$$

$$R_3 = 50.9\text{K}\Omega$$

To get an out-of-notch gain of 10, R_b must be 10 times R_a, or

$$R_b = 100\text{K}\Omega$$

USING THE NOTCH FILTER FOR DISTORTION MEASUREMENT

Since the notch filter shown in Figure 7–27 provides a band-pass response at point **B** and a notch at point **C**, it can be used to detect amplifier-created distortion in a sine wave.

Let us assume we are checking an amplifier for distortion by passing a 1KHz sine wave oscillator signal through the amplifier. If we set up the notch filter for a center frequency of 1KHz and feed the output of the amplifier into the filter, then any signal that appears at the output of the notch filter is distortion in the 1KHz waveform.

To verify that the notch is at 1KHz, the 1KHz oscillator signal peak can be checked by shifting the frequency slightly while viewing the band-pass output at point **B**. If the gain of the band-pass and summing circuit is unity (1), then the percent distortion can be determined by dividing the distortion voltage measured at point **C** by the signal voltage measured at point **B** and multiplying by 100, or

$$\%\ \text{Distortion} = \frac{V_C}{V_B} \times 100\%$$

CURRENT-TO-VOLTAGE CONVERTER

Since the output voltage from a closed loop op-amp is equal to the current through the feedback resistor times its value, we have current-to-voltage conversion. Most op-amp configurations we have studied have an input voltage signal that is converted to current by R_1 and then back to an output voltage with the drop across R_f. There are situations where the actual signal source is a current, and this is converted directly to an output voltage. An example of such a source is a photodiode, where current is a function of light falling on the diode.

A current-to-voltage converter circuit using a photodiode is shown in Figure 7–28. The increased reverse leakage current which occurs when light energy falls on the junction generates a positive voltage change at the output of the op-amp.

The amount of current increase we get with a given light power level (P_o) is determined by the responsivity (S) of the photodiode, which is measured in amps/watt. Typically, the optical power levels are in the microwatt range, and therefore currents are very low. To determine the change in voltage (ΔV_o) at the output of the op-amp, we first need to find the photodiode current.

$$I = P_o \times S$$

and

$$V_o = I \times R_f$$

so

$$V_o = (P_o)(S)(R_f) \qquad \textbf{(equation 7–22)}$$

EXAMPLE

A 1ms pulse of light at a power level of 100µW falls on a photodiode that has a responsivity of 0.3A/W. Describe the signal at the output of the op-amp if R_f is 100KΩ.

SOLUTION

From equation 7–22, the voltage change at the output of the op amp is

$$\Delta V_o = (100\mu W)(0.3A/W)(100K\Omega)$$

$$\Delta V_o = 3V$$

So we have a +3V pulse, 1ms wide, at the output of the op-amp.

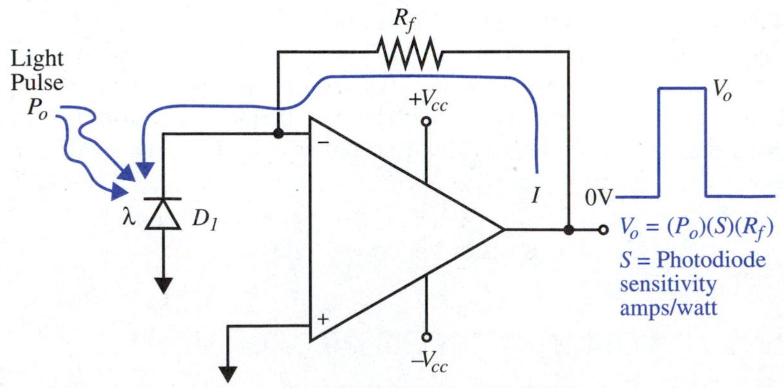

FIGURE 7–28 Photodiode current-to-voltage converter

SUMMARY

- The virtual ground on the inverting input is the reason an inverting adding circuit works.
- We can analyze an inverting adding circuit by using the R_1 value associated with a particular input along with the common R_f resistor to compute the gain.
- The expected output from an adding circuit should never be greater than about 2V less than the supply voltage.
- We can convert an inverting adding circuit into an averaging circuit by dividing R_f by the number of inputs.
- We can convert a noninverting averaging circuit into an adding circuit by adding gain equal to the number of inputs.
- A subtracting circuit can be used to find the difference between a signal on one input and a signal on the other input.
- An integrator circuit gives an output that is equal to the product of the input voltage and the time from when the signal was first applied.
- The output from a differentiator is proportional to the slope of the input waveform.
- An instrumentation amplifier circuit is used because of its noise immunity.
- Common mode rejection is how well an amplifier circuit rejects signals common to both inputs.
- For an oscillator to work, the fed-back signal must be fed back to the noninverting input to the amplifier, large enough to maintain the oscillation, and in phase with the output.
- The Wien bridge oscillator is a common low-frequency sine wave generator.
- A square wave generator relies on the delay in charging a capacitor to generate an output waveform.
- We can change a square wave generator into a pulse generator by charging the capacitor through one resistor and discharging it through one of a different value.
- An ideal diode circuit overcomes the 0.6V forward drop of the diode.
- A peak detector circuit stores the maximum voltage value on a capacitor.
- The output signal from an absolute-value circuit is of one polarity independent of the polarity of the input signal.
- A DC restorer circuit adds a DC component to an AC signal.
- Output signals are prevented from exceeding a predetermined value (clamped) by using a limiting circuit.
- A signal compression circuit restricts the dynamic range of the output signal.
- Because of their high input resistance and low output resistance, op-amps are a good choice for use in filter circuits.
- A first-order filter has a roll-off slope of −6db/octave (−20db/decade); a second-order filter has a slope of −12db/octave (−40db/decade).

- We can use a band-pass filter and an adding circuit to create a notch filter.
- A current-to-voltage converter uses an input current to develop a voltage drop across R_f, which becomes the output signal.

EXERCISE PROBLEMS

1. Draw the circuit for a three-input inverting adding circuit, where at the output, we require

$$V_o = -(V_a + V_b + 3V_c)$$

2. Draw the circuit for a four-input inverting averaging circuit.
3. A three-input noninverting averaging circuit has the following resistor values:

$$R_a = R_b = 2\text{K}\Omega$$
$$R_c = 1\text{K}\Omega$$

If V_a is 2V, V_b is 3V, and V_c is -3V, find the output voltage.

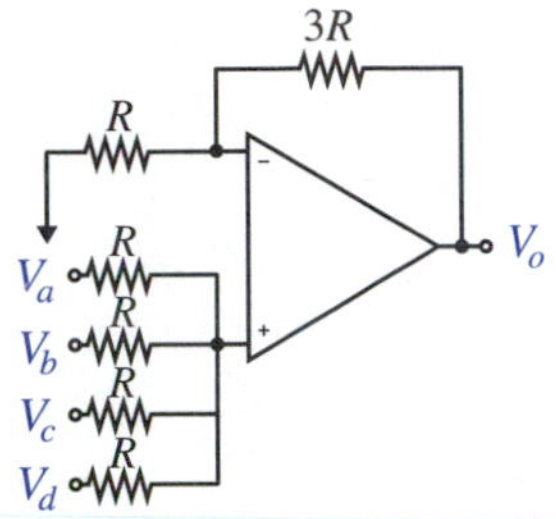

FIGURE 7–A (text figure 7–5)

4. Draw the circuit for a five-input noninverting adding circuit.
5. What is the output voltage equation if the R_1 resistor is open in Figure 7–A?
6. Show the circuit for a subtracting circuit with a gain of 7; let R_1 equal 10KΩ.
7. Show a circuit to satisfy the equation

$$V_o = 2V_a + 3V_b - V_c$$

where V_a, V_b, and V_c can be independently positive or negative. (*Hint:* Use two op-amps.)

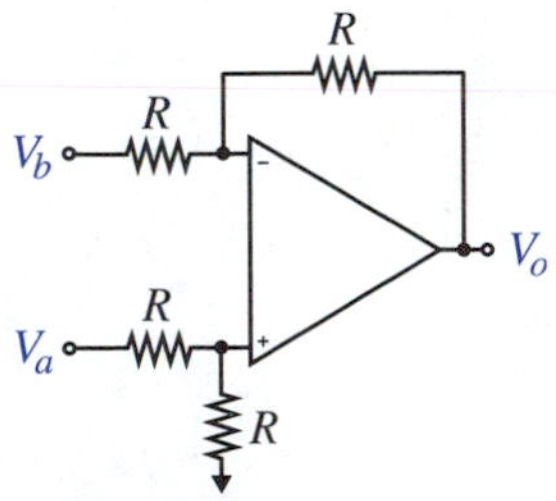

FIGURE 7–B (text figure 7–6)

8. What is the output voltage equation if the resistor from the noninverting input to ground is open in Figure 7–B?
9. The input signal to an integrator circuit is a 4V, 10ms pulse. If R is 1MΩ and C is 0.1μF, sketch and label the output voltage waveform. (See Figure 7–C.)
10. Show the circuit and define the input signal, if we desire a 100ms duration ramp output waveform that ramps at the rate of +50V/s with R of 1MΩ and C of 0.1μF.

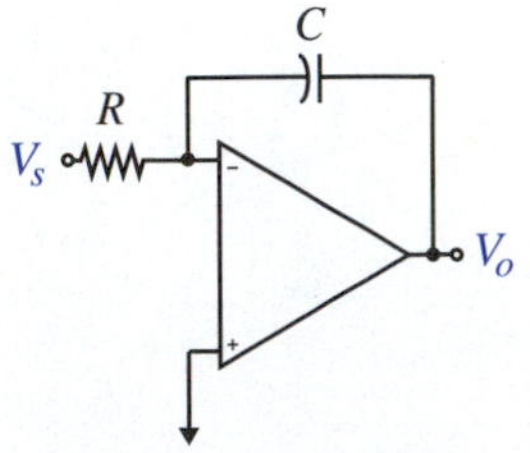

FIGURE 7–C (text figure 7–7)

11. A waveform is applied to a differentiator circuit with the following characteristics: a positive slope of 30V/s for 10ms followed by a negative slope of 40V/s for 20ms. If R is 100KΩ and C is 0.1μF, show the circuit and sketch the output voltage waveform. (See Figure 7–9.)
12. Which circuit would you use to obtain an exact +90° phase shift of a sine wave?
13. Show the circuit for an instrumentation amplifier with R_1 of 10KΩ and a required gain of −50.
14. What is the equation for the voltage gain (A_v) for the circuit of Figure 7–D if the R_1 resistor is open?

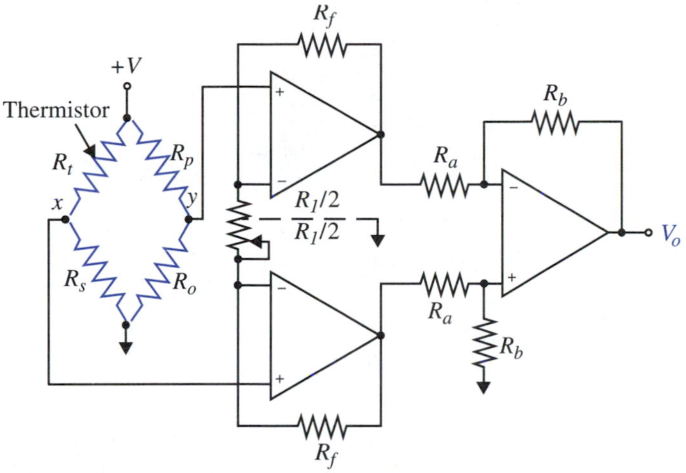

FIGURE 7–D (text figure 7–12)

15. Compute the signal and noise output voltages for the circuit of problem 13 if V_s is 50mV, V_n is 1V, and the common mode rejection is 90db. (*Hint:* CMRR = Inv log 4.5)

16. Compute the output signal level from the high-resistance instrumentation amplifier circuit of Figure 7–D if the input signal is 15mV, R_1 is 2KΩ, R_f is 20KΩ, R_a is 10KΩ, and R_b is 50KΩ.

17. The desired output frequency for a Wien bridge oscillator circuit is 6KHz. If C is 0.01μF, find R.

18. What will the output signal be if R_f in Figure 7–E is open?

19. We need a square wave generator with a period of 4ms. If C is 0.1μF find R_a.

20. Compute the values for R_a and R_b if we need a pulse generator with a positive pulse width of 1.2ms at a frequency of 200Hz. Let C equal 0.22μF.

21. Sketch the output waveform from the ideal diode circuit of Figure 7–F if R_1 is 3KΩ, R_f is 60KΩ, and the input is a 50mV peak-to-peak sine wave.

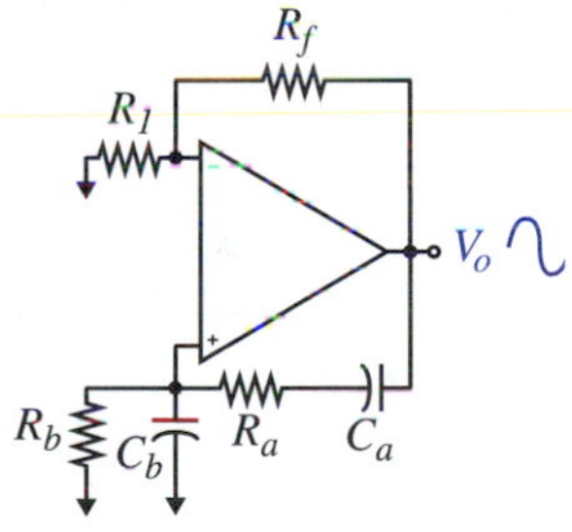

FIGURE 7–E (text figure 7–13a)

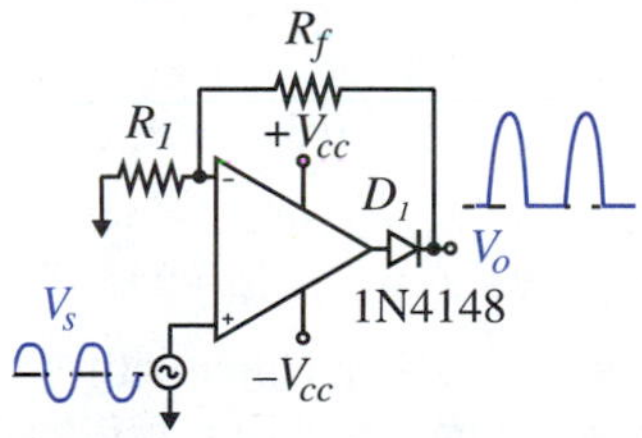

FIGURE 7–F (text figure 7–16b)

22. What would be the output, with a sine wave input, if the D_2 diode in the inverting ideal diode circuit of Figure 7–G became an open circuit?

23. For the absolute value circuit of Figure 7–H, describe the waveforms at points **A**, **B**, and **C** if the input voltage is a 200mV peak-to-peak sine wave and both D_1 and D_2 are reversed.

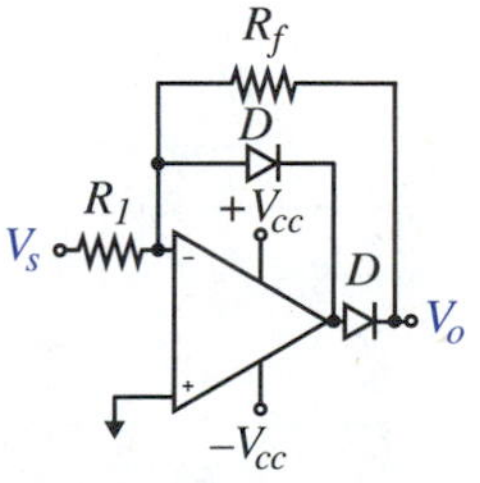

FIGURE 7–G (text figure 7–17)

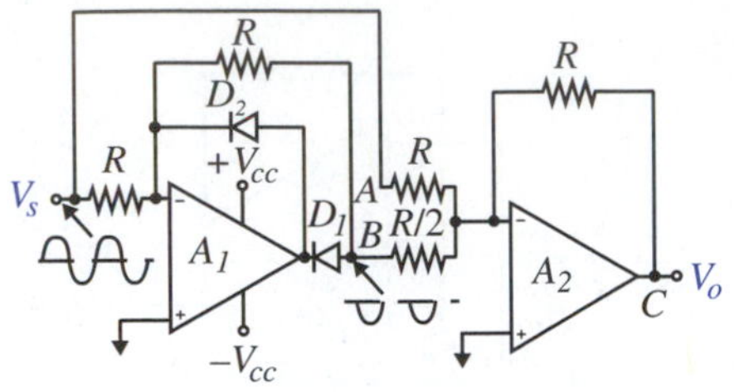

FIGURE 7–H (text figure 7–14)

24. What will the output signal be for the circuit of Figure 7–H if the input signal is a 300mV peak-to-peak sine wave, and the resistor connected to point **A** becomes an open circuit?

25. Sketch the output voltage waveform for the DC restorer circuit of Figure 7–J if the input signal is a 2V peak-to-peak sine wave and V_r is +2V.

26. Describe the output waveform relative to the input if the diode D_1 in Figure 7–I becomes an open circuit.

27. A double-ended limiter circuit has an R_1 of 10KΩ, R_f of 100KΩ, V_z of 5.1V, and r_z of 10Ω. If the input signal is a 2V peak-to-peak sine wave, determine the gain before limiting and after limiting and sketch the output waveform.

28. Sketch the output waveform for the circuit of problem 27 if the diode D_1 of Figure 7–K becomes shorted.

29. A unity gain signal compression circuit with R_f of 56KΩ has a sine wave input voltage of 6V peak to peak. Find the output p-p signal amplitude. Repeat for an input signal of $0.6V_{p-p}$.

30. Sketch the output waveforms for the circuit of problem 29 with a sine wave input if the diode D_1 in Figure 7–L becomes an open circuit.

31. How many filter sections are required with a low-pass filter to obtain a roll-off beyond the 0.707 point of −24db/octave?

32. Why are op-amps a good choice for filter circuits?

33. Specify the required values of R_1, R_2, and R_3 for a second-order Butterworth low-pass filter if A_v is 8, f_c is 4KHz, and C is 0.01μF.

34. Determine the values of R_1, R_2, and C_3 for a second-order Butterworth high-pass filter if A_v is 2, f_c is 1KHz, and C is 0.047μF.

35. It is desired to have a band-pass filter with a center frequency of 500Hz, a bandwidth of 100Hz, and a gain of 4. If C is 0.1μF, find R_1, R_2, and R_3.

36. We need a notch filter with a center frequency of 1KHz, a Q of 8, and an out-of-notch gain of 5. If C is 0.01μF and R_b is 100KΩ, find R_1, R_2, R_3, R_{a1} and R_{a2}.

37. If the band-pass filter portion of the notch filter of Figure 7–M has a gain of 3, show how the summing portion of the circuit needs to be changed for an overall gain of unity.

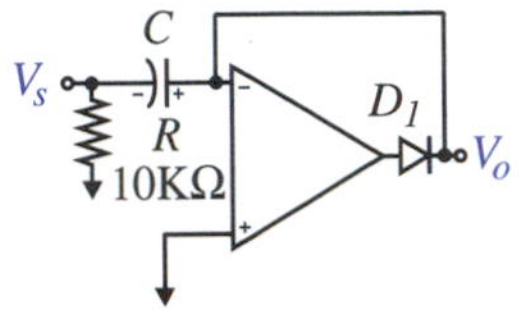

FIGURE 7–I (text figure 7–20a)

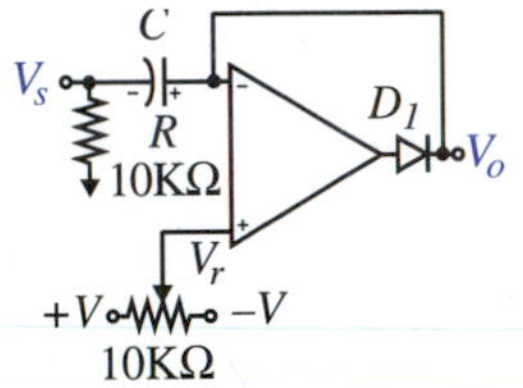

FIGURE 7–J (text figure 7–20b)

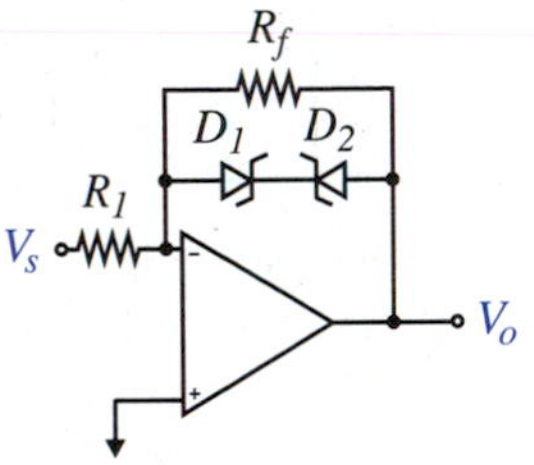

FIGURE 7–K (text figure 7–21)

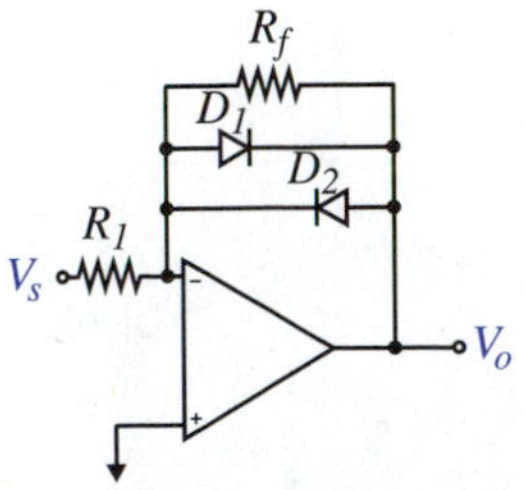

FIGURE 7–L (text figure 7–22

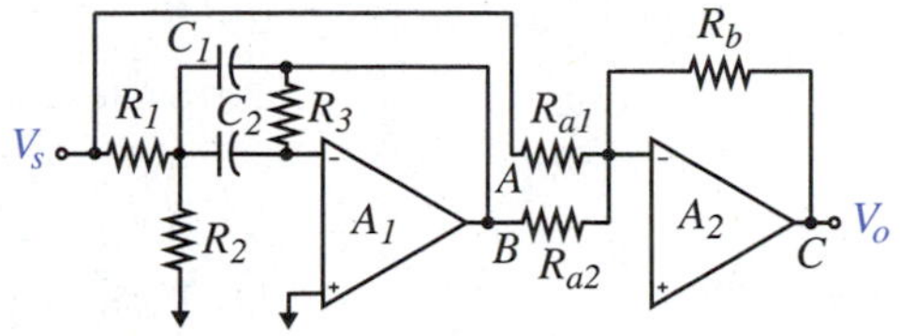

FIGURE 7–M (text figure 7–27)

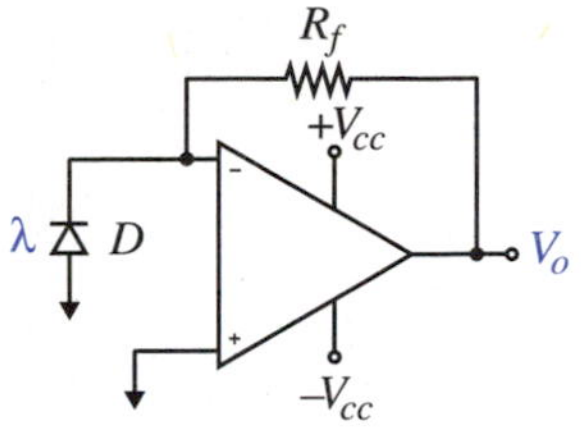

FIGURE 7–N (text figure 7–28)

38. Determine the light power level if a 2V change occurred at the output of the photodiode circuit of Figure 7–N with S of 0.5A/W and R_f of 56KΩ.

ABBREVIATIONS

I_n = Norton's current generator
V_{oa} = Output from V_a input signal
V_{ob} = Output from V_b input signal
V_m = Magnetic induced noise voltage
V_e = Electrostatic coupled noise voltage
V_n = Combined magnetic and electrostatic noise
V_{on} = Output combined noise signal
V_p = Peak signal voltage
V_r = Reference voltage
t_p = Pulse time duration
t_d = Time between pulses
f_p = Pulse repetition frequency
f_o = Center frequency
f_1 = Lower cutoff frequency
f_2 = Upper cutoff frequency
BW = Bandwidth—band of passed signals ($f_2 - f_1$)
Q = Peaking factor (f_o/BW)
r_d = Diode (zener) forward resistance
r_z = Zener diode reverse breakdown resistance

ANSWERS TO PROBLEMS

1. Use inverting summer.
2. See text.
3. $V_o = -0.25V$
4. Modify Figure 5–5 and make R_f equal to $4R$.
5. $V_o = (V_a + V_b + V_c + V_d)/4$
6. Use R_f = 7R in Figure 7–6.
7. Use two inverting summers.
8. $2V_a - V_b$
9. See Figure 7–8; negative ramp final voltage −400mV

10. Input signal −5V pulse of 100ms duration
11. −300mV for 10ms and +400mV for 20ms
12. Integrator circuit
13. Use Figure 7–11 with matched 500KΩ resistors for R_f.
14. $A_v = -R_b/R_a$
15. $V_{os} = -2.5V$; $V_{on} = -1.58mV$
16. $V_{os} = -1.575V$
17. $R = 2.65K\Omega$
18. Square wave, because there is no negative feedback
19. $R_a = 100K\Omega$
20. $R_a = 86.4K\Omega$; $R_b = 27.3K\Omega$
21. $V_{op} = 0.525V$
22. With positive input, D_1 reverse biases, and part of V_s appears at output via R_f and R_1.
23. See text.
24. Half-wave type output with amplitude $2V_p$
25. $V_{o(max)} = +4V$
26. Same as input with no DC restoration
27. $A_v = -10$; $A_{vl} = 4.66 \times 10^{-2}$
28. Positive part of waveform will clip at 0.6V.
29. $V_o = 1.16V_{p-p}$; $V_o = 1.05V_{p-p}$
30. Negative output not compressed
31. Four sections
32. High input resistance of the non-inverting doesn't load filter circuit.
33. $R_1 = 352\Omega$; $R_2 = 313\Omega$; $R_3 = 2.82K\Omega$; $C_2 = 180nF$
34. $R_1 = 1.92K\Omega$; $R_2 = 11.97K\Omega$; $C_3 = 23.5nF$
35. $R_1 = 3.98K\Omega$; $R_2 = 346\Omega$; $R_3 = 31.8K\Omega$
36. $R_1 = 127K\Omega$; $R_2 = 1K\Omega$; $R_3 = 255K\Omega$; $R_{a1} = R_{a2} = 20K\Omega$
37. Increase R_{a2} by a factor of 3.
38. $P_L = 71.4\mu W$

CHAPTER 8

Op-Amp Design Considerations

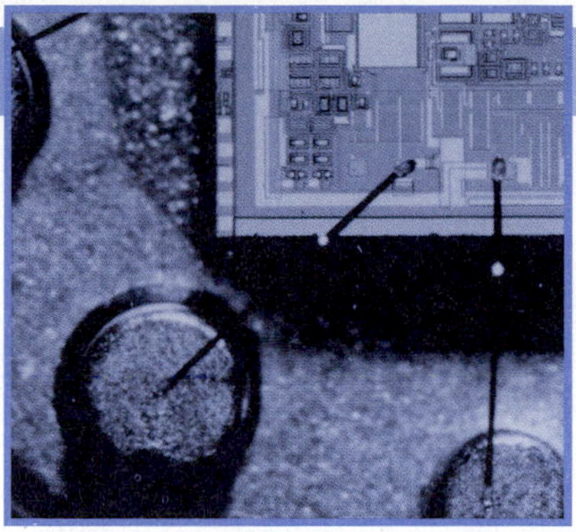

CHAPTER OBJECTIVES

After completing this chapter, you should be able to describe the following op-amp design concepts:

- The effect of offset voltage and offset current
- Effect of changes in the power supply voltages
- Frequency response of an op-amp
- Non-ideal op-amp gain equations
- Op-amp slew rate
- Op-amp rise time
- Noise effects
- How to select an op-amp

INTRODUCTION

Up to this point, our study of the op-amp has pictured the op-amp as an ideal device with few limitations. This is fine for those who wish to have just enough knowledge for basic troubleshooting. This chapter is intended for those persons who require a more in-depth understanding of the op-amp in order to optimize their troubleshooting skills or to perform actual circuit design.

We shall begin with a look at the effect on the op-amp of the bias current required for the input transistors in the op-amp differential amplifier. Differences of base-to-emitter voltage drop for these transistors will also be considered. The effect of op-amp frequency response on gain will be studied along with op-amp slew rate. Also, the impact of these factors on previously examined op-amp circuits will be pointed out. Finally, we shall look at available op-amps and compare their features for various applications.

OP-AMP BIAS CURRENT EFFECTS

Recall that the input circuitry to the op-amp consists of a pair of transistors connected as a differential amplifier, as shown in Figure 8–1.

The biasing of Q_1 and Q_2 requires that DC current flows into the base leads ($-V_{in}$ and $+V_{in}$). Because of the high β of these transistors, the current is rather small (600nA max

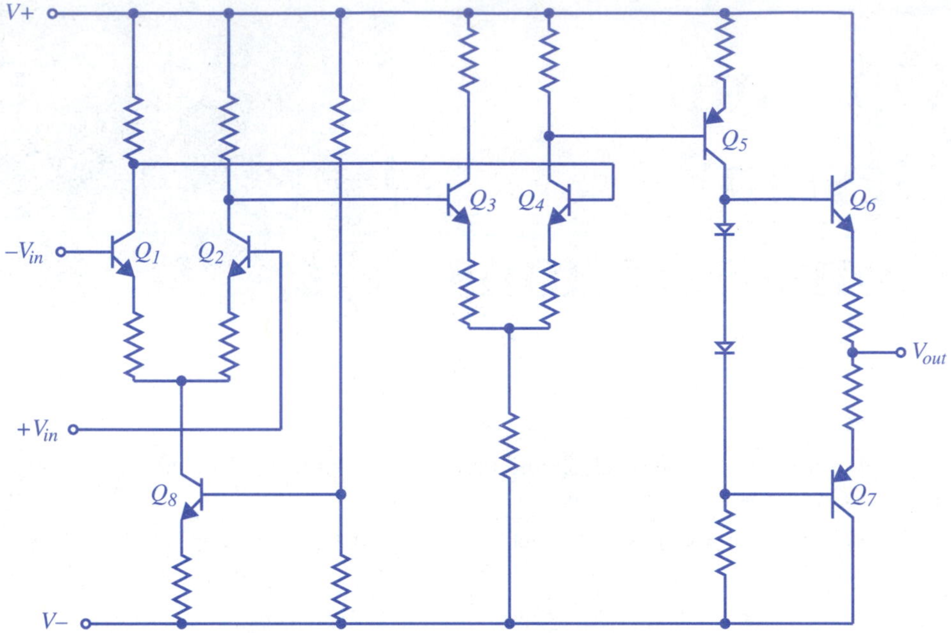

FIGURE 8–1 Operational amplifier schematic

for a 741 op-amp), but it does have an effect on circuit operation. The effect is a DC output voltage, even though there is no difference signal between the two op-amp inputs. This output voltage is called a **voltage offset,** and if it is too large, it can cause errors when using the op-amp to amplify DC signals, or it can shift the output voltage sufficiently off ground to prevent equal voltage swing (positive and negative) when amplifying AC signals.

We can determine the effect of current offset by analyzing the effect of the two input bias currents, I_{ba} and I_{bb}, with reference to Figure 8–2.

For convenience, we shall assume that both inputs are grounded. Notice that I_{ba} consists of currents I_1 and I_f, which flow respectively through resistors R_1 and R_f on the way to the base. These currents create a voltage (V_a) at the inverting input, which is a little negative with respect to ground. On the noninverting side, current I_{bb} flows directly from ground, so this terminal stays at ground potential (0V). We see at the output, then, a positive voltage V_{oa} resulting from V_a being amplified by the open loop gain A_{ol}, or

$$V_{oa} = (V_a)(A_{ol}) \qquad \textbf{(equation 8–1)}$$

From Figure 8–2a,

$$I_{ba} = I_1 + I_f$$

but

$$I_1 = V_1/R_1$$

and

$$V_1 = V_a$$

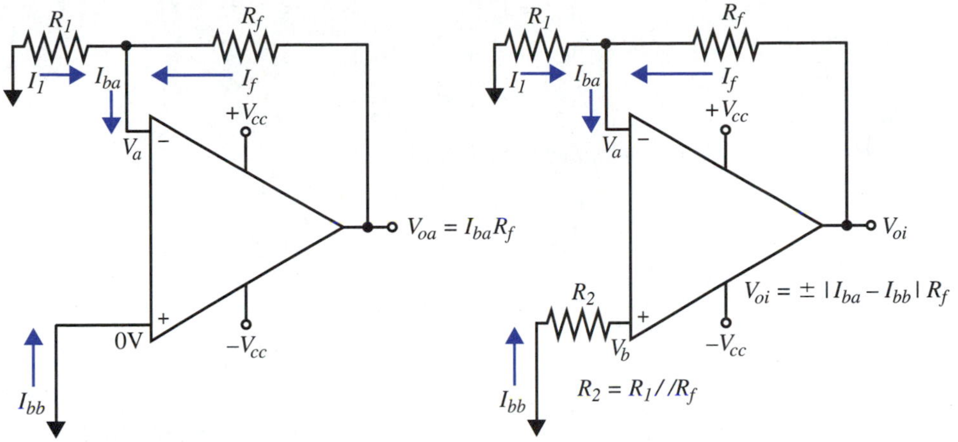

FIGURE 8–2 Determination of current offset effect

So

$$I_{ba} = \frac{V_a}{R_1} + I_f$$

Solving for V_a in equation 8–1 and substituting,

$$I_{ba} = \frac{V_{oa}}{A_{ol}R_1} + I_f \qquad \textbf{(equation 8–2)}$$

If A_{ol} is large, then

$$I_{ba} \approx I_f$$

and

$$V_{oa} = I_f R_f$$
$$V_{oa} = I_{ba} R_f \qquad \textbf{(equation 8–3)}$$

The significance of equation 8–3 is that the bias current I_{ba} flowing to the inverting input terminal flows through R_f and creates a DC voltage V_{oa} at the output of the op-amp.

EXAMPLE

Compute the output offset voltage V_{oa} for an amplifier if I_{ba} is 500nA, R_1 is 1KΩ, and R_f is 100KΩ. Also if A_{ol} is 100,000, find the magnitude of the I_1 relative to I_{ba}.

SOLUTION

Using equation 8–3,

$$V_{oa} = I_{ba}R_f$$

$$V_{oa} = 500\text{nA} \times 100\text{K}\Omega$$

$$V_{oa} = 50\text{mV}$$

Referring back to the development of equation 8–2

$$I_1 = V_{oa}/(A_{ol} \times R_1)$$

So

$$I_1 = \frac{50\text{mV}}{(100{,}000)(1\text{K}\Omega)}$$

$$I_1 = 0.50\text{nA}$$

This makes I_{ba} a thousand times larger than I_1 and justifies the dropping of I_1 in the derivation of equation 8–3 where we assumed

$$I_1 \ll I_{ba}$$

How can we reduce the output offset voltage V_{oa}? We could keep R_f as low as possible and/or add a resistor (R_2) in series with the noninverting input to generate a voltage V_b as a result of I_{bb} flowing through R_2, which counteracts V_a. This technique is shown in Figure 8–2b.

The noninverting input bias current I_{bb} now flows through the resistor R_2. This develops a negative voltage V_b at the noninverting input.

$$V_b = I_{bb}R_2 \qquad \textbf{(equation 8–4)}$$

V_b is amplified by the noninverting gain of the amplifier to provide a negative output voltage V_{ob} (remember that V_{oa} is positive).

$$V_{ob} = V_b\left(\frac{R_f}{R_1} + 1\right)$$

Substituting equation 8–4 for V_b;

$$V_{ob} = I_{bb}R_2\left(\frac{R_f}{R_1} + 1\right) \qquad \textbf{(equation 8–5)}$$

If we assume that I_{ba} equals I_{bb} and want no output offset voltage, then we must have V_{oa} equal V_{ob}. Equating equations 8–3 and 8–5

$$V_{oa} = V_{ob}$$

$$I_{ba} \times R_f = I_{bb}R_2\left(\frac{R_f}{R_1} + 1\right)$$

Remembering that $I_{ba} = I_{bb}$ and solving for R_2,

$$R_2 = \frac{(R_1)(R_f)}{(R_1 + R_f)} \qquad \textbf{(equation 8–6)}$$

What this result means is that both I_{ba} and I_{bb} flow through the same value of resistance, and if the two currents are of equal value, the voltage drops will be the same, and their effect will cancel at the output of the op-amp.

In practice, I_{ba} will not equal I_{bb} because the current requirements of the two differential amplifier transistors are different. However, equation 8–6 still provides the best compromise value for R_2.

With R_2 in the circuit, the output offset voltage V_{oi} due to the difference in the two biasing currents is

$$V_{oi} = (I_{ba} - I_{bb})R_f$$

The absolute difference between I_{ba} and I_{bb}—i.e., $|I_{ba} - I_{bb}|$—is called the **offset current** (I_{os}) in op-amp data sheets, while the average of I_{ba} and I_{bb} is called the **bias current** (I_b). So with R_2 in the circuit

$$V_{oi} = \pm I_{os}R_f$$

or

$$V_{oi} = (I_{ba} - I_{bb})R_f \qquad \textbf{(equation 8–7)}$$

EXAMPLE

The specification sheets for a 741 op-amp lists a maximum value of offset current of 200nA and bias current 500nA at 25°C. If R_f is 200KΩ and R_1 is 5KΩ, find the value of R_2 required for minimum output offset voltage and then find the worst-case voltage offset at the output of the amplifier with and without R_2.

SOLUTION

Using equation 8–6,

$$R_2 = R_1/R_f$$

$$R_2 = \frac{(200\text{K}\Omega)(5\text{K}\Omega)}{(200\text{K}\Omega + 5\text{K}\Omega)}$$

$$R_2 = 4.88\text{K}\Omega \quad (\text{use } 4.7\text{K}\Omega)$$

To find the output voltage offset, we must first find the values of I_{ba} and I_{bb}. The maximum 741 spec. value for I_{os} is 200 nA. If R_2 is connected in the circuit, the output offset is simply from equation 8–7.

$$V_{oi} = \pm I_{os}R_f$$

$$V_{oi} = \pm 200\text{nA} \times 200\text{K}\Omega$$

$$V_{oi} = \pm 40\text{mV}$$

The + or − means that we don't know whether I_{ba} or I_{bb} is larger! The maximum 741 spec. value for I_b is 500nA. But

$$I_b = \frac{(I_{ba} + I_{bb})}{2} \quad (\text{average})$$

and

$$I_{os} = |I_{ba} - I_{bb}| \quad \text{(absolute difference)}$$

Solving for I_{ba} and I_{bb}, using the values of 500nA for I_b and 200nA for I_{os}, we get

$$I_{ba} = 600\text{nA}$$

$$I_{bb} = 400\text{nA}$$

or

$$I_{ba} = 400\text{nA}$$

$$I_{bb} = 600\text{nA}$$

Without R_2, we shall assume I_{ba} has its worst-case value of 600nA. From equation 8–3,

$$V_{oa} = I_{ba}R_f$$

$$V_{oa} = 600\text{nA} \times 200\text{K}\Omega$$

$$V_{oa} = 120\text{mV}$$

We can see, then, that the offset voltage at the output can be three times larger without R_2.

CIRCUIT EFFECTS OF BIAS CURRENT

In addition to the offset voltage created at the output of the op-amp, operation of some circuits can be impaired by the effect of bias current, e.g., the integrator circuit shown in Figure 8–3.

With reference to Figure 8–3a, the I_{ba} bias current flowing from the integrator capacitor C (with no input signal level) causes it to charge to a positive voltage on the op-amp output side. The charging will continue until the op-amp output reaches the positive saturation level.

Solutions to this problem are shown in Figure 8–3b. Since an op-amp with an FET differential pair at the input has extremely small bias current (in the pA range), FET types are recommended along with making C as large as possible (reduce R). Also, keeping the capacitor shorted (with the MOSFET across the capacitor turned on) until the integrator is to be used, prevents charge buildup on the capacitor.

VOLTAGE OFFSET

The two transistors used in the op-amp differential amplifier can have a slight difference between their base emitter drops. For the 741 op-amp, this can be as much as 6mV. Although this value seems small, it does get amplified by the closed-loop gain of the op-amp.

We can analyze how this offset voltage is amplified with reference to Figure 8–4.

As Figure 8–4a shows, the input voltage offset V_{iv} can be represented by a battery (of unknown polarity) inside the op-amp between the two input terminals. This battery can

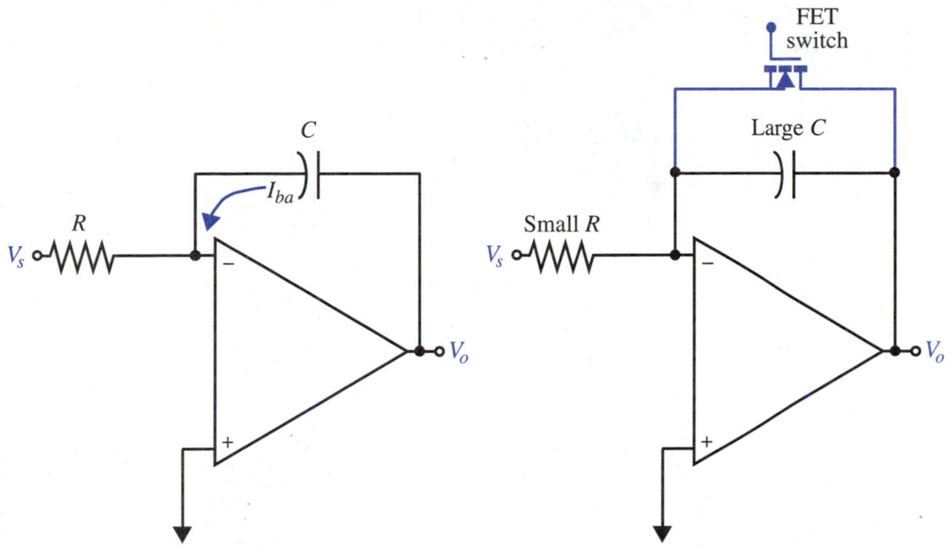

a) Basic circuit b) Improved circuit

FIGURE 8–3 Bias current effect on an integrator circuit

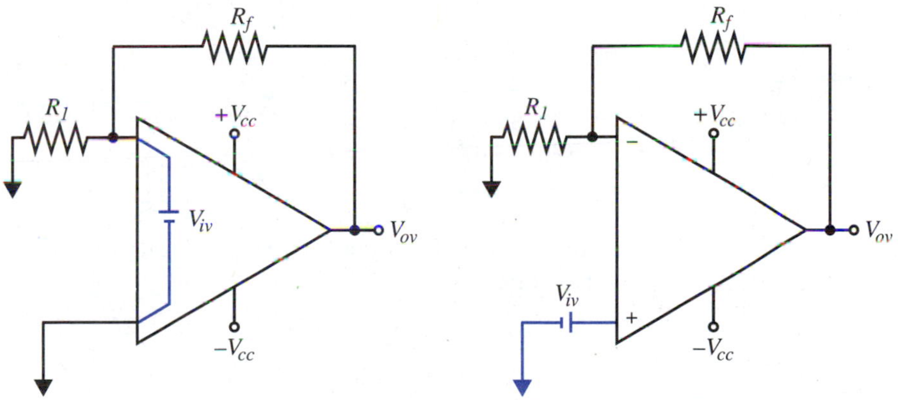

a) Internal offset voltage b) Offset voltage moved outside

FIGURE 8–4 Effect of op-amp voltage offset

be moved outside the op-amp noninverting input terminal, as shown in Figure 8–4b. (It can't be moved outside the inverting terminal because of the junction of R_1 and R_f.) Now it is apparent, from the position of the battery, that input offset voltage is amplified by the noninverting gain and causes a voltage (V_{ov}) at the output of the op-amp.

$$V_{ov} = \pm V_{iv}\left(\frac{R_f}{R_1} + 1\right) \qquad \textbf{(equation 8–8)}$$

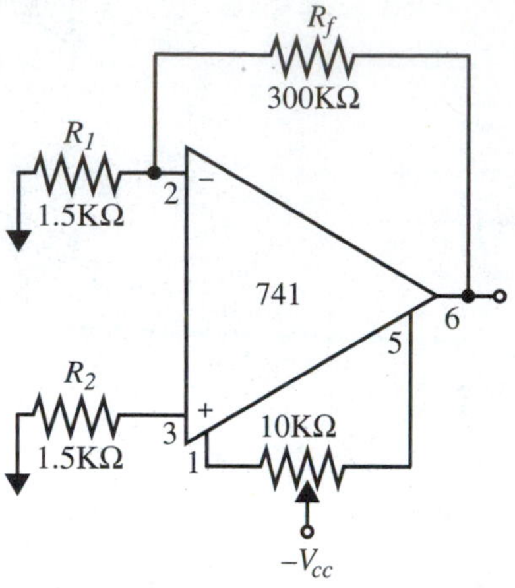

FIGURE 8–5 Combined input offset current and voltage

Both the voltage offset caused by current and that caused by voltage can be made to equal zero volts at the output of the op-amp by adjusting the potentiometer shown in Figure 8–5. This pot introduces an offset into the amplifier that cancels out the offset caused by the input transistors. However, the offset at the op-amp output can return if the temperature changes.

EXAMPLE

An inverting amplifier with an R_f of 240KΩ and R_1 of 3KΩ has an input offset voltage V_{iv} of 10mV. Find the output offset voltage V_{ov}.

SOLUTION

Even though we have an inverting amplifier, we still have to use the noninverting gain to compute the output offset voltage.

From equation 8–8,

$$V_{ov} = +V_{iv}\left(\frac{R_f}{R_1} + 1\right)$$

$$V_{ov} = 10\text{mV}\left(\frac{240\text{K}\Omega}{3\text{K}\Omega} + 1\right)$$

$$V_{ov} = +810\text{mV}$$

EXAMPLE

Find the total output offset voltage due to bias currents and input offset voltage for the circuit shown in Figure 8–5. The 741 op-amp has an offset current of 200 nA and input offset voltage of 6mV.

SOLUTION

Using equation 8–7 for the offset current $|I_{ba} - I_{bb}|$ effect,

$$V_{oi} = \pm I_{os}R_f$$

$$V_{oi} = \pm 200\text{nA} \times 300\text{K}\Omega$$

$$V_{oi} = \pm 60\text{mV}$$

From equation 8–8,

$$V_{ov} = \pm V_{iv}\left(\frac{R_f}{R_1} + 1\right)$$

$$V_{ov} = \pm 6\text{mV}\left(\frac{300\text{K}\Omega}{1.5\text{K}\Omega} + 1\right)$$

$$V_{ov} = \pm 1.206\text{V}$$

The combined output offset

$$V_{oc} = V_{oi} + V_{ov}$$

$$V_{oc} = \pm 1.266\text{V}$$

POWER SUPPLY VOLTAGE CHANGE REJECTION

The effect of voltage offset creates a DC voltage shift at the output of the op-amp. A DC shift can also occur at the output if the power supply voltages are changed. This is because the power supply voltage change creates an internal offset voltage within the op-amp. This change can be referenced to the input and handled like an input offset voltage. The output offset can then be calculated by multiplying this offset voltage by the noninverting gain of the amplifier.

Op-amp manufacturers can specify the power supply sensitivity in two ways. It can be specified as **power supply sensitivity** (PSS) in µV/V or as **power supply rejection ratio** (PSRR) or PSR in db.

The 741 op-amp power supply rejection, expressed as a ratio, has a minimum value of 7,000 (77db). What this means is that the actual equivalent input voltage offset is the total power supply voltage change ΔV_p (between positive and negative supplies) divided by 7,000. This equivalent offset voltage is multiplied by the noninverting gain to find the DC voltage shift at the output (V_{op}). In equation form,

$$V_{op} = \frac{\Delta V_p A_v}{\text{PSRR}} \qquad \textbf{(equation 8–9)}$$

EXAMPLE

Find the voltage shift at the output of a 741 op-amp with R_f of 68KΩ and R_1 of 2KΩ if both the positive and negative power supplies drop by one volt.

SOLUTION

With each power supply dropping one volt, ΔV_p is 2V. The noninverting gain is

$$68\text{K}\Omega/2\text{K}\Omega + 1 = 35$$

Using the worst-case PSRR for the 741 of 7,000 and putting these values in equation 8–9,

$$V_{op} = \frac{\Delta V_p A_v}{\text{PSRR}}$$

$$V_{op} = \frac{(2\text{V} \times 35)}{7000}$$

$$V_{op} = 10\text{mV}$$

We can see from this result that op-amps are not terribly sensitive to power supply changes.

OP-AMP FREQUENCY RESPONSE

All amplifiers have a frequency response, which means that they pass a given band of frequencies and reject all others. Most op-amps can pass frequencies on the low side down to DC, and so when we talk about op-amp frequency response, we really mean response to high frequencies. The op-amp, then, looks like a low-pass filter, and this is because, within the op-amp, there is capacitance to ground—consisting of stray and also the frequency-compensating capacitance (to be discussed later).

Let us review the operation of a low-pass filter, as shown in Figure 8–6.

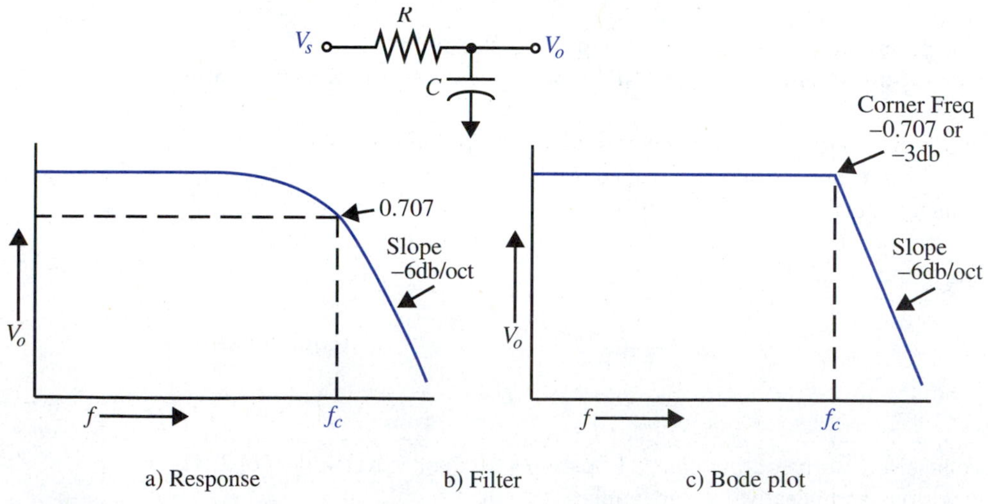

FIGURE 8–6 Low-pass filter

The bandwidth of a filter is defined as that frequency f_c where the output voltage is 0.707 (−3db) of the maximum amplitude. We can calculate this cutoff frequency because, at the 0.707 (half power) point, the reactance of C is equal to R.

$$R = X_c = \frac{1}{2\pi f_c C}$$

or

$$f_c = \frac{1}{2\pi RC}$$

At frequencies higher than f_c, the reactance of the C becomes lower than R, which causes the output signal to reduce. The reduction in the signal is proportional to the reactance of C. If we double the frequency, the reactance of C halves, and so does the output signal.

We can describe this again as a −6db per octave change (−6db means the signal voltage is halved and octave means we have doubled the frequency). Another way of expressing the change is −20db per decade (−20db means signal voltage is decreased to one-tenth and decade means the frequency is multiplied by ten).

We can draw the response of the low-pass filter as shown in Figure 8–6a, or we can approximate the response with straight lines as shown in Figure 8–6c. The straight line graph is called a **Bode plot**. Notice in the Bode plot of Figure 8–6c that the sloping portion has a −6db/octave slope starting at the cutoff frequency (corner frequency).

In addition to a drop in the output signal at the higher frequencies, a phase shift occurs between the input and output voltage. This phase (Ø) between input and output can be determined by the equation

$$Ø = \text{arc tan}\left(\frac{R}{X_c}\right)$$

as X_c gets smaller and smaller, Ø approaches 90°.

Within the op-amp, the various capacitors set up low-pass filters that reduce the open loop gain at the higher frequencies and also cause phase shifts. The overall effect can be a signal at the output that is shifted by 180°. Since in the closed loop configuration the output is fed back to the inverting input, it could end up in phase with the input signal. This causes the op-amp to break into oscillation—and it is no longer an effective amplifier.

The frequency response of an open loop hypothetical op-amp in Bode plot form is shown in Figure 8–7. Notice there are three internal RC filters, all with different corner frequencies.

It was noted previously that if we have two RC filters, we can have a maximum phase shift of 180° and the possibility of oscillation. With three filter sections, the chances for oscillation are even greater. We can determine the closed loop frequency response of an op-amp by drawing a horizontal line at the selected gain value as shown with the dotted line in Figure 8–7. If the intercept with this line and the sloping portion of the open loop graph occurs on a slope no greater than −20db/octave (one RC section), the amplifier will be stable (will not oscillate).

It can be seen from the graph for our hypothetical amplifier that circuits, using this amplifier, will be stable if the closed loop gain is greater than 80db (10,000).

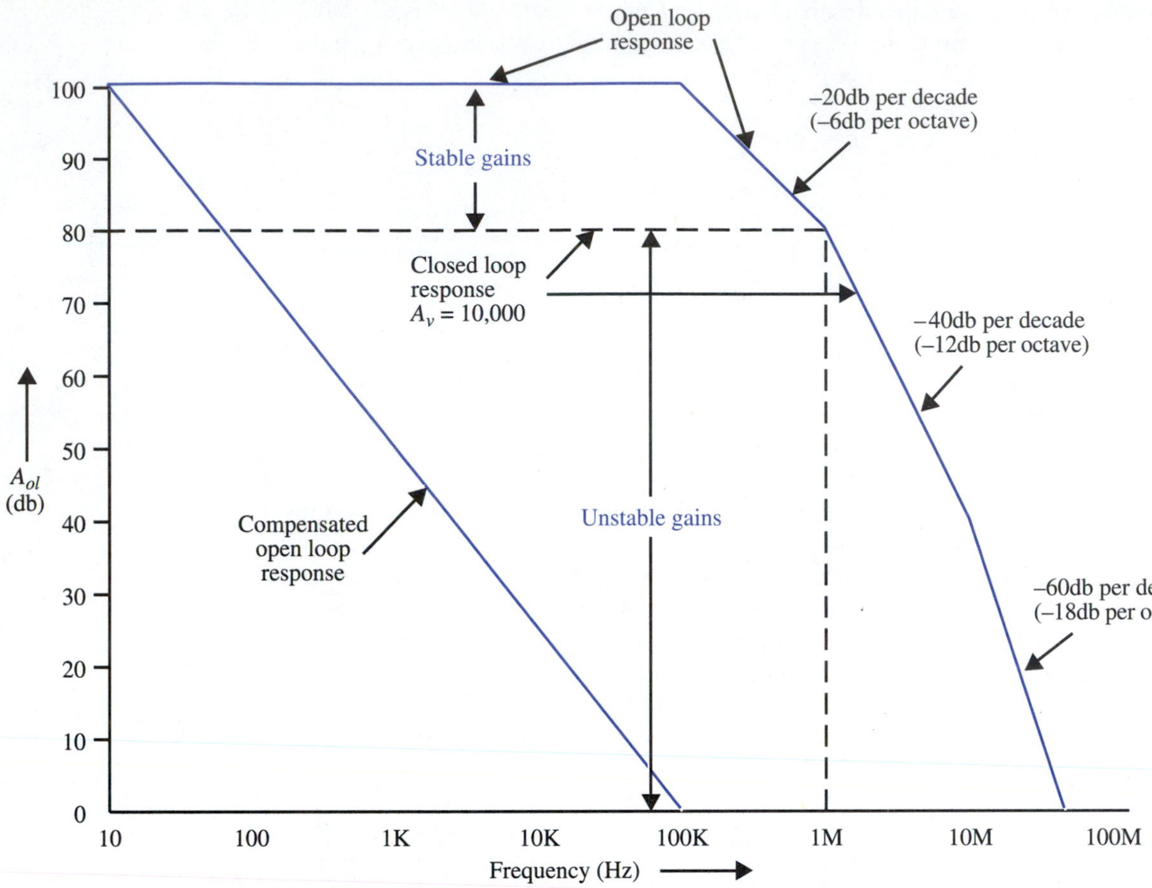

FIGURE 8–7 Frequency response of hypothetical op-amp

EXAMPLE

Determine whether an inverting amplifier with R_f of 200KΩ and R_1 of 1KΩ will be stable if the open loop response is as shown in Figure 8–7.

SOLUTION

The closed loop gain will be

$$A_v = -\frac{R_f}{R_1}$$

$$A_v = -\frac{200\text{K}\Omega}{1\text{K}\Omega}$$

$$A_v = -200$$

which in db = 20 log 200 = 46db.

Drawing a horizontal line on the graph at this gain, we see an intercept with the open loop graph at a slope greater than −20db/decade, so we conclude that the amplifier will be unstable. To prevent instability at the lower gain settings, a compensating capacitor is added internally (sometimes externally) to op-amps to ensure that the open loop gain never has a slope greater than −20db/decade. This results in a lower frequency response; so the trade-off becomes a reduction in amplifier bandwidth versus stability.

Figure 8–8 shows the open loop gain versus frequency curve for a 741 op-amp that uses an internal 30 pf compensating capacitor.

Note that the 741 response has the corner frequency point at 10Hz and the gain drops to 1 (0db) at a frequency of 1MHz. If we pick a closed loop gain and draw it on the graph, the intercept with the sloping portion gives us the bandwidth for that particular gain setting.

From the graph of Figure 8–8, we see that if we pick a closed loop gain of 10,000 (80db), it has a bandwidth of 100Hz; a closed loop gain of 10 (20db) has a bandwidth of 100KHz etc.

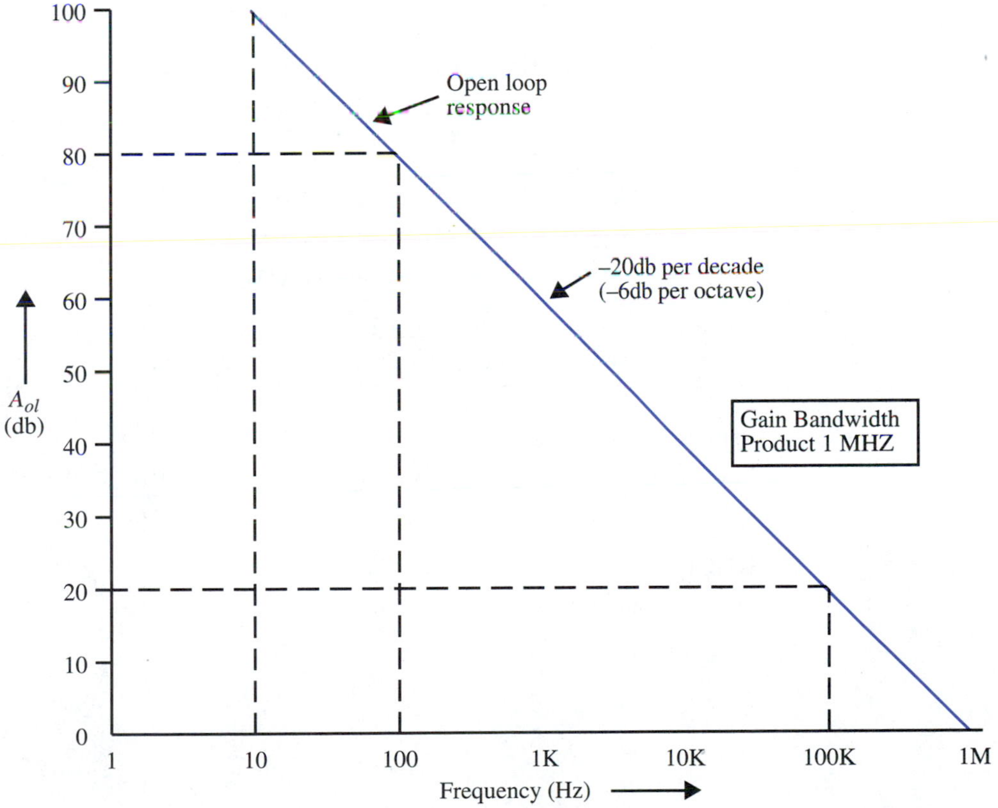

FIGURE 8–8 Open loop frequency response for a 741 op-amp

Multiplying the closed loop gain by the bandwidth gives the **gain-bandwidth product** (GBP). The result of these computations for the 741 op-amp gives a constant GBP of 1MHz. Knowing the GBP for an op-amp means that we can find bandwidth if we know the closed loop gain.

$$GBP = BW \times A_v$$

$$(A_v = A_{ol} \text{ in this equation}) \qquad \textbf{(equation 8–10)}$$

EXAMPLE

Find the bandwidth for an inverting amplifier using a 741 op-amp if R_f is 400KΩ and R_1 is 2KΩ.

SOLUTION

The closed loop gain

$$A_v = -\frac{R_f}{R_1}$$

$$A_v = -\frac{400K\Omega}{2K\Omega}$$

$$A_v = -200$$

Using the GBP of 1MHz for a 741 and rearranging equation 8–10,

$$BW = \frac{GBP}{A_v}$$

$$BW = \frac{1MHz}{200}$$

$$BW = 5KHz$$

Notice that we use the absolute value of the gain when determining the bandwidth, otherwise we would end up with a negative bandwidth.

EXAMPLE

An internally frequency compensated op-amp has a GBP of 20MHz. What is the maximum possible gain if the desired bandwidth is 100KHz?

SOLUTION

Using equation 8–10,

$$A_v = \frac{GBP}{BW}$$

$$A_v = \frac{20KHz}{100KHz}$$

$$A_v = 200$$

EFFECTS OF FREQUENCY RESPONSE

For all amplifiers, the GBP determines the frequency range over which we can get full gain. With op-amp filter circuits, we have to remember that the op-amp is inherently a low-pass filter, so when we make a high-pass filter, we really have a band-pass filter.

The differentiator circuit has a particular problem related to frequency response. The circuit is shown in Figure 8–9a, and its frequency response in Figure 8–9b. The response provides an increasing gain as frequency increases, since the reactance of C (acting like an R_1) decreases.

The problem is that where the rising response of this circuit meets with the open loop curve, it creates an intercept slope of −40db/decade—which means that the circuit will become unstable. However, this problem can be avoided by adding a resistor R_1 in series with C to make the gain curve horizontal before the intercept point is reached, as shown in Figure 8–10.

To determine the required value of R_1, we make the gain in the horizontal portion of the curve only a function of R_1, and at the break frequency f_c, we choose an R_1 of $10X_c$. To give a good margin of stability, we make f_c one-tenth of the intercept frequency.

At the intercept, then, from equation 8–10,

$$BW = \frac{GBP}{A_v}$$

and

$$A_v = \frac{R_f}{R_1}$$

$$BW = \frac{(GBP)R_1}{10R_f} \qquad \text{(equation 8–11)}$$

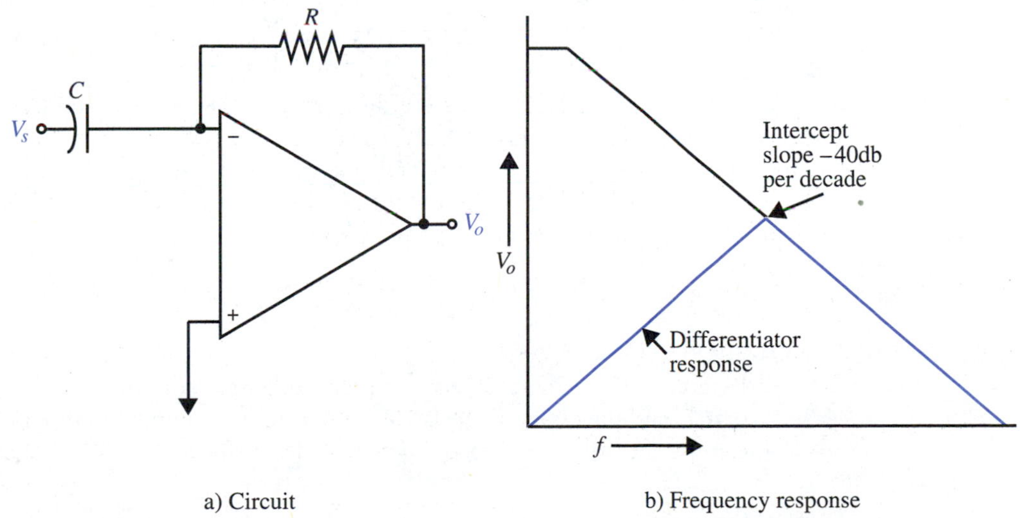

FIGURE 8–9 Differentiator stability

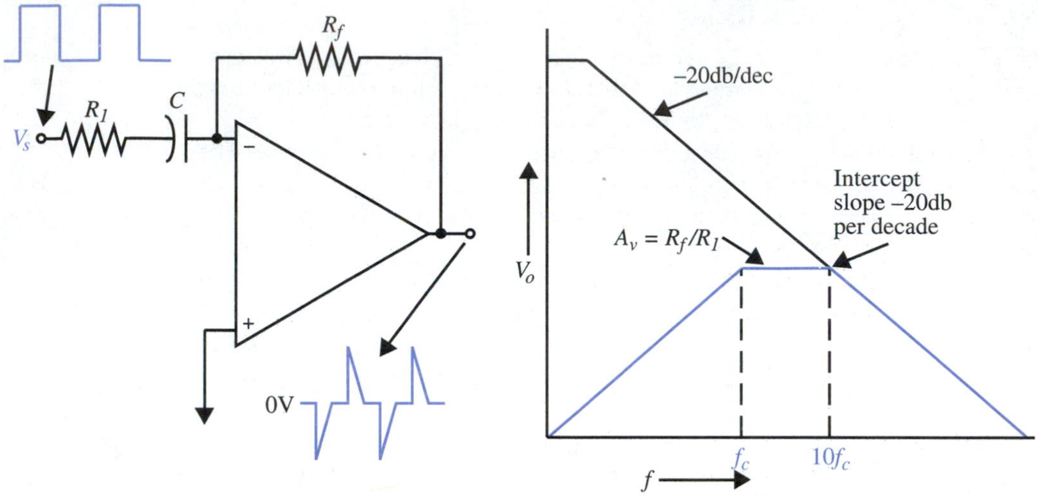

a) Improved circuit

b) Bode plot

FIGURE 8–10 Stabilized differentiator circuit

$$R_1 = 10X_c = \frac{10}{2\pi Cf_c} \qquad \textbf{(equation 8–12)}$$

Substituting for f_c from equation 8–11,

$$R_1 = \frac{10 \times R_f}{2\pi CR_1(\text{GBP})}$$

Solving for R_1

$$R_1 = \sqrt{\frac{100R_f}{2\pi C(\text{GBP})}} \qquad \textbf{(equation 8–13)}$$

Adding R_1 to the circuit makes the differentiator deviate from the true mathematical process. This is shown in Figure 8–10a, where the output waveform with a square wave input is not a sharp spike but has a trailing edge determined by the time constant R_1C. In order to get a good differentiated output, the input pulse width (t_p) to the differentiator should be at least ten times greater than R_1C, or

$$t_p = 10R_1C$$

EXAMPLE

Determine the value of R_1 required for a differentiator circuit if R_f is 100 KΩ, C is 0.1μF, and the op-amp GBP is 5MHz. Also find the minimum input pulse width we can use and still have a good differentiated waveform.

SOLUTION:

Using equation 8–13,

$$R_1 = \sqrt{\frac{100R_f}{2\pi C(\text{GBP})}}$$

$$R_1 = \sqrt{\frac{(100)(100\text{K}\Omega)}{(2\pi)(5\text{MHz})(0.1\mu\text{F})}}$$

$$R_1 = 1.78\text{K}\Omega$$

Using a t_p of $10R_1C$,

$$t_p = 10 \times 1.78\text{K}\Omega \times 0.1\mu\text{F}$$

$$t_p = 1.78\text{ms}$$

NON-IDEAL OP-AMP GAIN EQUATIONS

When we derived the equations for closed loop gain, we assumed the op-amp gain was infinite. However, we can see from the open loop response of an op-amp that at the higher frequencies, the open loop gain (without R_1 and R_f) can drop considerably; for example, the open loop gain for a 741 op-amp at 10KHz is just 100.

We shall now re-derive the closed loop gains for the inverting and noninverting amplifier to include consideration of the actual open loop gain—starting with the gain for the inverting amplifier (see Figure 8–11).

As seen in Figure 8–11, the voltage at the inverting input is V_{in} and the current i is

$$i = \frac{(V_s - V_{in})}{R_1}$$

but i also is

$$i = \frac{(V_{in} - V_o)}{R_f}$$

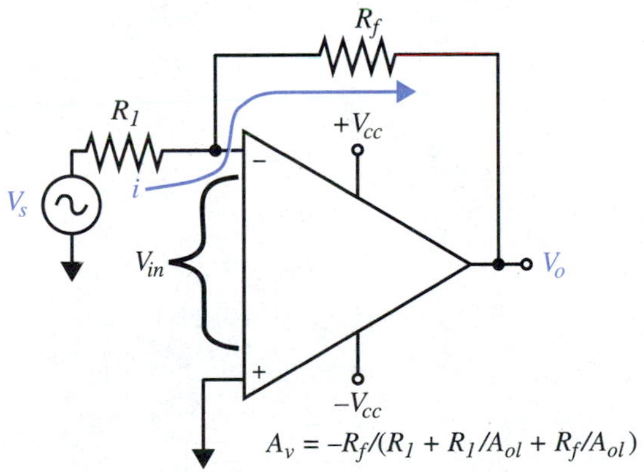

FIGURE 8–11 Non-ideal inverting amplifier gain

Combining these two equations,

$$R_f(V_s - V_{in}) = R_1(V_{in} - V_o)$$

but also

$$V_{in} = -\frac{V_o}{A_{ol}}$$

so substituting

$$R_f\left(V_s + \frac{V_o}{A_{ol}}\right) = R_1\left(-\frac{V_o}{A_{ol}} - V_o\right)$$

separating V_s and V_o terms,

$$R_f V_s = -V_o\left(\frac{R_f}{A_{ol}} + \frac{R_1}{A_{ol}} + R_1\right)$$

but since

$$A_v = \frac{V_o}{V_s}$$

The correct inverting gain equation is then

$$A_v = -\frac{R_f}{R_1 + \frac{R_1}{A_{ol}} + \frac{R_f}{A_{ol}}} \qquad \textbf{(equation 8–14)}$$

EXAMPLE

Find the gain of an inverting amplifier using a 741 op-amp if R_f is 100KΩ, R_1 is 1KΩ, and the operating frequency is 10KHz.

SOLUTION

Since the GBP for a 741 is 1MHz,

$$A_{ol} = \frac{\text{GBP}}{\text{BW}}$$

$$A_{ol} = \frac{1\text{MHz}}{10\text{KHz}}$$

$$A_{ol} = 100 \quad \text{(at 10KHz)}$$

Using equation 8–14,

$$A_v = -\frac{R_f}{R_1 + \frac{R_1}{A_{ol}} + \frac{R_f}{A_{ol}}}$$

$$A_v = -\frac{100\text{K}\Omega}{1\text{K}\Omega + \frac{1\text{K}\Omega}{100} + \frac{100\text{K}\Omega}{100}}$$

$$A_v = -49.8$$

which is half the gain we would expect using

$$A_v = \frac{R_f}{R_1}$$

Using a similar derivation for the non-ideal amplifier gain for a noninverting amplifier,

$$A_v = \frac{R_f + R_1}{R_1 + \frac{R_1}{A_{ol}} + \frac{R_f}{A_{ol}}} \qquad \textbf{(equation 8–15)}$$

To stay within 1% gain accuracy, these non-ideal amplifier gain equations should be used if the open loop gain of the op-amp is less than 100 times the expected closed gain obtained using the ideal gain equations.

OPEN LOOP GAIN EFFECT ON INPUT AND OUTPUT RESISTANCES

Since the input and output resistances of the op-amp are related to the open loop gain (equations 8–8 and 8–10) and the open loop gain depends on the operating frequency, the actual (not ideal) open loop gain should be calculated before using the input and output resistance equations.

EXAMPLE

Find the circuit output resistance R_o for an inverting amplifier using a 741 op-amp with R_f of 200KΩ, R_1 of 1KΩ at an operating frequency of 5KHz.

SOLUTION

The open loop output resistance r_o for a 741 is specified as 75Ω, and the open loop gain A_{ol} at 5KHz is

$$A_{ol} = \frac{\text{GBP}}{\text{BW}}$$

$$A_{ol} = \frac{1\text{MHz}}{5\text{KHz}}$$

$$A_{ol} = 200$$

From equation 6–11,

$$R_o = \frac{r_o}{1 + \frac{A_{ol}}{A_v}}$$

$$R_o = \frac{75\Omega}{1 + \frac{200}{100}}$$

$$R_o = 25\Omega$$

This is much greater than the output resistance we would get using an A_{ol} of 100,000. What this means is that at a higher operating frequency, we have a higher output resistance.

OP-AMP SLEW RATE

If we are trying to amplify a sharply rising waveform with an op-amp, we could find that the output cannot follow the input. This is because capacitors within the op-amp (mainly the frequency-compensating capacitor) have to be quickly charged to the input voltage. The problem is that the internal current sources within the op-amp can-not supply enough current to cause the required change in the capacitor voltage. In fact, these current sources reach a point where they limit and become constant current sources.

The equation for a capacitor being charged from a constant current source is

$$i = \frac{C\,\Delta V}{\Delta t}$$

If i is constant, $\Delta V/\Delta t$ will also be constant so we have a linear ramping voltage waveform across the capacitor.

The compensating capacitor inside a 741 op-amp has a value of 30pF, and the circuit that provides current to this capacitor can typically only provide 15μA. The maximum rate of change (slope) of the output voltage for the 741 is restricted, then, to

$$\frac{\Delta V}{\Delta t} = \frac{i}{C}$$

$$\frac{\Delta V}{\Delta t} = \frac{15\mu\text{A}}{30\text{pF}}$$

$$\frac{\Delta V}{\Delta t} = \frac{0.5\text{V}}{\mu\text{s}}$$

The term used for this constant rate of change of output voltage is **slew rate** (SR). So we can say that the 741 op-amp has a typical slew rate of 0.5V/μs.

The effect of slew rate on the op-amp response to waveforms is shown in Figure 8–12. If the voltage swing from the op-amp output is small, then the output will not be distorted by slew rate.

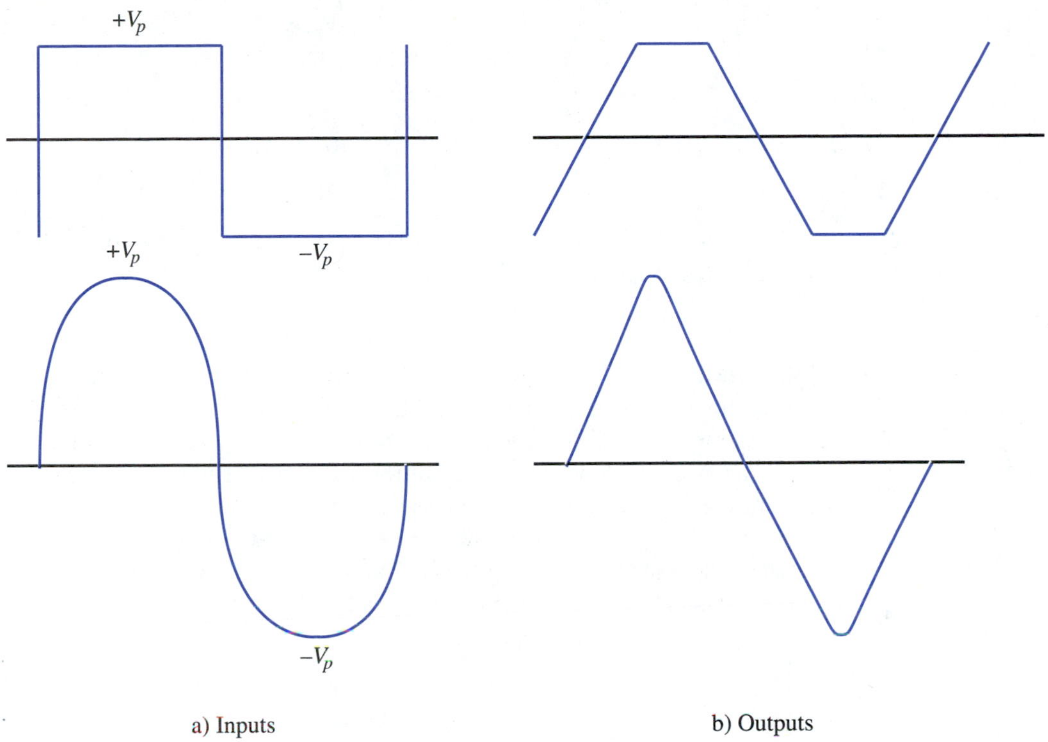

FIGURE 8–12 Slew rate limited waveforms

Notice that for square wave inputs, or sine wave inputs, the output can become triangular with slew rate limiting. This is more apparent on high-frequency signals because the slew rate generated slope becomes a greater portion of the waveform.

We can predict the slew rate limited output waveshape for square wave or pulse inputs, if we know the output voltage swing (V_{p-p}), by determining the time (t_{sr}) it takes the output to change the required amount from the minimum value on the waveform to the maximum value.

For step inputs

$$t_{sr} = \frac{V_{p-p}}{\text{SR}} \qquad \textbf{(equation 8–16)}$$

EXAMPLE

Describe the output waveshape from a 741 op-amp if a 25KHz square wave input signal of 10mV$_{p\text{-}p}$ is amplified by 1,000.

SOLUTION

The period of the 25KHz wave form is 1/25KHz or 40μs. A 10V$_{p-p}$ square wave should appear at the output; but the 741 op-amp can only respond at a 0.5V/μs rate. Using equation 8–16,

$$t_{sr} = \frac{V_{p-p}}{SR}$$

$$t_{sr} = \frac{10V}{0.5V/\mu s}$$

$$t_{sr} = 20\mu s$$

This is half the period of the 25KHz square wave so the output will be a $10V_{p-p}$, 25KHz triangle wave, rather than a $20V_{p-p}$ square wave (see Figure 8–12).

Since distortion is less tolerable in audio amplifiers, slew rate can be a problem with the amplification of high-level sine waves. The fastest changing portion of a sine wave is at the zero crossing point of the waveform, so the op-amp must have a slew rate fast enough to follow the steepest slope at this part of the waveform. The slope is steeper with higher-frequency and higher-amplitude sine wave signals.

The relationship between maximum sine wave frequency before slew rate distortion occurs (f_{sr}), the peak output amplitude (V_p), and op-amp slew rate factor (*SR*) can be derived using differential calculus (determines the point of maximum slope). The relationship is

Max sine wave frequency

$$f_{sr} = \frac{SR}{2\pi V_p} \qquad \textbf{(equation 8–17)}$$

EXAMPLE

If the input signal to a 741 op-amp with a voltage gain of 10 is a 0.5V peak sine wave, find the maximum input frequency (f_{sr}) before slew rate distortion occurs.

SOLUTION

The output signal will be 5V peak, and if we use the 0.5V/μs slew rate of the 741, then from equation 8–17,

$$f_{sr} = \frac{SR}{2\pi V_p}$$

$$f_{sr} = \frac{0.5V/\mu s}{2\pi(5V)}$$

$$f_{sr} = 15.9KHz$$

AMPLIFIER RISE TIME

A sharp-edged square wave consists of many sinusoidal harmonics of the fundamental square wave signal (reference Fourier analysis); the sharper the edges, the wider the amplifier bandwidth has to be to pass the square wave with good fidelity. The relationship between the rise time (t_{ra}) of the signal measured at the output of the amplifier and the amplifier bandwidth BW (out to the cutoff frequency f_c) is given by the equation

$$t_{ra} = \frac{0.35}{\text{BW}} \qquad \textbf{(equation 8–18)}$$

This equation assumes that the amplifier frequency roll-off is at −20db/decade (single low-pass filter) and is derived by finding the time t_{ra} (in terms of R and C) it takes the output waveform to go from 10% to 90% (rise time) of the square wave peak voltage. This yields a t_{ra} of 2.2RC. Using this equation to substitute for RC in the equation for the filter cutoff frequency ($f_c = 1/2\pi\text{RC}$ [0.707pt]) gives equation 8–18.

If the square wave input to the amplifier has a rise time t_{rs}, then the rise time at the output of the amplifier will be

$$t_{ra} = \sqrt{t_{rs}^2 + t_{ra}^2} \qquad \textbf{(equation 8–19)}$$

If the oscilloscope used to measure the rise time at the output of the amplifier has a rise time t_{ro}, then this term squared should also be added under the radical of equation 8–19.

EXAMPLE

Find the rise time and amplitude of a square wave at the output of a 741 op-amp amplifier circuit if the gain is set at 100 and the input square wave has a rise time of 10μs and an amplitude of 10mV_{p-p}.

SOLUTION

With a gain of 100, the 741 has a bandwidth of

$$\text{BW} = \frac{\text{GBP}}{A_v}$$

$$\text{BW} = \frac{1\text{MHz}}{100}$$

$$\text{BW} = 10\text{KHz}$$

Using this value in equation 8–18,

$$t_{ra} = \frac{0.35}{\text{BW}}$$

$$t_{ra} = \frac{0.35}{10\text{KHz}}$$

$$t_{ra} = 35\mu\text{s}$$

At the amplifier output, the rise time is from equation 8–19.

$$t_r = \sqrt{t_{rs}^2 + t_{ra}^2}$$

$$t_r = \sqrt{(10\mu\text{s})^2 + (35\mu\text{s})^2}$$

$$t_r = 36.4\mu\text{s}$$

and the amplitude

$$V_{os} = (A_v)(V_s)$$
$$V_{os} = (100)(10\text{mV}_{p-p})$$
$$V_{os} = 1\text{V}_{p-p}$$

The steepest portion of the exponential rising waveform at the output of an amplifier occurs on the initial slope. The question could be asked, "Does the op-amp have a fast enough slew rate to pass this portion of the signal?" If the slew rate is greater (more volts per microsecond) than the initial slope requires, the output waveform will be rise time limited; if less, slew rate limited.

The equation for this initial (maximum) slope of an exponential (rise time) waveform $\Delta V/\Delta t$ is

$$\frac{\Delta V}{\Delta t} = \frac{2.2(V_{p-p})}{t_r}$$

where V_{p-p} is the peak-to-peak amplifier output voltage.

The break point a between rise time limited waveform and slew rate limited is determined by setting this initial slope of the exponential waveform equal to the slew rate.

$$\frac{\Delta V}{\Delta t} = \text{SR}$$

Then the required slew rate for maximum waveform slope is

$$\text{SR} = \frac{2.2(V_{p-p})}{t_r} \qquad \textbf{(equation 8–20)}$$

Note: The t_r term in equation 8–20 should be left in microseconds because slew rate is given in volts per microsecond.

EXAMPLE

Find whether the output waveform from the previous example with an output of 1V_{p-p} and a rise time t_r of 36.4μs, is rise time limited or slew rate limited. Also determine with an output of 10V_{p-p} with the same rise time.

SOLUTION

Using equation 8–20 to find the required slew rate for this output,

$$\text{SR} = \frac{2.2(V_{p-p})}{t_r}$$

$$\text{SR} = \frac{2.2(1\text{V}_{p-p})}{3.64\mu\text{s}}$$

$$\text{SR} = \frac{0.06\text{V}}{\mu\text{s}}$$

This is much less than the 0.5V/μs capability for the 741, so the output is rise time limited. If the output from the op-amp is $10V_{p-p}$, the required slew rate at the output will be

$$SR = \frac{2.2(10V_{p-p})}{3.64\mu s}$$

$$SR = \frac{0.6V}{\mu s}$$

Now the output will be slew rate limited.

AMPLIFIER NOISE

How much can we amplify a signal? What is the weakest signal we can amplify? The answers to these questions depend on the amount of noise generated within the amplifier because the amplifier noise can mask out weak signals. If we amplify too much for weak input signals, the noise at the output of the amplifier will make it impossible to detect the wanted signal.

The most basic form of noise is thermal or Johnson noise. This noise exists in any component that has resistance and is caused by the random movement of electrons within the material. Increasing the temperature causes this type of noise to increase—thus the term thermal noise.

In transistor circuits, additional noise sources are flicker or $1/f$ noise, which has its highest amplitudes at low frequencies, and shot noise, which increases with the DC current and also increases with frequency.

We don't need to know the actual equations for the transistor noise sources because the manufacturers of op-amps provide graphs that give the equivalent input noise voltage and the noise current, like those shown in Figure 8–13 for the 741.

To use the graphs, we need to know the center operating frequency and also the bandwidth required for the signal. We read the value from the graph at the given center frequency and then multiply this value by the required bandwidth. The result is V_n^2 or I_n^2, and these values are used to find the amplifier output noise V_{on} using the equation

$$V_{on} = A_v\sqrt{(V_n)^2 + (I_n)^2(R_{eq})^2}$$
$$R_{eq} = R_1 \| R_f$$

(equation 8–21)

where R_{eq} is the equivalent resistance seen at the op-amp input terminals and is typically R_f in parallel with R_1. (See Figure 8–14.)

EXAMPLE

Find the noise voltage at the output of a 741 inverting op-amp with R_f of 100KΩ and R_1 of 1KΩ if the center frequency for the signal is 1KHz and the signal bandwidth 100Hz.

SOLUTION

From the voltage noise graph, at the center frequency of 1KHz, we read

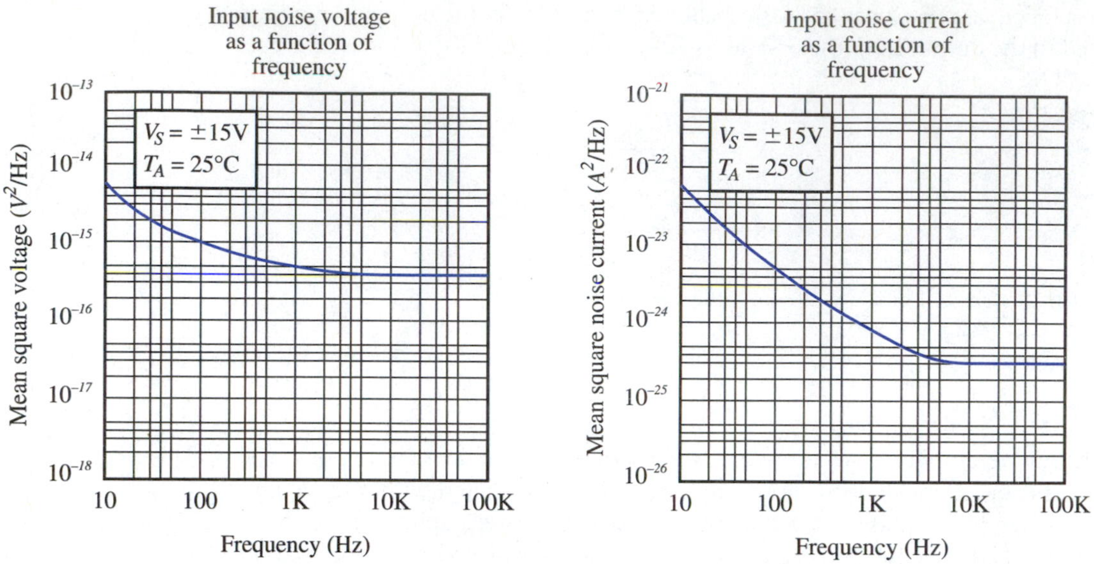

FIGURE 8–13 Noise graphs for the 741 op-amp. *Courtesy National Semiconductor.*

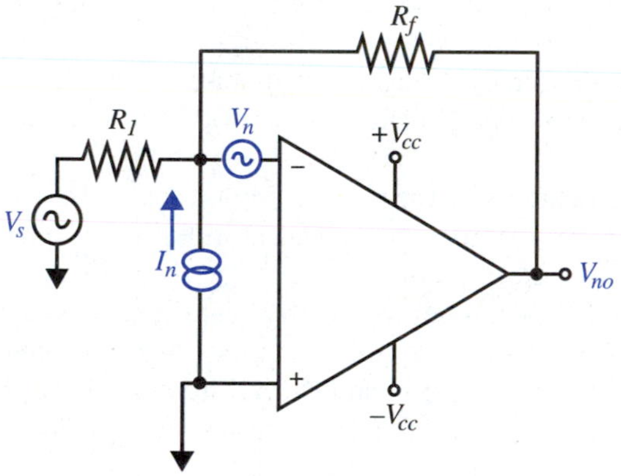

FIGURE 8–14 Equivalent noise circuit

$$V_n^2/\text{Hz} = 5 \times 10^{-16}$$

and if we multiply this number by the signal bandwidth, we get

$$V_n^2 = 5 \times 10^{-16} \times 100$$

$$V_n^2 = 5 \times 10^{-14}$$

From the current noise graph,

$$I_n^2 = 8 \times 10^{-25}$$

and if we multiply this number by the signal bandwidth, we get

$$I_n^2 = 8 \times 10^{-25}\,(100)$$
$$I_n^2 = 8 \times 10^{-23}$$

Now

$$R_{eq} = 100\text{K}\Omega / 1\text{K}\Omega$$
$$R_{eq} = 990\Omega$$
$$R_{eq}^2 = 9.8 \times 10^5$$

and

$$A_v = -\frac{R_f}{R_1}$$
$$A_v = -100$$

Putting these values in equation 8–21,

$$V_{on} = A_v\sqrt{(V_n)^2 + (I_n)^2(R_{eq})^2}$$
$$V_{on} = 100\sqrt{5 \times 10^{-14} + (8 \times 10^{-23})(9.8 \times 10^5)}$$
$$V_{on} = 22.38\mu\text{V}$$

We have looked at the noise generated by the op-amp but what about the thermal noise of R_f and R_1? The thermal noise voltage from an ideal resistor can be determined by the equation

$$V_{nr} = \sqrt{4KT(\text{BW})R} \qquad \textbf{(equation 8–22)}$$

where:

K is Boltzman's constant 1.38×10^{-23}J/°K

T is the absolute temperature in °K (degrees Celsius plus 273°)

BW is the bandwidth in Hz

R is the resistance in ohms and is the equivalent value of the R_f and R_1 resistors in parallel in our amplifier circuit.

Taking this resistor noise into account, equation 8–21 is modified to

$$V_{on} = A_v\sqrt{(V_n^2) + (I_n^2)(R_{eq}^2) + 4KT(\text{BW})R} \qquad \textbf{(equation 8–23)}$$

EXAMPLE

Find the effect on the result of adding the thermal resistance noise of R_f and R_1 to the previous example at 25°C.

SOLUTION

With the bandwidth of 100Hz, equivalent resistance of 990Ω, and temperature of 25°C, the noise term is

$$V_{nr}{}^2 = [4KT(\text{BW})R]$$

$$V_{nr}{}^2 = 4 \times 1.38 \times 10^{-23} \times (273 + 25) \times 100 \times 990$$

$$V_{nr}{}^2 = 1.623 \times 10^{-15}$$

Substituting this value, and the values from the previous example, in equation 8–23

$$V_{on} = A_v\sqrt{(V_n{}^2) + (I_n{}^2)(R_{eq}{}^2) + 4KT(\text{BW})R}$$

$$V_{on} = 100\sqrt{5 \times 10^{-14} + 8 \times 10^{-23} \times 9.8 \times 10^5 + 1.632 \times 10^{-15}}$$

$$V_{on} = 22.74\ \mu\text{V}$$

Comparing this result with that of the previous example, we see that the thermal resistance noise only increases the output noise by 1.6%.

Another noise consideration with resistors is that carbon resistors have intergranular contact, which causes additional noise above the thermal noise. So carbon resistors should not be used in the input stage of a low-noise amplifier. A better choice is to use metal film resistors.

AVAILABLE OP-AMPS

Up to this point, we have used the 741 op-amp as our example of the typical op-amp. Now we shall see that other available op-amps can give improved performance over the 741 in various applications.

Table 8–1 lists some representative op-amps along with their key parameters. Notice in the table that each op-amp is optimized for a particular parameter; for example, the

TABLE 8–1 Op-Amp Characteristics

Op-Amp	*A_{ol} (min)*	*GBP (MHz)*	*Slew Rate (V/μs)*	*Bias Cur. (nA)*	*Cur. Offst (nA)*	*Volt Offst (mV) max*	*CMR db*	*PSR db*	*Input Noise (V^2/Hz × 10^{-16})*	*Cost Fact*	*Key Features*
741C	20K	1.0	0.5	500	200	6.0	70	77	2.5	1.0	Low Cost
LF351	25K	4.0	13	0.1	0.025	10.0	70	77	2.5	1.4	FET Input.741 Pin-out
NE5534	25K	10.0	13	1500	300	4.0	70	80	0.4	1.8	Low Noise. 741 Pin-out
LP324	40K	0.1	0.05	20	4	9.0	75	74	—	2.6	Quad, Single supply. 125μA.
LM607	1500K	1.0	0.4	6.0	5.6	0.12	112	112	0.64	31.0	Precision.741 BM Pin-out
AD624A	1K (Av)	25.0	5.0	50	35	0.2	100	100	0.4	—	Instrumentation amp.
LMC660C	80K	1.4	0.7	0.002	0.001	6.3	62	62	4.8	5.5	C-MOS Quad
LM675	3.2K	5.5	8.0	2000	500	10.0	90	90	—	18.0	20 Watt Output
LM6365	5.0K	400	180	6000	1900	7.0	78	78	—	8.0	Wide BW 741 Pin-out

LMC660C has a very low bias and offset current but is not so good at common mode and power supply rejection (62db).

Let us look at each op-amp and find its most suited application based on key features.

- **741C** This op-amp has been used more than any other over the years because it is easy to use and has a very low cost.
- **LF351** Pin compatible with the 741 but has FET input circuitry, which accounts for the low bias and offset currents. Also has four times the GBP of the 741 and 26 times better slew rate. The low input current makes it a better choice than the 741 for integrator circuits, and the wider bandwidth and better slew rate improve audio frequency circuit performance.
- **NE5534** The key features of lower noise and greater GBP than a 741 makes this op-amp useful for amplifying low-level AC signals at frequencies up to about 100KHz. Pin compatible with 741.
- **LP324** This device contains four individual op-amps that operate from a single power supply ranging from +5 to +30V. The low power supply drain current maximum of 125μA compared with 2.8mA for a 741 makes this a good choice for battery-powered equipment.
- **LM607BM** High gain, low bias currents, input voltage offset, and noise, along with very high common mode and power supply rejection are the reasons this op-amp is called a precision device. Application would include low-level DC amplifiers.
- **AD624A** This device includes three op-amps in the high input resistance instrumentation configuration of Figure 9–12. Gain is programmed from 1 to 1,000 by using a single precision resistor between two of the terminals on the chip. It is a good choice for amplification in a noisy environment—which requires an instrumentation amplifier.
- **LMC660C** Has four C-MOS op-amps in one package that can operate from a power supply ranging from +5 to +15V. Unlike the 741 where output swing is about 2V less than the power supply, this op-amp can swing from the positive supply to ground (called **rail-to-rail swing**). This means that this op-amp should be used if the maximum output swing needs to be predictable. Input current offset is negligible (femto-amps), but the noise and common mode and power supply rejection are worse than the 741.
- **LM675** This is a power-type op-amp with a maximum output power of 20 watts. The high GBP means that it can easily operate beyond the audio frequency range.
- **LM6365** An extremely high GBP and slew rate are characteristic of this op-amp. These parameters mean sharp rise and fall times if this op-amp is used for a pulse amplifier or generator.

SUMMARY

- Biasing currents for the op-amp input transistors causes a voltage offset at the output of the op-amp.
- Output offset voltage creates errors when amplifying DC signals and causes clipping with large output AC signals

- Offset caused by I_{ba} is this current times R_f.
- If an R_2 compensating resistor is used, the output offset is the difference between I_{ba} and I_{bb} (I_{os}) times R_f.
- Input voltage offset is caused by the difference in base-emitter drop of the two input transistors of the op-amp.
- The output offset due to input voltage offset is the input voltage offset times the noninverting gain.
- A voltage offset can result from a change in the power supply voltage.
- Power supply rejection indicates how well an op-amp rejects changes in the power supply—the higher the rejection, the lower the generated offset voltage.
- Most amplifiers act like a low-pass filter.
- The Bode plot is a straight line approximation of the frequency response curve.
- A phase shift of 180° through a closed loop op-amp will cause oscillation.
- The intercept of the closed loop and the open loop response at a slope no greater than −20db/octave ensures stability.
- A frequency-compensated op-amp has the frequency response of a first-order low-pass filter and therefore is stable.
- The open loop frequency response curve for an op-amp provides the open loop gain at a given frequency.
- The gain-bandwidth product (GBP) can also be used to find the gain at a given frequency or the bandwidth given the gain.
- Reduction of open loop gain at the higher frequencies should be considered for the effect on closed loop gain, input resistance, and output resistance.
- Slew rate is the volt/μs slope capability of the op-amp.
- Slew rate is caused by limited current sources within the op-amp.
- Slew rate can cause square wave and sine wave inputs to become triangular at the output of the op-amp.
- The bandwidth of an amplifier controls the rise time of signals at the output of the op-amp.
- The output of an amplifier is rise time limited if it has a high enough slew rate to provide the greatest slope required by rise time.
- Amplification of weak signals is limited by the internal noise of the amplifier.
- Resistor thermal noise is usually lower than the internal amplifier noise.
- Carbon resistors are not a good choice for low-noise circuits.

EXERCISE PROBLEMS

1. Describe the effect of input bias currents on the voltage at the output of the op-amp.

2. How can the effect of input bias currents be reduced?
3. Describe the effect of input offset voltage on the voltage at the output of the op-amp.
4. An amplifier has an R_f of 470KΩ, R_1 of 3KΩ, and I_{ba} of 800nA. What is the output offset V_{oa}?
5. An amplifier has an R_f of 680KΩ, R_1 of 3KΩ, R_2 of 3 KΩ, I_{ba} of 700nA, and I_{bb} of 800nA. What is the output offset V_{oi}?
6. An amplifier has an R_f of 470KΩ, R_1 of 5KΩ, and V_{iv} of 5mV. What is the output offset V_{ov}?
7. Determine the value of R_2 required to minimize output offset voltage due to input bias currents if R_f is 56KΩ and R_1 is 4.7KΩ.
8. Find the total offset for a 741 op-amp due to both bias currents and input offset voltage if R_f is 200KΩ, R_1 is 1KΩ, and R_2 is 1KΩ. (Hint: Use maximum offset values from spec. sheet)
9. Determine the voltage offset at the output of the op-amp of problem 8 if the total power supply change is 3V and the PSR is 80db.
10. Compute the PSR in db for an amplifier if the change at the output of the op-amp is 10mV when the power supply changes by 2V. Let R_f be 270KΩ and R_1 be 10KΩ.
11. Determine the minimum stable closed loop gain for an op-amp if the open loop op-amp frequency response has its first corner frequency at 100Hz and the second at 10KHz. DC open loop gain is 120db.
12. A frequency-compensated op-amp has a GBP of 10MHz. If the gain is 100, what is the bandwidth?
13. A frequency-compensated op-amp has a GBP of 20MHz. If the bandwidth is 100KHz, what is the gain?
14. Determine the value of the R_1 compensating resistor required for a 741 differentiator circuit if R_f is 100KΩ and C is 1μF.
15. We wish to differentiate a $0.5V_{p-p}$ 1KHz square wave and have a peak output spike of 5V using a 741 op-amp. Find the required values of R_1 and R_f, and the maximum frequency if C is 10nF. (*Hint:* Use equation 8–13 to substitute for R_1 in the gain equation.)
16. Compute the gain for an inverting amplifier if the open loop gain is 500, R_f is 330KΩ, and R_1 is 1KΩ.
17. Compute the gain for a noninverting amplifier if the open loop gain is 1000, R_f is 220KΩ, and R_1 is 1KΩ.
18. If the output voltage from a 741 amplifier is a $16V_{p-p}$ square wave, compute the maximum input frequency before the amplitude of the output decreases due to the effect of slew rate. (*Hint:* Slew rate for 741 is 0.5V/μs.)
19. A 741 op-amp with a zero rise time input signal has a gain of 200 and a square wave output of $5V_{p-p}$. Determine whether the output signal is rise time limited or slew rate limited.

20. Determine the maximum voltage (p-p) output sine wave we can have before slew rate distortion occurs at the output of a 741 op-amp if the input frequency is 15KHz.
21. Find the rise time at the output of an amplifier with a gain of 100 and a GBP of 50MHz if the input signal has a rise time t_{rs} of 0.4μs.
22. Find the actual measured rise time on a 5MHz bandwidth oscilloscope for the output signal of problem 21.
23. Find the noise generated in a 10KΩ resistor over a bandwidth of 1KHz at 35°C.
24. A 741 op-amp has an inverting gain of 50 with R_f of 200KΩ. If the input signal center frequency is at 200Hz and requires a bandwidth of 50Hz, find the noise voltage at the output of the amplifier at 25°C.
25. Select an op-amp from Table 8–1 to be used to measure a weak signal in an electrically noisy environment.
26. Select an op-amp from Table 8–1 to be used in an integrator circuit.
27. Select an op-amp from Table 8–1 to be used as a battery-powered low-frequency amplifier.
28) Select an op-amp from Table 8–1 to be used in an application where +12V is available and an output voltage swing of ±6V is required.

ABBREVIATIONS

I_{ba} = Inverting input bias current
I_{bb} = Noninverting input bias current
I_{os} = Offset current $|I_{ba} - I_{bb}|$
V_{oa} = Output voltage from $I_{ba} \times R_f$
V_{ob} = Output voltage from $I_{bb} \times R_f$
V_{oi} = Output voltage offset due to input current offset ($I_{os} \times R_f$)
V_{iv} = Input offset voltage
V_{ov} = Output offset voltage due to input voltage offset ($A_v \times V_{iv}$)
V_{oc} = Combined output voltage from both bias current and offset voltage
V_{op} = Output voltage offset due to changes in the power supply voltages
V_{on} = Output noise voltage due to internal amp noise
V_{nr} = Noise voltage from resistor noise (thermal)
PSRR = Power supply rejection ratio
PSR = Power supply rejection in db
GBP = Gain-bandwidth product
SR = Slew rate
t_{sr} = Time to slew a given voltage
f_{sr} = Slew rate limited frequency

t_{ra} = Amplifier rise time
t_{rs} = Input signal rise time
t_r = Measured rise time—combines input and amplifier rise times

ANSWERS TO PROBLEMS

1. Input bias current can create an output offset voltage.
2. Reduce R_f and use a compensating resistor R_2.
3. Input offset voltage can create an output offset voltage.
4. $V_{oa} = 0.376V$
5. $V_{oi} = -68mV$
6. $V_{ov} = \pm 475mV$
7. $R_2 = 4.34K\Omega$
8. $V_{oi} = \pm 40mV$; $V_{ov} = \pm 1.206V$; $V_{oc} = \pm 1.246V$
9. $V_{op} = \pm 60.3mV$
10. PSRR = 5,600 PSR = 75 db
11. $A_{v(min)} = 80db$
12. BW = 100KHz
13. $A_v = 200$
14. $R_1 = 1.26K\Omega$
15. $R_1 = 1.59K\Omega$; $R_f = 15.9K\Omega$; $f_{max} = 314KHz$
16. $A_v = 199$
17. $A_v = 181$
18. $f_{max} = 9.9KHz$
19. Rise time limited
20. $V_{p-p} = 10.6V$
21. $t_{ra} = 0.806\mu s$
22. $t_r = 0.809\mu s$
23. $V_{nr} = 0.412\mu V$
24. $V_{no} = 10.7\mu V$
25. AD624A
26. LF351, because of low input current offset
27. LP 324, low current drain
28. LM660C

CHAPTER 9

Op-Amp Related Devices

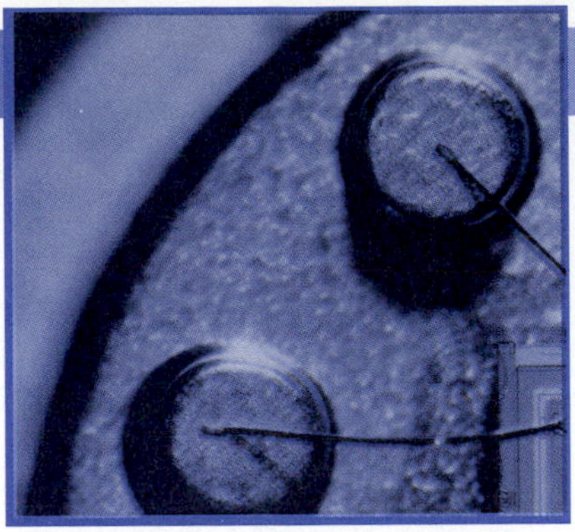

CHAPTER OBJECTIVES

After completing this chapter, you should be able to explain the following op-amp related device concepts:

- The four amplifier transfer functions
- The advantage of a current mode (transimpedance) amplifier
- Purpose of a voltage comparator
- Differences between a voltage comparator and an operational amplifier
- Zero crossing detector
- Schmitt trigger circuit
- Level detector circuit
- Window comparator
- Basic 555 timer operation
- The 555 oscillator circuit
- Output circuit for a 555 timer
- Differences and advantages of the 558 and 3905 timers over a 555 timer

INTRODUCTION

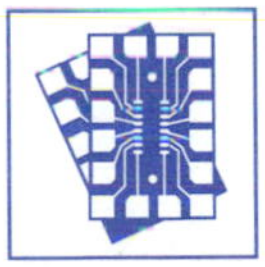

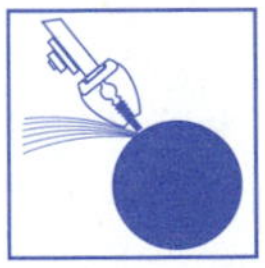

Now that we have a reasonably good background in operational amplifiers, we shall look at devices that are related to the op-amp or use the op-amp concepts.

We shall start by considering the transfer function of the four basic types of amplifiers and then look at one of these configurations—the transimpedance amplifier. Next we shall introduce the comparator and give examples of applications for this device. Finally we shall study the electronic timer, which uses the comparator technology.

AMPLIFIER TRANSFER FUNCTIONS

As indicated in Figure 9–1, amplifiers can be described by four different transfer relationships of voltage and current. The first type, the current amplifier, is the operating mode of a bipolar transistor with a transfer function of I_c/I_b or β. The second configuration (transconductance amplifier) can be illustrated by the FET, with a transfer function of I_d/V_g or g_m. The third version (voltage amplifier) represents the op-amp we have studied and has a transfer function of V_o/V_{in} or A_{ol}. Finally, we have a transfer function of V_o/I_s which, up

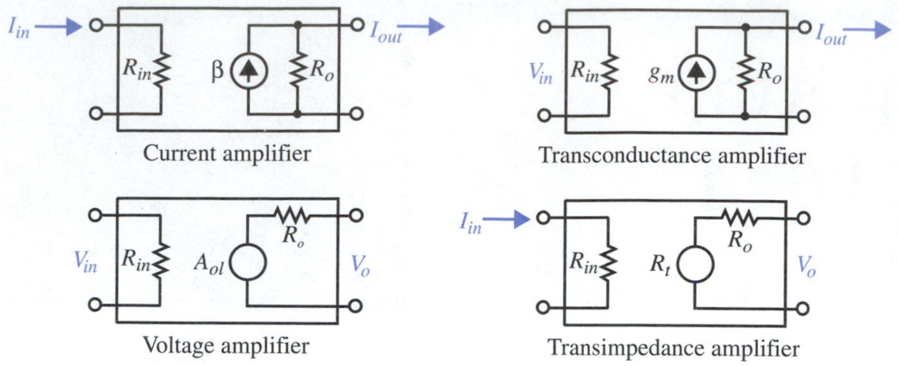

FIGURE 9–1 Four possible amplifier transfer functions

to this point, we have not studied. This last configuration is called a **transimpedance amplifier**, which we shall now consider.

TRANSIMPEDANCE AMPLIFIER

The transimpedance amplifier is the latest op-amp related device that is coming into general usage. Again, the basic difference between an op-amp and a transimpedance amplifier is that the transimpedance device is current controlled rather than voltage controlled. But this is not all, for if we compare the frequency-dependent voltage gain of the two devices, we find that transimpedance (current feedback) amplifier voltage gain-bandwidth product is not fixed—rather the bandwidth is almost independent of gain (see Figure 9–2b).

This frequency response difference in the two types of amplifiers is caused by the fact that with a conventional op-amp the open loop gain is fixed by the internal components

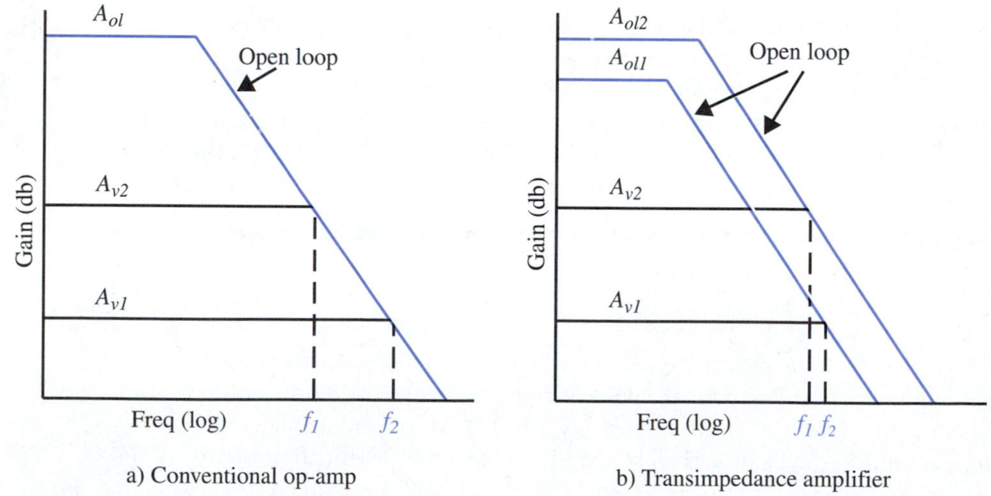

FIGURE 9–2 Comparison of op-amp and transimpedance frequency response

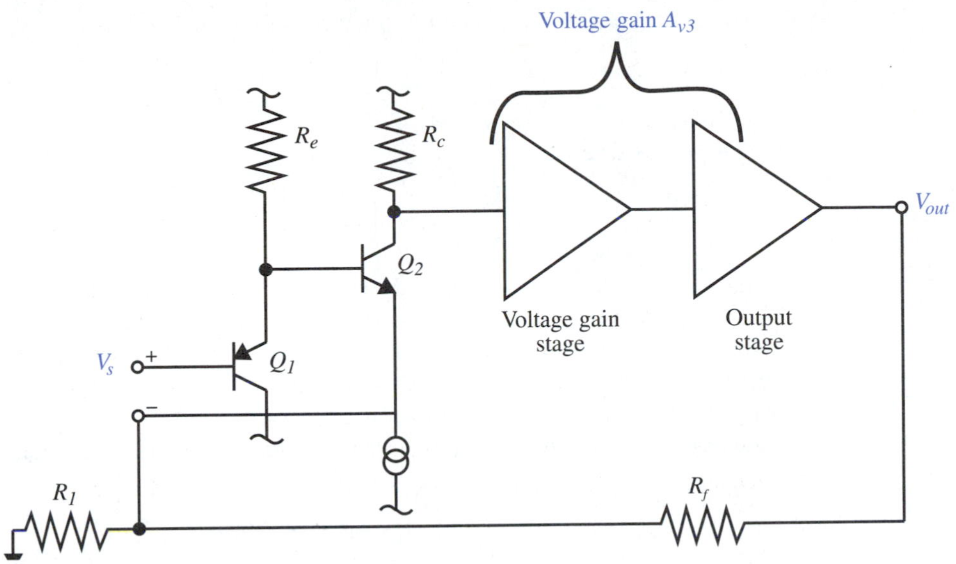

FIGURE 9–3 Transimpedance amplifier input circuit

and rolls off at high frequencies, as shown in Figure 9–2a. But the transimpedance amplifier open loop gain is a function of the value of the R_1 resistor—the lower the value of R_1 (higher gain), the higher the open loop gain. This is reflected in Figure 9–2b with the higher closed loop gain (A_{v2}) giving a higher open loop gain (A_{ol2}).

Why the open loop gain for the transimpedance amplifier is dependent on the value of R_1 will be shown with reference to Figure 9–3.

Figure 9–3 shows a noninverting amplifier configuration. Notice that the R_1 resistor is in the emitter circuit of Q_2 and provides the signal current to this stage via R_f. If we assume that the resistance of R_1 is much less than the high resistance of the emitter constant current source and R_f, then the voltage gain A_{v2} of the Q_2 (common base) stage is

$$A_{v2} = \frac{R_c}{R_1}$$

This equation verifies that the gain is a function of R_1. The overall open loop gain of the amplifier is from the figure,

$$A_{ol} = A_{v1} \times A_{v2} \times A_{v3}$$

The noninverting input signal passes through Q_1 a unity voltage gain follower so $A_{v1} = 1$, which doesn't add to the overall gain. Substituting $A_{v1} = 1$, and $A_{v2} = R_c/R_1$

$$A_{ol} = (R_c/R_1) \times A_{v3} \qquad \textbf{(equation 9–1)}$$

Let us go back to the non-ideal closed loop gain equation for the conventional op-amp noninverting amplifier (eq. 8–15):

$$A_v = \frac{R_f + R_1}{R_1 + R_1/A_{ol} + R_f/A_{ol}}$$

This equation can be inverted and rewritten (assuming $R_f > R_1$) as

$$\frac{1}{A_v} = \frac{1}{A_{vi}} + \frac{1}{A_{ol}} + \frac{1}{A_{ol}A_{vi}}$$

where A_{vi} is the ideal gain

$$\left(\frac{R_f + R_1}{R_1}\right)$$

What this equation shows is that if A_{ol} is large relative to A_{vi}, then A_v is equal to A_{vi}. But we know that A_{ol} is frequency dependent, and at higher frequencies, A_{ol} reduces and is not much larger than A_{vi}. So, the closed loop gain A_v is reduced (beyond f_1 or f_2, as shown in Figure 9–2a). If A_{vi} is increased by, say, reducing R_1, then the amplifier bandwidth is proportionally reduced (constant gain-bandwidth product).

Now if we consider the non-ideal closed loop gain equation for the noninverting transimpedance amplifier,

$$1/A_v = 1/A_{vi} + 1/A_{ol}$$

Substituting from equation 9–1,

$$\frac{1}{A_v} = \frac{R_1}{(R_1 + R_f)} + \frac{R_1}{(R_c A_{v3})} \qquad \textbf{(equation 9–2)}$$

The last term is again frequency dependent because of stray capacitance in the amplifier.

Examining this equation reveals that if we increase the gain by reducing R_1, we also increase the open loop gain of the amplifier ($R_c A_{v3}/R_1$). This means that A_v is not reduced as much as it is with the conventional op-amp, and so the bandwidth is less a function of gain.

EXAMPLE

Find the change in the open loop gain for a noninverting transimpedance amplifier if R_1 is doubled. Also find the closed loop gain if R_f is 1KΩ, R_1 is 200Ω, R_c is 1KΩ, and A_{v3} is 1000.

SOLUTION

From equation 9–2, the relationship for last term A_{ol} is

$$A_{ol} = \frac{R_c A_{v3}}{R_1}$$

We can see from this equation that doubling R_1 will halve A_{ol}.

Using equation 9–2 to solve for the closed loop gain,

$$\frac{1}{A_v} = \frac{R_1}{(R_1 + R_f)} + \frac{R_1}{(R_c A_{v3})}$$

$$\frac{1}{A_v} = \frac{200}{(200\Omega + 1K\Omega)} + \frac{200\Omega}{(1K\Omega \times 1000)}$$

Solving for A_v,

$$A_v = +5.99$$

We can see from this result that the closed loop gain is pretty much independent of A_{ol} and is about the same as we get by just using the noninverting gain equation $(R_f + R_1)/R_1$.

AN ACTUAL TRANSIMPEDANCE AMPLIFIER

A state-of-the-art transimpedance amplifier is the National Semiconductor LM6181. This device has a typical transimpedance of 1MΩ, a bandwidth of 100MHz with a closed loop gain of 2, and a bandwidth of 80MHz with a gain of 10.

Slew rate of conventional op-amps is determined by the size of the internal frequency compensating capacitor and the current available to charge this capacitance. Since the LM6181 doesn't use a compensating capacitor, it has a fast slew rate—with a typical over-driven output slew rate of 2000V/μs. The stated typical rise and fall times are a short 5ns, and this, along with a 100mA output drive capability, makes this device an excellent choice as a pulse driver for the low resistance encountered with data transmission lines.

To maintain the LM6181 high-frequency response, it is suggested that both R_f and R_1 be low values resistors with a typical value of 820Ω for R_f, and R_1 selected for the desired gain.

This concludes our coverage of the transimpedance amplifier.

THE VOLTAGE COMPARATOR

A voltage **comparator** is used to "compare" a varying input signal against a reference voltage. The comparator, like an op-amp, has two input terminals (inverting and noninverting) and one output terminal. When the varying input voltage changes polarity relative to the reference voltage (becomes greater or less than the reference), the output of the comparator changes state. This can be clarified by the circuit shown in Figure 9–4.

Following the convention of an open loop operational amplifier, when the varying input becomes slightly (less than a millivolt) more positive than the +5V reference, the output will swing from the high state to the low state (more positive to less positive). As soon as the varying input becomes less than the reference, the output will swing back from the low state to the high state.

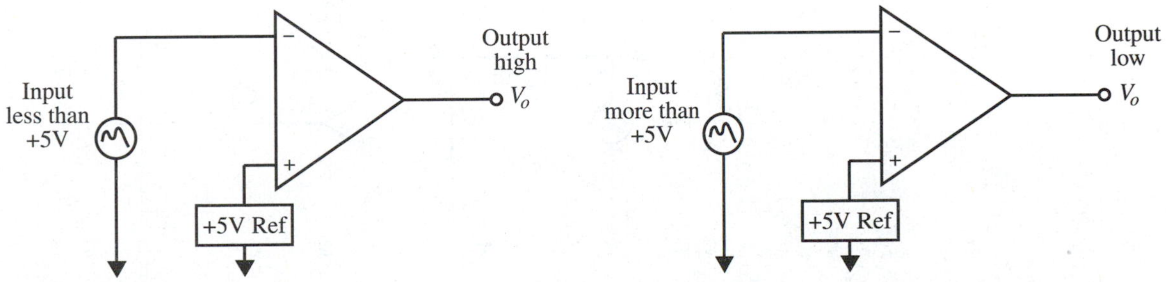

FIGURE 9–4 Voltage comparator

EXAMPLE

Determine the state of the output of the comparator circuit of Figure 9–4 if the reference is changed to −2V and the input voltage is at +1V.

SOLUTION

Since the inverting input is more positive than the noninverting input, the output will be low.

VOLTAGE COMPARATORS VERSUS OP-AMPS

The circuit of Figure 9–4 can use a conventional operational amplifier. However, voltage comparator integrated circuits are specifically made for this application with fast output voltage switching capability, which allows the comparator to be used as an interface device with digital logic circuits without creating excessive delay in circuit operation. This delay results from the requirement to reach a particular voltage level at the logic gate before it can operate (+2V for TTL logic). With a slow changing output from the comparator, an appreciable time can elapse from when the comparator changes state to when the logic is triggered. This creates what is called a **propagation delay**! If we were using a 741 op-amp with its 0.5V/μs slew rate, it would take 4μs to reach the 2V TTL trip point. A propagation delay of this magnitude could be intolerable in a given application.

Another advantage of the comparator over the conventional (741) type of op-amp is a lower offset current. This means the comparator will trigger closer to the desired reference voltage.

ZERO CROSSING DETECTOR

The **zero crossing detector** is a special case of the comparator with a grounded reference. It can be used to sense the zero voltage crossings of a sine wave. At the zero crossing point, the zero crossing detector creates an output with a well-defined transition for both the positive and negative input slopes. The fast rise time of the comparator output can quickly trigger digital circuits, which allows measurement of the frequency or phase of the sine wave. Figure 9–5 illustrates a typical zero crossing detector circuit. Input voltage offset can cause a slight error in the crossing point. If this is critical, an offset pot should be used.

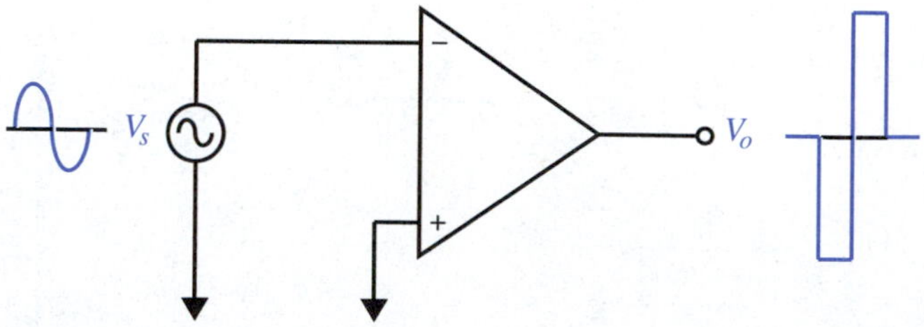

FIGURE 9–5 Zero crossing detector

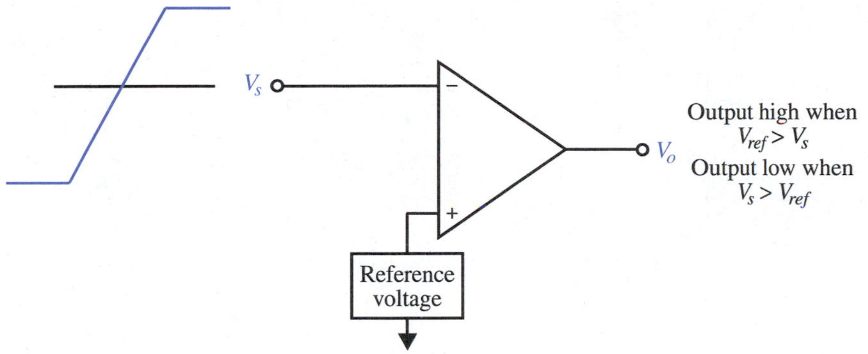

FIGURE 9–6 Level-sensing circuit

LEVEL-SENSING CIRCUIT

A **level-sensing circuit** determines when a particular predetermined voltage level is present at the input. In the circuit shown in Figure 9–6, the output changes from a high level to a low level when the input signal exceeds the reference voltage. If the input terminals are reversed, with the reference on the inverting terminal and the input on the noninverting terminal, the output will go from low to high when the input exceeds the reference.

The input voltage, for example, could be from a liquid-level sensor in a water tank, with the voltage from the sensor indicating the level in the tank. When the input voltage exceeds the reference, the change in the output signal from the comparator shuts off power to the motor that pumps water into the tank. Thus, by adjusting the reference voltage, we can change the water level in the tank (higher reference voltage means more water in the tank).

VOLTAGE COMPARATOR AS A SCHMITT TRIGGER

If noise is present on the input signal to a comparator, the output will "jitter" when the input signal is close to the reference value. This can be a problem if the output transitions are used to measure the frequency (zero crossings) of the input signal. A circuit that eliminates this problem is the Schmitt trigger circuit, shown in Figure 9–7.

Notice that this is a comparator circuit with positive feedback from the output to the noninverting input. Let us now look at the detailed operation of this circuit.

We will assume that the output will normally be high at +10V, and with the voltage divider shown, the noninverting input is then biased to +1V.

When the input exceeds +1V, the output will switch to −10V, and the noninverting input will go to −1V.

Now in order for the output to return to +10V, the input signal has to drop below −1V. So, there is a 2V difference between the trip points. This is called the circuit **hysteresis**.

To show how hysteresis increases noise immunity, consider the noisy waveform in Figure 9–8. If a zero crossing detector is used, an output transition occurs each time the waveform passes through zero (six times in the waveform shown in Figure 9–8); but with the Schmitt trigger circuit, once the input waveform exceeded the +1V upper trip point, a noise

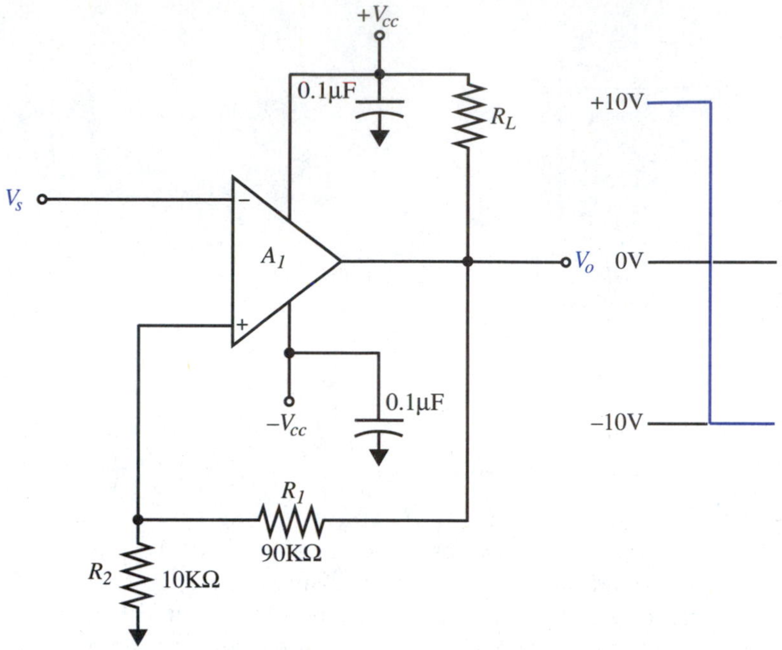

FIGURE 9–7 Schmitt trigger

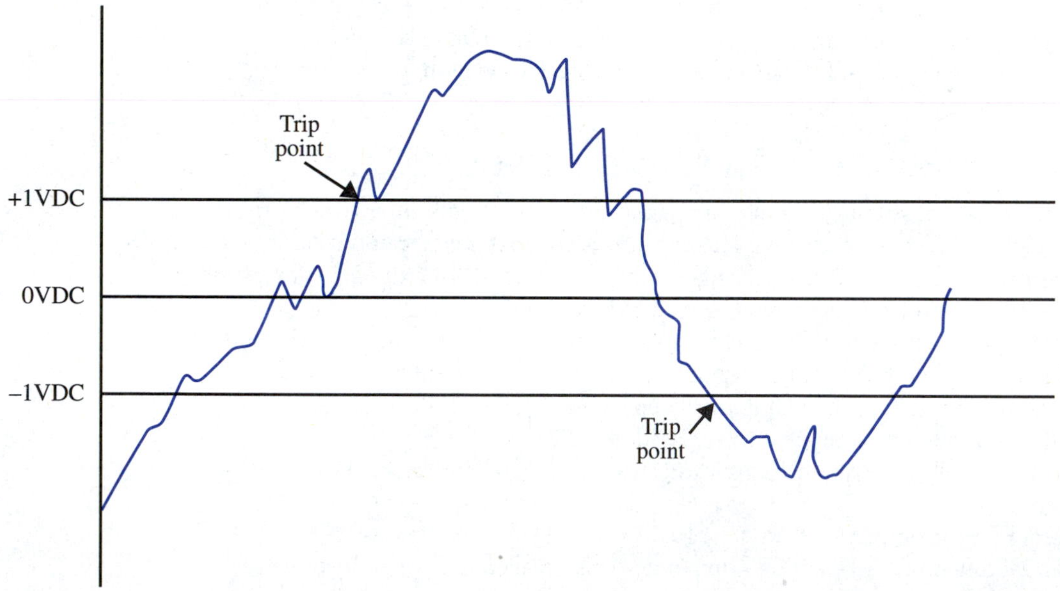

FIGURE 9–8 Noise waveform

spike has to be 2V in order to reach the lower −1V trip point. The amount of hysteresis can be varied by changing the ratio of the voltage divider resistors.

If the anticipated noise is low (millivolts), the divider ratio is increased to provide a lower offset on the noninverting input terminal. Note that the resistor R_L becomes part of the voltage divider when the output is high—so this resistor should be much smaller (one-hundredth) than the value of R_1.

EXAMPLE

Determine all circuit values for a Schmitt trigger circuit with a total hysteresis of 40mV using a LM311 comparator operating off ±10V supplies. Load to draw a maximum current of 10mA.

SOLUTION

With the collector of the comparator output transistor at −10V, the maximum current of 10mA will flow through the load resistor. The value of the load resistor will be:

$$R_L = \frac{+10\text{V} - (-10\text{V})}{10\text{mA}} = 2\text{K}\Omega$$

The resistance of R_1 and R_2 in series must be at least a hundred times larger than R_L or 200KΩ. These two resistors must also divide down the output voltage to provide 40mV/2 or 20mV when the output is at 10V. This requires a 500 : 1 divider, which means that R_1 must be 499 times larger than R_2. Choosing a R_1 value of 200KΩ sets the value of R_2 at:

$$R_2 = \frac{200\text{K}\Omega}{499} = 4.008\text{K}\Omega$$

WINDOW COMPARATOR

Sometimes we need to sense when a voltage is between two limits—or when it is above an upper limit or below a lower limit. An example is a voltage that represents a gauging operation where the measured part's dimensions must fall within two limits to be acceptable. A circuit that performs this function is called a **window comparator**, a version of which is shown in Figure 9–9.

The output circuits use a light-emitting diode (LED) to indicate when the input voltage is above the upper limit (V_H) or below the lower limit (V_L). The outputs of the comparators have open collectors that draw currents through the LEDs when the outputs go low. We could have used a relay or other device for the output indication, because the actual comparison is independent of the output circuit.

We will now study the detailed circuit operation.

When the input voltage exceeds the upper limit, the U_1 output ($V_{o\,HIGH}$) goes low to ground potential, which turns on the high LED. With this high input condition, the U_2 output ($V_{o\,LOW}$) is +12V, and the low LED is turned off.

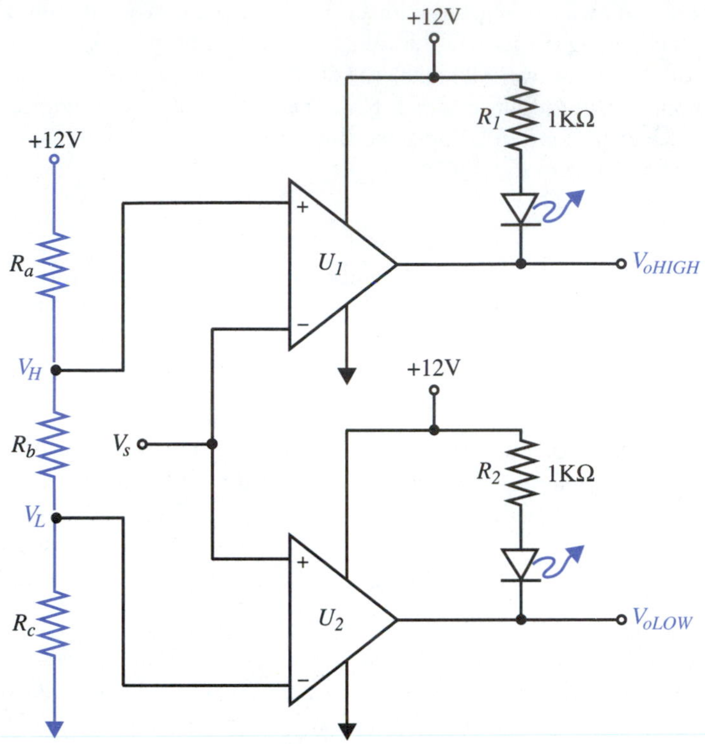

FIGURE 9–9 Window comparator

If the input voltage is now lowered so that it is between the upper and lower limits, both U_1 and U_2 outputs will be high (their output transistors *off*), and no current will flow in either LED.

The final situation is with the input voltage below the lower limit. This causes the U_1 output to be $+12V$ and the U_2 output to be ground potential, which forward biases the low LED.

The supply voltage used for the LEDs can assume any value providing that it stays within the maximum voltage value of the comparator and that the series resistor is changed to give the correct LED current.

A window comparator circuit, which uses a single LED to indicate when the input is within a given range, can be constructed by adding a transistor, as shown in Figure 9–10. With the input voltage within the upper and lower limits, both comparators internal output transistors are turned off. This allows current to flow into the base of Q_1, turning on the LED.

If the input voltage is outside of the limits, one comparator output transistor turns on and the base of Q_1 drops to about 0.2V. Since this voltage is not great enough to bias Q_1, the LED is turned off.

A TYPICAL VOLTAGE COMPARATOR

An actual integrated circuit comparator is the LM311. This device has a high input impedance with a maximum input bias current of 275nA and a maximum input voltage offset of 7.5mV.

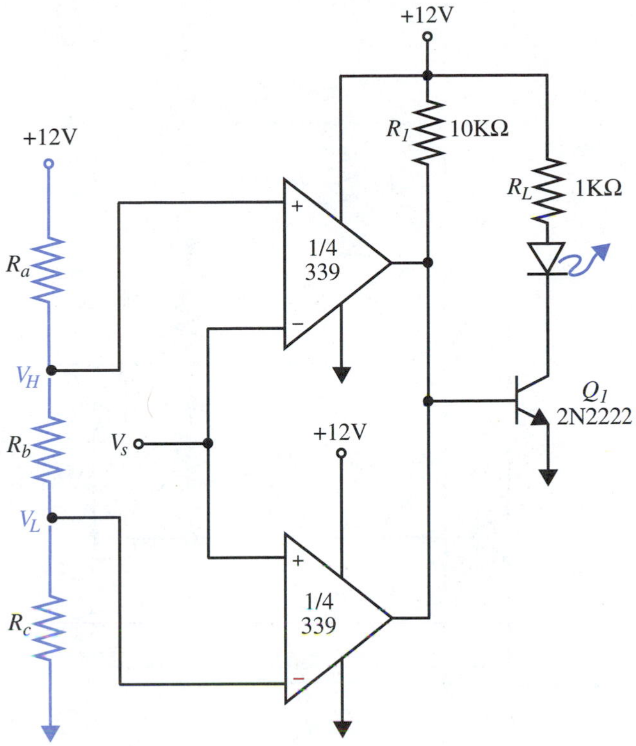

FIGURE 9–10 Single LED window comparator

The voltage on the input leads can be the same as the maximum supply voltages (+15V), which accommodates a large input voltage range. Both the input and the output circuits can be isolated from system ground, and the output circuits can drive loads connected to ground, to the positive supply, or to the negative supply.

The partial input/output schematic of Figure 9–11 illustrates these features. Notice that the output (Pin 7) is uncommitted (open collector), and we have the option of connecting the GND (Pin I) to system ground or to the negative supply. We can, therefore, connect the LM311 output load resistor in any of the three configurations shown in Figure 9–12. A disadvantage of the open collector is the longer rise time of output signals when compared to the totem pole output.

In Figure 9–12a we have an output voltage swing from $+V$ to $-V$ or ground. The output voltage swing in Figure 9–12b is from $+V$ to ground, and in Figure 9–12c it is from ground to $-V$.

We further have the option of connecting the load resistor (R_L) to a positive supply different from $+V$, providing we don't exceed 40V more positive than $-V$. This gives us tremendous flexibility with our output signal—which is good because comparators are required to interface with many different types of circuits.

If we don't connect Pin 1 to system ground, then the inputs are isolated from this connection. This can eliminate noise from appearing at the input via a common ground lead.

The typical switching time for the LM311 is 200ns. If faster speed is required, the LM160 with a maximum switching time of 20ns could be used. However, the faster comparators

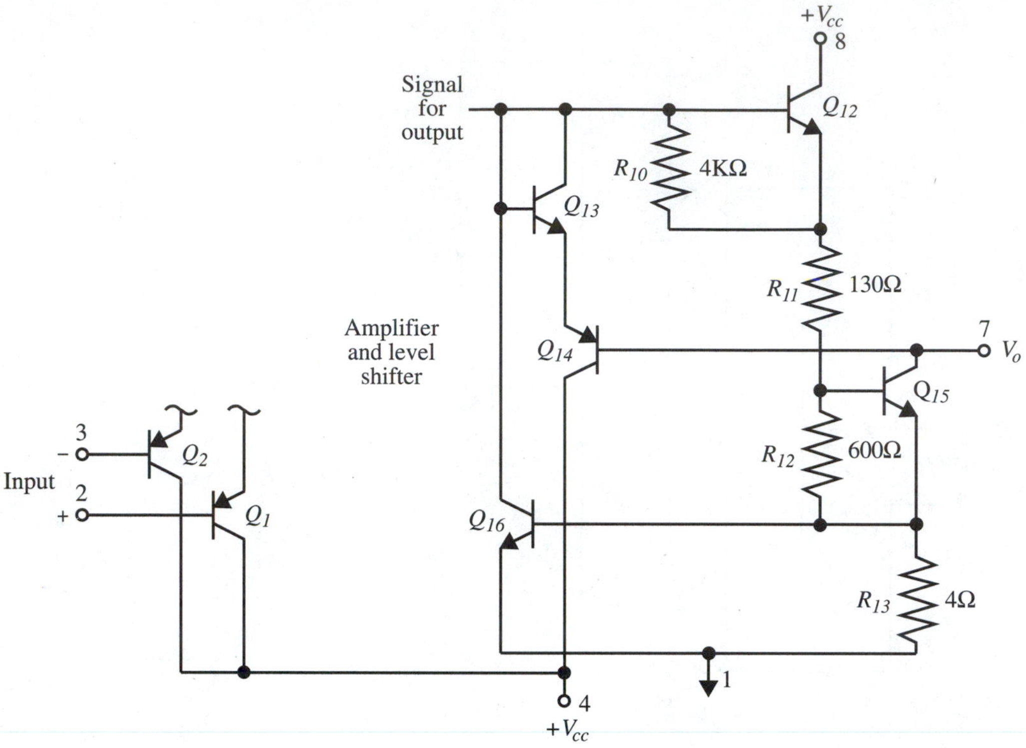

FIGURE 9–11 LM311 input/output circuit

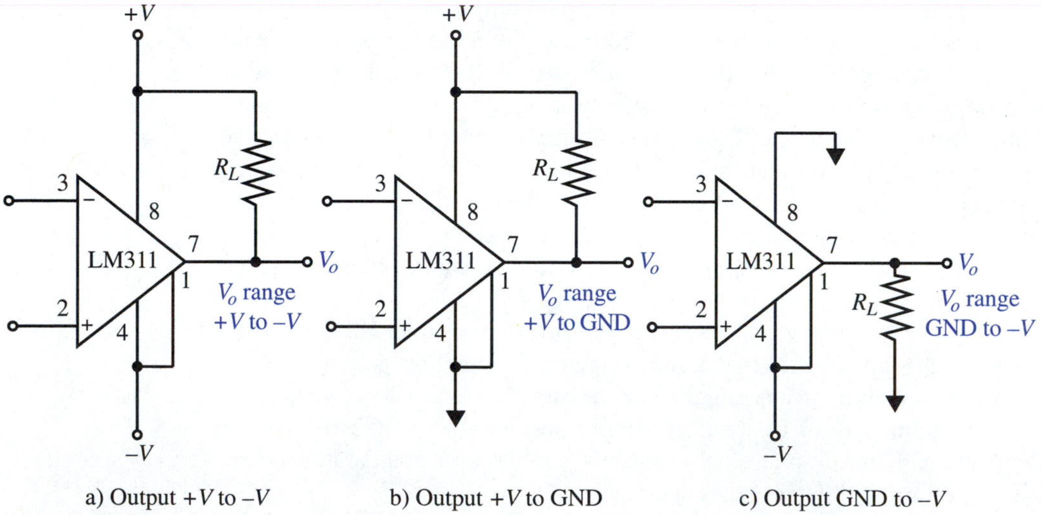

FIGURE 9–12 LM311 load resistor configurations

require care in circuit layout because of the tendency to oscillate; it is a good idea to use 0.1μF bypass capacitors right at the chip supply pins.

A much used comparator is the LM339 (or the low-power LP339), which consists of four comparators on a single chip. This device has a rather slow switching time of 1.3μs, but is a good choice in many low-frequency applications. Like the LM311, the LM339 has open collector outputs. We shall now look at other applications for comparators.

FLUID-LEVEL INDICATOR

A bar display can use light-emitting diodes (LED) to indicate the level of an input voltage. The actual display resembles a thermometer, with all lights mounted vertically (or horizontally) and the number of LEDs illuminated corresponds to the level of the input voltage. Figure 9–13 shows a circuit for a gasoline-level indicator.

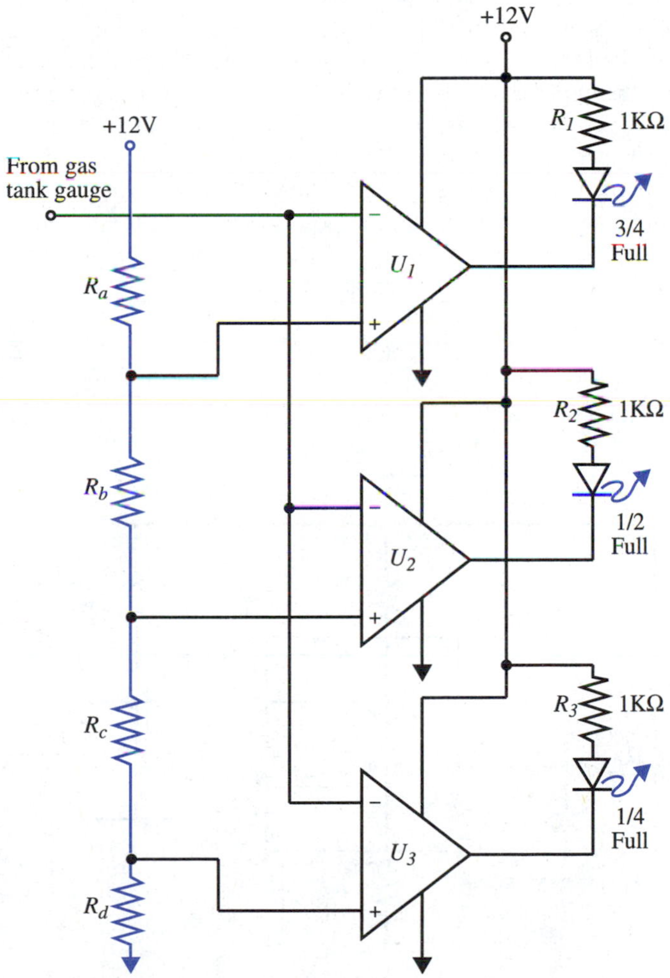

FIGURE 9–13 LED gas gauge

The circuit assumes that +12V on the fuel gauge corresponds to a full tank and 0V to an empty tank (not the case with an actual fuel gauge). If the tank is greater than three-quarters full, the voltage is greater than 9V and all comparators have a "low state" on the output, which will cause all LEDs to illuminate. When the tank drops below three-quarters full, the top LED will extinguish. Below a quarter of a tank, all LEDs extinguish.

In an actual situation, one comparator circuit could be used for each gallon of fuel, e.g., a 10-gallon tank would use 10 circuits and the LEDs could be marked in gallons. For this application, the LM3914 dot/bar display device, which contains 10 comparators, could be used.

AC LINE MONITOR CIRCUIT

Another application for the voltage comparator is in a line voltage monitor circuit, which is used to disconnect a piece of equipment from the AC line if the line voltage falls outside of a predetermined range of values. Figure 9–14 shows such a circuit, which operates if the line voltage differs by more than 10 percent from the nominal value.

In operation, the LMI36A-5 is a 5V voltage reference device with a tolerance of ±1%, which provides the upper limit voltage at the inverting input or U_2 and the lower limit of +4V via the voltage dividers R_4 and R_5 to the noninverting input of U_1.

The voltage divider R_1 and R_2 is set by adjusting R_1 to provide +4.5V at point **A** with the nominal line voltage. Variations in the AC line will cause point **A** to change. If the line voltage increases above 10%, the voltage at point **A** will exceed +5V, and comparator U_2 output will switch high deactivating relay K_2, which will open the circuit to the load. With the line voltage lower than 10%, comparator U_1 will switch and relay K_1 will open the load circuit.

The normally open contact (NO) relays used (K_1 and K_2) should have a coil pull in current of less than 20mA at a voltage of about 14V (low line). All resistors except R_s, R_2, and R_3 should be metal film, with R_4 and R_5, 1% or better tolerance.

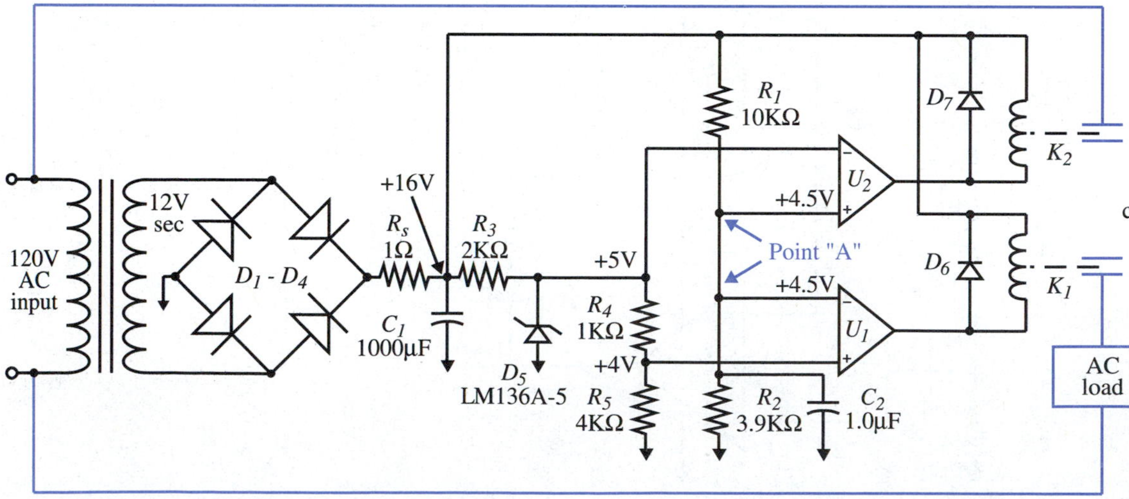

FIGURE 9–14 Under-over voltage protection circuit

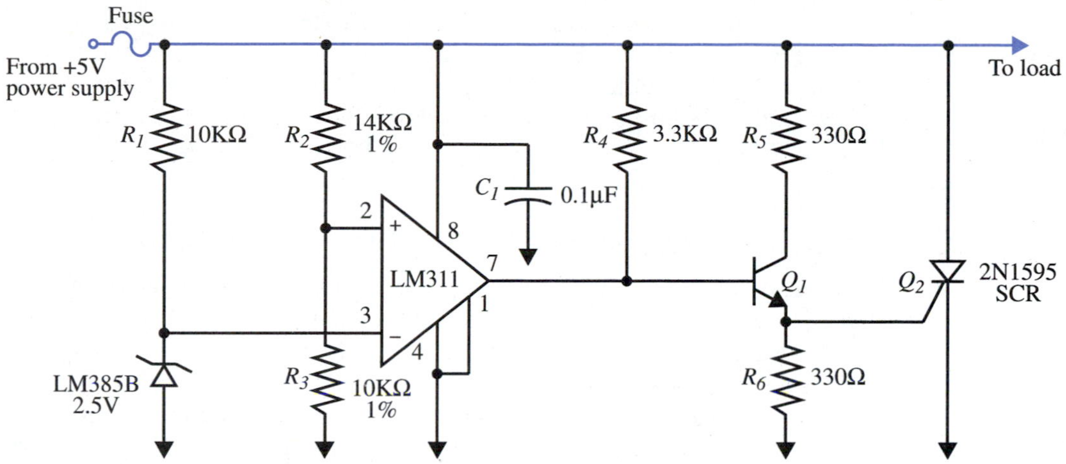

FIGURE 9–15 TTL power supply crowbar circuit

C_2 is selected to reduce the ripple or voltage transients by a factor of 100 from the voltage appearing across C_1. In turn, C_1 should be large enough to allow less than 1V ripple at the junction of R_5 and R_3.

CROWBAR CIRCUIT

A final comparator application is a **crowbar circuit**, which protects against an over-voltage condition causing damage to circuit components. The term *crowbar* refers to a protection approach which consists of throwing a short circuit across the power supply terminals when an over-voltage occurs. Such a circuit—using a comparator, voltage reference, and a silicon controlled rectifier (SCR—described in Chapter 13)—is shown in Figure 9–15.

The circuit is designed to prevent more than 6V from appearing on the transistor transistor logic (TTL) logic chips with a failure of power supply regulator. It does this by throwing a short across the supply using the Silicon Controlled Rectifier (SCR) and blowing the fuse. Maximum power supply voltage for the TTL is 7V—so we are allowing a 1V safety margin.

In operation, the LM385B provides 2.5V ±1.5% at the inverting terminal of the comparator. With the power supply at +5V, R_2 and R_3 establish +2V at the comparator noninverting terminal. This voltage makes the comparator's output low, preventing forward biasing of the base Q_1. Thus, Q_1 and Q_2 are normally off.

If the supply voltage increases to just greater than +6V, the voltage at the noninverting terminal exceeds the reference voltage and turns off the comparator output transistor. This allows the base of Q_1 to rise, turning on this transistor. When the voltage on the emitter of Q_1 reaches the gate trip point of *SCR* Q_2 (maximum 3V), the SCR fires. The SCR becomes a short across the power supply, when it fires, causing the fuse to blow. Rating of the fuse is determined by the normal maximum circuit current flow. A fuse size twice this current should suffice—but the fuse should be a "fast blow" type.

Considering the tolerance of the reference voltage and R_2 and R_3, the firing point of the circuit can be off by 3.5% or 0.21V. This means the circuit trip point could be as low as 5.79V or as high as 6.21V. If the power supply regulator is 5V ±4% (LM323 or LM309), the output could be as high as 5.2V, which gives a margin of (5.79V − 5.2V) or 0.59V.

TROUBLESHOOTING COMPARATOR CIRCUITS

Troubleshooting comparator circuits is similar to troubleshooting op-amp circuits. If the output is low, we can determine whether the comparator is good by jumpering the inverting input to the most negative voltage (ground or negative supply) and the output transistor should cut off causing the output to go high. With the output initially high, jumpering the inverting input to the positive supply should cause the output transistor to turn on and the output to go low.

Remember, a comparator will not operate correctly without a load resistor connected from the output transistor collector to a positive (relative to the transistor emitter) power source.

ELECTRONIC TIMERS

There are many applications that require a delay from the occurrence of one event until the start of another, e.g., the various cycles (rinse, wash, spin, etc.) of a modern washing machine, a traffic signal, etc. Electronic timers can provide this delay with times ranging from microseconds to hours. In other applications, electronic timers can provide a time window, during which a certain event is allowed to occur. This could be part of a burglar alarm sequence that allows a person to leave a building for a short period of time after the alarm is set. The timer would disable the alarm for a period of approximately 30 seconds. Timers also can be connected so the timing cycle continuously repeats. This is the astable or oscillator mode of operation. Applications for timers are almost as varied as those for op-amps. This section should give the reader a basic understanding of these devices sufficient to follow most timer configurations—and, if necessary, to improve upon existing designs.

LM555 ELECTRONIC TIMER

One of the more common types, and the most basic electronic timer, is the LM555. Figure 9–16 shows a functional block diagram of this timer in a one-shot circuit configuration. One-shot means that one output pulse is generated each time the circuit is triggered.

TIMER CIRCUIT OPERATION

The overall circuit operation starts when a negative trigger pulse is applied to Pin 2 in Figure 9–16, causing a positive pulse to occur at the output, which stays high for a period determined by the values of R_t and C_t.

Let us now look at the detailed steps of the timer circuit operation:

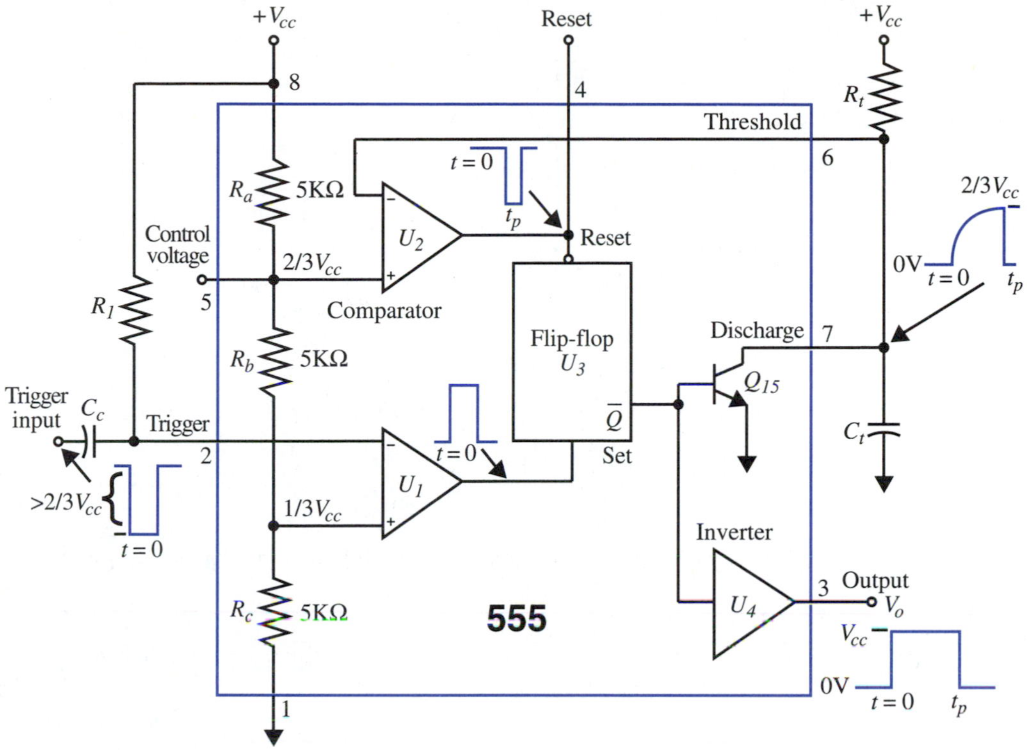

FIGURE 9–16 LM555 basic timer circuit. *Courtesy National Semiconductor.*

- If a negative pulse of sufficient amplitude is applied to the trigger input, it will drive Pin 2 from its normal value of V_{cc} to less than the internal voltage divider voltage of $\frac{1}{3}V_{cc}$, causing comparator U_1 output to go high.
- U_1 output going high sets the flip-flop, causing the "Q not" output to go low, which turns off Q_1 and also causes the timer output to go high ($+V_{cc}$).
- With Q_1 turned off, timing capacitor C_t is allowed to charge via the external timing resistor R_t toward V_{cc}.
- When the voltage of C_t exceeds $\frac{2}{3}V_{cc}$, comparator U_2 output goes low and resets the flip-flop. This causes the flip-flop "Q not" output to go high, dropping the timer output low (0V) and turning on Q_1.
- The Q_1 turn-on abruptly discharges C_t and holds the voltage across it at the Q_1 saturation voltage (about 0.2V).

This completes the total cycle, which caused the output to stay in the high state for the amount of time it took to charge the capacitor to $\frac{2}{3}V_{cc}$. So, by varying the time constant R_tC_t, we can vary the time interval the timer output is held in the high state.

Let us now determine the actual relationship between R_t, C_t, and the total time t_p that the output is high (actual timer interval). The instantaneous voltage on a capacitor (v_c) during charging is given by

$$v_c = V_{cc}(I - \varepsilon^{-t/RC})$$

But the time interval ends when

$$v_c = \tfrac{2}{3}V_{cc}\ (U_2 \text{ trip point})$$

So, at the completion of the timing cycle

$$\tfrac{2}{3}V_{cc} = V_{cc}(1 - \varepsilon^{-t/RC})$$

and rearranging,

$$\varepsilon^{-t/RC} = \tfrac{1}{3}$$

Taking the natural logarithm of both sides,

$$-t_p/R_tC_t = l_n(\tfrac{1}{3})$$

Solving for t_p, the output pulse width

$$t_p = 1.1R_tC_t \qquad \textbf{(equation 9–3)}$$

So we can vary the output pulse, then, by simply changing either R_t or C_t.

EXAMPLE

Compute the time the output from a 555 timer is in the high state (timer interval) if R_t is 470KΩ and C_t is 10μF.

SOLUTION

Using equation 9–3,

$$t_p = 1.1(R_t)(C_t)$$

$$t_p = 1.1(470\text{K}\Omega)(10\mu\text{F})$$

$$t_p = 5.17 \text{ seconds}$$

If we connect an external voltage at the control voltage terminal (Pin 5), we can change the divider voltages. This changes the timer cycle time and also the required trigger voltage. Raising the control voltage increases the voltage to the noninverting input of U_2 to above the $\tfrac{2}{3}V_{cc}$ value, which means that C_t has to charge longer to reach the new voltage—so, the timer interval is increased. The noninverting input to U_1 will also be raised, which means the required negative input trigger level is reduced. Connecting the center tap of a pot to Pin 5 allows the timing period to be accurately set to compensate for tolerance variations of R_t and C_t. One end of the pot should be connected to V_{cc} and the other end to ground. A multiturn 1KΩ wire-wound pot would be a good choice for this application.

The on time for the timer of $1.1R_tC_t$ could lead us to believe that we can get any time delay we want just by increasing the value of R_t and/or C_t. But there are practical limitations. The combined internal resistance at terminals 6 and 7 is about 20MΩ. This resistance sets up a voltage divider with R_t as shown in Figure 9–17.

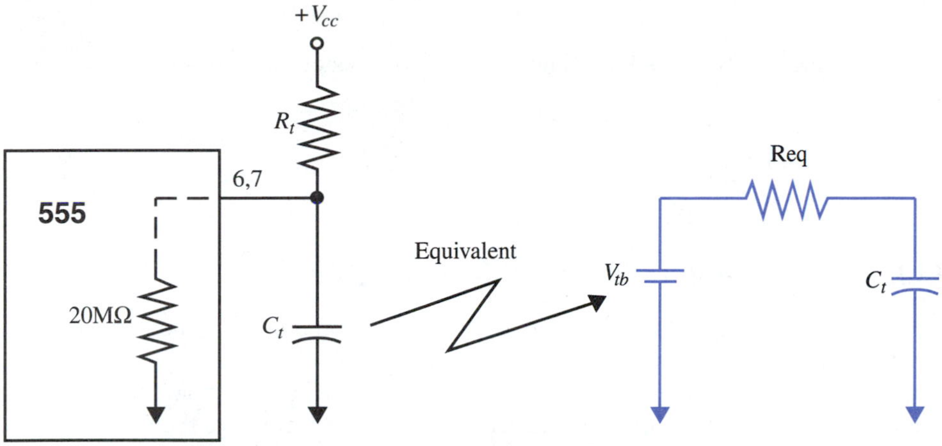

FIGURE 9–17 Voltage divider action by internal leakage resistance

The equivalent for the timing circuit shows that the internal resistance lowers both the effective supply voltage and the effective timing resistor. If, for example, we used an R_t of 20MΩ, the effective supply voltage for charging the capacitor would be halved—which means that the capacitor could never charge to $\frac{2}{3}V_{CC}$ and the timer would never shut off. Lower values of R_t could allow the capacitor to reach $\frac{2}{3}V_{CC}$, but the timing accuracy is still affected until we lower R_t to about 1MΩ. So a conservative approach is to not exceed 1MΩ for R_t.

The other way to increase the time interval is to raise the value of C_t. Increasing the size of this capacitor takes us into electrolytic types, which have lower leakage resistance and looser tolerance. The leakage resistance of the capacitor, again, sets up a voltage divider with R_t. These considerations give us conservative practical maximum timing interval for the 555 of about 15 minutes.

The trigger input signal should not be present when the timer turns off, otherwise the next cycle will begin automatically. A rule of thumb is to have the trigger pulse width less than one-tenth of the required timer interval. The trigger pulse should also be negative and have an amplitude greater than $\frac{2}{3}V_{CC}$. The trigger input (Pin 2) should be tied through a pull-up resistor to V_{CC} (R_1 in Figure 9–16) to prevent inadvertent noise triggering. A 10KΩ resistor is a good choice for R_1.

Manual triggering can be accomplished by momentarily grounding Pin 2 via a push-button switch.

EXAMPLE

Find the input trigger pulse width if the required time interval is 10 seconds and the power supply V_{CC} is 12V.

SOLUTION

Since the rule of thumb is that the trigger pulse be less than one-tenth of the time interval,

$$t_p = 10\text{sec}/10 = 1 \text{ second or less}$$

The negative trigger pulse amplitude should be greater than $\frac{2}{3}$ of V_{cc}, or

$$V_p = 12\text{V} \times \tfrac{2}{3} = 8\text{V}$$

A little bit larger than 8V, or up to 12V, is a good range.

ELIMINATION OF INITIAL TIMING CYCLE

When power is first applied to the timer, a timing cycle could begin without an input trigger. We can see why this can happen with reference to Figure 9–16.

Comparator U_1 inverting input is connected to V_{cc} via R_1 and its noninverting input is tied to $\frac{1}{3}V_{cc}$—so the output of the comparator that is the "set input" to the flip-flop U_3 will be low (not set). At the instant just after power is applied, the voltage across C_t will be zero (capacitor has had no time to charge), which holds the inverting input to U_2 low. The noninverting input to U_2 will be at $\frac{2}{3}V_{cc}$, causing the "active low" reset input to the flip-flop to be high (not reset). With these input states to the flip-flop, its "Q not" output can be high or low (indeterminate). If it is high, Q_1 will turn on, and C_t will be shorted out. However, if the flip-flop "Q not" output is low, a timing cycle will begin.

This inadvertent turn on can be undesirable in many applications because a momentary interruption of power caused, for example, by a lightning strike could cause an alarm to sound or equipment to start operating. A solution to this problem is to connect the timing capacitor C_t as shown in Figure 9–18.

The capacitor has been connected in parallel with R_t rather than in series, but the timer operation is essentially the same. Now, when power is first applied, the inverting input to U_2 in Figure 9–16 is at V_{cc}, causing the reset input to the flip-flop to be in its active low state. The flip-flop output "Q not" will be forced high, turning on Q_1 and grounding pins 6 and 7, and there is no false timing cycle.

The actual circuit operation is as follows (refer to Figures 9–16 and 9–18):

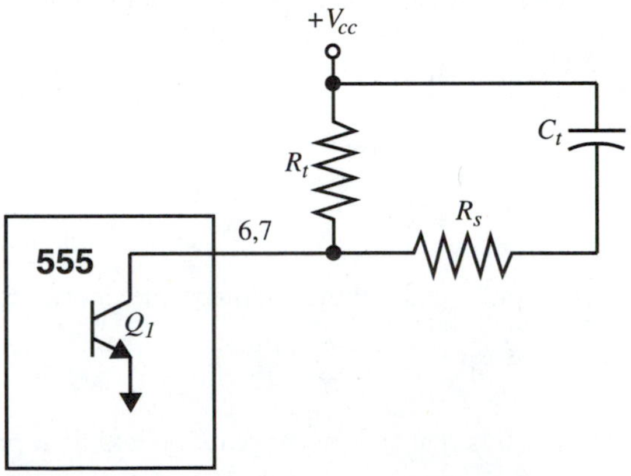

FIGURE 9–18 Circuit to prevent initial timing cycle

- With Q_1 on, C_t charges quickly through the low resistance of R_s (selected to limit the initial surge current to less than 100mA) to V_{CC}.
- An input trigger pulse initiates the timing cycle by turning off Q_1.
- With Q_1 off, the voltage (V_{CC}) across C_t now reduces as the capacitor discharges through R_s and R_t until the capacitor voltage is just less than $\frac{1}{3}V_{CC}$.
- With the capacitor voltage just less than $\frac{1}{3}V_{CC}$, the voltage from Pins 6 and 7 to ground is a little greater than $\frac{2}{3}V_{CC}$, so the inverting input to U_2 being more positive than the noninverting input causes this comparator's output to go low.
- With the inverting input to U_2 low, the flip-flop is reset and turns on Q_1, which completes the timing cycle.

If we neglect the resistance of R_s (much smaller than R_t), the timing cycle is still $1.1R_tC_t$.

EXAMPLE

If the power source for a 555 timer is 15V and R_t is 15KΩ, find the required value for R_s and the effect of this resistor on the timing interval.

SOLUTION

As previously stated, R_s should be selected to limit the surge current through Q_1 to less than 100mA. C_t is initially a short circuit when Q_1 turns on so the full V_{CC} voltage is across R_s. So

$$R_s = V_{CC}/100\text{mA}$$

$$R_s = 15\text{V}/100\text{mA}$$

$$R_s = 150\Omega$$

Since the ratio of R_s to R_t is 150Ω/15KΩ or 0.01, the effect of R_s in series with R_t during capacitor discharge is to increase the timing interval by 1 percent.

ELECTRONIC TIMER AS AN OSCILLATOR

The monostable or one-shot operation described earlier generates an output pulse for a time determined by R_t and C_t each time the timer is triggered by a negative pulse. When Q_1 is turned on at the completion of the timing cycle, a negative pulse is generated across capacitor C_t (capacitor discharges). If this pulse is coupled to the trigger input, the timer will recycle continuously. This is referred to as astable or free-running operation.

The 555 timer can be operated as an oscillator with the addition of a resistor R_l between Pin 7 and $+V_{CC}$, a resistor R_2 between Pin 7 and Pin 6, and the timing capacitor C_t between Pin 6 and ground. A shorting jumper is connected between Pins 2 and 6 to enable astable operation. Figure 9–19 shows the LM555 in an oscillator circuit configuration.

LM555 OSCILLATOR OPERATION

The operational steps are as follows:

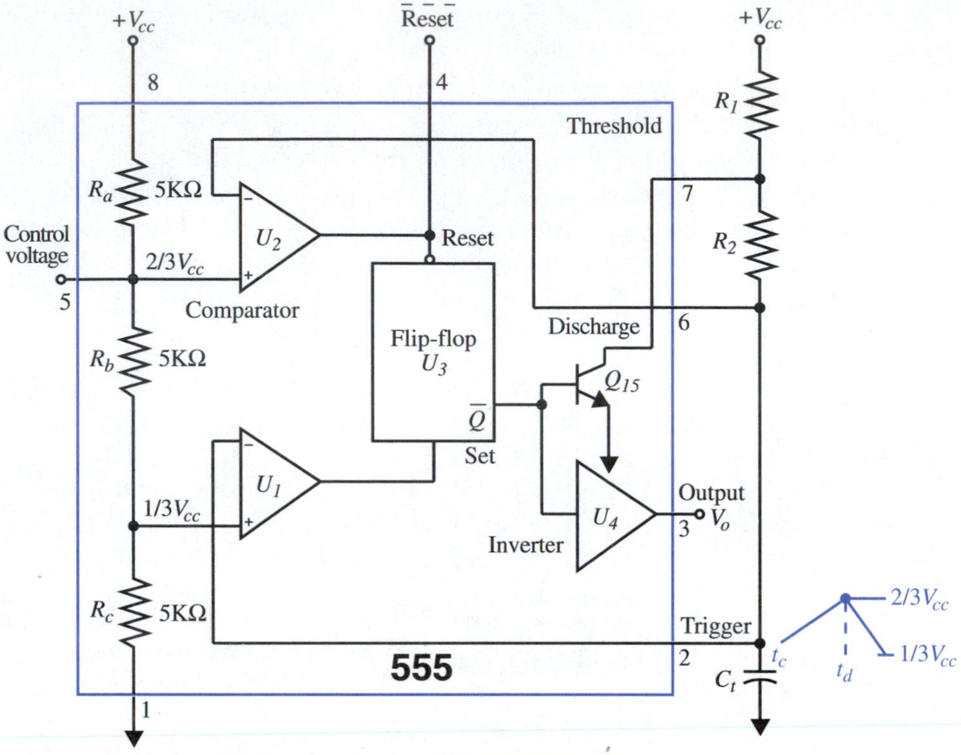

FIGURE 9–19 LM555 oscillator circuit

- Capacitor C_t charges to $\frac{2}{3}V_{CC}$, which causes the comparator U_2 output to change state.
- U_2 output going low resets the flip-flop U_3, which turns on Q_1 and effectively grounds Pin 7 and one end of resistor R_2.
- Capacitor C_t now starts to discharge toward ground through R_2. When the capacitor, and therefore the U_1 inverting input, voltage drops to $\frac{1}{3}V_{CC}$, comparator U_1 output changes state because its noninverting input is connected to $\frac{1}{3}V_{CC}$.
- U_1 output going high sets flip-flop U_3 and turns off Q_1. Capacitor C_t starts to charge via R_1 and R_2 in series toward V_{CC}, and the cycle repeats.

Using a derivation similar to that used to generate equation 9–3 but now with a capacitor voltage change of $\frac{1}{3}V_{CC}$ rather than $\frac{2}{3}V_{CC}$, and an effective power supply of $\frac{2}{3}V_{CC}$ (initial $V_{CC}/3$ charge on C_t), the LM555 charge time is

$$t_c = 0.693(R_1 + R_2)C_t \qquad \textbf{(equation 9–4)}$$

The LM555 discharge time is

$$t_d = 0.693R_2C_t \qquad \textbf{(equation 9–5)}$$

The total period is

$$T = t_c + t_d$$

Combining equations 7–1 and 7–2,

$$T = 0.693(R_1 + 2R_2)C_t \quad \textbf{(equation 9–6)}$$

The LM555 frequency of oscillation is

$$f = 1/T$$

$$f = \frac{1.44}{(R_1 + 2R_2)C_t} \quad \textbf{(equation 9-7)}$$

Notice that the timer "on" and "off" time, and frequency, are independent of long-term power supply variations. The waveforms generated in the free-running mode of operation are provided in Figure 9–20.

EXAMPLE

It is desired to have a 555 oscillator running at a frequency of 1KHz with the output high (time t_c) 70 percent of the period. If C_t is 10nF, find the required values for R_1 and R_2.

SOLUTION

The oscillator period T is

$$T = 1/f$$

$$T = 1/1\text{KHz}$$

$$T = 1\text{ms}$$

The output should be high for 70 percent of this time, or 700μs.

So t_c is 700μs and t_d is (1ms − 700μs) or 300μs.

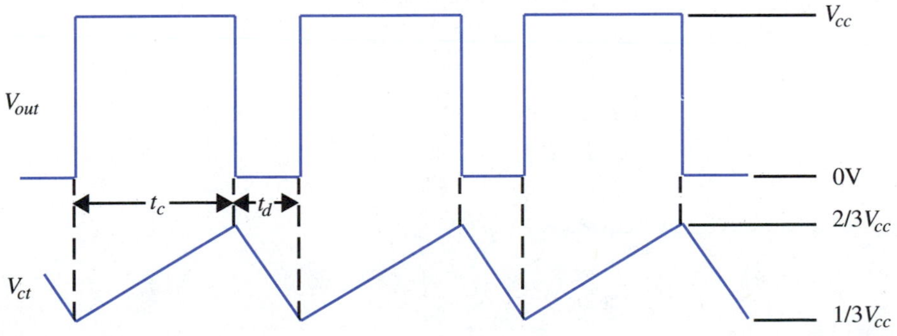

FIGURE 9–20 LM555 oscillator waveforms

Since it has just one variable, it is simpler to first use equation 9–5 and solve for R_2.

$$R_2 = \frac{t_d}{0.693C_t}$$

$$R_2 = \frac{300\mu s}{0.693 \times 10^{-8}F}$$

$$R_2 = 43.3K\Omega$$

Now using equation 9–4 to solve for R_1,

$$R_1 = \frac{t_c}{0.693C_t} - R_2$$

$$R_1 = \frac{700\mu s}{0.693 \times 10^{-8}F} - 43.3K\Omega$$

$$R_1 = 57.7K\Omega$$

Varying the voltage at the control voltage terminal (Pin 5) will cause the oscillator frequency to change proportionally to this voltage. This is called frequency modulation (FM) of the output waveform.

LM555 SQUARE WAVE GENERATION

If we make R_2 much greater than R_1, the output approaches a symmetrical square wave (equal on and off time). We can get a square wave by connecting a diode in series with R_1 and one in series with R_2 and by setting $R_1 = R_2 = R$, as shown in Figure 9–21. This connection causes the charge and discharge time constants to be the same, since during charging, current flows through R_1 and D_1, and when the capacitor is discharging, the flow is through R_2 and D_2. So the equal charge and discharge times ($t_c = t_d$) generate a square wave.

Determining the frequency of the square wave generator is not too straightforward because of the diode voltage drop in series with the timing resistors. Introducing the 0.6V diode drop into the charging equation gives a value for t_c of

$$t_c = -RC_t\left\{\left[\ln \frac{(1 - 1.2V/V_{cc})}{(2 - 1.2V/V_{cc})}\right]\right\} \qquad \textbf{(equation 9–8)}$$

Since

$$t_c = t_d$$

with the square wave, the frequency

$$f_{sq} = \tfrac{1}{2}t_c$$

EXAMPLE

It is desired to build a 10KHz square wave generator as shown in Figure 9–21 and powered from a +12V supply. If C_t is 10nF, find R ($R_1 = R_2$).

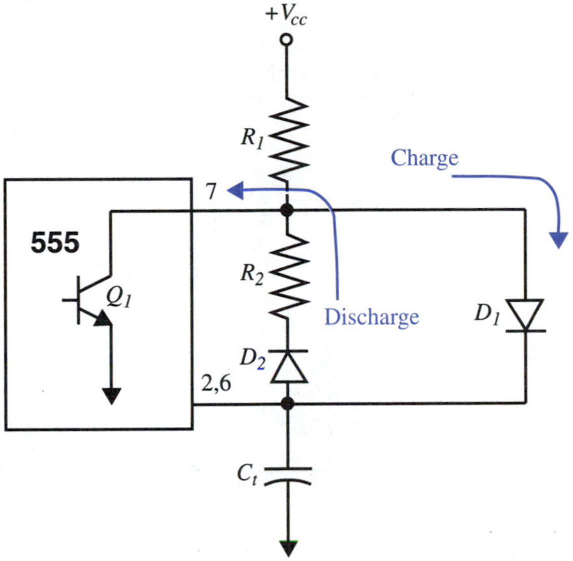

FIGURE 9–21 Timing circuit for square wave output

SOLUTION

With a 10KHz square wave, we can find t_c by finding half the period,

$$t_c = \frac{1}{2(10\text{KHz})}$$

$$t_c = 50\mu\text{s}$$

Using equation 9–8 to solve for R

$$R = \frac{t_c}{C_t K}$$

where

$$K = -\ln\left[\frac{(1 - 1.2\text{V}/V_{cc})}{(2 - 1.2\text{V}/V_{cc})}\right]$$

$$R = \frac{50\mu\text{s}}{(10^{-8}\text{F})(0.747)}$$

$$R = 6.69\text{K}\Omega$$

LM555 OUTPUT CIRCUIT

The internal output circuit of the 555 timer is shown in Figure 9–22. It is a "totem pole" type output capable of sourcing or sinking 200mA.

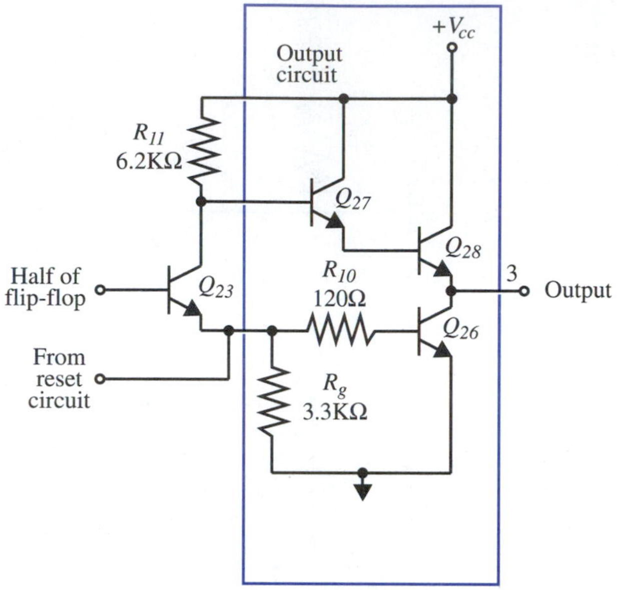

FIGURE 9–22 LM555 timer internal output circuitry

In operation, Q_{28} is turned on during the timing cycle and Q_{26} is off. At the completion of the timing cycle, Q_{26} turns on and Q_{28} turns off. This totem-pole connection charges or discharges any stray capacitance at the output quickly through the low on resistance of the transistors, thus giving sharp rise and fall times at the output (typically 100ns).

During the timing interval, the base of transistor Q_{23} is switched low, cutting off transistors Q_{23} and Q_{26}. With Q_{23} off, the collector of Q_{23} goes high, turning on transistors Q_{27} and Q_{28}, which are a Darlington pair. The output is pulled high to V_{cc}, minus the two base-to-emitter drops (1.2V) and the drop across the 6.2KΩ resistor (R_{11}).

With a 15V supply and a load of 200mA, the data sheets indicate an output voltage of about 12.5V. For the load to draw current during the timing cycle, the load should be connected from output to ground. This is called a **source connection**.

When the timing cycle is complete, Q_{23} and Q_{26} are turned on, and Q_{27} and Q_{28} are turned off. A load connected from the output to the supply voltage would now provide current into the collector of Q_{26}—which would sink this current. With a sink connected 200mA load, the collector voltage of Q_{26} would be at about 2.5V. Again, with a 15V supply, this would leave 12.5V across the load.

In summary, loads connected to ground are conducting current during the timing cycle, while those connected to V_{cc} are conducting when the timing cycle is off.

If a relay load (inductive) is connected to the 555 output, the manufacturer suggests both a diode in series with the relay as well as one in parallel (reversed).

LM555 RESET CIRCUIT

The reset input can be used to prevent the timing from starting or to terminate the timing cycle before the capacitor reaches the $\frac{2}{3}V_{cc}$ value. This is accomplished by lowering the voltage on the reset Pin 4 to approximately 0.5V.

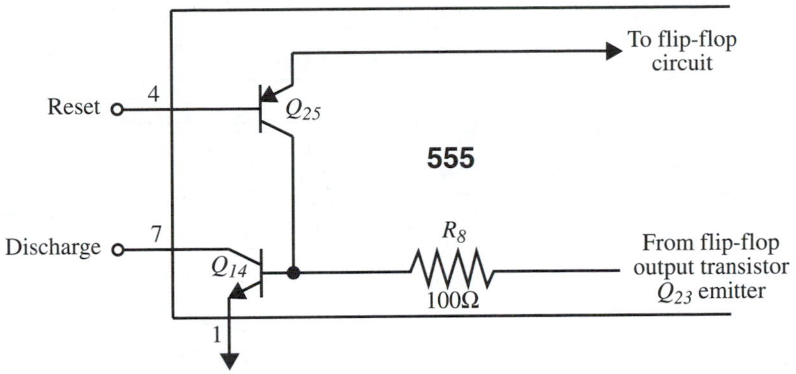

FIGURE 9–23 Reset circuit

With reference to Figure 9–23, transistor Q_{25} turns on when a low voltage or ground is applied to the reset Pin 4. This will supply current to the base of transistor Q_{14} (Q_1 previously), turning on this transistor and discharging the external timing capacitor. The flip-flop is also reset by the current flow through the Q_{25} emitter. But when the reset signal is removed, the timing cycle does not restart.

If the trigger pulse is present when the reset signal is applied, the external timing capacitor will discharge, but the flip-flop will not be reset. This allows the timing capacitor to start charging again (new timing cycle) when the reset signal is removed.

Applications for the 555 timer can fill a whole text. We will briefly look at a few applications that will illustrate the capabilities of the device.

BATTERY SAVER CIRCUIT

It is desired to have a battery saver circuit for a portable digital multimeter (DMM) that will automatically shut off the power after one minute of display. A possible circuit is shown in Figure 9–24.

The meter "on" timer circuit is activated by momentarily depressing the push button to apply power to the 555. Since capacitor C_1 cannot charge instantaneously, the trigger input Pin 2 is held low, which causes the timer cycle to initiate. The output Pin 3 now goes high, turning on transistor Q_1 and effectively grounding its collector. With Q_1 collector close to ground, the circuit for the DMM is completed and the meter turns on. Lowering the collector of Q_1 also turns on transistor Q_2, which shorts out the push-button switch and provides power to the circuit when the push button is released.

At the completion of the timing cycle (determined by R_t and C_t), the timer output goes low and Q_1 is turned off. The rising collector on this transistor turns off Q_2, and power is removed from the circuit.

R_2 allows the 555 output to drop to close to ground and turn off Q_1, and R_3 limits the current into the base of Q_1 when the 555 output goes high.

So, current is only drawn from the battery during the timing cycle, and the problem of battery drain, if the meter power switch is left on, is eliminated. All of the components shown in Figure 9–24 are mounted within the meter case.

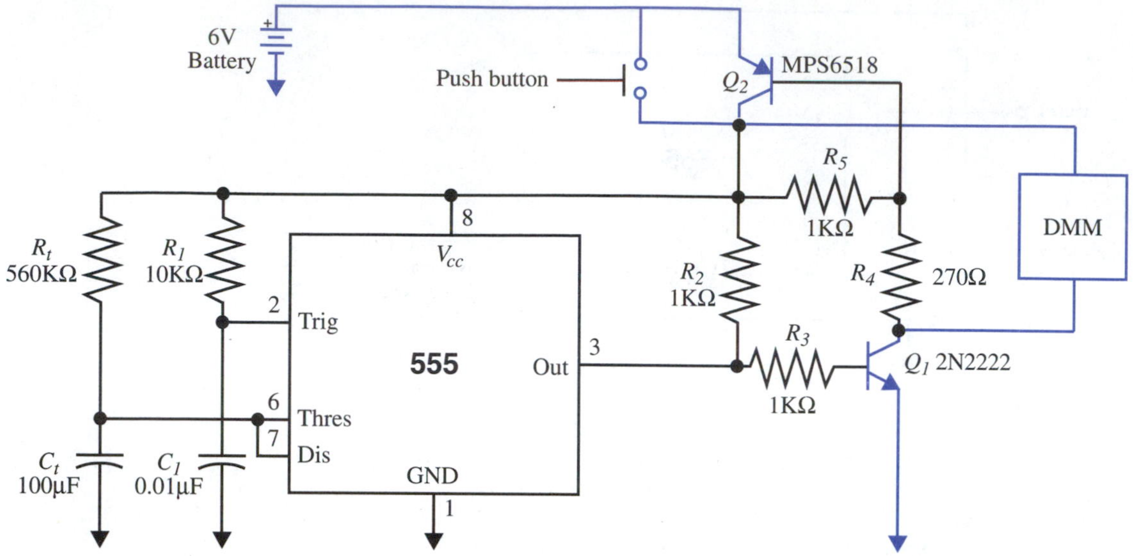

FIGURE 9–24 Battery saver circuit

LM555 OSCILLATOR APPLICATION

An example of an oscillator using a 555 timer is the power saver emergency flasher, shown in Figure 9–25.

The circuit components are selected so that the lamp is turned on for 0.1s and off for 1s. A conventional flasher is on and off for equal time intervals, resulting in greater battery drain.

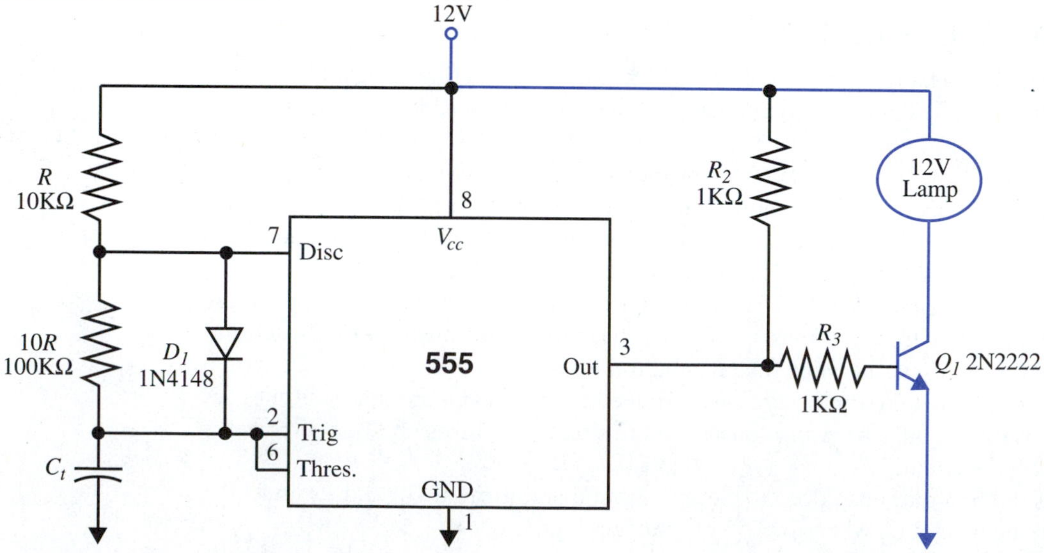

FIGURE 9–25 Power saver emergency flasher

You will recall that the output of the oscillator is high when the timing capacitor C_t is charging and low when it is discharging. Since the capacitor is charged through resistor R and the diode (diode shorts out resistor $10R$) and discharges through $10R$ (diode reverse biased), we get the required shorted on and longer off time—remembering that transistor Q_1 turns on when the output is high.

BURGLAR ALARM SYSTEM

Assume that a burglar alarm system is required, consisting of entry switches and an alarm siren. If a switch is opened, the alarm should sound for two minutes and then shut off to take care of false-alarm situations.

It is also required that upon activation of a disable switch, the alarm system not be active for a period of 45 seconds. This allows entry or exit of authorized persons.

The required 2-minute and 45-second time intervals can be achieved with monostable operation of a timer. Let us develop the circuit, using 555 timers as shown in Figure 9–26.

The normal circuit operation is controlled by the alarm switches. If one switch opens, indicating that access is being attempted, the alarm will sound.

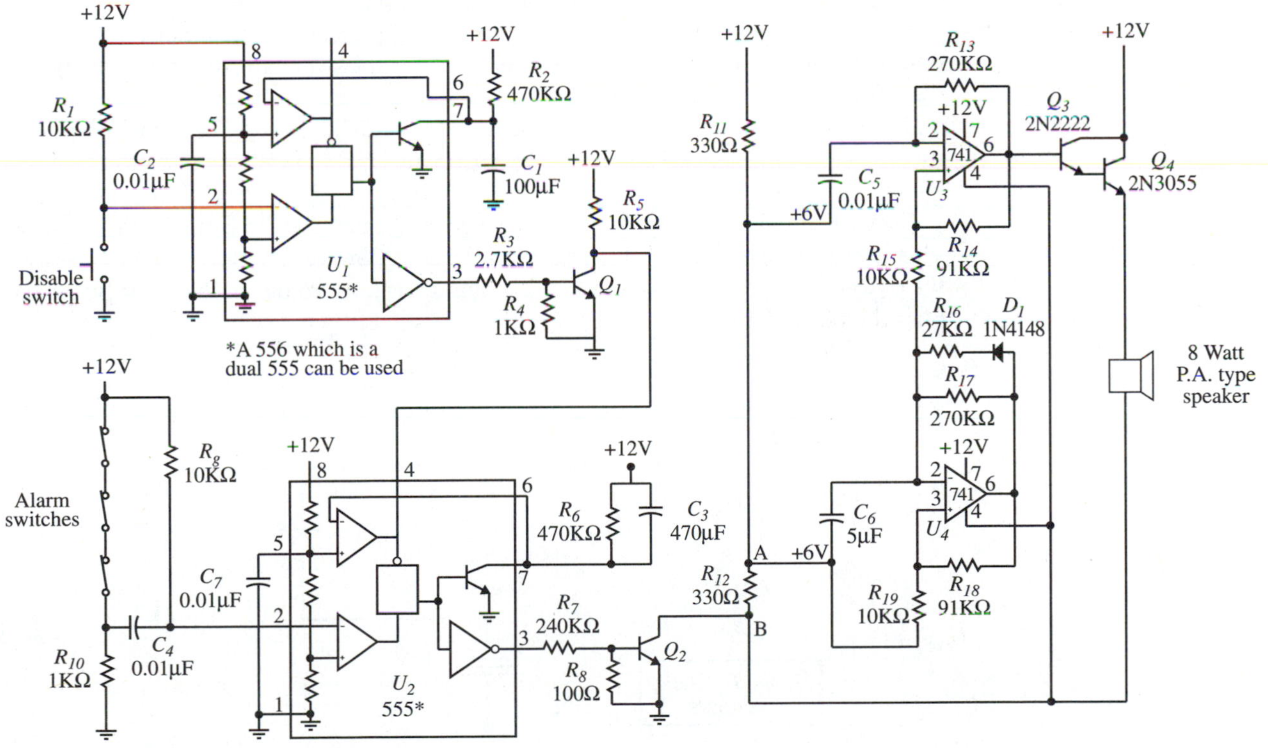

FIGURE 9–26 Burglar alarm circuit

- With a switch open, the alarm timer (U_2) is triggered because the top end of R_{10} goes to ground, causing Pin 2 of U_2 to go low.
- The output of U_2 now goes high, turning on Q_2, which provides a ground connection at point **B** for the alarm circuit.

ALARM CIRCUIT

The alarm consists of two square wave generators: U_3, which oscillates around 1KHz, and U_4, which is set for about 2Hz. Note that the frequency of U_3 is caused to shift (which creates the siren effect) by connecting the bottom of its voltage divider (R_{14} and R_{15}) to the U_2 timing capacitor (C_6). This causes the voltage at the noninverting input of U_3 to change, which in turn means that the voltage to which C_5 charges, moves, and the frequency of U_1 shifts.

The output of U_3 is connected to the Darlington circuit (Q_3 and Q_4), which provides the current gain to drive the speaker. Use of a negative supply voltage for the op-amps is avoided by use of the voltage divider R_{11} and R_{12}, which effectively provides +6V.

If the disable push button is depressed, U_1 is triggered and its output goes high, turning on Q_1. This brings the U_2 reset line to ground, which disables U_2. Now, building entry or exit is possible for the 45-second duration of U_1.

Activation of the alarm with power turn-on is avoided by connecting the timing capacitor for U_2 to the positive supply rather than to ground.

Alarm switches normally used are magnetic reed switches. A small magnet is mounted on the door (or window), which causes the leaves of the switch to come together and close the circuit.

The power source for the alarm should be maintained even if the main AC power is turned off. This can be accomplished by floating a 12V "Gel Cell"—solid electrolyte-type battery, as shown in Figure 9–27.

Notice that the power supply output voltage has been increased from +12V to +13.6V. This keeps the battery on trickle charge with 10mA flowing through the 100Ω resistor. If the power is lost, rectifier D_1 reverse biases and D_2 is forward biased. Now the battery supplies the power for the alarm.

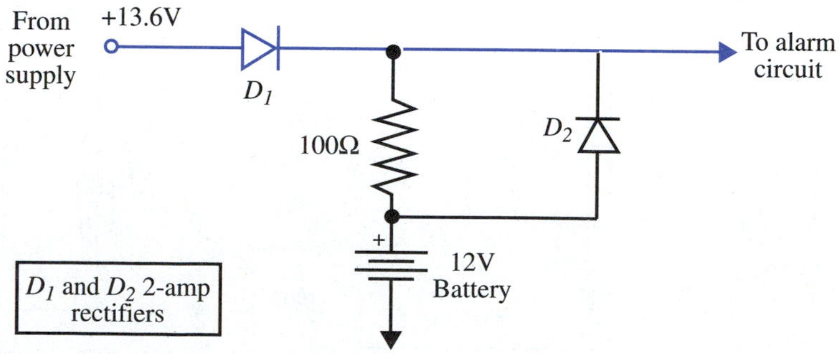

FIGURE 9–27 Standby power source

PRECISION TIMER

Another type of timer that has somewhat different capabilities than the 555 is the LM3905 precision timer. This timer offers greater versatility with better temperature stability (typically 0.003%/°C) than the 555 timer (typically 0.005%/°C). The LM3905 precision timer can operate with an unregulated power supply and is less sensitive to power-supply and trigger-voltage changes during the timing interval. Also, it has a logic capability for changing the sense of the output, i.e., high or low during the timing interval, and has an open collector and emitter on the output transistor.

Because of lower internal leakage currents (2.5nA versus 350nA for the 555), the LM3905 is capable of providing constant timing periods from microseconds to hours. A functional block diagram of this timer is shown in Figure 9–28.

When power is initially applied, the reset transistor Q_1 is turned on, timing capacitor C_t is discharged, and latching buffer U_2 output V_1 goes high. The detailed circuit operation is described in the following steps:

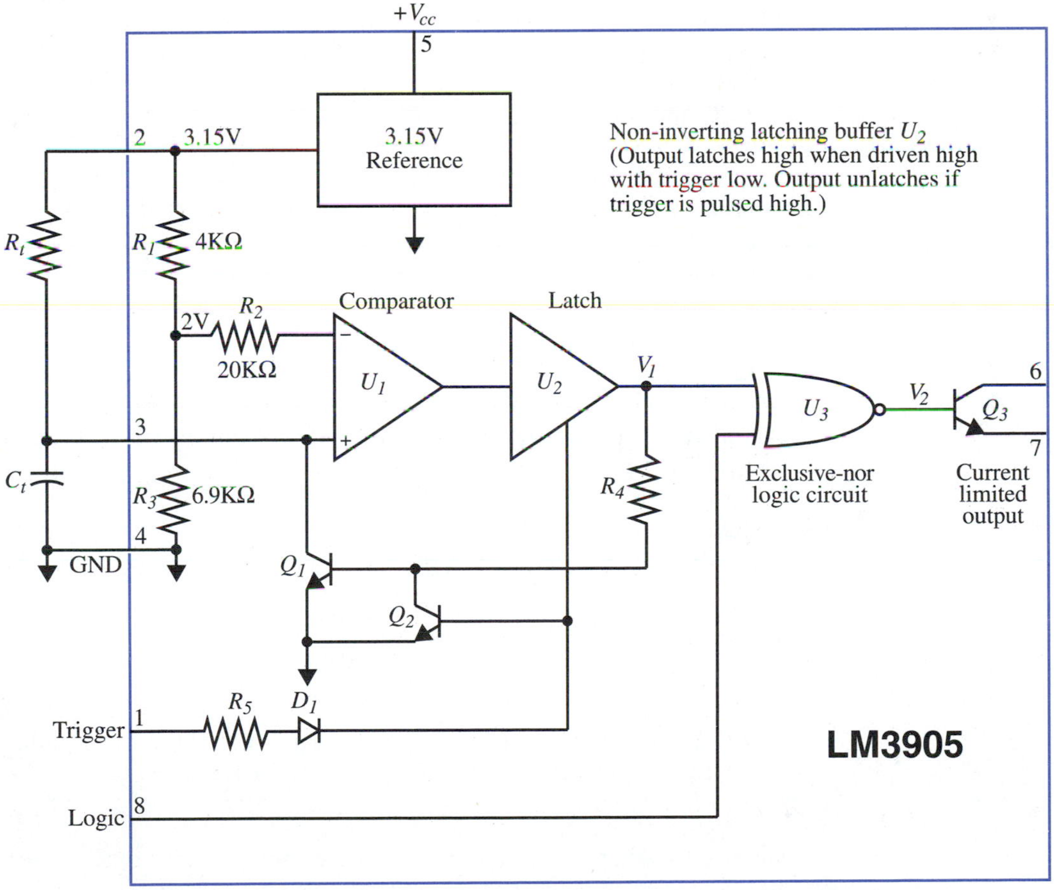

FIGURE 9–28 LM3905 functional block diagram. *Courtesy National Semiconductor.*

- Circuit operation is initiated when a positive pulse is applied to the trigger input Pin 1, which starts the timing cycle.
- When the trigger input pulse exceeds +1.6V, it unlatches buffer U_2, causing its output to go low and turn off Q_1.
- Timing capacitor C_t now starts to charge from V_{ref} via timing resistor R_t.
- Resistors R_1 and R_3 form a voltage divider which establishes a +2V input threshold level for comparator U_1.
- When the charging voltage on C_t exceeds the U_1 2V input threshold level, U_1 toggles, latching U_2, which returns V_1 to its initial high state.
- With V_1 high, it turns on Q_1, which discharges C_t and ends the timing cycle.

The LM3905 timing cycle is determined simply by the product R_tC_t without the 1.1 factor as required for the LM555 timer.

If the voltage on the trigger input (Q_2 base) is kept high when the timing cycle ends, transistor Q_2—which now has a collector supply voltage (since V_1 is high)—is turned on. This ensures that the base voltage of Q_1 is low, thus maintaining Q_1 off. With Q_1 off, the timing capacitor will continue to charge toward the reference voltage V_{ref} (+3.15V). During this period, V_1 will stay high, but buffer U_2 will not be latched. This means that the output pulse, unlike that of the 555, is independent of the width of the input trigger pulse.

The logic input Pin 8 controls the sense of the output signal. If the logic input is high or open, the output transistor Q_3 will be turned off during the timing cycle. (The Exclusive-nor U_3 has a low output when both inputs are different, and a high output when both inputs are the same.) With the logic input (Pin 8) grounded or low, the output transistor is turned on during the timing cycle. The sense of the output also can be changed, depending on whether the output is taken from the collector or emitter of Q_3. With Q_3 turned on, the emitter will go positive and the collector will go negative.

EXAMPLE

Find the width and polarity of the pulse seen at an emitter resistor connected to the output transistor Q_3 of a LM3905 timer, if R_t is 1MΩ, C_t is 100μF, and the logic pin is grounded.

SOLUTION

Since the output pulse width t_w for a LM3905 timer is simply R_tC_t,

$$t_w = (1\text{M}\Omega)(100\mu\text{F})$$

$$t_w = 100 \text{ seconds}$$

With the logic pin grounded and the V_1 low during the timing cycle, the output of the exclusive-nor circuit will be high, turning on Q_3. This means emitter current will flow, providing a positive pulse at the emitter for 100 seconds.

Voltages and currents for the LM3905 output transistor (Q_3) must be consistent with the maximum device dissipation of 500 mW at 25°C. Output currents can be about 50mA; while the maximum collector voltage is 40V. Since the collector is open, the collector supply voltage can be different from the voltage ($+V_{cc}$) used for the timing circuit. As with the 555 timer, connecting C_t in parallel with R_t prevents the timer from cycling when power is applied.

FREQUENCY REDUCTION CIRCUIT

Suppose we wish to derive a 60Hz TTL clock signal from a 1MHz crystal oscillator and we want the accuracy of the clock signal to be the same as the oscillator over a 20°C range.

We could use a digital dividing circuit, but another approach is to use two LM322 timers (3905 timers with a trim adjustment terminal). Consider the circuit shown in Figure 9–29.

The required frequency reduction is 1MHz/60Hz or 16,666 (recurring) to 1. To achieve this reduction, we will have each timer count down by a factor of 129 (the square root of 16,666). Since the 1MHz clock has a period of 1μs, U_1 will be set up for a time just less than 129μs. Choosing values of 0.01μF for C_1 and 13KΩ for R_1 will put us in the right timing range. By adjusting R_2, we can get the required time interval.

When the 129th clock pulse is received, U_1 will retrigger and the emitter of the output transistor (Pin 1) will go high, triggering U_2. This timer is set for a time interval of just less than the required 16,666μs (within 129 μs). Using a 1μF for C_2 and a 16KΩ resistor for R_4 will put us in the ballpark. Fine adjustment will be accomplished by adjusting R_5.

The parts chosen must be consistent with an accuracy of one part in 129, or 0.77%, over the required temperature change of 20°C. This amounts to a temperature coefficient of 38.8 parts per million per degree Celsius. The LM322 has a coefficient of 30 parts, and using a

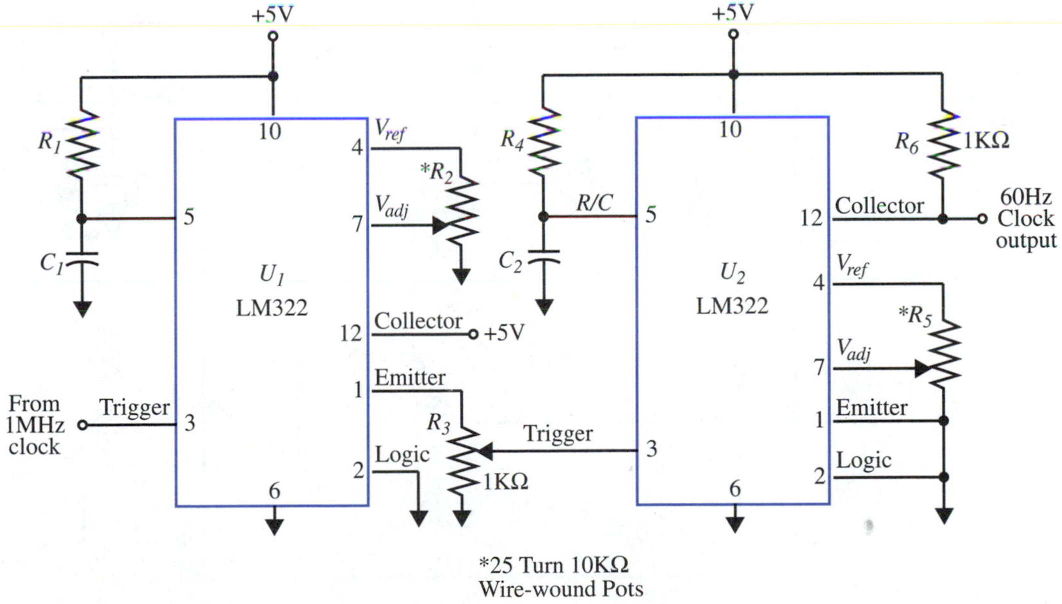

FIGURE 9–29 Frequency reduction circuit

ceramic capacitor for C_1 and C_2 and a metal film resistor for R_1, R_4 will provide the required stability.

SOME OTHER TIMERS

A dual version of the 555 timer exists; it is called the LM556. This 14-pin chip uses a common power source for the two timers, but all other signal pins are independent. Operation is identical to the 555 version.

A quad timer is available; it is called the LM558. This device is slightly different from the 555, as shown in Figure 9–30.

Differences are the use of one comparator rather than two, a negative edge triggered flip-flop, a divider set for $0.632V_{CC}$ rather than $\frac{2}{3}V_{CC}$, and an open collector (100mA) output.

Use of one comparator prevents the timer from being used as a stable oscillator (using two timer sections of the chip still provides this capability).

The negative edge triggered flip-flop means that the timer triggers on the positive-to-negative transition of the input waveform. An input coupling capacitor is not required, since if the input stays low after the timer is triggered, it has no effect on the timing cycle.

A $0.632V_{CC}$ divider causes the timing equation to be simply *RC* rather than *1.1RC.*

The open collector output allows greater flexibility with load resistor power sources, but the output rise time will be slower than the totem pole output of the 555.

The 558 has about one-sixth of the leakage current (threshold current) of the 555, allowing timing resistor values to be increased (up to six times larger). Also, temperature sta-

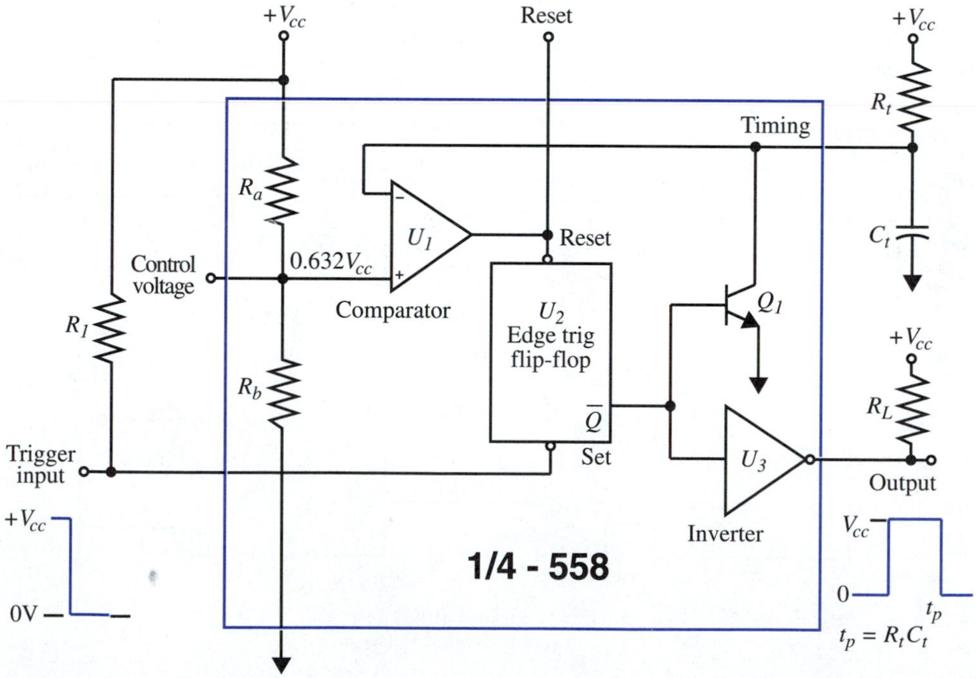

FIGURE 9–30 The LM558 timer block diagram

TABLE 9–1 555 One-Shot Normal Pin Conditions

Pin Number	*DC w/o Trigger*	*Waveform with Trigger*
1	0V	None
2	+15V	Negative pulse $>\frac{2}{3}V_{CC}$ (10V)
3	0V	Positive pulse for timer interval
4	Normally tied high to V_{CC}	None
5	$\frac{2}{3}V_{CC}$ (10V)	None
6	0V	Exponential 0V to 10V
7	0V	Exponential 0V to 10V
8	+15V	None

bility is about five times better than the 555. The 558 operates with a supply voltage range from 4.5V to 16V.

TROUBLESHOOTING TIMER CIRCUITS

The approach to troubleshooting timer circuits diagrammed here is one that will also be used for the more-complex integrated circuits covered in subsequent chapters.

When dealing with integrated circuit chips like the timer—which have more involved internal circuitry—we need to be aware of the expected DC voltages or waveforms at every pin. Since most of the signals are time varying, this requires the use of an oscilloscope to determine whether we have the correct signal.

Let us consider troubleshooting a 555 timer working from a 15V supply and used in a one-shot application (refer to Figure 9–16). With no trigger applied, we should measure the DC voltages listed in Table 9–1 at the various chip pins. If the circuit is being triggered in a repetitive fashion, we should see the waveforms indicated in Table 9–1, using an oscilloscope.

We need to be aware of the relationships between these waveforms. If, for example, the input trigger voltage swing is not large enough to initiate a timer cycle, we should not expect to see signal waveforms at pins 3, 6, or 7.

SUMMARY

- There are four transfer functions for the different types of amplifiers.
- A transimpedance amplifier has a current input and a voltage output.
- Transimpedance amplifiers do not have a fixed gain-bandwidth product because the open loop gain changes with R_1.
- Advantages of transimpedance amplifiers are wide bandwidth and high slew rate.
- A comparator compares two inputs, and the comparator output changes state when one input exceeds the other input.
- An op-amp can be used as a comparator if speed of response is not a problem.
- The output of a zero crossing detector changes state as the input signal passes through zero volts.
- Noise triggering of a comparator is avoided by using a Schmitt trigger circuit.

- A window comparator senses when the input is within a given voltage range.
- IC comparators usually have an open collector output stage, which allows for interface with various circuit devices, e.g., relays, lamps, etc.
- A timer can cause circuit activation (or deactivation) for a predetermined amount of time.
- Timers can be connected in a continuous or oscillator mode.
- The most common timer is the 555 device with a timing equation of $1.1R_tC_t$.
- Triggering of the 555 requires a negative pulse of amplitude greater than $\frac{2}{3}V_{CC}$.
- The limit to how long the timer cycle can be depends on the internal leakage resistance of the timer and the leakage resistance of the timing capacitor (C_7).
- Maximum conservative time period for the 555 is about 15 minutes, and the maximum value of the timing resistor (R_t) should be no greater than 1MΩ.
- The control voltage terminal allows for "fine tuning" of the timer period or for frequency modulation when the timer is connected as an oscillator.
- Use of a "totem pole" output for the 555 provides fast rise and fall times.
- The reset terminal can be used to inhibit the start of a timing cycle or shorten an in-process timing period.
- The 3905-type timer has the following advantages over the 555:
  1) Better temperature stability
  2) Timing cycle independent of trigger
  3) Less sensitivity to power supply changes
  4) Uncommitted output transistor
  5) Sense of output can be changed by a logic level input
  6) Less leakage, which means longer timer intervals
- The advantages of the SE558 over the 555 are:
  1) Edge triggering (output independent of trigger width)
  2) Four timers in one package
  3) Simpler timing equation
  4) Higher-value timing resistors possible
  5) Greater temperature stability

EXERCISE PROBLEMS

1. Describe the four types of amplifier transfer functions.
2. State the advantages of the transimpedance amplifier over the conventional op-amp.
3. Find the closed loop gain A_v for a noninverting transimpedance amplifier if R_1 is 100Ω, R_f is 820Ω, R_c is 1KΩ, and A_{v1} is 800.
4. Give two advantages of a comparator *IC* over an op-amp when used in a comparator application.

5. What is the main reason for using a Schmitt trigger circuit?
6. Give an advantage and a disadvantage for the open collector output used on a comparator.
7. What is a window comparator?
8. Calculate the values for R_a, R_b, and R_c for the window comparator circuit (Figure 9–A) if the LED is to illuminate with an input voltage range from 4V to 4.1V and the voltage divider current is 1mA.

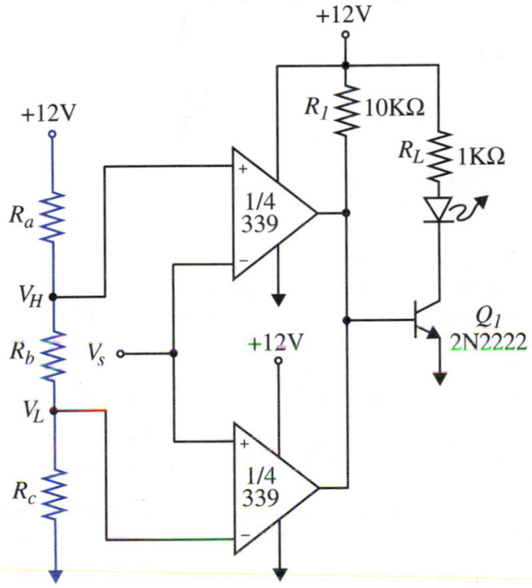

FIGURE 9–A (text figure 9–10)

9. Indicate the comparator output circuit connections in order to achieve a 0–2mA current swing through a 15KΩ resistor. Power supply voltages of ±15V are available.
10. Show a comparator circuit that has a high output when the input voltage exceeds +3V.
11. Show a comparator circuit that has a low output when the input voltage exceeds −3V.
12. Determine all circuit values for a Schmitt trigger circuit with a total hysteresis of 200mV, using an LM311 comparator operating off ±10V supplies. Load to be 20mA.
13. Design a window comparator circuit using an LM339 that activates a relay when the input voltage is between 2V and 3V. Use ±12V supplies and 1mA divider. Specify all component values (relay resistance 1KΩ, pull in current 20mA).
14. Determine the total DC current drawn from the power supply for the under-over voltage circuit in Figure 9–14 if the relays draw 15mA each, and the comparators are LM339 (draw 1mA total).

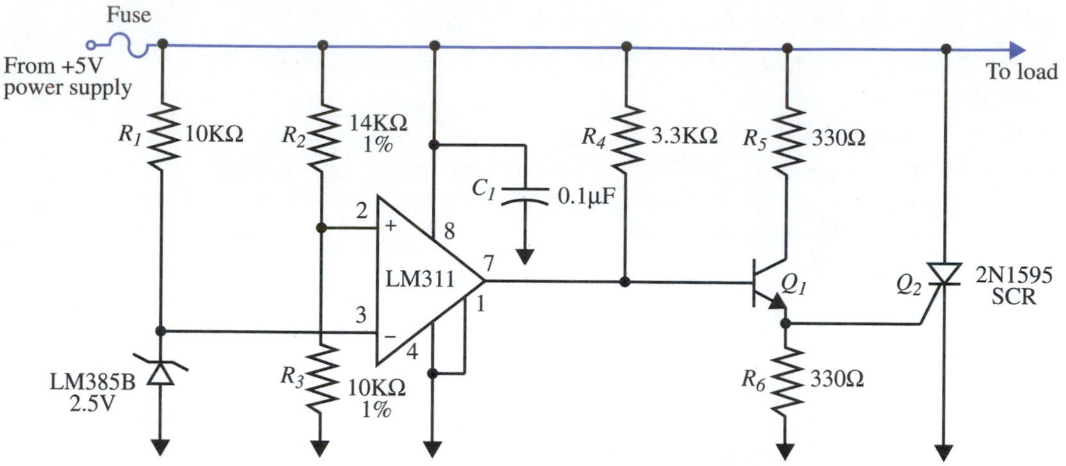

FIGURE 9–B (text figure 9–15)

15. Redesign the crowbar circuit of Figure 9–B for a +15V supply. Specify all component values. Use a 5V voltage reference and make the nominal trip voltage 16V.
16. Determine the required input trigger waveform (width and amplitude) for a 555 timer operating from a +15V power supply. Trigger pin is connected to the supply through a 10KΩ resistor. Timer is set for 10ms.
17. Show the complete circuit for a 555 timer that will turn on a 6V, 50mA lamp for 50 seconds. Circuit to be activated with a push-button switch; a +12V supply is available.
18. Determine the output voltage swing for a 555 timer with a source load of 150Ω and a power supply voltage of +15V. (*Hint:* See data sheets.)
19. Design a 5KHz oscillator circuit using a 555 timer. Output signal across 1KΩ load resistor is to be high 60 percent of the time. Power supply voltage is +15V. Use 10nF for C_t.
20. Find the value of R ($R_1 = R_2$) for a 1KHz square oscillator circuit using a 555 timer. Power supply voltage is +15 V. Use 22nF for C_t.
21. Using a 555 timer, design a 20sec battery saver for a digital voltmeter operating from +6V with a current drain of 200mA. Circuit is to be activated by the momentary depression of a push-button switch. Let C_t equal 100μF; determine all other component values.
22. With reference to Figure 9–C, determine all component values for an emergency flasher with 0.1sec on time and 1sec off time. Use a C_t of 1μF.
23. Show a circuit using an LM3905 timer to turn on a green (2V), 20mA LED for 1 minute. Also show the circuit change to have the LED normally on and turned off for 30 seconds. Use a +12V power supply and a C_t of 100μF.
24. Design a zero power dissipation timer using an LM3905 to turn on a relay for 30 seconds. Timer to be activated by a push-button switch. Relay operates on +6V drawing 100mA.

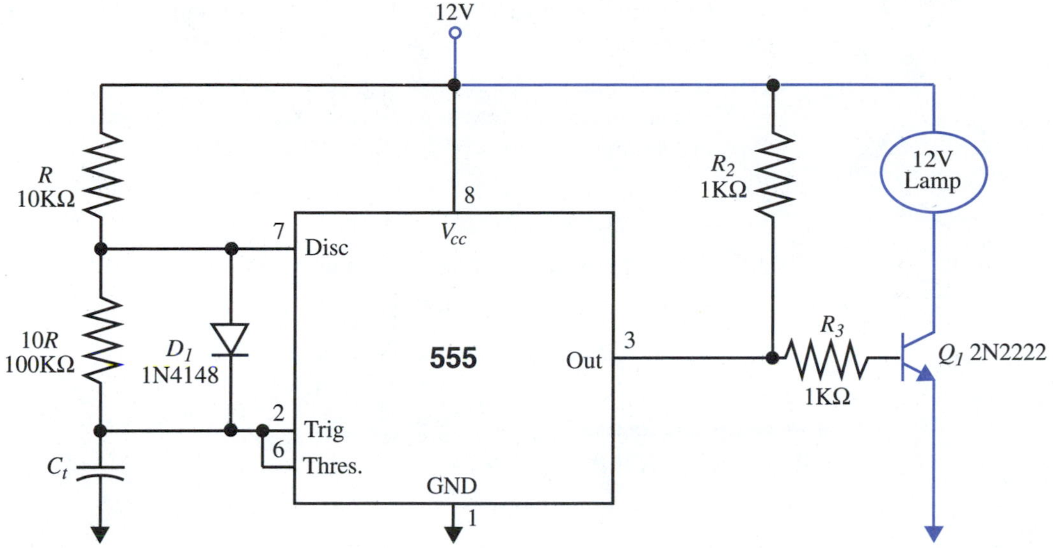

FIGURE 9–C (text figure 9–25)

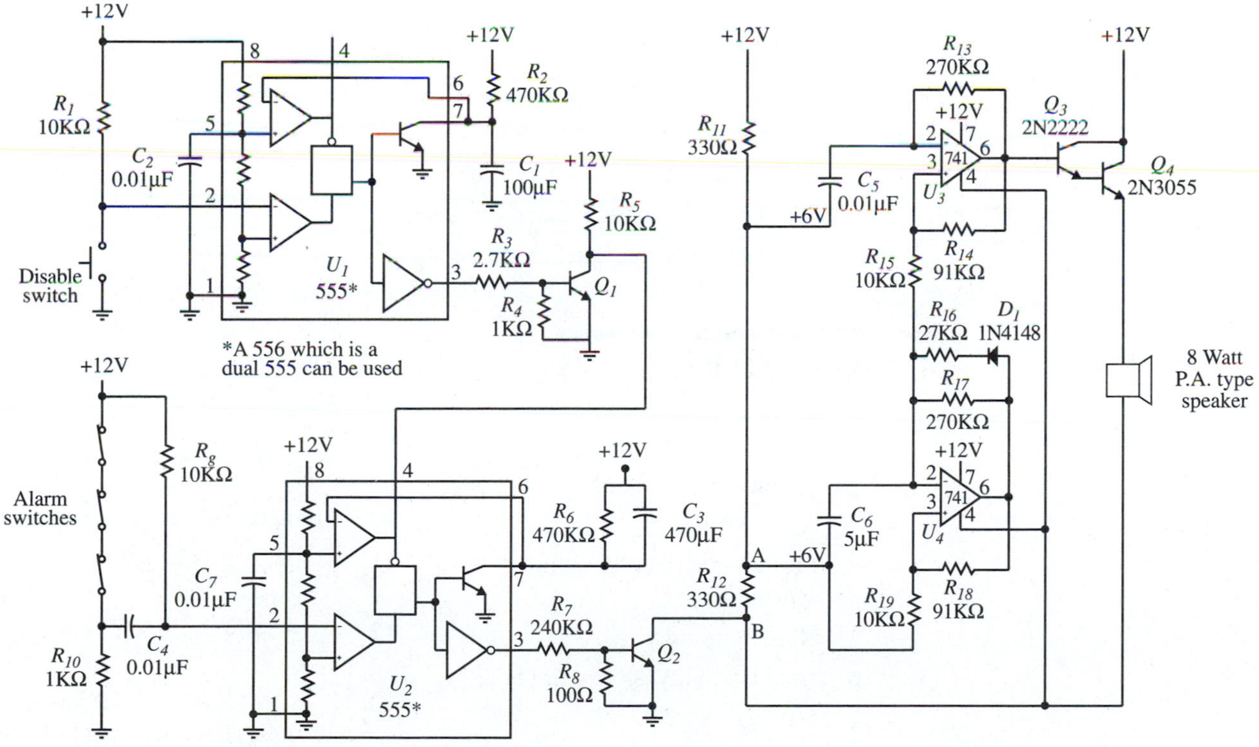

FIGURE 9–D (text figure 9–26)

25. Redesign the timer portion of the circuit in Figure 9–D, using a 3905 timer rather than a 555.
26. Using the device specifications, generate a table comparing the trigger requirements, maximum timer interval, output drive current capability, and possible load connections for a 555, 558, and 3905.
27. Determine the value of a resistor R_2 connected from Pin 5 of a 555 timer to ground that will provide a timer interval of RC rather than $1.1RC$.
28. Why is a "totem pole" output used on a 555 timer?
29. List the advantages of a 3905 and of a 558 timer over a 555.
30. List the disadvantages of the 558 timer.

ABBREVIATIONS

A_{ol} = Open loop gain
A_v = Closed loop gain
A_{vi} = Ideal closed loop gain
A_{v1} = Gain of all stages other than first for a transimpedance amplifier
C_t = Timing capacitor
R_t = Timing resistor
t_c = Capacitor charge time
t_d = Capacitor discharge time
t_p = Time duration of output pulse
TTL = Digital Transistor Transistor Logic
U = Designator for integrated circuits

ANSWERS TO PROBLEMS

1. Current, voltage, transconductance, and transimpedance
2. Bandwidth less a function of gain
3. $A_v = +9.19$
4. Less offset and faster output switching
5. Noise immunity
6. Load flexibility; slower rise time
7. See text.
8. $R_a = 7.9\text{K}\Omega$; $R_b = 100\Omega$; $R_c = 4\text{K}\Omega$.
9. Connect top of 15KΩ load to +15V, connect output emitter to −15V.
10. Connect +3V to inverting input.
11. Connect −3V to noninverting input.
12. $R_L = 1\text{K}\Omega$; $R_1 = 99\text{K}\Omega$; $R_2 = 1\text{K}\Omega$
13. $R_a = 9\text{K}\Omega$; $R_b = 1\text{K}\Omega$; $R_c = 2\text{K}\Omega$. Replace 1KΩ and LED with relay (see Figure 9–A).

14. $I_T = 38.2mA$.
15. Change R_2 to 22KΩ, R_5 to 1.5KΩ.
16. Negative >10V trigger pulse, 1ms wide
17. Refer to Figure 9–16. Connect push button between trig in and GND. Lamp and 120Ω series resistor from output to GND.
18. V_o swing 0V to 12.75V (from spec sheets)
19. $R_2 = 11.5K\Omega$; $R_1 = 5.78K\Omega$.
20. $R = 30.9K\Omega$
21. See Figure 9–24. $R_t = 182K\Omega$.
22. $R = 144K\Omega$; $10R = 1.44M\Omega$
23. (a) $R_t = 600K\Omega$; (b) $R_t =$ 300KΩ with logic pin high
24. Refer to Figure 9–24. Use +6V supply, reverse R_1 and C_1, ground pin 7, connect relay between +6V and pin 6.
25. See Figure 9–E.
26. See Specifications.
27. $R_2 = 61K\Omega$
28. Faster rise time
29. See summary.
30. See text.

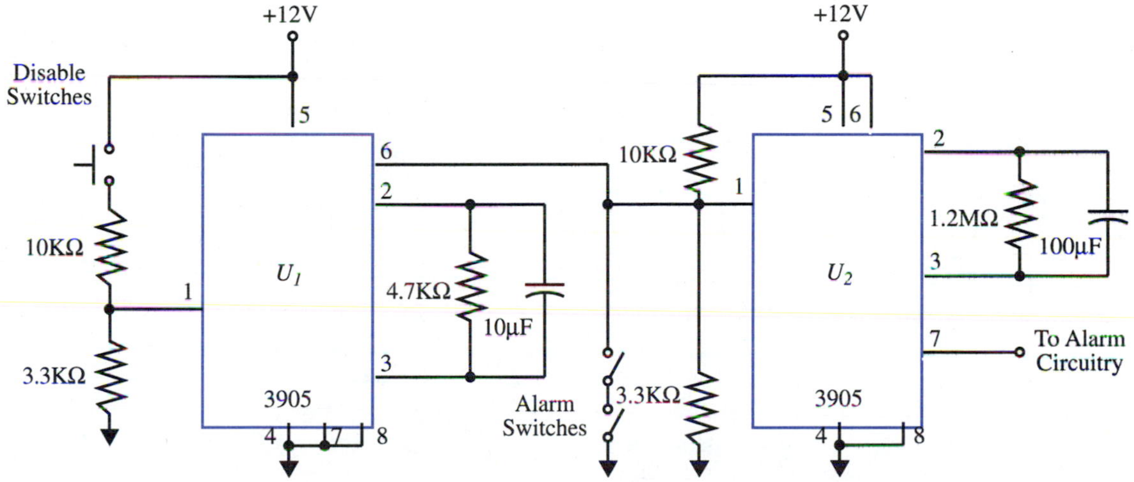

FIGURE 9–E

CHAPTER 10

Power Supplies

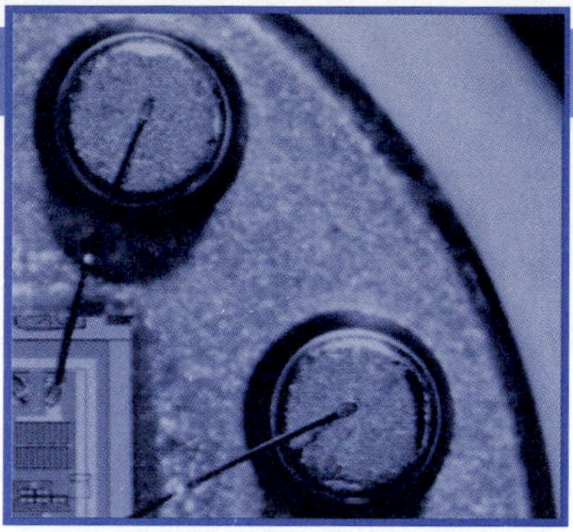

CHAPTER OBJECTIVES

After completing this chapter, you should be able to describe the following power supply concepts:

- Why a power supply is needed
- Half-wave rectification
- Full-wave rectification
- Bridge rectification
- Ripple filtering
- Need for power supply regulation
- Zener-diode regulator
- Transistor regulator
- Voltage doubling
- Voltage tripling
- Troubleshooting power supplies

INTRODUCTION

Most electronic circuits require a DC power source in order to operate. As we demonstrated with transistor circuits, the source of power for the circuit is DC, and this is used to set up the DC bias levels. The DC power source could be a battery—but batteries do have the disadvantage of discharging, which lowers the available DC voltage and requires replacement or recharge.

For most electronic equipment, it is better to be able to develop a DC power source from the sine wave provided by the AC mains. This chapter will be concerned with this method of DC power generation.

HOW AC IS CONVERTED TO DC

The basic difference between AC and DC is that with DC, electron current flows in just one direction; while AC is bidirectional. We can convert from AC to pulsating DC by restricting the AC power source to providing current flow through the circuit on only one-half (one polarity) of the sine wave, as shown in Figure 10–1.

The method to limit current to one direction, shown in Figure 10–1, uses the characteristic of a diode that permits current to pass only one way. This is called **half-wave rectification** of the input sine wave since conduction occurs on just one alternation (half cycle).

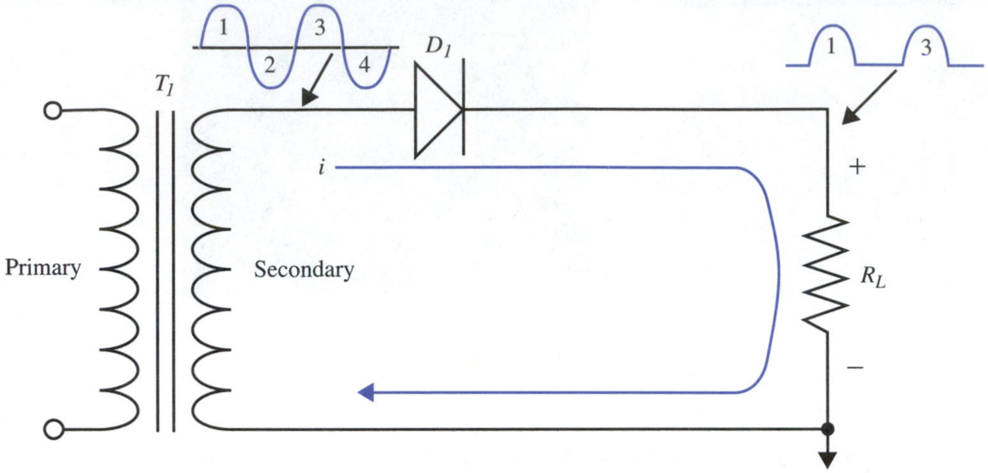

FIGURE 10–1 Conversion from AC to DC (half-wave rectification)

The circuit provides current in one direction through the load resistor, which increases and decreases with the input AC sine wave voltage. The transformer provides the ability to step the voltage down (or up) to the desired value required by the circuit; it also gives isolation (no common lead) from the AC mains.

FULL-WAVE RECTIFICATION

A circuit that allows current through the load on both alternations of the input sine wave is illustrated in Figure 10–2.

On one alternation (half cycle) of the input AC sine wave, the top end on the transformer has a positive voltage and the bottom end is negative relative to the center tap (common)

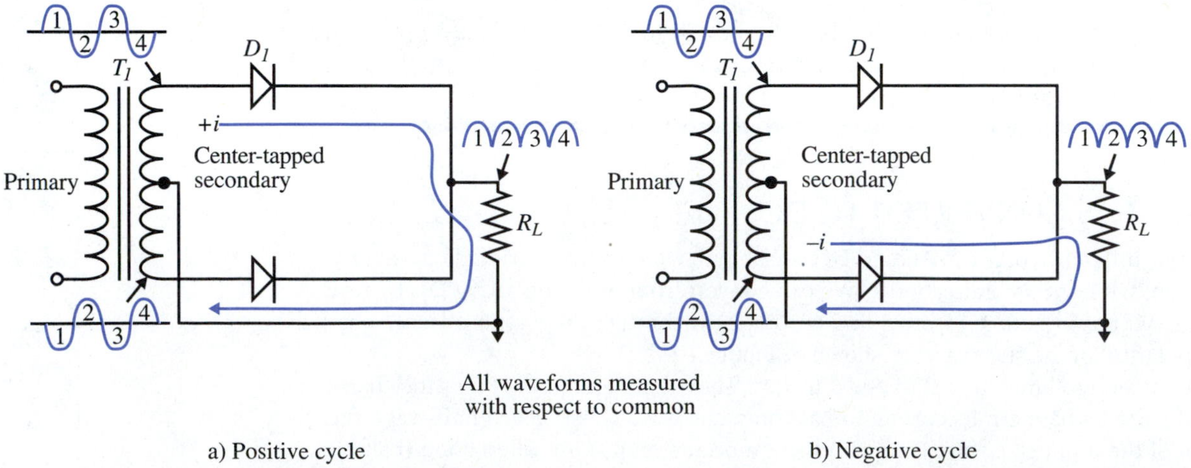

FIGURE 10–2 Conversion of AC to DC (full-wave rectification)

of the transformer. On the next alternation, the top of the transformer becomes negative and the bottom end goes positive.

When the voltage at the top of the transformer is positive, diode D_1 conducts and provides current through the load resistor and back through the center tap of the transformer. At this time, however, the bottom end of the transformer is negative, and diode D_2 is reverse biased and contributes no current to the load. On the next alternation, since the bottom end of the transformer is positive, D_2 conducts providing load current and D_1 is now reverse biased. Both rectified alternations appear across the load as pulsating DC.

The current path, when each diode conducts, is through the diode, then the load, and back through the center tap of the transformer. Notice that only one-half of the transformer is used for each alternation and therefore the voltage developed across the load is one-half the total transformer secondary peak voltage minus one diode drop.

BRIDGE RECTIFICATION

A **full-wave rectification** method that provides the total transformer voltage across the load and also eliminates the need for a center-tapped transformer is shown in Figure 10–3.

Four diodes are used in this configuration connected as a "bridge" (e.g., Wheatstone bridge). When the top end of the transformer is positive, D_1 conducts and current ($+i$) flows through the load and back through D_3 to the bottom end of the transformer. On the next alternation, the bottom end of the transformer is positive and current ($-i$) flows through D_2, the load, and through D_4 back to the top end of the transformer.

Because it provides current to the load on both alternations of the input sine wave, and doesn't need a center-tapped transformer, the bridge type is the most common method of power supply rectification. The four bridge diodes are usually sealed in a single package with four connecting leads.

RIPPLE FILTERING

The current, and therefore the voltage, developed across the load in the rectification circuits we have just studied, is not DC as provided by a battery, but a time-varying unipolar (one

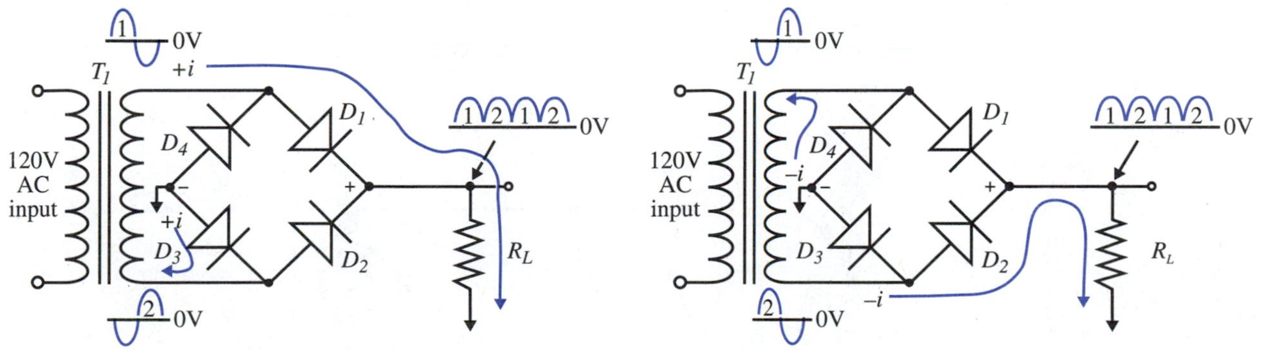

a) Positive cycle b) Negative cycle

FIGURE 10–3 Conversion of AC to DC (bridge)

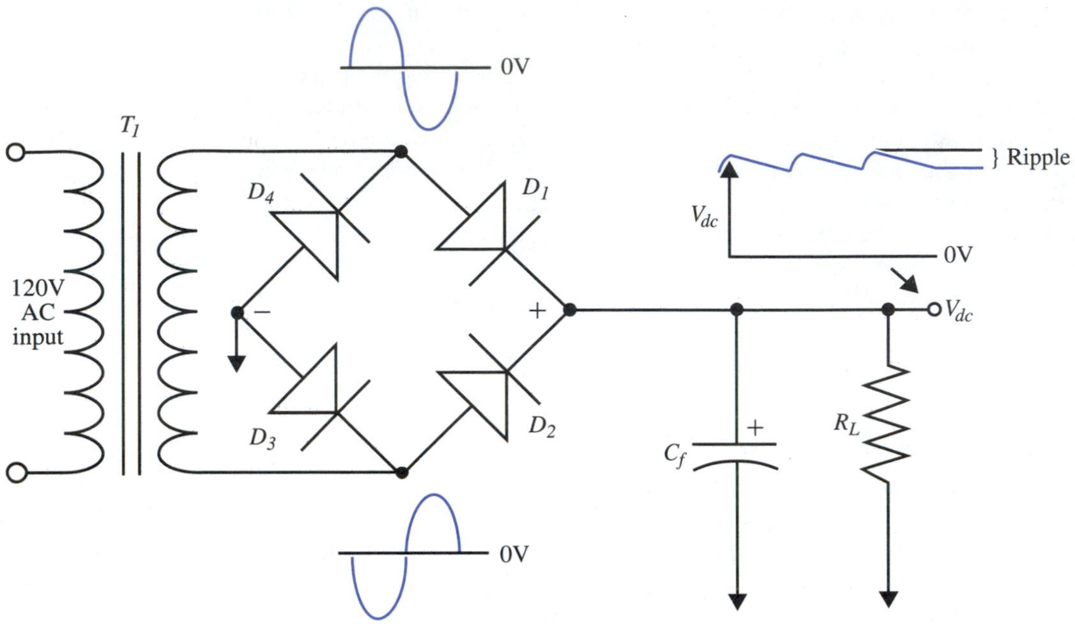

FIGURE 10–4 Capacitive filtering

polarity) voltage. There is a DC component of this waveform, which is 0.637 of the peak value, as well as a large AC component we call **ripple**.

To make a DC power supply more like a battery requires that we reduce the ripple component. We can do this with a **filter capacitor** (C_f in Figure 10–4).

This capacitor will maintain the output voltage near the peak value even when the input voltage drops to zero. It does this by storing charge and providing the load current when the AC input voltage from the transformer is less than the required DC output.

We can get a feel for the operation of a filter capacitor by analyzing the circuit of Figure 10–5. For simplicity, a half-wave rectifier is shown—but the same theory applies to a full-wave rectifier.

In the figure, the input sine wave has two vertical lines indicating the time interval the diode is conducting and recharging the capacitor. The capacitor charges to the peak of the input wave form minus the diode 0.6V drop.

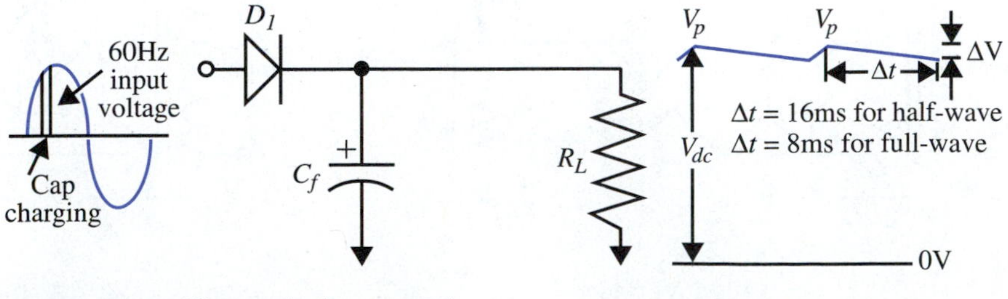

FIGURE 10–5 Filter capacitor operation

The voltage waveform across the capacitor shows the charging and discharging of the capacitor.

Discharging results from load current flowing from the capacitor during the time the diode is not conducting, and this causes the capacitor voltage to drop by an amount ΔV. But, we shall assume that the peak voltage across the capacitor is equal to the DC output voltage ($V_{dc} = V_p$). This assumption is good because the ripple voltage ΔV is usually much smaller than the V_p; since the actual V_{dc} is ($V_p - \Delta V/2$).

Ideally, to have pure DC at the load resistor, ΔV would be zero. But this would require an infinite size filter capacitor. Let us determine the relationship between the ripple ΔV and capacitor size.

The basic capacitor equation is

$$q = CV$$

where q is the charge on, and V the voltage across, the capacitor.

During discharge, let us assume the capacitor loses an amount of charge Δq during the increment of time Δt as a result of current flowing into the load. Further, assume the load current is constant over this time (the DC voltage change ΔV is too small to affect the current). We can determine the voltage change ΔV (ripple) on the capacitor by dividing the basic capacitor equation by the increment of time Δt.

$$\frac{\Delta q}{\Delta t} = C\frac{\Delta V}{\Delta t}$$

but

$$\frac{\Delta q}{\Delta t} = \text{I (current)}$$

so

$$I = C\frac{\Delta V}{\Delta t}$$

Solving for ΔV which is the peak-to-peak ripple,

$$\Delta V = \frac{I\,\Delta t}{C} \qquad \textbf{(equation 10–1)}$$

Since the current I is the independent variable determined by the load resistor, the ripple ΔV is inversely proportional to the size of the capacitor. The capacitor discharge time Δt is roughly 16ms for the half-wave rectification circuit of Figure 10–5. Actually, the ripple repeats at 16.67ms or 1/60Hz (we can assume the capacitor charges during the 0.67ms). For full-wave rectification, Δt is 8ms (1/120Hz).

EXAMPLE

Find the ripple voltage ΔV for a half-wave rectified power supply if I is 100mA and C is 4,000µF. Also, find the ripple if the capacitor size is increased to 8,000µF.

SOLUTION

Using equation 10–1,

$$\Delta V = \frac{i\,\Delta t}{C}$$

(Δt = 16ms for half-wave)

$$\Delta V = \frac{100\text{mA} \times 16\text{ms}}{4000\mu\text{F}}$$

$$\Delta V = 0.4\text{V}$$

With C of 8000μF,

$$\Delta V = \frac{100\text{mA} \times 16\text{ms}}{8000\mu\text{F}}$$

$$\Delta V = 0.2\text{V}$$

We can see from this example that doubling the capacitor size halves the ripple voltage—and the lower ripple is better!

Since for full-wave rectification the capacitor discharges only for 8ms before it is recharged, the capacitor size required for a given load current and ripple voltage is one-half the size of that required for half-wave.

EXAMPLE

Find the ripple voltage ΔV for a full-wave rectified power supply if I is 100mA and C is 2000μF.

SOLUTION

Using equation 10–1,

$$\Delta V = \frac{I\,\Delta t}{C}$$

$$\Delta V = \frac{100\text{mA} \times 8\text{ms}}{2000\mu\text{F}}$$

$$\Delta V = 0.4\text{V}$$

This is the same ripple voltage we got with a half-wave rectified power supply when we used twice the capacitor size (4000μF).

These results show why half-wave rectification is rarely used—it requires twice the capacitor size to obtain the same ripple voltage.

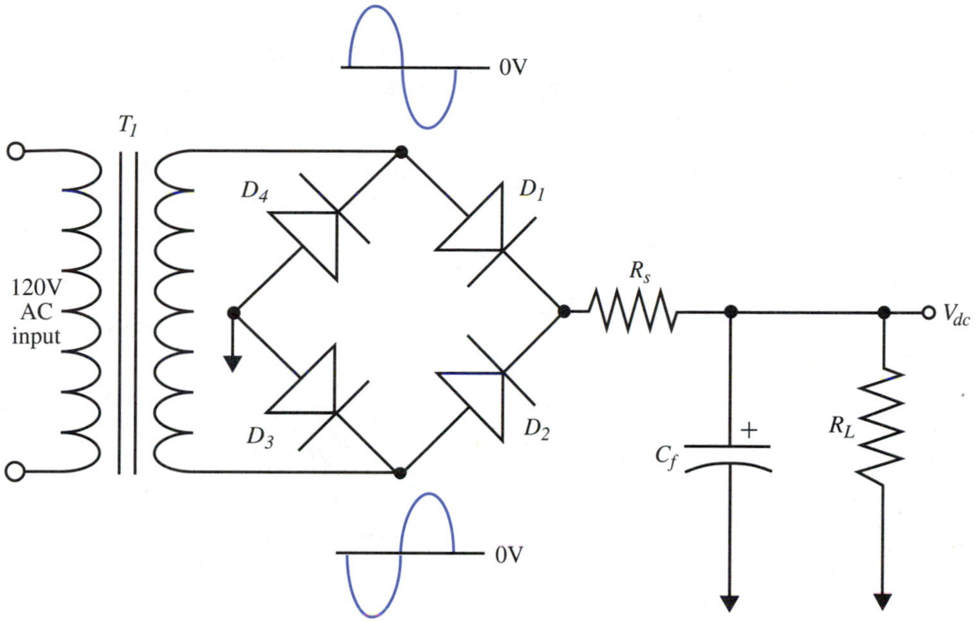

FIGURE 10–6 Complete power supply

COMPLETE POWER SUPPLY

We have looked at the basic elements of a power supply. Now we shall consider a practical power supply circuit. Figure 10–6 shows the power supply configuration we shall study. Notice that a full-wave bridge configuration is used for rectification. Again, a full-wave rectification circuit has the advantage of less discharge time for the filter capacitor, i.e., smaller capacitor; also, we don't need a center-tapped transformer with the bridge configuration.

The DC output voltage (V_{dc}) for this bridge circuit is the peak value of the input voltage V_{ac} minus two diode drops of 0.6V, or

$$V_{dc} \approx (V_{ac} \times \sqrt{2}) - 1.2\text{V} \qquad \textbf{(equation 10–2)}$$

A resistor R_s has been added between the diodes and the filter capacitor. The purpose of this resistor is to limit the diode surge current to a safe value when power is first turned on, since the capacitor, having no charge on it, looks like a short circuit.

R_s is selected to keep the surge current (I_s) below the one cycle surge current rating of the diode. For a diode that has a maximum forward current rating of 1A, the surge current rating is 30A; a 2A diode has a surge rating of 60A and so on.

It is possible that the DC resistance of the transformer, which is the secondary winding resistance and also the reflected primary resistance (changed by the square of the turns ratio), is sufficient to limit the current to below the diode surge current value. This must be taken into account when R_s is computed.

EXAMPLE

Find the required value of R_s for a power supply if the peak capacitor voltage V_c is 20V, the effective DC resistance of the transformer R_x is 0.07Ω, and a 3A bridge is used.

SOLUTION

$$R_s + R_x = \frac{V_c}{I_s}$$

For a 3A bridge rectifier, I_s is 90A.

$$R_s + R_x = \frac{20\text{V}}{90\text{A}}$$

$$R_s + R_x = 0.222\Omega$$

$$R_s = 0.222\Omega - 0.07\Omega$$

$$R_s = 0.15\Omega$$

RECTIFIER DIODE REVERSE VOLTAGE RATING

If we assume that the filter capacitor in Figure 10–6 charges to the positive peak of the transformer secondary AC voltage waveform (neglecting the 0.6V diode forward drops), then when the input waveform swings to the negative peak, there will be a reverse voltage of $2V_p$ across the rectifier diode, i.e., $+V_p$ on the diode cathode and $-V_p$ on the diode anode.

The rectifier diode should have a high enough peak reverse voltage (PRV) rating to withstand the reverse voltage it will encounter in the power supply circuit. To be conservative and accommodate AC line voltage spikes that can pass through the transformer, it is a good idea to use a diode with a PRV rating of twice the value of $2V_p$, or in other words $4V_p$.

EXAMPLE

Find the required peak reverse voltage rating for the diode bridge of Figure 10–6 if the capacitor voltage V_c is 25V.

SOLUTION

If we neglect the rectifier voltage drops and assume that V_c is equal to V_p, then the input waveform can swing to -25V. This means the rectifier diodes can see a peak reverse voltage rating of (2×25V) or 50V. To be conservative, we would use a bridge with a PRV of 100V.

EXAMPLE

For the power supply of Figure 10–6, we require DC output voltage of 16V, maximum current 500mA, and maximum 5% ripple. Find the transformer rating, diode rating, R_s if the effective transformer DC resistance R_x is 0.1Ω, and the required capacitor size.

SOLUTION

If we rearrange equation 10–2 to solve for V_{ac},

$$V_{ac} = \frac{(V_{dc} + 1.2V)}{\sqrt{2}}$$

$$V_{ac} = 12.16V \text{ (RMS)}$$

Use a 12V transformer.

To be conservative, we shall double our 500mA load current and use a 1A transformer and also a 1A bridge rectifier rated with a minimum PRV of four times V_c, or 64V (we shall use a standard size 100V PRV bridge).

When power is first turned on, we could have the maximum voltage of 16V driving current into the uncharged filter capacitor, which appears as a short circuit. With a 1A bridge, the specified surge current I_s is 30A.

So

$$R_s + R_X = \frac{V_{dc}}{I_s}$$

$$R_s + R_X = \frac{16V}{30A}$$

$$R_s + R_X = 0.53\Omega$$

Since R_X is given as 0.1Ω, R_s is 0.43Ω.

The ripple voltage is 5% of 16V or 0.8V, so the required filter capacitor is, from rearrangement of equation 10–1,

$$C = \frac{I\,\Delta t}{\Delta V}$$

$$C = \frac{500mA \times 8ms}{0.8V}$$

$$C = 5000\mu F$$

NEED FOR POWER SUPPLY REGULATION

If the input line voltage to our power supply of Figure 10–6 is lowered, the DC output voltage would also drop proportionally (refer to equation 10–2). The problem with this simple type of power supply is that the output DC will change with both the input line voltage and the load. Increasing the load current will cause a slight increase in voltage drop across the diode rectifiers; but a more pronounced voltage drop can occur from the DC resistance of the transformer and the surge resistor R_S.

EXAMPLE

Calculate the output voltage from a nominal 18V DC power supply with $R_x + R_s = 1.5\Omega$ if the load current changes from 200mA to 500mA. Also find the new output voltage with a load current held at 200mA, but the line voltage increases by 10 percent.

SOLUTION

Neglecting the change in the diode drop, the voltage drop ΔV for the 300mA increase in current ΔI is

$$\Delta V = \Delta I(R_s + R_x)$$

$$\Delta V = 300\text{mA} \times 1.5\Omega$$

$$\Delta V = 0.45\text{V}$$

The output voltage with the 500mA load is now

$$V_{dc} = 18\text{V} - 0.45\text{V}$$

$$V_{dc} = 17.55\text{V}$$

The change ΔV from our nominal 18V output if the line voltage increases by 10% is

$$\Delta V = 18\text{V} \times 0.1$$

$$\Delta V = 1.8\text{V}$$

Our new output voltage is

$$V_{dc} = 18\text{V} + 1.8\text{V}$$

$$V_{dc} = 19.8\text{V}$$

We could have obtained the same result by multiplying our nominal value by 1.1. For 10% low line we multiply V_{dc} by 0.9.

DEFINITION OF POWER SUPPLY REGULATION

Regulation of a power supply is defined as the change in output voltage (ΔV_{dc}) due to changes in the load current, or line voltage, divided by the nominal output voltage (V_{dc}), or

$$\text{Power Supply Regulation} = \frac{\Delta V_{dc}}{V_{dc}} \qquad \textbf{(equation 10–3)}$$

This is usually expressed in % regulation.

$$\%\ \text{Reg.} = \frac{\Delta V_{dc}}{V_{dc}} \times 100 \qquad \textbf{(equation 10–4)}$$

The lower the % regulation, the better the power supply.

Load regulation is defined in terms of the voltage change that occurs at the power supply output when the load current goes from no load (zero amps) to the full load condition. (Actually, regulator manufacturers specify some minimum load current to ensure that transistors are at their operating bias point.)

Line regulation is defined in terms of the voltage change that occurs at the power supply output when the input AC line voltage is increased or decreased by 10 percent.

EXAMPLE

Find the percent load regulation for a nominal 15V output power supply if the output voltage changes by 100mV when the load current is increased from no load to full load.

SOLUTION

From equation 10–4,

$$\% \text{ Load Reg.} = \frac{100\text{mV}}{15\text{V}} \times 100\%$$

$$\% \text{ Load Reg.} = 0.67\%$$

EXAMPLE

Find the percent line regulation for a nominal 12V output power supply if the output voltage drops 50mV when the input line voltage drops 10 percent.

SOLUTION

From equation 10–4,

$$\% \text{ Line Reg.} = \frac{50\text{mV}}{12\text{V}} \times 100\%$$

$$\% \text{ Line Reg.} = 0.417\ \%$$

MEASUREMENT OF POWER SUPPLY REGULATION

Power supply regulation was defined as the change in output voltage divided by the nominal value. Very little change in output voltage (millivolts) will occur with a well-regulated supply, so we can't measure the change by connecting a voltmeter from output to ground. For example, if the nominal output voltage of 10V changes by only 10mV, we won't see the change on, say, a 3 1/2 digit meter.

We need to use a **differential measurement technique** that will "buck out" the nominal output voltage and just show the change in output voltage.

Figure 10–7 shows the use of a second stable voltage source that is used to remove the nominal voltage from measurement.

Note that the voltmeter is not connected to ground—but instead is connected between the power supply output and the voltage reference.

The procedure for making the measurement is as follows:

1) The power supply output voltage is adjusted to the required value.
2) The DMM is connected between the output of the power supply and the voltage reference.
3) The reference voltage is adjusted until the meter reads 0V ($V_{ref} = V_{dc}$).

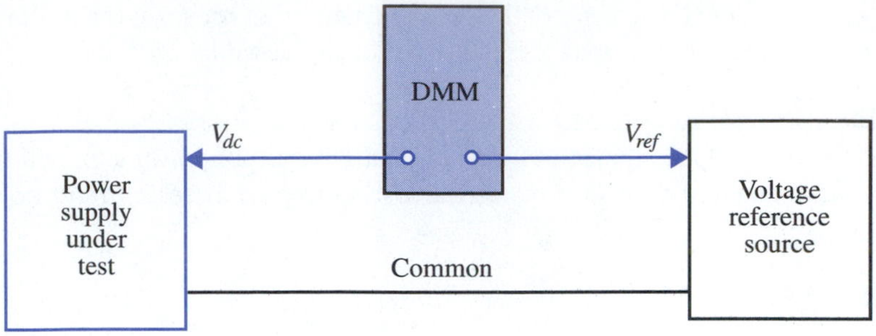

FIGURE 10–7 Differential voltage measurement

4) The DMM is set to its most sensitive voltage scale, and the reference is "tweaked" again until the meter reads close to 0V.
5) The power supply line voltage or load resistor is varied and change in the reading on the DMM is noted. This is the required power supply voltage change measurement.
6) The voltage regulation of the power supply is this change divided by the nominal power supply output voltage.

VOLTAGE REGULATOR

A device added to a power supply to reduce the effect of line and load variations on the output voltage is called a **power supply regulator**.

The simplest type of regulator uses a zener diode, as shown in Figure 10–8. This type of regulator uses the fact that the voltage across a zener diode is fairly constant—even with current changes through the device.

Notice the regulator is added to the output of our basic supply and consists of the zener diode and a voltage dropping resistor R_d.

Selection of the voltage dropping resistor R_d is based on providing enough current for the zener to operate (I_{zmin}) under all line voltage and load conditions.

The least current the zener will conduct is at the 10% low AC line (*LL*) voltage and with full load current (I_{Lmax}). Under these conditions, the current through R_d will be $I_{zmin} + I_{Lmax}$.

So

$$R_d = \frac{V_{cLL} - V_{dc}}{I_{zmin} + I_{Lmax}} \qquad \textbf{(equation 10–5)}$$

The zener voltage is picked to equal the desired output voltage from the power supply and so

$$V_z = V_{dc}$$

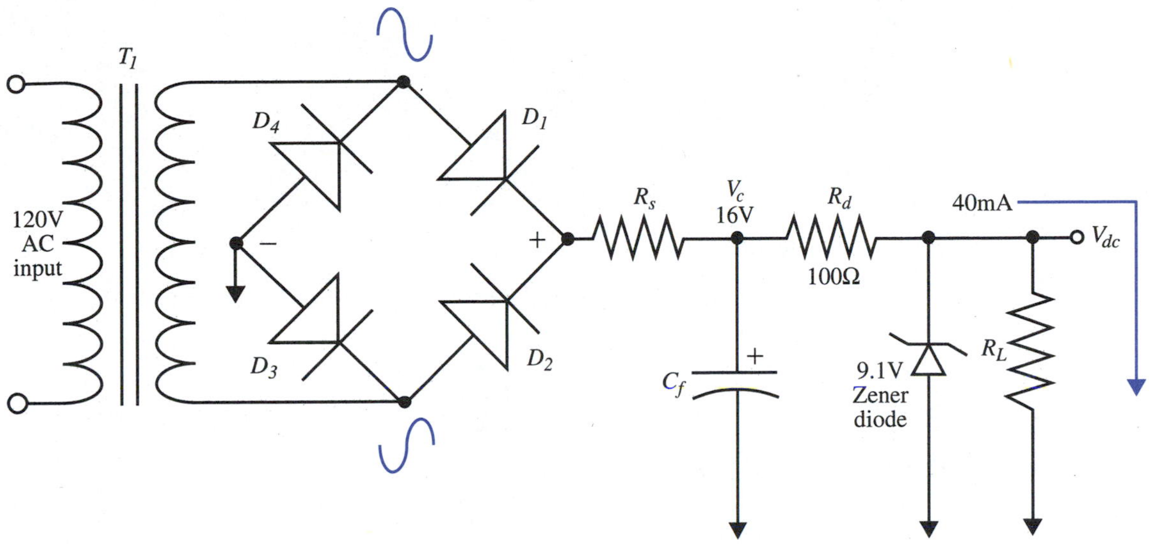

FIGURE 10–8 Zener diode regulator

The power rating of the zener is selected to handle the current at the 10% high line (*HL*) voltage condition with the load disconnected. (*Note:* I_{zmax} occurs with the load disconnected because all the current through R_d must now go through the zener.)

In equation form,

$$P_z = (V_{dc})(I_{zmax})$$

but

$$I_{zmax} = \frac{(V_{cHL} - V_{dc})}{R_d}$$

Therefore

$$P_z = (V_{dc})\frac{(V_{cHL} - V_{dc})}{R_d} \qquad \textbf{(equation 10–6)}$$

EXAMPLE

The required output from a zener-regulated power supply is 9V with a maximum load current of 40mA. If the nominal voltage at the filter capacitor V_c is 16V, and the minimum zener current I_{zmin} is 10mA, find the voltage and power rating of the zener, the resistance R_d value, and also the power rating of R_d.

SOLUTION

First, we can find the value of R_d by using equation 10–5 and remembering that V_{cLL} is the capacitor voltage at low line voltage and is equal to $0.9 \times V_c$.

$$R_d = \frac{V_{cLL} - V_{dc}}{I_{zmin} + I_{max}}$$

$$R_d = \frac{(16V \times 0.9) - 9V}{10mA + 40mA}$$

$$R_d = 108\Omega$$

(Use 100Ω standard R value—a lower value is more conservative.)

For the power rating P_r of the resistor R_d, we need to find the total current (I_{tHL}) at high line.

$$I_{tHL} = \frac{V_{cHL} - V_z}{R_d}$$

$$I_{tHL} = \frac{(16V \times 1.1) - 9V}{100\Omega}$$

$$I_{tHL} = 86mA$$

$$P_{rd} = I_{tHL}{}^2 R_d$$

$$P_{rd} = (86mA)^2 \times 100\Omega$$

$$P_{rd} = 0.74W$$

(To be conservative, use 2W resistor.)

The voltage rating of the zener is simply the required output voltage of 9V (standard zener value is 9.1V). For the power rating of the zener, we simply multiply the maximum current I_{tHL} through R_d (at high line) by the zener voltage rating or use equation 10–6.

$$P_z = 9V \times 86mA$$

$$P_z = 0.774W$$

(To be conservative, use a 2-watt zener.)

The component values for this example are shown in Figure 10–8.

As was indicated in Chapter 1, the voltage across a zener changes slightly with current through the device because of the zener resistance r_z.

The actual voltage rating for a zener is specified at a given test current and consists of the zener breakdown voltage plus the voltage drop across the zener impedance.

EXAMPLE

Determine the zener current I_z at maximum load and the % load regulation for the regulator in Figure 10–8. Use a 1N4739 (9.1V) zener that has a zener impedance r_z of 5Ω at a test current of 28mA.

SOLUTION

The current I_d through R_d is

$$I_d = \frac{(16\text{V} - 9.1\text{V})}{100\Omega}$$

$$I_d = 69\text{mA}$$

With a maximum load current of 40mA, the zener current I_z is

$$I_z = I_d - I_L$$

$$I_z = 69\text{mA} - 40\text{mA}$$

$$I_z = 29\text{mA}$$

The zener voltage (V_z) with this current, which is close to the specified value of 28ma, is 9.1V. If the load current goes to zero, all 69mA (I_d) will flow through the zener. The new zener voltage V_{znl} will be

$$V_{znl} = V_z + (r_z \times \Delta I_d)$$

$$V_{znl} = 9.1\text{V} + 5\Omega\ (69\text{mA} - 28\text{mA})$$

$$V_{znl} = 9.305\text{V}$$

The zener voltage change ΔV_z from no load to the maximum load condition is

$$\Delta V_z = 9.305\text{V} - 9.1\text{V}$$

$$\Delta V_z = 0.205\text{V}$$

The percent load regulation from equation 10–4 is

$$\%\ \text{Load Reg.} = \frac{\Delta V_z}{V_z} \times 100\%$$

$$\%\ \text{Load Reg.} = \frac{0.205\text{V}}{9.1\text{V}} \times 100\%$$

$$\%\ \text{Load Reg.} = 2.25\%$$

EXAMPLE

Find the percent line regulation (*%LR*) if the input line voltage increases by 10 percent for the circuit of Figure 10–8.

SOLUTION

Increasing the input line voltage by 10 percent means the voltage across the filter capacitor is going to increase by 10 percent to

$$16\text{V} \times 1.1 = 17.6\text{V}$$

This is an increase of 1.6V over the 16V value at normal line voltage. If we assume the zener voltage stays constant, then this 1.6V increase must all appear across R_d. This will increase the current through R_d by ΔI_{Rd}

where

$$\Delta I_{Rd} = \frac{1.6\text{V}}{100\Omega}$$

$$\Delta I_{Rd} = 16\text{mA}$$

Since the load current is fixed at 40mA, this additional current must all flow through the zener diode. Remembering that the specified zener resistance is 5**Ω**, then the voltage change (ΔV_z) across the zener (and also the load) will be

$$\Delta V_z = 16\text{mA} \times 5\Omega$$

$$\Delta V_z = 80\text{mV} = \Delta V_{dc}$$

The percent line regulation is

$$\frac{\Delta V_{dc}}{V_{dc}} \times 100\%$$

So

$$\% \text{ LR} = \frac{80\text{mV}}{9.1\text{V}} \times 100\%$$

$$\% \text{ LR} = 0.879\%$$

Compare this percent with a 10-percent change in the input.

RIPPLE REDUCTION EFFECT OF ZENER IMPEDANCE

One additional advantage of using a zener regulator is that it further reduces the ripple voltage. Since the AC resistance of the zener is simply r_z (see Figure 10–9), the ripple that appears at the filter capacitor ΔV_c is reduced at the output (across the zener) to ΔV_o because of the voltage divider action of R_d and r_z.

From Figure 10–9, we can see that the output ripple ΔV_o is

$$\Delta V_o = \frac{(\Delta V_c)(r_z)}{(R_d + r_z)} \quad \textbf{(equation 10–7)}$$

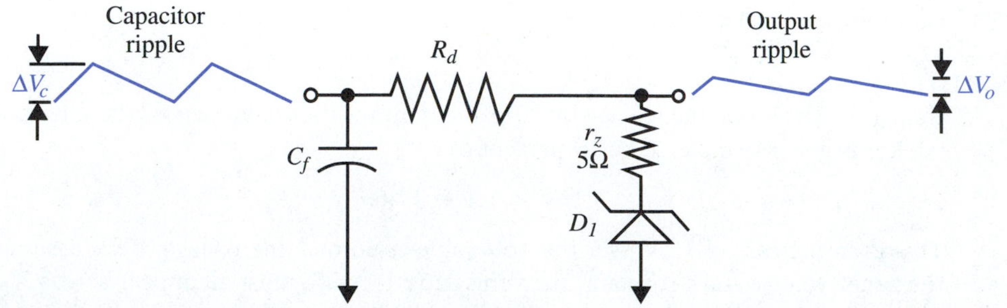

FIGURE 10–9 Ripple reduction of a zener regulator

EXAMPLE

The ripple on the filter capacitor ΔV_c is 1V; find the ripple ΔV_o at the power supply output if R_d is 100Ω and r_z is 5Ω.

SOLUTION

Using equation 10–7,

$$\Delta V_o = \frac{\Delta V_c r_z}{R_d + r_z}$$

$$\Delta V_o = \frac{1\text{V} \times 5\Omega}{100\Omega + 5\Omega}$$

$$\Delta V_o = 47.6\text{mV}$$

So the zener has provided about a 21-to-1 reduction in the ripple voltage!

TRANSISTOR REGULATORS

Disadvantages of the zener diode regulator are: The output voltage changes when the load current changes (zener impedance effect); the current drawn by the power supply is not reduced when the load current drops.

One way around the problems of the zener regulator is to add a transistor follower, as shown in Figure 10–10. This circuit is also called a **series regulator** because the transistor is in series with the main current path to the load.

What this circuit does is to make the current through the zener lower by the factor of the β of the transistor. The zener current change, when the load is applied, is reduced by the β of the transistor, so there is less zener resistance effect.

Notice in the circuit that the output voltage V_{dc} is equal to the zener voltage minus the 0.6V base-emitter voltage drop.

EXAMPLE

Calculate the change in the output voltage for the circuit of Figure 10–10 if the load current changes from 0mA to 40mA. Assume V_c is 16V, V_z is 10V, R_d is 1KΩ, r_z is 5Ω, and the β of the transistor Q_1 is 100.

SOLUTION

The output voltage will change only due to changes in the current flowing through the zener. So we need to find this current change.

The base current with a 40mA load current is

$$I_b = \frac{40\text{mA}}{100}$$

$$I_b = 400\mu\text{A}$$

With no load, the base current is zero.

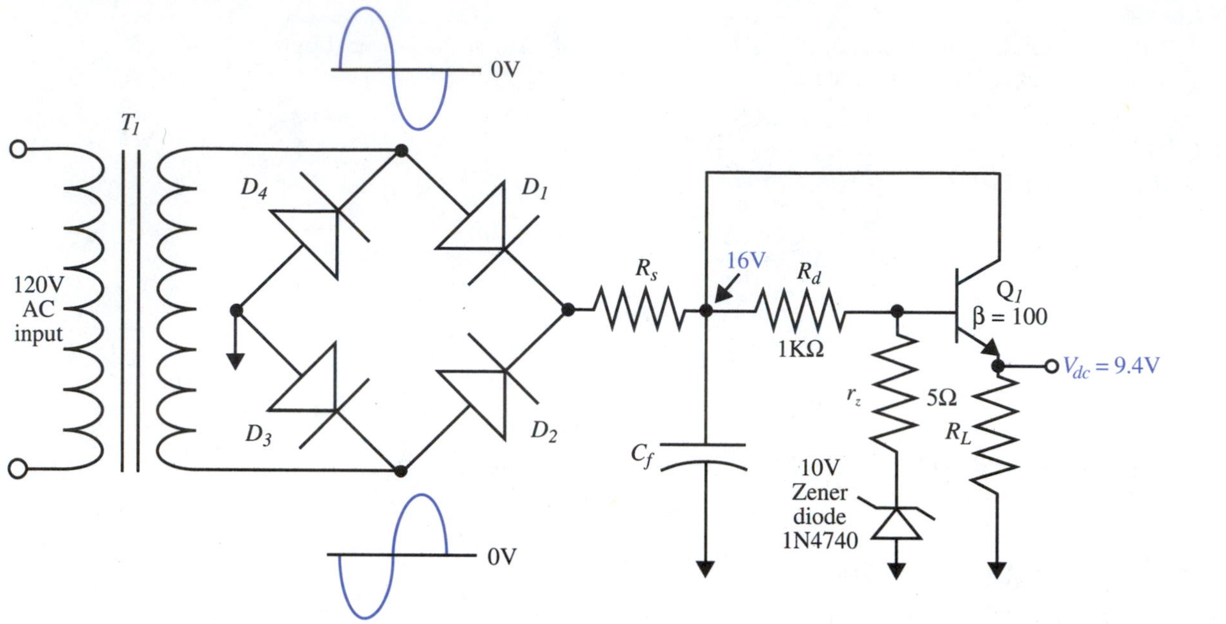

FIGURE 10–10 Follower (series) regulator

The difference between the base current and the current through R_d must flow into the zener. The actual change in current through the zener ΔI_z is the change in the base current of 0μA to 400μA, or 400μA. This current change times the zener impedance gives us the change in the output voltage ΔV_{dc}.

$$\Delta V_{dc} = \Delta I_z \times r_z$$

$$\Delta V_{dc} = 400\mu A \times 5\Omega$$

$$\Delta V_{dc} = 2mV$$

Without the transistor, the output voltage change would have been 40mA × 5**Ω**, which is 200mV or a hundred (β) times larger. (This is the result we got with the simple zener regulator in the earlier example.)

We conclude then that adding the transistor improves the load regulation by the transistor β factor.

The selection of the transistor for the follower regulator circuit is determined from

1) Load current
2) Output voltage
3) Power dissipation

The first two items are specified by the circuit requirements, and the transistor power dissipation (P_t) is the maximum load current (I_{Lmax}) times the voltage difference between the filter capacitor voltage V_c and V_{dc} (remember, $V_{dc} = V_z - 0.6V$).

$$P_t = I_{Lmax}(V_c - V_{dc}) \qquad \textbf{(equation 10–8)}$$

EXAMPLE

Specify the maximum power dissipation for a transistor if the load current varies from 5mA to 100mA with an output voltage V_{dc} of 15V and a filter capacitor voltage V_c of 19V.

SOLUTION

From equation 10–8,

$$P_t = I_{Lmax}(V_c - V_{dc})$$

$$P_t = 100\text{mA} \times (19\text{V} - 15\text{V})$$

$$P_t = 400\text{mW}$$

We used the 100mA load current value because this gives us the maximum transistor power dissipation.

The follower regulator does a good job of maintaining the output voltage constant with load changes, but it is a little sensitive to input line voltage changes.

EXAMPLE

Compute the change in the output voltage for a follower regulator if the input line voltage increases by 10 percent. Assume a nominal V_c of 16V, V_z of 10V (output 9.4V), R_d of 1KΩ, and r_z of 5Ω.

SOLUTION

If the line voltage increases by 10 percent, V_c becomes *16V × 1.1*, or 17.6V. This is an increase of 1.6V across R_d, which gives a current increase ΔI_z through this resistor of

$$\Delta I_z = \frac{1.6\text{V}}{1\text{K}\Omega}$$

$$\Delta I_z = 1.6\text{mA}$$

This current will cause an increase in voltage drop ΔV_z across the zener (and also the output) of

$$\Delta V_z = (\Delta I_z)(r_z)$$

$$\Delta V_z = (1.6\text{mA})(5\Omega)$$

$$\Delta V_z = 8\text{mV}$$

This change of 8mV in the nominal output voltage of 9.4V represents a 0.085% change—which is low when we consider that the line voltage changed by 10%.

ACTIVE REGULATOR

By adding another transistor, as shown in Figure 10–11, we can make an active regulator that is even less affected by line voltage variations and changes in the base-emitter voltage of Q_1.

In this new circuit, the main current through the zener is provided from the regulated output voltage.

In operation, Q_2 acts as an error amplifier to correct for changes in the output voltage. It does this by comparing the divided down (R_3, R_4, and R_5) output voltage with the zener voltage. R_4 allows for precise setting of the output voltage.

If the output voltage is too high, the base of Q_2 is high, causing the Q_2 collector current to increase. This increases the voltage drop across R_1 and lowers the base voltage of Q_1. Since Q_1 is a follower, the output voltage will drop back to the required value.

If the output voltage is too low, the lowered base voltage of Q_2 will cause the collector current to decrease, causing the voltage on the base of Q_1 to rise, which will restore the output voltage.

The design of the circuit shown in Figure 10–11 is based on a nominal output voltage of 9V and a maximum load current of 100mA.

The transformer will provide about 16V DC at the filter cap at normal line voltage. We shall assume a β for the Q_1 power transistor of 40. R_2 is selected to provide a zener current of 10.4mA or (9V − 6.2V)/270Ω, which is about 10 percent of the load current. The voltage divider at the output will draw about 1mA and is set up to provide 9V at the output with 6.8V at the base of Q_2 (Zener voltage + 0.6V).

Adding these three currents together, we find that Q_1 will be providing a maximum current I_{max} of

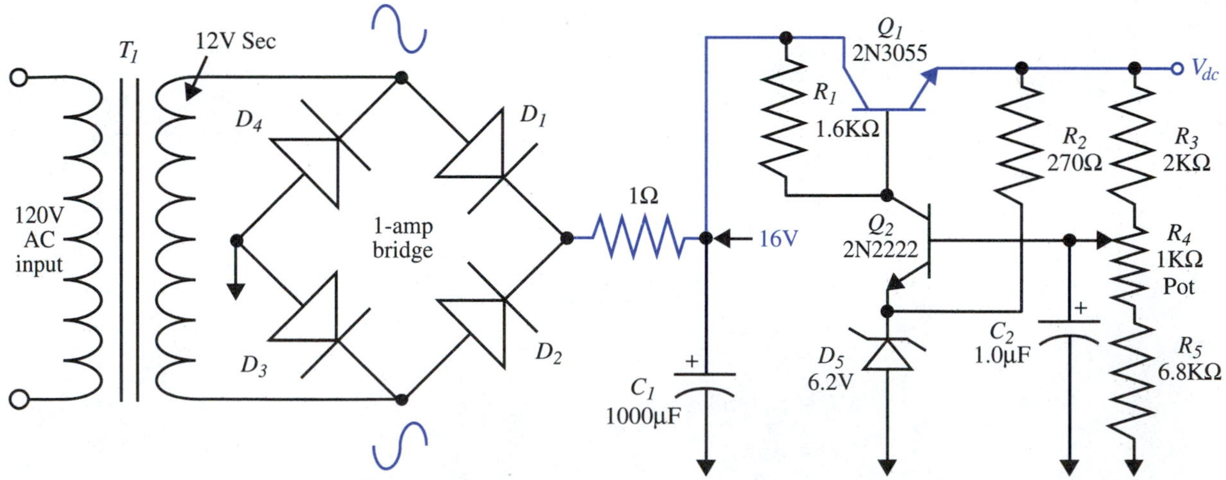

FIGURE 10–11 Active power supply regulator circuit

$$I_{max} = (100 + 10.4 + 1)\text{mA}$$

$$I_{max} = 111.4\text{mA}$$

With a β of 40, the maximum Q_1 base current will be

$$I_b = \frac{111.4\text{mA}}{40}$$

$$I_b = 2.785\text{mA}$$

The maximum current through R_1 will be this base current plus the collector current of Q_2. At 10 percent low line voltage, the voltage at the filter capacitor will be

$$V_c = 16\text{V} \times 0.9$$

$$V_c = 14.4\text{V}$$

If we assume that the Q_2 collector current is a minimum of 100μA at low (−10 percent) input line voltage, then the total current through R_1 at low line is this current plus the Q_1 base current, or

$$I_1 = 2.785\text{mA} + 100\mu\text{A}$$

$$I_1 = 2.885\text{mA}$$

R_1 is determined by realizing the base of Q_1 is 0.6V greater than the output voltage or 9.6V. The other end of R_1 is at 14.4V (we are computing with low line voltage), so the value of R_1 is

$$R_1 = \frac{14.4\text{V} - 9.6\text{V}}{2.885\text{mA}}$$

$$R_1 = 1.66\text{K}\Omega \quad \text{(use 1.6K}\Omega\text{ standard value)}$$

EXAMPLE

Find the change in the output voltage of the active regulator of Figure 10–11 if the input line voltage increases by 10 percent. Assume the zener impedance is 5Ω.

SOLUTION

If the line voltage increases by 10 percent, the filter capacitor voltage V_c (16V) will also increase by 10 percent (ΔV). This will increase the voltage at the top end of R_1 by 1.6V, which in turn will increase the current through this resistor by ΔI_1,

where

$$\Delta I_1 = \frac{\Delta V}{R_1}$$

$$\Delta I_1 = \frac{1.6\text{V}}{1.6\text{K}\Omega}$$

$$\Delta I_1 = 1\text{mA}$$

Because the load current is considered constant, the current into the base of Q_1 stays fixed; this increased current ΔI_1 must flow down into the Q_2 collector and through the zener diode. The voltage change across the zener ΔV_z due to this increased current is

$$\Delta V_z = 1\text{mA} \times 5\Omega$$

$$\Delta V_z = 5\text{mV}$$

Since the base-emitter voltage of Q_2 is fixed at 0.6V, this 5mV increase will also occur at the base of Q_2. This change in voltage is also across R_5 (and part of R_4) and, because R_3, R_4, and R_5 form a 6.8V to 9V voltage divider, this will cause the output to increase by ΔV_{dc} or

$$\Delta V_{dc} = \frac{(5\text{mV} \times 9\text{V})}{6.8\text{V}}$$

$$\Delta V_{dc} = 6.6\text{mV}$$

This change of 6.6mV represents a 0.07 percent change in the output voltage and is a slight improvement over the output voltage change of 8mV we saw with the same line voltage change for the follower regulator.

ACTIVE REGULATOR LOAD REGULATION

We now need to consider the effect of varying the load on the output voltage of an active regulator.

Any change that occurs in the output voltage of the regulator with load current changes must be due to the change in voltage drop across the internal output resistance of the regulator. The derivation of this output resistance will be more apparent if we restudy the output resistance of an operational amplifier as described in Chapter 6 (see equation 6–11).

If we look back into the emitter of Q_1 in Figure 10–11, we shall see an output resistance of R_1/β_1; but this resistance is lowered, because of the feedback (via the output divider and Q_2) in the circuit, by the loop gain factor A_L. This factor is determined from

$$A_L = \frac{A_2}{A_f}$$

where A_2 is the voltage gain of Q_2 and is equal to

$$A_2 = \frac{R_1 \| R_t}{r_{e2}}$$

(R_t is the input resistance seen at the base of Q_1.)

If we assume the wiper on R_4 is at mid-range,

$$A_f = \left(\frac{R_3 + \dfrac{R_4}{2}}{R_5 + \dfrac{R_4}{2}}\right) + 1$$

Combining these two relationships, we get

$$A_L = \frac{(R_1 \| R_t)\left(R_5 + \frac{R_4}{2}\right)}{r_{e2}\left(R_3 + \frac{R_4}{2}\right)} + 1$$

Finally, the regulator output resistance R_o is

$$R_o = \frac{R_1}{\beta_1 A_L}$$

$$R_o = R_1\left(\frac{2r_{e2}R_3 + r_{e2}R_4 + 2R_5 + R_4}{\beta_1(R_1 \| R_t)(2R_5 + R_4)}\right) \qquad \textbf{(equation 10–9)}$$

EXAMPLE

Compute the output resistance for the regulator circuit of Figure 10–11 with a 100mA load.

SOLUTION

In order to use equation 10–9, we must first determine the values of r_{e2} and R_t. The value of r_{e2} is determined from the emitter current of Q_2 (r_e is 26mV/I_e). With a nominal voltage at the filter cap of 16V and the base of Q_1 (also collector of Q_2) at 9.6V (since we have 9V at output), the current through R_1 is

$$I_1 = \frac{(16 - 9.6)\text{V}}{1.6\text{K}\Omega}$$

$$I_1 = 4\text{mA}$$

Since from our previous calculations, 2.785mA of this current flows into the base of Q_1, then

$$I_2 = 4\text{mA} - 2.785\text{mA}$$

$$I_2 = 1.215\text{mA}$$

So

$$r_{e2} = \frac{26\text{mV}}{1.215\text{mA}}$$

$$r_{e2} = 21.4\Omega$$

To determine R_t, we need to translate the Q_1 emitter current of 111.4mA to an equivalent R_e resistor.

$$R_e = \frac{9\text{V}}{111.4\text{mA}}$$

$$R_e = 81\Omega$$

And

$$R_t = \beta_1 \times R_e$$

$$R_t = 40 \times 81\Omega$$

$$R_t = 3.24\text{K}\Omega$$

With all the known values substituted in equation 10–9,

$$R_o = \frac{1.33 \times 10^4}{4 \times 10^4}$$

$$R_o = 0.33\Omega$$

We can now find the load regulation of the active regulator, from no load to full load, by multiplying the load current change by this output resistance.

EXAMPLE

Find the percent load regulation from no load to a full load current change (ΔI_L) of 100mA for the 9V active regulator in Figure 10–11.

SOLUTION

The voltage change at the output of the regulator is

$$\Delta V_o = (\Delta I_L)(R_o)$$

$$\Delta V_o = 100\text{mA} \times 0.33\Omega$$

$$\Delta V_o = 33\text{mV}$$

$$\%\ \text{regulation} = \frac{\Delta V_o}{V_o} \times 100\%$$

$$\%\ \text{regulation} = \frac{33\text{mV}}{9\text{V}} \times 100\%$$

$$\%\ \text{regulation} = 0.37\%$$

VOLTAGE DOUBLER

Sometimes we require an output DC voltage that is greater than what the transformer secondary can supply. To achieve this higher voltage, we can use a rectifying circuit (see Figure 10–12) that will double the effective input voltage.

On the positive alternation of the input sine wave ($i+$), D_1 conducts and D_2 is reverse biased. C_1 charges to the peak of the input voltage, as shown in Figure 10–12a. With the negative alternation ($i-$), shown in Figure 10–12b, D_1 is reverse biased and D_2 forward biased. The charge on C_1 causes the capacitor to be like a battery V_p in series with the input voltage, and so C_2 charges to twice V_p and the DC output voltage is twice the peak of the input voltage.

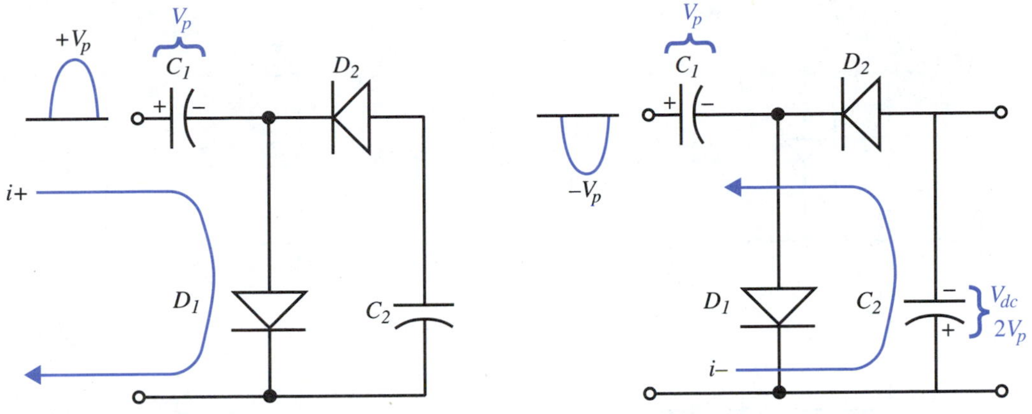

a) Positive alternation b) Negative alternation

FIGURE 10–12 Half-wave voltage doubler

This circuit is called a **half-wave voltage doubler** since the output capacitor (C_2) is charged once every cycle of the input waveform.

A **full-wave voltage doubler** is shown in Figure 10–13. This circuit has the advantage that the ripple on the output capacitors is twice the frequency of the half-wave doubler and therefore easier to filter; but a disadvantage is that both output lines are not common to the input as with the half-wave doubler.

In the circuit, C_1 charges on the positive alternation of the input sine wave ($i+$); while C_2 charges on the negative alternation ($i-$). The output voltage is taken across both capacitors and is equal to $2V_p$. R_s has been included to limit the surge current during power turn on.

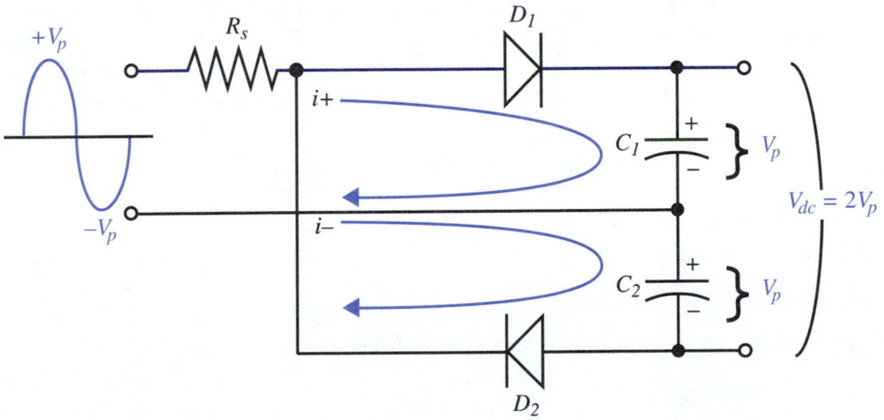

FIGURE 10–13 Full-wave voltage doubler

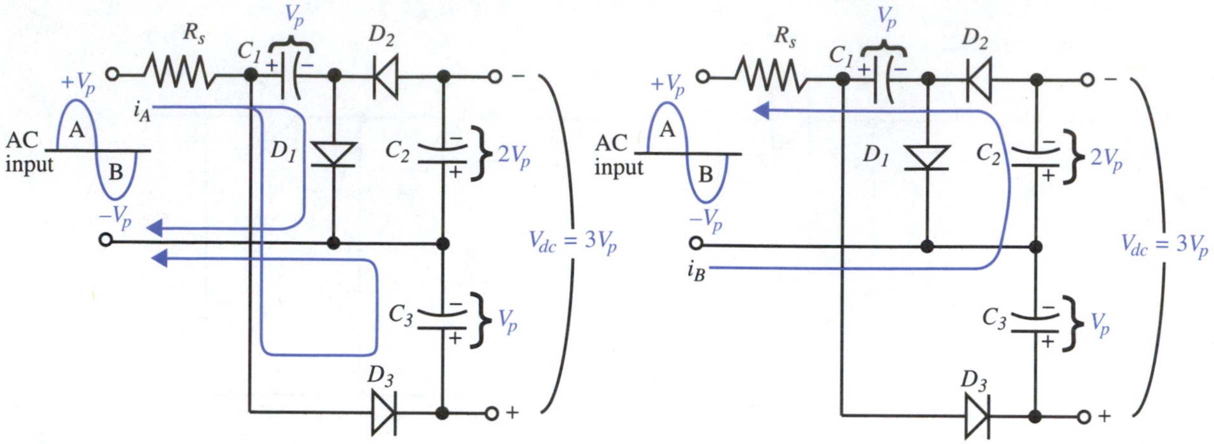

FIGURE 10–14 Voltage tripler (full-wave)

VOLTAGE TRIPLER

The voltage tripler is used when we want to increase the input voltage by a factor of three. Triplers are used in televisions and monitors to generate the high voltage required for the cathode ray tube. Figure 10–14 shows the circuit for a **full-wave tripler**. This circuit is similar to the half-wave doubler of Figure 10–12 but has an additional diode (D_3) and capacitor (C_3).

As with the half-wave doubler, C_1 charges to V_p on the first positive alternation because D_1 conducts; but D_3 also conducts and charges C_3 to V_p. On the negative alternation, D_2 conducts and C_2 charges to $2V_p$—which is the same result we got with the half-wave doubler circuit.

The output voltage is taken across C_2 and C_3 and is equal to the required $3V_p$.

TROUBLESHOOTING POWER SUPPLY CIRCUITS

Let us first look at troubleshooting the unregulated power supply shown in Figure 10–15. With this circuit, we expect to have a DC output of about $\sqrt{2} \times V_{ac}$ with a ripple component that is a function of the effective load resistor, i.e., the lower the resistor, the higher the ripple voltage. With the full-wave bridge configuration, the ripple frequency should be twice the input line frequency.

The following problems could occur with this power supply:

1) No DC output
2) Low DC output
3) Ripple voltage too high

Causes of no DC output are one or more of the following:

1) Fuse blown
2) Open line cord
3) Open R_s

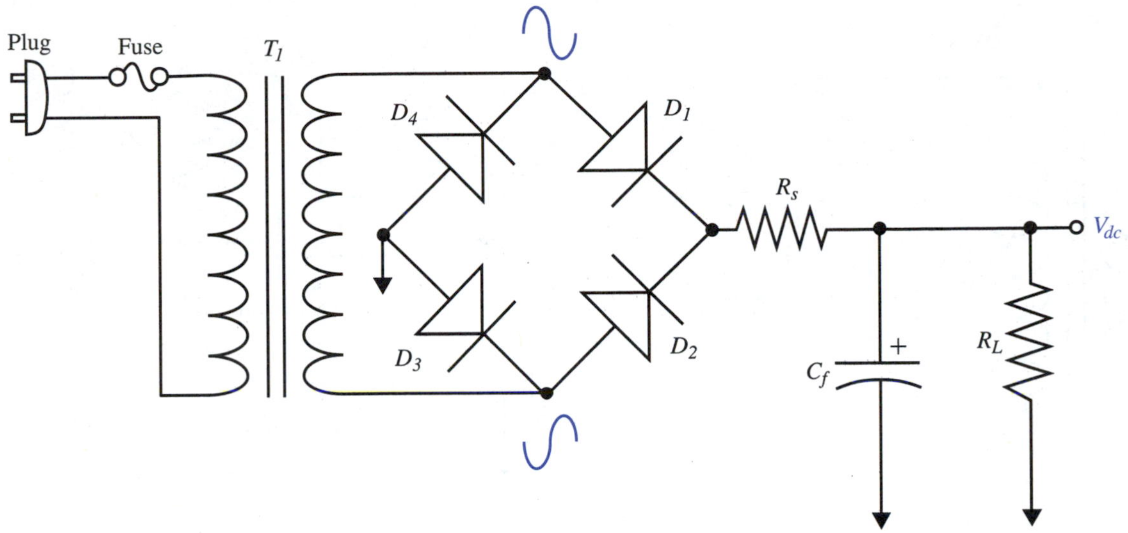

FIGURE 10–15 Troubleshooting the basic power supply

4) Open bridge
5) Open transformer secondary
6) Open transformer primary

To isolate the problem that causes no DC output voltage:

1) First check the fuse and condition of the line cord.
2) If the fuse has blown, check the power supply for a short circuit by removing the load and connecting a low-wattage 120V light bulb across the open fuse connection (in place of the fuse).
3) If the bulb lights brightly, the power supply is short-circuited. Disconnect the filter capacitor, and if the light bulb goes dull or out, the capacitor is shorted.
4) If the bulb stays bright, disconnect the bridge. A bright light now indicates a shorted transformer; a dull light or no light means a shorted bridge.
5) If line cord and the fuse were originally good (good fuse means no circuit short), measure the DC voltage to the left of R_S. If the DC voltage is present, R_S is open.
6) If no DC voltage appears to the left of R_S, measure the AC voltage on the transformer secondary. AC voltage at the secondary indicates an open bridge.
7) With an indication of no voltage at the transformer secondary, measure the primary voltage. If voltage is present at the primary, the transformer has an open circuit.
8) Low voltage at the output with a high percentage of ripple could mean that either the load is excessive (R_L too low) or one side of the bridge is open and the power supply is operating half-wave.

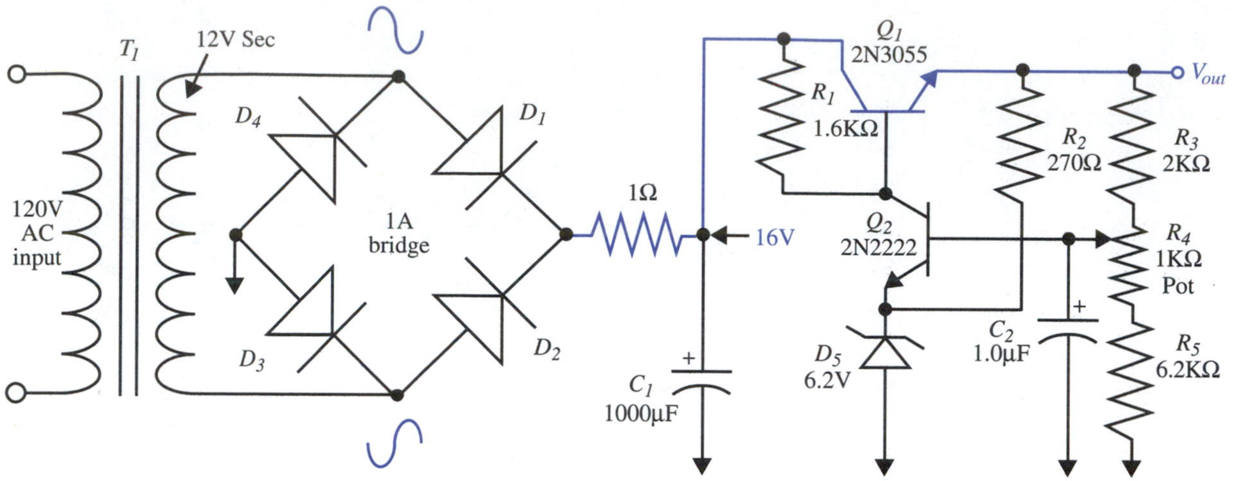

FIGURE 10–16 Troubleshooting an active regulator circuit

9) High ripple on the output of the power supply could be caused by an open filter capacitor. This would show as a rectified sine wave when viewed with an oscilloscope.

TROUBLESHOOTING AN ACTIVE REGULATOR

For our troubleshooting example of a regulated power supply, we shall use the active regulator of Figure 10–16. The techniques for troubleshooting the basic power supply portion have been covered—so we shall concentrate on the regulator portion of this supply by first assuming that the voltage at the filter capacitor is correct.

Three fault conditions could exist at the output of this power supply:

1) No output
2) Low output
3) High output

No output is caused by an open Q_1 or R_1 (assumes no open wires).

A low output can result from one of the following failures:

1) Zener shorted
2) R_5 open
3) Q_2 shorted
4) R_4 open below the wiper

The problem of an output voltage that is too high can be caused by one of the following failures:

1) Q_1 shorted
2) Q_2 open

3) Zener open
4) R_3, R_4 (above or at wiper) open
5) C_2 shorted
6) Wiper of R_4 open

The effect of an open R_2 would be slightly poorer regulation (higher r_z).

SUMMARY

- A power supply converts AC from the power lines to DC.
- Power supplies are preferred to batteries in nonportable applications since they don't have to be charged.
- AC is converted to DC by using diode rectification.
- Half-wave rectification is when current flows on one alternation of the input sine wave.
- Full-wave rectification is when current flows on both alternations of the input sine wave.
- A transformer is used in a power supply to provide the desired output voltage.
- Bridge-type full-wave rectification has the advantages of not requiring a center tap on the transformer and using the full transformer output voltage.
- Filter capacitors are used to remove the ripple voltage that results from rectification.
- The larger the capacitor, the less ripple voltage present on the power supply output.
- It takes twice the capacitance to achieve the same ripple reduction with half-wave rectification as it does with full-wave.
- Current flows into the filter capacitor only when the input voltage is greater than the capacitor voltage.
- The DC voltage on the filter capacitor is equal to the peak of the input AC waveform minus two diode drops (full-wave) of 0.6V.
- A regulator is used to keep the output voltage from the power supply constant despite changes in the line voltage and load current.
- A zener diode regulator works on the concept that voltage across a zener diode is relatively constant when current changes through the device.
- The current through the dropping resistor R_d in series with the zener is constant with load changes. Current shifts between the zener and the load.
- Zener impedance is the equivalent DC resistance inside the device.
- The voltage across a zener is the zener breakdown voltage plus the voltage drop generated by current through the zener impedance.
- The zener impedance works with R_d to divide down the ripple voltage.
- A follower regulator reduces the current through the zener and therefore provides a more constant output voltage with line and load changes.

- The active regulator has a more stable output with line voltage changes because zener current is provided from the regulated output voltage.
- The divided down output voltage is compared against a zener reference in the active regulator, and the circuit corrects for output voltage changes.
- Power supply regulation is defined as the change in output voltage divided by the nominal output voltage.
- A voltage doubler provides a DC output that is twice the peak of the input sine wave.
- A half-wave voltage doubler has half the ripple frequency and therefore twice the ripple voltage of a full-wave doubler.

EXERCISE PROBLEMS

1. State the advantage of a DC power supply over a battery.
2. How is AC converted to DC in a power supply?
3. If the input line frequency is 60Hz, what is the ripple frequency for a half-wave rectifier? a full-wave rectifier?
4. Which requires a larger filter capacitor: half-wave or full-wave?
5. Determine the DC voltage across the filter capacitor for a bridge rectifier power supply if the AC input is 18V RMS.
6. Determine the DC voltage across the filter capacitor for a half-wave rectifier power supply if the AC input is 18V RMS.
7. Find the required capacitor size to limit the ripple voltage to 50mV for a full-wave power supply if the load current is 200mA.
8. Find the required capacitor size to limit the ripple voltage to 50mV for a half-wave power supply if the load current is 200mA.
9. A power supply using 2A rectifiers has a maximum peak input voltage of 24V. Find the surge-limiting resistance if the transformer R_x is 0.1Ω.
10. What is the minimum PRV rating for the diodes of problem 9?
11. Why is power supply regulation required?
12. A zener diode develops 10V with a test current of 30mA and a zener impedance of 8Ω. Find the voltage across the zener if the current is reduced to 5mA.
13. Find the voltage across the zener of problem 12 when used in a zener diode regulator if the current through R_d is 50mA and the load current is 30mA.
14. The required output from a zener regulated power supply is 15V with a maximum load current of 50mA. If the nominal voltage at the filter capacitor is 22V, and the minimum zener current is 2mA, find the voltage and power rating of the zener diode and the resistance value and power rating of R_d.
15. The required output from a zener regulated power supply is 12V with a maximum load current of 80mA. If the nominal voltage at the filter capac-

itor is 16V, and the minimum zener current is 3mA, find the voltage and power rating of the zener diode and the resistance value and power rating of R_d.

16. Find the ripple voltage developed across the zener of problem 15 if the ripple on the filter capacitor is 1.5V and r_z is 2.5Ω.

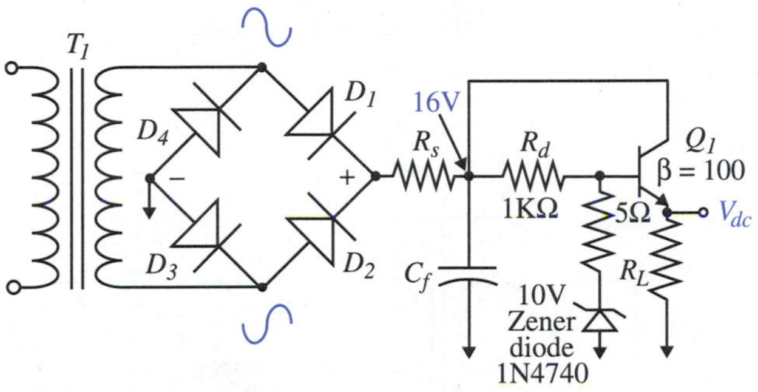

FIGURE 10–A (text figure 10–10)

17. Calculate the change in the output voltage for the circuit of Figure 10–A if the load current changes from 3mA to 60mA. Assume that V_c is 18V, V_z is 12V, R_d is 820Ω, r_z is 6Ω, and the β of the transistor is 100.

18. Determine the maximum power dissipation for the transistor of problem 17.

19. Compute the change in the output voltage for the circuit of problem 17 if the line voltage drops by 10 percent.

20. Give an advantage of the active regulator over the follower regulator.

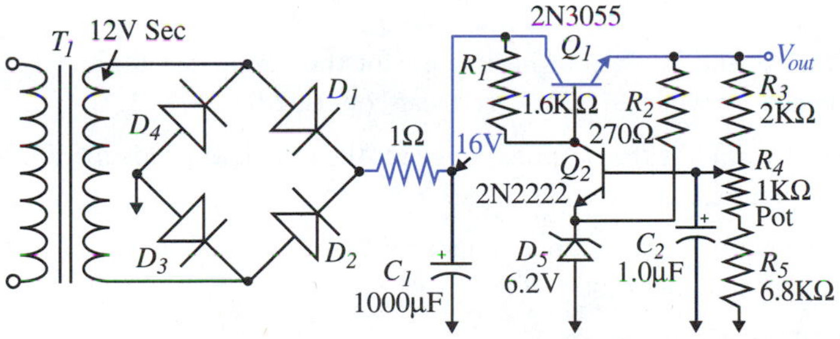

FIGURE 10–B (text figure 10–11)

21. Determine the change in the output voltage for the circuit of Figure 10–B if the line voltage drops by 5 percent and r_z is 5Ω.

22. Find the percent load regulation for the circuit of problem 17.

23. Find the percent line regulation for the circuit of problem 19.

24. State an advantage of the full-wave doubler over the half-wave.

25. What could cause the ripple frequency seen at the filter capacitor of a bridge rectifier power supply to be 60Hz?

26. What effect would you predict if the filter capacitor

    a) open-circuited?

    b) short-circuited?

27. What would be the effect of an open rectifier diode in a full-wave center-tapped transformer supply? (See Figure 10–C.)

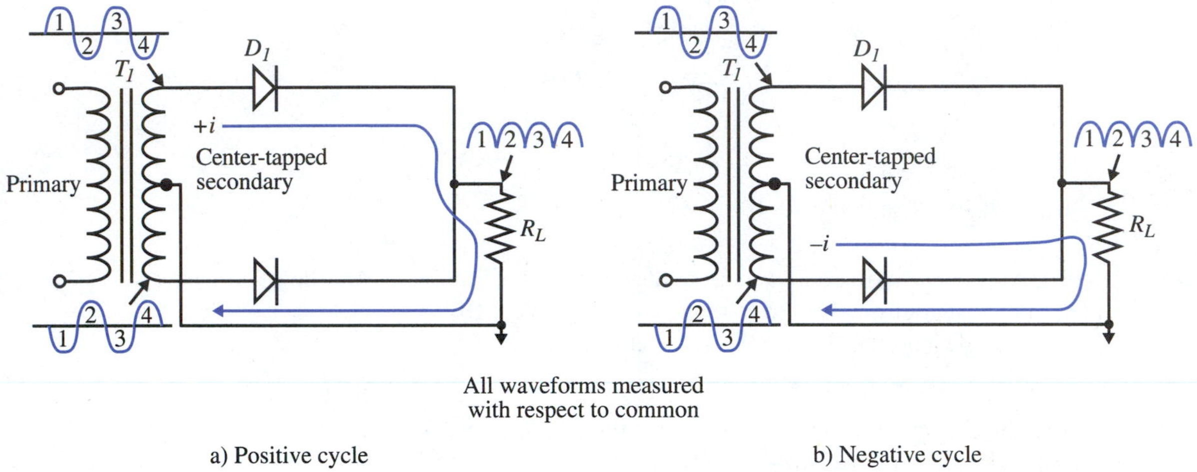

FIGURE 10–C (text figure 10–2)

28. What would be the output voltage for the circuit of problem 15 if the zener opened up while an 80mA load was connected?

29. What would be the output voltage for the circuit of problem 17 if the collector of Q_1 shorted to the emitter?

30. What would be the output voltage for the circuit of problem 17, with the 60mA load, if the zener diode

    a) open-circuited?

    b) short-circuited?

31. Which component could cause the output voltage of the power supply of Figure 10–C to drop to 5.6V?

32. Find the failed component that could cause the output voltage of the power supply of Figure 10–C to drop to 0.8V.

33. A transistor radio is operated from a power supply shown in Figure 10–D. A strong 120Hz tone is heard on the speaker. What is the problem?

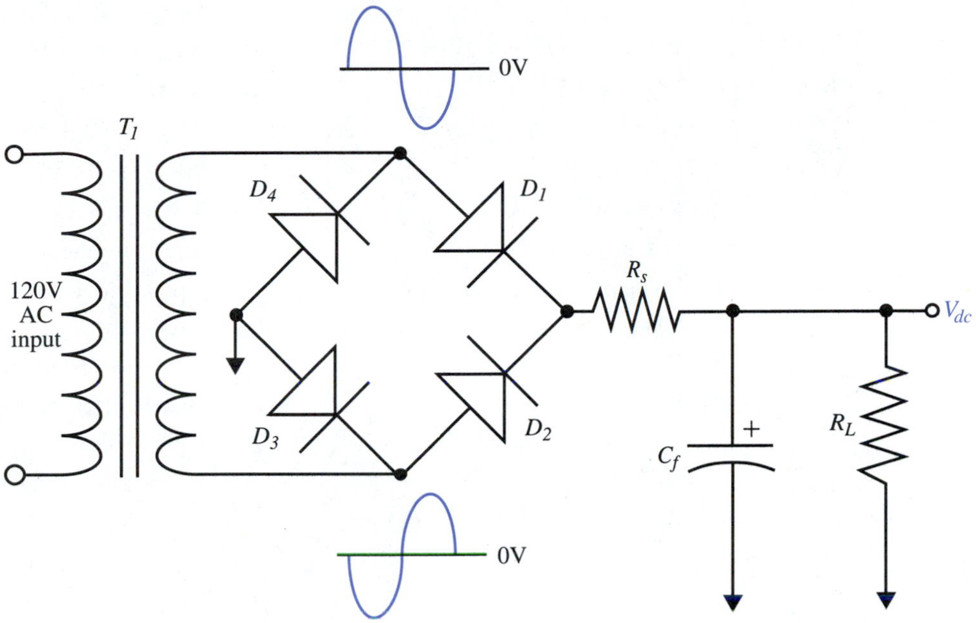

FIGURE 10–D (text figure 10–6)

ABBREVIATIONS

R_s = Diode surge current limit resistor
R_x = Transformer DC resistance
V_{cf} = Filter capacitor DC voltage
V_{ac} = Transformer secondary RMS voltage
V_{dc} = Voltage at output of power supply
ΔV_c = Ripple voltage on filter cap
ΔV_o = Ripple voltage at output of power supply
ΔV_z = Voltage drop across zener resistance (r_z)
PRV = Maximum reverse voltage diode can withstand
r_z = Zener internal resistance
R_d = Zener current-limiting resistor
V_{cLL} = voltage at low line
V_{cHL} = voltage at high line

ANSWERS TO PROBLEMS

1. Doesn't discharge
2. By rectification and filtering
3. 60Hz; 120Hz
4. Half-wave

5. $V_c = (18V\sqrt{2}) - 1.2V = 24.3V$
6. $V_c = 24.9V$
7. $C_f = 32mF$
8. $C_f = 64mF$
9. $R_s = 0.3\Omega$
10. 48V use 100V PRV
11. To maintain a constant output voltage
12. $V_z = 9.8V$
13. $V_z = 9.92V$
14. $V_z = 15V$; $P_z = 1.5W$; $R_d = 92\Omega$; $P_{Rd} = 920mW$
15. $R_d = 28.9\Omega$; $P_r = 1.09W$; $V_z = 12V$; $P_z = 2.33W$
16. $\Delta V_o = 119mV$
17. $\Delta V_{dc} = -3.42mV$
18. $P_{tmax} = 396mW$
19. $\Delta V_{dc} = -13.2mV$
20. See text.
21. $\Delta V_{dc} = -3.41mV$
22. %Reg. = 0.03%
23. %Reg. = 0.116%
24. Ripple frequency easier to filter
25. Open bridge diode
26. Open—high ripple; Short—blown bridge or fuse
27. Higher ripple at 60Hz (half-wave)
28. $V_{dc} = 13.4V$
29. $V_{dc} = 18V$
30. Open—16.5V; Short—0V
31. Q_2 collector-emitter short
32. Zener diode shorted
33. Open filter capacitor

CHAPTER 11

Linear Regulators

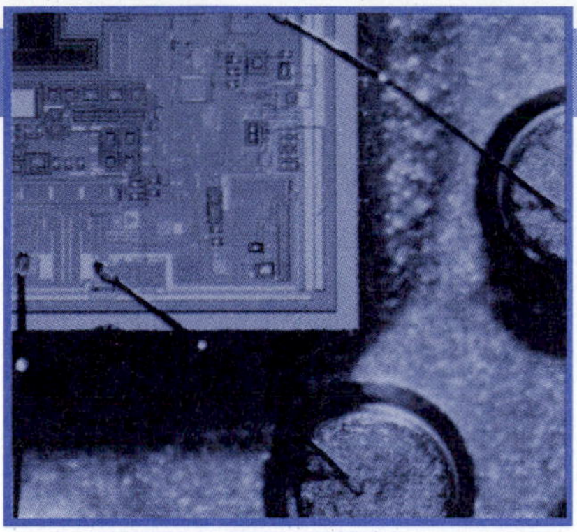

CHAPTER OBJECTIVES

After completing this chapter, you should be able to explain the following basic concepts of the voltage regulator, voltage reference, and current reference:

- Need for a regulator
- Fixed and variable regulators
- Short-circuit current protection
- Power dissipation calculations
- Ripple rejection
- Fold-back current limiting
- Thermal protection
- Regulation computation
- Tracking regulators
- Purpose of a voltage reference
- Advantages of a voltage reference over a zener diode
- Applications of a voltage reference
- Generation of a constant current source
- Applications for a constant current source

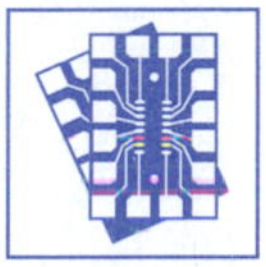

INTRODUCTION

This chapter will cover the integrated circuit linear regulator, which is derived from the transistor regulators of the previous chapter but provides better regulation in a more compact package. We shall also look at precise voltage and current references.

INTEGRATED CIRCUIT REGULATORS

As discussed previously, voltage regulators are used to maintain a predetermined output voltage of a power supply, despite variations in the input AC line voltage and in the output load. The voltage regulators are inserted between the output of an unregulated power supply and the input to the load, as shown in Figure 11–1.

The integrated circuit (*IC*) voltage regulators considerably simplify power supply design by replacing discrete components, such as transistors and vacuum tubes, that were used in early power supplies. However, we will find that power transistors are still sometimes used

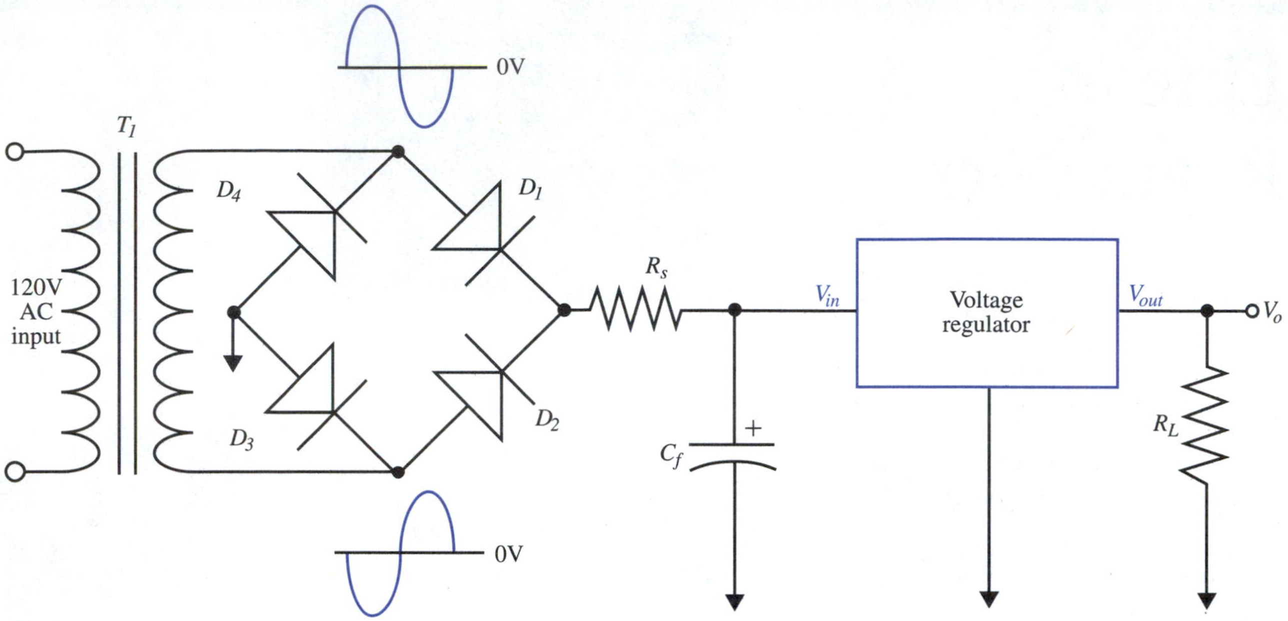

FIGURE 11–1 Location of regulator in a power supply

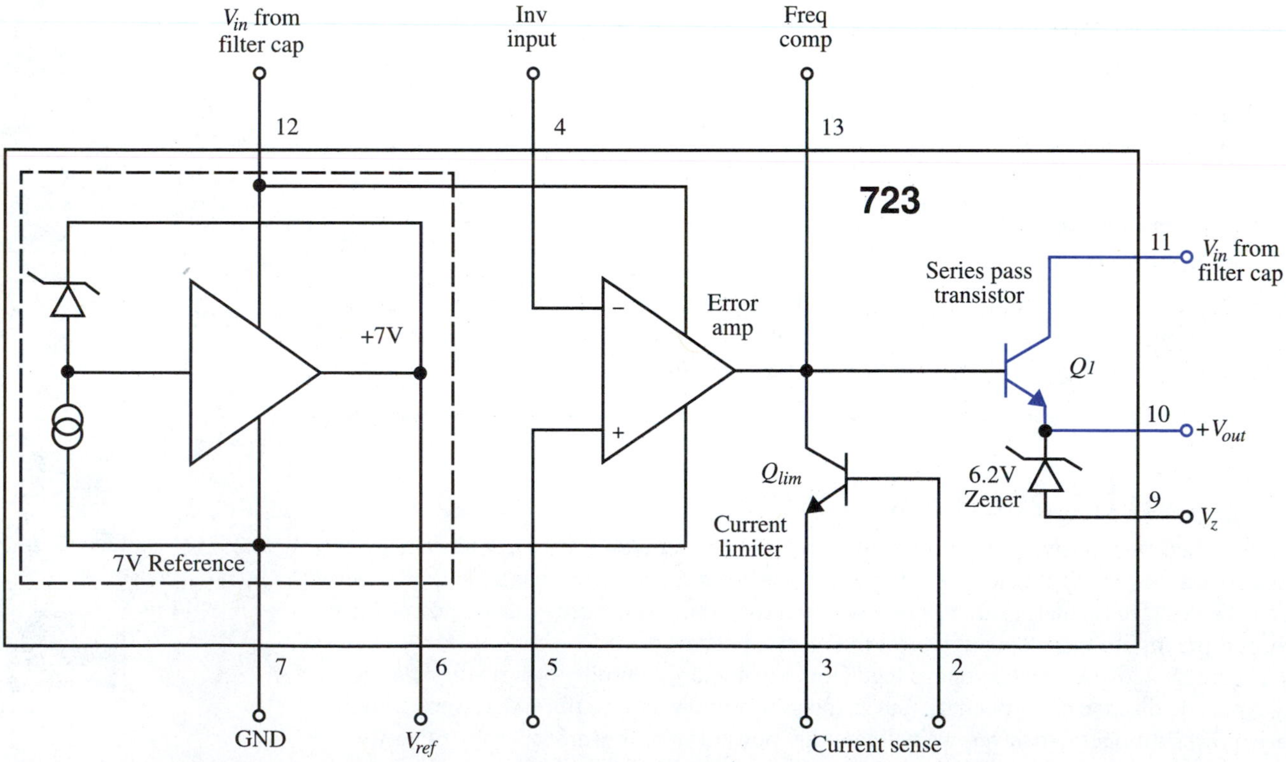

FIGURE 11–2 LM723 block diagram. *Courtesy National Semiconductor.*

in conjunction with the linear voltage regulators to control the higher currents associated with the larger power supplies.

We will consider the two main types of voltage regulators: the linear voltage regulator and the switching voltage regulator. The linear voltage regulator will be described in this chapter, and the switching voltage regulator in Chapter 12.

LINEAR VOLTAGE REGULATORS

One of the most basic types of linear regulator is the LM723. This device is built from components we have studied (op-amps, transistors, etc.) and is basically a refined version of the active regulator we examined in Chapter 10.

The internal block diagram for the LM723 output voltage regulator is shown in Fig. 11–2. The LM723 includes a +7V voltage reference source that can be used directly, or attenuated, to provide an input reference to the noninverting side of an error amplifier. A sample of the power supply output voltage is applied to the inverting input of this amplifier. These two voltages are compared by the error amplifier to produce a voltage at the base of the Q_1 series pass transistor.

When incorporated into an actual power supply, the series pass transistor's collector is connected to the filter capacitor and its emitter to the load resistor in a voltage follower configuration, as shown in Figure 11–3. The load current flows through Q_1 and then through the load resistor and develops the output voltage (V_o).

The 723 has built-in current limiting to protect the load and/or regulator from excessive power dissipation. Current limiting is achieved by sampling the load current through the resistor (R_{cl}) connected between the output of the regulator (Q_1 emitter) and the load. If the current increases too much, the voltage across this resistor reaches 0.6V and the current limiting transistor (Q_{lim}) is biased on. This draws current away from the base of Q_1, which reduces the emitter current of the series pass transistor and therefore prevents a further increase in the load current. To find R_{cl} for a given application, we divide the desired limit current into 0.6V, or

$$R_{cl} = \frac{0.6\text{V}}{I_{cl}} \qquad \textbf{(equation 11–1)}$$

EXAMPLE

Find the value of R_{cl} if the desired current limit I_{cl} is 30mA.

SOLUTION

From equation 11–1,

$$R_{cl} = \frac{0.6\text{V}}{I_{cl}}$$

$$R_{cl} = \frac{0.6\text{V}}{30\text{mA}}$$

$$R_{cl} = 20\Omega$$

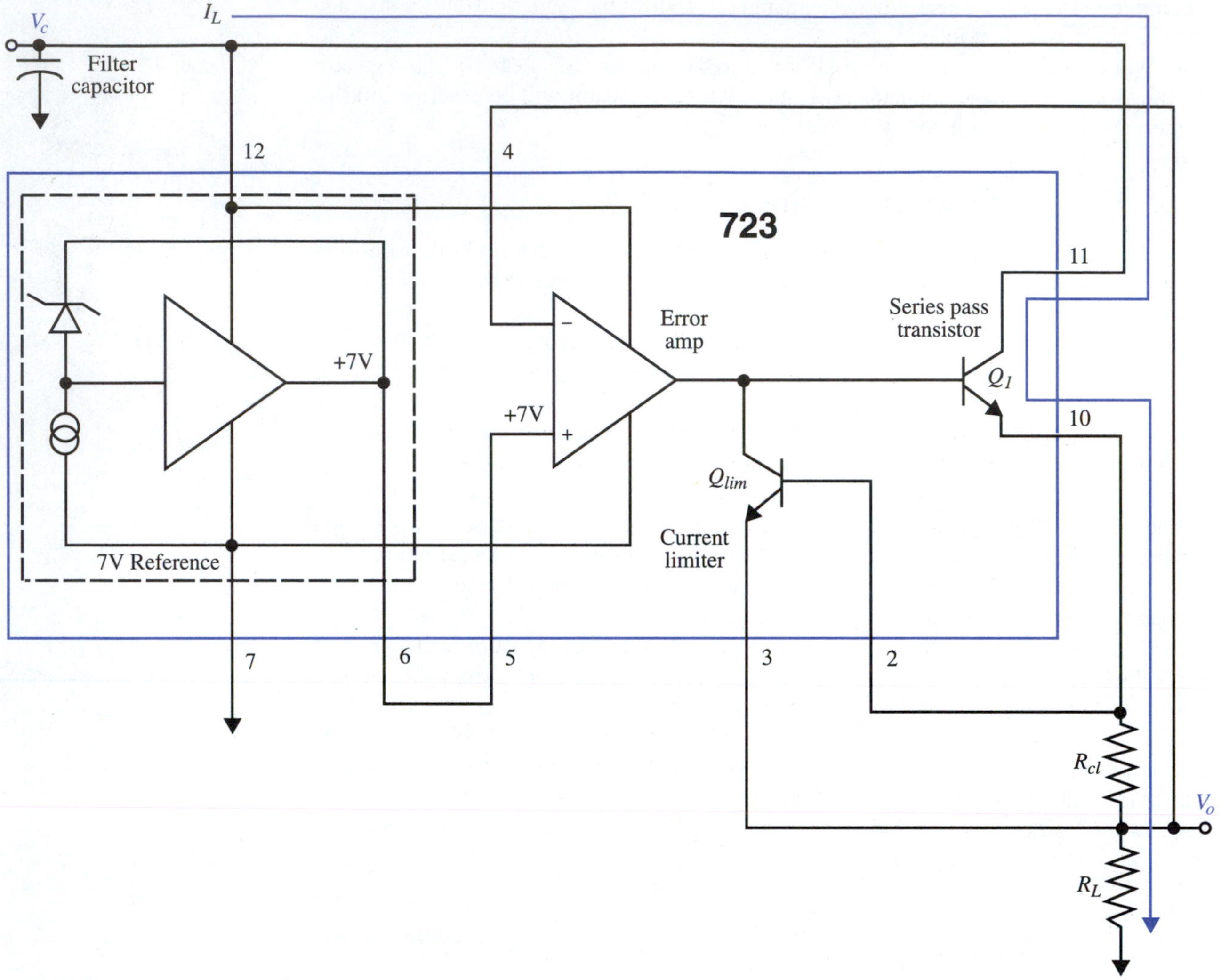

FIGURE 11–3 LM723 circuit connections

By tracing the connections to the op-amp in Figure 11–3, you will see that the inverting input is at +7V, and since the op-amp is connected as a follower, the output voltage across R_L is also +7V.

LOW-VOLTAGE VARIABLE REGULATOR

A typical application using a 723 regulator is shown in Figure 11–4, which provides a variable +2VDC to +7VDC power supply. We shall pick the unregulated input voltage to be +12V at the filter capacitor, and current limiting is selected for 50mA.

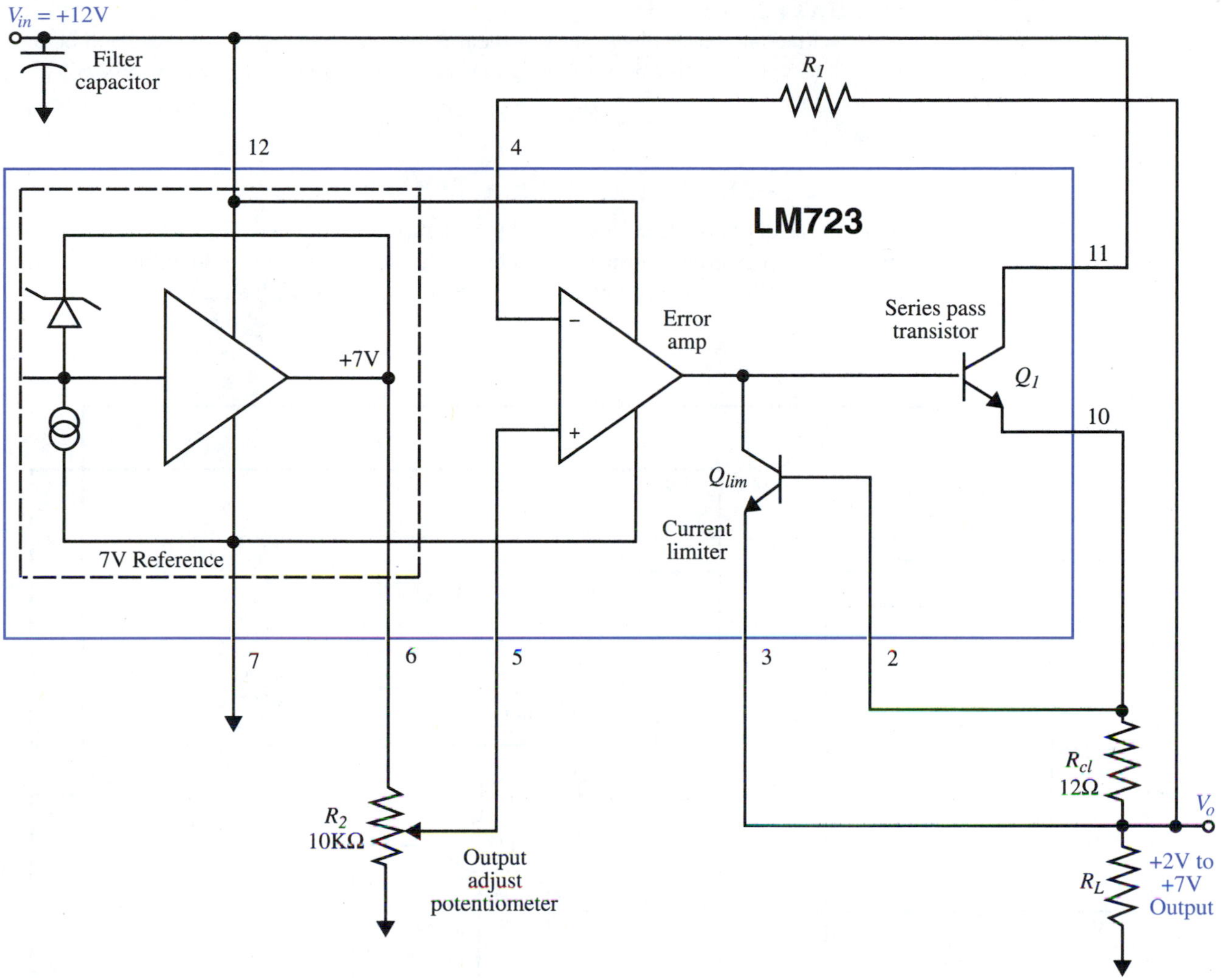

FIGURE 11–4 +2VDC to +7VDC variable 50mA power supply

The error amplifier is connected as a voltage follower, and therefore its output voltage is the same as the voltage at the center of the "output adjust" potentiometer R_2. The output cannot go lower than about +2V because of the required internal operating voltages for the 723.

Resistor R_1, in series with the amplifier inverting input, is used to minimize bias current offset. Ideally the resistance of R_1 should equal the equivalent resistance of the potentiometer R_2. The current-limiting resistor R_{cl} is determined by dividing 0.6V (current limit transistor turn-on voltage) by the specified limiting current (I_{cl}) of 50mA.

EXAMPLE

Find the output voltage from the regulator circuit of Figure 11–4 if the output adjustment potentiometer (R_2) is set at half range.

SOLUTION

With a reference voltage of 7V at the top of R_2, the voltage at half range will be 3.5V. This voltage is the input to the follower, so the output voltage will also be 3.5V.

REGULATOR WITH CURRENT BOOST

An example of a variable +7V to +34V power supply is shown in Figure 11–5. This power supply has an external transistor Q_2 to boost the output current up to 400mA. The internal error amplifier is connected as a noninverting amplifier with a gain A_v.

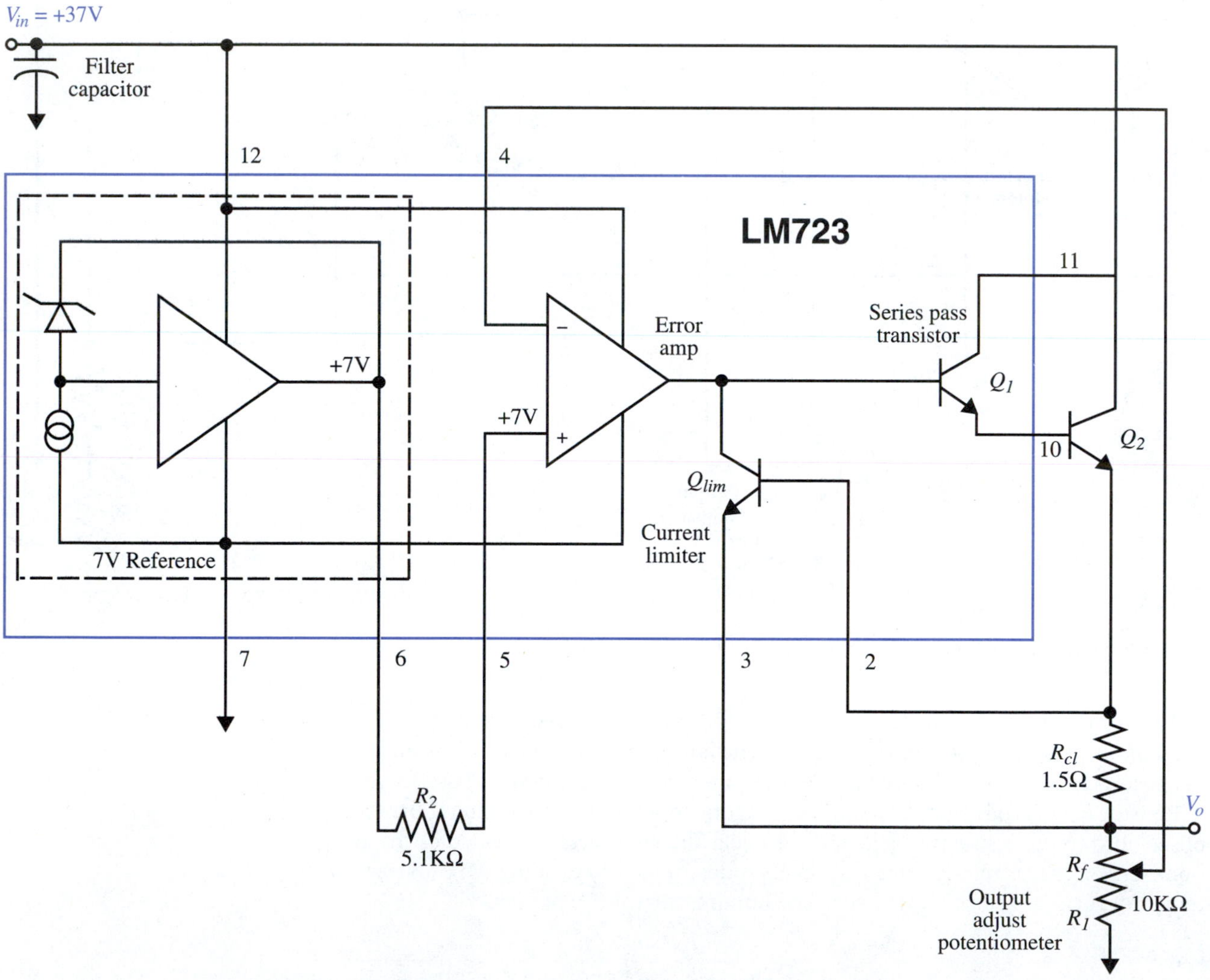

FIGURE 11–5 +7VDC to +34VDC variable high-current power supply

$$A_v = \frac{R_f}{R_1} + 1 \qquad \textbf{(equation 11–2)}$$

Resistors R_1 and R_f values are set by the wiper position of the output adjust potentiometer as indicated in Figure 11–5. Changing the potentiometer position changes the ratio of R_f to R_1 and therefore the gain. The output voltage V_o is simply the +7V reference source voltage multiplied by this closed loop gain A_v.

Notice that Q_2 is included in the feedback path, so we don't have to worry about its base-to-emitter 0.6V drop.

Resistor R_2 is used for current offset and should have an ohmic value of one-half of the output adjust potentiometer.

The required value of the current-limiting resistor R_{cl} is from equation 11–1.

$$R_{cl} = \frac{0.6\text{V}}{I_{cl}}$$

$$R_{cl} = \frac{0.6\text{V}}{400\text{mA}}$$

$$R_{cl} = 1.5\Omega$$

CURRENT BOOST TRANSISTOR CONSIDERATIONS

In order to determine the requirements for the external current boost transistor Q_2, we assume that we want to limit the internal power dissipation P_{id} in the regulator to 500mW, which is the 723 power rating at 50°C. If the regulator input voltage is +37V and the output is shorted to ground (worst-case load), then the maximum voltage drop V_{rm} across the 723 regulator is 37V.

This gives a maximum allowable regulator current $I_{R(max)}$ of

$$I_{R(max)} = \frac{P_{id(max)}}{V_{rm}}$$

$$I_{R(max)} = \frac{500\text{mW}}{37\text{V}} \quad \text{(at 50°C)}$$

$$I_{R(max)} = 13.5\text{mA}$$

We can determine the minimum β for Q_2 by assuming that

$$I_c \approx I_e \approx I_{out}$$

and then

$$\beta = \frac{I_{out(max)}}{I_{R\,(max)}}$$

$$\beta = \frac{400\text{mA}}{13.5\text{mA}}$$

$$\beta \approx 30 \qquad \textbf{(equation 11–3)}$$

Q_2 also requires a collector-to-emitter voltage rating of at least 37V.

If the β of the Q_2 is higher than 30 at 400mA, then less current is drawn from the regulator and the regulator power dissipation drops.

VOLTAGE REGULATOR INPUT VOLTAGE CONSIDERATIONS

We shall now consider the requirements for the regulator input power source. The limits of the input voltage are determined by the internal circuitry of the chip. If the input voltage is raised, the internal power dissipation increases, since power dissipated P_d is

$$P_d = (V_{in} - V_{out})I_{out} \qquad \textbf{(equation 11–4)}$$

where V_{out} is the regulator output voltage and I_{out} is the load current. Raising the input voltage too high will cause the internal semiconductors of the regulator to break down, and the chip will be damaged. If the input voltage is too low, the semiconductors within the regulator will have too little voltage across them to operate correctly.

The LM723 regulator has a maximum input voltage of 40V and a minimum input of 9.5V (required for the 7V internal reference). A minimum voltage of 3V is required across the regulator ($V_{in} - V_{out}$) in order to stay in regulation.

EXAMPLE

Find the minimum input voltage to a 723 regulator if the required output voltage is +15V, and also +5V.

SOLUTION

To maintain a minimum voltage across the 723 regulator of 3V, the input voltage for 15V out must be at least 18V.

For 5V out, we could assume that the required input voltage is 8V; but this is less than the minimum regulator input of 9.5V, so we must use 9.5V.

DETERMINING POWER SUPPLY COMPONENT VALUES

The selection of the power supply transformer, diodes, and filter capacitor is determined by the load voltage and current and the regulator characteristics.

The voltage at the filter capacitor must never drop below the required minimum input voltage to the regulator or exceed the maximum input voltage to the regulator.

When considering the minimum input voltage to the regulator, we must take into account the voltage drop that occurs when the filter capacitor is discharging (ripple). If we allow the ripple to cause the voltage to drop below the regulator minimum input value, a large ripple component will appear at the output of the regulator. So the allowable input voltage ripple ΔV_c to the regulator is

$$\Delta V_c = V_c - V_m \qquad \textbf{(equation 11–5)}$$

where V_c is the DC peak voltage on the filter capacitor and V_m the minimum allowable input voltage to the regulator for the required output voltage.

I think it is clear from equation 11–5 that the allowable ripple should be computed at low line voltage conditions because this provides the lowest value for V_c.

EXAMPLE

Find the maximum allowable ripple voltage at the filter capacitor for a bridge-type power supply using a 723 regulator if the regulator output voltage is 10V and the nominal 12V transformer secondary voltage is 10.8V (rms) at low line voltage.

SOLUTION

First we must determine the value of V_{cLL} at low line, using equation 10–2 from the previous chapter.

$$V_{cLL} = (V_{ac} \times \sqrt{2}) - 1.2\text{V}$$

$$V_{cLL} = (10.8\text{V} \times \sqrt{2}) - 1.2\text{V}$$

$$V_{cLL} = 14.07\text{V}$$

For 10V out, the minimum input voltage to the regulator

$$V_m = (10\text{V} + 3\text{V}) = 13\text{V}$$

Now using equation 11–5 to find the maximum allowable ripple voltage,

$$\Delta V_c = V_{cLL} - V_m$$

$$\Delta V_c = 14.07\text{V} - 13\text{V}$$

$$\Delta V_c = 1.07\text{V}$$

The actual capacitor value to provide the desired ripple voltage is a function of both the ripple voltage and the current drawn from the capacitor. Again from the previous chapter,

$$C = \frac{(I)(\Delta t)}{\Delta V_c} \qquad \textbf{(equation 11–6)}$$

The current (I) is the load current plus the current required for the 723 regulator to operate—which is a maximum of 4mA for the 723.

EXAMPLE

Find the value of the filter capacitor for the power supply of the previous example if the load current is 40mA.

SOLUTION

The total current is the load current of 40mA plus the required regulator current of 4mA, for a total current of 44mA.

Since we have a bridge-type power supply, the time the capacitor discharges is 8ms. The maximum allowable ripple voltage from the previous example is 1.07V. So, using these values in equation 11–6,

$$C = \frac{(I)(\Delta t)}{\Delta V_c}$$

$$C = \frac{(\text{rrmA} \times 8\text{ms})}{1.07\text{V}}$$

$$C = 329\mu\text{F} \quad (\text{use } 330\mu\text{F.})$$

REGULATOR RIPPLE REDUCTION

Because a regulator is very effective at reducing ripple, the ripple amplitude seen across the filter capacitor will not appear at the output.

The 723 specifications indicate that the ripple rejection is typically 74db (ratio 5011:1). This means that ripple across the filter capacitor (ΔV_c) is reduced by this factor at the output. The output ripple ΔV_o for 723 is

$$\Delta V_o = \frac{\Delta V_c}{5011} \qquad \textbf{(equation 11–7)}$$

If we compare a simple power supply with one using a 723 regulator, we find that to get the same ripple across the load, the simple power supply needs a filter capacitor 5011 times larger than the power supply using a 723 regulator.

EXAMPLE

Find the expected ripple ΔV_o at the regulator output (across the load) for the previous example with ΔV_c of 1.07V.

SOLUTION

Using equation 11–7,

$$V_o = \frac{\Delta V_c}{5011}$$

Solving for ΔV_o,

$$\Delta V_o = \frac{1.07\text{V}}{5011}$$

$$\Delta V_o = 214\mu\text{V}$$

COMPLETE REGULATED POWER SUPPLY

It will help to tie together the integrated regulator concepts we have studied if we go through an example of a complete power supply from input AC power to the output voltage.

Let us assume the following requirements:

- Load voltage, V_{out} is 10V ±0.5% (under line and load variations).
- Load current max ($I_{out(max)}$) is 50mA.
- Max peak-to-peak output ripple voltage (ΔV_o) at full load is 10mV.

- AC input line voltage variation is ±10%.
- Maximum operating temperature is 50°C.

To meet these requirements, we shall select an LM723 voltage regulator. This device has the following specified characteristics:

- Line regulation is 0.1%.
- Load regulation is 0.2%.
- Power supply ripple rejection is 74dB (5011:1).
- Minimum in-out voltage differential is 3V (min. input voltage 9.5V).
- Maximum input voltage differential is 38V.
- Max reg. power dissipation at 25°C is 660mW (500mW at 50°C).

The regulator input voltage at the filter capacitor can range from the required 10V output voltage plus the minimum output-input voltage differential of 3V, or 13V, to a maximum value of 40V.

With a load current of 50mA, plus 4mA idle current for the regulator, the maximum allowable voltage across the regulator V_{rm} with 500mW (P_d) internal power dissipation is

$$V_{rm} = \frac{P_d}{I_{out}}$$

$$V_{rm} = \frac{500\text{mW}}{54\text{mA}}$$

$$V_{rm} = 9.2\text{V} \qquad \textbf{(equation 11–8)}$$

With 10V at the output, the maximum regulator input voltage then is (9.2V + 10V) or 19.2V at 50°C, which is considerably less than the specified maximum value of 40V.

If a transformer with a nominal secondary voltage of 12V rms is available, the peak voltage across the secondary is approximately 17V (which is $12 \times \sqrt{2}$). The filter capacitor voltage is then 15.8V (17V minus two 0.6V diode drops).

Under 10 percent high input line voltage conditions, we have 17.5V at the capacitor, and with 10 percent low line, 14V. Note that the high line condition of 17.5V is less than the maximum input of 19.2V (determined above, from the maximum power dissipation).

Under low input AC line voltage conditions, the peak DC voltage across the filter capacitor is 14V. Remembering that the minimum input to the regulator is 13V for 10V out, the allowable ripple voltage

$$\Delta V_c = (14\text{V} - 13\text{V}) = 1\text{V}$$

We can determine the required size of the filter capacitor using equation 11–6.

$$C = \frac{(I)(\Delta t)}{\Delta V_c}$$

$$C = \frac{(54\text{mA})(8\text{ms})}{1\text{V}}$$

$$C = 432\mu\text{F} \quad (\text{use } 500\mu\text{F})$$

where I is the load current of 50mA plus the idle current of 4mA, and Δt is the capacitor discharge time (approximately 8ms for full-wave rectification and 16ms for half-wave rectification on a 60Hz AC line). A practical selection for the capacitor would be 500μF at 25V.

Again, the surge resistor R_s in series with the filter capacitor along with the transformer DC resistance R_t limits the initial turn-on surge current. If we assume we are going to use a 1A bridge (about the smallest size available), then we have a maximum one cycle surge current (I_s) rating of typically 30A. We must assume that the power is turned on at the worst possible time—a high line voltage and at the peak of the input sine wave. So

$$R_s + R_t = \frac{V_{c(max)}}{I_s}$$

$$R_s + R_t = \frac{17.5\text{V}}{30\text{A}}$$

$$R_s + R_t = 0.58\Omega \qquad \textbf{(equation 11–9)}$$

Using a transformer with measured dc resistance (R_t) of 0.08Ω requires that R_s has to be at least 0.5Ω.

This completes the selection of components. Now the question is whether the required output characteristics have been met.

With 1V peak-to-peak ripple voltage (ΔV_c) on the capacitor, and dividing this value by the 74db (ratio 5011) ripple rejection at the regulator, gives us an output ripple ΔV_o of

$$\Delta V_o = \frac{\Delta V_c}{5011}$$

$$\Delta V_o = \frac{1\text{V}}{5011}$$

$$\Delta V_o = 0.2\text{mV}$$

This is well within our desired 10mV value.

Going back to the 723 specifications, which state: A 10% change in the AC line voltage yields a 0.01% change in output voltage (10% × 0.1%). It is also stated that a load change from 5mA to 50mA will give a 0.2% output voltage change. Adding these two values gives a total of 0.21%, which again is well within the ±0.5% or 1% total required.

FOLD-BACK CURRENT LIMITING

The method of current limiting previously described limits the maximum current that can be drawn from the power supply to that value that develops 0.6V across the current-sensing resistor R_{cl}. If the output of the power supply is shorted to ground, then the full input voltage is dropped across the regulator, and the regulator power dissipation is the product of this voltage and the short-circuit current. With high input voltages, this could result in too much device dissipation and burnout of the regulator.

A technique to avoid the shorted output problem is to use a fold-back current-limiting technique (shown in Figure 11–6), which reduces the current drawn from the regulator

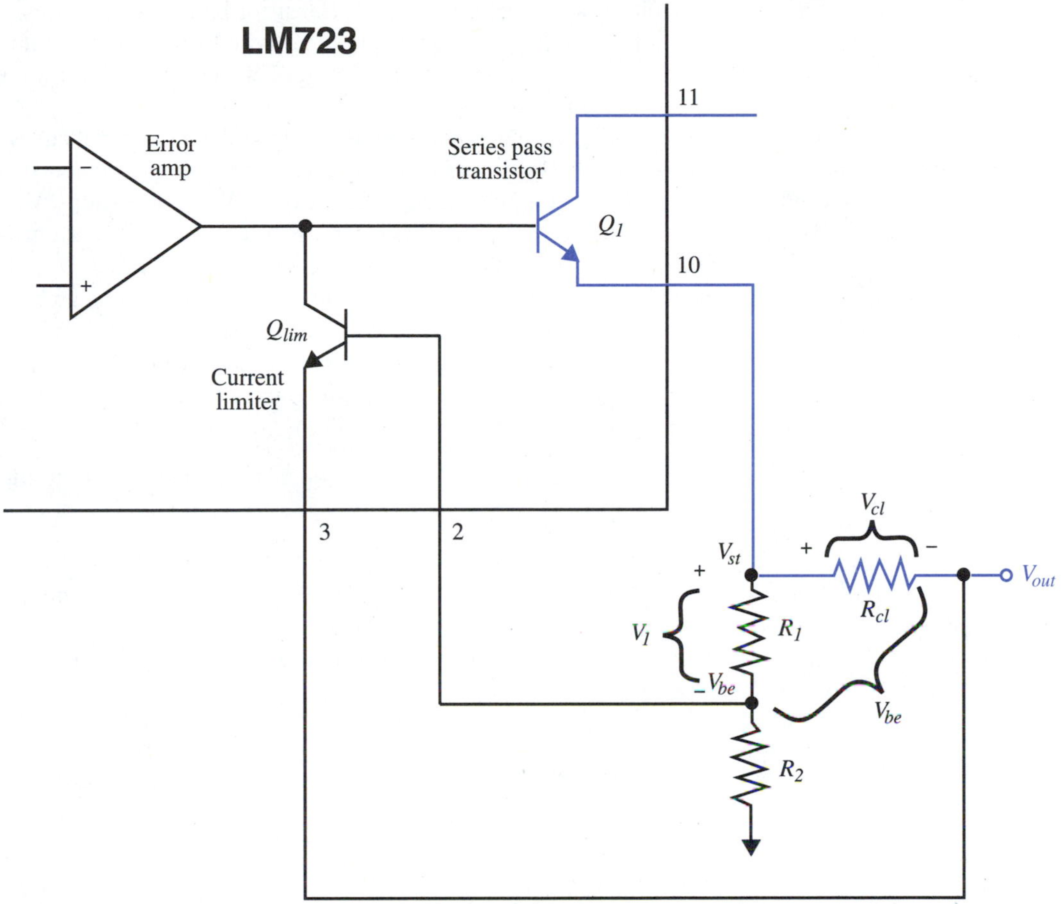

FIGURE 11–6 Fold-back current limiting circuit configuration

when a short is applied across the output. This circuit differs from the fixed current limiter in that the base of the current limit transistor Q_{lim} is connected to a voltage divider (R_1 and R_2) rather than to the output of the series regulator transistor Q_1.

Let us consider the circuit action between the voltage divider and the current-sensing resistor R_{cl}. The voltage from the base to emitter of Q_{lim} is

$$V_{be} = (V_{cl} - V_1)$$

V_1 is negative because V_b is closer to ground than V_{st}.

If no current is being drawn from the regulator (no load), then $V_{cl} = 0$ and the base-to-emitter junction of Q_{lim} is reverse biased because

$$V_{be} = -V_1$$

With an increasing current from the regulator, the voltage V_{cl} increases, and the negative bias on Q_{lim} is reduced. Further increase in load current will cause V_{cl} to cancel V_1, resulting in a positive voltage across the current limit transistor base-to-emitter junction. When this voltage V_{be} reaches 0.6V, Q_{lim} turns on, and the output load current will not be able to increase beyond the value that developed this voltage. This is referred to as the current limit point, and the limited load current is I_{cl}.

The operation described to this point is similar to that of conventional current limiting. However, if we reduce the load resistor (increasing load current) below the value that caused current limiting, the output voltage will now drop. When V_{out} drops, V_{st} also drops, and therefore V_1 is reduced. This reduction of V_1 means less load current is required through R_{cl} in order to maintain V_{be} at 0.6V. So the load current drops.

The relationship of the output voltage versus load current is shown in Figure 11–7. When the load resistance drops below R_{min}, the output voltage starts to drop, and current fold-back starts to occur. The current with a short on the output is called I_s.

With reference to Figure 11–6, a short across the output causes V_{out} to go to zero and so V_{st} equals V_{cl}. With this condition,

$$V_{be} = V_{st} - V_1 \quad \text{(with short on output)} \qquad \textbf{(equation 11–10)}$$

But

$$V_1 = \frac{V_{st}R_1}{(R_1 + R_2)} \qquad \textbf{(equation 11–11)}$$

At the current limit point V_{be} is 0.6V; substituting this value and equation 11–11 in equation 11–10,

$$0.6\text{V} = V_{st} - \frac{V_{st}R_1}{(R_1 + R_2)}$$

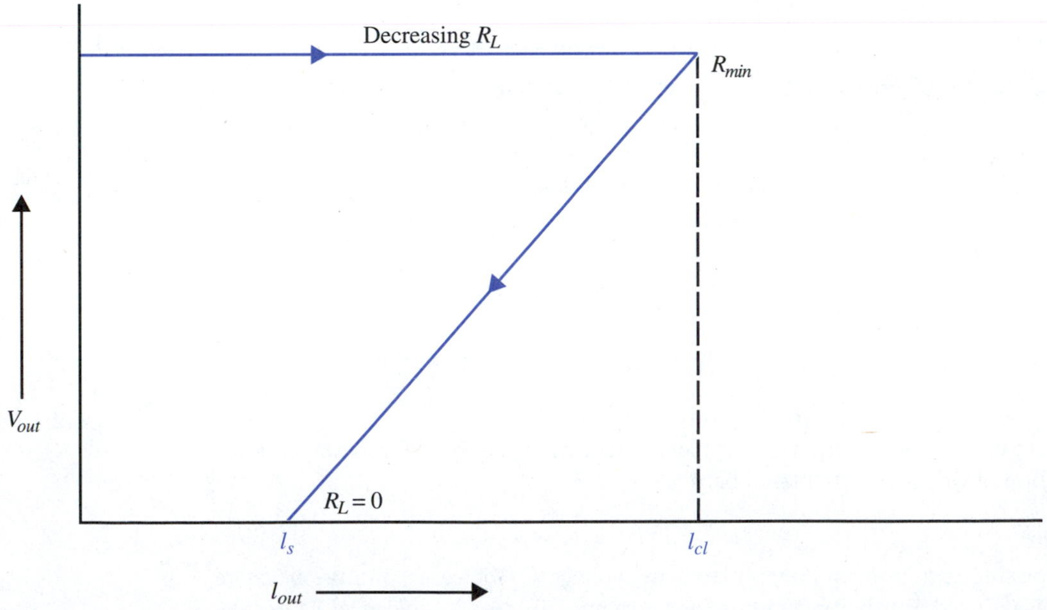

FIGURE 11–7 Fold-back current, output voltage versus load current

Solving for V_{cl} (since $V_{st} = V_{cl}$),

$$V_{cl} = 0.6\text{V}\left(\frac{R_1 + R_2}{R_2}\right) \qquad \textbf{(equation 11–12)}$$

but V_{cl} is also equal to $(I_s)(R_{cl})$, where I_s (determined from P_{max}/V_{in}) is the load current with a short on the output. Substituting this relationship in equation 11–12 and solving for R_{cl},

$$\mathbf{R_{cl} = \frac{0.6V}{I_s}\left(\frac{R_1 + R_2}{R_2}\right)} \qquad \textbf{(equation 11–13)}$$

Returning to the situation with a load resistor that just causes current limiting, then

$$V_{be} = 0.6\text{V} = V_{cl} - V_1 \qquad \textbf{(equation 11–14)}$$

and again,

$$V_1 = \frac{V_{st}R_1}{(R_1 + R_2)} \qquad \textbf{(equation 11–15)}$$

From Figure 11–6,

$$V_{st} = V_{cl} + V_{out} \qquad \textbf{(equation 11–16)}$$

Substituting equation 11–16 into 11–15, and the resultant V_1 term into equation 11–14,

$$0.6\text{V} = V_{cl} - \frac{(V_{out} + V_{cl})R_1}{R_1 + R_2}$$

Solving for V_{cl},

$$V_{cl} = \frac{0.6\text{V}(R_1 + R_2) + V_{out}R_1}{R_2} \qquad \textbf{(equation 11–17)}$$

But V_{cl} is also equal to $(I_{cl})(R_{cl})$ with I_{cl} the required non-shorted current limit (maximum current with no reduction in the output voltage). Substituting this relationship into equation 11–17 and solving for I_{cl},

$$I_{cl} = \frac{0.6\text{V}(R_1 + R_2) + V_{out}R_1}{R_2R_{cl}} \qquad \textbf{(equation 11–18)}$$

Let us now define the fold-back factor K as the ratio of I_{cl} divided by I_s.

$$\mathbf{K = \frac{I_{cl}}{I_s}} \qquad \textbf{(equation 11–19)}$$

Substituting for I_{cl} using equation 11–18, and solving for the short-circuit current I_s in equation 11–13 and substituting,

$$K = 1 + \frac{V_{out}R_1}{0.6\text{V}(R_1 + R_2)} \qquad \textbf{(equation 11–20)}$$

The fold-back factor K indicates how much the current will be reduced when the output is shorted.

EXAMPLE

Assume that the input voltage to a 723 regulator is 10V, the desired output voltage from the regulator is 5V, and the current limit (I_{cl}) is set at 80mA. Find the K factor and values of R_1, R_2, and R_{cl} if the maximum dissipation with a short on the output should not exceed 500mW.

SOLUTION

With a short on the output, the power dissipation without fold-back limiting is *10V × 80mA*, or 800mW. To limit the power dissipation to 500mW requires that the current I_s be limited to

$$I_s = \frac{500\text{mW}}{10\text{V}}$$

$$I_s = 50\text{mA}$$

Using equation 11–19,

$$K = \frac{I_{cl}}{I_s}$$

$$K = \frac{80\text{mA}}{50\text{mA}}$$

$$K = 1.6$$

(This is the minimum value for K.)

Let us choose R_1 equal to 1KΩ, then R_2 can be determined by rearranging equation 11–20.

$$\mathbf{R_2 = \frac{V_{out}R_1}{0.6V(K - 1)} - R_1} \qquad \textbf{(equation 11–21)}$$

$$R_2 = \frac{5\text{V} \times 1\text{K}\Omega}{0.6\text{V}(1.6 - 1)} - 1\text{K}\Omega$$

$$R_2 = 12.9\text{K}\Omega$$

A 13KΩ resistor could be used for R_2 since it is the closest standard 5% value. Finally, to find R_{cl}, we use equation 11–13.

$$R_{cl} = \frac{0.6\text{V}}{I_s}\left(\frac{R_1 + R_2}{R_2}\right)$$

$$R_{cl} = \frac{0.6\text{V}}{50\text{mA}}\left(\frac{1\text{K}\Omega + 13\text{K}\Omega}{13\text{K}\Omega}\right)$$

$$R_{cl} = 13\Omega$$

To summarize fold-back current limiting, it allows us to draw a greater current from the regulator under normal conditions while protecting the regulator from getting too much power dissipation with a short at the output.

We can see from the previous example that the important equations to determine component values for fold-back current limiting are 11–19, 11–21, and 11–13

THERMALLY PROTECTED REGULATORS

Some regulators, like the LMI40 series, have built-in thermal overload protection circuits that will shut down the regulator if the internal dissipation becomes too great. This prevents the regulator from being damaged from a combination of too large a voltage drop across it and too much current through it.

Figure 11–8 shows a simplified typical circuit used within the regulator for thermal protection. Referring to the figure, the combination of the voltage reference source and resistors R_1 and R_2 sets up a fixed voltage on the base of Q_1, which is below the transistor turn-on voltage. Q_1 base-to-emitter junction is temperature sensitive (typically −2.2mV/°C), and when the temperature of the base-to-emitter junction increases, the required turn-on voltage decreases.

Transistor Q_1 is physically located within the device next to the output regulator transistor (Q_2) for maximum thermal conduction. As power is dissipated by Q_2, Q_1 heats up and its turn-on voltage decreases. When the power dissipated by Q_2 is great enough, Q_1 will turn on and Q_2 base current I_b will be shunted to ground ($I_s = I_b$), causing Q_2 to turn off.

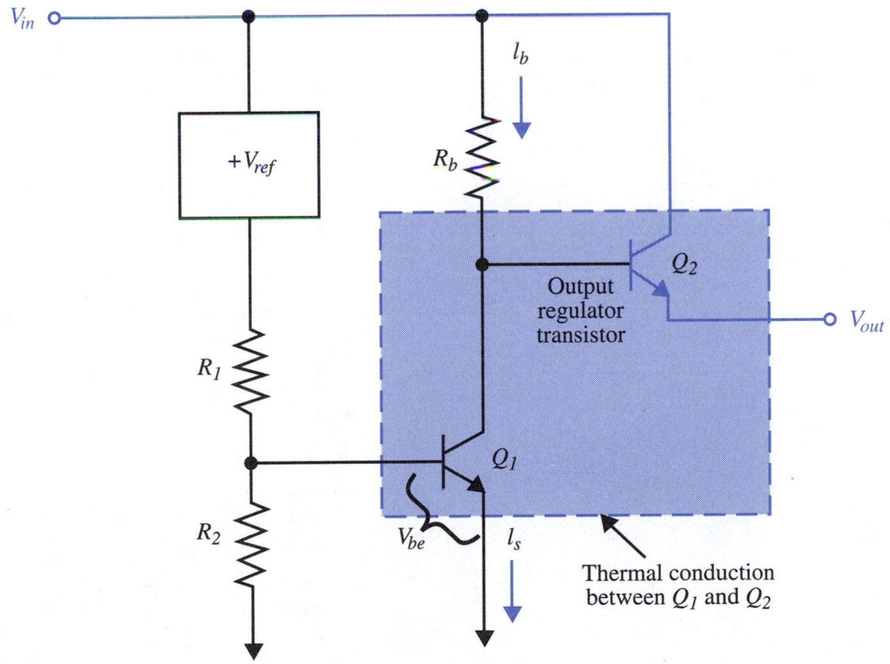

FIGURE 11–8 Thermal overload protection circuit

This causes the regulator output voltage and current to drop to zero, thus preventing the regulator from being damaged. Thermal turn-off usually occurs when the junction temperature is in the range of 150°C to 190°C.

Regulators with built-in thermal protection do not need fold-back current limiting for protection against a short circuit on the regulator output. However, conventional current limiting is sometimes still internally provided to protect against too much dissipation in the load circuits. Since thermally protected regulators operate on the basis of device temperature and less on the power controlled by the device, the amount of output current is a function of the device's heat sinking. The better the regulator is heat-sink mounted, the greater the possible output current.

THREE-TERMINAL ADJUSTABLE REGULATORS

The 723 regulator previously described was one of the first variable-voltage-type linear regulators. It has been replaced in new applications requiring more current by the three-terminal LM117 types (LM217, LM317). The LM117K can, for example, provide 1.5A and dissipate 20W in its TO-3 (TO–204AE) package. The LM317T can provide 1.5A and dissipate 15W in its TO–220 package.

These regulators have internal thermal overload protection as described in the previous section. Output voltage on the LM117 series is controlled by the resistors and an internal 1.25V reference, as shown in Fig. 11–9. Regulation is achieved by the device maintaining the 1.25V reference voltage across R_1 under all possible normal load conditions. The output voltage is

$$V_{out} = V_{ref} + (I_a + I_1)R_2$$

where I_1 is equal to V_{ref}/R_1 and I_a is the bias current from the adjust terminal with a maximum value of 100μA. If we neglect I_a, then

$$V_{out} = V_{ref} + V_{ref}R_2/R_1$$

$$V_{out} = V_{ref}(1 + R_2/R_1) \quad \textbf{(equation 11–22)}$$

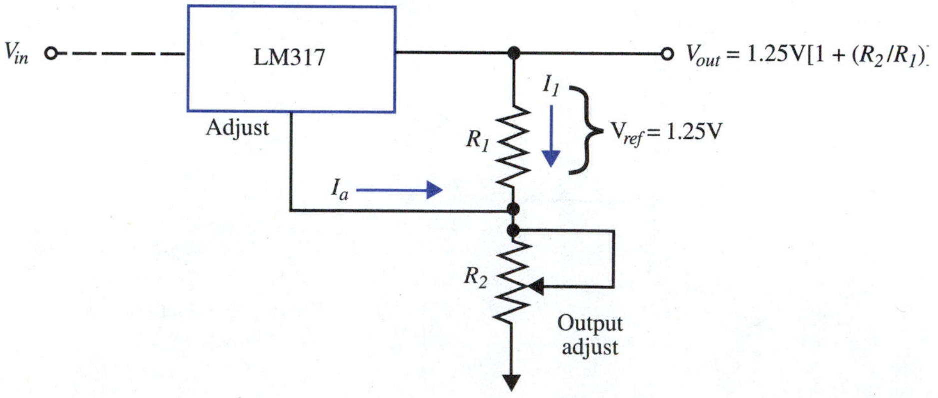

FIGURE 11–9 Output voltage adjust circuit for LM317

This equation indicates that if we decrease R_2 to zero, the output voltage is simply the internal reference voltage. Increasing R_2 can raise the output voltage to a desired maximum value of about 2V (regulator voltage drop) less than the regulator input voltage.

With a maximum input voltage specification for the LM317 of 40V, the maximum output then is 38V. It is not suggested, however, that the regulator be operated at the absolute maximum input level of 40V, since an increase in power line voltage caused by transients could destroy the device. A maximum input voltage of 35V is more conservative and yields an output voltage as high as 33V.

EXAMPLE

Assume we have the following power supply requirements:

Output voltage range—at least 2V to 20V

Output current capability—1.0 A

Line voltage variation—10%

Maximum output ripple—10mV

Determine the required circuit components.

SOLUTION

We shall use an LM317T regulator. For the output voltage of 20V, we need to determine values for R_1 and the potentiometer used for R_2. Let us first find the value of R_1. Remember, we neglected the 100μA of bias current when we derived equation 11–22. In order to do this, we should make the current (I_1) through R_1 100 times larger than I_a, or

$$I_1 = 100 \times I_a$$

$$I_1 = 10\text{mA}$$

From Ohm's law,

$$R_1 = \frac{V_{ref}}{I_1}$$

$$R_1 = \frac{1.25\text{V}}{10\text{mA}}$$

$$R_1 = 125\Omega$$

Let us use a standard value of 120Ω (5%).

V_{ref} has a nominal value of 1.25V, but from the specifications sheets, it can be anywhere in the range of 1.2V to 1.3V for a given device. The maximum value of R_2 occurs with a reference voltage of 1.2V. Using this value and solving for R_2 in equation 11–22,

$$R_2 = R_1\left(\frac{V_{out} - V_{ref}}{V_{ref}}\right)$$

$$R_2 = 120\left(\frac{20\text{V} - 1.2\text{V}}{1.2\text{V}}\right)$$

$$R_2 = 1.88\text{K}\Omega$$

If we pick a standard 2KΩ potentiometer, this value will accommodate the tolerance variations of R_1. For 20V out, we must have an input voltage under low line voltage conditions of no less than 22V. Selecting a standard 20V (RMS) transformer provides about 24V (V_{in}) at the input to the regulator under 10% low line conditions.

$$V_{in} = [(\sqrt{2} \times 20 \times 0.9) - 1.2]\ \text{V}$$

$$V_{in} \approx 24\text{V (low line)}$$

The 1.2V in the equation corresponds to the two diode drops in a bridge rectifier, and the 0.9 factor reduces the nominal voltage by 10 percent. Under high line voltage conditions, the voltage input to the regulator will be

$$V_{in} = [(\sqrt{2} \times 20 \times 1.1) - 1.2]\ \text{V}$$

$$V_{in} \approx 30\text{V (high line)}$$

This is all right because it is below our chosen maximum value of 35V.

FIXED VOLTAGE THREE-TERMINAL REGULATORS

Regulators used in electronic equipment are mostly the fixed voltage type. TTL logic, for example, requires a fixed voltage of +5V supply, and op-amps typically ±12V.

Fixed voltage regulators are simpler to use since they are usually three-terminal devices. The input terminal is connected to the power supply filter capacitor, the common terminal is connected to ground, and the output terminal is connected to the load.

The same adjustable regulator features of internal current limiting and thermal overload protection are common to fixed regulators.

A popular fixed regulator is the LM340. This device comes in three versions, with output voltages of +5V, +12V, or +15V and maximum output currents from 0.1A to 1.5A (depending on package configuration). Line regulation ranges from 0.2 percent for the higher quality units to 1 percent for the commercial versions. (*Note:* Line regulation is listed with a range of DC input voltages to the regulator rather than AC line variations.) Load regulation can be as low as 0.5 percent or as high as 1 percent, and ripple rejection ranges from 60dB to 80dB. As always, the higher the price you are willing to pay, the better the regulator.

Table 11–1 lists the key characteristics for several common types of fixed regulators, for both positive and negative output voltages. Also listed are some variable regulators.

The maximum current listed for these regulators is at a power dissipation level below thermal shutoff, and as we shall see in Chapter 19, these current values are considerably lower at the higher power dissipation levels.

Note that both fixed and variable negative voltage regulators are available that complement the positive voltage devices. Also see from the table that the maximum output voltages are quite optimistic for the variable regulators when one considers the maximum

TABLE 11–1 Common Voltage Regulators

Type	*Output Voltage*	*Max Input Voltage*	*Max Current*	*Line* Regul(%)*	*Load* Regul (%)*	*Ripple Rej (db)*	*Max Junct Temp (°C)*	*Thermal Resistance Junction to Case (°C/Watt)*
LM340	+5, +12, +15V	+35V	1.5A	1	1	72	150	4 (TO–3)
LM78XX	+5, +12, +15V	+35V	1A	1	1	72	150	4 (TO–3)
LM79XX	−5, −12, −15V	−35V (−5V), −40V (−12, −15) −15)	1.5A	1	2	70	150	5 (TO–3)
LM2940	+5, +8, +10V	+26V	1A	1	1	66	150	3 (TO–3)
LM323	+5V	+20V	3A	0.5	2	60	125	2 (TO–3)
LM117	+1.25 to +37V	+40V	1.5A	0.7 (+10V)	1.5	65	125	3 (TO–3)
LM117HV	+1.25 to +37V	+60V	1.5A	0.7 (+10V)	1.5	65	125	3 (TO–3)
LM337	−1.25V to −37V	−40V	1.5A	0.7 (−10V)	1.5	60	125	3 (TO–3)
LM337HV	−1.25 to −37V	−60V	1.5A	0.7 (−10V)	1.5	60	125	3 (TO–3)
LM396	+1.25 to +15V	+20V	10A	0.2 (+10V)	0.1 (+10V)	74	175	1.2 (TO–3)

*Check specifications for line and load variations.

possible input voltages and allows for line voltage variations. The LM117HV and the LM137HV provide for more conservative designs if higher output voltages are required.

LOW DROP-OUT VOLTAGE REGULATOR

Although not indicated as such in Table 1–1, the LM2940 is a low drop-out voltage regulator—which means it is able to operate with a low voltage drop from input to output. For the LM2940, drop-out voltage is typically 0.5V when 1A is flowing (LM340 has 2V at 1A). The significance of this is that with less voltage drop across the regulator, there is less power dissipation. The possible +8V output for this regulator might sound like an odd value, but this voltage is becoming a standard for automobile control signals.

LOAD AND LINE REGULATION OF LINEAR REGULATORS

The stated purpose of a regulator is to maintain the power supply output voltage constant with variations of line voltage and load. Let us now look at some examples of line and load regulation, using devices listed in Table 11–1.

EXAMPLE

Find the predicted variation in the output voltage (ΔV) for an LM2940–8 (8V) regulator if the input is from a car battery with a variation of +10V to +16V. Specifications for the regulator list a typical variation of 20mV with an input range from +10V to +26V.

SOLUTION

If we assume the output variation is proportional to the input voltage change, we can set up the following ratio:

$$\frac{(\Delta V)}{(16\text{V} - 10\text{V})} = \frac{(20\text{mV})}{(26\text{V} - 10\text{V})}$$

Solving for ΔV,

$$(\Delta V) = 7.5\text{mV}$$

EXAMPLE

Find the predicted variation in the output voltage (ΔV) for an LM317 regulator adjusted for +10V if the output current changes from 10mA to 500mA.

The LM317 specifications state that a 1.5 percent change occurs in the output voltage if the load current changes from 10mA to 1.5A.

SOLUTION

The 1.5 percent change in +10V is (10V × 0.015) or 150mV. As before, let us assume that the change in output voltage is proportional to the output current change and set up the following ratio:

$$\frac{(\Delta V)}{(500\text{mA} - 10\text{mA})} = \frac{(150\text{mV})}{(1.5\text{A} - 10\text{mA})}$$

$$(\Delta V) = 49.3\text{mV}$$

PREVENTING REGULATOR OSCILLATION

All of the regulators described in this chapter use high-gain amplifiers to maintain good regulation. This high gain can cause oscillations to appear at the output of the regulator. These oscillations can be reduced by adding a high-frequency bypass capacitor at the input and also at the output of the regulator. A good choice is a 0.1μF ceramic or mica capacitor connected as close as possible to the regulator terminals.

TRACKING POWER SUPPLIES

When both positive and negative power supplies are used, it is sometimes required that if one of the supplies has a change in output voltage, the other supply also must change by the same amount. This is called power supply tracking and is used to make the operation of circuits using dual supplies more predictable. One circuit used to accomplish tracking is shown in Figure 11–10 using positive (LM340) and negative (LM7912) voltage regulators.

The input voltages (V_{in}) to the regulators are chosen to exceed the minimum voltage drop across the regulators (2V for the LM340 and 1.1V for the LM7912).

As previously discussed, the filter capacitors (C_f) must keep the 120Hz input ripple from dropping the regulator input voltages below the minimum value.

Tracking is accomplished by the op-amp and the resistive divider. If both outputs are exactly 12V, the center tap of the divider (Pin 2 of 741) will be at the common (ground) po-

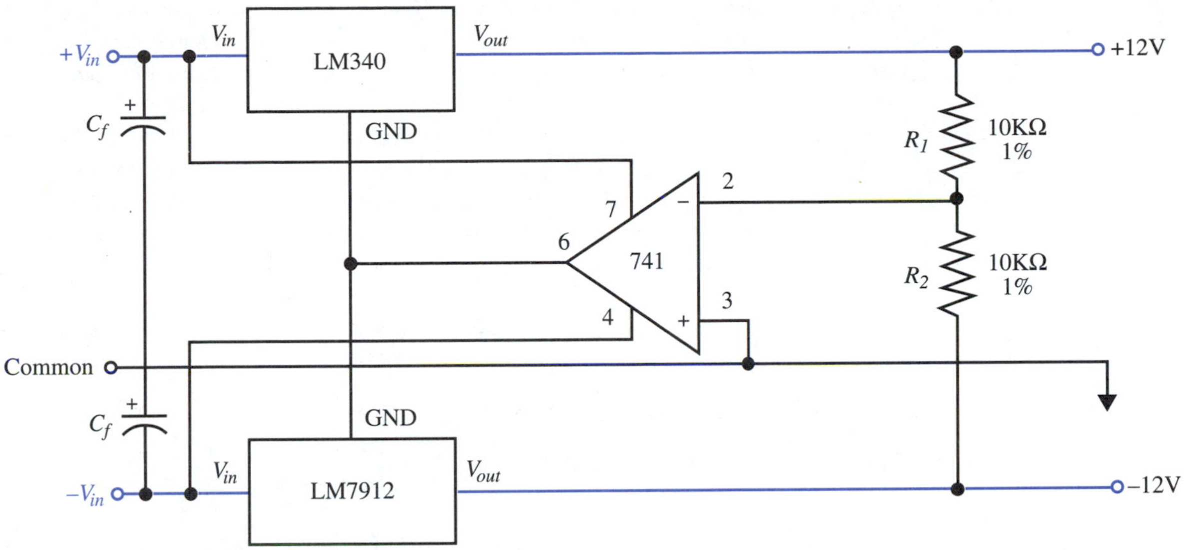

FIGURE 11–10 Dual tracking ±12V power supplies

tential. This means there will be no difference in voltage between the op-amp input leads, and the op-amp output will also be at the common potential.

Let us now assume that a load on the +12V supply causes this supply to drop by 100mV. The voltage across the R_1,R_2 divider is now 23.9V rather than 24V, and the voltage drop across each resistor will be 11.95V. This means the voltage at the center of the divider will be −50mV. The inverting input to the op-amp is now more negative (by 50mV) than the noninverting input, which causes the op-amp output to go positive.

Because of the load—the positive regulator has a fixed voltage of 11.9V between the output and common terminal—raising the voltage on its common terminal will cause the output voltage to increase. The positive output voltage will increase until the center tap of the divider is back to the common potential. This occurs when the positive output voltage is increased by 50mV. To achieve this, the op-amp output has to increase by this amount. But since the common terminal of the negative regulator is also raised, the negative supply decreases by 50mV. The final result is matched output voltages of ±11.95V.

We must have a slight difference in the op-amp inputs in order to have +50mV at the output. However, with the 741 op-amp open loop gain of 100,000, the center tap of the resistors (inverting input) will only be 0.5μV above the common potential. This represents the error in tracking of the output voltages. A similar analysis shows that increases or decreases in both the positive and negative supplies will be tracked.

TROUBLESHOOTING VOLTAGE REGULATORS

The approach to troubleshooting voltage regulators is a function of whether we are dealing with a multipin device like the LM723 or a three-terminal device, i.e., the LM317. Let us first look at troubleshooting the 723.

To troubleshoot the LM723, we must consider each functional block within the device. We can break the device down into a reference source (+7V), error op-amp, pass transistor, and current limit transistor. For the device to work, all these functional blocks must be operational.

We can check the voltage reference by looking for +7V (6.8V to 7.5V) at Pin 6.

If the error amp is working, we should see the same DC voltage at Pin 4 as we do at Pin 5.

The voltage measured at the emitter of the series pass transistor (output Pin 10) should be the value predicted from the gain of the error amp times the voltage measured at Pin 5 input. Finally, since we have access to the base, emitter, and collector of the current limit transistor (Pins 2, 3, and 13), we can check the junctions of this transistor with the power removed.

The three-terminal regulators are rather simple to troubleshoot. Three basic problems can occur:

1) The output can be too high.
2) The output can be too low.
3) The regulator can be oscillating.

With the output too high, the problem is most likely a shorted series pass transistor. But if the regulator is the adjustable type, check components in the adjust circuitry.

If the output is low, it can be a defective regulator, or the regulator could be in a current limit mode because the load current is too high or the ambient temperature is too high. For the latter situations, remove the load and/or check the component temperature before replacing the regulator.

Oscillations can be a difficult problem to isolate. But, check the filter capacitors at the regulator input and output. Are the capacitor leads short enough?

VOLTAGE AND CURRENT REFERENCES

Whenever there is a need for a voltage or current to some prescribed accuracy, there has to be some absolute reference device that can be used for comparison. With the 723 regulator, the output voltage was compared with an internal reference voltage of 7V. In the next few sections, we will look at integrated circuit references, which are available for use in equipment where a standard voltage reference or current source is required. But first we will consider the zener diode.

ZENER DIODE AS A VOLTAGE REFERENCE

The most common reference voltage source is a zener diode, a device that relies on the constant voltage characteristic of reverse breakdown. However, as discussed previously, variations in the zener voltage can occur when current is changed through the diode. This again is due to the internal resistance (zener resistance) of the diode.

EXAMPLE

A particular zener has a breakdown voltage of 12V at 50mA and a zener resistance of 10Ω. If the current through the zener is changed to 100mA, what is the new zener voltage?

SOLUTION

The new zener voltage is

$$V_z = 12\text{V} + (100\text{mA} - 50\text{mA}) \times 10\Omega$$

$$V_z = 12.5\text{V}$$

We see that the zener voltage has changed by 0.5V with increased current through the device.

The higher the zener resistance, the greater the voltage change for a given current. With a given zener, we can eliminate the voltage change by keeping the current constant through the device.

VOLTAGE REFERENCE DEVICES

A voltage reference semiconductor consists of an integrated circuit with a zener diode and active circuits that keep the current through the zener relatively constant. This gives a very small voltage change with load current changes, and therefore the voltage reference device has a low effective zener impedance—typically less than 1Ω. Figure 11–11 shows the typical internal circuit of the device.

In operation, the + input terminal is equivalent to the cathode terminal of a zener diode and is connected in series with a voltage dropping resistor R_d. If the + input terminal goes more positive by an amount ΔV, the change is transferred by the zener diode to the base of Q_1. This causes a more positive voltage on the base of Q_1, and this transistor is turned on more, resulting in more collector current and a drop in collector voltage. Since the collector of Q_1 is connected directly to the base of Q_2 (PNP), this transistor is also turned on more, causing greater current to flow through the external resistor R_L. The net effect is that the original positive increase is negated by the additional volt drop across R_L caused by the increased current through Q_2. The gain of Q_1 and Q_2 enables the circuit to correct for changes in the input voltage, and this lowers the effective zener resistance of the device.

Since the components are on one IC chip, greater temperature stability is achieved with the voltage reference by use of internal compensation. Being a two-terminal device like a zener, the voltage reference is used in circuit applications the same way a zener is used. However, the range of available voltages for the voltage reference devices is not as extensive as can be obtained with zener diodes.

TYPICAL VOLTAGE REFERENCES

One particular voltage reference is the LM329. This device has a nominal voltage of 6.9V and has an operating current range of 1mA to 15mA. The zener resistance is about 1Ω so that we can expect only 14mV change in the reference voltage over the complete current excursion. The temperature coefficient K_t for the LM329B is a maximum of 50 parts per million (5×10^{-5}) per degree centigrade.

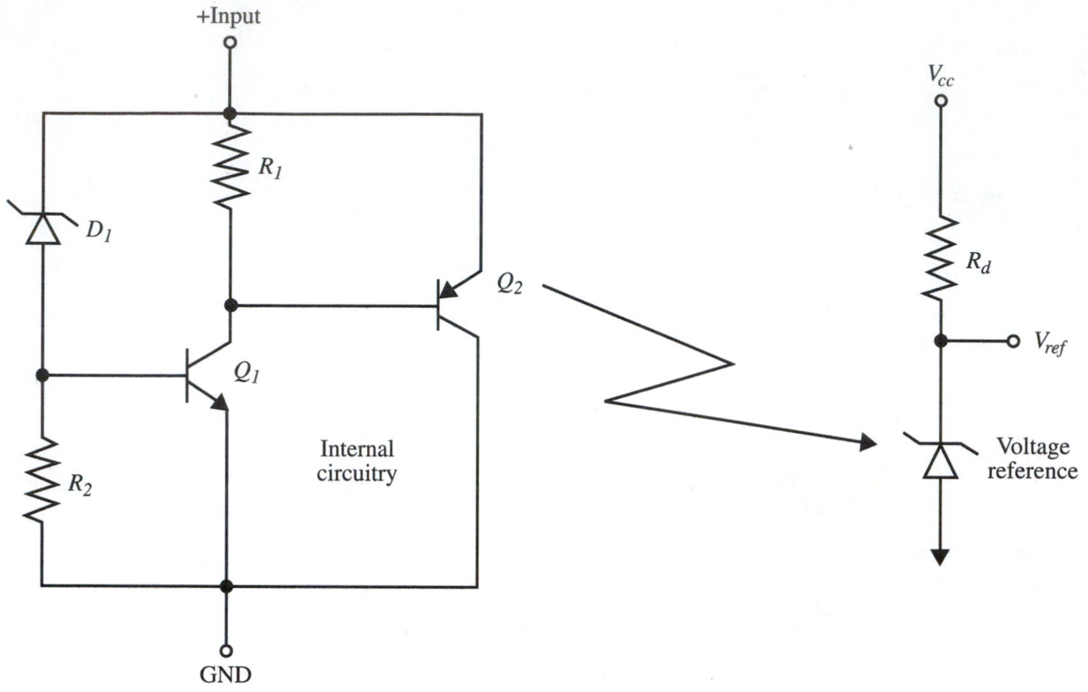

FIGURE 11–11 Voltage reference equivalent circuit

EXAMPLE

If the ambient temperature changes from 25°C to 50°C, find the new reference voltage for an LM329B.

SOLUTION

The reference voltage V_{ref} at 50°C will be

$$V_{ref} = 6.9V + 6.9V(50°C - 25°C)5 \times 10^{-5}/°C$$

$$V_{ref} = 6.9V + 0.008625V$$

$$V_{ref} = 6.908625V$$

Or a change of 8.625mV over the temperature range.

Other voltage references are shown in Table 11–2. The table indicates a reference voltage range from 1.22V to 30V and initial tolerances from 0.05% to 3%. Zener resistances are from 0.2Ω to 1Ω, and temperature drift is from 0.5 parts per million to 0.5% over the full temperature range. The LP2951 and LP2950 devices are really precision voltage regulators (the LP2951 is adjustable) that can provide currents up to 100mA.

TABLE 11–2 Some Available Voltage References

Voltage V_r	*Device*	*Tolerance*	*ppm/°C*	*Temp Range*	*Current*	*r_z(typ)*
1.22V	LM113–2	±1%	50 (typ)	−55°C to +125°C	500μA to 20mA	0.8Ω
1.235V	LM185–1.2	±1%	150	−40°C to +85°C	10μA to 20mA	1.0Ω
1.24V–5.3V	LM285BX	±1%	30	−40°C to +85°C	10μA to 20mA	0.3Ω
1.24V–6.3V	LM611C	±2%	150	0°C to 70°C	16μA to 10mA	0.27Ω
2.49V	LM136A	±1%	72	−55°C to+125°C	400μA to 10mA	0.4Ω
2.5V	LM385–2.5	±3%	150	0°C to 70°C	20μA to 20mA	1.0Ω
5.0V	LM136A	±1%	72	−55°C to +125°C	400μA to 10mA	0.8Ω
6.95V	LM199A	±2%	0.5	−55°C to +125°C	500μA to 10mA	0.5Ω
10.24V	LH0071–2	±0.05%	8	−40°C to + 85°C	0mA to 5mA	0.2Ω
1.235 to 30V Adj.	LP2951C	±1%	±1% over temp	− 40°C to +125°C	120μA to 100mA	100ppm per mA
5V	LP2950AC	±0.5%	±0.5% over temp	−40°C to +125°C	120μA to 100mA	100ppm per mA

VOLTAGE REFERENCE APPLICATIONS

In operation, the two terminals of the voltage reference are used like a zener diode. In fact, as indicated in Figure 11–11, the schematic symbol for a voltage reference is the same as that of the zener. However, the voltage reference can be recognized on a schematic by its **LM** prefix rather than a **1N** designator.

The voltage reference provides a more precise voltage than a zener diode but typically is able to pass less current. If the application requires the precision of a voltage reference and the extra current capability of a zener, then a voltage reference is used with current-boost circuitry.

Let us now look at some application examples for a voltage reference.

DIFFERENTIAL VOLTAGE MEASUREMENT

In the previous chapter, we saw how differential voltage measurement is used to measure the small change that occurs in the output voltage for a power supply during load and line regulation measurements. Figure 11–12 shows how an LM329 can be used for regulation measurements of a +5V power supply

Again, what the reference circuit does is to null out the 5V output so we are able to detect only the changes (ΔV) in the output voltage on a more sensitive scale of the DMM.

POSITIVE AND NEGATIVE VOLTAGE REFERENCE

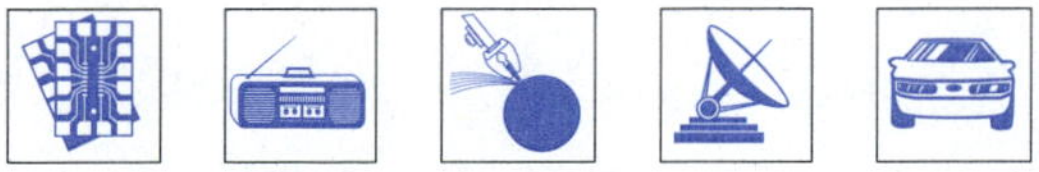

Let us say that we would like to have a variable voltage reference that can be adjusted to both positive and negative levels. Consider the circuit of Figure 11–13.

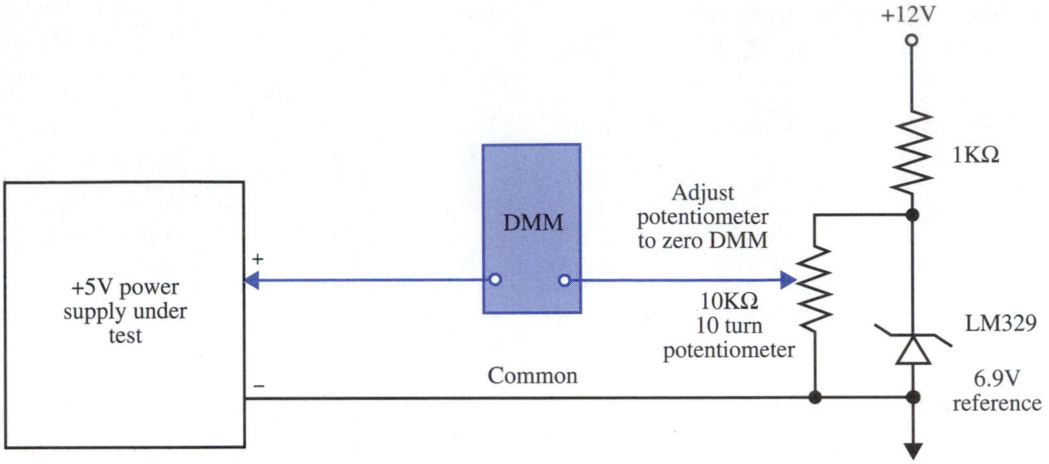

FIGURE 11–12 Differential measurement using a voltage reference

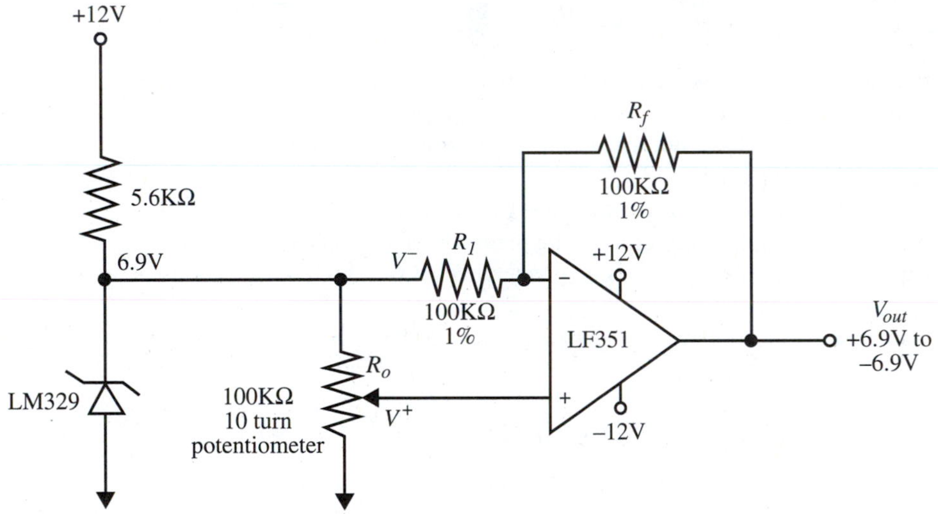

FIGURE 11–13 Positive and negative voltage reference

Using an LM329, which is a 6.9V reference, the output voltage for this circuit is

$$V_{out} = V^-(-R_f/R_1) + V^+(R_f/R_1 + 1)$$

$$V_{out} = 6.9\text{V}(-100\text{K}\Omega/100\text{K}\Omega) + V^+(100\text{K}\Omega/100\text{K}\Omega + 1)$$

$$V_{out} = -6.9\text{V} + 2(V^+)$$

From the circuit, we see that V^- is +6.9V and V^+ ranges from 0V to +6.9V. With V^+ of 0V, V_{out} is −6.9V, and with V^+ of +6.9V, V_{out} is +6.9V. The output can be adjusted to any value between these limits with 0V output when the wiper is in the middle of the potentiometer.

The LF351 FET input op-amp used for this circuit has a low bias and offset current so that temperature drift of these parameters is negligible.

ADJUSTING THE LM317 OUTPUT TO 0V

Remember that the LM317 regulator has a minimum output voltage of +1.25V. But, by biasing the lower terminal of the voltage adjust potentiometer negative, we are able to provide a zero output voltage capability.

As the circuit in Figure 11–14 shows, the LM385–2.5 provides a stable −2.5V at the bottom of R_2. Using −2.5V rather than −1.25V ensures that the output will reach zero volts. Since the output voltage (and current flow) is taken from the V_{out} terminal to the GND terminal, the output voltage cannot go negative.

With the adjust potentiometer shorted, the adjust terminal is at −2.5V, and the output is at 0V. With the full 5KΩ potentiometer resistance in the circuit, voltage drop across the series combination of R_1 and R_2 is 32.5V, and the output is at +30V.

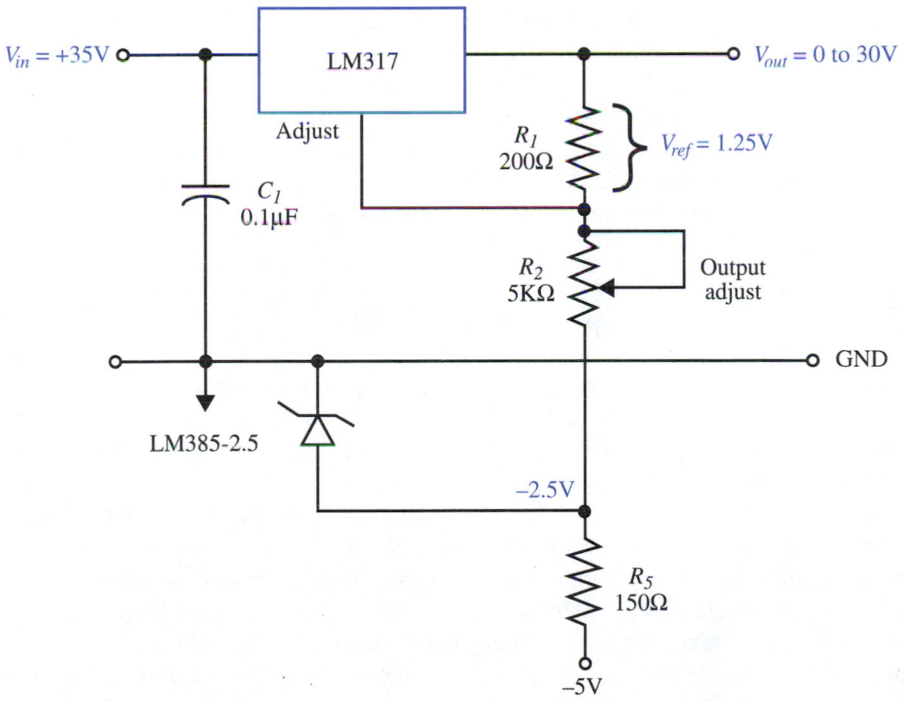

FIGURE 11–14 0V to +30VDC power supply

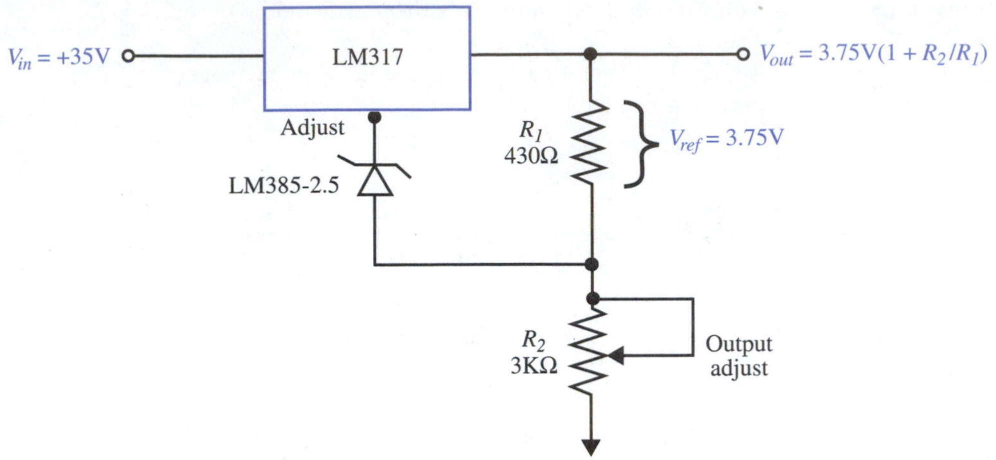

FIGURE 11–15 Improved LM317 regulator

IMPROVING THE REGULATION OF AN LM317

Another application of the voltage reference is to improve the regulation of an adjustable regulator like the LM317. The circuit is shown in Figure 11–15. Recall that the LM317 has an internal reference of 1.25V and that the device regulates by maintaining this voltage.

Adding the LM385–2.5 in series with the adjustment terminal effectively increases the reference to (1.25V + 2.5V), or 3.75V—which means a greater portion of the output voltage is regulated. Actually, the regulator is improved by the factor of three that the reference voltage is increased. The disadvantage of this circuit is that the minimum output voltage is now 3.75V rather than 1.25V. However, by adding a negative reference similar to that shown in Figure 11–14 , the output voltage can be brought back to zero.

ELECTRONIC STANDARD CELL

Let us now look at an application of a voltage reference as a calibration standard. A much used standard for voltage calibration is a **standard cell**—which is a battery with a precise output voltage. This standard can be replaced, however, with the circuit given in Figure 11–16, which uses an LM199 precision reference.

If we want extreme (standard cell) accuracy with a voltage reference, we must keep the temperature of the device constant. The LM199 (299, 399) comes in a ceramic package that contains a small oven and a built-in, temperature-controlled heater. By maintaining the temperature of the reference at a value greater than the ambient (surrounding) temperature, the temperature of the reference is essentially independent of ambient temperature changes. That is, if the ambient temperature rises, the internal heater output is reduced—which maintains a constant temperature at the reference.

A standard cell has an output voltage of 1.0192V with a temperature coefficient of −10μV/°C. Let us now determine whether the circuit of Figure 11–16 meets these requirements.

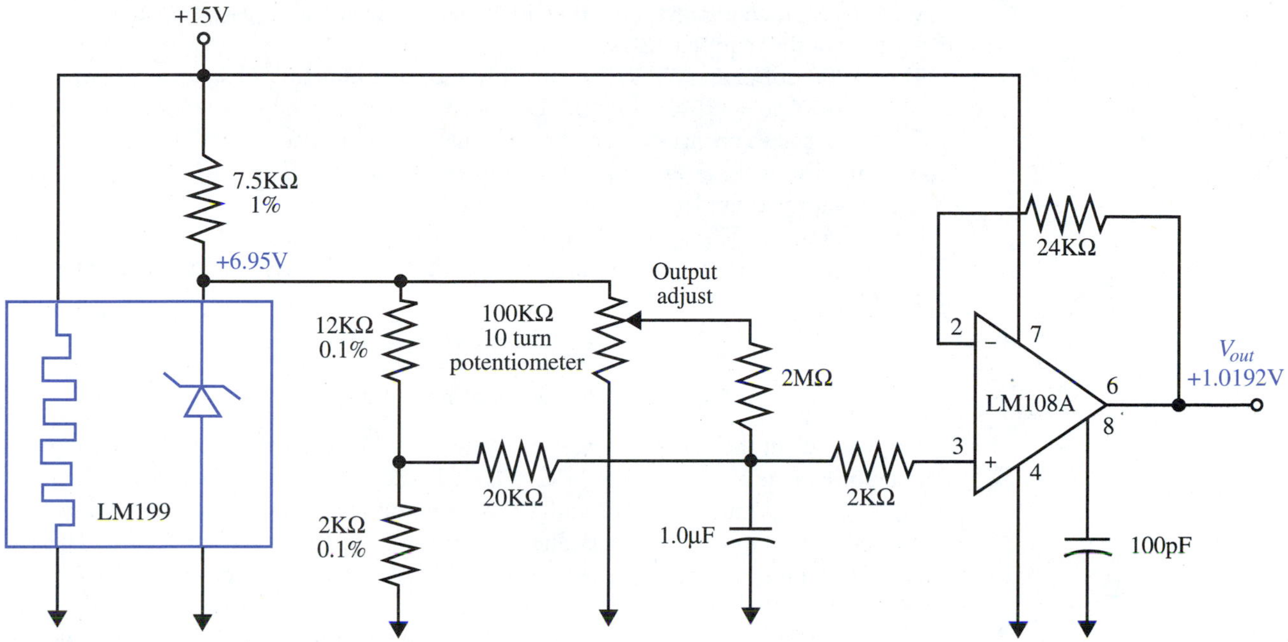

FIGURE 11–16 Standard cell replacement circuit

In the schematic, the 6.95V reference is reduced by the 12KΩ and 2KΩ voltage divider to provide a voltage V_D.

$$V_D = \frac{6.95\text{V} \times 2\text{K}\Omega}{(2\text{K}\Omega + 12\text{K}\Omega)}$$

$$V_D = 0.9929\text{V}$$

The top portion (above wiper) of the 100KΩ potentiometer in series with the 2MΩ resistor forms another voltage divider with the 20KΩ resistor to provide the input to the noninverting side of the follower. Let us assume that the wiper on the pot is all the way to the top, then the input voltage (V_{ni}) to the noninverting input to the follower is

$$\text{max. } V_{ni} = \frac{(6.95 - 0.9929)\text{V} \times 20\text{K}\Omega}{(20\text{K}\Omega + 2\text{M}\Omega)} + 0.9929\text{V}$$

$$\text{max. } V_{ni} = 1.0519\text{V}$$

With the wiper on the potentiometer at the bottom of its range, the top end of the 2MΩ resistor goes to ground and forms a voltage divider with the 20KΩ resistor, which divides down the 0.9929V to

$$\text{min. } V_{ni} = \frac{0.9929\text{V} \times 2\text{M}\Omega}{(20\text{K}\Omega + 2\text{M}\Omega)}$$

$$\text{min. } V_{ni} = 0.9831\text{V}$$

We can see, then, that with the potentiometer, we have the adjustment capability to set the voltage at the required 1.0192V.

The LM108A op-amp used in this circuit has a maximum 1mV offset over a temperature range of −55°C to +125°C. Since this can be adjusted out with the potentiometer, the critical factor becomes changes of the offset voltage and offset current with temperature. The maximum offset voltage variation with temperature of 5μV/°C for this op-amp compares favorably with the standard cell's value of 10μV.

Considering the offset current change over the temperature range, the coefficient is 2.5pA/°C. The change in output voltage (ΔV_o) due to this current change and with the R_f of 24KΩ is

$$\Delta V_o = (2.5 \times 10^{-12}\text{A/°C})(2.4 \times 10^{4}\Omega)(180°\text{C})$$

$$\Delta V_o = 9.9\mu\text{V} \quad \text{(which is negligible!)}$$

But what about the voltage reference? The temperature coefficient for the LM199 over a temperature range of −55°C to +85°C is 0.0001%/°C. This is equivalent to a coefficient of 6.95μV/°C. Combining this value with the op-amp offset voltage coefficient of 5μV/°C, we get a total of 11.95μV/°C, which is slightly over the 10 μV/°C stated for the standard cell.

Over the temperature range from 85°C to 125°C, the LM199 degrades to 15μV/°C, which takes it beyond the standard cell variation. However, since standard cells are mainly used in calibration labs, which are temperature controlled at around 25°C, this should not create a problem.

CONSTANT CURRENT SOURCES

A current source is generated by putting a very high resistance (R_s) in series with the output terminals of a voltage source. If R_s is at least 100 times greater than the maximum load resistor, we can say that the current is constant, regardless of changes in the load resistor. Current sources are useful for calibrating meters or for generating linear ramp-type waveforms. The equation for charging a capacitor is

$$i = C\frac{\Delta V}{\Delta t}$$

If the current is kept constant, the slope $\Delta V/\Delta t$ is constant. A linear ramp generated this way can be used for the horizontal and vertical deflection circuits used with a cathode ray tube.

ADJUSTABLE CURRENT SOURCE

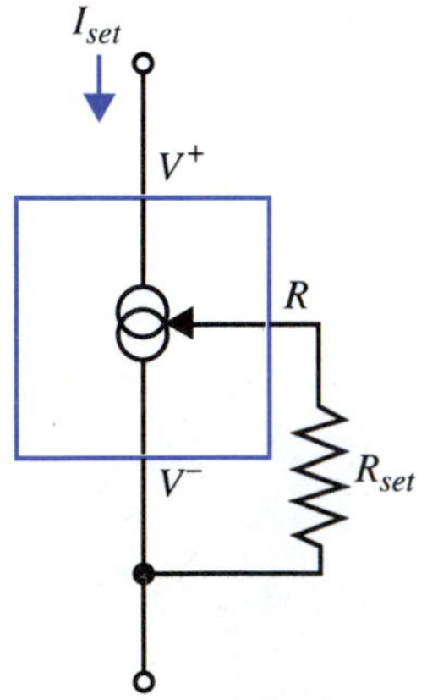

FIGURE 11–17 LM234

The LM234 is a three-terminal adjustable current source with a 10,000-to-1 range of current adjustment. The current is set by selecting the value of a set resistor (see Figure 11–17).

The relationship between the I_{set} current and the value of R_s and the temperature is

$$I_{set} = \frac{V_{set}}{R_{set}} = \frac{(215\mu\text{V/°K})(\text{T})}{R_{set}} \qquad \textbf{(equation 11–23)}$$

where T is the temperature in degrees Kelvin (°C + 273°).

At 25°C, the numerator term is

$$V_{set} = (215\mu V/°K)(25°C + 273°)$$

$$V_{set} = 64mV$$

This is the voltage across the R_{set} resistor.

EXAMPLE

Find the voltage across the R_{set} resistor used with an LM234 at 75°C.

SOLUTION

$$V_{set} = 215\mu V/°K \times (273 + 75)°K$$

$$V_{set} = 74.8mV$$

EXAMPLE

Find the value of R_{set} required for an LM234 if the desired I_{set} is 1mA at 75°C.

SOLUTION

Rearranging equation 11–23,

$$R_{set} = \frac{(215\mu V/°K \times T)}{I_{set}}$$

$$R_{set} = \frac{215\mu V/°K(273 + 75)°K}{1mA}$$

$$R_{set} = 74.8\Omega$$

The fact that the LM334 current I_{set} varies with temperature suggests that the device can be used in a thermometer.

Figure 11–18 shows an LM234 used in a temperature-sensing circuit. The components are chosen to provide a voltage at the output that indicates the temperature in degrees Fahrenheit. The calibration provides 10mV per °C at the output with 0V at 0°C.

CURRENT SOURCE FROM A VOLTAGE SOURCE

We can generate a precise current source by using a voltage source. Look at the precision 1mA current source shown in Figure 11–19, which uses a voltage reference.

Operation of the circuit is as follows:

The voltage at the noninverting input is the supply voltage of 15V minus the 2.5V of the LM385–2.5 reference, or 12.5V. Since there can be no difference in the voltage between the two inputs of an op-amp, the voltage at the bottom of the 2.5KΩ resistor (R_5) must also be at 12.5V. This means that the voltage drop across the 2.5KΩ resistor is the

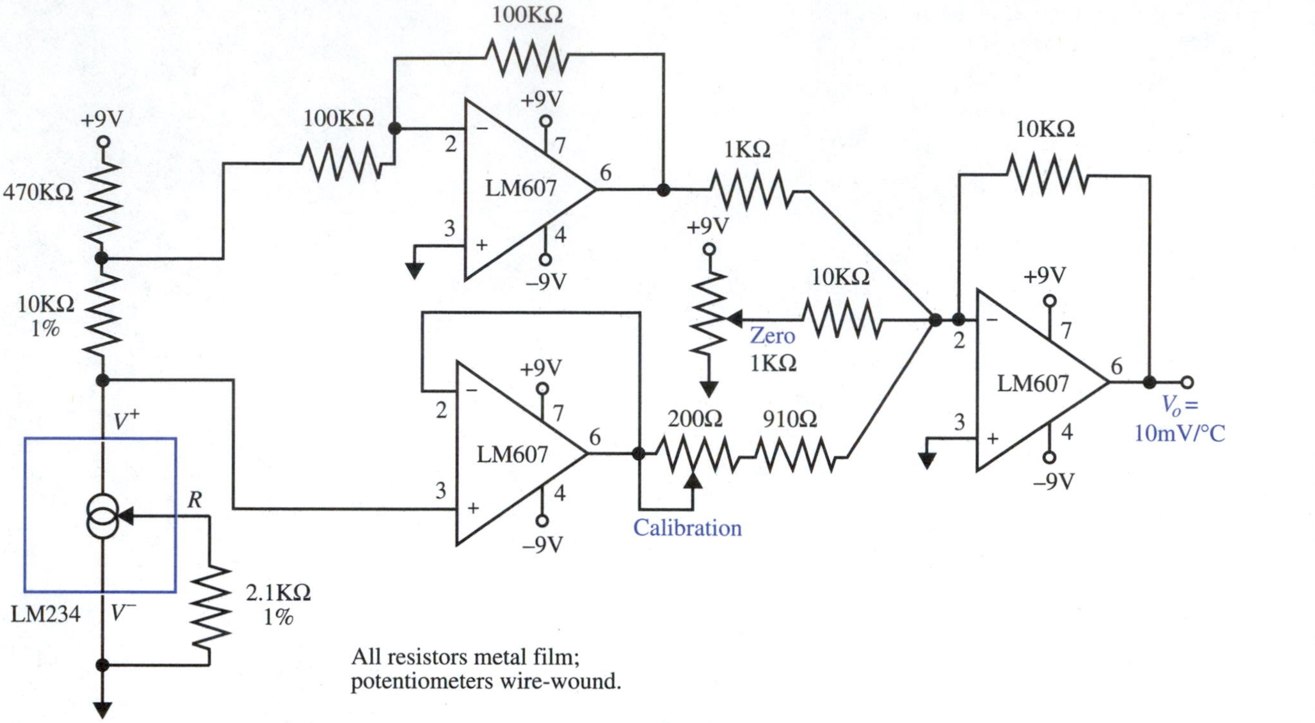

FIGURE 11–18 Temperature sensor

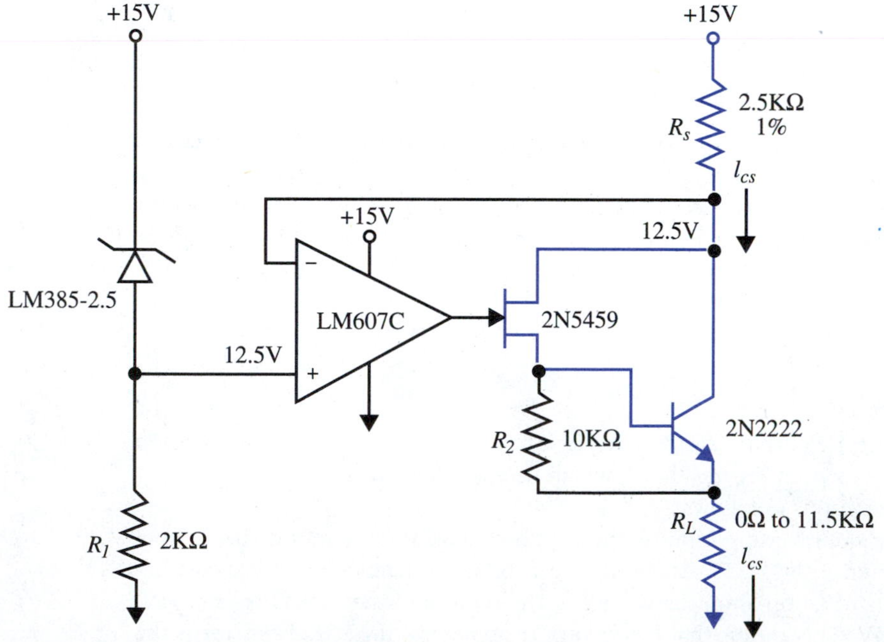

FIGURE 11–19 Precision one-milliamp current source

same as the 2.5V voltage reference, and the current flowing through this resistor is (2.5V/2.5KΩ), or 1mA.

Since no current flows into the op-amp inverting terminal, the full output 1mA is divided between the field effect transistor and the bipolar transistor. If we assume that the FET gate current is zero, then this same 1mA is flowing out of the junction of the 10KΩ resistor and the transistor emitter and into the load (R_L).

The current from this circuit is precise because it is controlled only by the voltage reference and the 2.5KΩ, 1% resistor. If we desire more current, then the 2.5KΩ resistor should be reduced.

Limits on the size of the load resistor (R_L) range from 0Ω to 11.5KΩ The upper load resistor limit is determined by allowing a minimum 1V drop across the transistors to maintain their correct operation.

In a similar fashion, we can generate a precision 1mA current "sink" (current flows in rather than out) circuit as indicated in Figure 11–20 by reversing the position of both the reference and the 2.5KΩ resistor.

As before, we have the 2.5V across the 2.5KΩ resistor, and the sink current is 1mA. The allowable range of load resistors is from 0Ω to a value determined by the maximum collector/drain voltage of 30V. If we assume this maximum value of 30V with a 1V drop across the 2N2222 and the 2.5V drop across R_S, the maximum load resistor R_L is

$$R_L = \frac{(30\text{V} - 1\text{V} - 2.5\text{V})}{1\text{mA}}$$

$$R_L = 26.5\text{K}\Omega$$

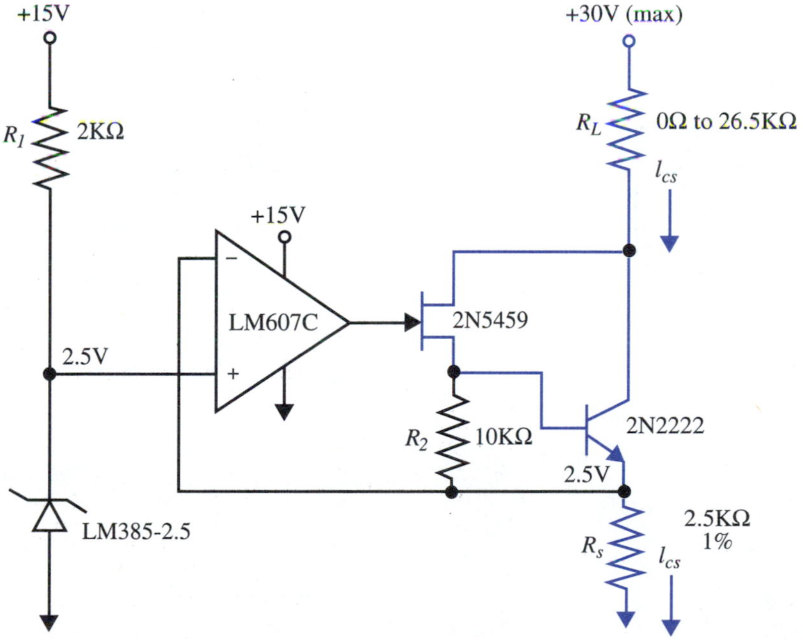

FIGURE 11–20 Precision one-milliamp current sink

The current source and sink circuits described have an output current capability limited only by the power dissipation of the 2N2222 output transistors. At 25°C, this transistor can dissipate 0.8W. For example, if we want an output current of 50mA, then the maximum voltage drop across the transistor can be no greater than 0.8W/50mA, or 16V.

When the required current is less than 1mA, the circuits can be simplified by deletion of the output transistors, as shown in Figure 11–21. Since the voltage at the input terminals of a closed-loop op-amp (terminals 2 and 3 in Figure 11–21) must be the same, the voltage drops across R_1 and the voltage reference are identical ($V_{Ref} = V_{R_1}$). We can obtain the required current I_{cs} by choosing R_1 to satisfy the relationship

$$R_1 = \frac{V_{ref}}{I_{cs}} \qquad \textbf{(equation 11–24)}$$

Notice that in the circuits of Figure 11–21, the voltage across the load resistor R_L appears at the noninverting input to the follower. As the load voltage increases, there is less voltage drop across the 10KΩ resistor, and therefore there is a current change ΔI through the reference.

This current change causes a slight voltage change (ΔV) across the reference of

$$\Delta V = \Delta I \times r_z$$

where r_z is the dynamic (zener) resistance of the reference.

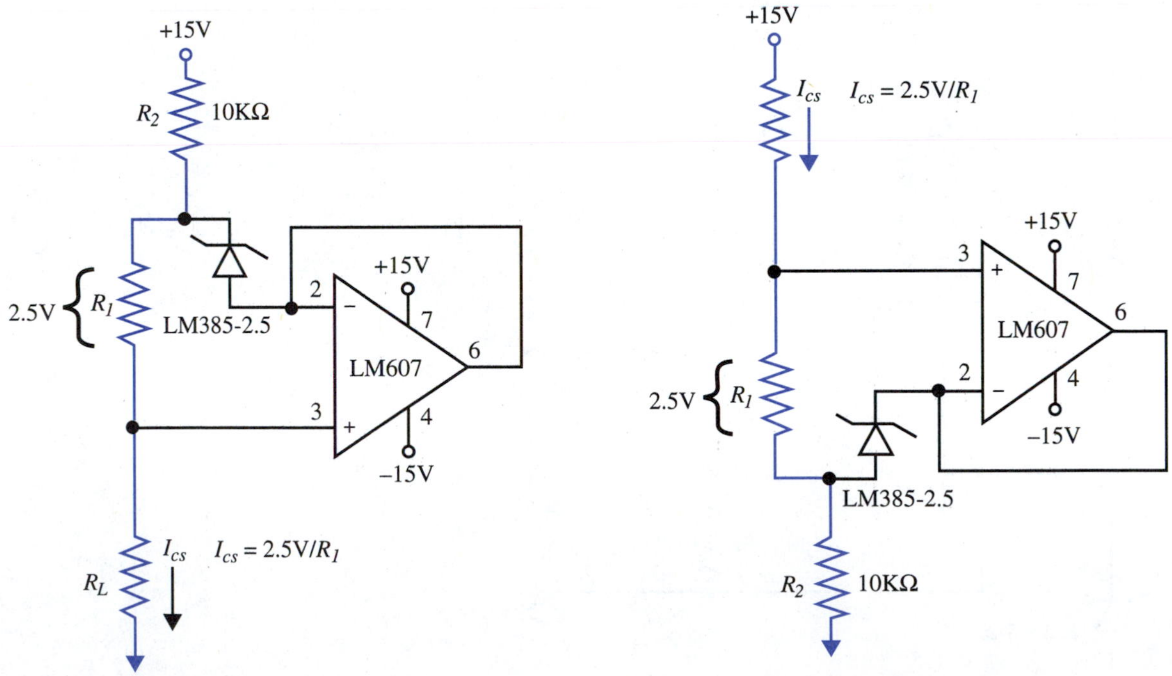

a) Current source circuit b) Current sink circuit

FIGURE 11–21 Low current source and sink circuit

EXAMPLE

Find the required value of R_1 for the circuit of Figure 11–21 if the desired source current is $80\mu A$.

SOLUTION

Using equation 11–24,

$$R_1 = \frac{V_{ref}}{I_{cs}}$$

$$R_1 = \frac{2.5V}{80\mu A}$$

$$R_1 = 31.25K\Omega$$

EXAMPLE

Determine the change in voltage across the LM385–2.5 in Figure 11–21 if the value of R_1 is $31.25K\Omega$ (I_{cs} is $80\mu A$) and the load resistor R_L is increased from $10K\Omega$ to $100K\Omega$. Assume that r_z is 1Ω.

SOLUTION

We need to find the voltage change across R_2, which will enable us to determine the change in current through the voltage reference. Any increase in voltage across R_L will decrease the voltage across R_2. The change in voltage across R_2 is

$$\Delta V_{R2} = 80\mu A(100K\Omega - 10K\Omega)$$

$$\Delta V_{R2} = 7.2V$$

The current change ΔI_{R2} through R_2 is

$$\Delta I_{R2} = \frac{7.2V}{10K\Omega}$$

$$\Delta I_{R2} = 0.72mA$$

Since the current through R_1 is fixed at $80\mu A$, this change (decrease) in current must occur in the reference. So, the change in voltage across the reference is

$$\Delta V_{ref} = 0.72mA \times 1\Omega$$

$$\Delta V_{ref} = 0.72mV$$

Actually the small voltage change will also cause a slight change in the current through R_1.

TROUBLESHOOTING VOLTAGE REFERENCES

Being a two-terminal—or at most a three-terminal—device, the voltage reference is a relatively easy device to troubleshoot. With power applied, we should measure the predicted output voltage—taking into account the component tolerances. Again we need to be aware that a failure in the load circuitry for the reference could be the reason the output is too low. An output that is too high typically means a failure within the reference device.

SUMMARY

- A regulator is used between the power supply filter capacitor and the load to keep the load voltage constant with variations in the input line voltage and load current.
- All integrated circuit regulators contain the following internal elements: a reference voltage, an error amplifier, and a means for protecting against excessive current flow.
- The LM723 regulator is a low-power adjustable regulator that provides load currents up to 150mA and has an output voltage range from +2V to about +35V (conservative).
- Current limiting can protect both the regulator and the load under normal operating conditions.
- Fold-back current limiting reduces the current as the load resistance decreases to abnormally low values (possible short circuit) and protects the regulator from excessive power dissipation.
- The fold-back factor K is found by dividing the maximum output current, with full output voltage, by the current obtained with a short on the output.
- Regulators with internal thermal overload protection do not require fold-back current limiting.
- The LM317 regulator is a high-current (1.5A) regulator with an output voltage range from +2V to +35V.
- Since there is less voltage drop across an adjustable regulator when the output voltage is high, the output current can be greater for the same power dissipation.
- Allowable ripple on a filter capacitor is the difference between the low line input voltage and the minimum permissible input voltage to the regulator. The larger the allowable ripple, the smaller the capacitance value.
- A voltage reference provides a constant voltage for power supplies and calibration circuits.
- Comparing a voltage reference with a zener diode, the reference has a lower dynamic resistance and has better temperature capability.
- Initial tolerances on voltage references range from 0.05% to 5%, and dynamic resistance is typically less than 1Ω.
- A voltage reference can be used to increase the regulation of an *IC* voltage regulator.
- Temperature stability is so good for the temperature-stabilized LM199-type reference that it can replace a standard cell.

- Current sources provide a current that is independent of the value of the load resistor.
- A precision current source can be made from a voltage reference and an op-amp.
- Current flows out of a current source, and current flows into a current sink.
- The LM334 can be used as both a temperature sensor and a variable current source.

EXERCISE PROBLEMS

1. Find the power dissipation in the transistor Q_2 in Figure 11–A with a short on the regulator output.

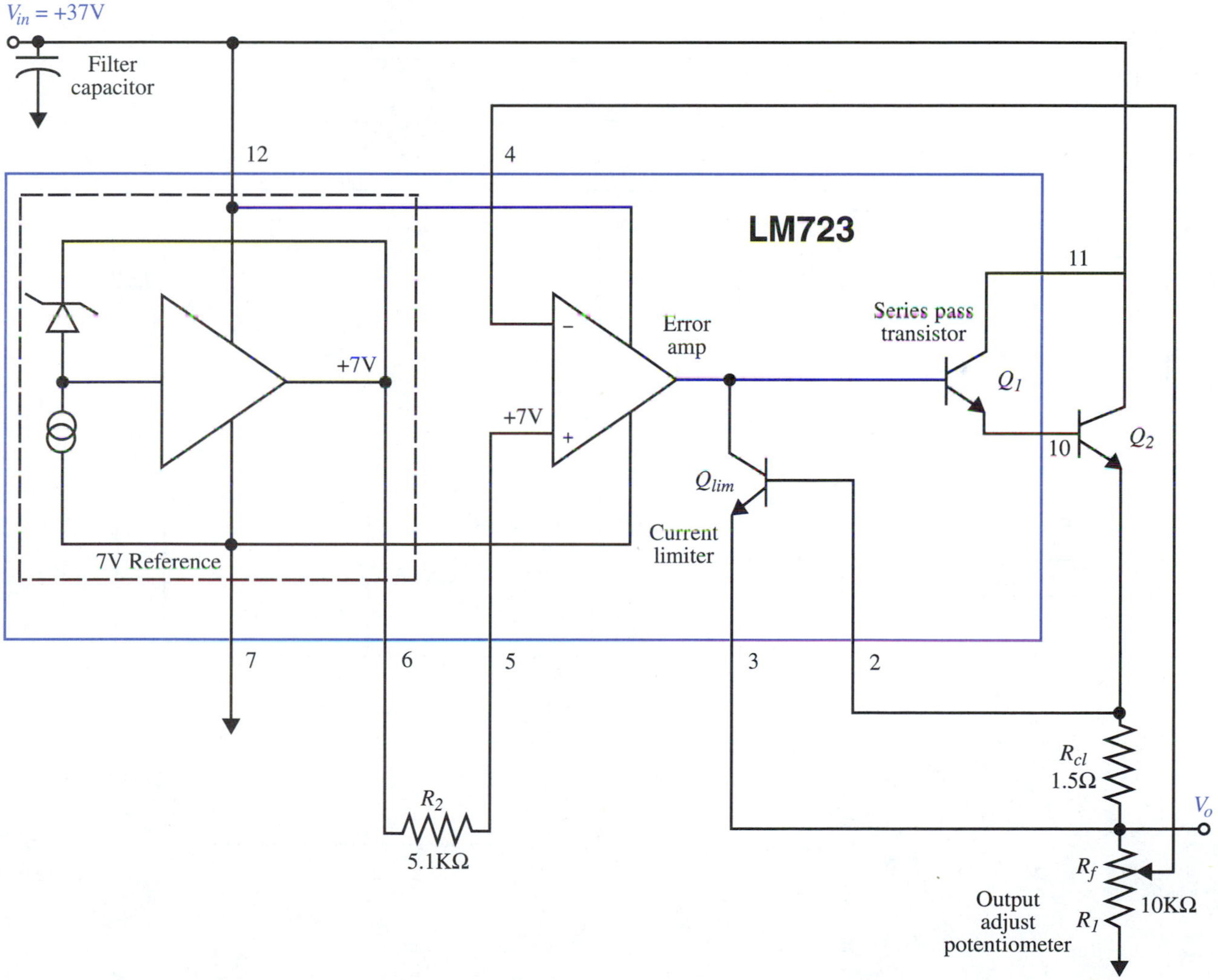

FIGURE 11–A (text figure 11–5)

2. If the output voltage of a 723 regulator is 10V, and the input voltage to the regulator is 15V at low line, find the maximum allowable ripple on the filter capacitor.
3. Design a regulated power supply using an LM723 voltage regulator (Figure 11–A without boost transistor) to provide an output of +9V up to 25mA (current limit point) with an output ripple voltage of less than 10mV peak to peak. Specify all component values and power ratings, from transformer to output. Use 12V (RMS) transformer.
4. Design a regulated power supply to provide an output of +5V at 20mA with less than 25mV peak-to-peak ripple. Use an LM723 voltage regulator (see Figure 11–B) and specify all component values from transformer to output. Use 10V (RMS) transformer.

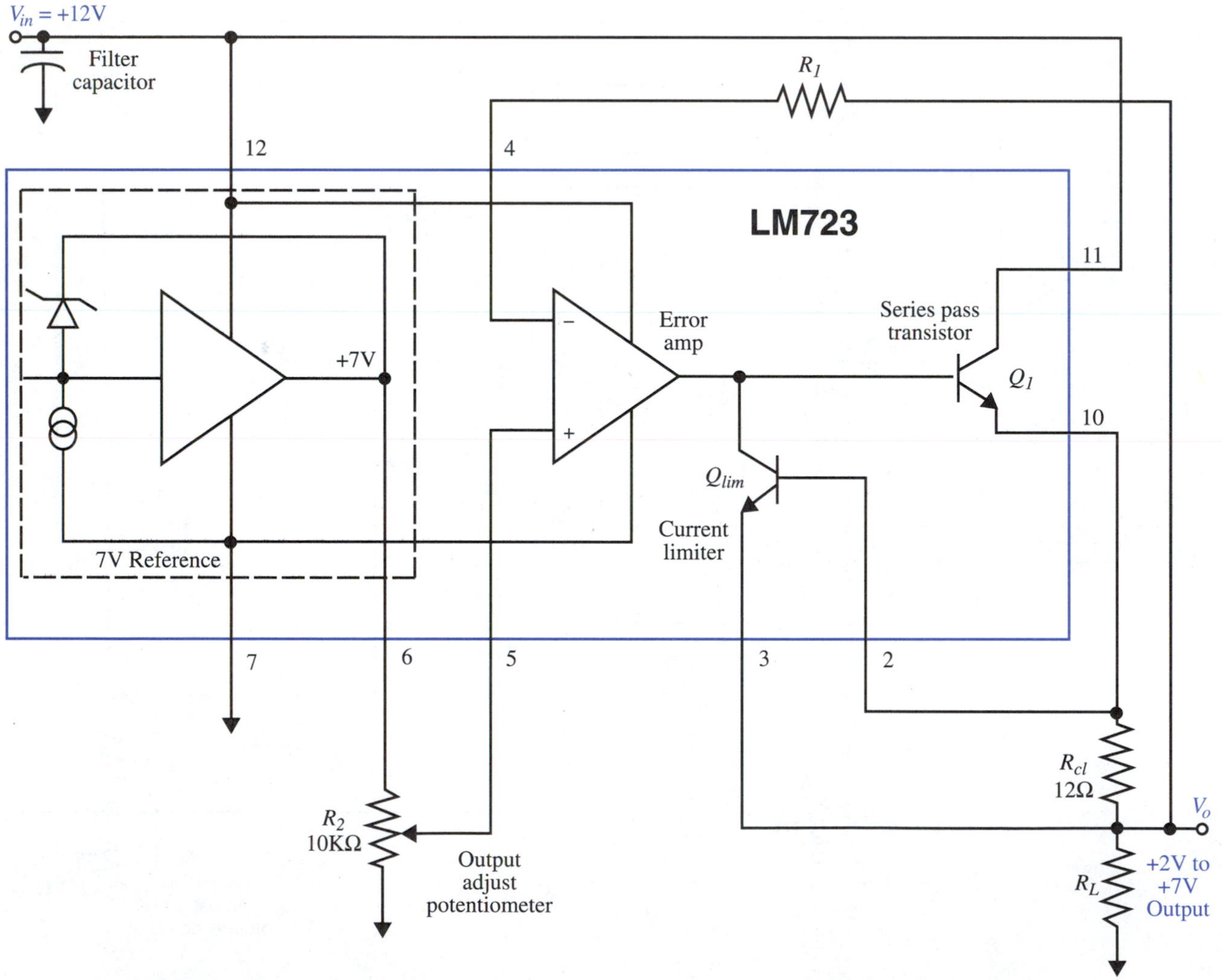

FIGURE 11–B (text figure 11–4)

5. Provide a circuit and specify parts for a regulated power supply with an output voltage variable from +7V to +25V with a maximum load current of 10mA using an LM723 regulator. Use a 24V(RMS) transformer.
6. A power supply is required with an output of +15V at 150mA. Input voltage to the regulator is +20V. Specify all component values for the regulator circuit. (Assume maximum temperature of 50°C and use an LM723 regulator.)
7. Design a fold-back current-limiting circuit for an LM723 regulator with an input voltage of +15V and output of +5V at 50mA. Select a K factor to protect the device at 50°C when the output is shorted. (Let $R_l = 1\text{K}\Omega$.)
8. Specify the output transistor and design a fold-back current-limiting circuit for a +10V, 1A power supply using an LM723. Input voltage to the LM723 regulator is +15 V. Circuit should be able to sustain a continuous short on the output at 50°C. (Let $R_l = 1\text{K}\Omega$; $\beta = 20$.)
9. Using an LM317K regulator, show the complete circuit (from transformer to output) for a variable power supply capable of providing 0.5A from 1.25V to 18V. Maximum output ripple is to be 10mV. Use 18V (RMS) transformer.
10. Explain why a low drop-out regulator is desirable.
11. Why are tracking power supplies used?
12. Compute the output voltages for the tracking power supply circuit of Figure 11–C if reducing the load causes the negative supply to initially increase by 40mV.

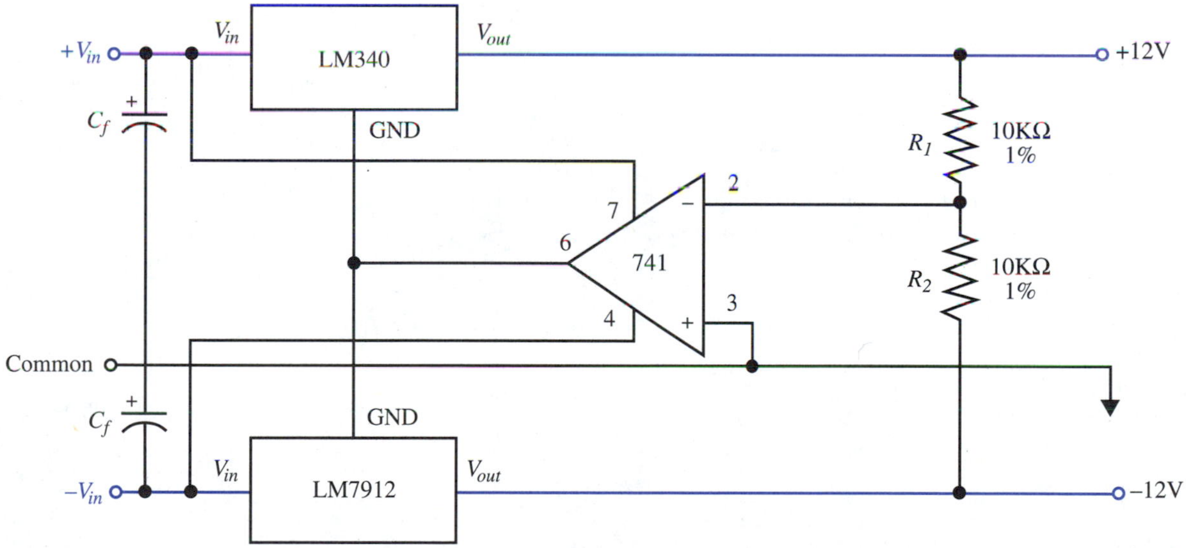

FIGURE 11–C (text figure 11–10)

13. Determine the change in voltage across an LM329 voltage reference if the current through the device changes from 2mA to 14mA. Assume a zener impedance of 1.6Ω.

14. Determine the maximum voltage change across an LM329B voltage reference caused by the temperature changing from 25°C to 65°C. Assume an initial voltage of 6.9V.
15. State two advantages of a voltage reference over a zener diode.
16. Explain how the internal circuitry of a voltage reference lowers the zener impedance.
17. An LM329 voltage reference is connected in series with a 1.6KΩ resistor. If the input voltage (V_{cc}) changes from 15V to 18V, find the voltage change ΔV across the reference.
18. Select a 6.95V reference from Table 11–2 that will not vary by more than 0.01% over a temperature range from −30°C to +60°C and has an initial tolerance of 2% or less.
19. Show a circuit that will increase the percent regulation capability of an LM317 regulator by a factor of 5. What is the minimum output voltage?
20. For the standard cell circuit of Figure 11–D, find the change in output voltage if the input supply voltage changes by 1 percent ($r_{z(max)}$ is 1Ω).

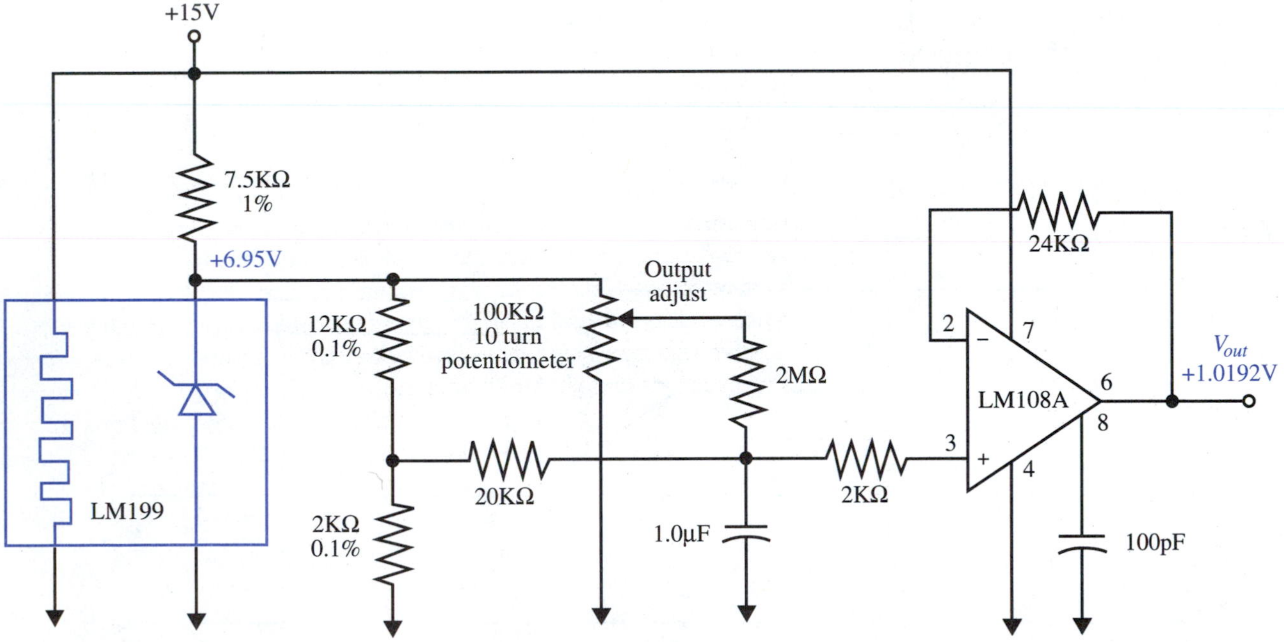

FIGURE 11–D (text figure 11–16)

21. Assume the supply voltage in the circuit of Figure 11–E is +15 V, R_1 is 100KΩ, and r_z is 1.5Ω. Find the change in the current source if the load resistor changes from 10KΩ to 200KΩ.

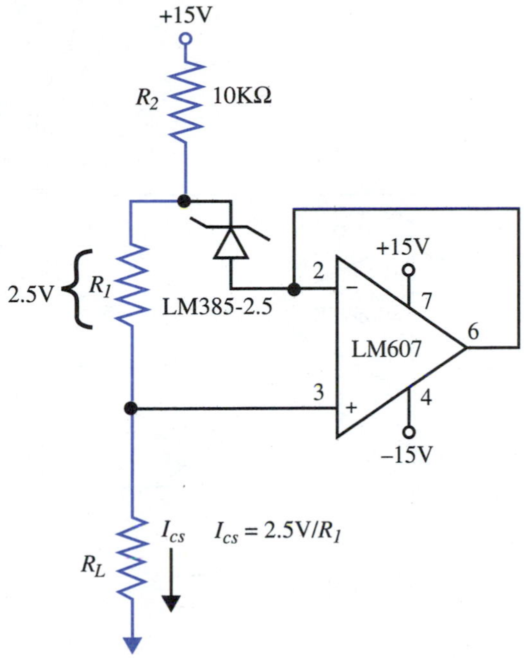

FIGURE 11–E (text figure 11–21a)

22. What is the required temperature-stability of the 100KΩ potentiometer used in the standard cell circuit in Figure 11–D?

23. What effect does the initial accuracy of the LM329 reference, used in the differential measuring circuit of Figure 11–F, have on the measurement accuracy?

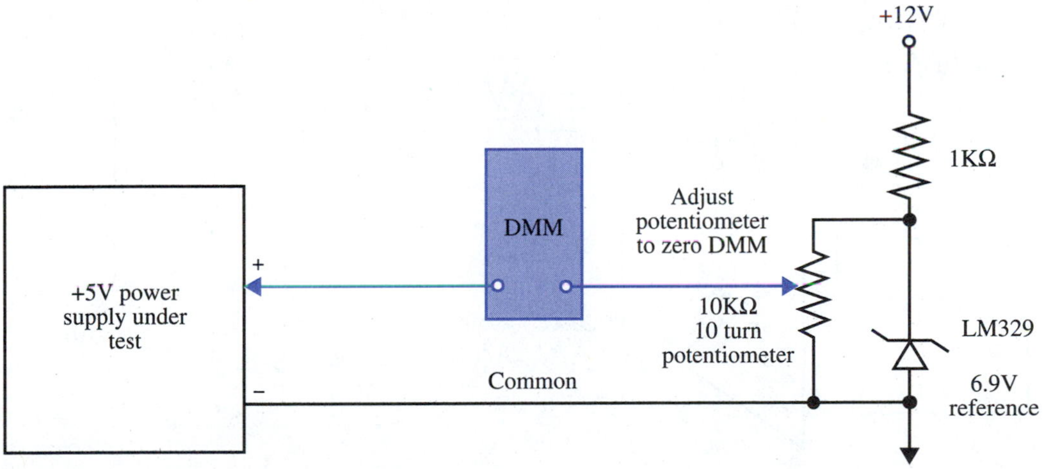

FIGURE 11–F (text figure 11–12)

24. How much error (%) does the maximum voltage offset of the LM607C create in the precision current source of Figure 11–G?

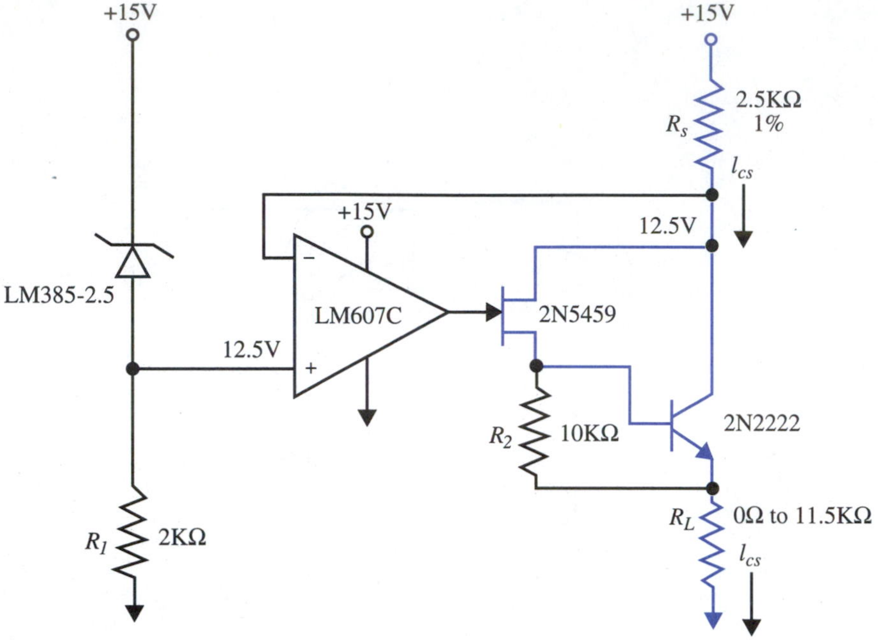

FIGURE 11–G (text figure 11–19)

25. Determine the R_{set} value to obtain an I_{set} current of 50μA at 50°C for an LM334.
26. Determine the voltage drop across the 10KΩ resistor in Figure 11–H at 20°C.

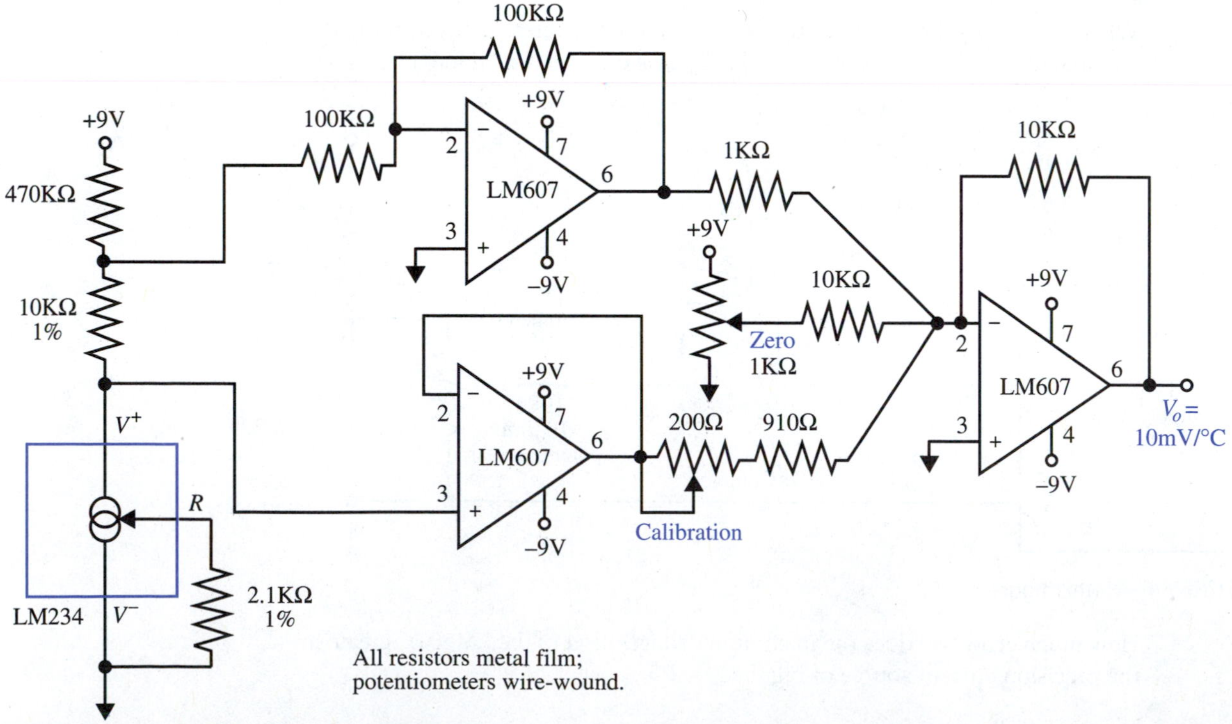

FIGURE 11–H (text figure 11–18)

ABBREVIATIONS

I_{cl} = Current when current limiting occurs
I_{cs} = Current generated by a constant source/sink circuit
I_s = Current with a short on the output
I_{set} = Current flowing into LM234
I_R = Current flowing through regulator
K = Ratio of I_{cl}/I_s
K_t = Temperature coefficient
P_d = Internal power dissipation in regulator
r_z = Zener resistance
R_{cl} = Current-sensing resistor
R_s = Surge current limiting resistor
R_{set} = Resistor that "sets" the LM234 set current
R_t = Resistance of transformer winding
ΔV_c = Allowable ripple at input to regulator
ΔV_o = Ripple at output of regulator
V_m = Minimum allowable voltage at input to regulator
V_{ref} = Reference voltage
V_c = Peak voltage on filter capacitor

ANSWERS TO PROBLEMS

1. $P_d = 14.6\text{W}$
2. $\Delta V_c = 2\text{V}$
3. 1A; 100V diode bridge; $R_s + R_x = 0.58\Omega$; $C_f = 110\mu\text{F}$ (use 150μF, 25V), $R_1 + R_f = 10\text{K}\Omega$ potentiometer; $R_{cl} = 24\Omega$; $V_r = 0.42$ mV
4. 1A; 100V diode bridge; $R_s + R_t = 0.48\Omega$; use $C_f = 100\mu\text{F}$ (25V); $R_{cl} = 30\Omega$
5. 1A; 200V (doubled for safety) diode bridge; $R_s + R_t = 1.2\Omega$; use $C_f = 100\mu\text{F}$ (50V); $R_1 + R_f = 10\text{K}\Omega$ potentiometer, $R_{cl} = 60\Omega$
6. $R_{cl} = 4\Omega$; $I_{R(max)} = 25\text{mA}$; $I_L = 21\text{mA}$; $\beta_{(min)} = 7.5$
7. $I_s = 33.3\text{mA}$; $K = 1.6$; $R_1 = 1\text{K}\Omega$; $R_2 = 12.4\text{K}\Omega$; $R_{cl} = 20.8\Omega$
8. $I_{cl} = 50\text{mA}$; $I_s = 34.2\text{mA}$; $K = 1.46$; $R_2 = 35.2\text{K}\Omega$; $R_{cl} = 18\Omega$
9. 1A; 100V diode bridge; $R_s + R_t = 0.89\Omega$; use $C_f = 2500\mu\text{F}$ (35mV); $R_1 = 120\Omega$; $R_2 = 1.675\text{K}\Omega$; use 2KΩ potentiometer
10. Less regulator power dissipation
11. To maintain correct circuit operation
12. ±12.02V
13. $\Delta V = 19.2\text{mV}$
14. $\Delta V = 13.8\text{mV}$

15. Lower zener resistance and better temperature stability
16. See text.
17. $\Delta V = 1.875\text{mV}$
18. LM199
19. See Figure 11–15. Use LM136–5 reference.
20. $\Delta V = 2.93\mu\text{V}$
21. $\Delta I_{cs} = -4.75\text{nA}$
22. Since the potentiometer is used as a voltage divider, temperature stability is not a problem.
23. None; 10KΩ potentiometer compensates.
24. Error = 0.01%
25. $R_{set} = 1.389\text{K}\Omega$
26. $V = 0.2999\text{V}$

CHAPTER 12

Switching Regulators

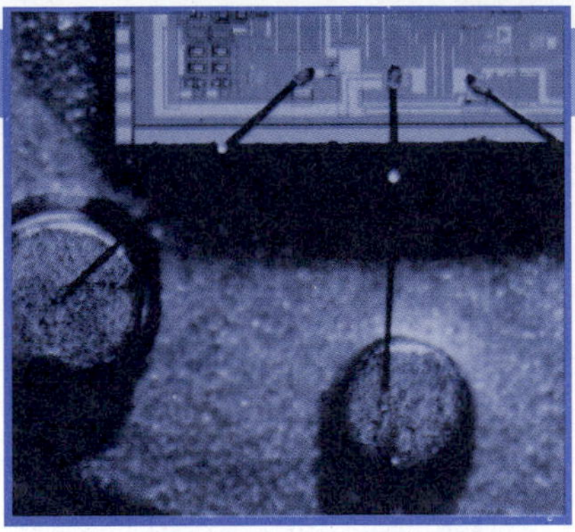

CHAPTER OBJECTIVES

After completing this chapter, you should be able to explain the following basic concepts of the switching regulator:

- Reason for using a switching regulator
- Relative efficiencies for a switching and a linear regulator
- Step-down (bucking) switching regulator operation
- Step-up (boosting) switching regulator operation
- Fly-back switching regulator operation
- Determination of regulator filter components
- Regulator stability criteria
- Switched capacitor-type voltage converter.
- Series resonant power supplies
- Advantages of SRPS

INTRODUCTION

This chapter will first describe the basic concepts of a switching regulator used in a switched mode power supply (SMPS) and explain why it is taking the place of the linear regulator in many applications. Examples of switching regulator circuits then will be covered, including the voltage step-down, voltage step-up, and voltage inverting. Finally, self-contained switching regulator chips will be examined.

WHY A SWITCHING REGULATOR IS USED

The linear regulator can be considered as a variable voltage dropping resistor in series with the load resistor. When the load current changes, or the power supply input voltage changes, this "variable resistor" automatically adjusts to keep the output voltage constant. Thinking of the regulator as a resistor makes us aware of the fact that the combination of current through and voltage across the device results in a power loss.

EXAMPLE

What is the efficiency of a linear regulator if the input voltage to the regulator is 15V and the output voltage is 5V.

SOLUTION

If we assume the same current (I_L) flows through both the regulator and the load, the input power is

$$P_{in} = 15V \times I_L$$

The load power is

$$P_L = 5V \times I_L$$

Efficiency is the ratio of these two powers.

$$\%Eff = \left(\frac{P_L}{P_{in}}\right) \times 100\%$$

$$\%Eff = \left(\frac{5V \times I_L}{15V \times I_L}\right) \times 100\%$$

$$\%Eff = 33.3\%$$

With a given output voltage, the greater the input voltage, the lower the efficiency of a linear regulator.

The switching regulator improves power supply efficiency at the expense of increased complexity. An ideal switch has no power loss associated with its operation. There is either full voltage across the switch and no current (switch open) or full current and no voltage (switch closed)—in both cases, no power ($V \times I$) loss. We shall see that the switching regulator uses this concept to increase efficiency.

Let us now look at the three basic types of switching regulators: the **step-down switching regulator, the step-up switching regulator, the fly-back switching regulator.**

STEP-DOWN SWITCHING REGULATOR

A step-down switching regulator reduces an input DC voltage to a lower DC output level in a much more efficient manner than the voltage divider approach of the linear regulator.

This method of converting one DC voltage to another is like having a DC transformer.

The switching method is compared with the linear method of voltage conversion in Figure 12–1. In Figure 12–1a, the control element is shown as a variable resistor, and the power loss in the linear regulator is $V_r \times I_L$. In Figure 12–1b, the regulator is shown as a switch with the power loss equal to zero.

The output voltage of the switching regulator is proportional to the time that the switch is closed. Without filtering, the output voltage for this circuit would be a series of pulses corresponding to the time the switch is closed and with an amplitude equal to the input voltage (V_{in}). Filtering is used to convert this signal to the required DC output.

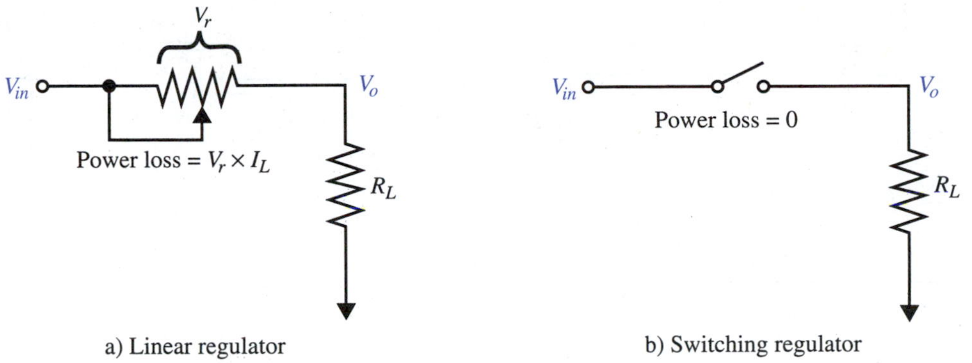

FIGURE 12–1 Comparison of linear and switching regulation

A basic step-down switching regulator, which includes an inductor and capacitor filter, along with the control element is shown in Figure 12–2.

Operation of the circuit can be described by first assuming that the transistor switch (Q_1) is closed in response to sensing that the output voltage is too low. With Q_1 closed, current will flow through L_1 building up a magnetic field in the inductor, charging the capacitor, and flowing through the load resistor (see Figure 12–3a).

If the transistor switch is now opened, L_1 tries to keep the current flowing by developing a voltage (from its collapsing magnetic field) that is positive at the load side and negative at the switch side. This voltage forward biases D_1 and completes the current path (see Figure 12–3b).

The filtering action of L_1 provides a resistance to current change and therefore a more constant current to the load; however, there is some change in current through the inductor (ΔI_L), otherwise there would be no voltage generated by the inductor's changing magnetic field.

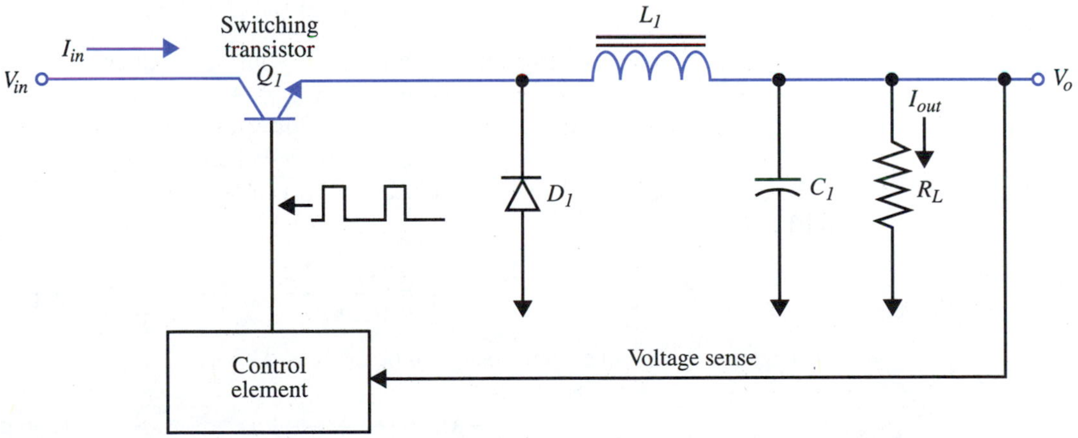

FIGURE 12–2 Step-down switching regulator

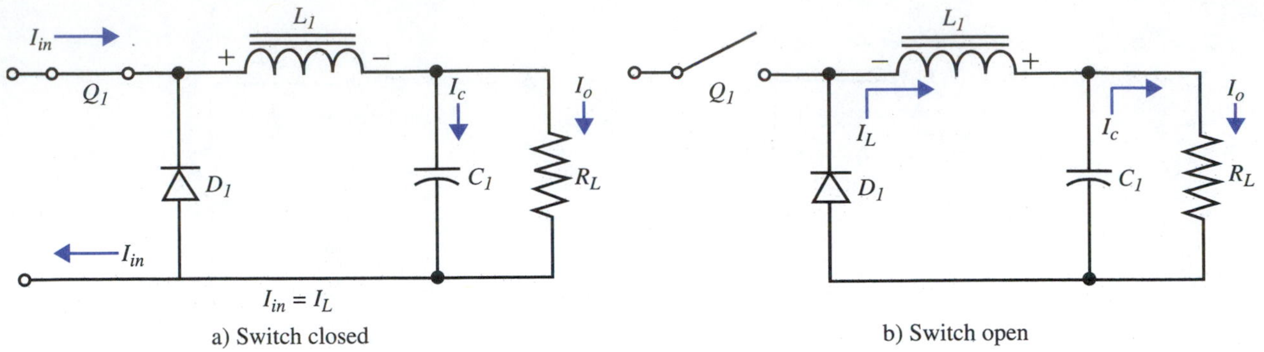

FIGURE 12–3 Current paths for step-down switching regulator

Capacitor C_1 provides current to the load when the current through the inductor drops. This maintains the current to the load very constant, which means that the ripple voltage across the load resistor will be small.

Let us now look at the equations for the step-down switching regulator, starting with the equation for the voltage generated across L_1.

$$V_L = L_1\left(\frac{\Delta I_L}{\Delta t}\right)$$

Solving for ΔI_L,

$$\Delta I_L = \left(\frac{V_L\,\Delta t}{L_1}\right)$$

This changing current in the inductor must not exceed twice the output DC load current (I_o), or current through the inductor will be interrupted and the output could become unstable.

$$\Delta I_{L(p-p)} < 2I_o$$

Figure 12–4 shows the variation of I_{L1} relative to I_o during the times when the transistor switch is closed (t_{on}) and when the switch is open (t_{off}). Also indicated in Figure 12–4 are the times when the capacitor C_1 is being charged (inductor current greater than load current) and when C_1 is discharging (inductor current less than load current).

The positive change in the inductor current $+\Delta I_L$ can be expressed as (see Figures 12–2 and 12–4)

$$+\Delta I_L = \left(\frac{V_{in} - V_o}{L_1}\right)t_{on} \qquad \textbf{(equation 12–1)}$$

The negative change in current $-\Delta I_L$ can be expressed as

$$-\Delta I_L = \left(\frac{V_o + V_{D1}}{L_1}\right)t_{off} \qquad \textbf{(equation 12–2)}$$

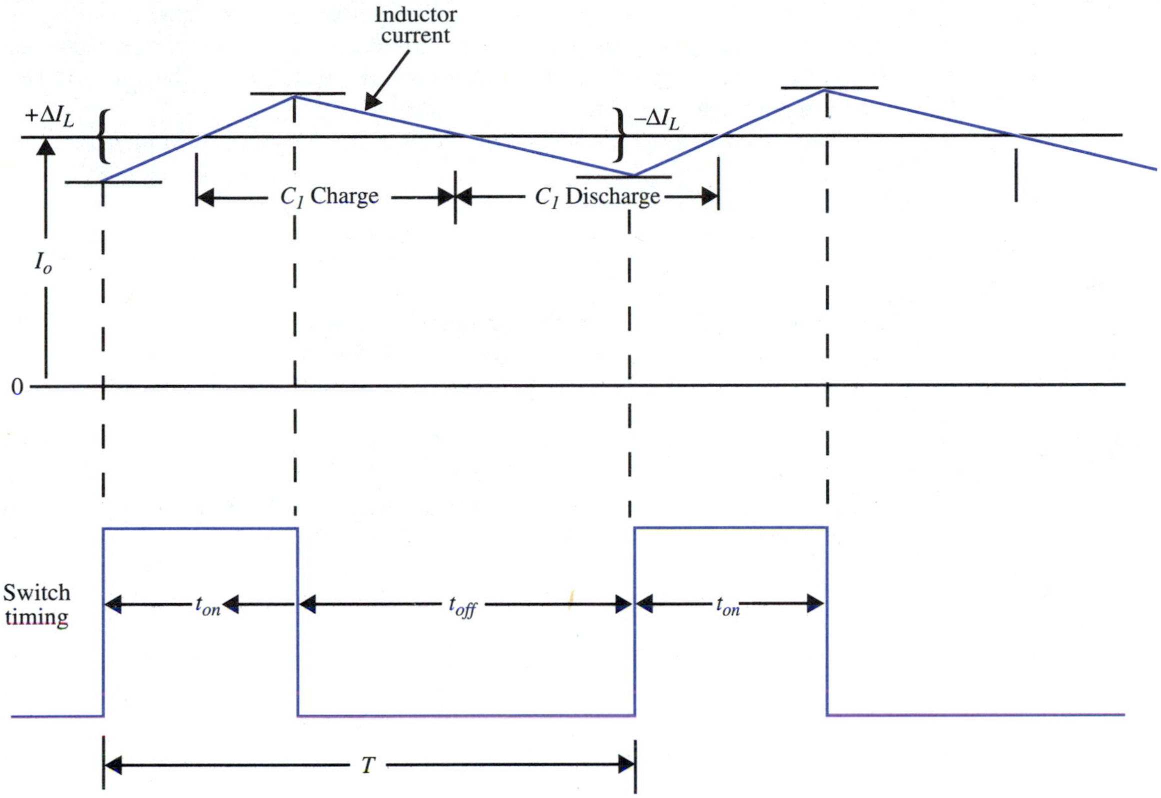

FIGURE 12–4 Switching regulator current waveforms

where V_{D1} is the forward drop across the diode when the switch is open. From Figure 12–4,

$$+\Delta I_L = -\Delta I_L$$

so

$$\left(\frac{V_{in} - V_o}{L_1}\right)t_{on} = \left(\frac{V_o + V_{D1}}{L_1}\right)t_{off}$$

Neglecting the small voltage drop across D_1 and solving for V_o,

$$V_o = V_{in}\left(\frac{t_{on}}{T}\right) \qquad \textbf{(equation 12–3)}$$

where

$$T = (t_{on} + t_{off})$$

What equation 12–3 indicates is that the input voltage is stepped down by the factor (t_{on}/T) at the output—again, much like the transformer turns ratio factor. The equation indicates, as expected, that if the switching transistor is turned on for the full period ($t_{on} = T$), the output voltage is equal to the input voltage.

As with a transformer, if the voltage is reduced, the current must increase by the same factor, or

$$I_o = I_{in}\left(\frac{T}{t_{on}}\right) \qquad \textbf{(equation 12–4)}$$

If we assume a 1V drop across the on transistor switch Q_1 and a 1V drop across the diode D_1, the ratio of power out (P_o) over power in (P_{in}), or the efficiency, is

$$\text{Eff} = \frac{V_o}{V_o + 1} \qquad \textbf{(equation 12–5)}$$

Equation 12–5 indicates that the higher the output voltage, the higher the step-down regulator efficiency.

EXAMPLE

Find the output voltage and efficiency for a step-down switching regulator if the period of the switching waveform is 20μs, the switch-on time is 12μs, and the input voltage is 15V.

SOLUTION

We can use equation 12–3 to find the output voltage.

$$V_o = V_{in}\left(\frac{t_{on}}{T}\right)$$

$$V_o = 15\text{V}\left(\frac{12\mu\text{s}}{20\mu\text{s}}\right)$$

$$V_o = 9\text{V}$$

Using equation 12–5 to find the efficiency,

$$\text{Eff} = \frac{V_o}{V_o + 1}$$

$$\text{Eff} = \frac{9\text{V}}{(9\text{V} + 1)}$$

$$\text{Eff} = 0.9 \text{ or } 90\%$$

Efficiency can be reduced if we have switching losses in the switching transistor Q_1 (see Chapter 5).

DETERMINING SWITCHING REGULATOR VALUES FOR L_1 AND C_1

Remember that the purpose of L_1 and C_1 is to filter out the variations in current that occur because of the use of switching to control the amplitude of the output voltage.

The equation for determining the value of the inductor L_1 can be derived from equations 12–1 and 12–3 by assuming that ΔI_{L1} is equal to $K \times I_o$ (usually the current change factor K is 0.3), and the frequency f (switching oscillator) is $1/T$. With these assumptions,

$$L_1 = \frac{V_o(V_{in} - V_o)}{KI_oV_{in}f} \qquad \textbf{(equation 12–6)}$$

To determine the equation for the capacitor C_1, we need to recognize from Figure 12–4 that the peak current flowing out of the capacitor is equal to $\Delta I_{L1}/2$ during t_{off} (capacitor supplies load current when inductive current drops), and this current flows through the **equivalent series resistance** (ESR) within the capacitor C_1. This resistance ranges from about 0.03Ω to 0.3Ω for electrolytic capacitors (usually specified at 120Hz but is lower at the higher switching frequencies), and the current flowing out of the capacitor develops a voltage drop across this resistance, which increases the output ripple voltage.

Using these facts, we are able to provide the following equation for the minimum value of C_1:

$$C_{1(min)} = \frac{KI_o}{2f[\Delta V_o - (2KI_o \times \text{ESR})]} \qquad \textbf{(equation 12–7)}$$

where ΔV_o is the desired ripple across the load resistor.

EXAMPLE

Determine the value of L_1 and C_1 if the maximum input voltage $V_{in(max)}$ to a step-down switching regulator is 13V, the output voltage is 8V, the load current is 2A, the desired maximum output ripple voltage is 120mV, the ESR is 0.09Ω, the ratio of inductor current to load current K is 0.3, and the switching frequency is 52KHz.

SOLUTION

First using equation 12–6 to find L_1,

$$L_1 = \frac{V_o(V_{in} - V_o)}{KI_oV_{in}f}$$

$$L_1 = \frac{8\text{V}(13\text{V} - 8\text{V})}{0.3 \times 2\text{A} \times 13\text{V} \times 52\text{KHz}}$$

$$L_1 = 98.6\mu\text{H}$$

(Use 100μH.)

Now using equation 12–7 to find C_1,

$$C_{1(min)} = \frac{KI_o}{2f[\Delta V_o - (2KI_o \times ESR)]}$$

$$C_1 = \frac{0.3 \times 2A}{2(52KHz)[0.12V - (2 \times 0.3 \times 2A \times 0.09\Omega)]}$$

$$C_1 = 481\mu F$$

(Use 500μF.)

THE STEP-DOWN SWITCHING CONTROLLER

Up to this point we have assumed that the switching transistor Q_1 was turned on and off without considering how this was accomplished. In Figure 12–2, a control element is shown with an output line that drives the base of the switching transistor and an input line that senses the load voltage. The purpose of the controller is to control the transistor on-time so that the load voltage is kept constant, i.e., if the load voltage drops, the on-time is increased. This is called **pulse width modulation**.

A typical control device for a step-down switching regulator is the National Semiconductor LM2576 "Simple Switcher." This device can control load currents up to 3A, accommodate input voltages from 4V to 40V, and provide output voltages from 1.23V to 37V. An internal switching oscillator runs at a fixed frequency of 52KHz, giving a period T of about 20μs. A control pin is provided that allows remote turn-on of the regulator. Figure 12–5 is a functional block diagram of this device.

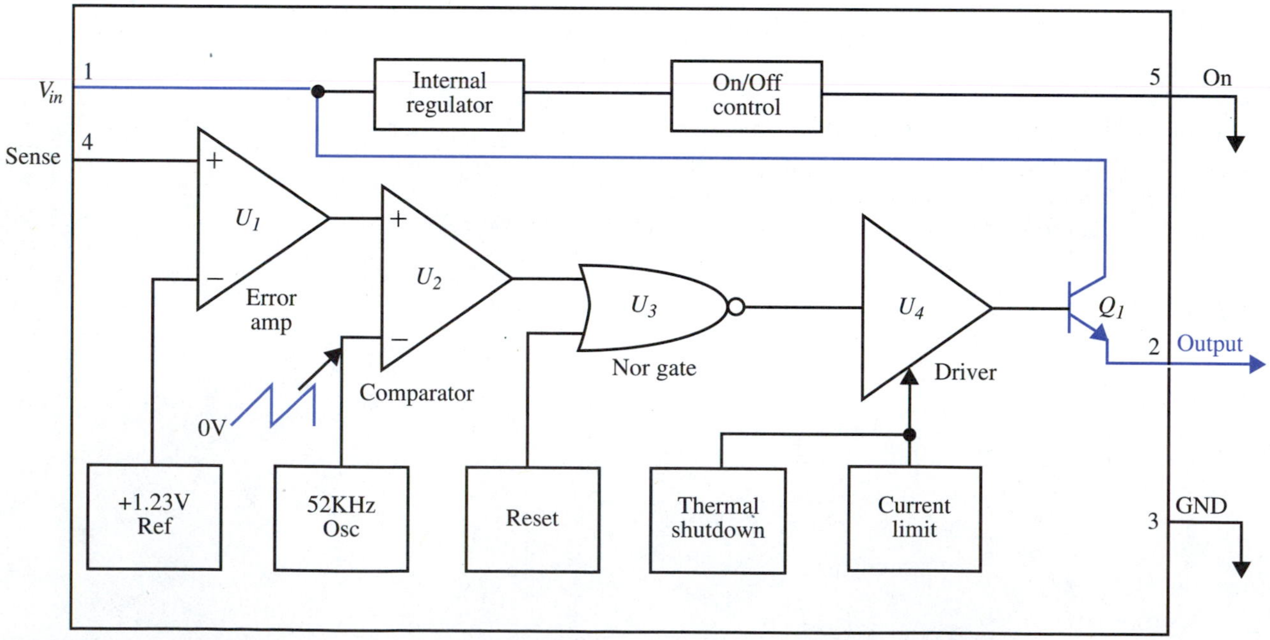

FIGURE 12–5 Block diagram of the LM2576 switching regulator. *Courtesy National Semiconductor.*

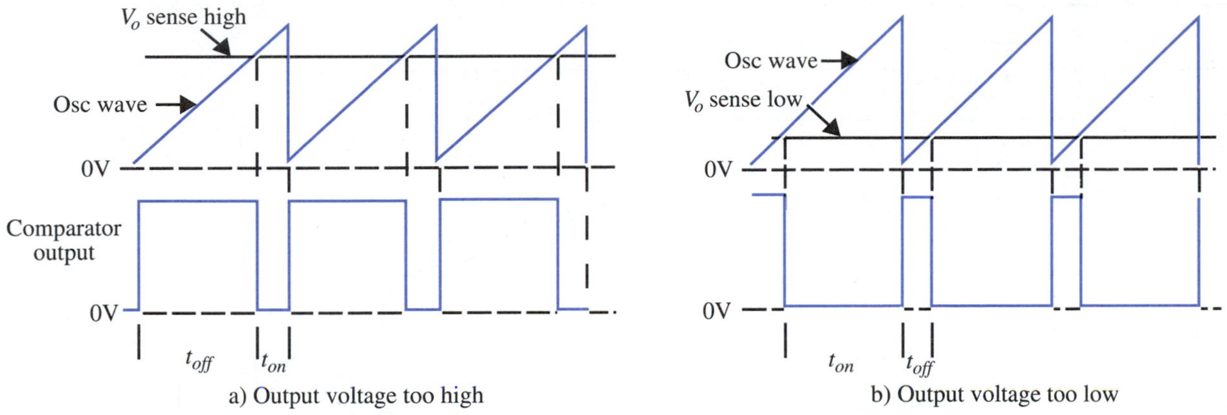

FIGURE 12–6 LM2576 internal comparator waveforms

We can begin our description of the operation of the LM2576 by assuming the condition that the divided down output voltage, provided to the sense input (Pin 4) of the chip, now indicates the output voltage is too high.

With a high input (>+1.23V), the noninverting input to the op-amp will be greater than the inverting input. This means the voltage output of the error amp will be more positive. With this positive input to the noninverting side of the comparator, and the oscillator sawtooth waveform on the inverting input, the comparator output will spend more time in the high state.

With the input to the nor gate more often high, the gate output will spend more time low; which means the on time (t_{on}) of Q_1 will be reduced. Reduction of t_{on} will cause less current to be provided to the load, and the desired reduction in output voltage will occur.

The operation of the comparator can be better understood by viewing the waveforms in Figure 12–6.

In Figure 12–6a, the output voltage is too high, and the comparator output spends more time in the high state—which means a shorter t_{on}. Figure 12–6b shows the case with the output voltage too low, and the comparator spends less time in the high state—which means a longer t_{on}.

AN ACTUAL STEP-DOWN SWITCHING REGULATOR CIRCUIT

We shall now look at a complete step-down switching regulator circuit, using an LM2576 as shown in Figure 12–7.

The unregulated DC input is connected to Pin 1 of the chip, and Pin 4 senses the divided down output voltage. Selection of the resistor values for the $R_f : R_1$ divider can be obtained by thinking of the regulator as a noninverting amplifier with +1.23V on the noninverting input and the divided down output at the inverting input. We pick R_f and R_1, then, for the desired DC output voltage from the power supply using the equation for a noninverting amplifier.

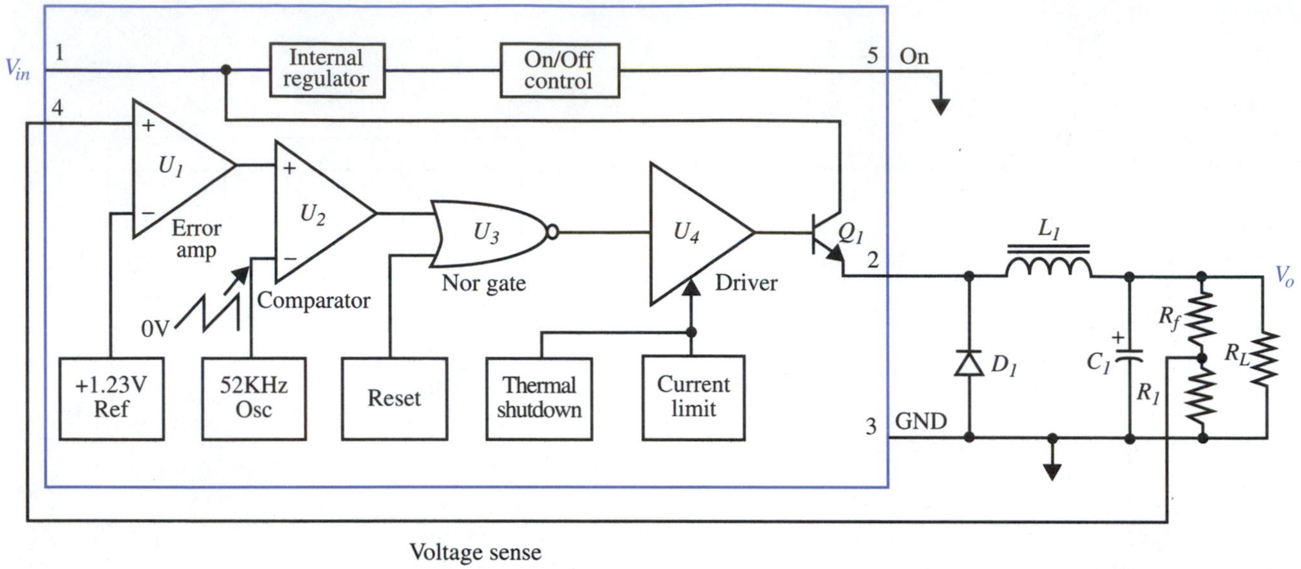

FIGURE 12–7 Complete step-down switching regulator

$$V_o = 1.23V\left(\frac{R_f}{R_1} + 1\right) \qquad \textbf{(equation 12–8)}$$

EXAMPLE

Find the value of R_f for the circuit of Figure 12–7 if the desired output voltage is 8V and R_1 is 1KΩ.

SOLUTION

Rearranging equation 12–8 to solve for R_f,

$$R_f = R_1\frac{(V_o - 1.23V)}{1.23V}$$

$$R_f = 1K\Omega\frac{(8V - 1.23V)}{1.23V}$$

$$R_f = 5.5K\Omega$$

Notice that the ON Pin 5 of the LM2576 is tied to ground. This 0V input (or <0.8V) enables the regulator to operate. Disconnecting an input from this pin or taking it "high" (>2.4V), will cause the regulator (and the power supply) to shut down.

SWITCHING REGULATOR DESIGN COMPUTER PROGRAM

National Semiconductor provides a "Simple Switcher" computer program that allows design of a power supply regulator using the LM2576. Based on input specs for the regulator, the program selects components considering circuit stability and temperature. Table 12–1 shows the readout from the program after the specs (left column) are entered.

Notice that the input specs for the table are the same as those used in a previous example; however, the component values obtained are somewhat different.

Using equation 12–6, we got an L_1 inductance of 98.6μH rather than the 99μH, from the computer program

The value of C_1 is quite different, with 481μF obtained in the previous example, since the computer program provided a value of 330μF—but the ESP was not specified. The manufacturer provides an additional equation for the minimum value of C_1 required to ensure frequency stability of the circuit.

$$C_1 > 13{,}300 \frac{V_{in(max)}}{V_o \times L(\mu H)} (\mu F) \qquad \textbf{(equation 12–9)}$$

Using this equation, we obtain 216μF, which is lower than the two previous values, but ripple considerations require using the higher value.

The values of R_f and R_1 are both twice the value we obtained in a previous example, so we get the same output voltage of 8V using equation 12–8.

C_{in} is specified at 100μF and was not considered in the previous examples because it is required by the LM2576 for stability. This capacitor should be mounted as close as possible to the terminals of the LM2576.

The 5A diode selected for D_1 is a Schottky-type for low voltage drop (less power loss) and has a 2.5 multiplier over the maximum current of 2A to accommodate the inductive current surges.

After the components are selected, the program asks "Is there a need for a temperature analysis?" If the answer is yes, the regulator package choice is requested, and data are provided on the required heat sink along with the temperature of the regulator junction (refer to Chapter 19).

TABLE 12–1 Computer Based Switching Regulator Design

Input Specs	*Limit Values*	*Component Values*
$V_{in(min)}$ = 12–00V	L > 99.0μH	L = 100μH
$V_{in(max)}$ = 13.00V	Mode = Continuous	
$T_{a(max)}$ = 50.00°C	C_1 > 330μF	C_1 = 330μF
$T_{a(min)}$ = −20.00°C	$ESR_{(max)}$ < 96.85mΩ	
V_{out} = 8.00V	$ESR_{(min)}$ > 48.42mΩ	
$I_{L(max)}$ = 2.00A	R_1 = 2.00KΩ	R_1 = 2.00KΩ
Diode = Schottky	R_f = 10.96KΩ	R_f = 11.00KΩ
	C_{in} > 100μF	C_{in} = 100μF
	ΔV_o = 0.12V	D_1 = 5.00A

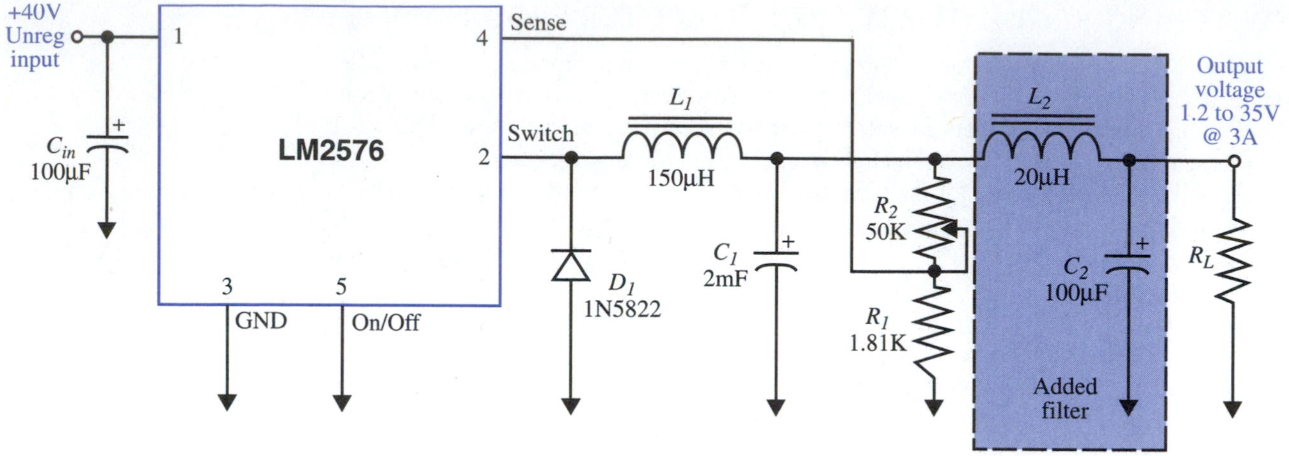

FIGURE 12–8 Variable step-down with added ripple filter. *Courtesy National Semiconductor.*

Finally, the program provides a regulator schematic and parts list.

FURTHER REDUCTION OF SWITCHING REGULATOR OUTPUT RIPPLE

Output ripple voltage can be further reduced by adding an *L* section filter to the output.

In Figure 12–8, this added filter is shown attached to the output of a variable power supply (circuit provided by National Semiconductor).

The computed ripple reduction is better than 10:1 with this added filter, and the predicted ripple, at the output of the circuit shown, is about 5mV.

STEP-UP SWITCHING REGULATOR

Sometimes a low-voltage, high-current, DC power supply is used to provide a higher DC voltage, lower-current supply. For example, a +12V power source could be required and a high-current +5V source is available. A step-up switching regulator can provide the required voltage conversion.

The fundamental circuit for a step-up switching regulator is given in Figure 12–9.

The step-up regulator differs from the step-down regulator in that the inductor is connected to ground when Q_1 is turned on, rather than being directly in series with the load. This allows energy to be stored in the magnetic field of the inductor.

When Q_1 is turned off, the decrease of current through the inductor causes the lower end of the inductor to swing to a high positive voltage. This causes the diode D_1 to forward bias, which provides current to the load and to also charge C_1.

Figure 12–10 shows the current paths with the transistor (Q_1) on and with it off.

With the transistor turned on, the current flows through the inductor L_1 and through the transistor to ground. When the transistor turns off, the left side of D_1 goes positive, and the diode is forward biased. Current now flows into the load and into C_1.

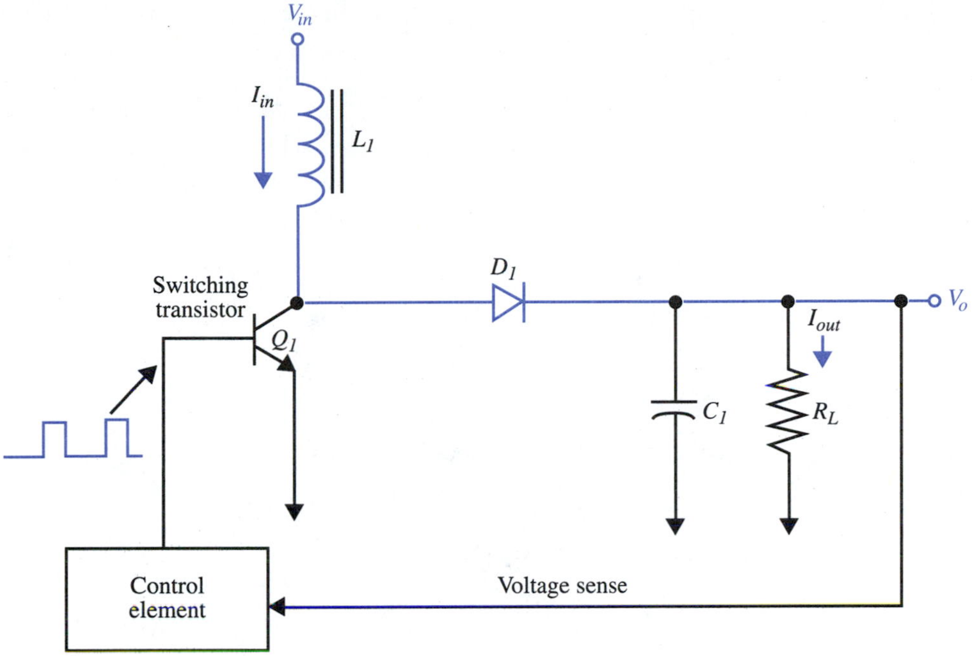

FIGURE 12–9 Step-up switching regulator

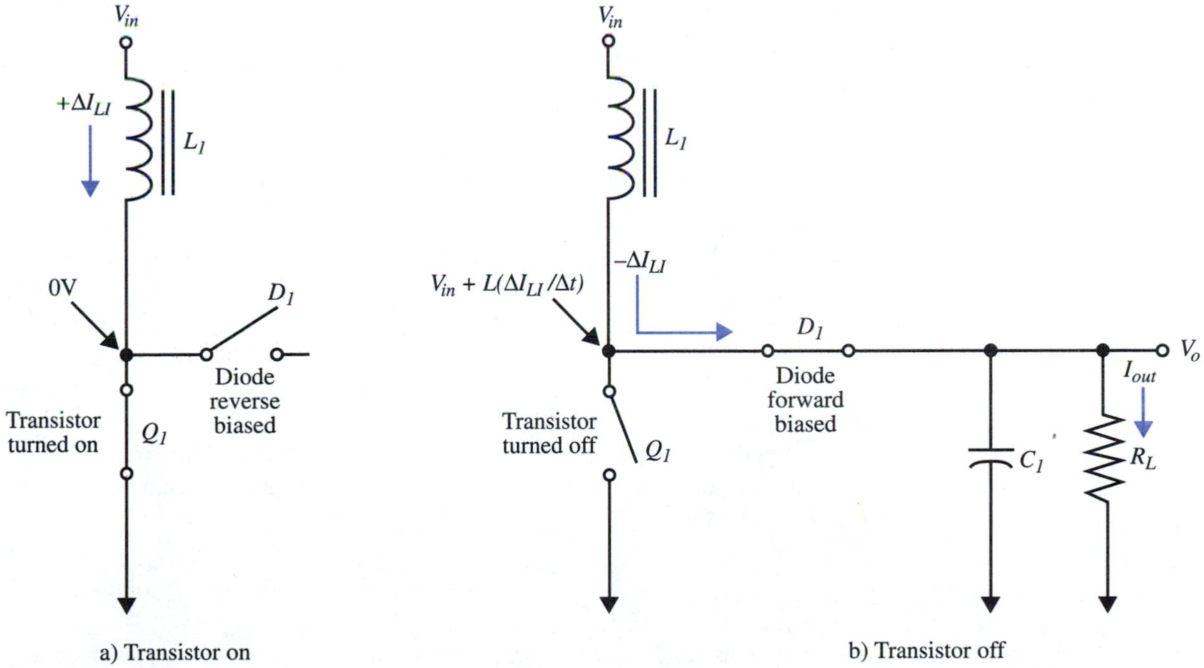

FIGURE 12–10 Current paths for step-up regulator

Notice that the driving voltage at the left side of the diode, when the transistor turns off, is

$$V_{in} + L\frac{\Delta i}{\Delta t}$$

We will now develop the equations for the step-up regulator.
With the switching transistor on, the change in inductor current is

$$+\Delta I_{L1} = \frac{V_{in}t_{on}}{L_1} \qquad \textbf{(equation 12–10)}$$

With the switching transistor off, the change in inductor current is

$$-\Delta I_{L1} = \frac{(V_o - V_{in})t_{off}}{L_1} \qquad \textbf{(equation 12–11)}$$

Since the inductor current doesn't go to zero,

$$+\Delta I_{L1} = -\Delta I_{L1}$$

Substituting equations 12–10 and 12–11,

$$\frac{V_{in}T_{on}}{L_1} = \frac{(V_o - V_{in})t_{off}}{L_1}$$

Solving for V_o,

$$V_o = V_{in}\left(\frac{T}{t_{off}}\right) \qquad \textbf{(equation 12–12)}$$

This equation shows that the output voltage is always equal to or greater than the input voltage because

$$\left(\frac{T}{t_{off}}\right) \geq 1$$

Neglecting losses in Q_1 and D_1, and setting power out equal to power in, we can solve for the output current.

$$I_o = I_{in}\left(\frac{t_{off}}{T}\right) \qquad \textbf{(equation 12–13)}$$

Again we see the transformer effect—the output voltage is higher than the input voltage, and the output current is less than the input current.

The efficiency for the step-up power supply is the ratio of actual power out to power in. As before, if we assume a 1V drop across Q_1 and D_1, we arrive at the following equation for step-up regulator efficiency:

$$\text{Eff} = \frac{V_{in}}{(V_{in} + 1)} \qquad \textbf{(equation 12–14)}$$

This equation assumes only DC losses. Efficiency can be reduced by switching losses in Q_1 and D_1 (see Chapter 5).

EXAMPLE

Find the output voltage and efficiency for a step-up switching regulator if the input voltage is +5V, the oscillator frequency is 50KHz, and t_{on} is 12μs.

SOLUTION

First we need to find the oscillator period T.

$$T = 1/f$$
$$T = 1/50\text{KHz}$$
$$T = 20\mu\text{s}$$

Solving for t_{off},

$$t_{off} = T - t_{on}$$
$$t_{off} = 20\mu\text{s} - 12\mu\text{s}$$
$$t_{off} = 8\mu\text{s}$$

Now we can solve for V_o, using equation 12–12,

$$V_o = V_{in}\left(\frac{T}{t_{off}}\right)$$
$$V_o = 5\text{V}\left(\frac{20\mu\text{s}}{8\mu\text{s}}\right)$$
$$V_o = 12.5\text{V}$$

For efficiency, we use equation 12–14.

$$\text{Eff} = \frac{V_{in}}{(V_{in} + 1)}$$
$$\text{Eff} = \frac{5\text{V}}{(5\text{V} + 1)}$$
$$\text{Eff} = 0.833 \quad (83.3\%)$$

DETERMINATION OF C_1 AND L_1

Only the output capacitor C_1 provides current to the load during t_{on}. The voltage change on C_1 during this time will be the ripple voltage ΔV_o.

$$\Delta V_o = \frac{I_o t_{on}}{C_1}$$

Solving for C_1,

$$C_1 = \frac{I_o t_{on}}{\Delta V_o}$$

Using equation 12–12 and $t_{on} = (T - t_{off})$ to substitute for t_{on},

$$C_{1(min)} = \frac{I_o(V_o - V_{in})}{f[\Delta V_o - (I_o \times ESR)]V_o} \qquad \textbf{(equation 12–15)}$$

EXAMPLE

Find the minimum required value for C_1 if a step-up regulator has an input voltage of 5V, an output voltage of 15V, a load current of 1A, a switching frequency of 50KHz, and the maximum allowable ripple of 100mV (assume C_1 has ESR of 60mΩ).

SOLUTION

Using equation 12–15,

$$C_{1(min)} = \frac{I_o(V_o - V_{in})}{f[\Delta V_o - (I_o \times ESR)]V_o}$$

$$C_{1(min)} = \frac{1A(15V - 5V)}{50KHz[100mV - (1A \times 60m\Omega)]15V}$$

$$C_{1(min)} = 333\mu F$$

In deriving the equation for the inductor L_1 used with the step-up regulator, we recognize that the full input voltage is across the inductor during the time t_{on}.

$$L_1 = \frac{V_{in}t_{on}}{\Delta I_{L1}}$$

Using the relationships

$$\Delta I_{L1} = KI_{in}$$

$$I_{in} = I_o \times \frac{V_o}{V_{in}}$$

$$t_{on} = T\left(1 - \frac{V_{in}}{V_o}\right)$$

$$L_1 = \frac{(V_{in})^2(V_o - V_{in})}{KI_o f(V_o)^2} \qquad \textbf{(equation 12–16)}$$

EXAMPLE

Find the required value of L_1 for a step-up switching regulator if the input voltage is 5V, the output voltage is 15V, the load current is 1A, switching frequency is 50KHz, and the ratio of inductor current swing to input current K is 0.3.

SOLUTION

Using equation 12–16,

$$L_1 = \frac{(V_{in})^2(V_o - V_{in})}{KI_o f(V_o)^2}$$

$$L_1 = \frac{(5V)^2(15V - 5V)}{0.3 \times 1A \times 50KHz(15V)^2}$$

$$L_1 = 74\mu H$$

THE STEP-UP SWITCHING CONTROLLER

An available step-up regulator control circuit is National Semiconductor's LM2577, which can provide an output variable up to +60V, with load currents up to 3A. The device is also available in fixed output voltage versions (voltage sense divider is inside the chip).

A fixed +12V output voltage device is the LM2577–12. This device can operate with an input voltage as low as 3.5V. Figure 12–11 shows the internal block diagram for the LM2577–12.

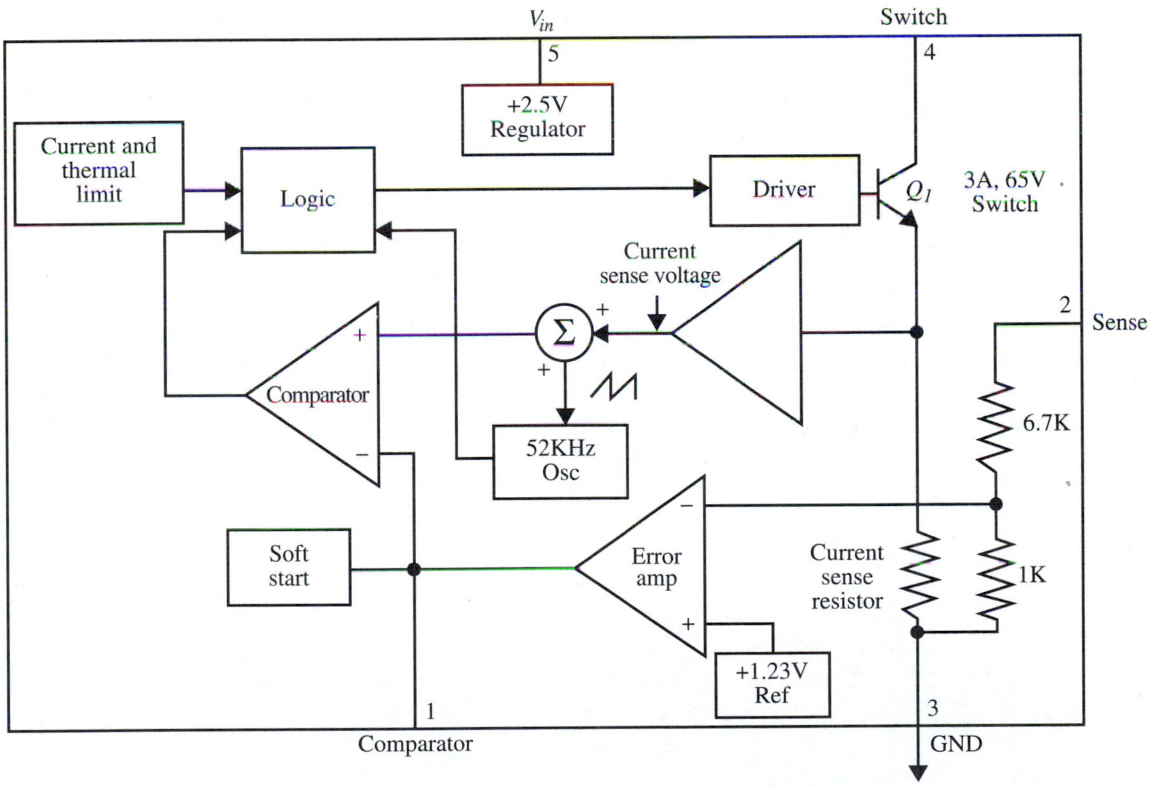

FIGURE 12–11 LM2577–12 step-up regulator circuit. *Courtesy National Semiconductor.*

This device is similar to the LM2576, but with the following differences:

1) It steps the voltage up rather than down.
2) It has a built-in soft-start circuit.
3) External frequency compensation is required.

Like the LM2576, the device maintains the output voltage by varying the time interval that the transistor switch is closed.

In operation, the sensed output voltage is divided down and compared with the reference in the error amplifier. The output from this amplifier is compared with the signal from the summing circuit (Σ) which is derived from the addition of the oscillator ramp and the voltage generated across the current sense resistor. The net result is that if the output voltage or current is too high, the width of the pulse supplied to the Q_1 switch is reduced.

The soft-start control delays the regulator turn-on (called soft-start) when power is first applied. This is done to prevent damaging current surges required to charge C_1 during start-up.

All the internal circuitry is operated from the 2.5V internal regulator—so this is why the input voltage can be as low as 3.5V.

AN ACTUAL STEP-UP SWITCHING REGULATOR CIRCUIT

Figure 12–12 provides the complete circuit for a step-up switching regulator power supply. The input voltage from the power supply filter is connected to the regulator input filter capacitor C_{in}, the inductor L_1, and the power supply pin (5) on the chip. The output voltage is sensed, via the line from the load, by Pin 2 on the chip. The frequency compensation circuit is connected from Pin 1 to ground.

We can select the value of L_1 by using equation 12–16, and C_1 by using equation 12–15.

A 0.1μF low ESR capacitor is suggested for C_{in}. If the leads from the regulator to the power supply are long, an additional 47μF capacitor in parallel with the 0.1μF is required.

Values for the compensation components R_c and C_c are computed using the following equations:

$$*R_{c(max)} \le \frac{750 \times I_{o(max)} \times (12V)^2}{V_{in(min)}^{\;2}} \qquad \textbf{(equation 12–17)}$$

$$*C_{c(min)} \ge \frac{58.5 \times (12)^2 \times C_1}{(R_c)^2 \times V_{in(min)}} \qquad \textbf{(equation 12–18)}$$

*Valid if ESR ≤ $V_{in}/115 \times I_o$, R_c **maximum value is 3KΩ.**

We can tie these equations for the step-up regulator together by working through an example of a step-up power supply.

EXAMPLE

It is desired to build a 5V to 12V step-up regulator circuit using an LM2577–12 capable of providing 1A of load current with a maximum 120mV of output ripple and

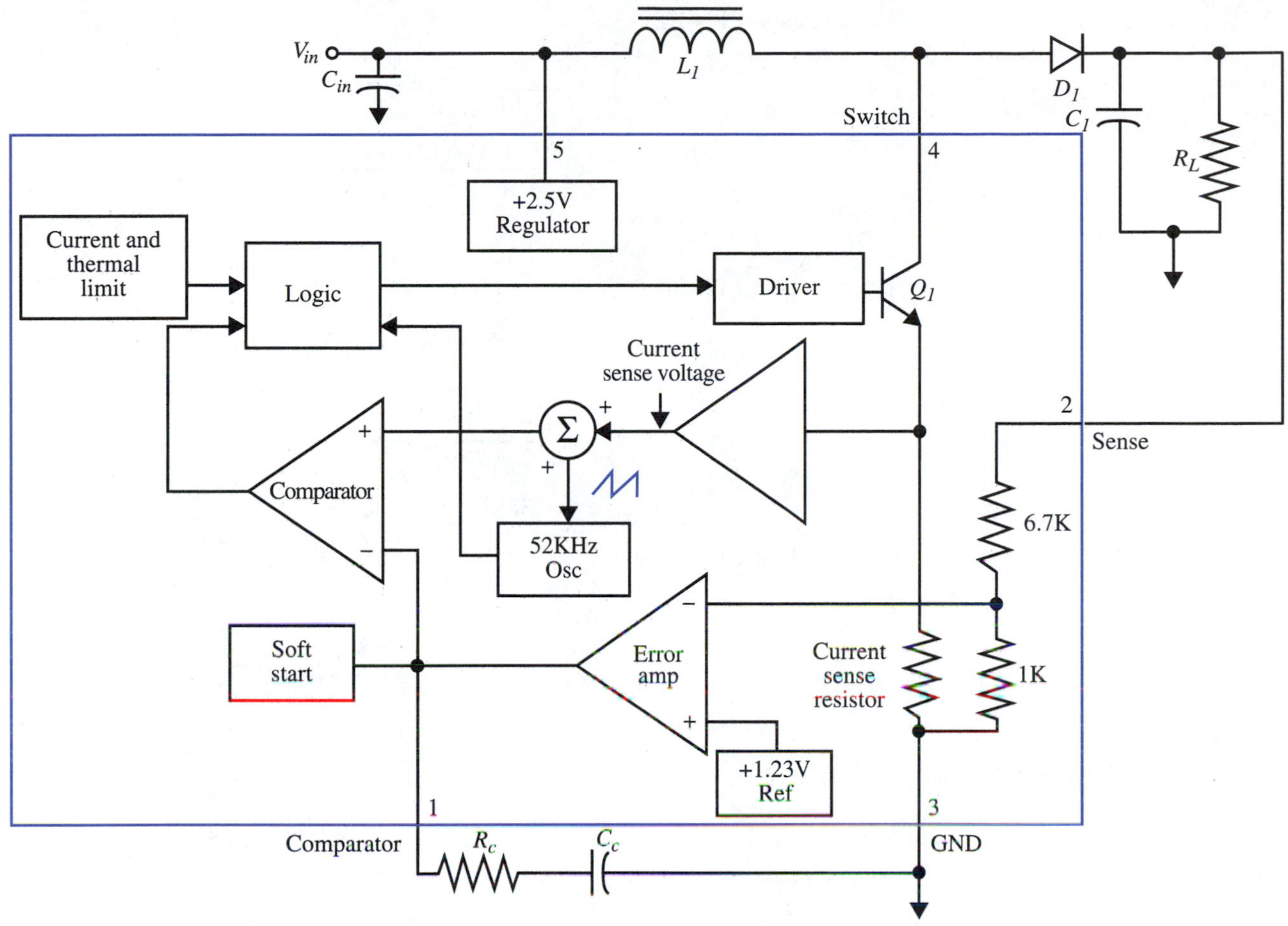

FIGURE 12–12 Complete step-up regulator circuit

with a current factor K of 0.3. Find the required values for C_1 (assume ESR is 40mΩ), L_1, R_c, and L_c.

SOLUTION

From equation 12–15,

$$C_{1(min)} = \frac{I_o(V_o - V_{in})}{f[\Delta V_o - (I_o \times \text{ESR})]V_o}$$

$$C_{1(min)} = \frac{1\text{A}(12\text{V} - 5\text{V})}{52\text{KHz}[120\text{mV} - (1\text{A} \times 40\text{m}\Omega)]12\text{V}}$$

$$C_{1(min)} = 140\mu\text{F}$$

From equation 12–16,

$$L_1 = \frac{(V_{in})^2(V_o - V_{in})}{KI_o f(V_o)^2}$$

$$L_1 = \frac{(5V)^2(12V - 5V)}{0.3 \times 1A \times 52KHz(12V)^2}$$

$$L_1 = 78\mu H$$

From equation 12–17,

$$R_{c(max)} = \frac{750 \times I_{o(max)} \times (12V)^2}{V_{in(min)}{}^2}$$

$$R_{c(max)} = \frac{750 \times 1A \times (12V)^2}{(5V)^2}$$

$$R_{c(max)} = 4.32K\Omega$$

(Use 3KΩ max.)

From equation 12–18,

$$C_{c(min)} \geq \frac{58.5 \times (12)^2 \times C_1}{(R_c)^2 \times V_{in(min)}}$$

$$C_{c(min)} \geq \frac{58.5 \times (12)^2 \times 140\mu F}{(3K\Omega)^2 \times 5V}$$

$$C_{c(min)} = 26.2nF$$

Checking the value of ESR,

$$ESR \leq \frac{V_{in}}{115 \times I_o}$$

$$ESR \leq \frac{5V}{115 \times 1A}$$

$$ESR \leq 43.5m\Omega$$

Since we specified a value of 40mΩ, the equations for C_c and R_c are valid.

FLY-BACK REGULATOR

The fly-back regulator stores energy in the primary winding of a transformer and transfers this energy to a circuit in the secondary. Unlike the previous switching regulators, the output voltage can be greater or less than the input voltage and of the same or opposite polarity.

A fly-back regulator using an LM2577 and capable of providing ±15V at 225mA from a +5V source is shown in Figure 12–13.

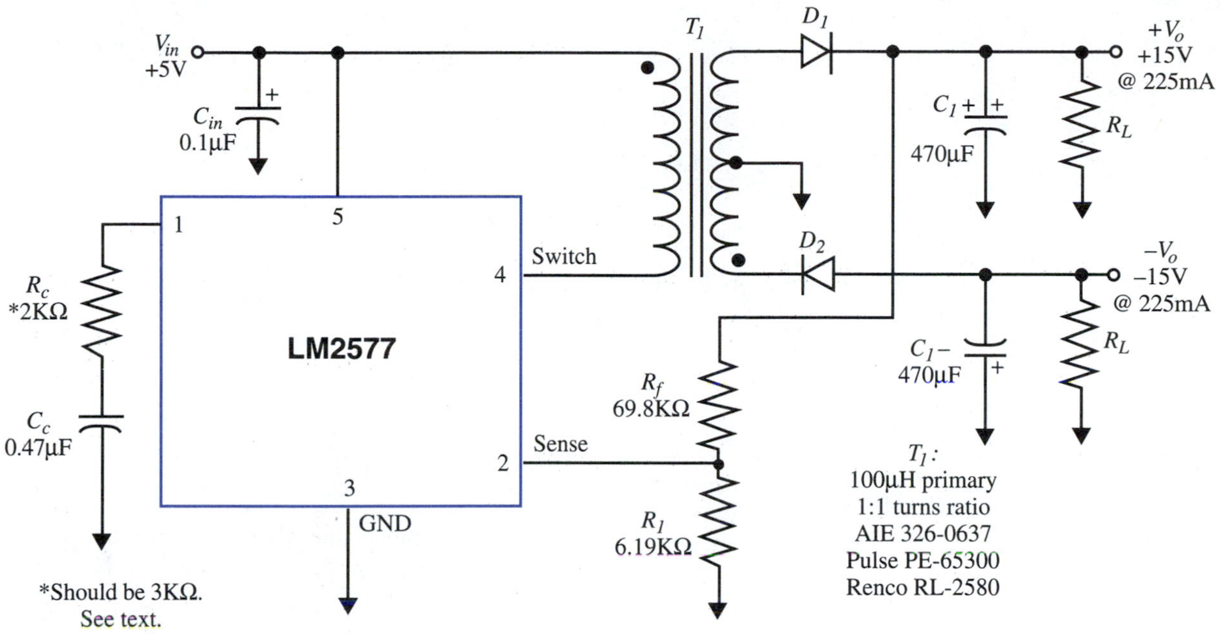

FIGURE 12–13 Fly-back regulator with positive and negative outputs

The amount of energy stored in the primary of T_1 is a function of how long the switch is turned on. Since the primary and secondary windings are out of phase, no energy is transferred to the secondary winding during the on time (diodes are reverse biased).

When the switch turns off, the collapsing magnetic field develops a voltage across the secondary, which forward biases D_1 and D_2 charging the output capacitors and supplying current to the load. The amount of energy stored, and therefore the secondary voltage, depends on the time the transistor switch is closed (t_{on}).

Let us look at the component selection for the circuit shown in Figure 12–13.

The ratio of R_f to R_1 is determined by the ratio of the desired 15V output voltage to the 1.23V reference.

Transformer selection (by vendor) is listed in National Semiconductor's data sheets for the particular input and output voltages.

The equations provided by National Semiconductor for C_1, R_c, and C_c are not the same as with the step-up regulator because of the discontinuous current in the fly-back circuit. Selection of components is interrelated, using the following equations:

$$R_c \leq \frac{750 \times \Sigma I_{o(max)} \times (V_o + V_{in(min)}N)^2}{V_{in(min)}{}^2} \qquad \textbf{(equation 12–19)}$$

Where ΣI_o is the sum of the output currents and N the transformer turns ratio. The value for R_c should be no greater than 3KΩ.

$$C_1 \geq \frac{0.19 \times R_c \times L_p \times \Sigma I_{o(max)}}{V_o \times V_{in(min)}} \qquad \textbf{(equation 12–20)}$$

and

$$C_1 \geq \frac{V_{in(min)} \times R_c \times N^2 \times [V_{in(min)} + (3.74 \times 105 \times L_p)]}{487{,}800 \times V_{out}^2 \times (V_{out} + V_{in(min)} \times N)}$$

with $C_1 = (C_1+) + (C_1-)$.

The larger value of C_1 from these two equations must be used to ensure regulator stability.

$$C_c \geq \frac{58.5 \times C_1 \times V_o \times [V_o + (V_{in(min)} \times N)]}{R_c^2 \times V_{in(min)} \times N} \qquad \textbf{(equation 12–21)}$$

EXAMPLE

Using equations 12–19, 12–20, and 12–21, determine whether the component values selected for the circuit of Figure 12–13 meet the criteria required by these equations. Assume $V_{in(min)}$ is 5V, N is 1, and L_p is 100μH.

SOLUTION

First checking R_c,

$$R_c \leq \frac{750 \times \Sigma I_{o(max)} \times (V_o + V_{in(min)}N)^2}{V_{in(min)}^2}$$

$$R_c \leq \frac{750 \times 450\text{mA} \times (15\text{V} + 5\text{V})^2}{(5\text{V})^2}$$

$$R_c \leq 5.4\text{K}\Omega \ (3\text{K}\Omega \text{ max})$$

R_c used is 2KΩ, so criteria are met.

Checking C_1,

$$C_1 \geq \frac{0.19 \times R_c \times L_p \times \Sigma I_{o(max)}}{V_o \times V_{in(min)}}$$

$$C_1 \geq \frac{0.19 \times 2\text{K}\Omega \times 100\mu\text{H} \times 450\text{mA}}{15\text{V} \times 5\text{V}}$$

$$C_1 \geq 228\mu\text{F}$$

and

$$C_1 \geq \frac{V_{in(min)} \times R_c \times N^2 \times [V_{in(min)} + (3.74 \times 105 \times L_p)]}{487{,}800 \times V_{out}^2 \times (V_{out} + V_{in(min)} \times N)}$$

$$C_1 \geq \frac{5\text{V} \times 2\text{K}\Omega \times 1 \times [5\text{V} + (3.74 \times 105 \times 100\mu\text{H})]}{487{,}800 \times (15\text{V})^2 \times (15\text{V} + 5\text{V})}$$

$$C_1 \geq 22.96\mu\text{F}$$

Circuit used $(C_1+) + (C_1-) = 900\mu F$, so both criteria are met. Finally, verifying C_c,

$$C_c \geq \frac{58.5 \times C_1 \times V_o \times [V_o + (V_{in(min)} \times N)]}{R_c^2 \times V_{in(min)} \times N}$$

$$C_c \geq \frac{58.5 \times 900\mu F \times 15V \times [15V + (5V \times 1)]}{(2K\Omega)^2 \times 5V \times 1}$$

$$C_c \geq 0.79\mu F$$

Circuit used $0.47\mu F$, which is too small.

Using the "Simple Switcher" computer program to verify the selected values in the schematic showed a difference in the selection of R_c with a value of $3K\Omega$ rather than $2K\Omega$. Substituting this value in the equation for C_c gives the result that the C_c used must be greater than $0.35\mu F$. The schematic value of $0.47\mu F$ is now correct. Using the new $3K\Omega$ for R_c in the equations for C_1 doesn't affect the result for this selection of this capacitor.

SWITCHED CAPACITOR VOLTAGE CONVERTERS

There is a different method available for stepping up DC voltages, which uses the basic voltage doubling technique described in Chapter 10. Figure 12–14 shows a simplified circuit.

The clock signal controls the switch positions with the inverter ensuring that S_1 and S_3 are closed when S_2 and S_4 are open, and vice versa. With S_1 and S_3 closed, the capacitor C_1 charges to V_{cc} with the polarity shown in Figure 12–14. When these switches open and S_2 and S_4 close, the voltage applied to C_2 plus the supply voltage for a total of $2V_{cc}$. So the output voltage across C_2 is double the supply voltage.

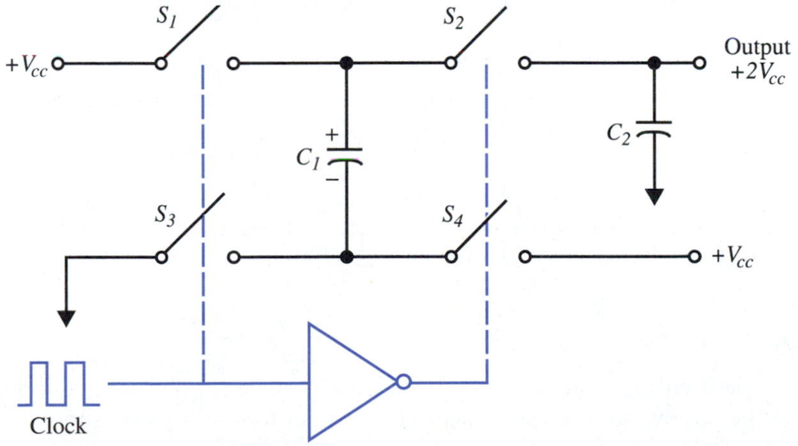

FIGURE 12–14 Switched capacitor voltage doubler

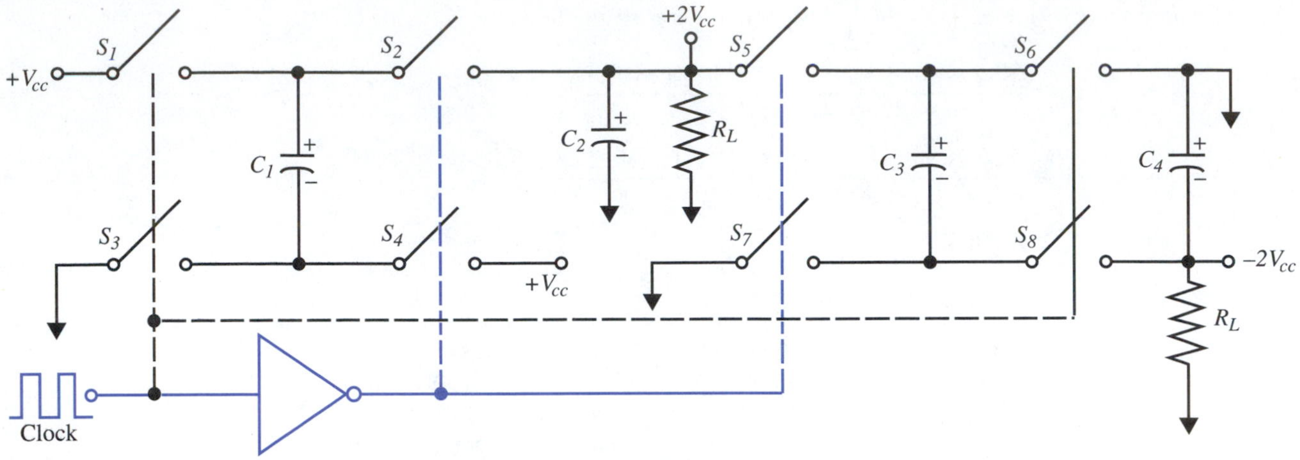

FIGURE 12–15 Dual switched power supply output $\pm 2V_{CC}$

The switches used are low voltage drop C-MOS power FETS. The switches and the clock can be part of an integrated circuit.

If we wish to have a negative as well as positive output voltage equal to $2V_{CC}$, we can use the circuit of Figure 12–15. In this circuit, the $+2V_{CC}$ output from the positive doubler (see Figure 12–14) charges the capacitor C_3 when switches S_5 and S_7 are closed. When S_5 and S_7 open and S_6 and S_8 close, the charge on C_3 is transferred to C_4. But since the top end of C_4 is grounded, the output voltage is $-2V_{CC}$.

An integrated circuit that basically contains the circuitry of Figure 12–15 is the Maxim MAX681. Input voltages as high as +6V can provide ±12V out of this chip, with currents up to about 10mA. An on-board clock runs at 8KHz to control the C-MOS switches.

The MAX681 is not a regulator (no feedback from output), and the output resistance for the positive output is about 150Ω and the negative output is 90Ω. Applications, then, are restricted to low-current C-MOS-type loads. However, the chip does provide a very simple approach to obtaining the higher positive and negative voltages required for op-amp types of circuits.

A device similar to the MAX681 is the MAX680, which requires external capacitors for C_1, C_2, C_3, and C_4. Though the added components add to the circuit's space requirements, the use of larger capacitors allows the output ripple voltage to be reduced.

EXAMPLE

Determine the voltage ripple at the positive and negative outputs from a Max680 switched capacitor voltage doubler if all capacitors are 10μF and the load currents are 10mA.

SOLUTION

First we need to determine the ripple at the positive supply (ΔV_o+), being aware that the positive supply also provides the current for the negative supply for a total of 20mA.

$$\Delta V_o+ = \frac{I_o \times \Delta t}{C_2}$$

where

$$\Delta t = \frac{1}{2f}$$

$$\Delta V_o+ = \frac{I_o}{2f \times C_2}$$

$$\Delta V_o+ = \frac{20\text{mA}}{2 \times 8\text{KHz} \times 10\mu\text{F}}$$

$$\Delta V_o+ = 125\text{mV}$$

The negative supply, having half the current, will have half the ripple.

$$\Delta V_o- = 62.5\text{mV}$$

SERIES RESONANT POWER SUPPLY

The latest switched power supply is the series resonant type (SRPS). It has the following advantages over the power supplies we have studied:

1) It has the highest efficiency.
2) It causes little conducted and radiated interference (little voltage across switch at turn-on).
3) It uses rather simple components.
4) It can continue to operate with the output open or shorted.

A disadvantage of the series resonant power supply is the complexity of its operation, which makes it a difficult circuit to analyze.

The basic circuit for a SRPS is shown in Figure 12–16.

The series resonant part of the circuit is L_1 in series with C_1 and C_o. A circuit requirement is that C_o be at least twice C_1. Also, L_s is selected to be at least ten times greater than L_1 so it has no effect on the resonant circuit operation.

The GTO is a gate-controlled switch that turns on and becomes a low resistance from anode (*A*) to cathode (*K*) when a positive voltage is applied to the gate (*G*). The GTO can be turned off by applying a negative voltage to the gate. (Refer to Chapter 13.)

If we assume initially that the GTO is not being turned on, then C_1 will be charged to the DC input voltage V_s. Now let us turn on the GTO just long enough to discharge C_1, and not long enough for the current through L_1 to change. If we now turn off the GTO, the parallel resonant (tank) circuit consisting of C_1, L_1, and C_o will oscillate at a frequency f_o.

$$f_o = \frac{1}{2\pi\sqrt{L_1 C_t}}$$

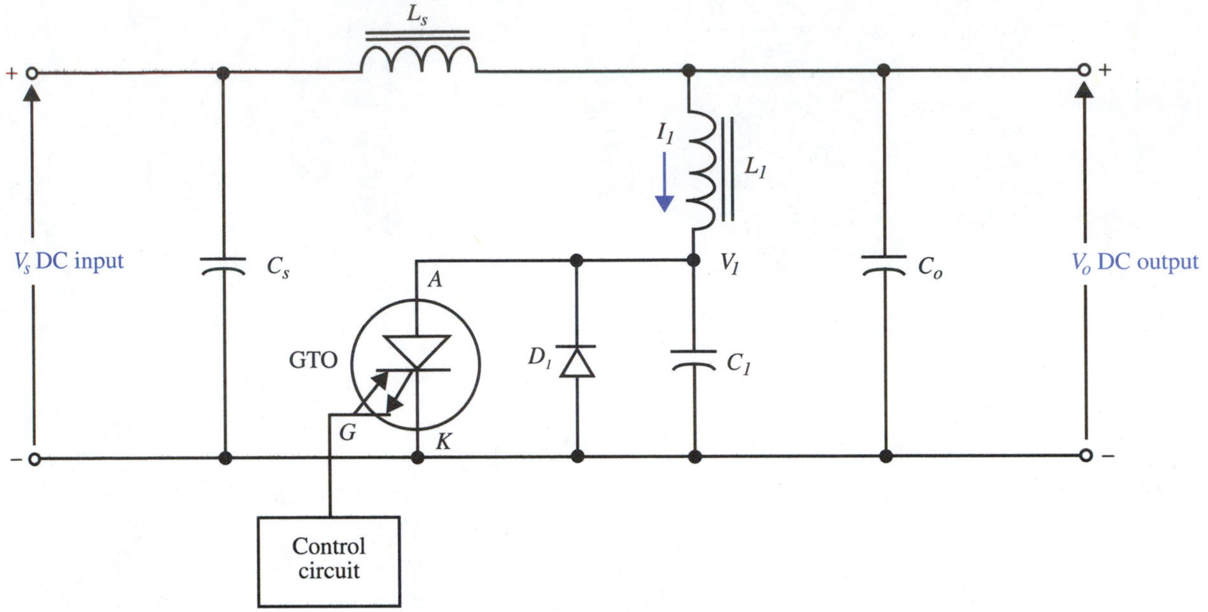

FIGURE 12–16 Basic SRPS circuit

where

$$C_t = \frac{C_1 C_o}{C_1 + C_o}$$

The peak current (I_{1p}) in the tank circuit at resonance is from basic circuit theory.

$$I_{1p} = V_s/Z_1$$

The instantaneous value of I_1 is

$$i_1 = V_s/Z_1 \sin 2\pi f_o t$$

For stable oscillation of the circuit, the voltage V_1 across C_1 must reach zero during each cycle, so the instantaneous value of V_1 is

$$V_1 = V_s(1 - \cos 2\pi f_o t) \qquad \textbf{(equation 12–22)}$$

What this equation indicates is that without oscillation ($f_o = 0$), the voltage across C_1 is the DC input voltage V_s, and with oscillation, the voltage across C_1 varies from 0V to $2V_s$. The change in voltage across C_o is in the ratio of C_1 to C_o, or

$$V_o = V_1 C_1/C_o \qquad \textbf{(equation 12–23)}$$

The peak AC value of the output voltage will be

$$V_{op} = V_s C_1/C_o \qquad \textbf{(equation 12–24)}$$

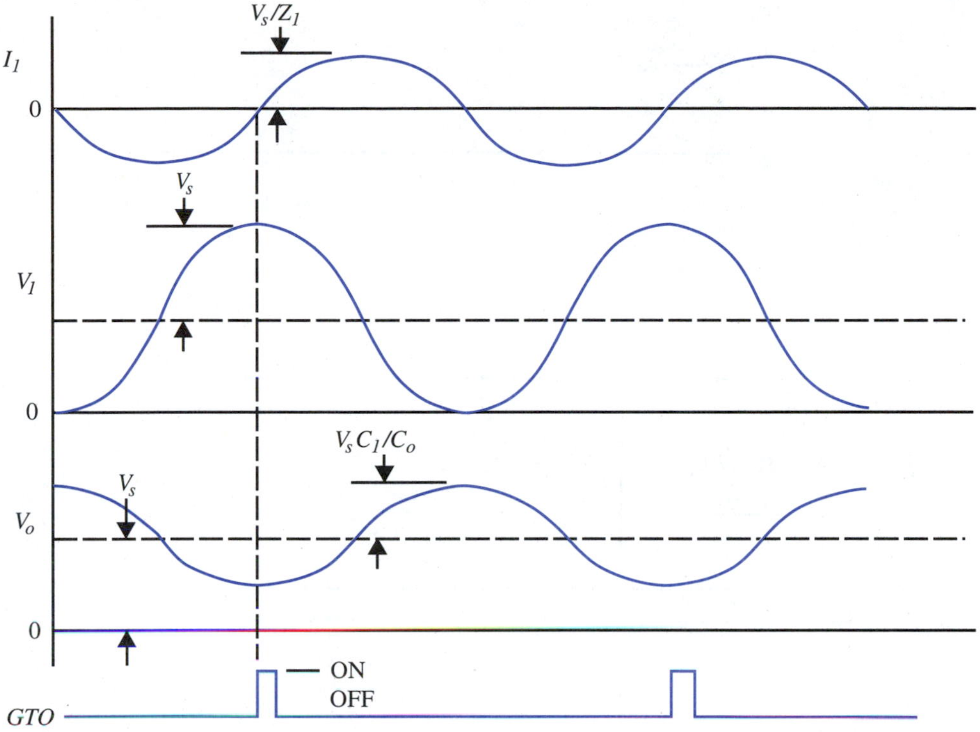

FIGURE 12–17 Waveforms for an SRPS with short GTO turn-on

If, as suggested,

$$C_o = 2C_1$$

then the voltage swing across C_o is half that across C_1, or $\pm V_s/2$. This change rides on the DC component V_s. Figure 12–17 shows the waveforms.

The first waveform shows the current I_1 through L_1 and C_1. Notice that the zero current point occurs at the time that the GTO turns on (C_1 discharged). The AC voltage across C_o is one half the amplitude of the voltage across C_1 (with C_o equal to $2C_1$) and 180° out of phase (current flows through the capacitors in the opposite direction relative to ground).

If we now increase the time the GTO is turned on, beyond the time necessary to discharge C_1, the capacitor C_o will start to discharge, and current will flow through L_1 from L_s. This will increase the tank circuit oscillating current and also the AC voltage developed across C_1 and C_o. If the current increase is by a factor M, then the new peak of the AC voltage across C_1 will be

$$V_{1p} = MV_s$$

and from equation 12–23,

$$V_{op} = MV_sC_1/C_o \qquad \textbf{(equation 12–25)}$$

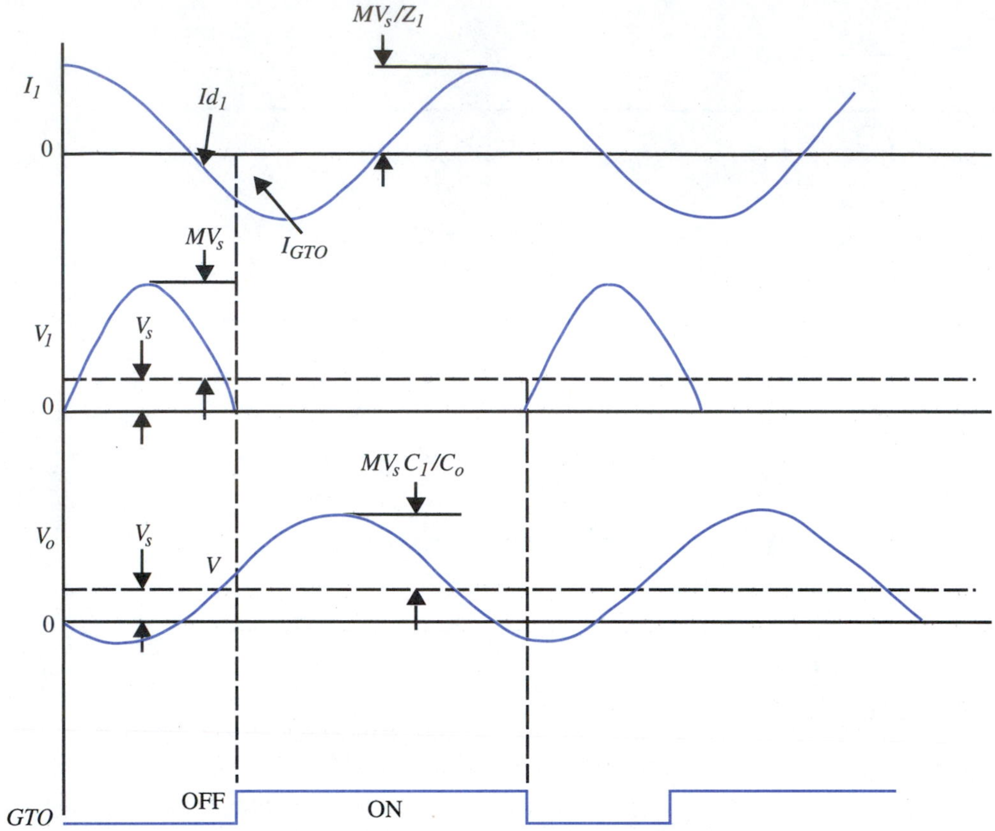

FIGURE 12–18 Waveforms for an SRPS with long GTO turn-on time

The significance of equation 12–25 is that by changing the on-time of the GTO, we can change the output voltage.

Figure 12–18 shows the new waveforms with the increased GTO turn-on time. Notice that V_o is increased by the factor M.

Since the voltage at the top side of C_1 now can go slightly negative, the diode D_1 conducts for a portion of the time that the GTO has a positive trigger on the gate. This is apparent from the I_1 current waveform.

VOLTAGE CONTROL RANGE OF THE SRPS

Assuming that V_s is the rectified line voltage peak of 170V ($120V \times \sqrt{2}$), and the maximum allowable voltage across the GTO is 1200V, we can determine from these voltages the maximum multiplier M that can be used. Since they are in parallel, the voltage across the GTO is the same as the voltage across C_1.

$$V_{GTO} = V_1 = V_s + MV_s$$

Solving for M,

$$M = \frac{V_{GTO} - V_s}{V_s}$$

$$M = \frac{1200\text{V} - 170\text{V}}{170\text{V}}$$

$$M = 6$$

Using this value of M, and again assuming that C_o is $2C_1$, the peak AC output voltage from equation 12–25 is

$$V_{op(max)} = MV_sC_1/C_o$$

$$V_{op(max)} = 6 \times 170\text{V} \times 1/2$$

$$V_{op(max)} = 510\text{V}$$

The minimum AC peak output voltage ($M = 1$) from equation 12–24 is

$$V_{op(min)} = V_sC_1/C_o$$

$$V_{op(min)} = 170 \times 1/2$$

$$V_{op(min)} = 85\text{V}$$

So the output voltage can be changed by the factor 510/85, or 6, by changing the on-time of the GTO. Looking at this another way, the circuit can accommodate 6 to 1 changes of the input voltage V_s and still keep the output at a constant voltage.

CONNECTING A LOAD TO THE SPRS

There are three basic ways a load can be connected to the output of an SRPS:

1) From a diode from the top of C_o to a filter capacitor. This will provide an output DC voltage of $V_s(1 + M)$.
2) From a capacitor from the top of C_o to the load. This will provide an AC output voltage with a peak voltage of $V_sM/2$.
3) By replacing L_s with a transformer and matching the load by selecting the correct turns ratio for the load on the transformer secondary winding.

These three load connections are shown in Figure 12–19.

SINGLE-CHIP POWER SUPPLY

The ideal power supply would include transformer, rectifiers, and regulator on one chip. With a few limitations, the Harris HV–2405E single-chip power supply can satisfy this need. It can provide an output voltage from 5V to 24V at a maximum current of 50mA with an input voltage range of 28V RMS to 264V RMS. The functional diagram of this device is shown in Figure 12–20.

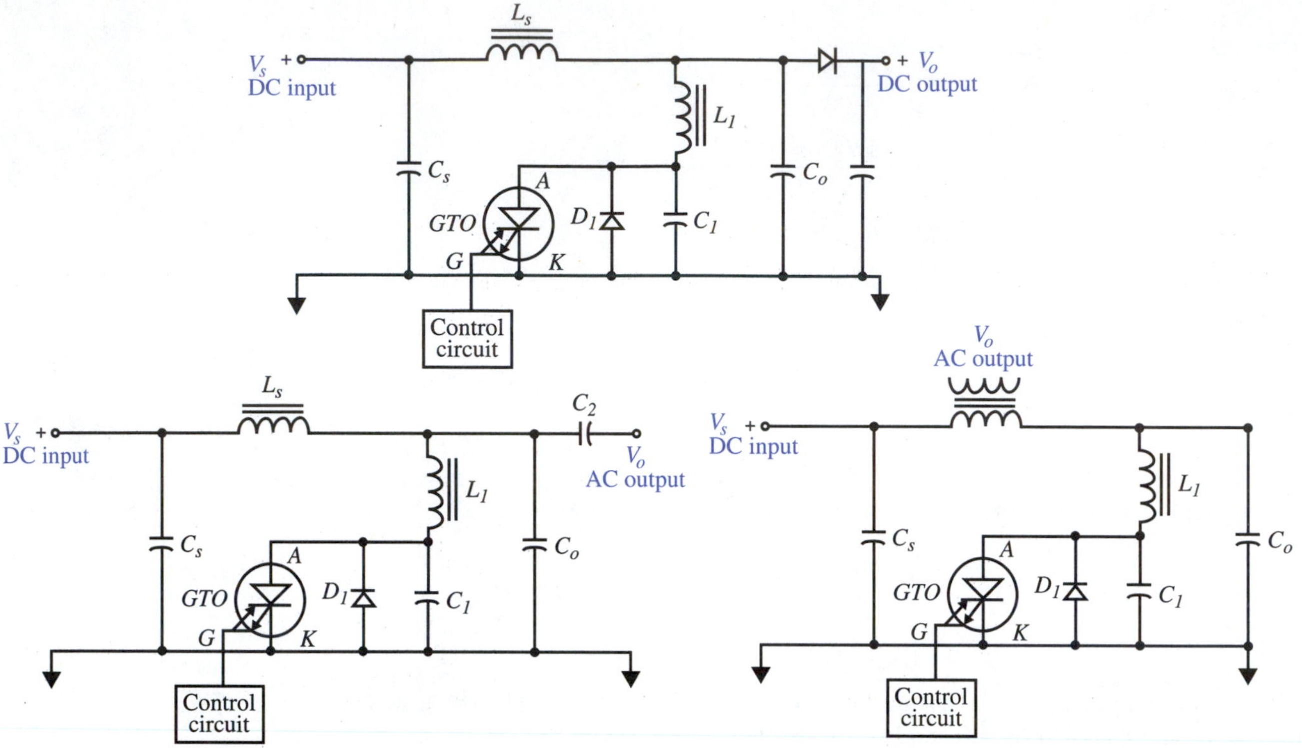

FIGURE 12–19 Load connections for SRPS

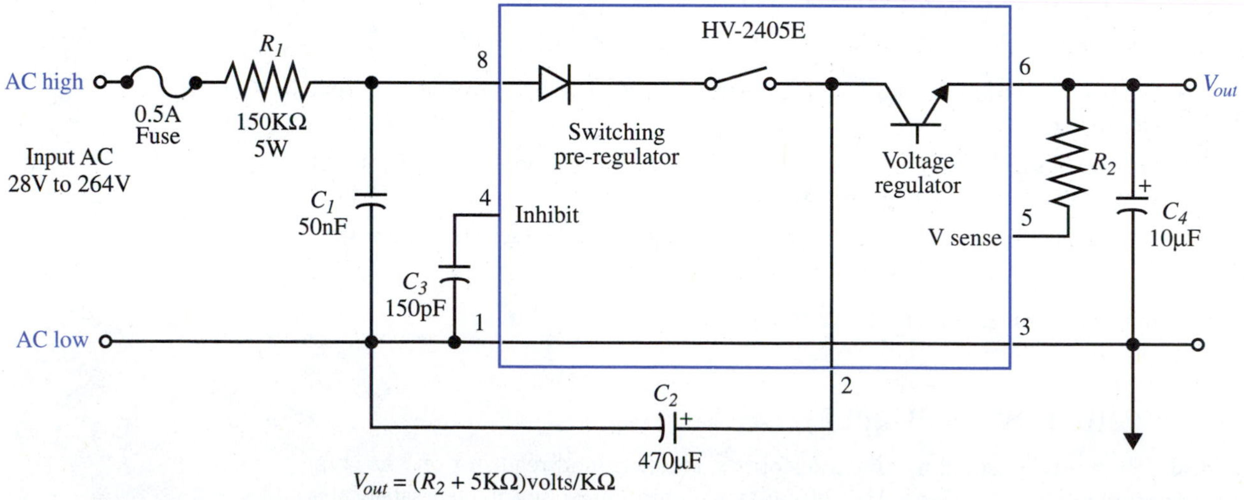

FIGURE 12–20 HV–2405E functional diagram

The HV–2405E contains an input switching regulator that connects the input voltage to the capacitor C_2 from when the input swings positive from the zero crossing of the input sine wave until the input reaches the desired output voltage plus 6V, or (V_{out} + 6V). This charges C_2 to the output voltage plus 6V, and this is the input to the linear regulator portion of the chip.

The linear regulator contains a 1.21V reference connected to the noninverting input of an op-amp. The inverting input is connected to the power supply output via R_2 plus an internal 3.79KΩ resistor. There is also an internal 1.21KΩ resistor from the inverting input to ground. We can determine the output then by using the noninverting op-amp equation

$$V_{out} = \left(\frac{R_2 + 3.79\text{K}\Omega}{1.21\text{K}\Omega} + 1\right)1.21\text{V}$$

Solving for R_2,

$$R_2 = V_{out}(\text{K}\Omega/\text{V}) - 5\text{K}\Omega \qquad \textbf{(equation 12–26)}$$

EXAMPLE

Find the value of R_2 required to provide 8V at the output of a HV–2405E power supply.

SOLUTION

Using equation 12–26,

$$R_2 = V_{out}(\text{K}\Omega/\text{V}) - 5\text{K}\Omega$$

$$R_2 = 8\text{V}(\text{K}\Omega/\text{V}) - 5\text{K}\Omega$$

$$R_2 = 3\text{K}\Omega$$

Varying the input from 80V RMS to 264V RMS will cause a maximum change in the output voltage of 20mV. Changing the load current from 5mA to 50mA will also cause a maximum change in the output voltage of 20mV.

The power dissipation P_d in the dropping resistor R_1 can be determined from the following equation:

$$P_d = 9.3\sqrt{\pi R_1}\, V_p(I_{out})^3 \qquad \textbf{(equation 12–27)}$$

where V_p is the peak of the AC input voltage and I_{out} the output load current.

EXAMPLE

Find the power dissipation in R_1 for the circuit of Figure 12–20 if the input voltage is 120V RMS and the load current is 50mA.

SOLUTION

Using equation 12–27,

$$P_d = 9.3\sqrt{\pi R_1}\, V_p(I_{out})^3$$

$$P_d = 9.3\sqrt{\pi \times 150\Omega} \times 120 \times \sqrt{2}\,(50\text{mA})^3$$

$$P_d = 4.28\text{W}$$

TROUBLESHOOTING SWITCHING REGULATORS

The approach to troubleshooting switching regulators is the same as with other linear integrated circuits. The operation of the regulator must be understood so that it can be determined whether the signal or DC voltage at each pin is correct. (Be sure to check right at the pin.)

Let us use the step-down regulator originally shown in Figure 12–7, and repeated in Figure 12–21 (with component values), as an example circuit.

The troubleshooting steps are as follows:

1) Verify that the input voltage at Pin 1 is +12V and that the ripple is consistent with the input rectification and filtering used (see Chapter 10).
2) Check if the output DC voltage and the ripple are as per power supply specifications.
   a) If the DC voltage is not present, check for a switched waveform at Pin 2 of the LM2576. If present, check continuity of the inductor and also look for a short on the output. If waveform is not present, verify that Pin 5 is grounded; if it is, replace the LM2576.

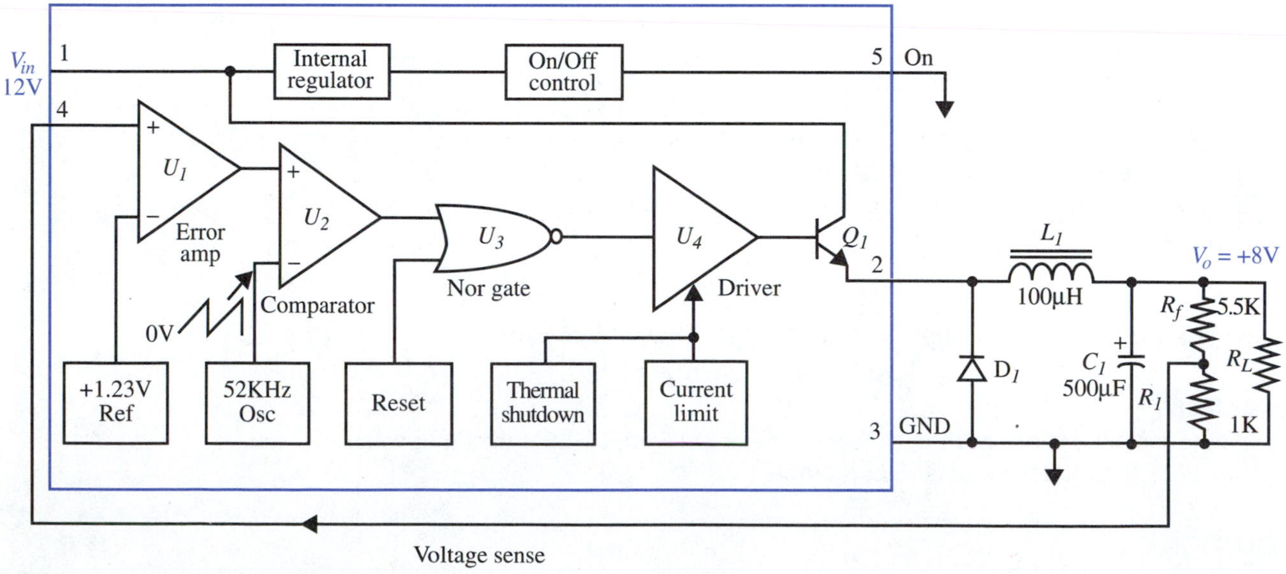

FIGURE 12–21 Troubleshooting example

b) If the DC output is too low, check to see whether R_1 is open-circuited.

c) If the output voltage is equal to the input voltage of 12V, check for an open ground or open R_f. If OK, replace the LM2576.

d) If the output ripple is too high, check for excessive load or open filter capacitor (C_1).

SUMMARY

- A linear regulator can be considered as a resistor in series with the load.
- The reasons for using a switching regulator are greater efficiency and smaller packaging.
- Greater efficiency results from using a switch as the regulating element.
- A disadvantage of the switching regulator is the greater circuit complexity.
- There are three basic types of switching regulators, and with reference to the output voltage, they can step down (buck), step up (boost), and fly back (invert).
- Switching regulators typically use pulse width variation to regulate the output voltage. A low output voltage requires a wider pulse to supply more current to the load.
- An inductor and a capacitor are used with a switching regulator to convert a pulsing voltage to a DC output.
- The efficiency of a step-down regulator increases with a higher output voltage.
- The efficiency of a step-up regulator increases with a higher input voltage.
- A low current voltage up-converter can be built using the voltage doubler concept with C-MOS switches.
- The series resonance power supply (SRPS) has the advantage of greater efficiency and less noise generation; but it is more difficult to analyze, so results are not as predictable.

EXERCISE PROBLEMS

1. State the advantages of a switching regulator over a linear regulator. Disadvantages?
2. Determine the output voltage from a step-down switching regulator if the input voltage is 12V and the switching transistor is on 40 percent of the time period.
3. Find the output from a step-up switching regulator if the conditions are the same as in problem 2.
4. Calculate the efficiency for a switching regulator with 5V input and 15V output. Neglect the switching losses.
5. Calculate the efficiency for a switching regulator with 15V input and 5V output. Neglect the switching losses.

6. Determine the value of L_1 and C_1 if the maximum input voltage $V_{in(max)}$ to a step-down switching regulator is 10V, the output voltage is 5V, the load current is 2A, the desired maximum output ripple voltage is 100mV, the capacitor ESR is 0.06Ω, the ratio of inductor current to load current K is 0.3, and the switching frequency is 52KHz.
7. Determine the value of L_1 and C_1 if the maximum input voltage $V_{in(max)}$ to a step-down switching regulator is 12V, the output voltage is 6V, the load current is 1A, the desired maximum output ripple voltage is 100mV, the capacitor ESR is 0.07Ω, the ratio of inductor current to load current K is 0.3, and the switching frequency 52KHz.
8. Find the value of R_f for an LM2576 regulator circuit if the desired output voltage is 5V and R_1 is 2KΩ.
9. Find the value of R_1 for an LM2576 regulator circuit if the desired output voltage is 6V and R_f is 12KΩ.
10. Find the minimum required value for C_1 if a step-up regulator has an input voltage of 8V, an output voltage of 16V, a load current of 1A, a switching frequency of 50KHz, and a maximum allowable ripple of 100mV. (Assume that C_1 has ESR of 50mΩ.)
11. Find the minimum required value for C_1 if a step-up regulator has an input voltage of 10V, an output voltage of 18V, a load current of 2A, a switching frequency of 50KHz, and a maximum allowable ripple of 100mV. (Assume that C_1 has ESR of 40mΩ.)
12. Find the required value of L_1 for a step-up switching regulator if the input voltage is 8V, the output voltage is 15V, the load current is 1A, the switching frequency is 50KHz, and the ratio of inductor current swing to input current K is 0.3.
13. Find the required value of L_1 for a step-up switching regulator if the input voltage is 12V, the output voltage is 18V, the load current 1.5A, the switching frequency 50KHz, and the ratio of inductor current swing to input current K is 0.3.
14. It is desired to build an 8V to 12V step-up regulator circuit using an LM2577–12 capable of providing 0.8A of load current with a maximum 100mV of output ripple and with K of 0.3. Find the required values for C_1 (assume that ESR is 43mΩ), L_1,R_c, and C_c.
15. It is desired to build a 6V to 12V step-up regulator circuit using an LM2577–12 capable of providing 0.5A of load current with a maximum 80mV of output ripple and with K of 0.3. Find the required values for C_1 (assume that ESR is 45mΩ), L_1, and R_c, C_c.
16. Determine the values for R_c, R_1,C_c, and C_1 for the fly-back regulator in Figure 12–A if the desired output voltages are ±12V with maximum currents of 200mA. Assume that $V_{in(min)}$ is 4.5V, L_1 is 100μH, R_2 is 10KΩ, and N is 1.
17. Compute the voltage drops for both the positive and negative output voltages in going from no load to 10mA for the MAX681 converter.
18. State the advantages of the SRPS.

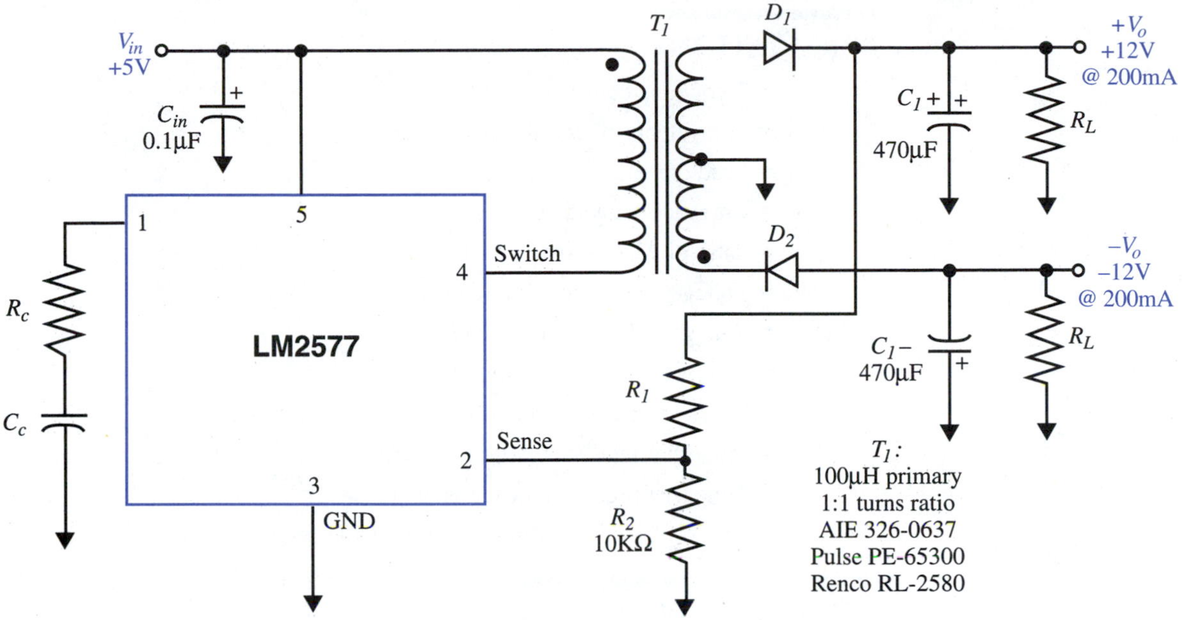

FIGURE 12–A

19. Determine the peak voltage across the GTO and across C_o in Figure 12–B if the input voltage (V_s) is the rectified line voltage, and the GTO is turned on just long enough to discharge C_1. Assume that C_o is $2C_1$.

20. Determine the peak and AC peak of the voltage across C_o in Figure 12–B if the GTO is turned on for a longer period to create an M of 3. Assume that V_s is the rectified input line voltage and that C_o is $2C_1$.

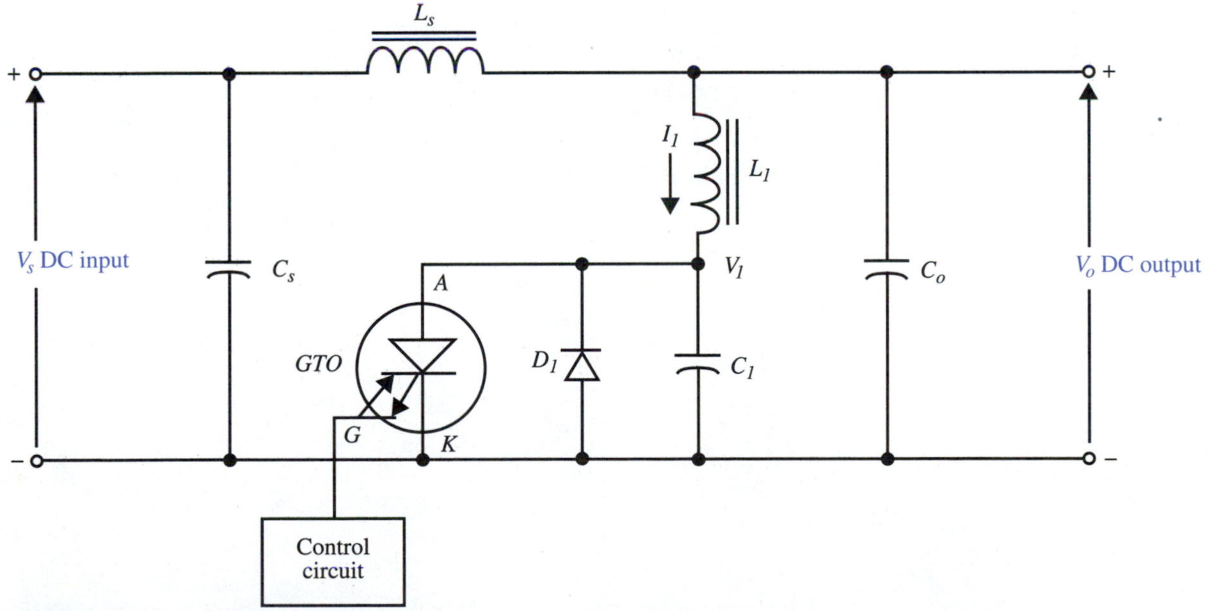

FIGURE 12–B (text figure 12–16)

ABBREVIATIONS

C_c = Frequency compensating capacitor
ESR = Equivalent series resistance of capacitor
I_L = Load current
ΔI_L = Current change through L_1
K = Ratio of $\Delta I_L/I_o$ (current factor)
V_r = Voltage drop across the regulator
V_{L1} = Voltage induced across the inductor L_1
R_c = Frequency-compensating resistor
SRPS = Series resonant power supply
t_{on} = Time that regulator switch is closed
t_{off} = Time that regulator switch is open
T = Period of oscillation in switching regulator ($1/f_{osc}$)

ANSWERS TO PROBLEMS

1. See text.
2. $V_o = 4.8V$
3. $V_o = 20V$
4. Eff = 83%
5. Eff = 83%
6. $L_1 = 80\mu H$; $C_1 = 206\mu F$
7. $L_1 = 192\mu H$, $C_1 = 49.7\mu F$
8. $R_f = 6.13K\Omega$
9. $R_1 = 3.09K\Omega$
10. $C_1 = 200\mu F$
11. $C_1 = 889\mu F$
12. $L_1 = 133\mu H$
13. $L_1 = 119\mu H$
14. $C_1 = 78\mu F$; $L_1 = 142\mu H$; $R_c = 1.35K\Omega$; $C_c = 45nF$
15. $C_1 = 83.6\mu F$; $L_1 = 192\mu H$; $R_c = 1.5K\Omega$; $C_c = 52nF$
16. $R_c = 4.03K\Omega$ (use $3K\Omega$); $C_1 = 422\mu F$; $C_c = 121nF$; $R_1 = 87.6K\Omega$
17. $\Delta V+ = 1.5V$; $\Delta V- = 0.9V$
18. See text.
19. $V_{c1} = 340V$; $V_{co} = 170V$
20. With reference to Fig 12–18: peak = 425V; AC peak 255V

CHAPTER 13

Power Control Devices

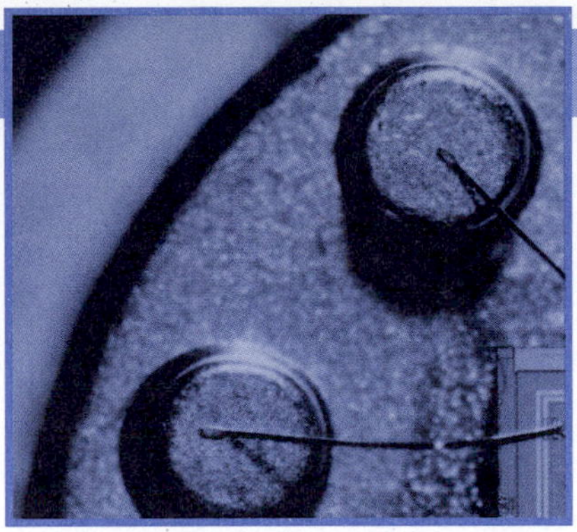

CHAPTER OBJECTIVES

After completing this chapter, you should be able to explain the following power control device concepts:

- How a bipolar transistor controls the application of power
- Power control with a transistor-relay combination
- How an FET controls the application of power
- Advantages of an IGBET
- Operation of a silicon controlled rectifier (SCR)
- DC power control with an SCR
- AC power control with an SCR
- Operation of a Shockley (four-layer) diode
- How a Shockley diode improves SCR operation
- Operation of a unijunction transistor
- Reason for triggering an SCR with a unijunction transistor
- SCS operation
- GTO characteristics
- Operation of a triac
- AC power control using a triac
- Operation of a diac
- Reason for triggering a triac with a diac
- Zero crossing switching
- Static switching of an SCR
- Optical power control devices
- Purpose of a Smart Switch
- Troubleshooting power control devices

INTRODUCTION

As the name indicates, power control devices are used to control the application of power to a piece of equipment or to a device, e.g., motor, light, heater, fuel injector, etc. These devices are designed to perform with the highest efficiency, or in other words, with minimum power loss in the control device. Power control can be closing a switch to apply power to a device, or partial application of power as with a lamp dimmer.

This chapter will present the basic operation of these power control devices and their support components, and also show practical power circuit applications and troubleshooting techniques.

BIPOLAR TRANSISTOR POWER CONTROL

We saw in Chapter 5 that a signal level transistor can be used as a switch because of the low forward voltage drop when in the "on" state. Since we are typically dealing with high power in power control circuits, the question arises "Can we also achieve a low voltage drop with high power transistors?" When using a transistor for power control, we can, by operating it in the saturated mode (common emitter), minimize the "on" voltage drop (remember that a follower will not saturate).

If we are controlling the application of power from a positive power supply, an NPN transistor is connected on the power supply common side of the load (low side switch); whereas a PNP is connected on the power supply side of the load (high side switch), as shown in Figure 13–1.

Let us consider using the common 2N3055 NPN power transistor as a switch. This transistor can block up to 60V when off and conduct up to 15A when turned on. Saturated current gain (β or h_{fe}) ranges from 20 with a current of 1A down to a β of 3 at 10A. As a power control device, we are interested in the voltage drop across the device when conducting. With 1A flowing through the transistor, the drop is 0.1V; at 10A, it is a maximum of 3V.

EXAMPLE

We wish to use a 2N3055 to control power to a 28V, 1A, DC motor. Determine the required base current to the transistor, the power loss in the transistor, and the circuit efficiency. Also, determine the power loss and efficiency with the motor current increased to 10A.

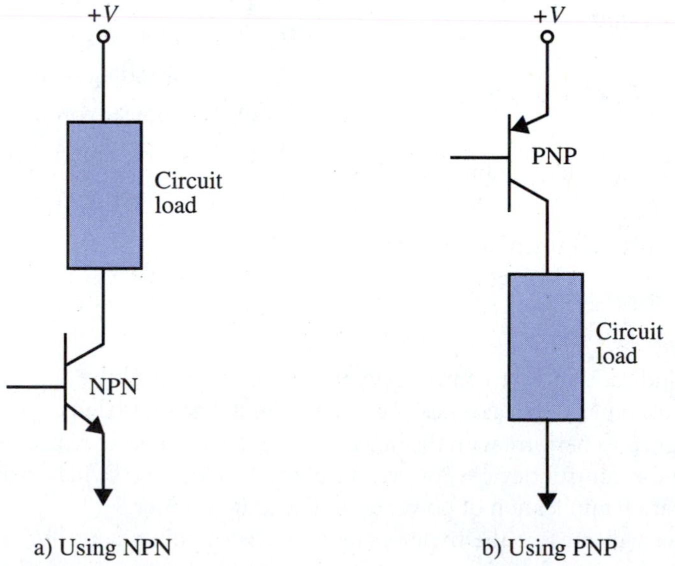

FIGURE 13–1 Transistor power control

SOLUTION

With a β of 20 at 1A,

$$I_b = \frac{I_c}{\beta}$$

$$I_b = \frac{1\text{A}}{20}$$

$$I_b = 50\text{mA}$$

Power loss (P_l) is the product of the collector current of 1A and the collector-emitter voltage drop of 0.1V, or

$$P_l = 0.1\text{V} \times 1\text{A}$$

$$P_l = 0.1\text{W}$$

Efficiency is the amount of power used by the motor (P_m) divided by the amount provided by the source (P_s), or

$$\text{Eff} = \frac{P_m}{P_s}$$

Since the transistor drops 0.1V, the voltage across the motor is $28V - 0.1V$, or 27.9V.

$$\%\text{Eff} = \left(\frac{P_m}{P_s}\right) \times 100$$

$$\%\text{Eff} = \left(\frac{27.9\text{V} \times 1\text{A}}{28\text{V} \times 1\text{A}}\right) \times 100$$

$$\%\text{Eff} = 99.6\%$$

If the motor is changed to one that draws 10A, then the drop across the transistor increases to 3V, and the required base current is ($10A/3$), or 3.3A. The new power loss is

$$P_l = 3\text{V} \times 10\text{A}$$

$$P_l = 30\text{W}$$

The voltage drop across the motor is now $28V - 3V$, or 25V, and the % efficiency is

$$\%\text{Eff} = \left(\frac{25\text{V} \times 10\text{A}}{28\text{V} \times 10\text{A}}\right) \times 100$$

$$\%\text{Eff} = 89\%$$

We can see from this example how the voltage drop across the control device can lower efficiency at higher currents. Efficiency is further lowered because the high base current causes additional input power loss ($I_b \times V_{be}$), which is $3.3A \times 0.6V$, or about 2W.

Base current to operate the control transistor can be reduced by using a power Darlington control transistor (the emitter of the first transistor is connected to the base of the second). The 2N6576 power Darlington requires 1mA rather than 50mA to saturate the transistor for 1A of collector current and 10mA rather than 3.3A at a collector current of 10A.

Collector-to-emitter saturation voltages for the 2N6576 transistor are 1V with a collector current of 1A and 2.8V with 10A. But, if we want to raise the control transistor efficiency at 10A, we need to go to a higher-power Darlington transistor. The Motorola MJ11028 (NPN) and MJ11029 (PNP) have 10A saturation voltages of 1V rather than 2.8V and so are able to provide higher-power switching efficiency.

TRANSISTOR-RELAY CONTROL

A transistor can be used to control power via a relay. Figure 13–2 shows a circuit using a low-power switching transistor to turn on a relay—which in turn, controls the main circuit power.

The advantages of this circuit are the low input power required to turn on the transistor and the isolation between the power source being controlled and the power source used to turn on the transistor (i.e., the power circuit commons can be isolated from one another, and both AC and DC power can be controlled). Disadvantages of using a relay are the speed, size, and the reliability of a mechanical device.

Notice that a protection diode D_1 must always be connected across the relay coil when it is being activated by a transistor. The diode suppresses the positive voltage spike from the coil's inductance generated by the collapsing magnetic field when the transistor turns off. This prevents the collector of the transistor from going more positive than 0.6V greater than the supply voltage (12.6V in Figure 13–2). Unsuppressed, this voltage spike could break down the collector-base junction and destroy the transistor.

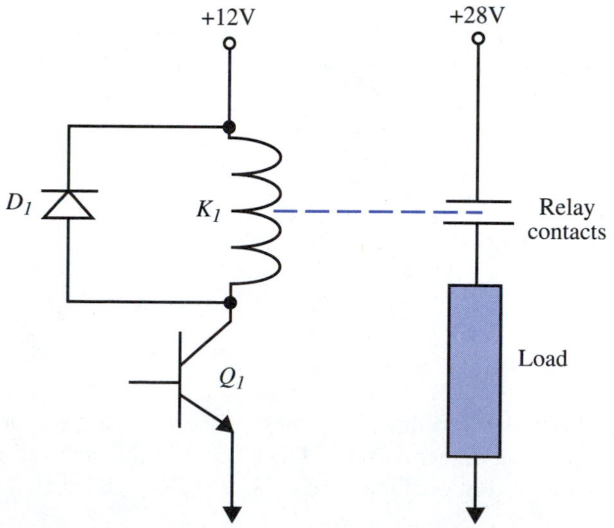

FIGURE 13–2 Transistor-relay power control

EXAMPLE

Find the base current required to activate the transistor in the transistor-relay power control circuit of Figure 13–2. Assume that the relay resistance is 100Ω and the β of the 2N2222 is 100. Also find the efficiency of the circuit if the relay contacts provide power to a 10A, 28V motor.

SOLUTION

When the relay is turned on, the transistor is saturated, so essentially all the power supply voltage is dropped across the relay resistance. The transistor current I_c is then

$$I_c = \frac{12V}{100}$$

$$I_c = 120mA$$

In our discussion of switching circuits in Chapter 5, the ground rule given was that to saturate a transistor, the base current should be one-tenth of the collector current, or

$$I_b = \frac{120mA}{10}$$

$$I_b = 12mA$$

To determine the circuit efficiency, we need to find the total power P_t used in the circuit and divide this value into the useful output power P_L $(28V \times 10A)$.

Neglecting the base drive, the power used to turn on the relay P_r is

$$P_r = 12V \times 120mA$$

$$P_r = 1.44W$$

The total source power P_s is P_r plus the load power P_L.

$$P_s = P_r + P_L$$

$$P_s = 1.44W + (28V \times 10A)$$

$$P_s = 281.44W$$

The efficiency is

$$\%Eff = \left(\frac{P_L}{P_s}\right) \times 100\%$$

$$\%Eff = 99.5\%$$

FET POWER CONTROL

We can use an FET rather than a bipolar transistor for power control. A power FET has the advantage of requiring negligible input power (drive) to operate the device.

The power MOSFET, which we will use for our motor power control example, is the N-channel enhancement (open circuit until biased) type MTP15N05E. This device has

TABLE 13–1 Bipolar versus FET for Power Control

Feature	*Bipolar*	*FET*
Price	Cost effective for voltages > 200V	Higher for high voltage
Drive	Complex	Simple
Speed	Storage limited	Fast
On-Voltage	Lower for hi-volt devices	Lower for lo-volt devices
Paralleling	More complex	Fairly simple
Thermal stability	Requires some precautions	Less susceptible
Efficiency	Requires drive power	More efficient

a continuous current rating of 15A and a maximum drain-source voltage of 50V. The "on" drain-source voltage drop with a drain current of 10A is 1V (with + 10V on the gate). Being a MOSFET, the gate current (and control input power) is negligible.

EXAMPLE

Calculate the percent efficiency of the MTP15N05E MOSFET when used to control the 10A, 28V motor.

SOLUTION

With the 1V loss across the FET with 10A, we have 27V across the motor, and

$$\%\ \text{Eff} = \left(\frac{P_m}{P_s}\right) \times 100$$

$$\%\ \text{Eff} = \left(\frac{27\text{V} \times 10\text{A}}{28\text{V} \times 10\text{A}}\right) \times 100$$

$$\%\ \text{Eff} = 96.4\%$$

This is better than a 7% improvement over the 2N3055; not considering the base drive power required for the bipolar transistor—which is critical, for example, when driving from a microprocessor output.

A comparison of bipolar and FET power control devices is shown in Table 13–1.

To summarize the table listings, bipolars are better when you are controlling high-voltage (>200V) power sources, but FETs are faster, simpler to drive, easier to operate in parallel, better for handling higher currents, and more efficient and stable with temperature change.

INSULATED GATE BIPOLAR TRANSISTOR

A newer device developed for high-voltage power switching is the Insulated Gate Bipolar Transistor (IGBT). As the name suggests, this device is a combination of an FET and bipolar transistor with a gate, collector, and emitter. It is an enhancement-mode device which has a vertical power structure and an on-resistance of about one-half that of a conventional power MOSFET. Figure 13–3 shows the schematic symbol for an IGBT.

As an example of the IGBT specifications, the Harris Semiconductor HGTP10N50C1 can operate with collector-emitter voltages up to 500V and has a maximum collector-emitter voltage drop of 2.5V when conducting 10A. Since the device is controlled from an FET gate, the input drive for turn-on is negligible so it can be operated from a low-powered integrated circuit (about +6V is required for turn-on).

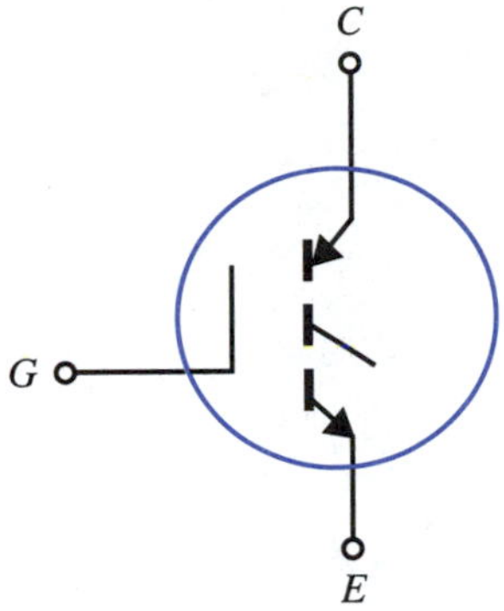

FIGURE 13–3 IGBT symbol

EXAMPLE

Assuming that a 500V power source is used, find the maximum efficiency for the HGTP10N50C1 IGBT with a load current of 10A.

SOLUTION

The efficiency is again determined from the output power divided by the input power.

$$P_{in} = 500V \times 10A$$

$$P_{in} = 5000W$$

$$P_{out} = (500V - 2.5V) \times 10A$$

$$P_{out} = 4975W$$

$$\%Eff = \frac{P_{out}}{P_{in}} \times 100\%$$

$$\%Eff = \frac{4975W}{5000W} \times 100\%$$

$$\%Eff = 99.5\%$$

This high efficiency results from the great difference between the high supply voltage used and the low voltage drop across the IGBET.

The turn-on time for an IGBT is shorter than for a conventional power MOSFET—100ns versus 200ns; but the turn-off time is about twice as long—1μs vs 0.5μs.

THYRISTORS

Thyristors are semiconductor devices that can be triggered from a nonconducting (off) to a conducting (on) state by the application of an electrical signal. These devices come in various configurations. Some are triggered by a signal applied to a gate; others trigger when a particular voltage across the device is exceeded. There are thyristors that will conduct current in one direction when triggered; others will conduct in both directions for AC power control applications.

SILICON CONTROL RECTIFIER

We shall look at various forms of thyristors in our study of power control circuits, starting with the silicon controlled rectifier—basically a diode that doesn't forward conduct until triggered.

Before we use a silicon control rectifier (SCR) in a power control application, we need to understand how the device operates. Figure 13–4a shows the bipolar transistor equivalent of the SCR and Figure 13–4b gives the schematic symbol.

With reference to the equivalent circuit of Figure 13–4a, forward biasing the base-emitter (gate-cathode) junction of Q_1 causes this transistor to conduct. Current flow (I_1) in the collector of Q_1 forward biases Q_2, which then provides, via its collector, a bias current (I_2) for Q_1. So, if we initially forward bias the base of Q_1 from an external source, we can then remove this source because Q_2 provides the necessary bias current.

With both Q_1 and Q_2 conducting and reinforcing one another, both transistors go into saturation. The current through the transistors in this saturated condition is determined by the external load resistor. The voltage across the SCR from anode to cathode when conducting becomes one base-emitter drop (0.6V) plus one saturated transistor drop (0.2V), or 0.8V. At high currents, the drop will be higher because of the internal bulk resistance.

So, by momentarily providing a turn-on pulse to trigger the SCR, we cause it to conduct and allow current through the load resistor. But how do we turn the SCR off? We must interrupt the load current or drop it down below what is called the SCR holding current.

Holding current (I_h) is the minimum current necessary to keep the transistors conducting, because below this value of current, the transistor βs are too low to maintain the current feedback values required for saturation.

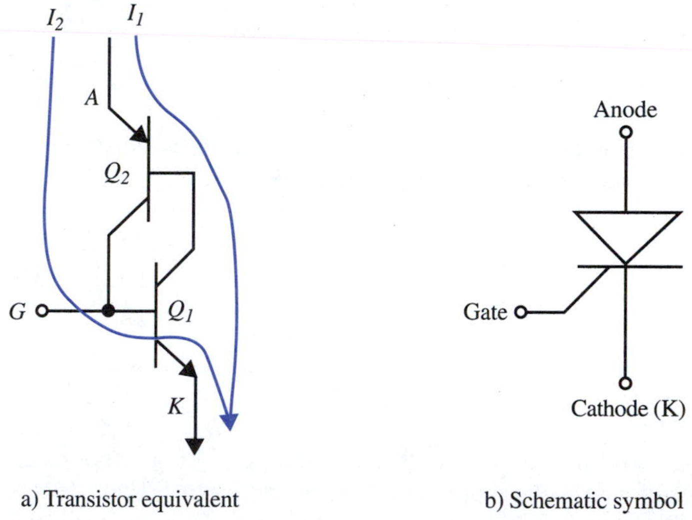

FIGURE 13–4 Transistor equivalent of an SCR

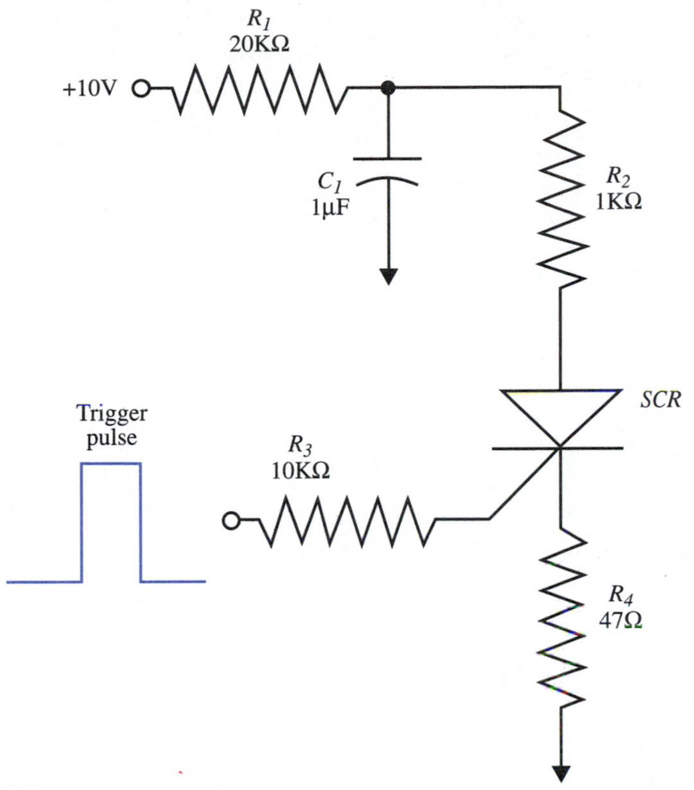

FIGURE 13–5 SCR test circuit

A typical low-power SCR is the 2N5062. This SCR can sustain anode-to-cathode voltages up to 100V; during conduction, it handles up to 0.8A with 1.25V drop from anode to cathode. Voltage drop from anode to cathode with 10mA is given as 0.7V. Maximum holding current for the 2N5062 is 5mA. The maximum gate voltage to fire the SCR (VGT) is 0.8V, and the maximum gate current (IGT) is 200μA.

A simple SCR test circuit using the 2N5062 is shown in Figure 13–5.

In operation, a trigger pulse greater than the required level will turn on the SCR. There is a slight delay (about 1μs) in reaching the trigger point because of gate-to-cathode capacitance.

Resistor R_1 (20KΩ) limits the current through the SCR to about 0.5mA, but capacitor C_1 is selected to provide SCR current (up to 10mA, limited by R_2) that will decay down to the holding current. When C_1 discharges sufficiently, the current will drop below the holding current, and the SCR will turn off. A time equal to $5R_1C_1$ should be allowed for the capacitor to recharge before the SCR is retriggered. The SCR test circuit can be used to measure the voltage and current trip points and also the holding current.

EXAMPLE

We are required to measure the voltage and current trip points and the holding current for a given SCR.

SOLUTION

The circuit of Figure 13–5 can be used.

Voltage trip point can be measured by using a 10μs positive pulse (long enough to charge the gate capacitance) at a pulse repetition frequency of 10Hz ($1/5R_1C_1$). The voltage amplitude of the input pulse is slowly increased from zero until the SCR fires—this will be indicated by a slight jump in the gate pulse amplitude caused by the SCR internal feedback current.

Once the trigger voltage trip point is determined, the trigger current can be found by noting the difference in voltage at both ends of R_3 (using dual 'scope) and dividing this difference by the value of R_3. Holding current can be measured by measuring the voltage drop across R_4 at the point where the SCR turns off, i.e., at the minimum voltage across C_1.

DC POWER CONTROL USING AN SCR

Since the 2N5062 can only handle currents up to 0.8A, it could not be used to control power to the 10A motor used in previous examples.

A higher-power SCR is the 2N6400. This device can control voltages up to 50V, with a maximum continuous current of 16A. The maximum required trigger voltage is 1.5V; the maximum required trigger current is 30mA.

Holding current for the 2N6400 has a maximum value of 40mA, and the maximum voltage drop across the SCR with 10A of anode current is 1.2V.

We can use the 2N6400 to control the power for our 28VDC, 10A motor—but once we turn the motor on by firing the SCR, how do we turn it off? Again, we have to cause the SCR anode current to fall below its holding current value for SCR turn-off.

An SCR circuit that can control power to our motor and has the capability of turning off the SCR, is shown in Figure 13–6.

Let us look at the circuit operation. We shall assume an initial state with SCR_2 conducting (caused by a power-up circuit, which is not shown). This means the anode of SCR_2 is at about +1.2V, which lowers the voltage on the R_2 and R_3 divider allowing an SCR trigger pulse to pass when applied to the left side of the diode D_2 (since SCR_1 anode is at +28V, the R_6 and R_7 divider prevents conduction of D_3 and therefore triggering of SCR_2).

The capacitor C_1 (called a commutating capacitor) charges via the motor to (28V − 1.2V) or 26.8V. The first input pulse via D_2 turns on SCR_1 and the motor starts to run. The left side of C_1 will now be connected to ground (plus the SCR_2 anode-cathode drop). Remembering that the voltage across a capacitor cannot change instantaneously, the right side of the capacitor will be driven to −26.8V. This means the anode of SCR_2 will be driven negative by this amount of voltage. The result is that all of the R_1 current now

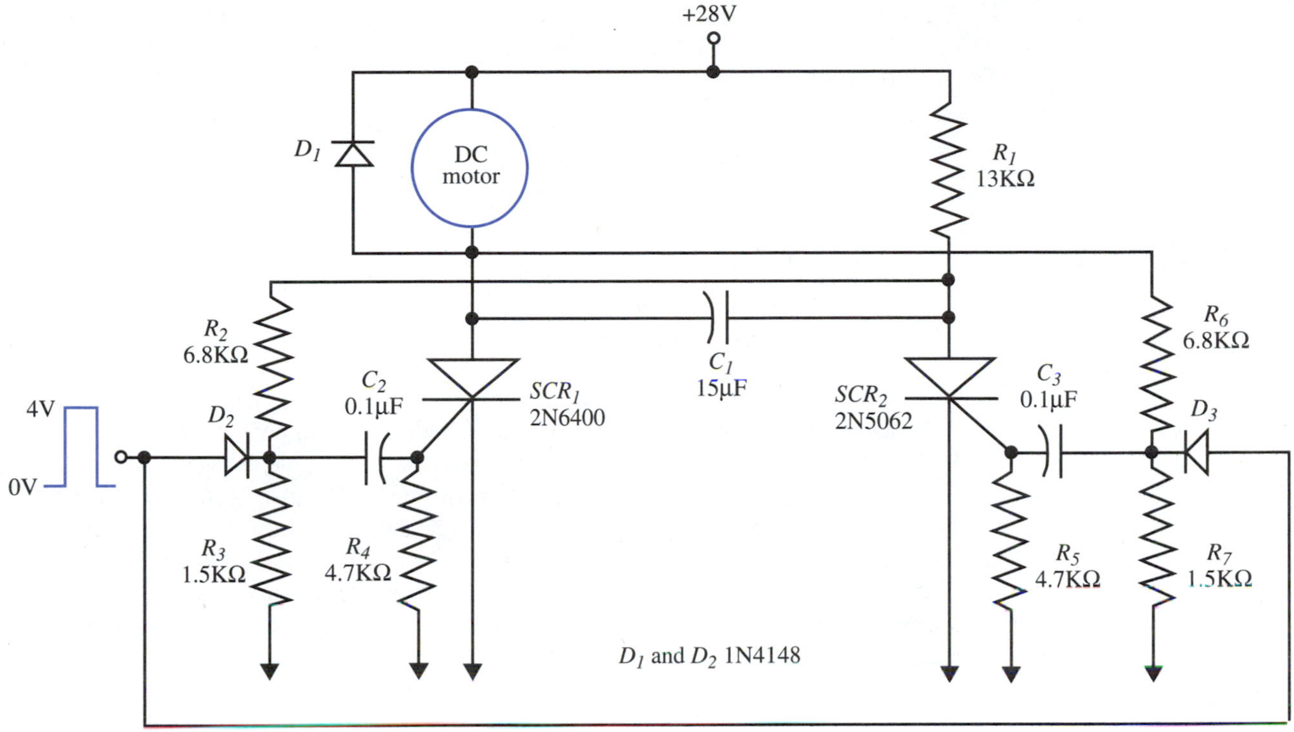

FIGURE 13–6 SCR power control

flows into the more negative C_1. Since the current in SCR_2 falls below the holding current, SCR_2 turns off.

If we now turn on SCR_2, the motor current will flow into the left side of C_1 rather than into SCR_2 and this capacitor will quickly charge to the +28V supply voltage. In order for the 2N6400 to turn off, the current flowing into this SCR must be taken below the holding current value for at least the specified 15μs. We can ensure this by not allowing the voltage on the SCR anode to exceed 0V for, say, 2 × 15μs or 30 μs.

If we assume the motor inductance maintains the 10A motor current flow while charging the capacitor C_1 from the negative voltage to zero volts, then we can determine the required value for C_1. The relationship to determine the size of a capacitor, given a constant charging current (i_c) and a required voltage change (ΔV_c) in a given time (Δt) is

$$C = i_c\left(\frac{\Delta t}{\Delta V_c}\right) \qquad \textbf{(equation 13–1)}$$

EXAMPLE

Determine the size of the commutating capacitor C_1 to ensure turn-off of the motor control SCR in the circuit of Figure 13–6.

SOLUTION

Using equation 13–1 and the specified values of motor current, voltage change, and turn-off time,

$$C_1 = 10\text{A} \times \frac{30\mu\text{s}}{(28\text{V} - 1.2\text{V})}$$

$$C_1 = 11.2\mu\text{F}$$

(Use 15μF.)

We note that this capacitor can be charged with either polarity and therefore must be a non-polarized electrolytic.

As stated before, when SCR_1 turns off, the left side of C_1 will now charge to +28V. With SCR_2 now on, the right side of this capacitor is at the "on" anode voltage of SCR_2 (<1V).

If we want the motor current to again flow through SCR $_1$, we need to retrigger SCR_1. When SCR_1 goes on, the right side of C_1 is forced negative, turning off SCR_2. What we have, then, is a flip-flop action between SCR_1 and SCR_2 with only one SCR being on at a time.

When SCR_1 just turns on and SCR_2 turns off, C_1 now starts charging through R_1. It takes a time equal to $5R_1C_1$ for charging C_1, so the value of R_1 is determined by the minimum time (t_m) we require the motor to be provided current by SCR_1, so

$$R_1 = \frac{t_m}{5C_1} \qquad \textbf{(equation 13–2)}$$

Since the current through SCR_2, when it is on, is determined by the value of R_1, this SCR's maximum continuous current rating is determined by R_1. However, we can assume that a motor surge current of 10A (just as SCR_1 turns off) will occur through SCR_2 during the time that C_2 (left side) is charging from zero volts to +28V, i.e., about 30μs. So we need to consider both the continuous and surge ratings of SCR_2.

EXAMPLE

Find the required value of R_1 and the continuous and surge current ratings of SCR_2 in Figure 13–6 if the minimum time the motor is turned on is 1 sec.

SOLUTION

From the previous example, we suggested a value for C_1 of 15 μF. Using equation 13–2,

$$R_1 = \frac{t_m}{5C_1}$$

$$R_1 = \frac{1\text{s}}{5(1.5 \times 10^{-5}\text{F})}$$

$$R_1 = 13.3\text{K}\Omega$$

(Use 13KΩ.)

If we assume the full 28V supply is across R_1 when SCR_2 is on, then the continuous current rating (I_m) of SCR_2 is

$$I_m = \frac{28V}{13K\Omega}$$

$$I_m = 2.15mA$$

The surge current rating is the 10A mentioned above and exists for about 30 μs.

Both current ratings are within the capability of the low-power 2N5062, since the continuous current rating is 0.8A and a sinusoidal pulse of 6A peak over 8.3ms can be accommodated (our 10A pulse, although greater, only occurs over 30μs).

The motor turn-on time of one second may seem short—but by pulsing the power to a DC motor, the average motor current can be changed and the speed controlled. This is much more efficient than controlling speed with a variable resistor in series with the power source. This method is used for power control of battery-powered fork lift trucks and electric cars.

The diode D_1 in parallel with the motor in Figure 13–6 (called a freewheeling diode) serves two purposes: It prevents the voltage on the anode of SCR_2 from becoming greater than the 28V supply voltage (plus one diode drop); it also allows a path for motor current when SCR_1 turns off. This current is driven by the voltage generated when the inductive motor field collapses, and we find that even if the motor is under pulse speed control, D_1 serves to keep the motor current nonpulsing or DC.

FIRING AN SCR WITH A UNIJUNCTION TRANSISTOR

If, for example, we need to fire an SCR at a given frequency for motor speed control, we can provide a solid trigger pulse and provide the required frequency by using a **unijunction transistor**.

The operation of a unijunction transistor is explained with reference to the functional equivalent bipolar transistor circuit shown in Figure 13–7.

The configuration of the transistors in Figure 13–7 is similar to that shown for the SCR in Figure 13–4, except an input signal is applied to the emitter of the PNP transistor rather than to the base of the NPN transistor.

If we ignore the voltage drop across R_3, because it is small, we can say the voltage divider consisting of R_1 and R_2 will provide a voltage of $0.75V_{cc}$ at the base of Q_1. Now if we input a voltage at the emitter of Q_1 and increase it so that it is equal to $0.75V_{cc} + 0.6V$, transistor Q_1 will start conducting and cause current to flow into the base of Q_2. Now Q_2 turns on and draws part of its collector current from the base of Q_1, and the remainder from R_1. This causes the base voltage of Q_1 to drop, turning Q_1 on harder and both transistors conduct heavily. The current of both transistors flowing through R_3 causes a voltage drop across this resistor, and the output voltage V_o goes much more positive.

If the initial voltage provided to the emitter of Q_1 is from a charged capacitor, the capacitor will discharge through the turned-on transistors. When the capacitor voltage drops,

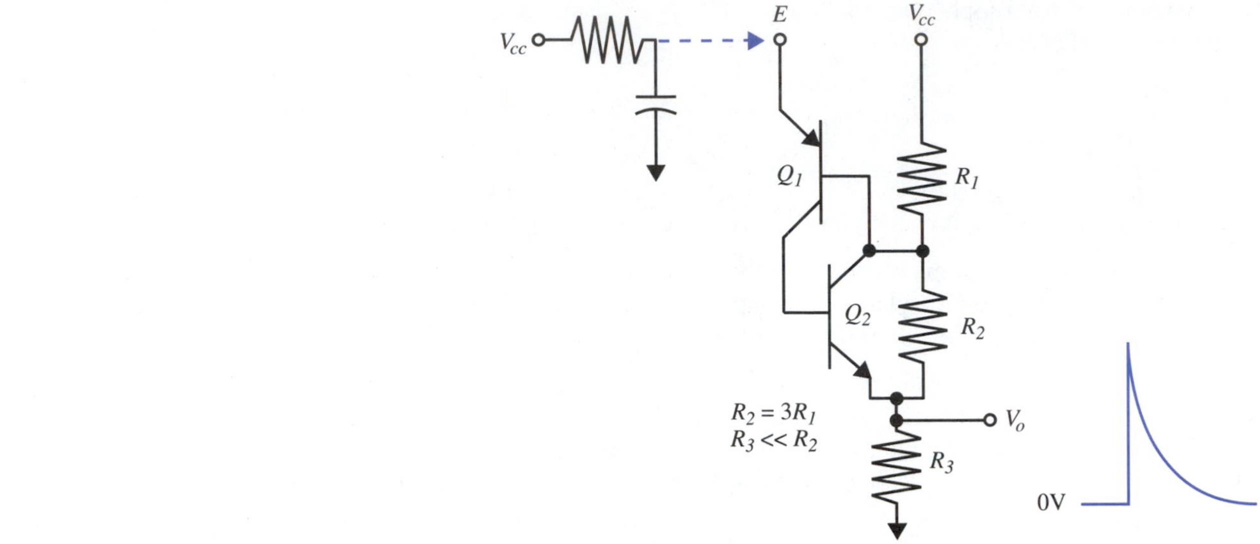

FIGURE 13–7 Functional equivalent of the unijunction transistor

the current through the transistors will fall low enough to turn the transistors off. The effect on V_o is a sharp positive jump when the transistors turn on, and then an exponential decay from V_o back to the original voltage (near 0V) as the capacitor discharges.

Although our transistor equivalent is a good model of a unijunction transistor (we can emulate a unijunction with an NPN, PNP transistor, and two resistors), the actual construction is somewhat different, as shown in Figure 13–8.

As can be seen from Figure 13–8a, the structure of a unijunction is rather similar to that of a JFET. There is an N-type channel (bar) with a small P-type region diffused into the channel. Connections to the unijunction are labeled "base 1," "base 2," and "emitter." In operation, base 2 is connected to the positive power supply, the emitter is typically connected to a timing capacitor, and base 1 is connected via the load resistor to ground.

When the voltage on the emitter (V_e) is sufficient to forward bias the PN diode, electrons flow from base 1 to the emitter. The material of the channel is such that an increase of electrons causes a decrease in the resistance of the channel between emitter and base 1. This decrease in resistance allows more current to flow, which further decreases resistance. The result is that the unijunction timing capacitor is quickly discharged.

Since the current drops to a low value when the capacitor is almost discharged, the resistance between the emitter and base 1 now increases—further reducing the current. The final result is that the emitter-base 1 resistance reverts to its high-resistance state.

The forward conduction characteristics of the unijunction transistor are illustrated in Figure 13–9.

The **valley point**, on the curve, is the point at which the current causes lowest resistance between emitter and base 1 and also the least voltage drop (V_v) between these terminals.

Further increase in current beyond this point causes an increase in emitter-base 1 voltage drop (saturation or constant resistance region). The $I_{b2} = 0$ curve is with base 2 disconnected and just shows the forward conduction of the emitter-base 1 diode.

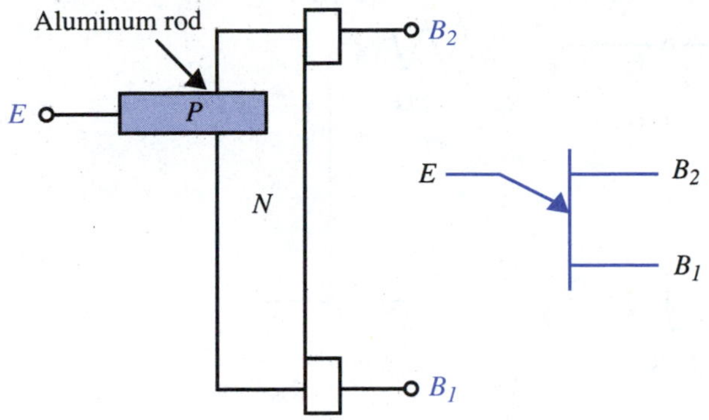

FIGURE 13–8 Unijunction transistor

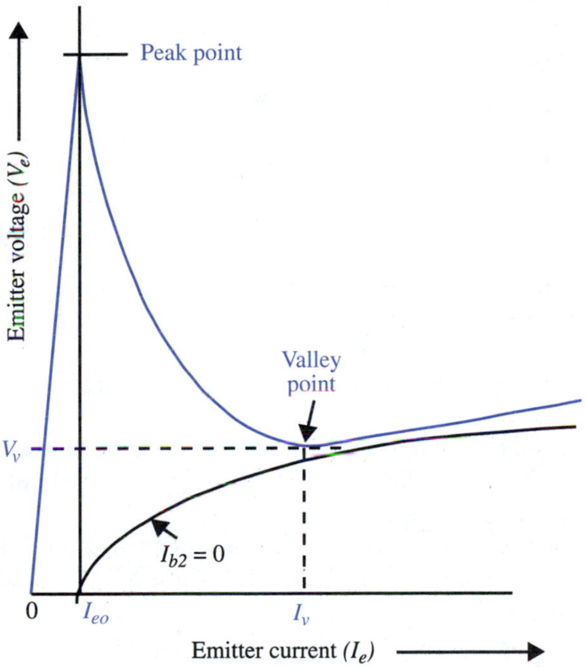

FIGURE 13–9 Unijunction characteristics

A typical unijunction is the 2N4871. Specifications for this device are as follows:

- Intrinsic standoff ratio (η) is 0.77 of the voltage from base 2 to base 1 (V_{12}).
- Peak point emitter current is 5μA.
- Valley point current is 5mA.

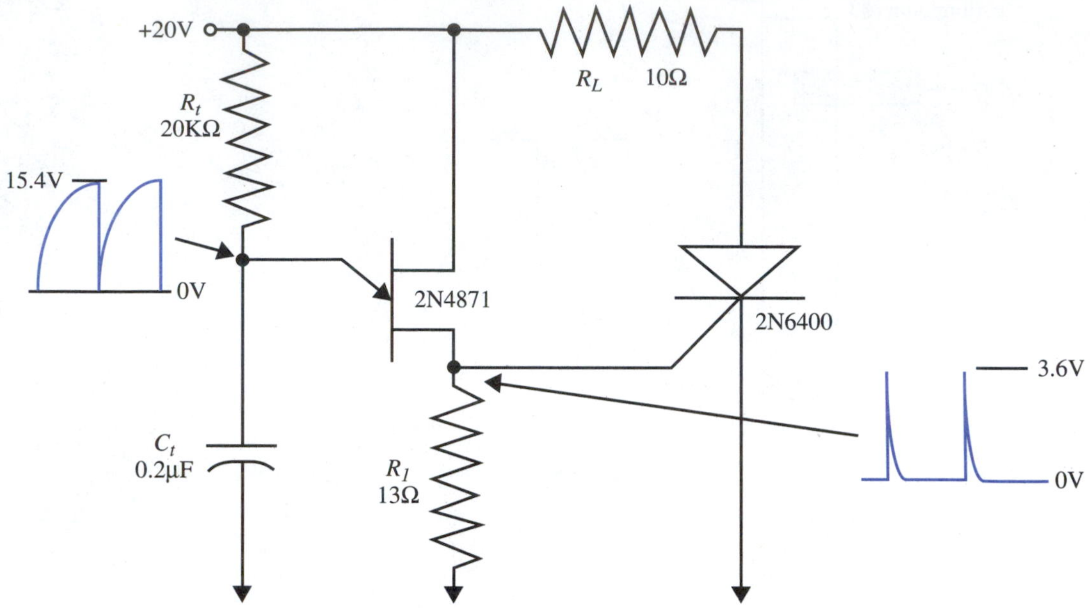

FIGURE 13–10 Unijunction in a firing circuit

The **intrinsic standoff ratio** defines where the emitter is connected on the interbase (B_2 to B_1) resistance and therefore determines the **peak point** firing voltage for the unijunction (the actual firing voltage is V_{12} + 0.6V).

An actual SCR firing circuit, using a 2N4871 unijunction transistor to fire a 2N6400 SCR, is shown in Figure 13–10.

After power is turned on, C_t starts to charge via R_t. When the voltage on C_t reaches peak point (ηV_{cc} + 0.6V), the unijunction conducts and provides a current pulse through R_1 that fires the SCR and discharges C_t.

With the capacitor discharged, the current from emitter to base 1 falls low enough to turn off the unijunction (capacitor voltage is about 1.5V at turn-off). The capacitor now starts to charge again and the cycle repeats.

The time to charge the capacitor is derived from the equation

$$v = (V_{cc} - 1.5\text{V})(1 - e^{-t/R_tC_t})$$

by making v equal to ($\eta V_{cc} + 0.6V$) and solving for t,

$$t = -R_tC_t \times ln\left[\frac{(V_{cc}(1 - \eta) - 2.1\text{V})}{(V_{cc} - 1.5\text{V})}\right] \qquad \textbf{(equation 13–3)}$$

where *ln* is the natural logarithm.

The unijunction frequency (f_p) at which current pulses are provided to R_1 is

$$f_p = \frac{1}{t}$$

$$f_p = -\frac{1}{R_tC_t \times \ln\left[\frac{(V_{cc}(1-\eta) - 2.1V)}{(V_{cc} - 1.5V)}\right]} \qquad \textbf{(equation 13–4)}$$

EXAMPLE

Find the pulse frequency for the unijunction circuit of Figure 13–10.

SOLUTION

Remembering that the intrinsic ratio (η) for the 2N4871 is 0.77 and using equation 13–4,

$$f_p = -\frac{1}{20K\Omega \times 0.2\mu F \times \ln\left[\frac{20V(1-0.77) - 2.1V}{(20V - 1.5V)}\right]}$$

$$f_p = 125Hz$$

This frequency can be raised by lowering the value of R_t. Changing C_t will change the pulse frequency but it will also change the amount of current through (and voltage across) R_1. (See Figure 13–11.)

The maximum required input firing voltage for the 2N6400 is 1.5V, and the maximum current is 30mA. This looks like a worst-case input resistance of 1.5V/30mA or 50Ω.

The R_{b1} resistor (13Ω) is in parallel with this resistance, and the effective R_{b1} resistor is then 13‖50 or 10Ω.

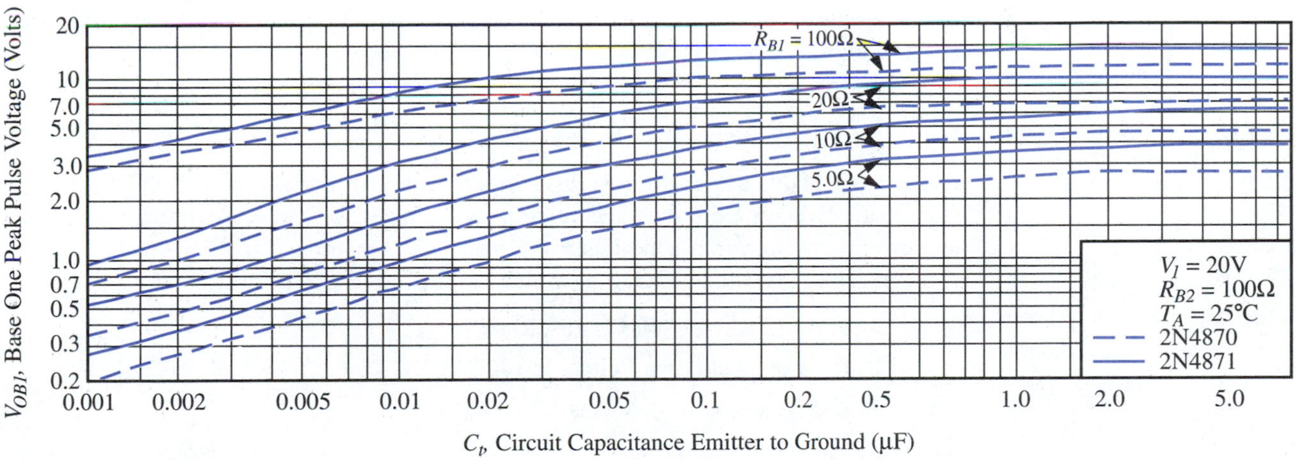

FIGURE 13–11 R_1 voltage versus C_t. *Courtesy Motorola Semiconductor.*

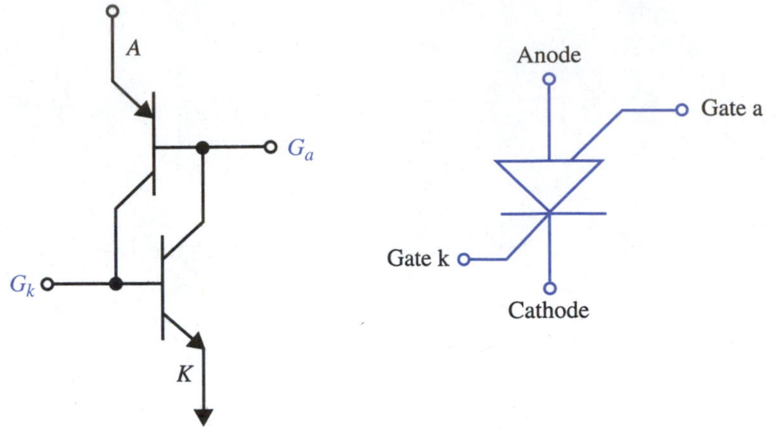

FIGURE 13–12 Silicon controlled switch

From Figure 13–11 we see that the voltage developed across an effective R_{b1} of 10Ω, when the unijunction conducts, is about 4V($C_t = 0.2\mu f$), which exceeds the required value of 1.5V and therefore guarantees the SCR will fire.

In summary, the unijunction is used with an SCR to provide a solid (guaranteed to fire) trigger pulse to the SCR, which insures that it fires at the desired time.

SILICON CONTROLLED SWITCH (SCS)

If we add another gate connection to the base of the PNP transistor in the SCR equivalent circuit of Figure 13–4, we have the circuit for an SCS that is shown in Figure 13–12.

The SCS can be fired by taking the gate G_k voltage high, as with the SCR, or taking the gate G_a voltage low, which will cause the PNP transistor to conduct. This device has the additional feature of allowing turn-off by taking G_k voltage low or G_a voltage high.

The disadvantage of the SCS is that it conducts a maximum current in the high mA range rather than the amps possible with an SCR.

GATE TURN-OFF (GTO) THYRISTOR

As the name of this device implies, this thyristor can be turned off by reversing the voltage at the gate. It differs from the SCS in that it has a single gate and can conduct much heavier currents and sustain much higher voltages.

A typical GTO is the MGTO1000, which has the following specifications:

- Forward Current Max is 18A RMS.
- Reverse Breakdown Voltage is 1000V.
- Gate Firing Voltage Max is 1.5V.
- Gate Firing Current Max is 300mA.
- Gate turn-off voltage is 12V.
- Gate turn-off current peak is 5A.

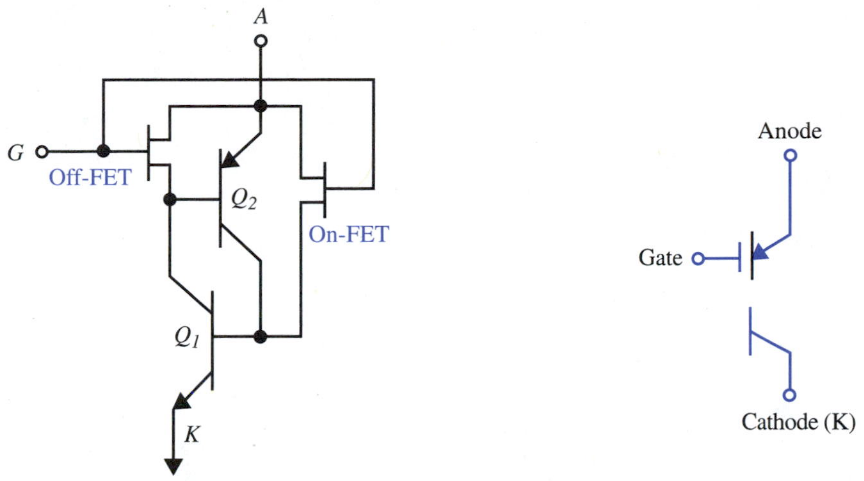

FIGURE 13–13 MCT equivalent circuit and symbol

Notice that the GTO gate trigger voltage and current are similar to the SCR; but the gate signal required to turn off the device is considerably greater than the turn-on value.

MOSFET CONTROLLED THYRISTOR

Derived by Harris Semiconductor from the technology of the insulated gate bipolar transistor (IGBT) is the newest power switching device, the MOSFET controlled thyristor (MCT). This device is an SCR that can be turned on or off via the gate of a dual MOSFET. The equivalent circuit and symbol are shown in Figure 13–13.

The on-FET shown is a P-type MOSFET; the off-FET is an N-type. Applying a negative voltage to the gate will turn on the on-FET and turn off the off-FET. This condition will turn on Q_1 and cause the SCR to latch on. Applying a positive voltage to the gate will cause conduction of the off-FET, which generates a short across the base-emitter of Q_2. This drops the SCR current to a value below the holding current, and the SCR turns off.

The advantage of the MCT is that it can control currents over 100A with very little voltage drop (about 1.5V at 100A) and withstand forward voltages over 1,000V.

Comparing the MCT with the IGBT, the MCT has about one-third the forward voltage drop at a given current. Losses per switching cycle are about twice as high with an MCT as with the IGBT—so the decision as to which is used is determined by the relative values of conduction losses to switching losses.

THE TRIAC

A device similar to an SCR but capable of conducting in both directions for AC power control is the **triac**. In simple terms, a triac can be considered as consisting of back-to-back SCRs, as shown in Figure 13–14.

Notice that the triac shown as two SCRs in Figure 13–14a has a common gate connection. A positive pulse relative to MT_1 on the gate will cause SCR_2 to conduct while a posi-

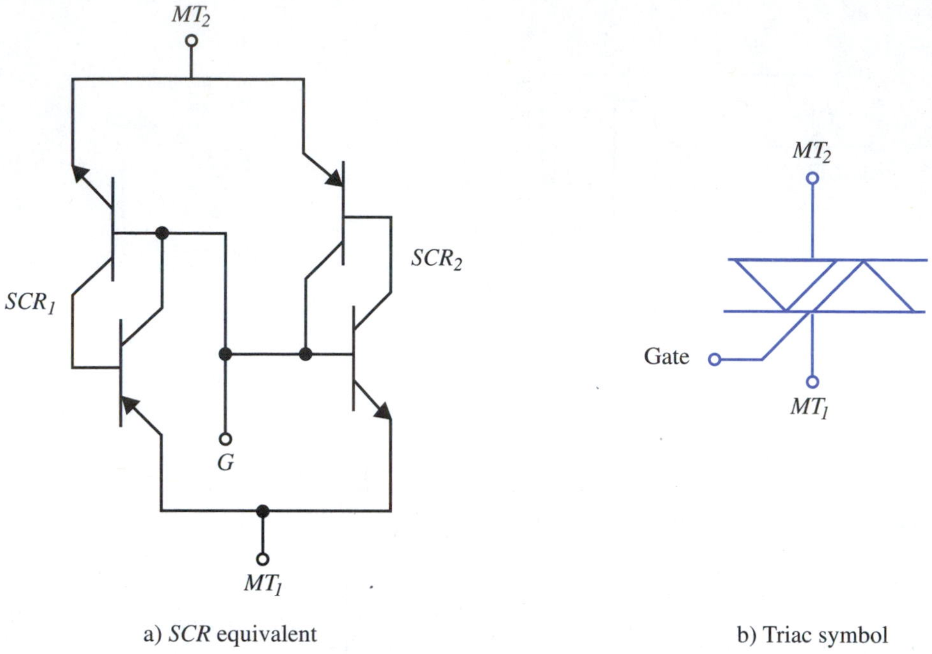

FIGURE 13–14 Triac/SCR equivalent circuit

tive pulse relative to MT_2 causes conduction of SCR_1. For conduction to occur, the anode of each SCR must be more positive than its cathode.

Actually, a triac can be fired four different ways, as indicated in Table 13–2.

The quadrants 1(+) and 3(−) operation can be understood by considering the triac as back-to-back SCRs. Quadrants 1(−) and 3(+) are the result of the actual configuration of the triac.

Figure 13–15 shows the actual semiconductor arrangement along with the equivalent circuits for 1(+) and 3(−). Since quadrants 1(−) and 3(+) are not often used, we shall not consider their operation.

With quadrant 1(+) operation, MT_2 is more positive than MT_1, and a positive pulse on the gate causes the P_2/N_3 junction to forward bias. This turns on Q_1 which then turns on Q_2, and the triac is conducting.

TABLE 13–2 Triac Triggering Modes

Gate to MT_1 Voltage	*MT_2 to MT_1 Voltage*	*Quadrant*
Positive	Positive	1(+)
Negative	Positive	1(−)
Positive	Negative	3(+)
Negative	Negative	3(−)

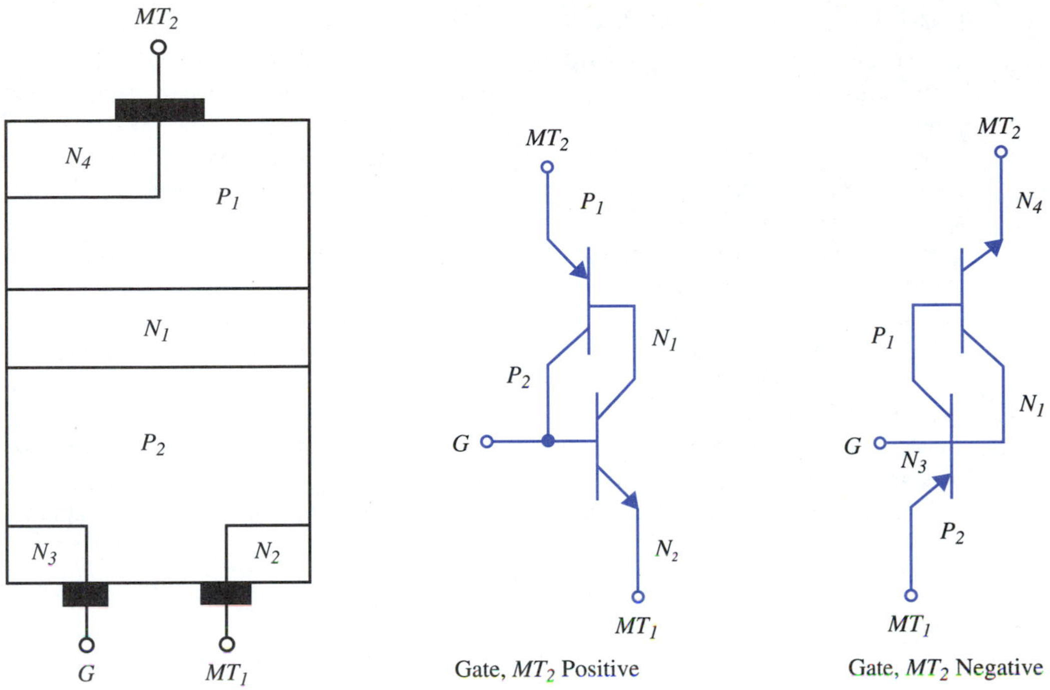

FIGURE 13–15 Actual triac configuration

In quadrant 3(−) operation, MT_1 is more positive than MT_2. Making the gate negative turns on Q_3 by forward biasing the P_2/N_4 junction. Q_4 is turned on from the collector current of Q_3, and the triac goes into conduction.

AN ACTUAL TRIAC

A typical triac is the Motorola MAC228A6 which has the following specifications:

- Triac RMS current rating is 8A.
- Triac Breakdown voltage is 400V.
- Triac Maximum required trigger current is 5mA [10mA for quadrant 3(+)].
- Triac Maximum required trigger voltage is 2V [2.5V for quadrant 3(+)].

The RMS current rating is the maximum continuous current that can be passed through the device without damage. If the breakdown voltage is exceeded, the device can turn on without a trigger being applied—with possible damage to the junctions.

To ensure that the triac will trigger symmetrically, both the maximum required trigger voltage and current must be exceeded.

CIRCUIT OPERATION OF THE TRIAC

The SCR is used for DC power control both as a switch or as a controllable rectifier to regulate the input AC voltage for a DC power supply. The triac was developed for bidirectional power control of AC circuits.

Although the triac can be used as a simple power on-off switch (called static operation) for AC circuits, its main use is to control the amount of effective (power-generating) voltage and current provided to a load.

Figure 13–16a shows an example of a triac circuit that controls the power to the load (R_L) by varying the firing time relative to the input AC waveform. R_1 and C_1 serve to delay (phase shift) the voltage applied to the gate of the triac. When the triac fires, the supply voltage will be dropped across the load resistor, and the voltage applied to the gate will go to zero. We don't have to worry about turning off a triac since automatic turn-off occurs whenever the input AC waveform swings through zero.

With the triac MT_2 terminal going positive, the voltage across capacitor C_1 will be positive, and when MT_2 is going negative, the capacitor will charge to a negative voltage.

The voltage (v_c) developed across the capacitor C_1 can be determined by the equation

$$v_c = V_p\left(\frac{X_c}{Z}\right)\sin\phi_g \qquad \textbf{(equation 13–5)}$$

where

$$X_c = \frac{1}{2\pi f C_1}$$
$$Z = \sqrt{(X_c)^2 + (R_1)^2}$$

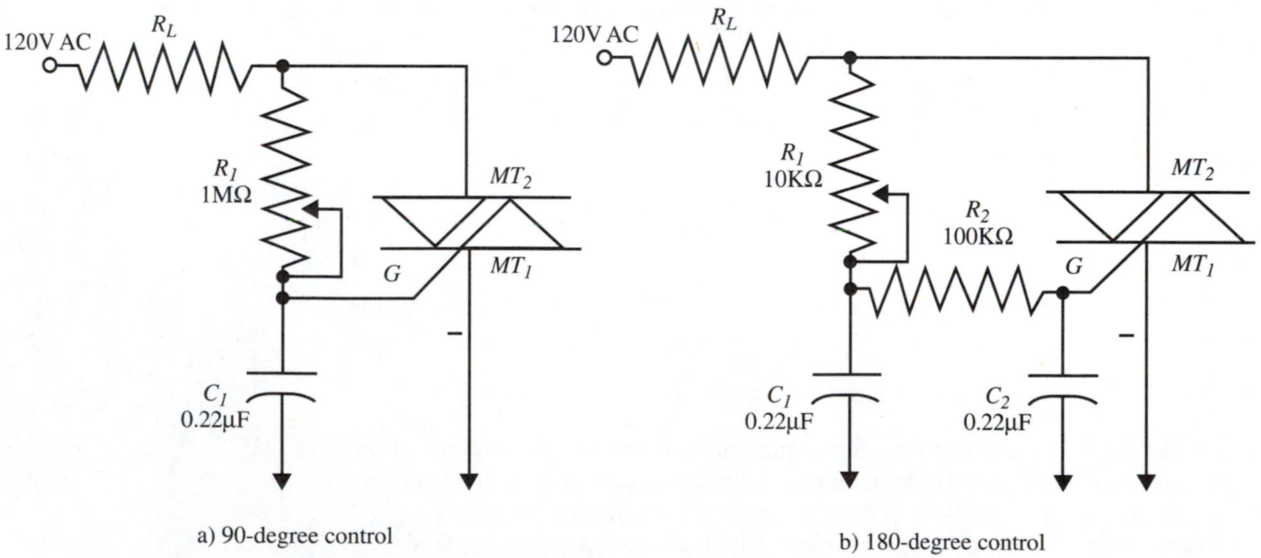

FIGURE 13–16 Triac AC power control

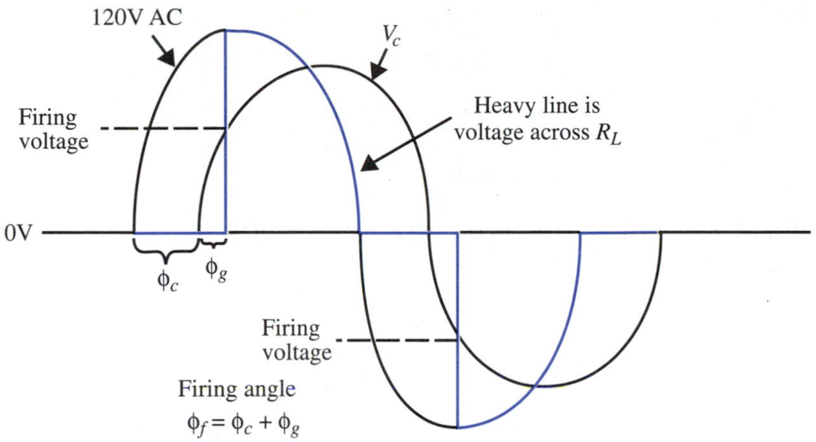

FIGURE 13–17 Triac waveforms

At the firing point, the capacitor voltage v_c must just exceed the required gate firing voltage for the triac. If we reference to the phase angle of the input power source waveform, we can see in Figure 13–17 that the angle at which the triac fires (firing angle) depends on when the gate firing voltage is reached.

With reference to Figure 13–17, the angle ϕ_g represents the phase change (delay) since the zero crossing point of the sine wave developed across the capacitor. The triac firing angle (ϕ_f) is defined as the change in phase from the zero crossing of the input AC to the phase angle on the input waveform when the triac fires. Its value consists of the phase shift across the capacitor (ϕ_c) plus the elapsed phase to the gate firing point, or

$$\phi_f = \phi_c + \phi_g \qquad \textbf{(equation 13–6)}$$

where

$$\phi_c = \text{arc tan } R_1/X_c$$

$$\phi_g = \text{arc sin}\left(\frac{v_c Z}{V_p X_c}\right)$$

EXAMPLE

Find the firing angle for the circuit of Figure 13–16a with a power source voltage of 120V ($V_p = 120\text{V} \times \sqrt{2}$), a frequency of 60Hz, and R_1 set for 5KΩ. Also find the angle when R_1 is changed to 800KΩ.

SOLUTION

With the MAC228A6, the maximum required firing voltage is 2.5V, so we set v_c equal to this value. The capacitive reactance X_c is

$$X_c = \frac{1}{(2\pi \times 60 \times 0.22\mu\text{F})}$$

$$X_c = 12.1\ \text{K}\Omega$$

And

$$Z = \sqrt{(X_c)^2 + (R_1)^2}$$

$$Z = \sqrt{(12.1\text{K}\Omega)^2 + (5\text{K}\Omega)^2}$$

$$Z = 13.1\text{K}\Omega$$

Using equation 13–6 to find for ϕ_f with R_1 of 5KΩ and v_c of 2.5V,

$$\phi_f = \phi_c + \phi_g$$

$$\phi_f = \text{arc tan}\frac{R_1}{X_c} + \text{arc sin}\left(\frac{v_c Z}{V_p X_c}\right)$$

$$\phi_f = \text{arc tan}\left(\frac{5\text{K}\Omega}{12.1\text{K}\Omega}\right) + \text{arc sin}\left(\frac{2.5\text{V} \times 13.1\text{K}\Omega}{170\text{V} \times 12.1\text{K}\Omega}\right)$$

$$\phi_f = 22.45^\circ + 0.91^\circ$$

$$\phi_f = 23.4^\circ$$

Changing R_1 to 800KΩ;

$$Z = \sqrt{(12.1\text{K}\Omega)^2 + (800\text{K}\Omega)^2}$$

$$Z = 800\Omega$$

$$\phi_f = \text{arc tan}\left(\frac{800\text{K}\Omega}{12.1\text{K}\Omega}\right) + \text{arc sin}\left(\frac{2.5\text{V} \times 800\text{K}\Omega}{170 \times 12.1\text{K}\Omega}\right)$$

$$\phi_f = 89.1^\circ + 76.5^\circ$$

$$\phi_f = 165.6^\circ$$

What these results mean is that the triac will turn on 23.4° (R_1 = 5KΩ) or 165.6° (R_1 = 800KΩ) after the input sine wave passes through zero volts (at 0 degrees). The longer the delay in turn-on, the less power delivered to the load resistor.

The maximum firing delay we can achieve with the circuit of Figure 13–16a is a little less than 180°, since the maximum phase shift across a single RC circuit is a little less than 90° and the peak of the sine wave occurs 90° later. Adding a second RC section, as shown in Figure 13–16b, will guarantee a delay out to 180°. The second RC components have a resistance (impedance) ten times greater than the first set to avoid loading and interaction between the two networks.

In our example of a firing control circuit for the triac, we used the worst-case gate firing voltage for the MAC228A6 of 2.5V. But what if the particular triac we use has a lower firing voltage! Then the firing angle will not be as we predicted.

EXAMPLE

Find the firing angle for the triac in Figure 13–16a if the firing voltage is 1.25V and R_1 is set for 800KΩ.

SOLUTION

This means v_c is 1.25V, and using the formulas and some results from the previous example, Z is 800KΩ, X_c is 12.1KΩ, and V_p is 170V.

$$\phi_f = \arctan\left(\frac{R_1}{X_c}\right) + \arcsin\left(\frac{v_c Z}{V_p X_c}\right)$$

$$\phi_f = \arctan\left(\frac{800\text{K}\Omega}{12.1\text{K}\Omega}\right) + \arcsin\left(\frac{1.25 \times 800\text{K}\Omega}{170 \times 12.1\text{K}\Omega}\right)$$

$$\phi_f = 89.1^\circ + 29.1^\circ$$

$$\phi_f = 118.2^\circ$$

This is 47.4° less than the firing angle with a gate firing voltage of 2.5V, which demonstrates firing angle sensitivity to gate firing voltage.

FIRING A TRIAC WITH A DIAC

A way to make the firing angle independent of the variable firing voltage of the triac is to fire the triac with a **diac**, as indicated in Figure 13–18.

The diac is like a triac without the gate connection. It conducts when the voltage across the device reaches the break-over voltage in either direction. Diacs come with different breakdown voltages, with values ranging from 20V to 36V and a tolerance as close as +2V. With a 20V unit, a tolerance of 2V is 10 percent; compare this with the variation in gate firing voltage for a triac, which is greater than 100 percent, and we can see why diacs are often used to fire triacs.

In the circuit, as soon as the diac breaks down, the voltage that was across the diac is transferred to the gate of the triac. Since the voltage now on the gate is much greater than the required firing voltage, the triac immediately fires. This makes the triac firing only a function of the diac breakdown voltage.

EXAMPLE

Find the range of firing angles for the triac in the circuit of Figure 13–18 if the diac breakdown voltage (V_d) is 28 ±2V and R_1 ranges from 1KΩ to 68KΩ.

SOLUTION

The minimum firing angle will occur with an R_1 of 1KΩ and the minimum value for V_d of 26V. Since the resistance of the diac before breakdown is very high, we shall consider that all of the capacitor voltage v_c appears across the diac. Then at breakdown

$$v_c = V_d = 26\text{V}$$

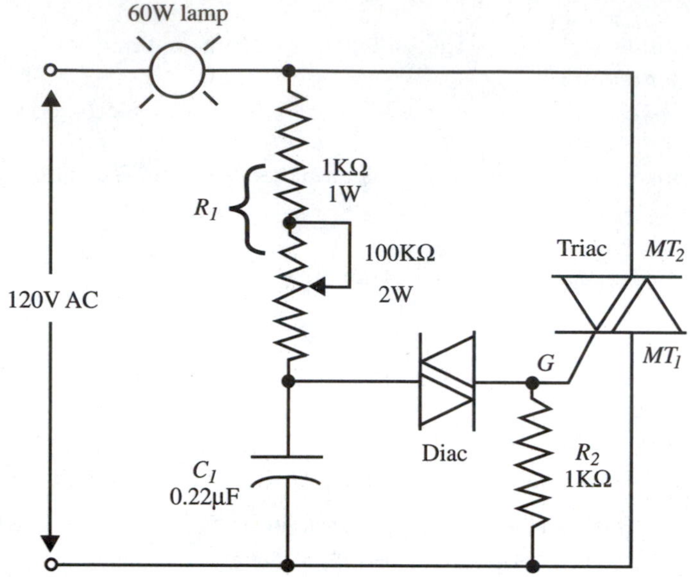

FIGURE 13–18 Firing a triac with a diac

From our previous results, X_c is 12.1KΩ, with R_1 of 1KΩ and Z of 12.14KΩ.

$$\phi_f = \text{arc tan}\left(\frac{R_1}{X_c}\right) + \text{arc sin}\left(\frac{v_c Z}{V_p X_c}\right)$$

$$\phi_f = \text{arc tan}\left(\frac{1\text{K}\Omega}{12.1\text{K}\Omega}\right) + \text{arc sin}\left(\frac{26\text{V} \times 12.14\text{K}\Omega}{170 \times 12.1\text{K}\Omega}\right)$$

$$\phi_f = 4.72° + 8.83°$$

$$\phi_f = 13.6°$$

For maximum phase delay,

$$R_1 = 68\text{K}\Omega$$

$$Z = 69\text{K}\Omega$$

$$v_c = V_d = 30\text{V}$$

$$\phi_f = \text{arc tan}\left(\frac{68\text{K}\Omega}{12.1\text{K}\Omega}\right) + \text{arc sin}\left(\frac{30\text{V} \times 69\text{K}\Omega}{170 \times 12.1\text{K}\Omega}\right)$$

$$\phi_f = 79.9° + 90°$$

$$\phi_f = 169.9°$$

The range then is 13.6° to 169.9° with this diac firing circuit. Notice that basically the same delay is achieved with 68KΩ in this circuit as was obtained with 800KΩ without the diac. The reason for this is the additional delay required to reach the higher firing voltage when a diac is used.

Note that if the 100KΩ potentiometer is shorted (R_1 of 1KΩ), and the triac doesn't fire because it is defective, the power dissipation in the firing circuit will be excessive, and R_1 and R_2 will burn out. This problem can be avoided by putting a 1/100 A fuse in series with R_1.

EFFECTIVE LOAD POWER IN A TRIAC CIRCUIT

We have studied how the triac trigger can be delayed to control the power supplied to the load, but what is the relationship between trigger delay angle and load power?

In order to find the relationship, we have to use integral calculus to determine the effective power (P_e) of a partial sine wave. The process involves squaring the sine wave voltage ($V_p sin\phi$) and taking the calculus integral over the angle when load current is flowing—which is called the **conduction angle**

$$\text{conduction angle} = 180° - \phi_f$$

The result is the following equation for effective power:

$$P_e = \frac{V_p^2}{R\pi}\left\{\frac{\pi}{2} - \left[\frac{\sin(2\phi_f) + \phi_f}{2}\right]\right\} \qquad \textbf{(equation 13–7)}$$

where ϕ_f is in **radians**, i.e., 180°/π or 57.3°, and R is the load resistance.

EXAMPLE

Find the effective load power for the firing angle delays of the previous example with an input voltage of 120V RMS (V_p is 170V) and a load resistance of 100Ω.

SOLUTION

Converting ϕ_f of 13.6° to radians,

$$\phi_f = \frac{13.6°}{57.3°} = 0.227 \text{ radians}$$

Substituting in equation 13–7,

$$P_e = \frac{(170)^2}{100\pi}\left\{\frac{\pi}{2} - \left[\sin(2 \times 0.227) + \frac{0.227}{2}\right]\right\}$$

$$P_e = 93.7\text{W}$$

With ϕ_f of 169.9°,

$$\phi_f = 169.9°/57.3° = 2.97 \text{ radians}$$

$$P_e = \frac{(170)^2}{100\pi}\left\{\frac{\pi}{2} - \left[\sin(2 \times 2.97) + \frac{2.97}{2}\right]\right\}$$

$$P_e = 38.85\text{W}$$

With ϕ_f of 0° (no delay), equation 13–7 yields the common equation $(V_p)^2/2R$, or 144.5W. With ϕ_f of 180° (π radians), equation 13–7 gives, as expected, 0W.

So, as stated before, the larger the firing angle, the smaller the power delivered to the load.

ZERO CROSSING POWER CONTROL

When a triac is turned on, the sharp increase in load current generates electromagnetic interference (EMI) and disruptive and possibly destructive transient voltages. This effect can be minimized by turning on the triac only when the voltage across the device is close to zero volts, or in other words,

$$\phi_f \approx 0°$$

Power control with this method is achieved by turning the triac on for a given number of cycles of the input sine wave and then off for another number of cycles. The average power P_a delivered to the load is determined from the number of "on" cycles divided by the total of "off" and "on" cycles.

EXAMPLE

If the power delivered when the triac is conducting is 200W, determine the average power (P_a) delivered to the load if the triac is on for one cycle and off for three.

SOLUTION

$$P_a = \frac{200\text{W} \times 1}{(1 + 3)}$$

$$P_a = 50\text{W}$$

The circuitry for zero crossing control can be contained within the triac, or it can be external of the device.

An example of a zero crossing trigger source is the Harris Semiconductor CA3059. This device generates an output trigger pulse at the zero crossing of the input sine wave when certain control conditions are satisfied. It is capable of providing 10mA at a voltage level of 3V during the zero crossing of the ac line. The internal schematic for this device is shown in Figure 13–19.

The CA3059 can operate directly from the ac line with a nominal input voltage range of 24V to 277V RMS, depending on the value of the dropping resistor R_s as indicated in Table 13–3.

OPERATION OF THE CA3059

We shall now look at the internal operation of the CA3059 with reference to Figure 13–19. The input AC current is limited by R_s, and the back-to-back input (zener) diodes limit the voltage input to the device. The clipped sine wave, resulting from the limiting, is half-wave rectified, and a peak voltage of about 6.4V is a available at Pin 2. Attaching a filter capacitor to this pin converts the waveform to a 6.4V DC output.

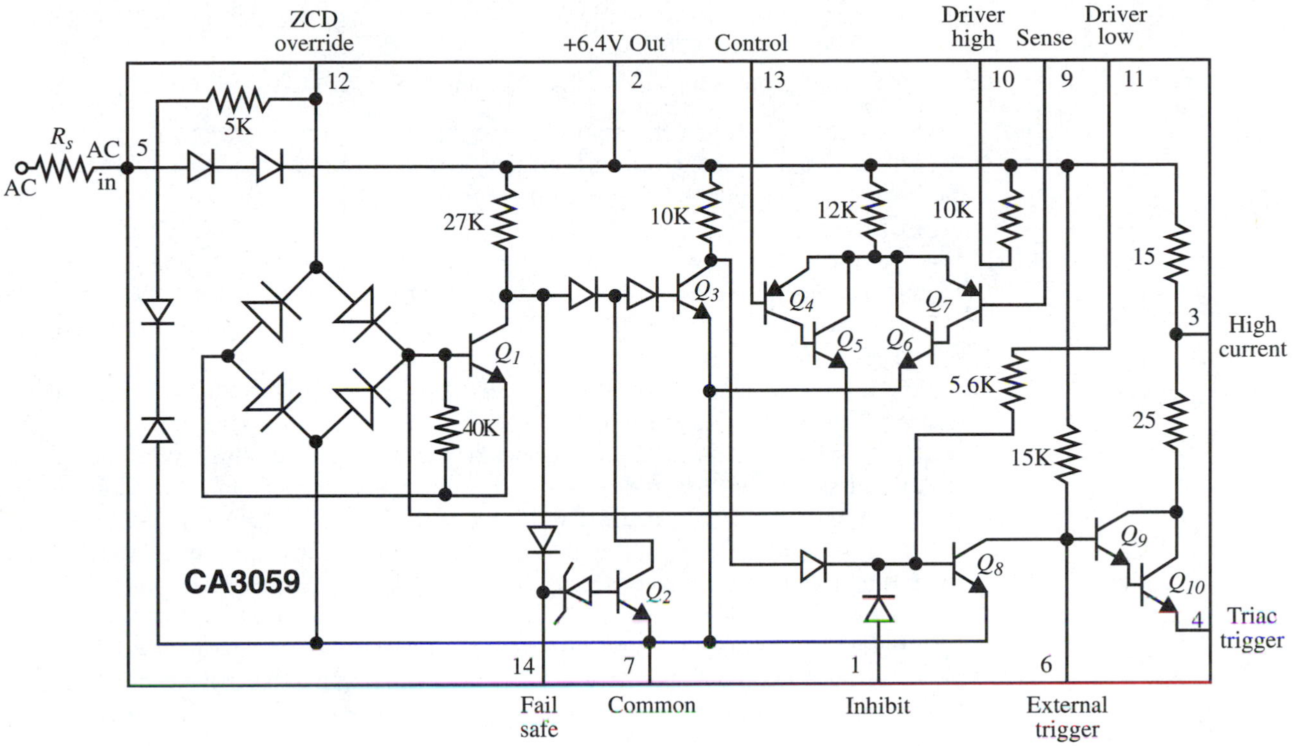

FIGURE 13–19 CA3059 zero crossing trigger source. *Courtesy Harris Semiconductor.*

The clipped sine wave is also passed through a full-wave bridge to provide a positive-going full-wave rectified signal at the base of Q_1. Let us first consider the case when this voltage is positive enough to turn on Q_1.

- With Q_1 conducting, its collector is low, turning off Q_3.
- With Q_3 off, its collector is high, which turns on Q_8, and its collector goes low.

TABLE 13–3 CA3059 Dropping Resistor (R_S) Values

AC Input Voltage (50/60 or 400Hz)	*Input Resistor (R_S)*	*Wattage Rating for R_S*
24V AC	2KΩ	0.5W
120V AC	10KΩ	2W
208/230V AC	20KΩ	4W
277V AC	25KΩ	5W

- A low on the collector of Q_8 is transferred to the output via the Darlington connection of Q_9 and Q_{10}.
- The output being low means that there is no signal to trigger a triac.

The other case is when the input AC waveform is approaching zero. Now the voltage will not be high enough to turn on Q_1. Following through the same analysis as we did before, we conclude that now the output will go high to a voltage level capable of firing a triac.

Let us now consider the control part of the circuit. First, to simplify the analysis, we shall assume that Pins 9, 10, and 11 are all jumpered together. With this connection, the voltage at the base of Q_7 will be at about +3V because of the divider action of the 10KΩ and 9.6KΩ resistors. The voltage at the emitter of Q_7 will be 0.6V higher or +3.6V.

Now if the voltage at the base of Q_4 (Pin 13) is made greater than +3V, Q_4 will be turned off along with Q_5. The emitter of Q_5 will float to the same signal as the base of Q_1, and the Q_1 circuit will operate in the normal manner as described earlier. With the voltage at Pin 13 less than +3V, both Q_4 and Q_5 transistors will turn on, and the emitter of Q_5 (and Q_1 base) will rise to a voltage about 1V less than the voltage at Pin 13. If the voltage at Pin 13 is in the range from +1V to +3V, the voltage at the emitter of Q_4 will be high enough to keep Q_1 in conduction, and the output can't provide a trigger pulse. In other words, it is inhibited.

So the state of the output is controlled by the relative voltages on Pins 9 and 13. It should be realized that Pin 9 does not have to be connected to the internal resistors but could be connected to an external voltage. The conditions of Pin 13 to allow a trigger pulse at the output can be described by the following expression:

$$1V < V_{13} < V_9. \quad \text{(trigger pulse inhibited)}$$

Lowering the voltage on Pin 9 allows a greater range of voltage on Pin 13 without inhibiting the output.

Connecting a positive voltage greater than about 1.2V to inhibit pin 1 will turn on Q_8, and this will also inhibit the output. Taking the Fail-Safe Pin 14 to ground will cause the collector of Q_1 to go low—again inhibiting the output.

A lower-cost zero crossing trigger is the CA3079. This device is similar to the CA3059 minus the inhibit functions. It has the same pin connections, but Pins 1, 6, and 14 are not used.

A lamp control circuit, using the CA3079 and a MAC228A6 sensitive gate triac, is shown in Figure 13–20. The circuit operation is as follows:

The unijunction transistor operates as a relaxation oscillator with a period of about 0.2 sec. Charging current for the 1μF capacitor is provided by the PNP transistor, which is set up as a constant current source to generate a linear ramp across this capacitor, which ramps from about 1V to 4V. This ramping voltage is applied to Pin 9, and a potentiometer-adjusted voltage is applied to Pin 13.

The relationship between the waveforms and the voltages is shown in Figure 13-21. In the figure, we can see that the triac trigger waveforms occur at the same time as the zero crossings of the input line ac line. They are present at Pin 4 of the CA3079 only when the voltage on Pin 13 is greater than the voltage on Pin 9 or the voltage on Pin 9 is less than about 1V.

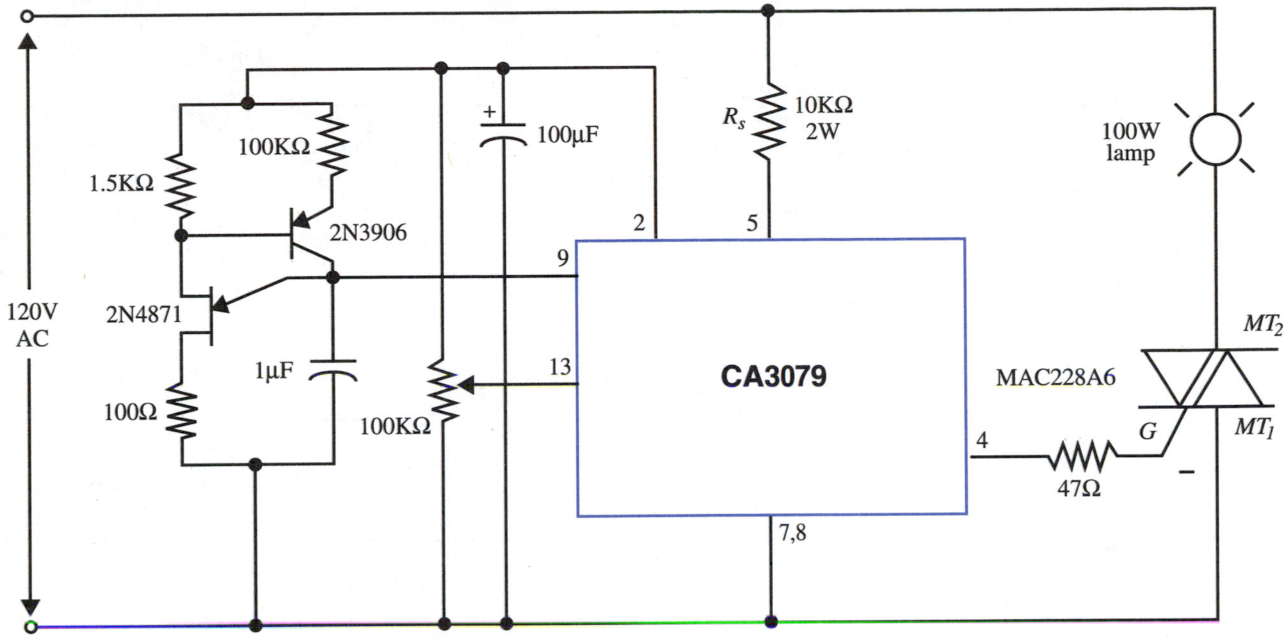

FIGURE 13–20 Low-noise lamp dimmer

STATIC SWITCHING

If we just want to use the triac as a simple switch that is either on or off (called **static switching**), we can use the circuit shown in Figure 13–22. All we need to do is provide a current to the gate sufficient to trigger the triac. When the trigger current is removed, the triac will turn off as the input sine wave next passes through zero volts.

Two methods to trigger the triac are shown in Figure 13–22:

1) The gate current is enabled by closing the low-power contacts of a switch or relay.
2) Light from an LED causes the resistance of a photoresistor to drop sufficiently to allow the required gate current.

The advantage of static switching is that we are able to control a large current by providing a small (gate) current.

OPTICAL POWER CONTROL

When we studied relay power control, we learned that the relay provides isolation between the power source used for control, and the power source being controlled. We can achieve the same sort of isolation, without the low reliability of a mechanically activated component, by using optical coupling (4N designation is used with optocouplers).

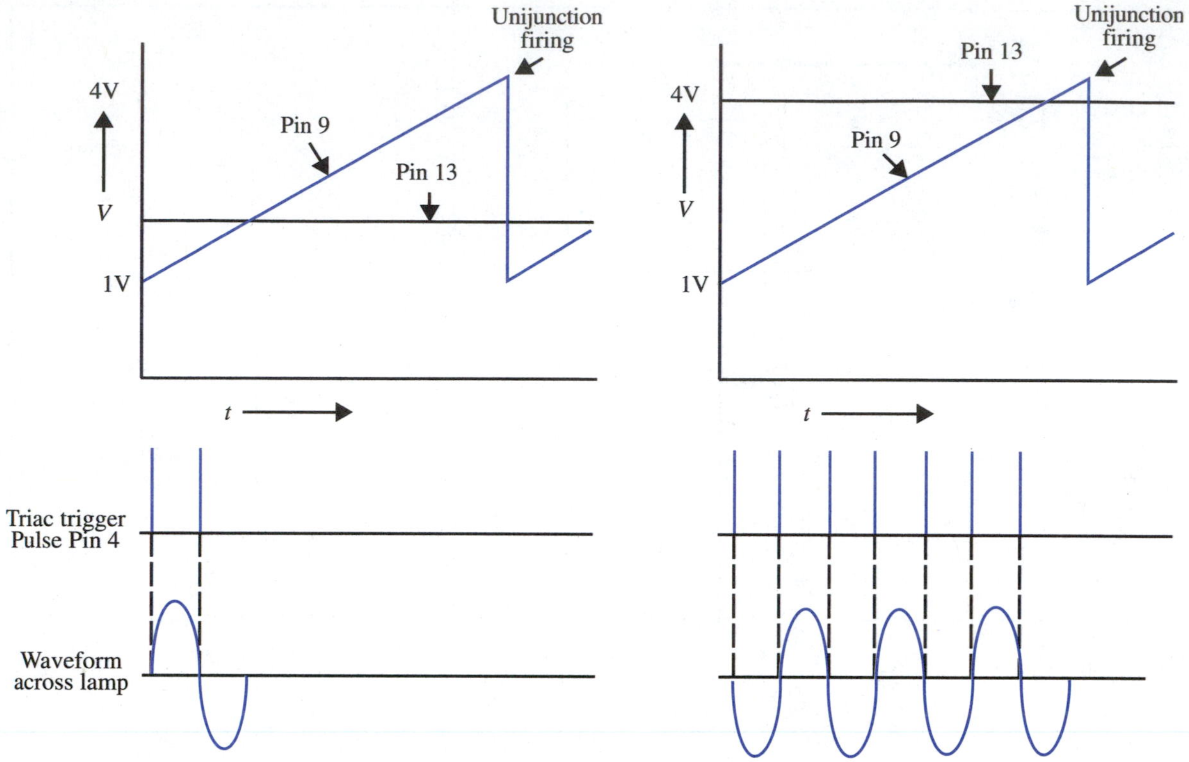

FIGURE 13–21 Low-noise lamp dimmer waveforms

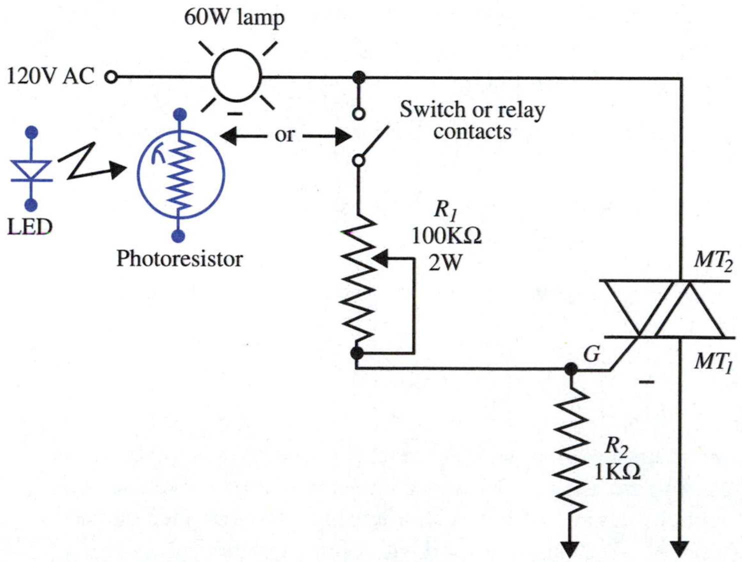

FIGURE 13–22 Triac static switching

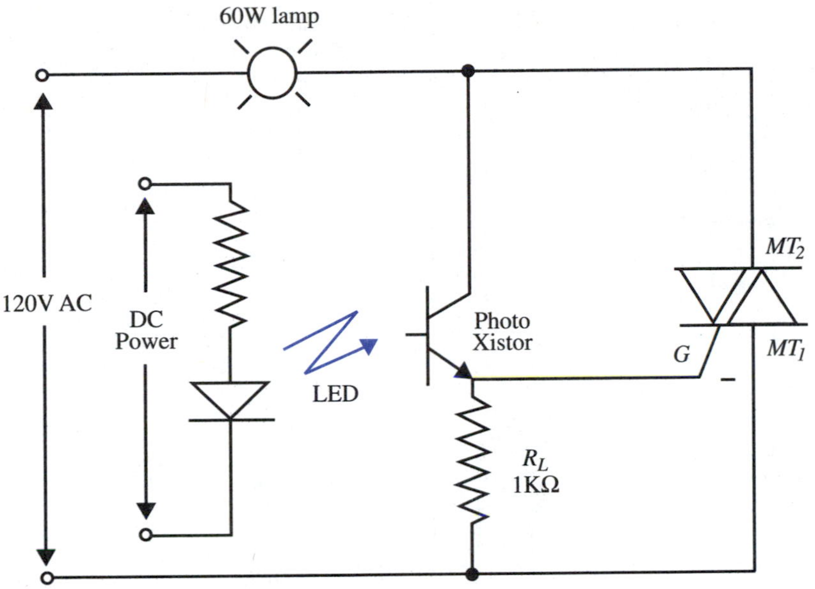

FIGURE 13–23 Phototriac equivalent circuit

The basic elements used in optical coupling are the light-emitting diode (LED) and the photodetector (photodiode, phototransistor, or photodiac). Operation of the LED and photodiode was described in Chapter 1. The phototransistor has an optical window that allows light to fall on the collector-base junction, which increases the minority (reverse) current. The base connection is left floating so the minority current passes through the base-emitter junction, causing transistor action, i.e., amplification of the minority current. The net result is that the collector current (and emitter current) increases in response to the incident light level.

If we connect an NPN phototransistor as a follower, with the load resistor between the emitter and ground—as shown in Figure 13–23—then the light from the LED will cause the voltage across the load resistor to go positive and fire the triac.

Optical couplers can also include triacs or diacs within the package. Phototriacs (or diacs) have the LED and detector built into the device. To fire the triac (or diac), we only need to provide the current (about 10mA) to activate the LED.

TRANSIENT BREAK-OVER ($\Delta V/\Delta t$)

All of the thyristors we have studied will turn on if voltage is applied to the device at too fast a rate. The reason for this is that current is required to charge the capacitance of the thyristor semiconductor junctions when the voltage across the device changes. If the voltage changes too fast, the resulting capacitor current will be large enough to turn on the thyristor. This is called **$\Delta V/\Delta t$ turn-on.** Typically, thyristors can tolerate $\Delta V/\Delta t$ rates up to about 200V/μs; but this lowers if the temperature is raised.

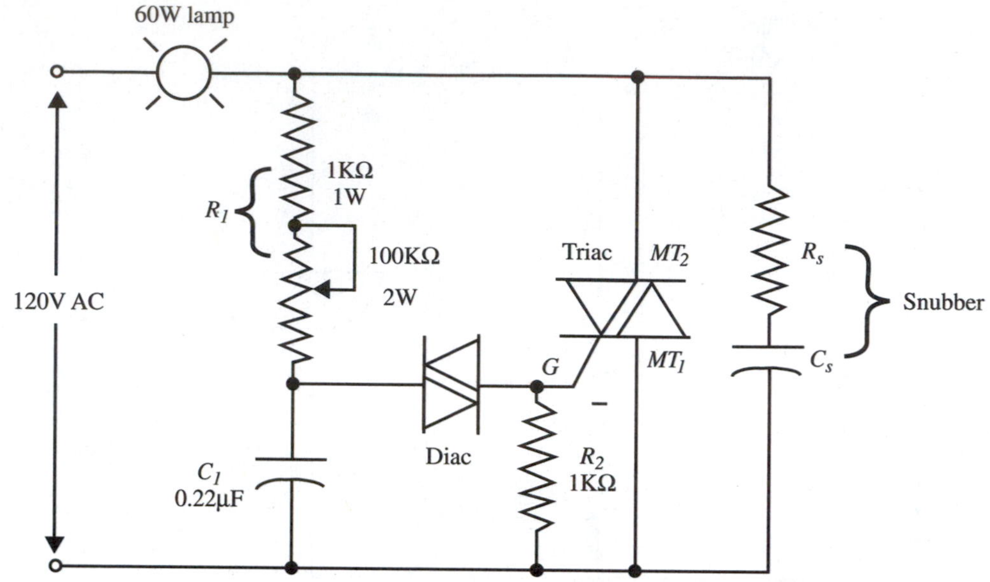

FIGURE 13–24 Thyristor $\Delta V/\Delta t$ suppression

We can prevent $\Delta V/\Delta t$ turn-on by connecting a resistor—capacitor combination (called a **snubber circuit**) across the thyristor, as shown in Figure 13–24. Charging of this added capacitor slows the voltage buildup across the device. The resistor is included in series with the capacitor to limit the discharge current from the capacitor when the thyristor is intentionally turned on. The resistor is selected to prevent the rated thyristor surge current from being exceeded, and the value usually is in the 50Ω–100Ω range.

SMART SWITCHES

A series of power control switches have been developed that provide status information on the load and the switch conditions. These switches are typically used under computer control, and the computer is able to make decisions based on the status provided from the switch.

An example of a **smart switch** is the LMD18400. This integrated circuit device has four separate "high side" switches, each of which can provide up to 3A (1A continuous) with an input voltage range from 6V to 28V. The maximum "on" resistance for each switch is 1.3Ω. Suggested uses are: relay and solenoid drivers, automotive fuel injectors, lamp driver, power supply switching, and dc motor drivers. The power-limiting feature of the chip allows a "soft turn-on" for lamp loads to restrict the surge current caused by the cold resistance of the light bulb (one-tenth of the hot resistance). This extends lamp life significantly.

The external connections for the LMD18400 are shown in Figure 13–25. The individual switches are turned on by providing a TTL logic 1 level to the associated Switch Select In-

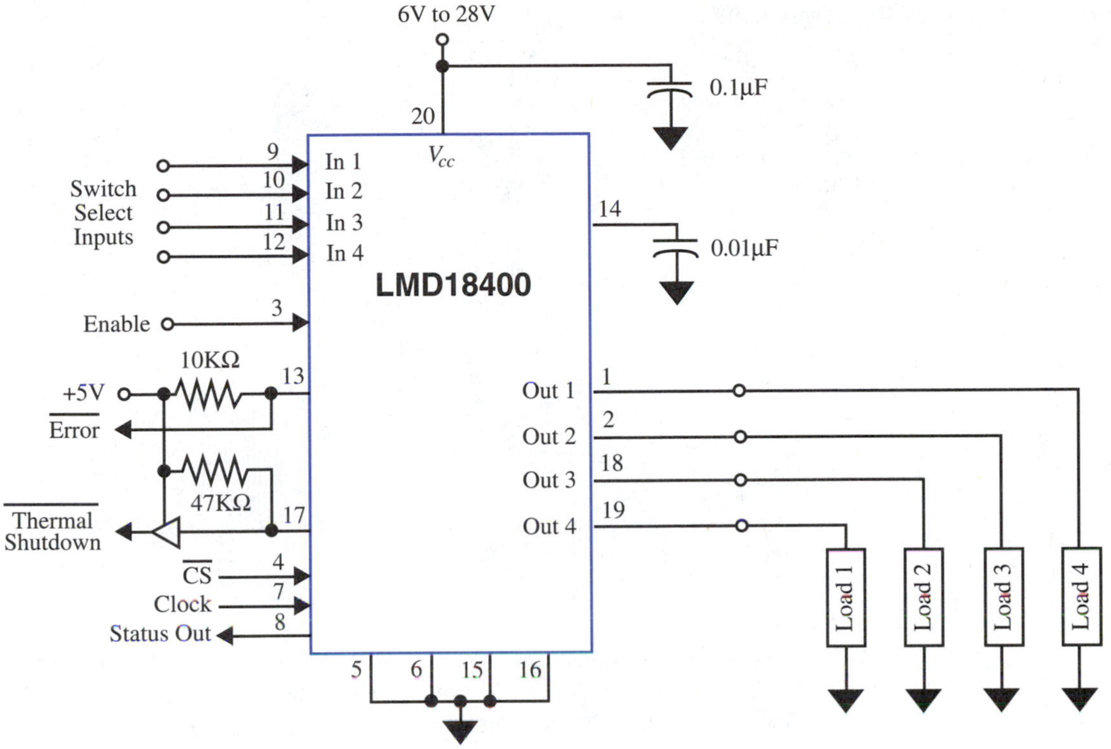

FIGURE 13–25 LMD18400 external connections

put (Pins 9 through 12). Loads are connected from the switch outputs to ground consistent with the high side switching.

The **Enable Input** makes the chip active when high and drops to the **sleep mode** when this line is taken low. In the sleep mode, the chip draws a maximum of only 10μA from the power source.

Error Output is an open collector output that indicates normal conditions when high, and an open, or short-circuited, load condition when low.

Thermal Shutdown output is in a low state if the device junction temperature exceeds 170°C. With this condition, *all* switches are turned off.

With the Chip Select (CS) low, and external Clock signal present, an eleven-bit serial data word is shifted out of the chip. The significance of each bit is as indicated in Table 13–4.

Open load detection is accomplished by internally providing a small current from V_{cc} via a 50KΩ to the load when the control switch is open. If the load resistance is high enough to develop a voltage greater than 4.1V, the load is considered open.

Short-circuit detection occurs when the control switch is closed. If the voltage across the load is less than 4.1V, the load is considered shorted.

TABLE 13–4 Status Bits from LMD18400

Bit#	*1* CH 1	*2* CH 2	*3* CH 3	*4* CH 4	*5* CH 1	*6* CH 2	*7* CH 3	*8* CH 4	*9*	*10*	*11*
	ERROR STATUS				ON/OFF SWITCH STATUS				T_j	T_j	V_{cc}
Logic 1	LOAD OK				SWITCH OFF				<145°C	<170°C	<35V
Logic 0	LOAD ERROR				SWITCH ON				>145°C	>170°C	>35V

Status bit 9 is an early warning flag that the chip is getting too hot. Bit 10 in the high state indicates that the junction is greater than 170°C, and all switches have been turned off. The switches will automatically turn back on when the junction cools to 165°C.

EXAMPLE

If the status message received by a computer from the LMD18400 is 11110101111, what is the status?

SOLUTION

The *0* in bit positions 5 and 7 indicates that switches 5 and 7 are closed, and that the rest are open. The remaining 1s indicate that the loads are OK, the chip junction temperature is below 145°C, and the supply voltage is less than 35V.

TROUBLESHOOTING POWER CONTROL CIRCUITS

We can determine the source of a problem in a bipolar transistor power control circuit by using the troubleshooting techniques suggested in Chapters 2 and 3. Since we are using the transistor as a switch, it should either be conducting (saturated) or off.

When conducting, there should be a low voltage between collector and emitter and 0.6V between base and emitter. If the load is connected and not open-circuited, but we have the supply voltage from collector to emitter with 0.6V base to emitter, we have an open transistor, or open connection.

If the base emitter voltage is 0V, but we have a low voltage (<2V) between collector and emitter, we have a shorted transistor.

With MOSFET power control, failures can include an open gate circuit, which will prevent power turn-on; or a shorted gate to channel, which prevents correct enhancement and again will result in no power turn-on.

Techniques for troubleshooting thyristor circuits depends on the type of thyristor in the circuit.

SCR Troubleshooting

For an SCR, we can have an anode-to-cathode short, which will apply power to the load at all times. An open circuit between these two terminals means we are unable to power the

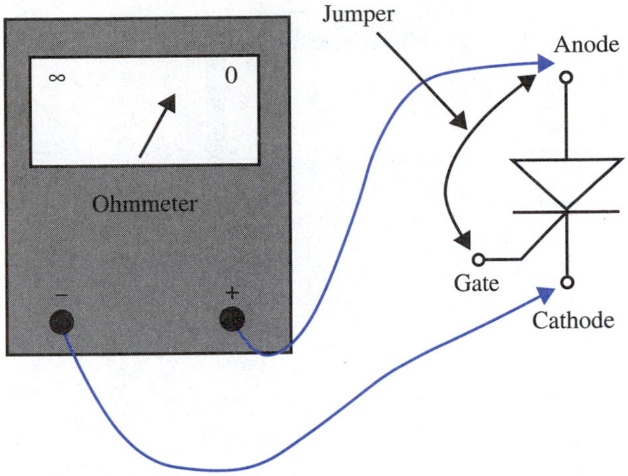

FIGURE 13–26 Checking an SCR out-of-circuit

load even with the correct gate signal. With an SCR out-of-circuit, we should see a high resistance (open circuit) when measuring the anode-to-gate resistance. The gate-to-cathode should appear as a diode with the P side at the gate.

We can check an SCR for trigger operation, using an ohmmeter, as shown in Figure 13–26.

The meter used for the SCR trigger test should have at least voltage 3V across the test leads (VOM), when measuring ohms (on high ohms scale), to provide sufficient gate-firing voltage. As connected in Figure 13–26, the SCR should turn on, and the meter should read a low resistance between anode and cathode.

Unijunction Troubleshooting

In circuit, a unijunction should fire when the emitter voltage reaches about 75 percent of the supply voltage. After firing, the device should turn off (indicated by no voltage across the R_{b1} resistor) when the emitter is shorted to ground. Out of circuit, a unijunction will read a relatively low resistance (k-ohms) in both directions between B_1 and B_2 and will act as a diode between the emitter and B_1 or B_2.

Triac Troubleshooting

A triac can fail shorted between MT_1 and MT_2, which will apply full power to the load. It can also fail open between these terminals, which means no power to the load. An open or shorted gate will prevent the triac from being fired. Out of circuit, a triac can be tested as shown in Figure 13–27.

Resistance should read high both ways between MT_1 and MT_2. With MT_2, along with the gate, connected to the positive lead of the ohmmeter and the negative lead to MT_1, a low resistance should be observed. Similarly, connecting MT_2 and the gate to the negative lead of the ohmmeter, and the positive lead to MT_1, should yield a low resistance.

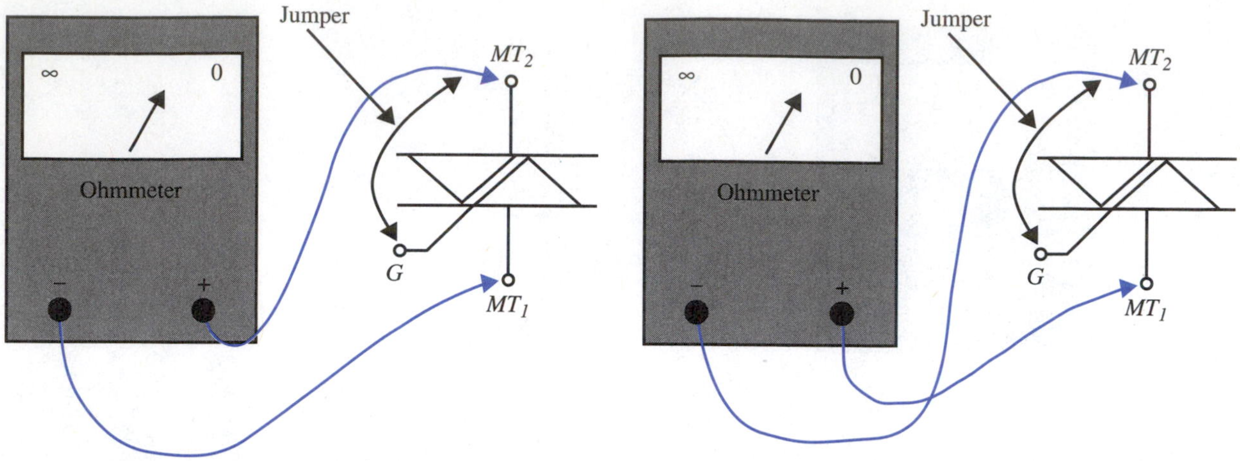

FIGURE 13–27 Checking a triac out-of-circuit

SUMMARY

- Power control devices are used to apply the primary power to a circuit.
- The efficiency of a power control device can be determined from the voltage drop across the device and the voltage drop across the load.
- Power can be applied to a circuit by driving a bipolar transistor into the saturated (common emitter) mode.
- With bipolar power control, a PNP transistor is connected on the positive voltage side of the load, but an NPN transistor is connected on the negative side.
- Base current required to operate (saturate) a bipolar transistor can be reduced by using a Darlington transistor.
- Power control using a transistor-activated relay has the advantage of isolation between the control power source and the power source being controlled.
- A disadvantage of using a relay is the inherent low reliability of a mechanically operated device.
- Power enhancement type MOSFETs can be used for power control.
- The advantages of the MOSFET over a bipolar for power control are the negligible input power required to operate the device, drive simplicity, speed, and temperature stability.
- The bipolar transistor is better suited for high-voltage (>200V) power control than the MOSFET.
- An IGBET has both the advantages of a power bipolar transistor and a power FET.
- A thyristor is a device that is triggered from a nonconducting state to a conducting state.

- Thyristors are of two basic types: those that are triggered into conduction by application of a trigger to a gate, and those that trigger when the voltage across the device is exceeded.
- A silicon controlled rectifier (SCR) is functionally a diode that conducts when its gate is triggered.
- SCRs are used for power control of DC circuits or for control of rectification in DC power supplies.
- The operation of an SCR can be explained by a totem pole configuration of a PNP and an NPN transistor.
- SCR holding current is the minimum current to maintain conduction.
- Once an SCR is conducting, the gate no longer has control, and turn-off can only be achieved by dropping the current through the device below the holding current value.
- A commutating circuit, using the energy stored in an capacitor, can be used to turn off an SCR.
- When controlling power to a DC motor by pulsing an SCR, a freewheeling diode connected across the motor maintains a nonpulsing motor current.
- A unijunction transistor is a trigger device that fires when the emitter voltage reaches $\eta VC + 0.6V$.
- Firing an SCR with a unijunction makes the firing independent of the gate characteristics of the particular SCR.
- A silicon controlled switch (SCS) has two gates and can be turned off as well as on via a gate.
- A disadvantage of an SCS is that it is a low-current device.
- A GTO is a medium-power device that can be turned on and off from a single gate.
- An MTC is a high-power device that can be turned on and off from a single gate.
- The MCT has the lowest forward voltage drop at a given current.
- A triac can be considered as back-to-back SCRs, which allows for power control of AC circuits.
- Triacs have three terminals: main terminal 1 (MT1), main terminal 2 (MT2), and gate terminal.
- Triacs can be fired with four combinations of polarity between the gate and the other two terminals.
- We don't have to worry about turning a triac off since the power source controlled (AC) swings through zero volts.
- Variable power control is achieved with a triac by delaying (phase lag) the signal applied to the gate until the source voltage has built up across the triac.
- A diac is a two-terminal thyristor that conducts when the voltage in either direction exceeds a specified value.
- Firing a triac with a diac makes the firing point independent of the triac trigger characteristics.

- Zero crossing power control means the triac is turned on when the voltage (either polarity) across the triac just passes through zero.
- Zero crossing power control minimizes the generation of electrical interference.
- Static switching of a triac means the gate is left in the trigger state for the amount of time we require full power to the load.
- Optical power control of a triac has the advantage of complete electrical isolation between the firing circuit and the power circuit being controlled.
- $\Delta V/\Delta t$ protection of a thyristor means we use a snubber circuit to prevent charging of the semiconductor junctions from turning on the device.
- A "smart" switch provides status information on the load and the switch conditions.

EXERCISE PROBLEMS

1. Define "power control circuit."
2. Why is a common emitter configuration used for bipolar transistor power control?
3. Determine the efficiency of controlling a 10A, 12V motor with a 2N3055.
4. Why is a Darlington transistor used for bipolar power control?
5. What are the advantages and disadvantages of using a relay for power control?
6. Why should a diode always be connected across a relay activated by a solid-state device?
7. Compute the % efficiency of the circuit of Figure 13–A if the relay has a 62Ω, 12V coil and the load is a 10A, 28V motor.

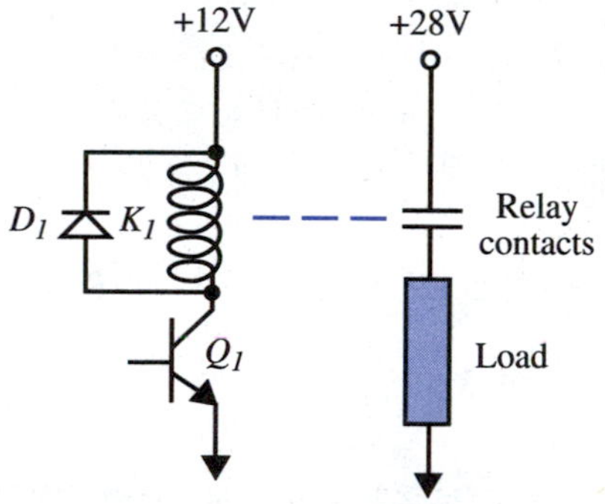

FIGURE 13–A (text figure 13–2)

8. Determine the efficiency of controlling a 10A, 12V motor with an MTP15N05E MOSFET.
9. Give the advantages for using a MOSFET rather than a bipolar for power control.
10. Define "thyristor."

11. Explain the operation of an SCR.
12. How is an SCR turned on, and how is it turned off?
13. What is the holding current for an SCR used in the test circuit of Figure 13–B if the voltage measured across R_4, when the voltage across C_2 is minimum, is 0.141V?

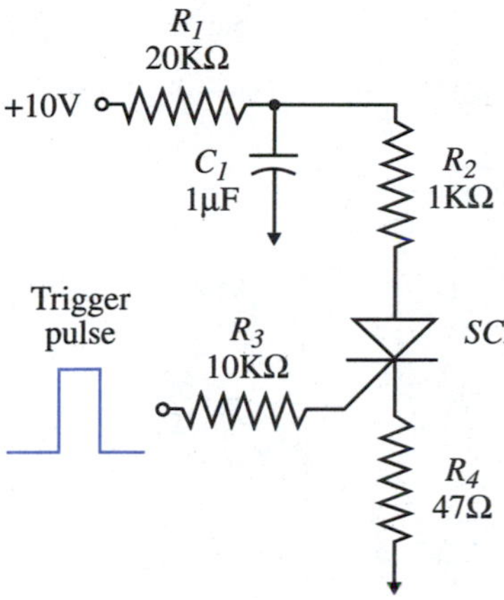

FIGURE 13–B (text figure 13–5)

14. Determine the trigger current for an SCR under test in the circuit of Figure 13–B if the voltage at the trigger source side of R_3 is 1.5V and on the gate side is 1.1V.
15. Explain the purpose of the capacitor C_1 in the circuit of Figure 13–C.

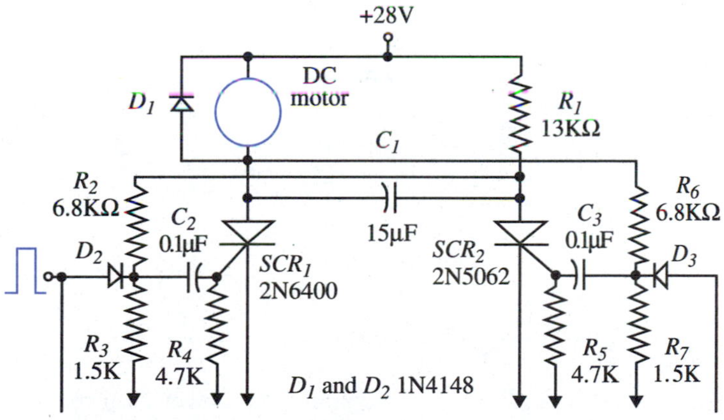

FIGURE 13–C (text figure 13–6)

16. Determine the size of the capacitor C_1 in Figure 13–C if the motor voltage is 28V and the motor current 5A.
17. Find the required value of R_1 in Figure 13–C if the minimum time the motor is turned on is 10 seconds.

18. Explain the operation of a unijunction transistor.
19. Determine the emitter firing voltage for a unijunction transistor if V_{cc} is 12V and the intrinsic standoff ratio is 0.68.
20. Find the pulse frequency for the circuit of Figure 13–D if R_t is reduced to 5.6KΩ.

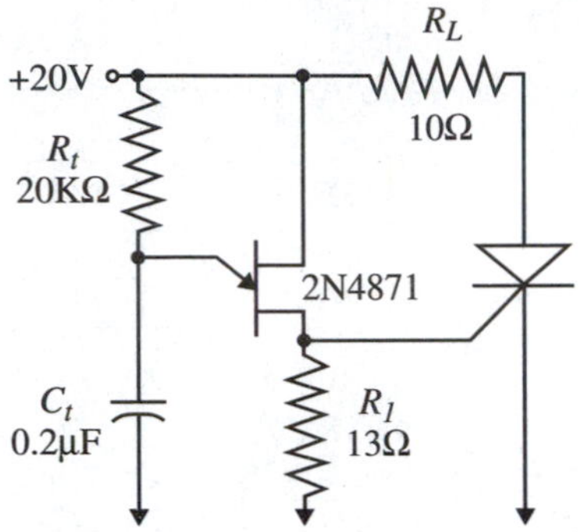

FIGURE 13–D (text figure 13–10)

21. What advantages does an SCS or a GTO have over the SCR? What disadvantages?
22. What advantages does an IGBT or an MCT have over the SCS or GTO?
23. What are the two most used ways, of the four possible, to fire a triac?
24. Find the firing angle for the triac of Figure 13–E if the source voltage is 120V, the frequency is 60Hz, the firing voltage is 2.5V, and R_1 is set for 100KΩ.

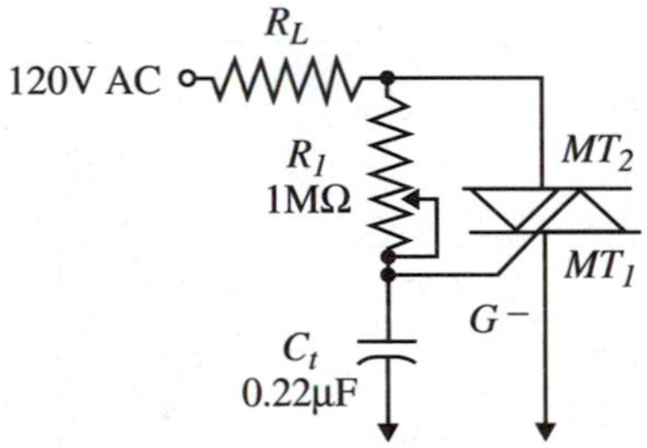

FIGURE 13–E (text figure 13–16a)

25. Find the firing angle for the conditions of problem 24 if the firing voltage is changed to 1.5V.
26. Why is a diac used to fire a triac?
27. Find the firing angle for the circuit of Figure 13–F if the diac measured firing voltage is 31V and R_1 is 27KΩ.
28. Determine the firing angle for the circuit of Figure 13–F if the diac measured firing voltage is 26V and R_1 is 47KΩ.

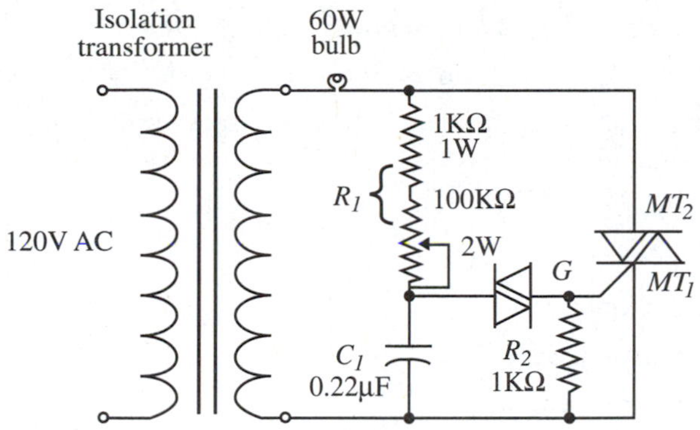

FIGURE 13–F (text figure 13–18)

29. Find the effective load power for the firing angle determined in problem 27 if the source voltage is 120V RMS and the effective load resistance is 20Ω.
30. Determine the effective load power for the firing angle determined in problem 28 if the source voltage is 120V RMS and the effective load resistance is 20Ω.
31. Explain why zero crossing power control is used.
32. Determine the average power delivered to the load with the triac on for three cycles and off for two cycles, if when conducting, the triac delivers 500W.
33. Find the number of consecutive "on" cycles for the low-noise lamp dimmer circuit of Figure 13–G if the voltage at Pin 13 is set at 3V. Assume the unijunction period is 0.2 seconds.

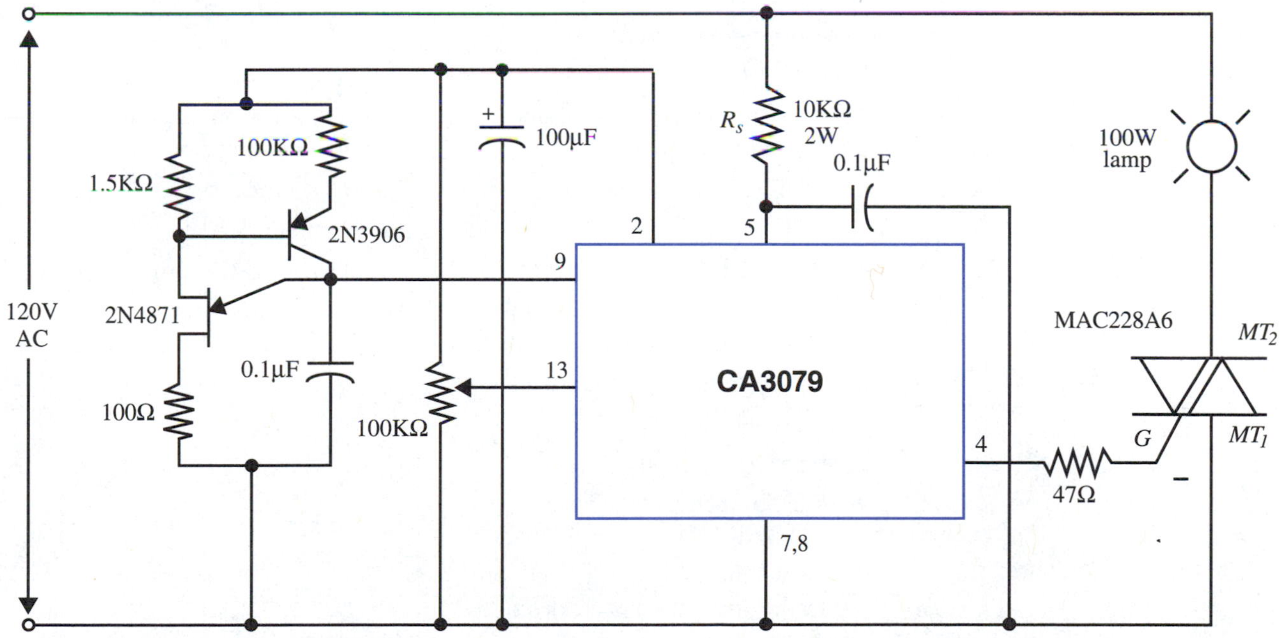

FIGURE 13–G (text figure 13–20)

34. State the advantage of optical power control.
35. Why is $\Delta V/\Delta t$ suppression used?
36. Determine the status from an LMD18400 smart switch if the status message is 01110111111.

ABBREVIATIONS

f_p = Pulse frequency
GTO = Gate turn-off SCR
I_m = Maximum current
P_a = Average power
P_e = Effective power
P_l = Power loss in switching device
P_m = Motor power
P_r = Relay power
SCR = Silicon controlled rectifier
SCS = Silicon controlled switch
V_v = Minimum voltage across an "on" unijunction
ϕ_c = Phase delay at phase shift capacitor
ϕ_g = Phase delay from zero crossing of waveform across capacitor to firing point
ϕ_f = Firing angle = $\phi_c + \phi_g$

ANSWERS TO PROBLEMS

1. A circuit that controls the application of power
2. Transistor can saturate.
3. %Eff = 75%
4. Reduced base current (drive)
5. Can handle large currents, isolation, but is unreliable
6. To suppress voltage transients
7. Eff = 99.2%
8. Eff = 91.7%
9. Very low activation power required
10. Device that can be "triggered" on by the application of an electrical signal
11. See text.
12. Triggered on; turns off going below holding current
13. $I_h = 3\text{mA}$
14. $I_t = 40\mu\text{A}$
15. Provides SCR turn-off
16. $C_1 = 5.6\mu\text{F}$
17. $R_1 = 133\text{K}\Omega$
18. See text.
19. $V_e = 8.76\text{V}$
20. $f_p = 446\text{Hz}$
21. See text.
22. See text.
23. See text.
24. $\phi_f = 90.1°$
25. $\phi_f = 87.3°$

26. Overcomes variable firing voltage of triac
27. $\phi_f = 92.4°$
28. $\phi_f = 113.4°$
29. $P_e = 389W$
30. $P_e = 600W$
31. To reduce noise transients
32. $P_a = 300W$
33. Eight cycles of line frequency
34. Isolation between control circuit and power circuit
35. To prevent inadvertent turn-on
36. Channel 1 switch is closed, but it has a load error.

CHAPTER 14

Audio Circuits

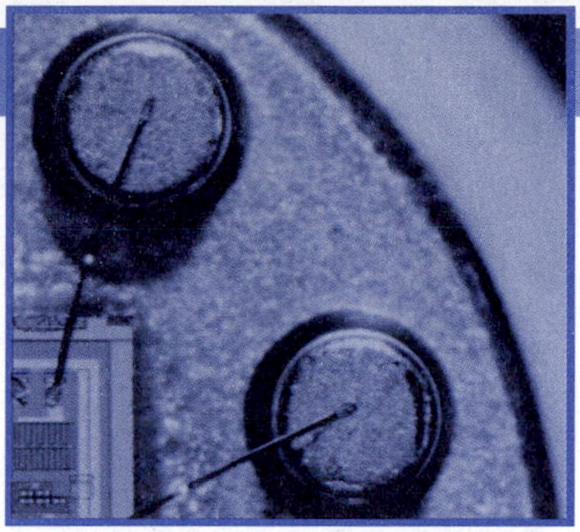

CHAPTER OBJECTIVES

After completing this chapter, you should be able to explain the following audio circuit concepts:

- Basics of hearing
- Psychoacoustic effects
- Why preamplifiers are used
- Why equalization is used
- How audio band noise can be reduced
- Dolby noise reduction
- How stereo is heard
- Dynamic range of audio
- Control of audio volume, tone, and balance
- Stereo enhancement
- Drivers
- Power amplifiers
- Bridge amplifiers
- Switching amplifiers
- Switched capacitor filter

INTRODUCTION

In this chapter, we will first review some of the basics of hearing, and then look at the electronic processing of audio signals provided by live and recorded sources. We will first study the different types of audio amplifiers; finally, we shall look at audio filter circuits.

BASICS OF HEARING

The process of hearing is not completely understood, and much of our knowledge of hearing is based on measured (empirical) data. This data is derived from tests, conducted on thousands of individuals, that measured hearing sensitivity, frequency response, distortion, and also various psychoacoustic effects.

The ear is an extremely sensitive device with a maximum hearing sensitivity of 10^{-16} W/cm^2 at 3KHz. This sound pressure level results in a movement of the eardrum of 10^{-9}cm, which is about one-tenth the diameter of a hydrogen molecule. The response of

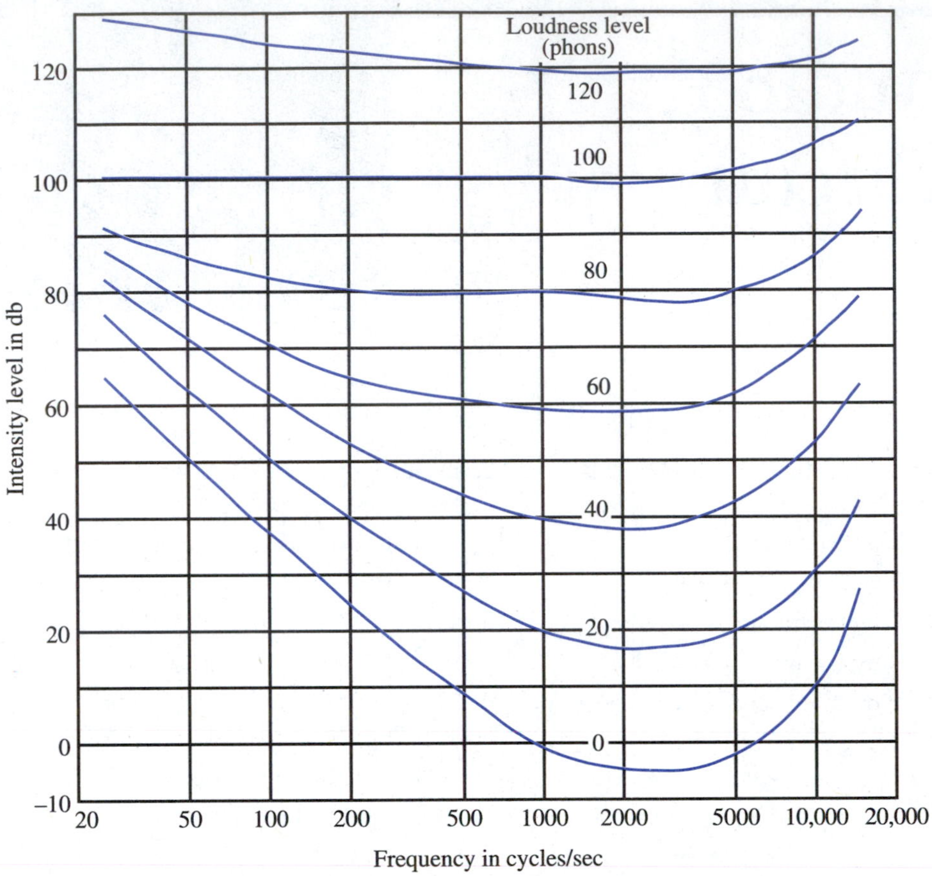

FIGURE 14–1 Equal loudness curves

the ear is logarithmic, and that means the sound power must increase by a factor of ten before the ear senses a doubling of volume.

As shown in the equal loudness curves (Fletcher-Munson curves) of Figure 14–1, the sensitivity of the ear changes with frequency. The ear's most sensitive frequency band is from about 1KHz to 5KHz. The loudness is measured in **phons**, which is the sound intensity relative to the level at a reference frequency of 1KHz. Notice in the figure that the relative loudness (in phons), at a particular frequency, changes with the frequency of the sound. The effect is greater at the lower sound levels.

The typical audio range is from 20Hz to 20KHz for music and 100Hz to 5KHz for voice. Devices (voice, instruments) provide more output power at the low frequencies with about a 6db per octave roll-off of the higher frequencies.

The frequency response of the ear varies with the individual and with the person's age. The high-frequency response drops off with age, and a response drop to 8KHz is typical with persons older than 60 years. Female high-frequency response is typically greater than that of the male. Low-frequency response is less subject to the individual or age.

The ear's response to sound is nonlinear—meaning it creates harmonic distortion. If we hear a 200Hz and a 604Hz tone, the sound level will vary at a frequency of 4Hz. This results from the third harmonic of 200Hz (600Hz) created by the ear's nonlinearity beating with the 604Hz tone. Test results indicate that the tones above 500Hz at sound levels below 40 phons do not generate harmonics of "significant" magnitude. What this is saying is that those high-level signals, particularly at low frequency, overload the ear and create distortion, i.e., new tones that were not present at the entrance to the ear.

Psychoacoustic effects refers to hearing effects that cannot be explained by physical phenomena. A typical example is synthetic bass—the ear senses the presence of a low-frequency tone when only the harmonics of the tone are applied to the ear. Another example is the addition of white (wide band) noise to music, with poor high-frequency response causing the ear to extend the frequency range. Still other examples are the audio masking of one tone by a another higher-level tone (important with noise reduction), and the phantom center image formed when listening to stereo speakers or headphones.

ELECTRONIC PROCESSING OF AUDIO SIGNALS

Audio signals direct from the voice or music source are processed by electronic equipment for amplification and/or recording. The processing can involve reshaping of the frequency spectrum (equalizing) and noise reduction techniques.

Amplification of audio signals is required, for example, to bring the low-power level at the output of a microphone to the several watts (thousands?) required to drive a loudspeaker.

Equalization can be used to boost the low-amplitude high-frequency components of the audio signal so they are not lost in noise during the recording process. The NAB curve recommended for tape recorders has about a 30db boost from 100Hz to 2KHz. Another common use for equalizers is to compensate for loudspeaker deficiencies at low frequency by allowing a bass boost.

Audio noise reduction techniques are used to achieve a high signal-to-noise ratio that allows greater dynamic range (ratio of highest-level signal to lowest-level) of the audio signal. This is important with some types of music that have a dynamic range close to 100dB (100,000 to 1). Methods to reduce noise include equalization, Dolby-type signal processing, and—the ultimate approach—digital signal conversion of audio.

DIGITAL AUDIO

Digital audio could imply that the loudspeaker is assuming an on or off position. Actually, *digital audio* is a contradiction in terms because audio is an analog not a digital signal. *Digital audio* refers to the conversion of the analog audio signal to digital prior to recording (or transmitting) to maintain the signal-to-noise ratio (see Chapter 16). If we state that equipment is digital ready, it means it can handle the wide dynamic range of the audio released by the digital media.

STEREO AUDIO

Stereo audio is the result of humans' having two ears separated by about 20cm. The separation enables us to sense the direction of a sound by the difference in phase shift and/or

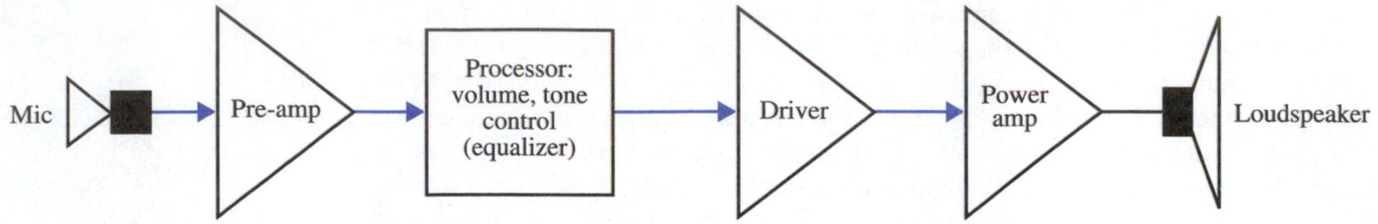

FIGURE 14–2 Audio system components

amplitude of the signal received at each ear. The stereo effect is not readily apparent below about 200Hz. Up to about 3KHz, phase of the signal is primarily used for sound direction; above 3KHz, relative intensity is used. Experiments have indicated a dropout of the stereo effect at frequencies close to 3KHz caused by inability of the ear to detect a phase shift (ear spacing is about a half wavelength) and by the directivity of the sound at this frequency not being sufficient to provide a significant amplitude difference.

AUDIO CIRCUITS

In the remainder of this chapter, we shall look at specific circuits used to implement the audio requirements. The emphasis will be on linear integrated circuits since they are becoming predominant in new audio designs—but we shall still see transistors used in some applications where they are appropriate.

We can see a block diagram of a typical audio system in Figure 14–2. Sound converted to an electrical signal by the microphone is amplified, processed, and amplified to a level to drive the loudspeaker.

The description of audio circuits will start with preamplifier circuits, progress through the various types of audio processing, then cover power amplifiers, and finally look at a speech synthesis system.

AUDIO PREAMPLIFIER

The purpose of a preamplifier is to bring the weak signal (could be as low as 20μV) from a microphone (or pickup) to a level suitable for processing or being applied to the power amplifier driver.

A preamplifier must be low noise so that it doesn't mask out the weak signals and has sufficient dynamic range to accommodate the 120dB variation of audio.

Figure 14–3 provides the circuit for a low-noise preamplifier using a low-noise transistor differential stage followed by an op-amp differential amplifier. The equivalent input noise for this circuit is $2.4nV/\sqrt{Hz}$, which means the total input noise level (V_{ni}) over the audio band of 20Hz to 20KHz is

$$V_{ni} = \frac{2.4\text{nV}}{\sqrt{\text{Hz}}} \times \sqrt{20\text{KHz} - 20\text{Hz})}$$

$$V_{ni} = 339\text{nV}$$

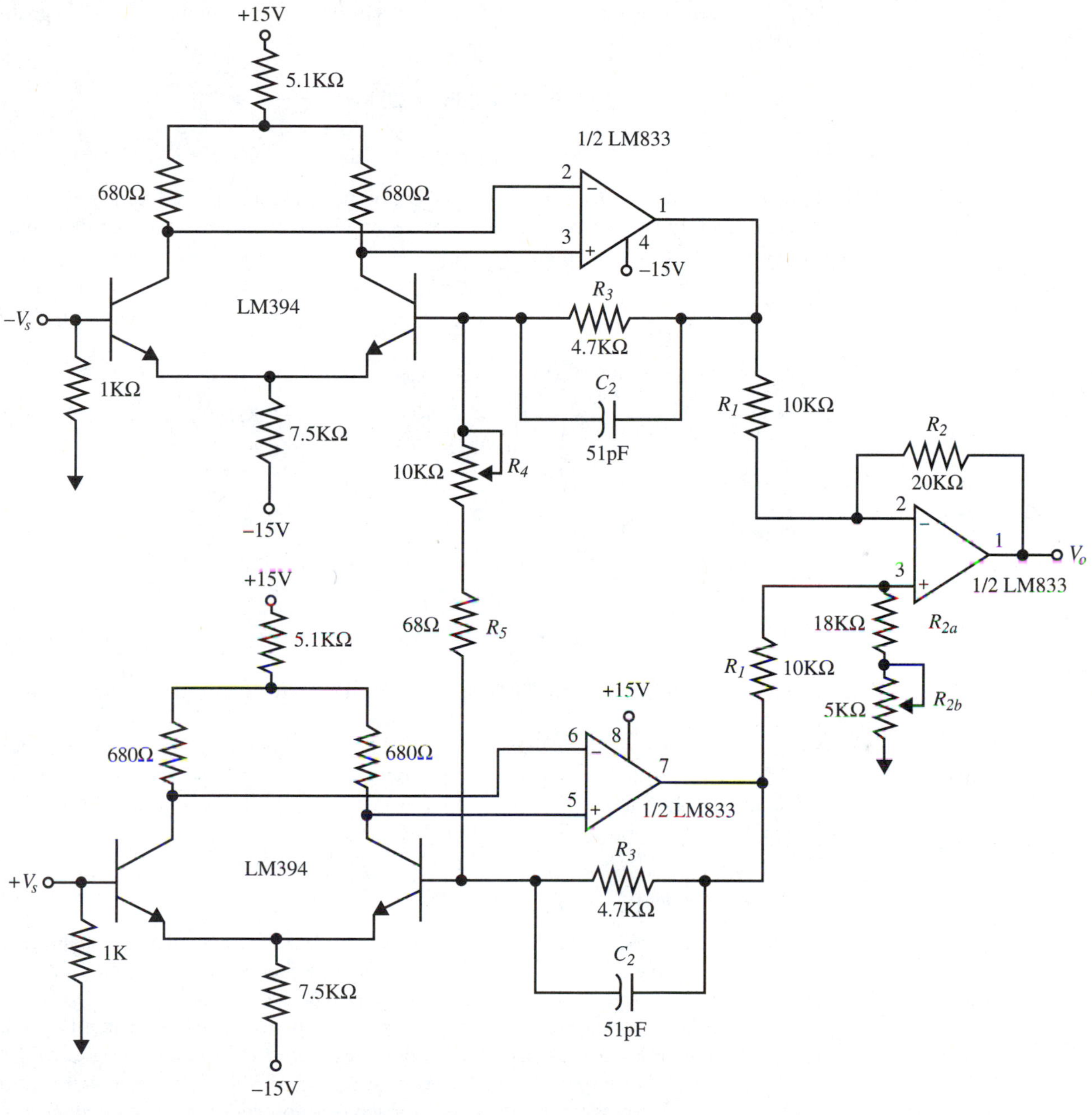

FIGURE 14–3 Low-noise preamplifier

This means the worst-case signal-to-noise ratio in db with the minimum input signal of 20μV is

$$V_s/V_{ni}(\text{db}) = 20 \log \frac{20\mu V}{339nV}$$

$$V_s/V_{ni}(\text{db}) = 35.4\text{dB}$$

In percentage, this means that just 1.7 percent of the weakest amplified signals are noise.

We can adjust the gain for the circuit by varying R_4, with the range of gain determined from the following equation:

$$A_v = \left(\frac{R_3}{(R_4 + R_5)/2} + 1\right)\left(\frac{R_2}{R_1}\right) \qquad \textbf{(equation 14–1)}$$

With R_4 at maximum value,

$$A_v = \left(\frac{4.7K\Omega}{(10K\Omega + 68\Omega)/2} + 1\right)\left(\frac{20K\Omega}{10K\Omega}\right)$$

$$A_v = 3.87$$

With R_4 at minimum value,

$$A_v = \left(\frac{4.7K\Omega}{(68\Omega)/2} + 1\right)\left(\frac{20K\Omega}{10K\Omega}\right)$$

$$A_v = 278$$

The total harmonic distortion (THD) specified for the circuit at this maximum gain is 0.01 percent.

The differential configuration of this preamplifier gives good rejection of common mode noise pickup. To ensure maximum rejection, the input leads should be twisted and shielded, and the value of R_{2b} should be adjusted for a zero signal output with the same (common mode) signal applied to both inputs.

AUDIO PROCESSING

The term **audio processing** means, for our discussion, frequency response shaping and noise reduction. We can accomplish frequency response shaping by use of tone control circuits or equalizers, and noise reduction with Dolby-type processors.

The audio frequency response of an amplifier can be changed to cause a flatter frequency response at the ear for low audio levels by boosting the high and low frequencies (see Figure 14–1). We can also change the frequency response to suit individual preferences (some people like accentuated bass) or to compensate for acoustic absorption of higher frequencies.

Let us now look at a tone control integrated circuit that also has the capability for volume, stereo balance, and stereo enhancement.

AN AUDIO CONTROL DEVICE

The IC audio control device we shall study is the LM1040, shown in block diagram form in Figure 14–4.

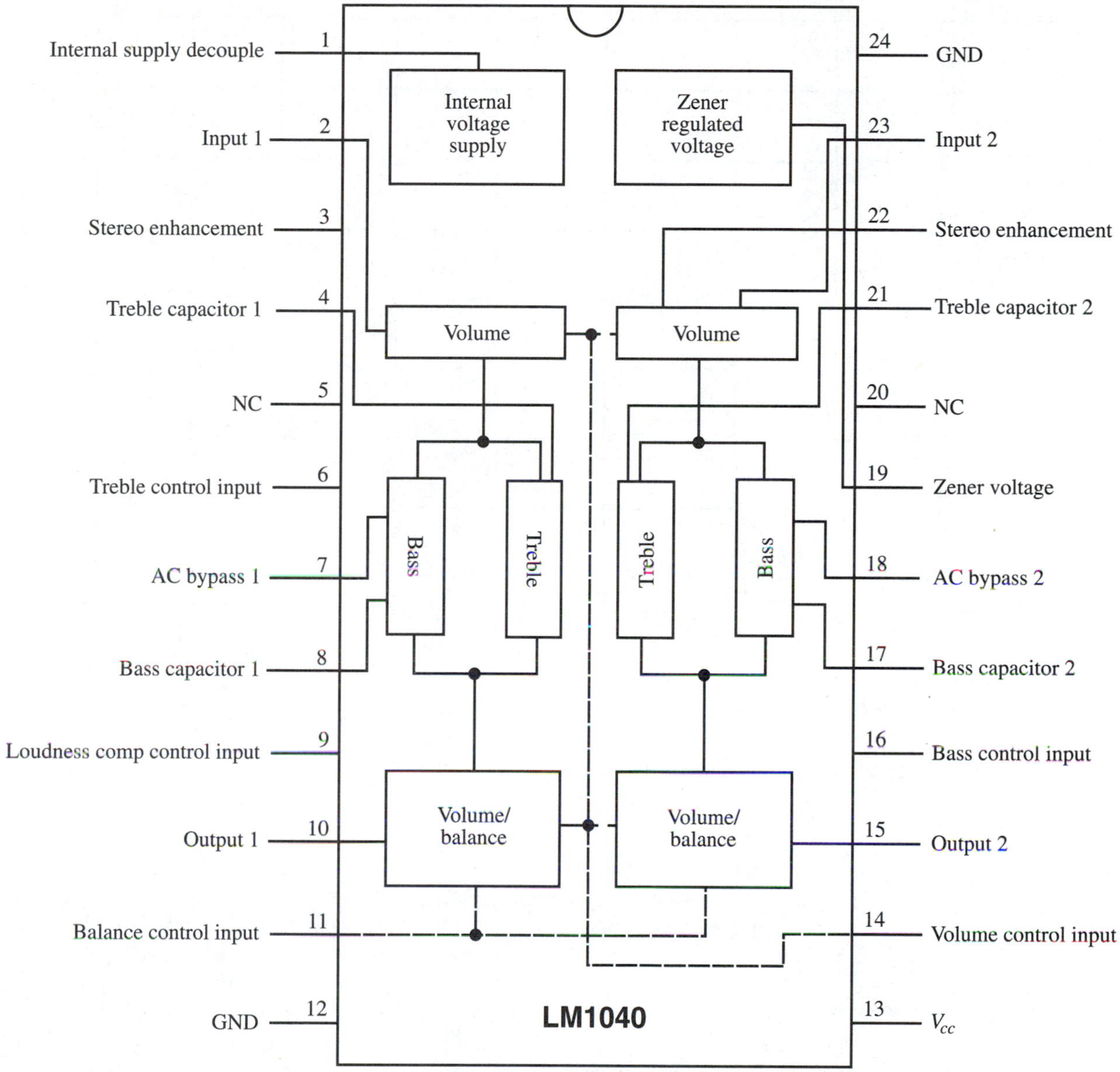

FIGURE 14–4 LM1040 Audio processing circuit. *Courtesy National Semiconductor.*

This chip uses a DC voltage controlled tone (bass/treble), volume, and balance circuit designed for stereo applications. A stereo enhancement feature is included to improve the apparent stereo separation for systems where the speaker's required mounting is too close.

Four control inputs are provided to control bass, treble, balance, and volume by the application of DC control voltages from either a remote location or from fixed voltages derived from the internal 5.4V zener diode.

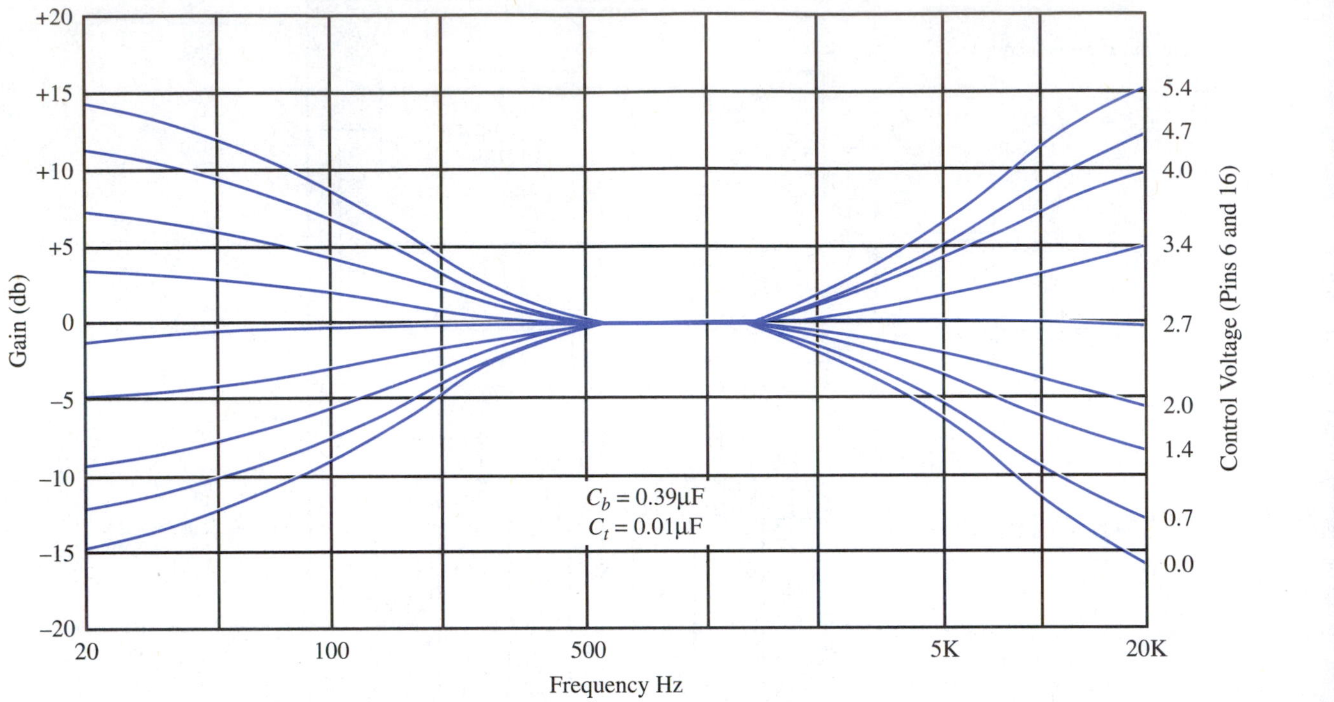

FIGURE 14–5 LM1040 control voltage versus frequency response

TONE RESPONSE

The specific tone response is determined by the selection of the DC voltage applied to Pins 6 and 16 in conjunction with the capacitors used for bass (Pins 8 and 17) and for treble (Pins 4 and 21).

Figure 14–5 shows the response as a function of the control voltage applied jointly to the bass and treble control pins—using a 0.39μF bass capacitor and a 0.01μF treble capacitor. With the control voltage at 2.7V, which is one half the zener reference voltage, there is no bass or treble boost, or in other words, flat response of the signal from the pre-amplifier. With a control voltage of 5.4V, maximum bass and treble boost occur, and with 0V, both high and low frequencies are cut (reduced). Since separate control pins are provided for bass and treble, the high and low frequency responses can be varied independently.

VOLUME AND LOUDNESS CONTROL

Volume is a change in the level of the audio signal; loudness changes both the level and the frequency response to compensate for the ear's drop-off at high and low frequencies with low sound levels (again, refer to Figure 14–1).

The volume can be varied over a control range of 70db by applying a DC voltage to Pin 14. The gain reduction is 70db (3162) with 0V on Pin 14 and 0db (gain of 1) with +5.4V on this pin.

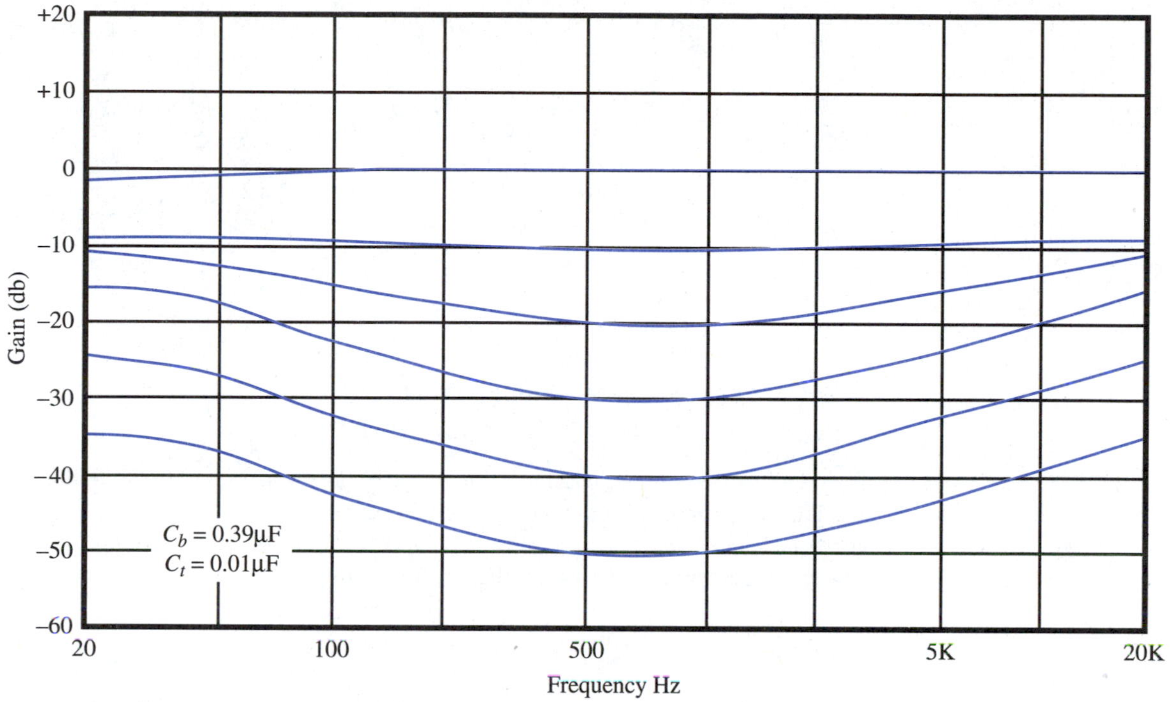

FIGURE 14–6 LM1040 loudness characteristics

By tying Pin 14 to Pin 9, the loudness is controlled rather than the volume. Now the high and low frequencies are not attenuated as much as midband frequencies when we lower the DC voltage. Figure 14–6 shows the loudness curves.

Notice the similarity of the shape of the loudness curves and the Fletcher-Munson curves of Figure 14–1.

STEREO BALANCE AND ENHANCEMENT

Stereo balance is the relative level of the two stereo channels. With a person located closer to one stereo speaker than the other, the sound level of that speaker is too high to maintain the stereo effect. A stereo balance control allows level compensation.

The LM1040 has a typical stereo balance control range of 26db. This is accomplished by applying a DC voltage to Pin 11. If we apply 0V, we attenuate stereo channel 1 by 26db and channel 1 by 0db. With 5.4V applied, we attenuate stereo channel 2 by 26db and channel 2 by 0db. If the control voltage is set at 2.7V, we attenuate neither channel.

As mentioned previously, we use stereo enhancement to compensate for speakers mounted too close to one another because of equipment and or cabinet limitations. The stereo effect can be enhanced by allowing some phase-reversed coupling between the two channels. This occurs within the LM1040, using the circuit of Figure 14–7.

The stereo channel 1 input is applied to Pin 2 and the channel 2 input to Pin 23. The cross-coupled signal is applied to the transistor emitter, and the amount of signal at the

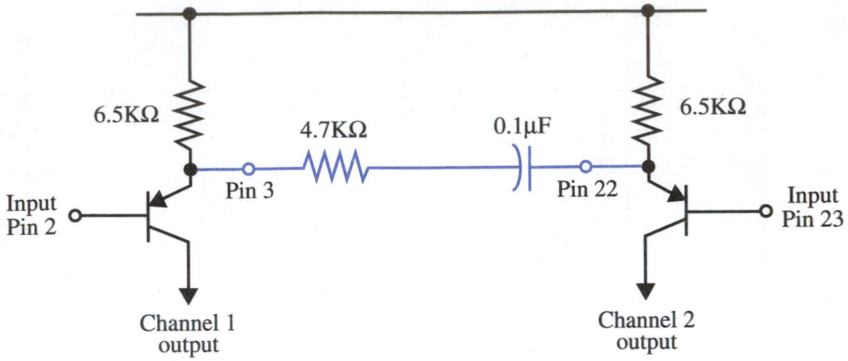

FIGURE 14–7 Stereo-enhancement circuit

emitter is a function of the dividing action of the 4.7KΩ resistor and the 6.5KΩ emitter resistor. This ratio provides about a 60 percent cross-coupling.

The 0.1μF capacitor in series with the 4.7KΩ resistor reduces the coupling at frequencies below 300Hz, where the stereo effect is minimal.

APPLICATION CIRCUIT FOR THE LM1040

Figure 14–8 provides an application circuit for the LM1040, showing the connection for the functions we have studied. Switches enable selection of stereo enhancement or loudness. Independent controls are provided to allow bass and treble to be individually adjusted.

The maximum input signal using a 12V supply voltage is 1.3V RMS; the maximum output voltage with the same supply voltage is 0.8V RMS.

AN EQUALIZER CIRCUIT

It was mentioned earlier that an audio amplifier's frequency response can be altered by a tone control circuit or by an equalizer. An equalizer differs from a tone control in that the tone control response at one frequency is related to that at another frequency. In contrast, the equalizer breaks the audio spectrum into frequency bands that can be independently adjusted.

An example of a graphic equalizer is the LMC835 integrated circuit. This chip is designed for digital computer control of 14 frequency bands with 25 amplitude steps. Each band is set up by reception of a unique eight-bit serial code. The data words are of two types: one selects the band (bit D7 = high) and the other sets the gain (bit D7 = low). The word formats are indicated as follows.

An external clock up to 500KHz is provided to shift in the serial data. The eight-bit control word is preserved in an internal latch when a strobe input is taken low. Conditions of each band are retained until changed by a new control word. The maximum boost (gain) or cut (attenuation) within the band is 6dB or 12dB, depending on the state of the D4 (B) and D5 (A) bits.

The frequency of each band is determined by external components connected to the LC pins. Figure 14–9 shows an application for a stereo seven-band equalizer.

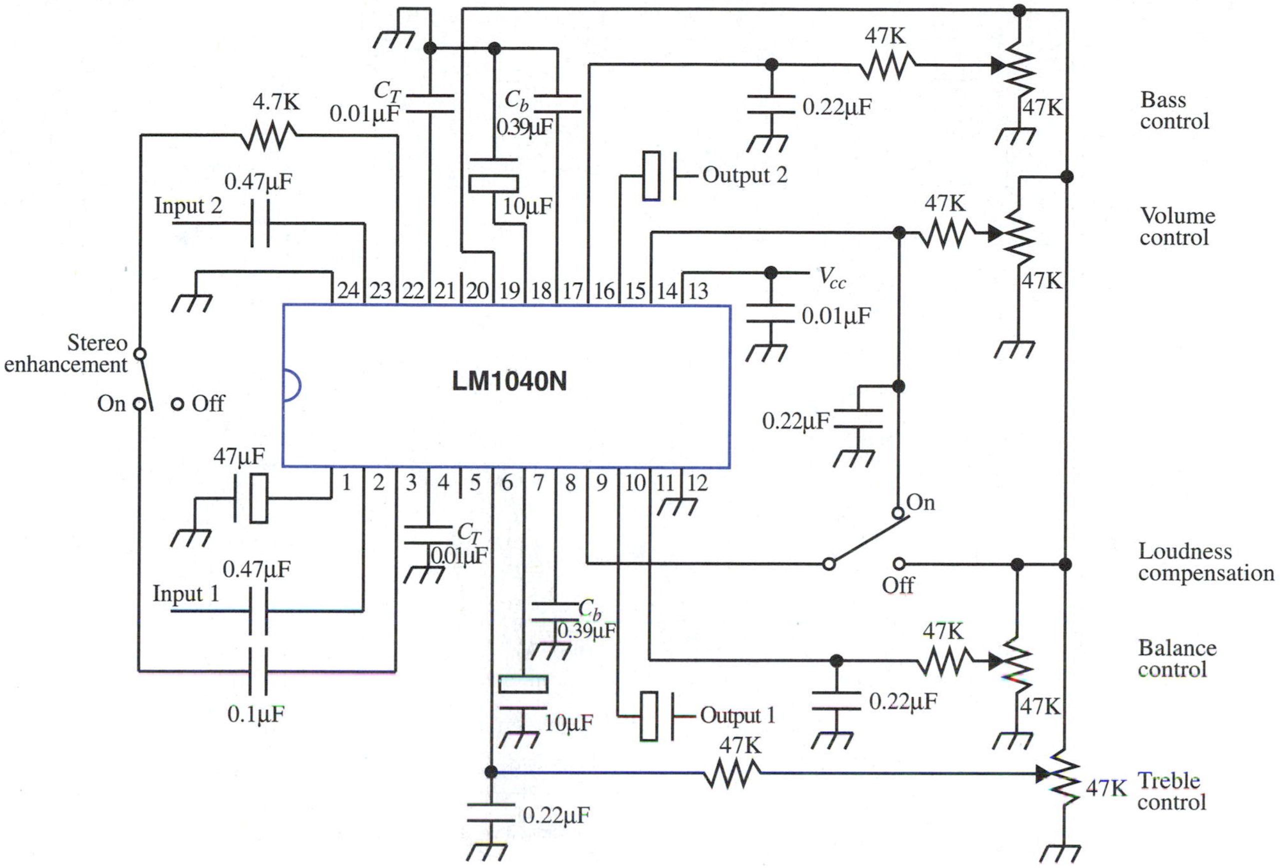

FIGURE 14–8 LM1040 Application circuit. *Courtesy National Semiconductor.*

Data Word I (Band Selection)							
D7	D6	D5	D4	D3	D2	D1	D0
H	X	S	S	S	S	S	S

Notes:
D0 → D3: selects band
D4: selects 12db range when "H" and 6db range when "L" for Channel B
D5: selects 12db range when "H" and 6db range when "L" for Channel A
D6: don't care
D7: band selection when "H"

Data Word II (Gain Selection)							
D7	D6	D5	D4	D3	D2	D1	D0
L	S	S	S	S	S	S	S

Notes:
D0 → D5: selects boost or cut level

D6: boost "H", cut "L"

D7: gain selection when "L"

The frequency bands are listed in the table along with the required value of capacitors and resistors to achieve these frequencies using the circuits in the "Z" blocks. This equalizer has independent boost/cut control range of either channel (A or B).

To keep within the 18V total power supply rating, the LMC835 in this circuit is powered from ±7.5V regulators derived from the ±15V supplies used for the op-amps.

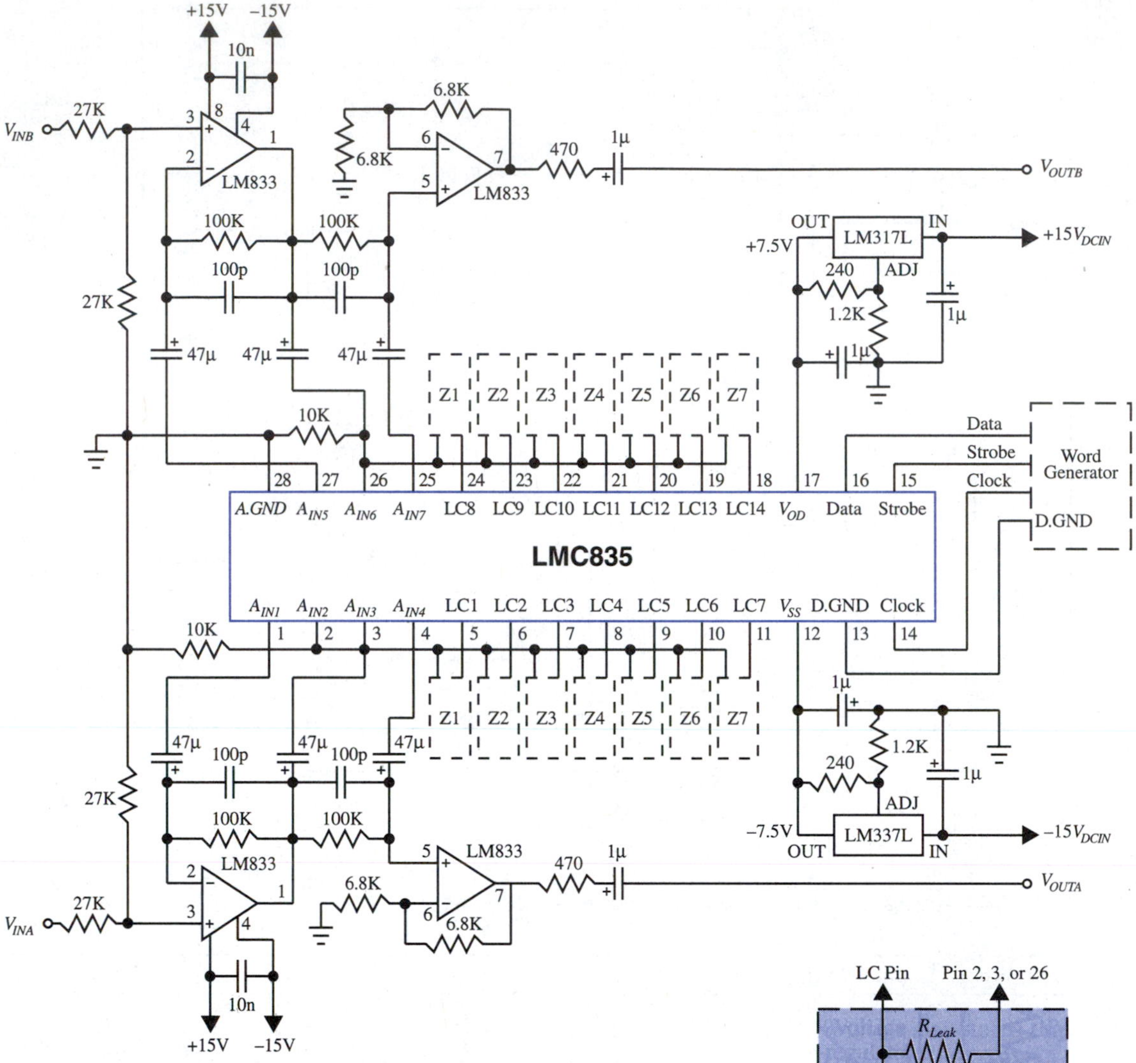

Z	f_o (Hz)	C_o (μF)	C_L (F)	R_L (Ω)	R_o (Ω)
Z_1	63	1.0	0.1	100K	680
Z_2	160	0.47	0.033	100K	680
Z_3	400	0.15	0.015	100K	680
Z_4	1K	0.068	0.0068	82K	680
Z_5	2.5K	0.022	0.0033	82K	680
Z_6	6.3K	0.01	0.0015	82K	680
Z_7	16K	0.0047	680p	47K	680

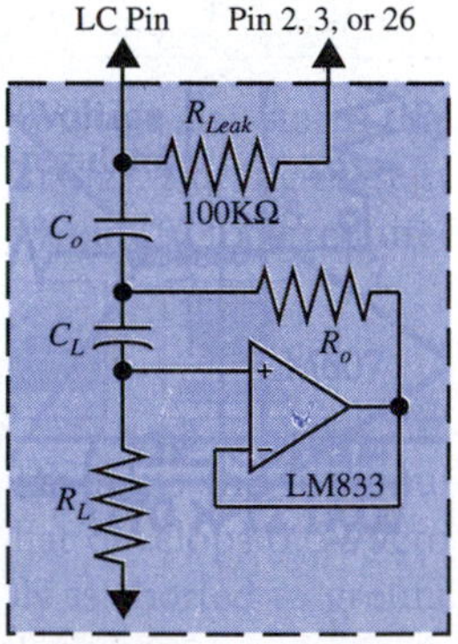

$$L_o = C_L R_L R_o \qquad f_o = \frac{1}{2\pi\sqrt{L_o C_o}} \qquad Q_o = \sqrt{\frac{L_o}{C_o R_o^2}}$$

$$Q_{12db} = \frac{R_o Q_o}{R_o + 1590} \qquad Q_o = 3.5, \qquad Q_{12db} = 1.05$$

FIGURE 14–9 LMC835 stereo-band equalizer. *Courtesy National Semiconductor.*

A possible application for this computer-controlled equalizer could be with a high-quality audio system. A calibration of frequency response could be as follows:

- A test microphone with a known frequency response is located near the audio listener.
- An audio sweep frequency generator is initially connected to the preamplifier input and to a frequency-to-voltage converter. The output of the converter is converted to digital via an analog-to-digital converter (ADC).
- The test microphone output is also converted to digital.
- The computer monitors the DC voltage from the frequency-to-voltage converter to determine the instantaneous sweep generator frequency and at the same time monitors the voltage at the test microphone.
- Based on its inputs, the computer adjusts the equalizer bands to generate a flat frequency response at the microphone.

This automatic adjustment of the equalizer settings would compensate for unevenness in the sound level caused by the loudspeaker response and the acoustics of the listening area.

AUDIO NOISE REDUCTION

To preserve the large dynamic range of audio, the noise level must be small relative to the smallest audio signals. Noise can be introduced during the transmission or recording of the analog audio signal. This has resulted in the development of various noise reduction systems—most of which process the signal prior to being exposed to the noise source.

The basic approach to noise reduction is to reduce the dynamic range of the audio by compressing (**companding**) the signal. Audio compression lowers the gain for high-level signals and raises the gain for low-level signals. The low-level signals now are less sensitive to noise introduced beyond the compressor. Another approach to noise reduction is to dynamically restrict the bandwidth to that required for the signal, and thereby rejecting out-of-band noise.

The most common form of noise reduction used are the Dolby systems. Dolby A-type is a dual path system where high-level signals pass through without processing, and low-level signal companding takes place in four independent frequency bands. The four bands allow a more selective control of the signal amplitude without interaction. Dobly A-type provides 10db of noise reduction from 30Hz to 5KHz, rising to 15db at 15KHz.

In Dolby B-type and C-type companding, the band of frequencies that are companded are always higher than the high-level signal frequency (higher-frequency signals are usually lower in amplitude and therefore more susceptible to noise). If the strong-signal frequency rises, then the companding band slides up out of its way.

In the newer Dolby S-type, a combination of fixed and sliding bands is used that provides a 10db noise reduction for low frequencies and a 24db reduction for the higher frequencies.

The Dolby-type companders are double ended—meaning that processing must occur before and also after transmission or recording. A noise reduction approach that is single ended, i.e., processing only after noise is introduced, is provided by the LM1894, shown in block diagram form in Figure 14–10.

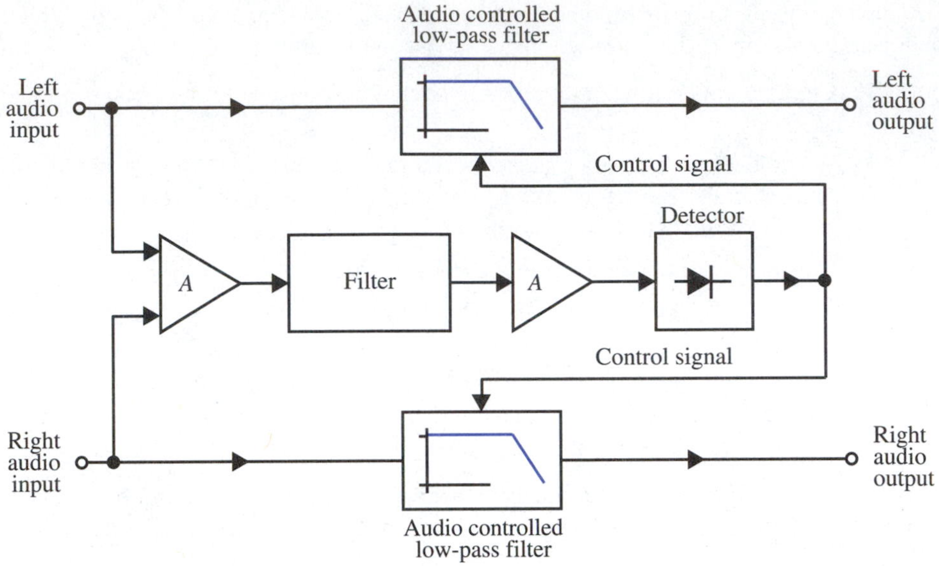

FIGURE 14–10 LM1894 stereo noise reduction system. *Courtesy National Semiconductor.*

The approach with this type of noise reduction is to reduce the bandwidth when the signal level is low and to increase the bandwidth when the signal strength is sufficient to mask out the noise. This is accomplished by controlling the bandwidth of a low-pass filter with a signal generated from the summing of the two stereo input signals. The bandwidth reduces to 1KHz when no signal is present and increases to 35KHz with maximum signal amplitude.

A high-pass filter is used in the control path to make the circuit more responsive to the amplitude of the higher-frequency signals (signals between 1KHz and 6KHz are of a lower amplitude and are therefore more susceptible to noise masking).

The recommended dynamic noise reduction (DNR) circuit using the LM1894 is shown in Figure 14–11. The circuit includes a 19KHz notch filter for preventing the pilot carrier, used in FM stereo, from activating the system. This filter can be bypassed in non-FM applications.

This circuit can give up to 14db effective noise reduction, and the THD is a maximum of 0.1% with an input signal of 300mV. The voltage gain is -1 with the peak detector disabled (Pin 9 grounded), and the bandwidth is typically 34KHz.

Let us now look at the equations that determine the operation of this noise reduction system.

The output current (I_o) from the transconductance (g_m) amplifier is

$$I_o = g_m V_{in}$$

This current charges the 3.3nF capacitor to provide an output voltage (V_o) of

$$V_o = \frac{I_o}{2\pi f C}$$

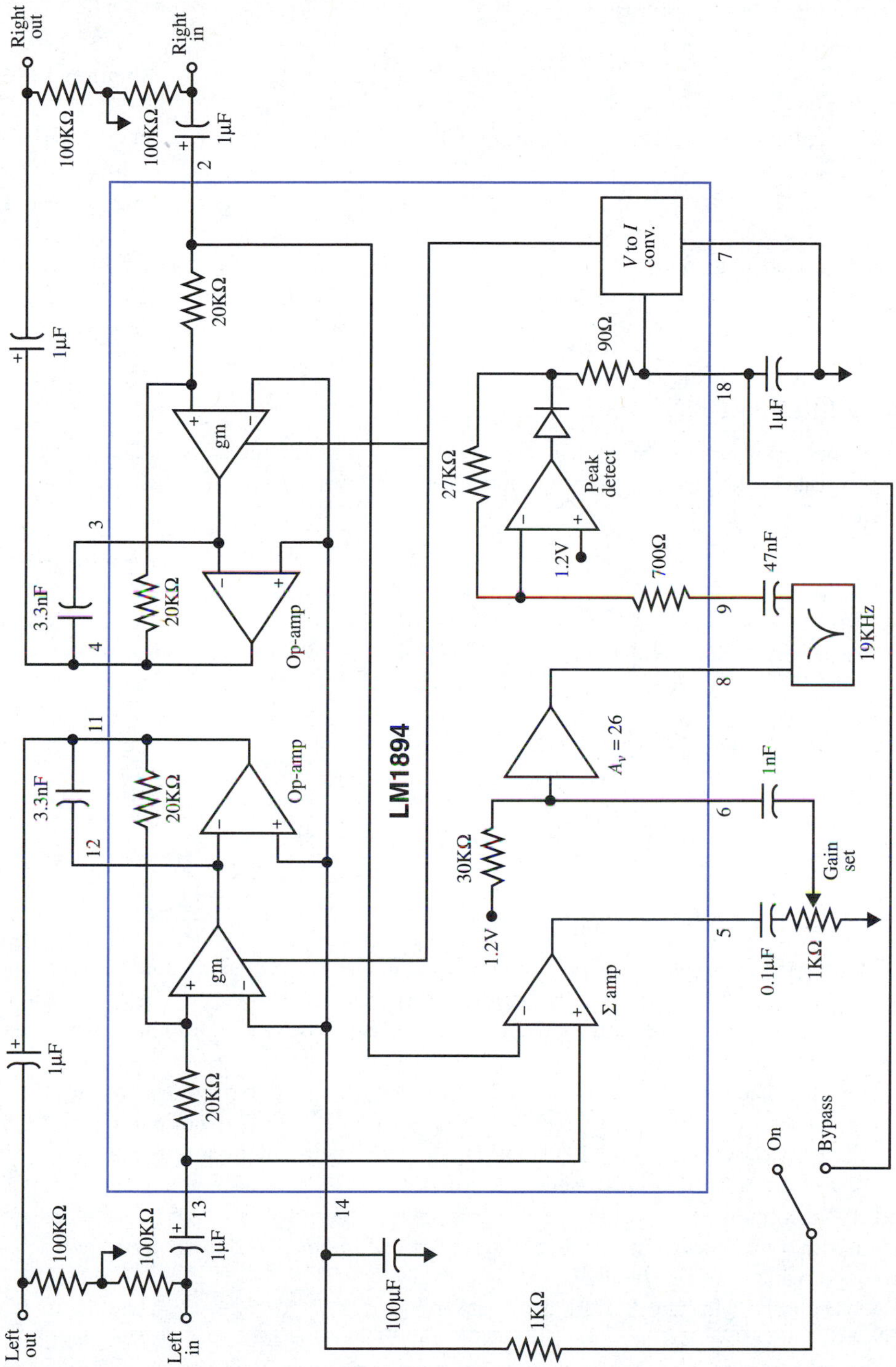

FIGURE 14–11 Complete stereo noise reduction circuit

Combining these two equations,

$$A_v = \frac{V_o}{V_{in}} = \frac{g_m}{2\pi fC} \qquad \textbf{(equation 14–2)}$$

At a high-enough frequency, the gain A_v will fall to unity, and this frequency (f_u) is

$$f_u = \frac{g_m}{2\pi C} \qquad \textbf{(equation 14–3)}$$

With a fixed value of C, the unity gain frequency (f_u), and therefore the bandwidth, will change with g_m. Since g_m increases with the peak detected input signal voltage, the bandwidth also increases with the input signal, and this provides the required noise reduction.

AUDIO DRIVER STAGE

The signal level from a preamplifier of an audio processing circuit is not sufficient to drive a loudspeaker. The relationship between the speaker power and the required peak-to-peak voltage is as follows:

$$V_{p-p} = 2\sqrt{2PR_s} \qquad \textbf{(equation 14–4)}$$

EXAMPLE

Find the peak-to-peak voltage required to provide 20W to an 8Ω speaker.

SOLUTION

Using equation 14–4,

$$V_{p-p} = 2\sqrt{2PR_s}$$

$$V_{p-p} = 2\sqrt{2 \times 20\text{W} \times 8\Omega}$$

$$V_{p-p} = 35.8\text{V}$$

To bring the voltage from the preamplifier to a sufficient level requires amplification, and this is performed by a stage called an **audio driver**. The driver precedes the power amplifier stage that provides the actual current to the speaker. The power amp has a small voltage gain but high current gain.

One integrated circuit driver is the LM391, which is designed to drive a power stage consisting of power transistors at power levels from 10W to 100W. The internal schematic for this driver is shown in Figure 14–12. This chip can operate with supply voltages up to ±50V and provide an output voltage swing of ±43V. The minimum voltage gain is 1000, and the maximum THD is 0.25%. Both a source (Pin 8) and sink (Pin 5) output are provided with a maximum current capability of 5mA. A remote shutdown capability is also provided, which can also be used to delay turn-on.

An application for the LM391 driver is provided in Figure 14–13. This circuit is a 40W-8Ω, or a 60W-4Ω, amplifier using NPN and PNP transistors for the power amplifier.

The voltage gain A_v of the circuit is determined using the 5.1KΩ (R_1) from Pin 2 to AC ground and the 100KΩ (R_f) resistor from the output (Pin 9) to the inverting input (Pin 2).

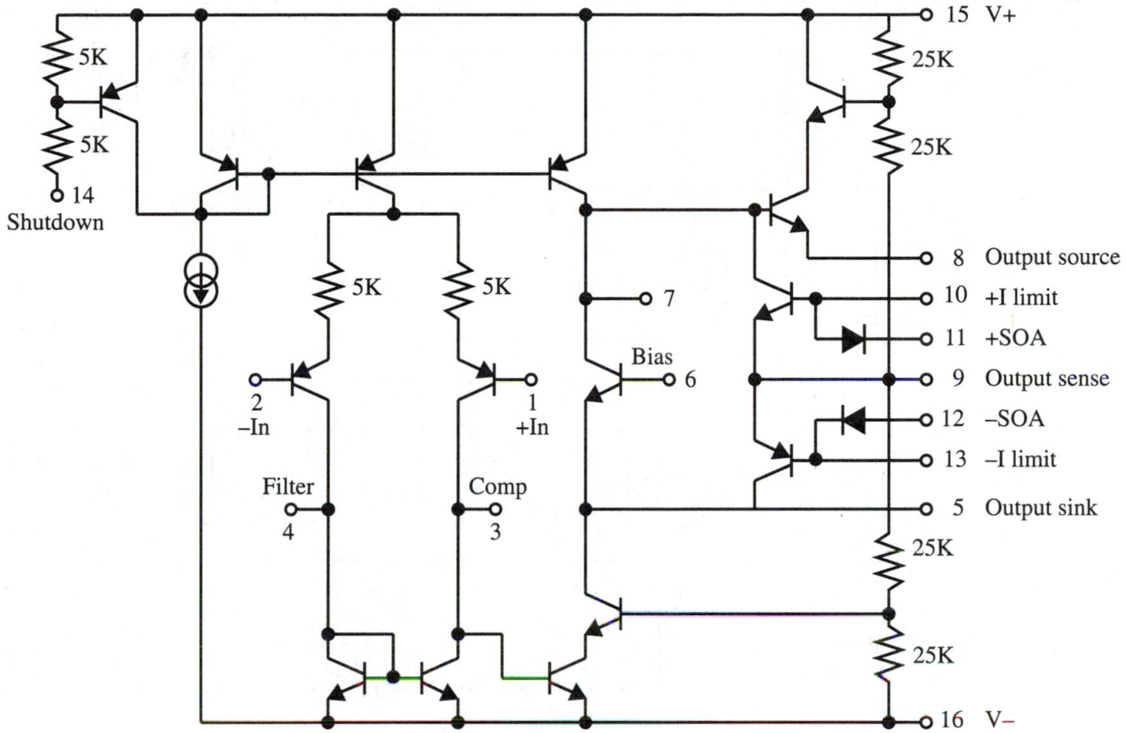

FIGURE 14–12 LM391 Audio driver schematic. *Courtesy National Semiconductor.*

$$A_v = \frac{R_f}{R_1} + 1$$

$$A_v = \frac{100\text{K}\Omega}{5.1\text{K}\Omega} + 1$$

$$A_v = 20.6$$

The input resistance to the circuit is the 100KΩ resistor from Pin 1 to ground.

EXAMPLE

Find the peak input voltage and current required for the circuit of Figure 14–13 to provide 40W into an 8Ω speaker.

SOLUTION

First we need to find the required voltage across the speaker at the 40W level. Using equation 14–4,

$$V_{p-p} = 2\sqrt{2PR_s}$$

$$V_{p-p} = 2\sqrt{2 \times 40\text{W} \times 8\Omega}$$

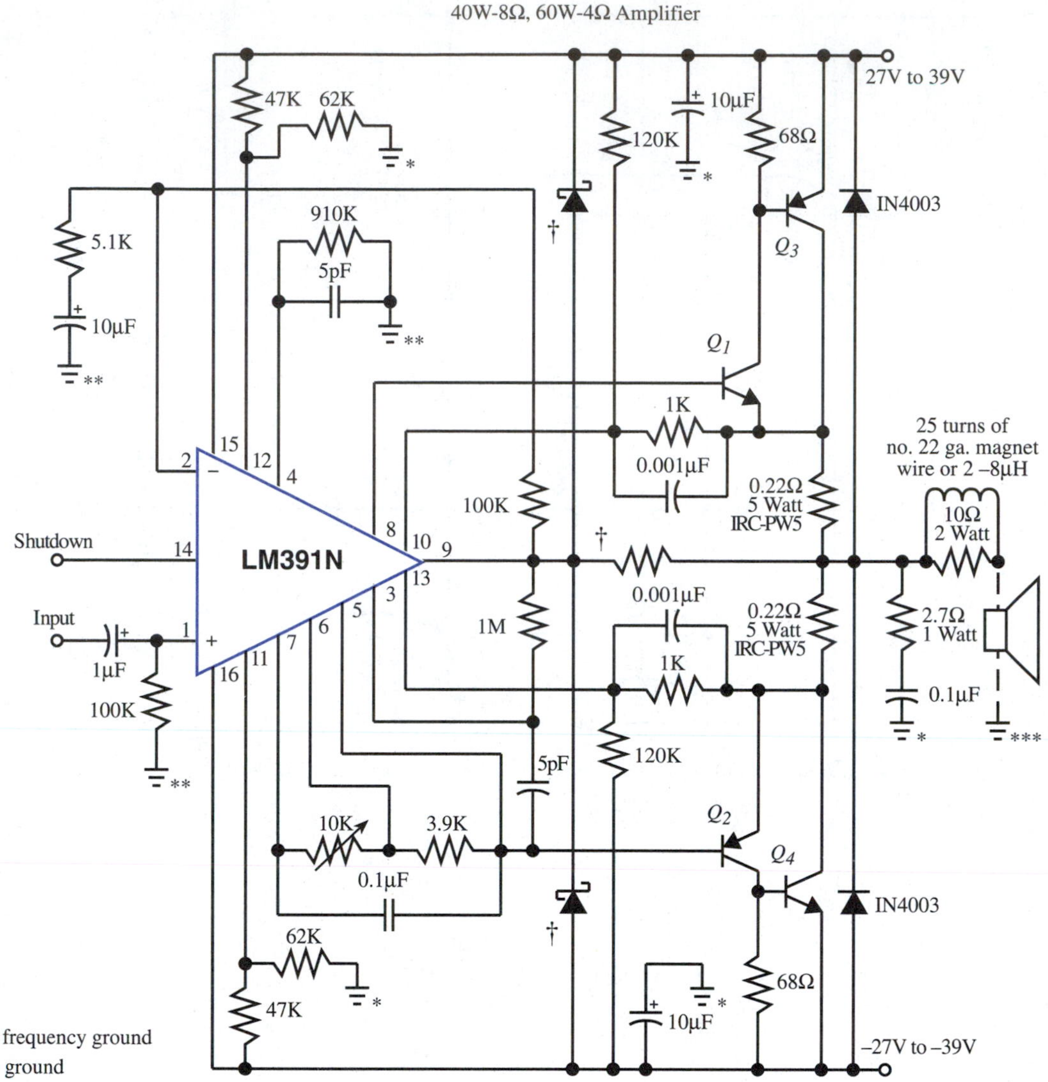

* High frequency ground
** Input ground
*** Speaker Ground
Note: All grounds should be tied together only at power supply ground
† Additional protection for LM391N; Schottky diodes and R ≅ 100Ω

FIGURE 14–13 LM391 used with a power amplifier

$$V_{p-p} = 50.6\text{V}$$
$$V_p = 25.3\text{V}$$

With the gain of 20.6, the required peak input voltage is

$$V_{in(p)} = \frac{25.3\text{V}}{20.6}$$
$$V_{in(p)} = 1.23\text{V} \quad (0.87\text{V RMS})$$

The peak input current is this voltage divided by the 100KΩ input resistance.

$$I_p = \frac{1.23\text{V}}{100\text{K}\Omega}$$

$$I_p = 12.28\mu\text{A}$$

Q_1 is an MJE722 (2N4923) NPN and Q_2 is an MJE712 (2N4920) PNP plastic power transistor with minimum β of 35 at 100mA. Q_3 is a 2N5880 PNP and Q_4 is a 2N5882 NPN TO-3 package transistor with minimum β of 30 at 3A. If we assume a maximum current through the speaker of 3A then the current provided by the LM391 must be

$$I_{(391)} = \frac{3\text{A}}{30 \times 35}$$

$$I_{(391)} = 2.86\text{mA}$$

This is within the maximum 5mA source/sink current capability of the LM391 driver.

The 0.22Ω resistors in the emitters of the NPN transistors serve to balance the outputs and also provide current sensing via Pins 10 and 13. It is assumed that current limiting will occur when the voltage across these resistors reaches 0.65V—which means a current of

$$I_{cl} = \frac{0.65\text{V}}{0.22\Omega}$$

$$I_{cl} = 2.95\text{A}$$

The remaining circuit components are selected to reduce distortion and maintain a safe operating area (SOA) for the power transistors.

THE AUDIO POWER AMPLIFIER

We have seen an example of a bipolar transistor power stage used with the LM391 driver. Now we shall look at an integrated circuit power amplifier. The amplifier we shall study is the LM12. This device has the specifications indicated in Table 14–1.

TABLE 14–1 LM12 Power Amplifier Characteristics

Parameter	*Conditions*	*Value*
Power output	$R_L = 4\Omega$	150W
Output voltage swing	$I_{out} = 10\text{A}$	±35V
Peak output current	$V_{out} = 0$	±13A
Bandwidth	$A_v = 1$	700KHz
Slew rate	$R_L = 4\Omega$	9V/μs
Total harmonic distortion	$R_L = 4\Omega$	0.01%
Input offset voltage	$V_{cm} = 0$	2mV
Input bias current	$V_{cm} = 0$	150nA
Voltage gain	$R_L = 4\Omega$	50,000
Continuous DC dissipation	Case temp 25°C	90W
Supply current	$I_{out} = 0$	60mA

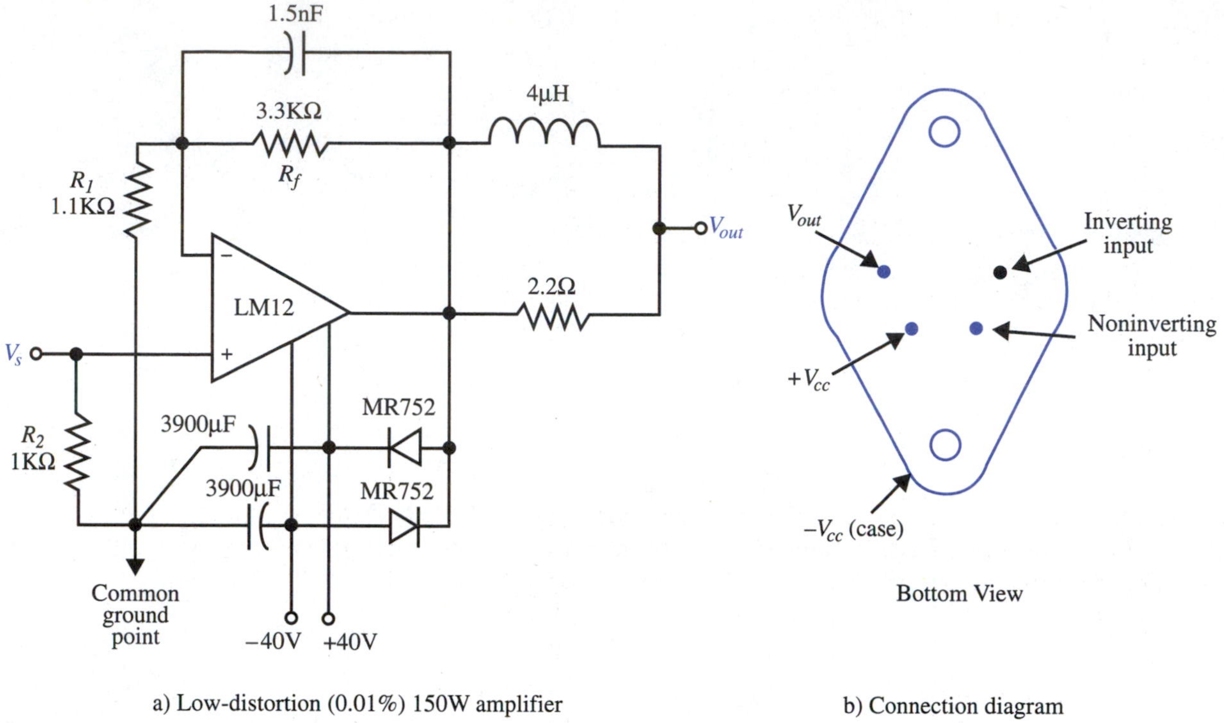

FIGURE 14–14 LM12 power amplifier

In addition to the listed specifications, the full power (150W) bandwidth extends to 60KHz—which is well beyond the audio band.

A power amplifier circuit using the LM12 is shown in Figure 14–14, along with the case terminals.

The amplifier is set up as a noninverting amplifier with a gain A_v of

$$A_v = \frac{R_f}{R_1} + 1$$

$$A_v = \frac{3.3\text{K}\Omega}{1.1\text{K}\Omega} + 1$$

$$A_v = 4$$

EXAMPLE

Find the input voltage and current required to provide a 150W output across a 4Ω load from the LM12 circuit of Figure 14–14.

SOLUTION

To get the required ±35V output voltage swing required for the 150W output requires an input voltage V_s of

$$V_s = \frac{V_o}{A_v}$$

$$V_s = \frac{\pm 35\text{V}}{4}$$

$$V_s = \pm 8.75\text{V}$$

The peak current required is this input voltage divided by the input resistance of 1KΩ—which is 8.75mA. This voltage and current drive could, for example, be supplied by an LM833 or LF451 op-amp operating from ±12V supplies.

The large current flowing in the LM12 requires care with power supply bypassing, lead inductance, and ground loops (these concepts are covered in greater detail in Chapter 18).

The power supply bypass capacitors (3900μF in Figure 14–14) should have short leads and be connected directly to the LM12 terminals.

The LM12 is sensitive to inductance of the power supply leads and also the output lead. Power supply leads should be kept as short as possible, and the feedback resistor (R_f) should be connected directly to the output terminal. An inductor is connected between the LM12 output and the load to isolate the amplifier from load capacitance, which can cause instability.

Ground loops can be avoided by single point grounding, as shown in Figure 14–14.

BRIDGE AMPLIFIER

To provide twice the output voltage swing across the load (power increases by 4), the power amp can be connected in a bridge connection, as shown in Figure 14–15.

One LM12 is connected as an inverting amplifier, and the other as a noninverting amplifier. Both are set up for a voltage gain of 4. If the input voltage reaches a peak of +8V, the output of the inverting amplifier (the left side of the speaker) will be −32V; while the output of the noninverting amplifier will be +32V (right side of speaker). This puts 64V

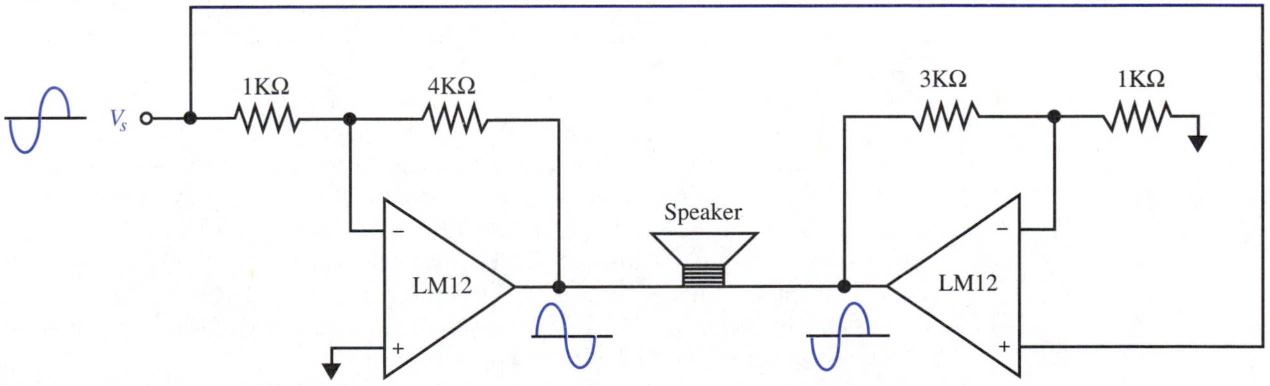

FIGURE 14–15 Bridge amplifier

across the speaker terminals in one direction. If the input voltage now swings to −8V, the voltage across the speaker will again be 64V but in the other direction. So, the peak-to-peak voltage swing across the speaker is 128V.

EXAMPLE

Find the power delivered to an 8Ω speaker if the voltage across its terminals is $128V_{p-p}$.

SOLUTION

Solving equation 14–4 for the speaker power,

$$P_s = \frac{(V_{p-p})^2}{8R_s}$$

$$P_s = \frac{(128\text{V})^2}{8 \times 8\Omega}$$

$$P_s = 256\text{W}$$

When the power amplifier is used in a bridge configuration, the same cautions have to be followed relative to lead length and power supply bypassing.

SWITCHING POWER AMPLIFIER (CLASS D)

A switching amplifier converts the input audio signal to a series of fixed amplitude pulses whose width varies with the sampled amplitude of the input signal. The pulses occur at a sampling frequency f_s that is much higher than the highest audio frequency, and their amplitude is equal to the power supply voltage (typically plus and minus supplies).

Although the switching or Class D amplifier is more complex than the conventional push-pull (class B) amplifiers, it has much higher efficiency at high-power levels. At output power levels in excess of 25W, switching amplifiers have efficiencies greater than 90 percent rather than the 78.5 percent limit for class B amplifiers (see Chapter 3). Above 100W output power rating, the switching amplifier costs less than a class B amplifier and requires a heat sink of about one-tenth the size.

A block diagram for a Class D amplifier is provided in Figure 14–16.

In the figure, we see that the input audio signal is added to the inverted and chopped output from the switching circuit and passed through an integrator. This operation ensures the fidelity of the chopped waveform. The integrator acts like a low-pass filter to eliminate the fast transitions in the chopped waveform during this signal comparison.

At the circuit output, the low-pass filter converts the variable width pulses back to the shape of the input audio by removing the switching frequency. The filter works, because the switching frequency is much higher than the audio frequencies. (The circuit described by Motorola in their AN1042 bulletin has an f_s of 120KHz.)

A more detailed Motorola circuit from the output of the chopper to the speaker appears in Figure 14–17.

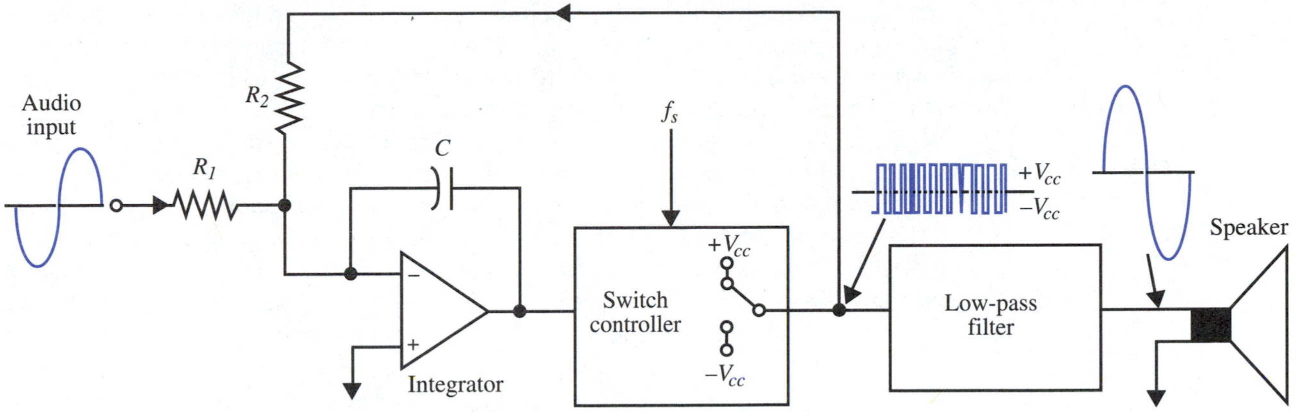

FIGURE 14–16 Block diagram of a switching amplifier

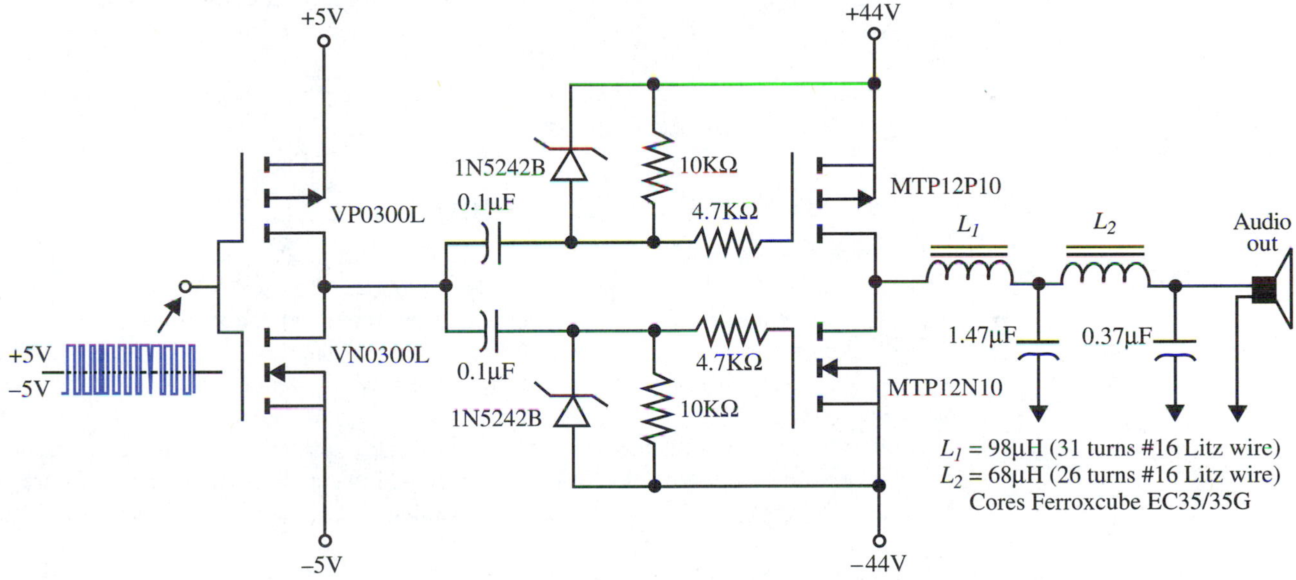

FIGURE 14–17 Switching amplifier output circuit. *Courtesy Motorola Semiconductor.*

MOSFETs are used in the output circuit because they are fast, and have low "on" resistance. The complementary arrangement (P and N MOSFETs) provides current to the speaker on both the positive and negative voltage swing.

LOUDSPEAKER CONSIDERATIONS

We have considered the power amplifier load to be a pure resistor. Actually a loudspeaker is a complex impedance consisting of inductance, capacitance, and resistance. Further, these components break down into both electrical and acoustic parts—for example, we have ohmic resistance of the speaker coil and also the acoustic radiation resistance. The result is that we have an equivalent electrical circuit representing the loudspeaker that does not have a flat frequency response and can oscillate (damped) when excited.

Cabinet design complements the speaker and can make the frequency response acceptable. Speaker resonances that cause a fuzziness, when there is a sharp change in sound level, can be reduced by the damping effect of the low output resistance of the power amplifier.

The nominal resistance (impedance) of a loudspeaker is specified at a frequency of 1KHz—but can vary over the frequency range. To get a more even distribution of power over the audio frequency range, speakers are designed to specifically operate over a limited frequency band. A woofer operates at the low frequencies, a midrange speaker handles the mid frequencies, and the tweeter the high frequencies.

To channel the frequency band to a particular speaker requires filter networks. A low-pass for the woofer, band-pass for the midrange, and high-pass for the tweeter. This filtering is usually accomplished at the output of the power amplifier with so called "crossover networks." The disadvantage of this approach is that the filter components must handle the high-level speaker current. With the lower cost of integrated circuit power amplifiers, it now makes more sense to do the filtering at the output of the preamplifier and using a separate power amplifier for each speaker. This is not a bad approach because most of the power is required to move the woofer, and the more efficient midrange and tweeter can operate from relatively low-power amplifiers.

THE SWITCHED CAPACITOR FILTER

I think that it is appropriate in a chapter concerned with audio circuits to introduce an integrated circuit that can perform audio signal filtering using a band-pass, notch, low-pass, or high-pass configuration. This device is called a **switched capacitor filter** (SCF).

We looked at various op-amp active filter networks in Chapter 7. The band-pass, notch, low-pass, and high-pass were all covered; in all these configurations an op-amp was used along with precision resistors and capacitors. Using precision components adds to the cost of equipment as well as making it more difficult to get replacements. This is particularly true with precision capacitors. Manufacturers of linear integrated circuits realized this problem and developed the switched capacitor active filter (SCF) chip, which does away with the need for external capacitors.

Since a key element of the SCF is the integrator, let us first look at how we can filter with this type of circuit. Figure 14–18a shows an integrator used for a low-pass filter. In operation, as the input frequency increases, the reactance of the feedback capacitor decreases, which reduces the gain, and the output signal drops. In Figure. 14–18b, we see that a high-pass filter can be generated by using the integrator along with a summing amplifier (A_2). Both the input signal and the output of the low-pass filter, which is shifted by 180°, are added together (actually subtracted because of the 180° phase shift) to generate a high-pass characteristic.

By rearranging the feedback and using a second integrator, a band-pass or notch filter can be generated. Filters that use integrators and summing amplifiers are called **state variable types**.

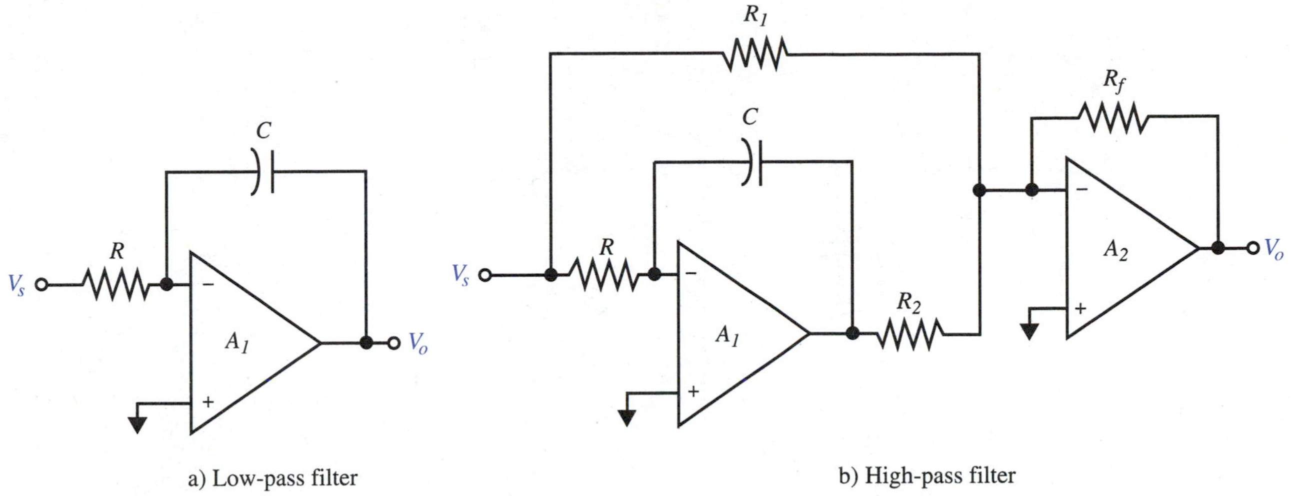

FIGURE 14–18 Integrator used as a filter

With a switched capacitor filter, the integrator input resistor is replaced by two switches and a capacitor, as shown in Figure 14–19.

The clock signal controls the switches S_1 and S_2. On a positive clock pulse S_1 closes and C_1 quickly charges to the input voltage V_s (assumes V_s is from a low-resistance source). When the clock signal goes low, S_1 opens and S_2 closes. The charge (q) on C_1, which is equal to V_sC_1, is now all quickly transferred to C_2. The voltage (and charge) on C_1 goes to zero because of the virtual ground at the op-amp input. This transfer of charge to C_2 on each clock cycle causes the voltage on C_2 to increase in steps. With a positive input voltage, the output will step to a more negative voltage.

Comparing the output signal to the conventional integrator, we get a stepped slope rather than a steady slope. But notice that if the clock frequency is increased (more charge transfers), the slope increases. The average current (I_a) that flows into C_1, and then into C_2 is

$$I_a = \frac{q}{t}$$

$$I_a = \frac{V_sC_1}{T}$$

$$I_a = V_sC_1f_{CLK}$$

where T is the clock period and f_{CLK} the clock frequency. This current flow can be thought of as resulting from a fictional resistor R_1 where

$$R_1 = \frac{V_s}{I}$$

$$R_1 = \frac{1}{C_1f_{CLK}} \qquad \textbf{(equation 14–5)}$$

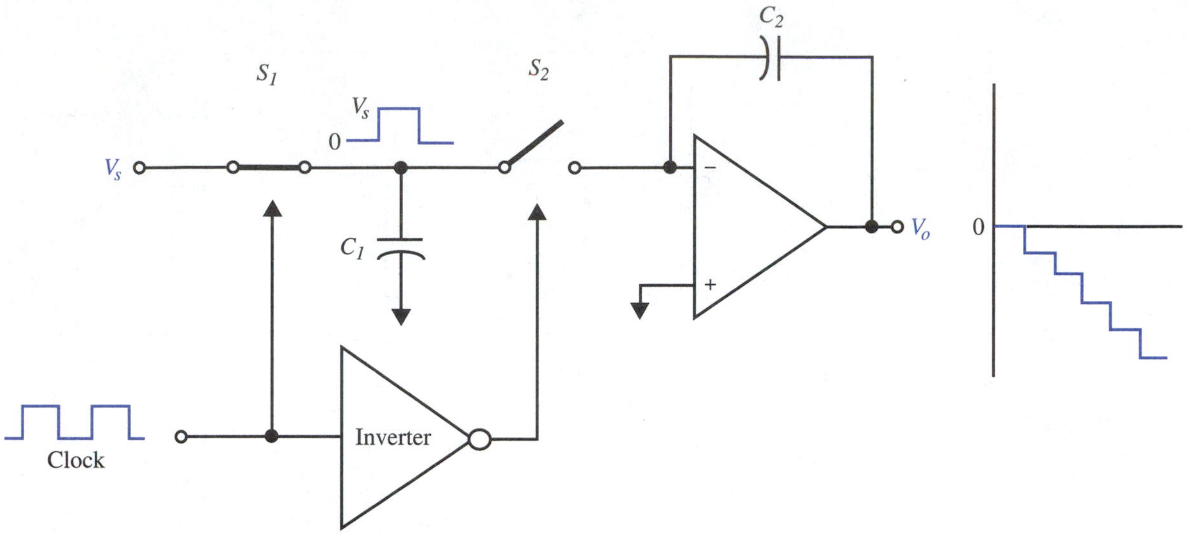

FIGURE 14–19 Basic switched capacitor circuit

This effective resistor represents the charging and discharging of C_1, and its value is controlled by the clock frequency. So we can change the response (slope) of the integrator by simply changing the clock frequency. The filter characteristics, then, are also under control of the clock.

In a switched capacitor filter chip, the capacitors C_1 and C_2 are formed in the semiconductor material and are therefore closely matched. There is no longer a concern, then, with the drift of external capacitors.

The actual integrator used in the SCF is a noninverting type. This is achieved by reversing the connection of C_1 when it is connected to C_2, as shown in Figure 14–20. Both sections of S_1 are closed first, charging C_1, but when S_{2a} and S_{2b} close, the polarity of C_1 is reversed and is connected to the op-amp input.

As before, S_1 and S_2 are controlled by opposite states of the input clock. The clock frequency is much higher than the signal frequency, so the steps in the output of the integrator can be filtered out with a simple RC low-pass filter.

A typical SCF is the LMF100 (an updated version of the original MF10). This device can be used as a filter for frequencies up to about 70KHz with a maximum clock frequency of 3.5MHz.

The filter center, or 3dB down, frequency is set by dividing the clock by 50 or 100 (depending on the voltage on a control pin). Filter accuracy is as good as the frequency stability of the clock. If we use a crystal-controlled clock, we will have a very accurate filter. The LMFl00 contains two separate second-order filters, each as shown in Figure 14–21. As can be seen from the figure, there are five different frequency responses available from three outputs, notch (N), all-pass (AP), high-pass (HP), band-pass (BP) and low-pass (LP).

Input signals can enter at Pin 4 for a inverted output or Pin 5 (S_1) for a noninverted output. When using Pin 5, the driving impedance should be less than 1KΩ. The signal passes through a summing junction where two of the inputs (−) are subtracted from a third (+).

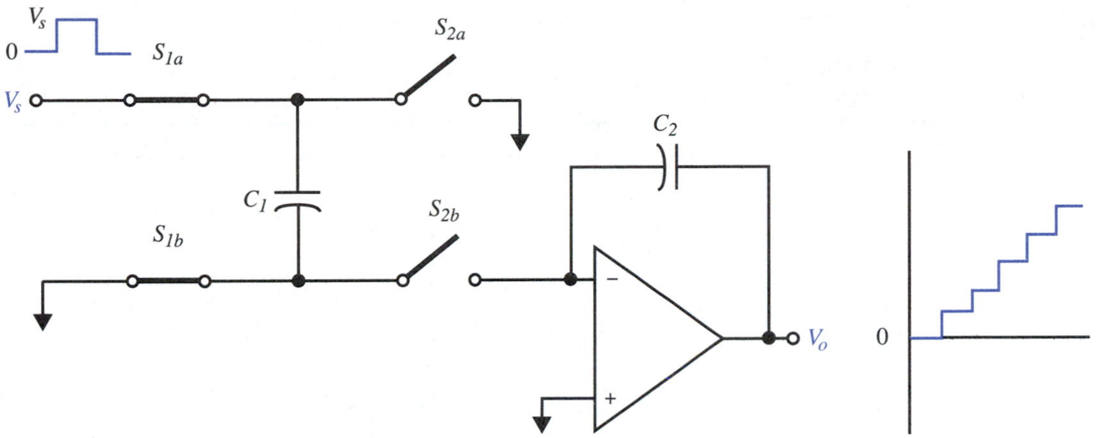

FIGURE 14–20 Non-inverting integrator

Taking Pin 6 (S_{AB}) low will cause the mode switch to move to a position that will ground the input to the summing junction. With Pin 6 high, the output of the second integrator will be connected to the summing junction. We will see later how this can change the operating mode.

The level shift (Pin 9) allows for various input clock levels. With a TTL input clock, this pin should be grounded, but if the input clock excursion swings to a negative supply value, Pin 9 should be tied to this supply.

There are several possible circuit configurations (modes) using the LMF100. We shall look at two that will give us the four filter characteristics. The first configuration shown in Figure 14–22 is called mode 1 by the manufacturer and provides notch, band-pass, and low-pass outputs by using three external resistors.

We need to define the filter responses in terms of in-band gain, cutoff frequency, center frequency, Q, etc. Figure 14–23 shows the responses along with the equations for the mode1 configurations.

Note that the equations for the low-pass are somewhat more involved than those of the notch and band-pass. Let us look at an example using these equations.

EXAMPLE

A band-pass filter is required with a center frequency (f_o) of 3KHz, a bandwidth of 300Hz ($Q = 10$), and a gain of -10. If R_2 is specified to be 10KΩ, find the required values of R_1, and R_3. Use a 100:1 clock frequency for least clock ripple in the output signal.

SOLUTION

First, we shall find the required clock frequency.

$$f_{CLK} = 100 f_o$$

$$f_{CLK} = 100 \times 3\text{KHz}$$

$$f_{CLK} = 300\text{KHz}$$

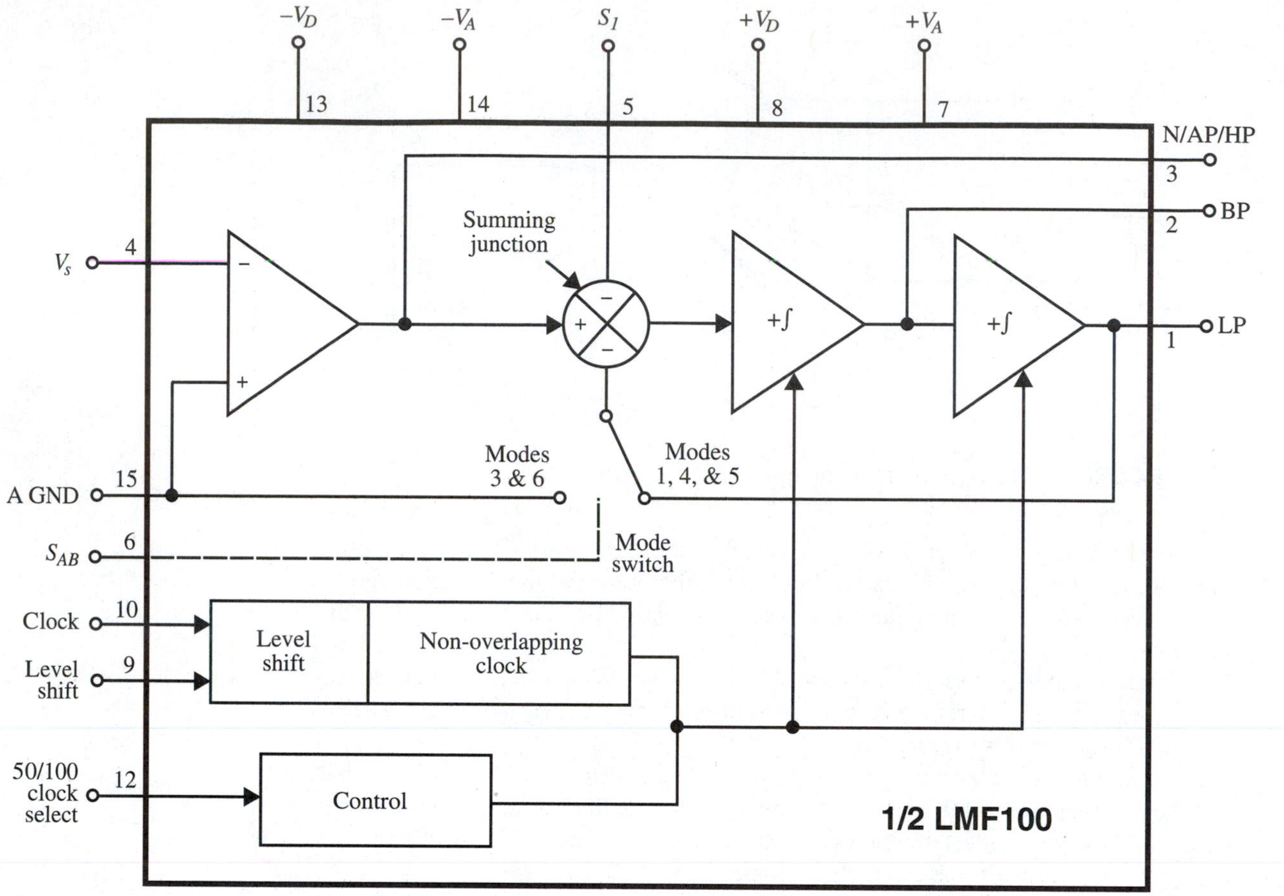

FIGURE 14–21 LMF100 block diagram (one half). *Courtesy National Semiconductor.*

The required bandwidth of 300Hz requires a Q of 10. Solving for R_3,

$$R_3 = QR_2$$

$$R_3 = 10 \times 10\text{K}\Omega$$

$$R_3 = 100\text{K}\Omega$$

Using the required gain (H_{obp}) of -10, we can now solve for R_1.

$$R_1 = \frac{-R_3}{H_{obp}}$$

$$R_1 = \frac{-100\text{K}\Omega}{-10}$$

$$R_1 = 10\text{K}\Omega$$

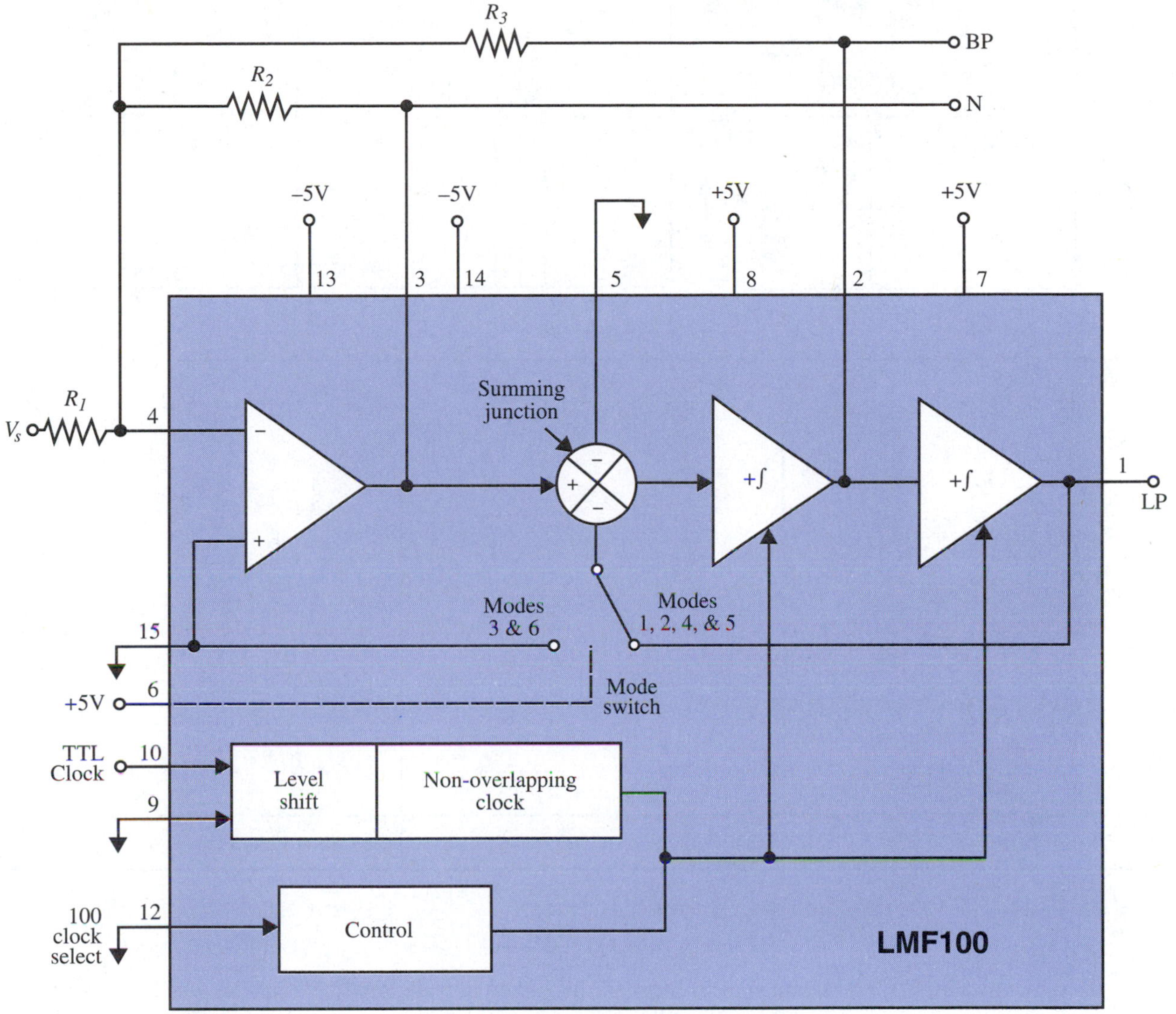

FIGURE 14–22 LMF100 mode 1 notch, band-pass, and low-pass filters

So, with just two 10KΩ resistors and one 100KΩ resistor, we have a second-order filter. The band-pass output is taken from Pin 2 (see Figure 14–22). A low-pass response is available at Pin 1, and a notch response at Pin 3.

The other configuration we shall study is mode 3. This mode can provide high-pass, low-pass and band-pass. The circuit arrangement for mode 3 is shown in Figure 14–24.

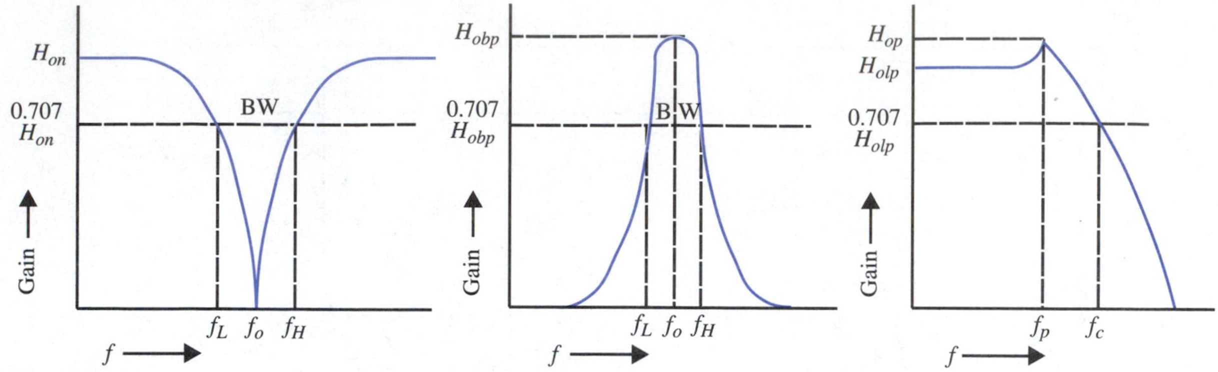

$f_o = \frac{f_{CLK}}{100}$ or $\frac{f_{CLK}}{50}$ $Q = \frac{f_o}{f_H - f_L} = \frac{R_3}{R_2}$ $H_{on} = \frac{R_2}{R_1}$	$f_o = \frac{f_{CLK}}{100}$ or $\frac{f_{CLK}}{50}$ $Q = \frac{f_o}{f_H - f_L} = \frac{R_3}{R_2}$ $H_{obp} = -\frac{R_3}{R_1}$	$f_o = \frac{f_{CLK}}{100}$ or $\frac{f_{CLK}}{50}$, $Q = \frac{R_3}{R_2}$ $H_{olp} = -\frac{R_2}{R_1}$, $H_{op} = \frac{H_{olp}Q}{\sqrt{1 - \frac{1}{4Q^2}}}$ $f_p = f_o\sqrt{1 - \frac{1}{2Q^2}}$ $f_c = f_o\sqrt{\left(1 - \frac{1}{2Q^2}\right) + \sqrt{\left(1 - \frac{1}{2Q^2}\right) + 1}}$
a) Notch Filter	**b) Band Pass Filter**	**c) Low Pass Filter**

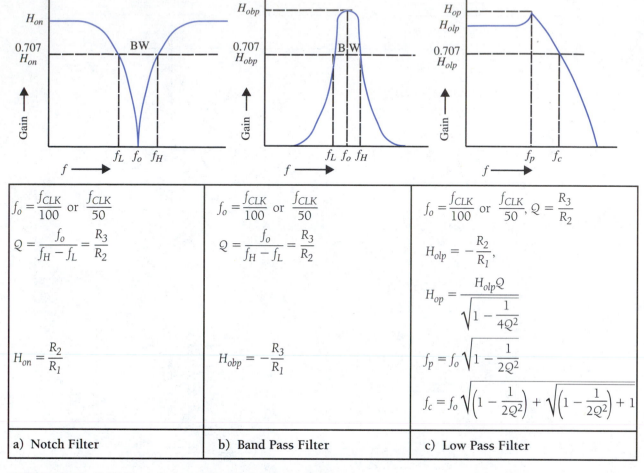

FIGURE 14–23 LMF100 mode 1 responses and design equations

The differences between this circuit and the mode 1 circuit is that Pin 6 is grounded rather than connected to the positive supply, and an extra resistor (R_4) is added. Since we had no high-pass response for mode 1, we shall show this response and the equations using mode 3 (see Figure 14–25).

EXAMPLE

A high-pass filter is required with a cutoff frequency (f_c) of 1KHz and a gain of 5 (H_{ohp}). We want little peaking of the response curve (H_{op} close to H_{ohp}), so we shall select Q equal to 1. Let R_1 be 5KΩ. Find the values of R_2, R_3, and R_4, and also the peak amplitude and the frequency where it occurs.

SOLUTION

First, solving for R_2,

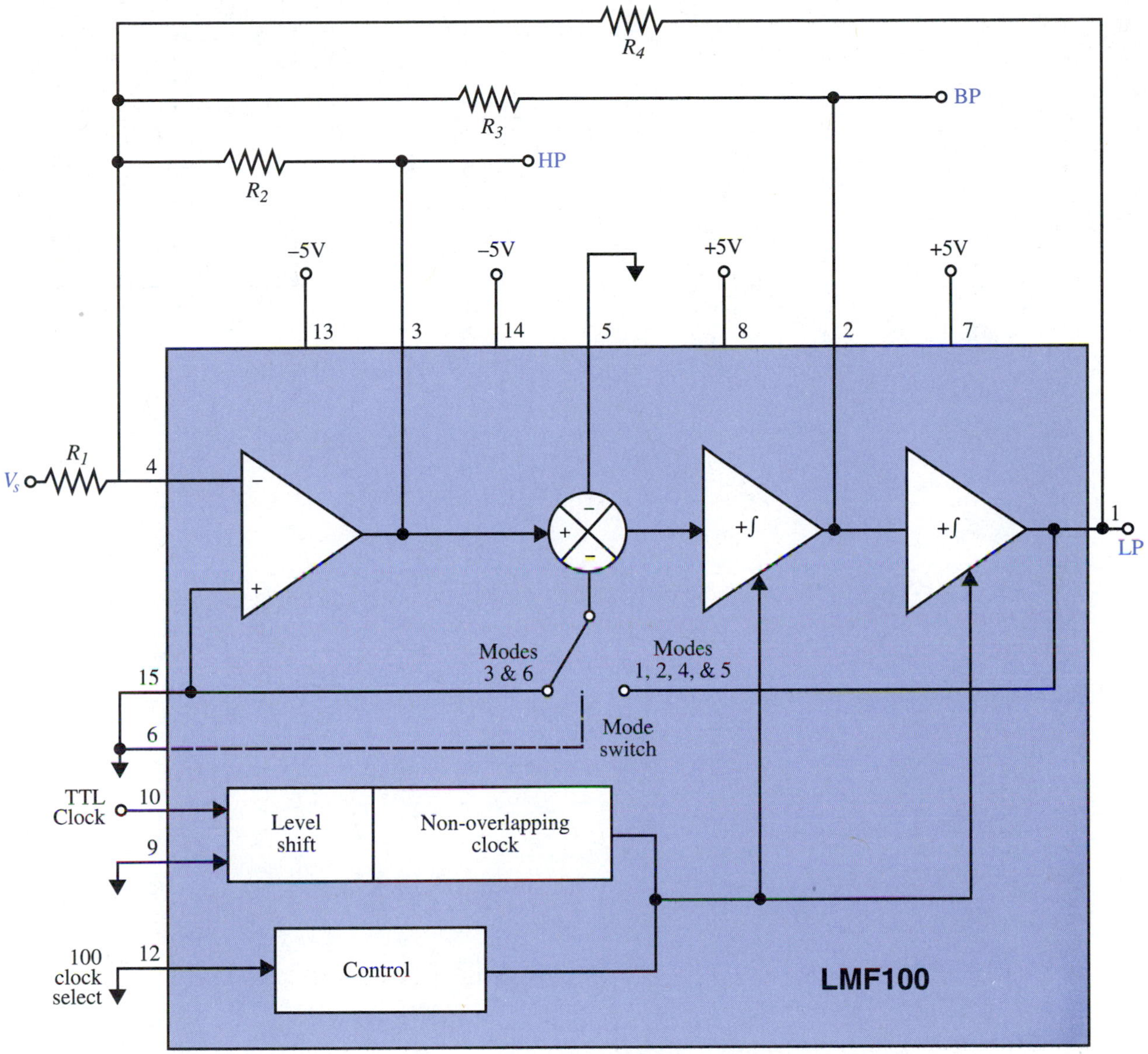

FIGURE 14–24 LMF100 mode 3 high-pass, low-pass, and band-pass. *Courtesy National Semiconductor.*

$$R_2 = H_{ohp} R_1$$

$$R_2 = 5 \times 5\text{K}\Omega$$

$$R_2 = 25\text{K}\Omega$$

Using Q of 1 and f_c of 1KHz and solving for f_o,

$$f_o = 1.272 \text{ KHz}$$

$$f_o = \frac{f_{CLK}}{100} \times \sqrt{\frac{R_3}{R_4}} \text{ or } \frac{f_{CLK}}{50} \times \sqrt{\frac{R_3}{R_4}}$$

$$H_{ohp} = -\frac{R_2}{R_1} \qquad Q = \frac{R_3}{R_2}\sqrt{\frac{R_2}{R_4}}$$

$$H_{op} = \frac{H_{ohp}Q}{\sqrt{1 - \frac{1}{4Q^2}}} \qquad f_p = f_o \frac{1}{\sqrt{1 - \frac{1}{2Q^2}}}$$

$$f_o = f_c\sqrt{\left(1 - \frac{1}{2Q^2}\right) + \sqrt{\left(1 - \frac{1}{2Q^2}\right)^2 + 1}}$$

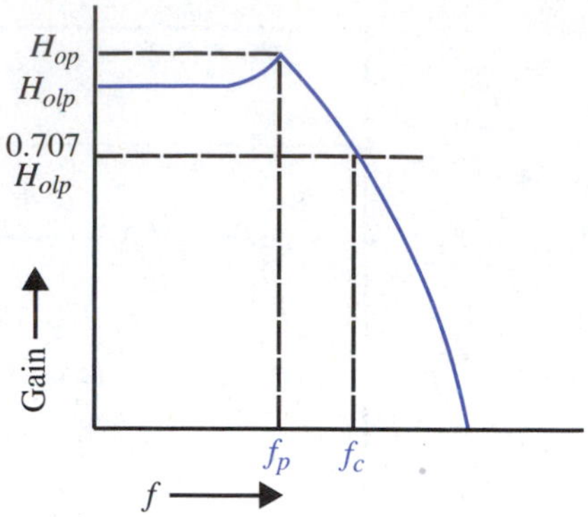

FIGURE 14–25 LMF100 mode 3 high-pass response and equations

If we want the clock frequency 100 times greater than f_o, then

$$f_o = \frac{f_{CLK}}{100} \times \sqrt{\frac{R_2}{R_4}}$$

$$f_{CLK} = 127.2 \text{ KHz} \quad \text{and} \quad \sqrt{\frac{R_2}{R_4}} = 1$$

So therefore

$$R_4 = R_2 = 25\text{K}\Omega$$

To find R_3 we can solve for this term in the Q equation.

$$Q = \frac{R_3}{R_2}\sqrt{\frac{R_2}{R_4}}$$

$$Q = \frac{R_3}{R_2} \times 1$$

$$R_3 = QR_2$$

$$R_3 = 1 \times 25\text{K}\Omega$$

$$R_3 = 25\text{K}\Omega$$

We can determine the peak gain H_{op}.

$$H_{op} = \frac{H_{ohp}Q}{\sqrt{\frac{1}{1 - 4Q^2}}}$$

$$H_{op} = \frac{5 \times 1}{\sqrt{1 - \frac{1}{4 \times 1}}}$$

$$H_{op} = 5.77$$

This amounts to a 15 percent peaking from the high-frequency gain (H_{ohp}).

Finally, we can find the frequency (f_p) at which the peak gain occurs.

$$f_p = f_o \frac{1}{\sqrt{1 - \frac{1}{2Q^2}}}$$

$$f_p = 1.272\text{KHz} \frac{1}{\sqrt{1 - \frac{1}{2 \times 1}}}$$

$$f_p = 1.8\text{KHz}$$

Being a second-order filter, the slope in the rejection region is 40dB per decade (12dB per octave).

COMPUTER CONTROL OF FILTER FREQUENCY

Since the LMF100 filter cutoff/center frequency is controlled by the clock frequency, we can move the filter frequency by simply changing the clock. Using a programmable counter between the clock oscillator and the filter gives us the capability of changing the filter frequency by varying the countdown factor (N) of the counter.

With the computer controlled system of Figure 14–26, we can select 16 different clock frequencies for the filter. If we add another programmable counter, we can expand to 256 frequencies. This gives us a tremendous versatility in filtering.

If we are using the filters for detecting control tones (train control systems use three of eight tones) and desire to change the frequencies for security reasons, we just program new data out of the microcomputer to the counter. A similar action would have to happen at the tone transmitters in order for the tones and filters to track.

TROUBLESHOOTING AUDIO CIRCUITS

In addition to the common circuit failures, audio circuits can exhibit oscillations and distortion.

Determining whether oscillations exist in an audio circuit requires the use of a wide-band oscilloscope (50MHz or greater)—otherwise many hours can be spent without finding the cause of circuit overheating and distortion.

Measuring the power capability of an audio amplifier can be accomplished by replacing the speaker with a load resistor of the same ohmic value. The peak-to-peak voltage across this resistor can now be viewed on the oscilloscope while the input to the amplifier is raised until distortion becomes apparent. Dropping the input just below the distortion level will provide the voltage used for the maximum power equation—which is

$$P_{max} = \frac{(V_{p-p})^2}{8R}$$

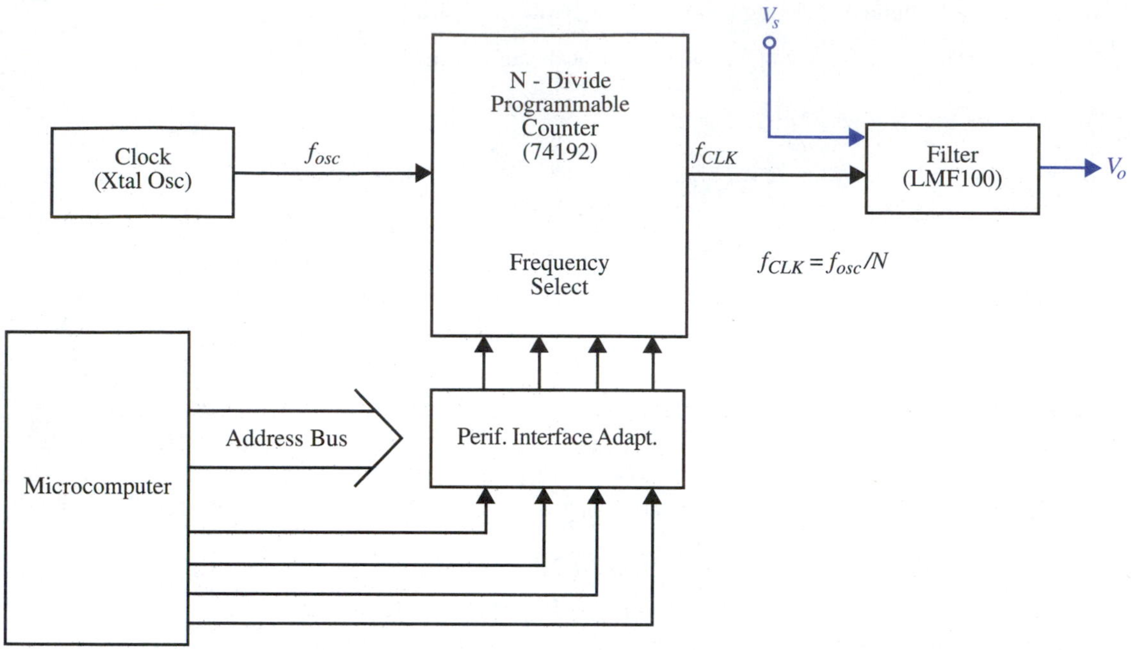

FIGURE 14–26 Computer selection of filter frequency

To determine the exact amount of distortion in the waveform requires the use of a distortion analyzer. Typically a low-distortion 1KHz sine wave is applied to the input of the amplifier under test, and the analyzer is connected across the load (resistor). The analyzer contains a variable-frequency high-Q band-pass filter that is peaked at the 1KHz frequency. Now the analyzer is switched to a high-Q notch filter, and the 1KHz signal is rejected. The remaining signal is the distortion plus noise. Changing the amplitude of the input signal allows measurement of the distortion at different power levels.

SUMMARY

- Our knowledge of hearing is based on measured (empirical) data.
- The ear is an extremely sensitive device with a maximum hearing sensitivity of 10^{-16}W/cm^2 at a frequency of 3KHz.
- The response of the ear is logarithmic, which means the sound power must increase by a factor of ten before the ear senses a doubling of volume.
- The ear's response to sound is nonlinear—meaning it creates harmonic and intermodulation distortion.
- Psychoacoustic effects refers to hearing effects that cannot be explained by physical phenomena.

- Amplification of audio signals is required, for example, to bring the milliwatt power levels at the output of a microphone to the several watts (thousands?) required to drive a loudspeaker.
- Equalization can be used to boost the low-amplitude high-frequency components of the audio signal so they are not lost in noise during the recording process.
- Digital audio refers to the conversion of the analog audio signal to digital prior to recording.
- The purpose of a preamplifier is to bring the weak signal (could be as low as 20μV) from a microphone (or pickup) to a level where processing can be performed, or can be applied to the power amplifier driver.
- A preamplifier must be low noise so that it doesn't mask out the weak signals and have sufficient dynamic range to accommodate the 100db variation of audio.
- The LM1040 chip uses a DC voltage controlled tone (bass/treble), volume, and balance circuit designed for stereo applications.
- Volume is a change in the level of the audio signal; loudness changes both the level and the frequency response to compensate for the ear's drop-off of high and low frequencies at low sound levels.
- An example of a graphic equalizer is the LMC835 integrated circuit. This chip is designed for digital computer control of 14 frequency bands with 25 amplitude steps.
- To preserve the large dynamic range of audio, the noise level must be small relative to the smallest audio signals.
- The most common forms of noise reduction used are the Dolby systems.
- A noise reduction approach that is single ended, i.e., processing only after noise is introduced, is provided by the LM1894.
- The LM1894 circuit can give up to 14db effective noise reduction.
- To bring the voltage from the preamplifier to a sufficient level requires amplification, and this is performed by a stage called an audio driver. The driver precedes the power amplifier stage that provides the actual current to the speaker.
- An integrated circuit driver is the LM391, which is designed to drive a power stage consisting of power transistors at power levels from 10W to 100W.
- The LM12 power amp has a full power (150W) bandwidth that extends to 60KHz.
- To provide twice the output voltage swing across the load (power increases by 4), the power amp can be connected in a bridge connection.
- A switching amplifier (class D) is the most efficient power amplifier.
- The switched capacitor filter can provide band-pass, notch, low-pass, and high-pass without the use of capacitors.
- The filter frequency of an SCF can be changed by simply changing the clock frequency.

EXERCISE PROBLEMS

1. How much would the sound volume increase if the speaker power level goes from 1W to 10W?
2. What is the most sensitive frequency for the ear at low sound levels?
3. What is the difference between volume and loudness?
4. The ear linear or nonlinear? What effect does this have on the sound detected by the ear?
5. What are psychoacoustic effects?
6. Why is equalization used prior to recording?
7. What is the dynamic range of music?
8. What does digital audio mean?
9. How do the ears detect stereo?
10. What is the total input noise level for an audio preamplifier if the equivalent input noise is $2.9nV/\sqrt{Hz}$?
11. Find the signal-to-noise ratio in db for the preamplifier of problem 10 if the minimum signal level is 20μV.
12. What is the gain at 100Hz for an LM1040 tone control circuit if the voltage applied to Pins 6 and 16 is +0.7V?
13. What is the gain at 5KHz for an LM1040 tone control circuit if the voltage applied to Pins 6 and 16 is +4.7V?
14. What amount of channel cross-coupling is used for stereo enhancement?
15. What is the significance of the D7 data bit input to the LMC835?
16. With the LMC835 D7 bit low, what does the D6 bit control?
17. List the LMC835 center frequencies when used as a stereo equalizer.
18. Describe the Dolby B-type noise reduction.
19. What amount of noise reduction is achieved by Dolby S-type?
20. What are the minimum and maximum bandwidths for the LM1894?
21. Find the peak-to-peak voltage required to provide 100W to an 8Ω speaker.
22. What is the purpose of a driver stage?
23. Find the peak input voltage and current required to provide a 100W output from the LM12 circuit of Figure 14–A.
24. Describe the operation of a switching amplifier and state why it is most efficient.
25. What is the purpose of the low-pass filter at the output of a switching amplifier?
26. What is the advantage of an SCF?
27. An SCF mode 1 band-pass filter is required with a center frequency (f_o) of 5KHz, a bandwidth of 500Hz (Q is 10), and a gain of −5. If R_2 is specified to be 10KΩ, find the required values of R_1 and R_3. Use a 100:1 clock frequency for least clock ripple in the output signal.

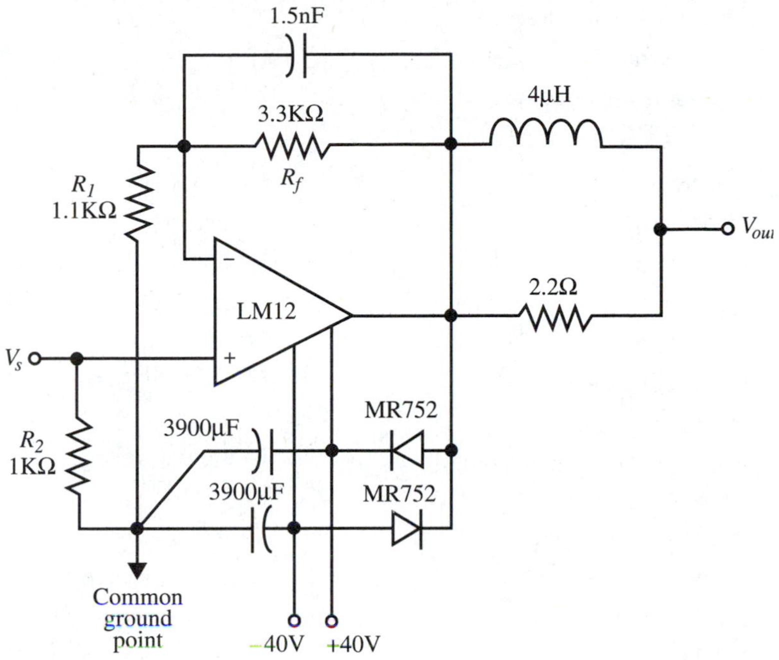

FIGURE 14–A

28. An SCF mode 3 high-pass filter is required with a cutoff frequency (f_c) of 3KHz and a gain of 10 (H_{ohp}). We want little peaking of the response curve (H_{op} close to H_{ohp}), so we shall select Q to be equal to 1. Also we choose f_{CLK} to be $100f_c$. Let R_1 be 10KΩ. Find the values of R_2, R_3, and R_4, and the peak gain and the frequency where it occurs.

ABBREVIATIONS

DNR = Dynamic noise reduction
f_u = Frequency at which the gain is unity
I_a = Average current
I_{cl} = Current at limit point
H = Gain term used with active filters
P_s = Power delivered to the speaker
R_s = Equivalent speaker resistance
SCF = Switched capacitor filter
SOA = Safe operating area (region) for a device
THD = Total harmonic distortion
$V_{in(p)}$ = Peak input voltage

V_s = Input signal voltage
V_{ni} = Input noise voltage

ANSWERS TO PROBLEMS

1. Doubles
2. 3KHz
3. Loudness compensates for ear.
4. Nonlinear; creates distortion
5. See text.
6. Noise reduction
7. 100db
8. Accommodates extended dynamic range
9. By phase and amplitude
10. 410nV
11. 33.76db
12. −7.5db
13. +5db
14. 60 percent
15. Determines band or gain selection
16. Boost or cut
17. See text.
18. See text.
19. 10db at low, 24db at high frequencies
20. 1KHz, 35KHz
21. $80V_{p-p}$
22. To provide voltage gain
23. $V_p = 7.07V$; $I_p = 7.07mA$
24. See text.
25. To remove switching frequency
26. No capacitors, easy frequency change
27. $R_1 = 20K\Omega$; $R_3 = 100K\Omega$
28. $R_2 = 100K\Omega$; $R_3 = 78.6K\Omega$; $R_4 = 61.8K\Omega$; $H_{op} = 11.55$; $f_p = 5.397KHz$

CHAPTER 15

Transducers

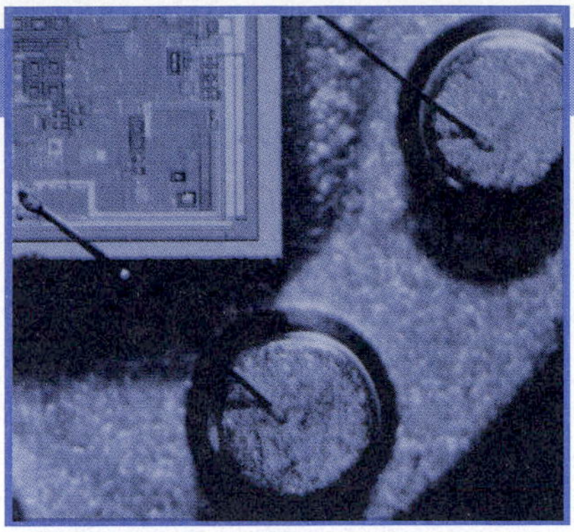

CHAPTER OBJECTIVES

After completing this chapter, you should be able to explain the operation of the following types of transducers:

- Thermocouple
- Resistance temperature device (RTD)
- Thermistor
- Semiconductor temperature sensor
- Resistance strain gauge
- Semiconductor strain gauge
- Piezoelectric force transducers
- Pressure drop flow sensors
- Vane flow sensors
- Turbine flow sensors
- Frequency-to-voltage conversion
- Hot wire flow measurement
- RTD fluid level sensor
- Ultrasonic level sensor
- Ultrasonic distance sensor
- Magnetic distance sensor (LVDT)
- Capacitive distance sensor
- Velocity sensors
- Acceleration sensors

INTRODUCTION

A **transducer** is defined as a device that converts other forms of energy into electrical energy. In this chapter, we shall look at transducers that are used to measure the following physical variables:

- Temperature
- Force
- Pressure
- Flow
- Level
- Distance or Presence
- Velocity
- Acceleration

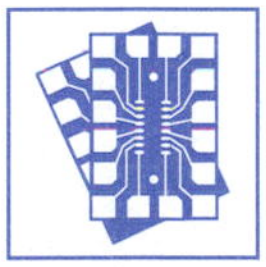

First we shall look at actual transducers used for measurement of these variables and then the circuits used to process the outputs from the transducers. We shall also study application examples.

TEMPERATURE MEASUREMENT

Temperature is an indication of the heat level in an object. To generate an electrical signal proportional to temperature, we can use one of the following heat transducers:

1) Thermocouple
2) Resistance temperature device (RTD)
3) Thermistor
4) Semiconductor

Each of these temperature transducers has its advantages and disadvantages for a given application.

THERMOCOUPLES

The number of free electrons within a material is proportional to both the temperature and the type of material. If two dissimilar metals are held in contact with one another, one will have more free electrons than the other, and this will create a voltage across the junction. Since the number of free electrons is proportional to temperature, the junction voltage is also proportional to temperature.

We conclude then that a thermocouple consists of two (couple) wires of different metals connected together that can be used to measure temperature. Table 15–1 lists typical wire combinations along with their temperature range and sensitivity.

The sensitivities listed in Table 15–1 are average because the curves of voltage versus temperature are not completely linear. If more-accurate data are required, extensive tables are obtainable from the manufacturers of thermocouples.

From Table 15–1, we see that the most sensitive thermocouple is the chromel-constantan, and that the tungsten-rhenium type is able to operate over the greatest temperature

TABLE 15–1 Thermocouple Characteristics

Junction Materials	*Temperature Range °C*	*Average Sensitivity* K_t *(μV/°C)*	*ANSI Designation*
Platinum-rhodium	38–1800	*7.8	B
Tungsten-rhenium	0–2300	16.1	C
Chromel-constantan	0–982	76.4	E
Iron-constantan	−184–760	53.0	J
Chromel-alumel	−184–1260	34.6	K
Plat.-plat.13% rhod	0–1593	11.7	R
Plat.-plat.10% rhod	0–1538	10.4	S
Copper-constantan	−184–400	44.5	T

*not linear below 1000°C

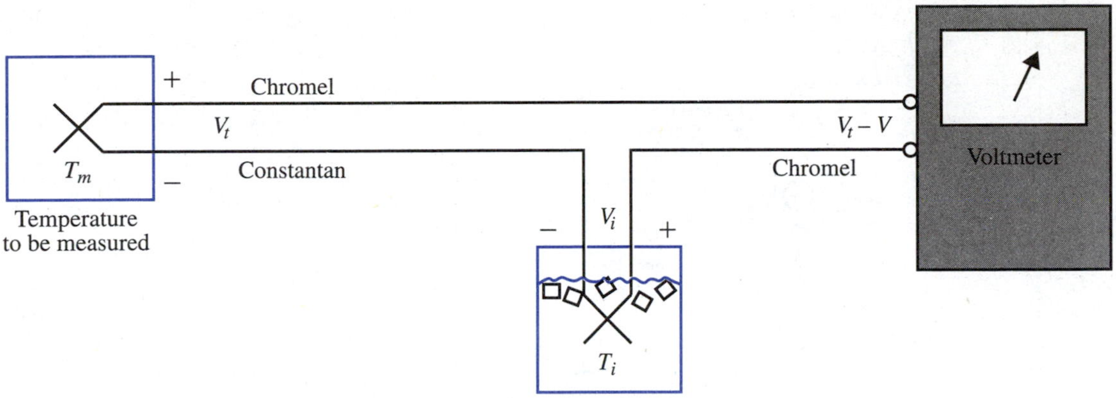

FIGURE 15–1 Thermocouple differential measurement

range. The table doesn't list the voltage developed at a particular temperature but rather the change in voltage that occurs for a one-degree-Celsius change. The reason for this is that thermocouples are usually used in a differential mode with two thermocouples in series (**bucking**)—with one thermocouple measuring the required temperature and the other held at a reference temperature, as shown in Figure 15–1.

Since the ice bath keeps the reference at 0°C, a voltage will appear at the meter if the measured temperature is different from 0°C. If the measured temperature is 0°C, then the voltage at the meter will be 0V. With the measured temperature above 0°C, the voltage at the meter will be positive. With the measured temperature below 0°C, the voltage at the meter will be negative. The actual temperature T_m is determined by the equation

$$T_m = \frac{V_t - V_i}{K_t} \qquad \textbf{(equation 15–1)}$$

where K_t is the thermocouple sensitivity in volts per degree Celsius (see Table 15–1).

EXAMPLE

Find the measured temperature for the configuration of Figure 15–1 if the voltage read at the meter is 16mV.

SOLUTION

Since $V_t - V_i$ is given as 16mV, we can use this value in equation 15–1 along with the K_t for chromel-constantan value of 76.4μV/°C from Table 15–1.

$$T_m = \frac{V_t - V_i}{K_t}$$

$$T_m = \frac{16\text{mV}}{76.4\mu\text{V/°C}}$$

$$T_m = 209.4\text{°C}$$

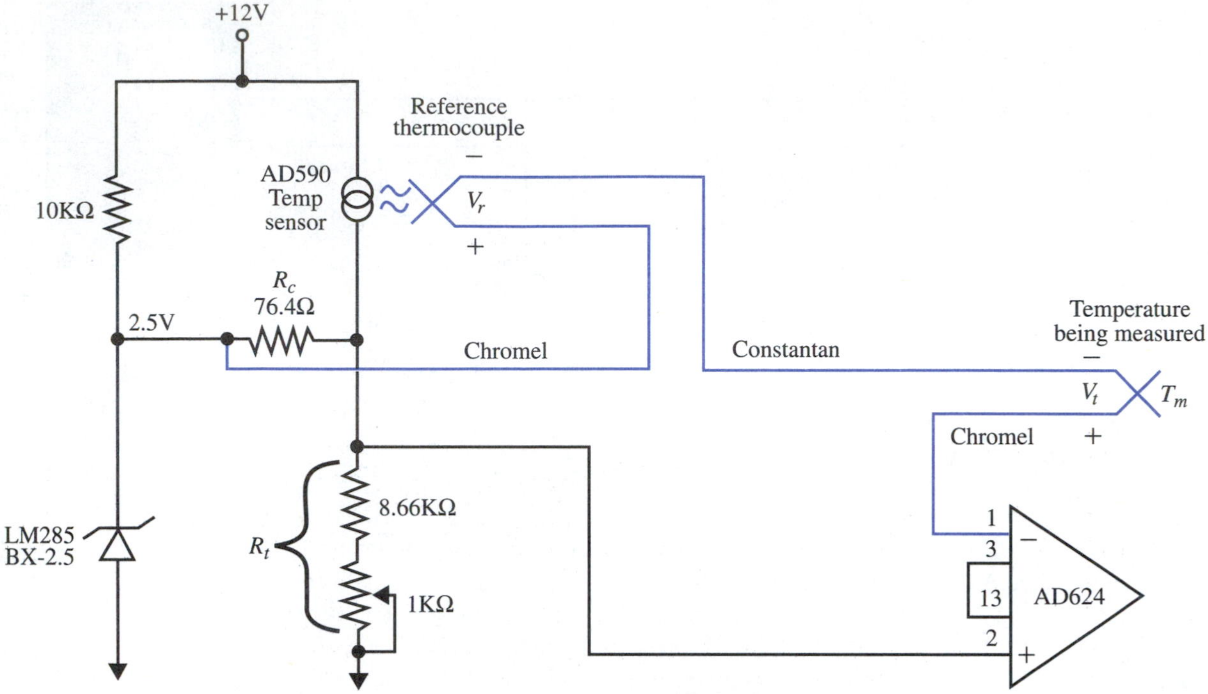

Note: $R_c = 76.4\Omega$ assumes a thermocouple temperature sensitivity K_t of $76.4\mu V/^\circ C$ over the operating range of this circuit (15–35 degrees Celsius).

FIGURE 15–2 Automatic cold junction compensation (E type)

The configuration shown in Figure 15–1 has the disadvantages of requiring an ice bath and having just a meter readout rather than a print-out or display. A circuit that overcomes these disadvantages is shown in Figure 15–2.

The circuit of Figure 15–2 provides automatic compensation of the reference junction by mounting it in contact with an AD590 temperature transducer chip. This transducer provides an additional 1μA of current for every increase of 1°K (Kelvin)—which means that it puts out 273μA at 0°C (273°K).

The basic circuit operation is that changes in the voltage across the reference thermocouple due to ambient temperature changes are canceled out by a voltage drop across R_c.

Let us initially assume that the ambient temperature is 0°C, the potentiometer in R_t is set at midrange, and all the AD590 current of 273μA flows down through R_t. With R_t at midrange, its value is 8.66KΩ + 1KΩ/2 = 9.16KΩ.

The voltage across R_t is then

$$R_t = 9.16K\Omega \times 273\mu A$$

$$R_t = 2.5V$$

This is the same voltage as the reference voltage at the left side of R_c, so no current flows through this resistor. With no voltage drop across R_c, the full reference thermocouple voltage V_r is subtracted from the temperature-measuring thermocouple voltage and the voltage at the op-amp input corresponds to the difference in temperature from 0°C.

Now let us assume the ambient air increases by 1°C—which means that an additional 1μA will flow out of the AD590 terminal. This additional current will flow through the low resistance of R_c and into the 2.5V reference source. The 1μA through the 76.4Ω resistor will provide a voltage drop across it of 76.4μV. But the reference thermocouple has a temperature coefficient of 76.4μV/°C, so its voltage will also increase by 76.4μV. These voltages are in opposite polarity, so they will cancel and the effective voltage provided by the reference thermocouple is still the 0°C value.

The potentiometer in R_t should be adjusted to calibrate the circuit. A calibration method would be to put the transducer, reference thermocouple, and measurement thermocouple in an ice bath and adjust the potentiometer for zero volts at the input to the op-amp.

The value of R_c used in the circuit is for the chromel-constantan type of thermocouple. If other thermocouples are used, the value of R_c should be selected in accordance with the equation

$$R_c = \frac{K_t}{\mu V/\Omega °C}$$

Components used in the circuit of Figure 15–2 should have low temperature drift, i.e., metal film resistors, etc. Remember that every pair of dissimilar metals forms a thermocouple—for example, a copper/solder junction has a temperature coefficient of 3μV/°C.

RESISTANCE TEMPERATURE DETECTORS (RTD)

An RTD is a metal temperature transducer that increases in resistance with temperature in a rather linear fashion. RTDs are available with resistances ranging from the tens of ohms to thousands of ohms. The most used RTDs are made of platinum, but other metals like nickel, copper, and nickel-iron are also used. Table 15–2 shows the characteristics of this type of temperature transducer.

Platinum is less sensitive to temperature than nickel or nickel-iron, but it operates over a greater temperature range, has a higher resistance, and has a very constant temperature coefficient. The higher resistance of platinum makes it easier to use this material in actual circuit application because, for a given current, more voltage is developed.

TABLE 15–2 Characteristics of RTDs

Material	*Temp Range °C*	*Temp Coeff K_t(%/°C) at 25°C*
Platinum	−200–+850	0.39
Nickel	−80–+320	0.67
Copper	−200–+260	0.38
Nickel–iron	−200–+260	0.46

When using an RTD, the current through the device must be low enough so that it does not cause appreciable power dissipation and therefore generate self-heating—otherwise the measured temperature will be in error.

A simple circuit configuration using a 100Ω (at 0°C) platinum RTD is shown in Figure 15–3. The LM285 sets up a 2.5V reference at the inverting input to the op-amp. With the 500Ω potentiometer adjusted to provide a total R_1 resistance of 9.399KΩ, the current flowing through R_1 (and also the RTD) is 2.5V/9.399KΩ or 0.266mA. At 0°C, this current flowing through the 100Ω resistance of the RTD causes the op-amp output to be +0.266V. However, if a voltage of −0.266V is applied to the noninverting input to the op-amp, the gain of approximately unity from this input (1 + 100Ω/9.399KΩ) will cause the output voltage to cancel and be zero—corresponding to 0°C.

If now the temperature increases to 266°C, the resistance of the RTD will increase to

$$R_{p200} = 100\Omega[1 + (0.39\%/°C \times 266°C)]$$

$$R_{p200} = 200\Omega$$

This is a doubling of resistance from the 0°C value, or an increase in resistance of 100Ω. With the same 0.266mA of current flowing through R_p, the output voltage will now increase from 0V to 100Ω × 0.266mA or 0.266V.

We can see from this result that with the component values used in Figure 15–3, the output voltage changes by 1mV from every degree-Celsius change in temperature.

The circuit is calibrated by setting the 500Ω span potentiometer at midrange with the RTD in an ice bath and then adjusting the 1KΩ offset potentiometer for 0V out. Next, the RTD is raised to the maximum temperature to be encountered (using a calibrated temperature-measuring device), and the scan potentiometer adjusted for the correct out-

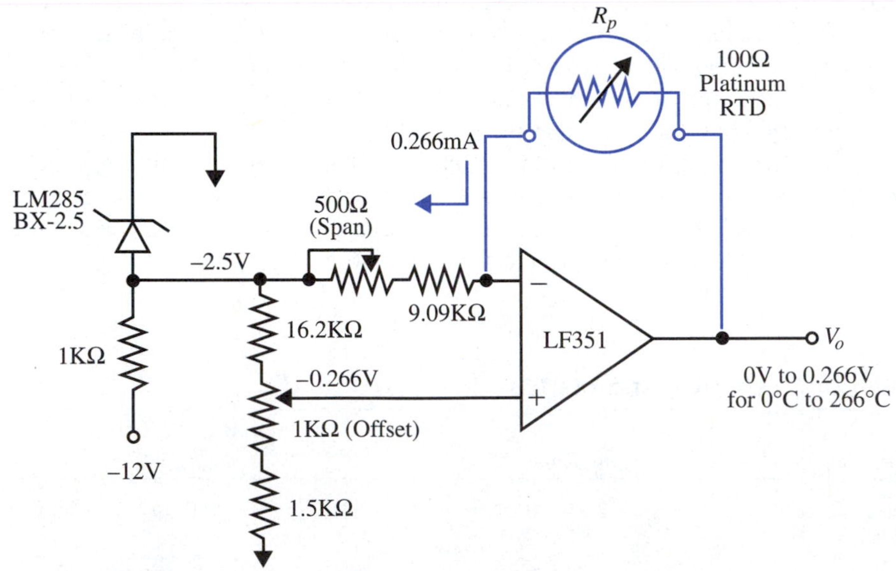

FIGURE 15–3 Low-cost RTD circuit

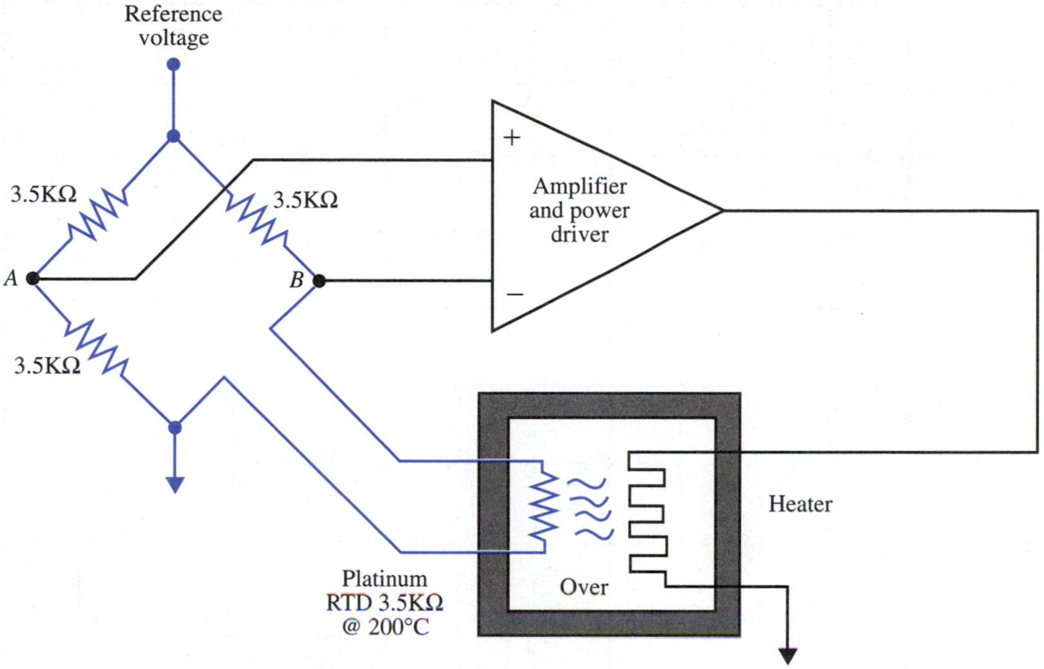

FIGURE 15–4 RTD bridge heater control

put voltage corresponding to this temperature. Putting the RTD back into the ice bath will verify that no interaction has occurred between the potentiometer settings—the offset potentiometer may need to be retweaked to restore the 0V output.

RTD PRECISION CONTROLLER

RTDs are often used in a bridge configuration. Figure 15–4 shows a portion of a circuit used to keep a small oven in a spacecraft at a temperature of 200°C ± 0.1°C for five years in a varying environment. If the temperature in the oven falls below 200°C because of an outside ambient temperature change, the resistance of the RTD decreases and causes the voltage at point B to be less than the voltage at point A. This condition causes the voltage applied to the heater to be increased, which raises the temperature back to the desired 200°C value.

THERMISTOR TEMPERATURE DETECTORS

A thermistor is made from a material that reduces in resistance as the temperature increases. The resistance change is very sensitive to temperature (about −4%/°C) but is very nonlinear following an exponential-type curve.

Figure 15–5 shows the typical temperature versus resistance characteristic for a thermistor.

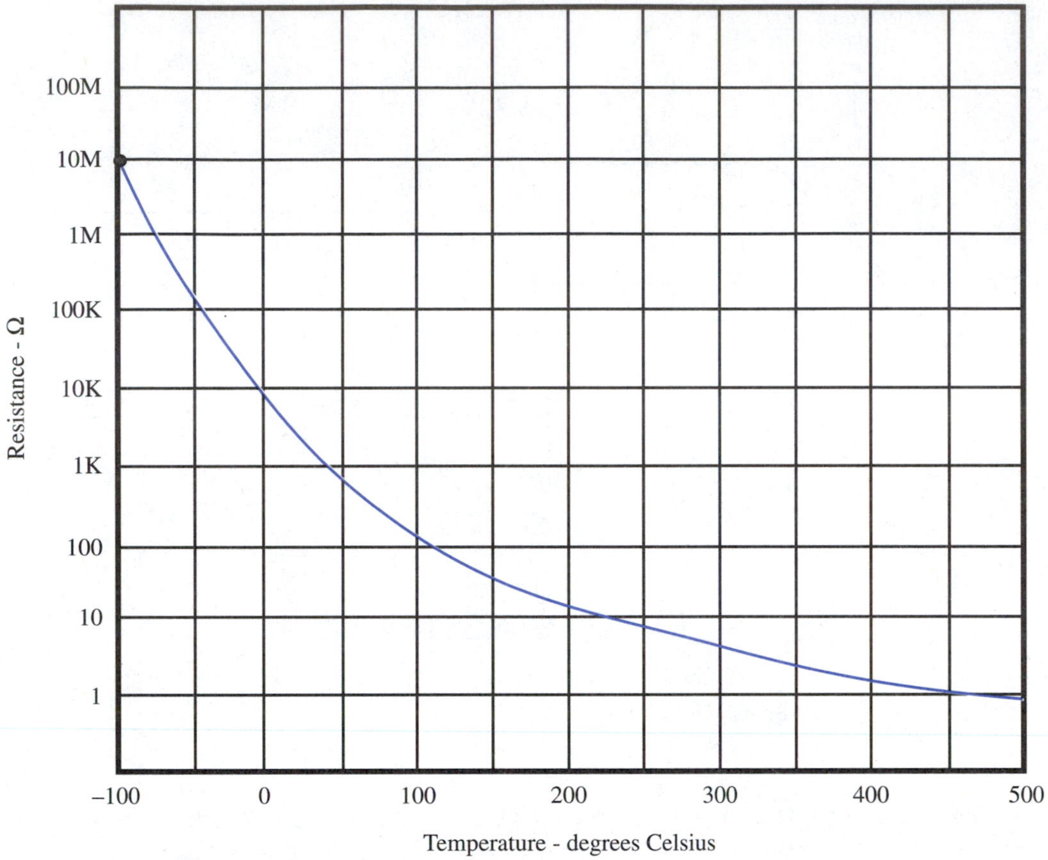

FIGURE 15–5 Thermistor characteristics

Linearizing networks are available that make the thermistor more linear at the expense of about a factor of ten reduction in sensitivity. Figure 15–6 provides the configuration for a Yellow Springs Inst. CO 44000 series linearized thermistor. This device can be set up as a voltage divider or a variable resistor, and has a uniformity and interchange ability to within 0.1% from −40°C to +100°C and 0.2% from −80°C to +150°C. The four resistors (two are thermistors) are provided under the part number for the network. They can be set up as a voltage divider or a variable series resistor, as shown in the figure. The circuit of Figure 15–7 uses the voltage divider configuration in a bridge circuit.

For calibration, the network is maintained at a temperature of 0°C and the Zero Adj potentiometer is turned until the output reads 0V. At this setting, the voltage V_b should ideally read 2.163V. With the network now raised to 100°C, R_o is adjusted for 1.00V at the output.

With the op-amp connections shown, the AD624 instrumentation amplifier is set up for a gain of unity.

Figure 15–8 shows the network, in a series configuration, as the feedback element for an op-amp. The 2.5V reference voltage across the 3.83KΩ resistor provides a current of

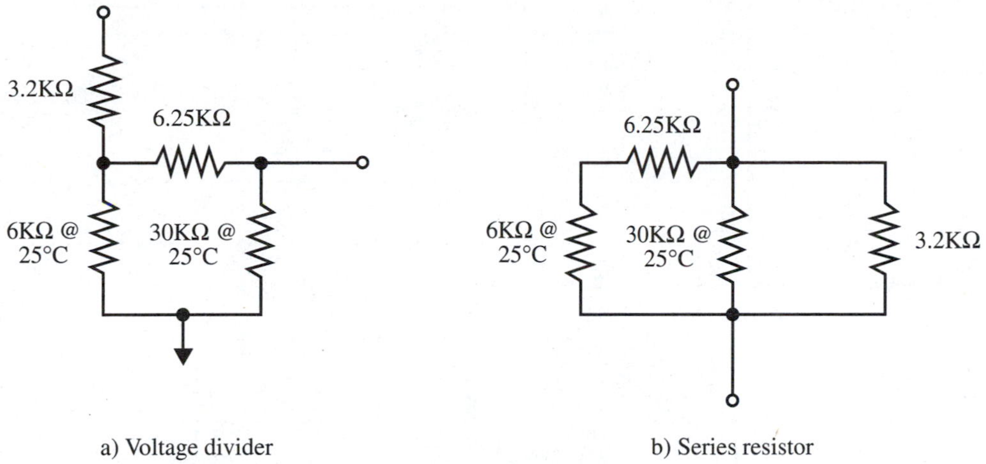

FIGURE 15–6 Linearized thermistor

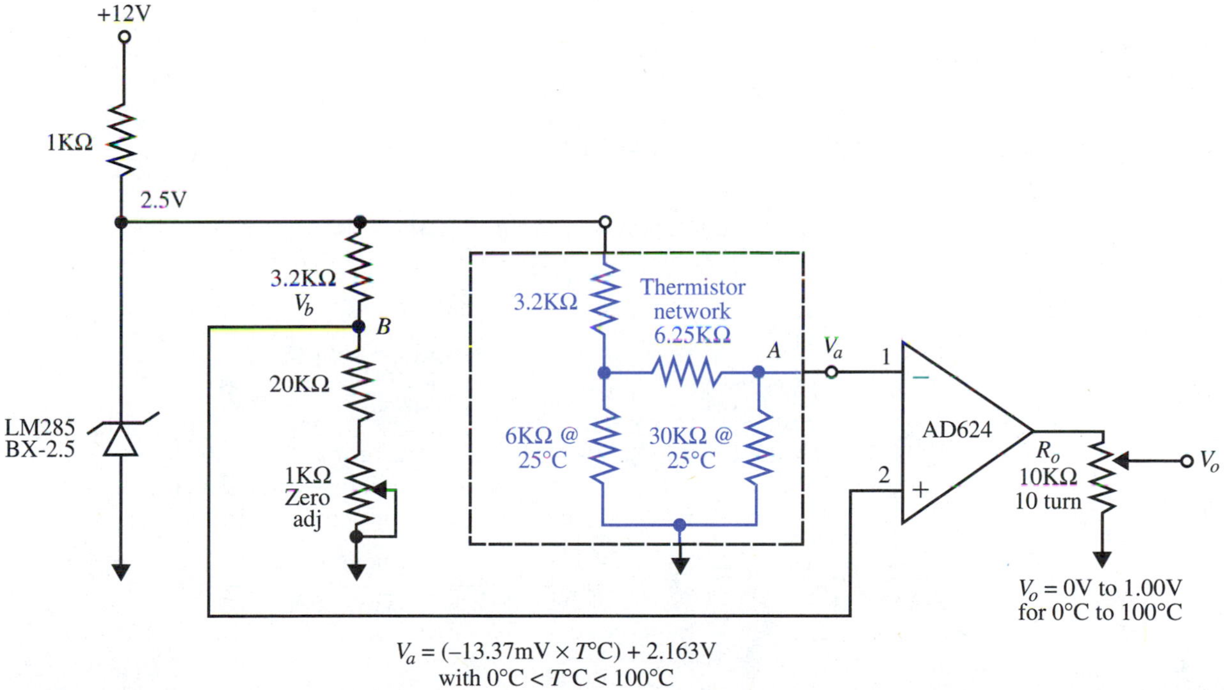

FIGURE 15–7 Temperature detector—thermistor bridge. *Courtesy Yellow Springs Instruments.*

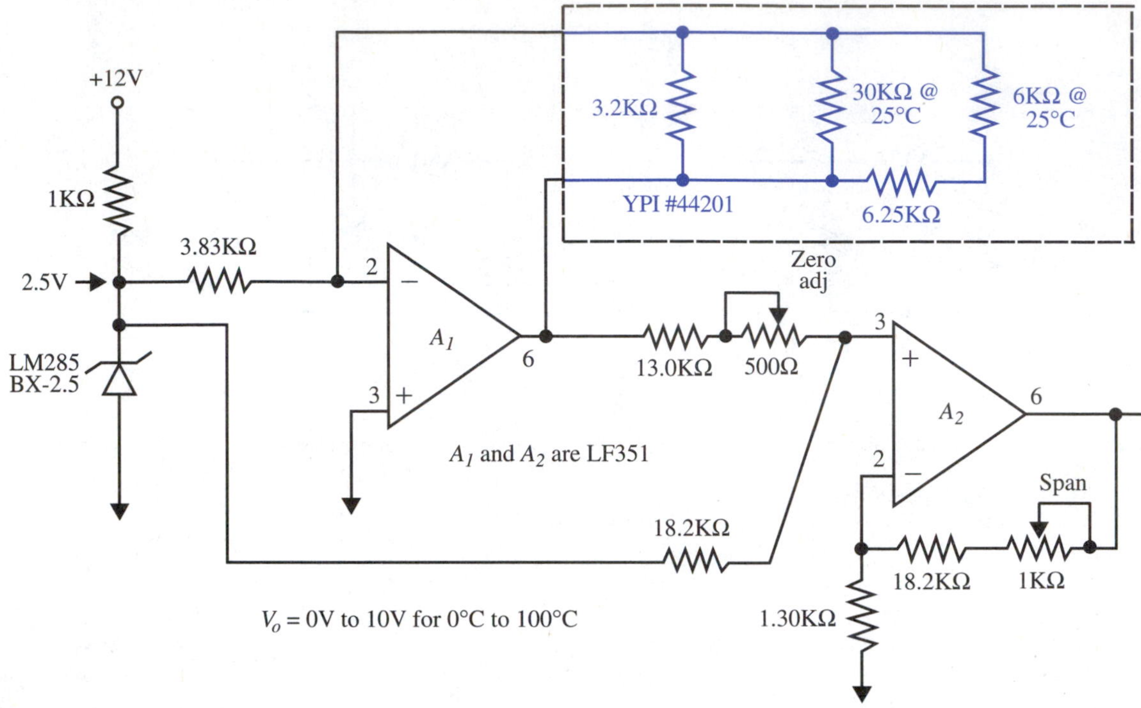

FIGURE 15–8 Temperature detector—series mode. *Courtesy Yellow Springs Instruments.*

0.653mA (2.5V/3.83KΩ) to the thermistor network. Since the network has a resistance of 2.768KΩ at 0°C, the voltage at the output of op-amp A_1 at this temperature will be

$$V_{o1} = -I_1 \times R_f$$

$$V_{o1} = -0.653\text{mA} \times 2.768\text{K}\Omega$$

$$V_{o1} = -1.8075\text{V (0°C)}$$

The 13KΩ resistor and Zero Adj potentiometer form a voltage divider with the 18.2KΩ resistor. So, the voltage V_{ni2} at the noninverting input to A_2, with the Zero Adj at midrange, is the 2.5V reference voltage minus the voltage drop across the 18.2KΩ resistor.

$$V_{ni2} = 2.5\text{V} - \left(\frac{2.5\text{V} - (-1.8068\text{V})}{18.2\text{K}\Omega + 13\text{K}\Omega + 250\Omega}\right)(18.2\text{K}\Omega)$$

$$V_{ni2} = 7.67\text{mV}$$

Decreasing the adjusted resistance of the Zero Adj potentiometer slightly will cause this voltage, and therefore the output voltage of A_2, to go to zero—corresponding to the temperature of 0°C.

When the thermistor network is at 100°C, its resistance is 1.057KΩ—which makes the voltage at the output of A_1

$$V_{o1} = -I_1 \times R_f$$

$$V_{o1} = -0.653\text{mA} \times 1.057\text{K}\Omega$$

$$V_{o1} = -0.69022\text{V} \quad (@100°\text{C})$$

The voltage at the noninverting input to op-amp A_2 will now be

$$V_{ni2} = 2.5\text{V} - \left(\frac{2.5\text{V} - (-0.69022\text{V})}{18.2\text{K}\Omega + 13\text{K}\Omega + 250\Omega}\right)(18.2\text{K}\Omega)$$

$$V_{ni2} = 0.6538\text{V}$$

The voltage gain of amplifier A_2 is

$$A_{v2} = 1 + \frac{18.2\text{K}\Omega + 1\text{K}\Omega/2}{1.3\text{K}\Omega}$$

$$A_{v2} = 15.385$$

The output voltage is then

$$V_{o2} = A_{v2} \times V_{i2}$$

$$V_{o2} = 15.385 \times 0.6538\text{V}$$

$$V_{o2} = 10.059\text{V}$$

A slight reduction in the adjusted resistance of the span (range) potentiometer will make this output voltage 10.0V corresponding to 100°C.

SEMICONDUCTOR TEMPERATURE TRANSDUCERS

We encountered the Analog Devices AD590 semiconductor temperature transducer when we studied the thermocouple cold junction compensating circuit (Figure 15–2). Again, this device is a temperature-controlled current source that provides an increase in output current of 1μA for every 1°K increase in temperature. The AD590 operates over a temperature range of −55°C to +150°C with an absolute error of ±3°C for the J version, and ±1°C for the more expensive M version.

A differential-temperature-measuring circuit using two AD590 chips is shown in Figure 15–9.

Each of the AD590s provides a constant current determined only by its temperature. Any difference in current between the two sensors caused by a difference in temperature must flow through R_f. The output voltage generated by this current flowing through R_f reflects the difference in temperature between the two temperature sensors.

The difference current flowing through R_f is the temperature difference in °C times 1μA/°C. The output voltage is this current times R_f, or

$$V_o = (T_2 - T_1)(1\mu\text{A/°C})(R_f) \qquad \textbf{(equation 15–2)}$$

$$V_o = (T_2 - T_1)(10\text{mV/°C})$$

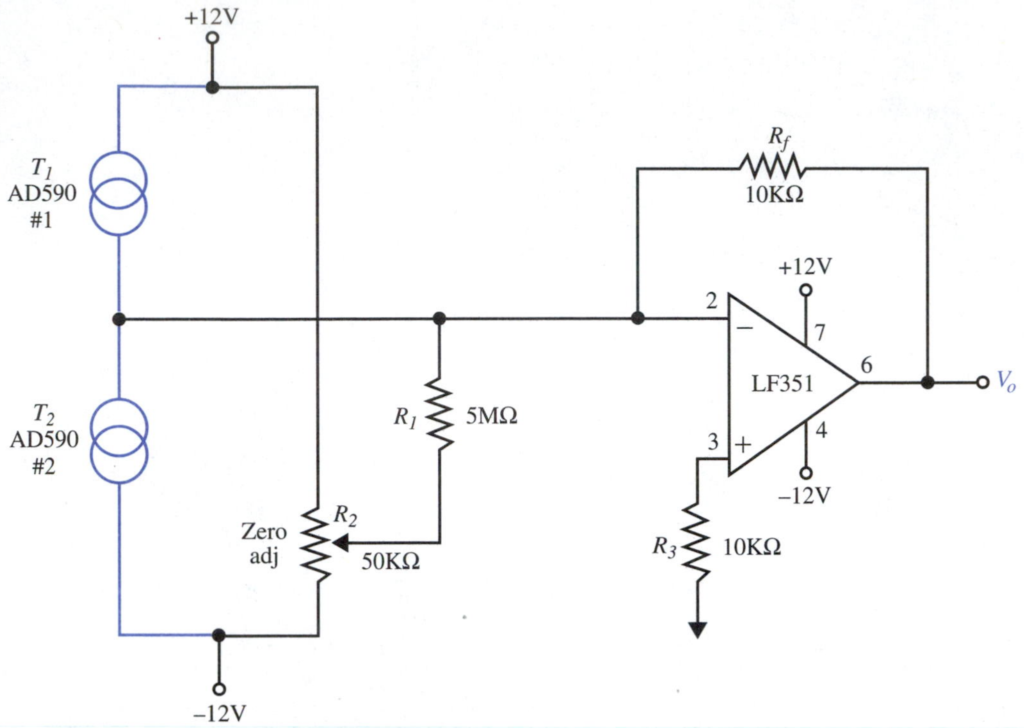

FIGURE 15–9 Differential temperature measurement

EXAMPLE

Find the output voltage for the circuit of Figure 15–9 if the #2 sensor has a temperature of 65°C and the #1 sensor a temperature of 25°C.

SOLUTION

Putting the temperature values in equation 15–2 with R_f of 10KΩ, we get an output voltage of

$$V_o = (T_2 - T_1)(1\mu A/°C)(R_f)$$

$$V_o = (65°C - 25°C)(1\mu A/°C)(10K\Omega)$$

$$V_o = 0.4V$$

OTHER SEMICONDUCTOR TEMPERATURE TRANSDUCERS

National Semiconductor makes two semiconductor temperature transducers that are very simple to use and require no offset to read from zero degrees.

The LM34 is a three-terminal Fahrenheit device that provides an output voltage of 10mV/°F over a range of −50°F to 300°F with an accuracy of ±2°F. It can operate from a power supply of 5V to 30V with current drain of 70μA. With a single positive supply, the chip can provide an output (10mV/°F) from +5°F to +300°F. Both a positive and negative supply are required to operate over the full temperature range.

The LM35 is a three-terminal Celsius device that provides an output voltage of 10mV/°C over a range of −55°C to 150°C with an accuracy of ±1°C. It can operate from a power supply of 4V to 30V with current drain of 60μA. With a single positive supply, the chip can provide an output (10mV/°C) from +2°C to +150°C. Again, both a positive and negative supply are required to operate over the full temperature range.

Both devices can be easily connected to an analog-to-digital converter to provide a digital readout.

FORCE MEASUREMENT

When a force is applied to an object, it causes a change in its shape. The amount of change is proportional to the applied force, and this is the basis for force measurement.

Devices used to measure force are resistance strain gauges, semiconductor strain gauges, and piezoelectric transducers; we shall study each of these in turn.

A resistance strain gauge consists of a metal strip or wire that changes in length, and therefore resistance, when a force is applied to the sample on which the gauge is mounted and causes stretching or compression of the sample (the gauge deforms in the same way as the sample). Typical configurations for resistive strain gauges are shown in Figure 15–10.

The bonded-wire gauge consists of a wire bonded to a base, and the foil-type is made by etching a foil or depositing on a substrate. The two-axis gauge can determine both the magnitude and direction of the applied force.

Strain gauges are often used as part of a load cell. The load cell consists of strain gauges mounted on a precisely machined spring element. The gauges are placed at various angles on the spring element so that the effects of temperature changes are nullified.

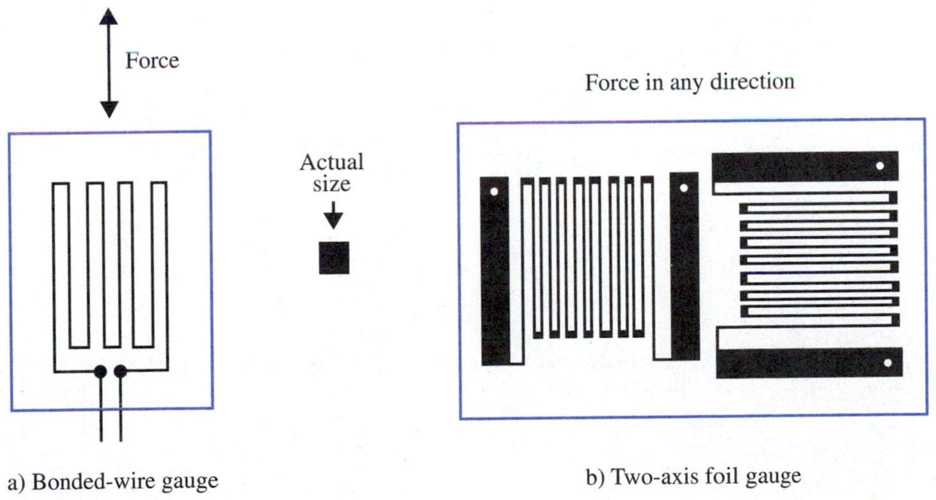

FIGURE 15–10 Resistive strain gauges

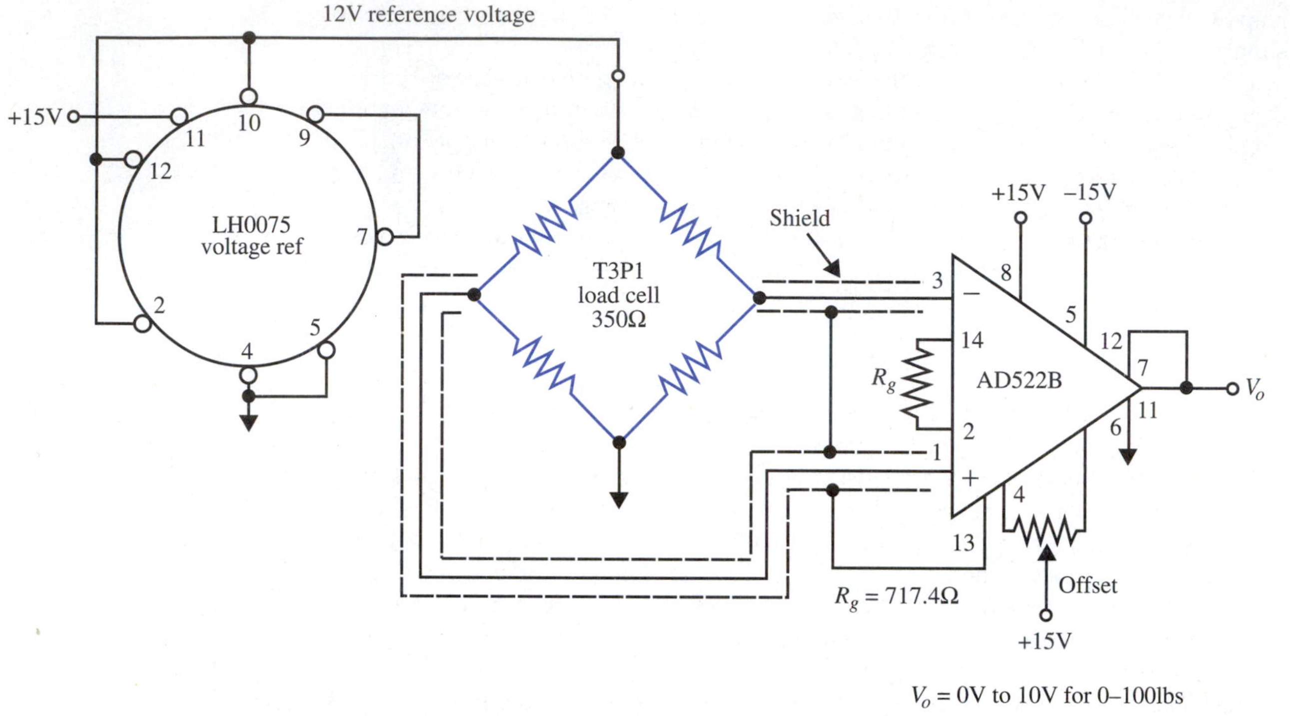

FIGURE 15–11 Strain gauge load cell. *Courtesy Analog Devices.*

To reduce temperature effects, strain gauges are typically used in a bridge configuration. Figure 15–11 shows a bridge arrangement for the 100lb full-scale, BLH Electronics T3P1 350Ω strain-gauge-based load cell.

The sensitivity of the T3P1 load cell is 3mV per volt applied across the cell. With the excitation voltage of 12V the full-scale output with a 100lb load is

$$3\text{mV/V} \times 12\text{V} = 36\text{mV}$$

If we require the full-scale voltage output from the AD522B instrumentation amplifier to be 10V, this requires an op-amp gain of

$$A_v = \frac{10\text{V}}{36\text{mV}}$$

$$A_v = 277.8$$

The AD522B has a capability of gain programming by selecting a single resistor (R_g) in accordance with the equation

$$A_v = 1 + \frac{2 \times 10^5\Omega}{R_g}$$

Solving for R_g and using the required value for A_v,

$$R_g = \frac{2 \times 10^5 \Omega}{(A_v - 1)}$$

$$R_g = \frac{2 \times 10^5 \Omega}{(277.8 - 1)}$$

$$R_g = 722.5 \Omega$$

A standard value 715Ω metal film resistor could be used with 1% error—or a fixed metal film resistor with a small trim potentiometer could also be used.

Notice that the leads from the bridge to the amplifier, which could be long, are shielded to reduce noise pickup typical in an industrial environment.

The reference voltage, for the bridge, is provided by an LH0075 precision programmable voltage regulator that is wired to provide 12V ± 0.1%. This chip can provide up to 200mA and, for other applications, voltages of 5V, 6V, 10V, and 15V can be selected by changing the wiring connections.

SEMICONDUCTOR STRAIN GAUGES

Semiconductor strain gauges have the advantage of greater resistance change with a deformation of the semiconductor material. In a bridge configuration, the semiconductor type is about 30 times more sensitive than a metallic type. But the semiconductor strain gauge is less accurate, much more temperature sensitive, and is harder to compensate.

Uses for the semiconductor strain gauges are limited to applications where the increased sensitivity is important and temperature variations are small. Suggested use is in null-bridge applications where the bridge is driven back into a null condition, and the output signal is the amount of drive required to achieve the null.

PIEZOELECTRIC FORCE TRANSDUCERS

The piezoelectric transducers are used to measure forces that are continuously changing over, perhaps, millisecond time intervals. The **piezoelectric effect** is the generation of a voltage when the material is deformed by an external force.

The output of a piezoelectric transducer can be represented by a voltage source in series with a small capacitor. The applied force results in an effective change in the capacitance. This can be sensed by using a charge amplifier, as shown in Figure 15–12.

Changes in the charge on C_p, caused by the applied force, are transferred (current flows) to the feedback capacitor C_f. Since the left side of C_f is held at ground by the op-amp feedback, the charge transferred to C_f develops an output voltage ΔV_o.

$$\Delta V_o = \frac{-\Delta Q}{C_f} \qquad \textbf{(equation 15–3)}$$

From this equation, we see that the amount of output voltage is inversely proportional to the size of the feedback capacitor.

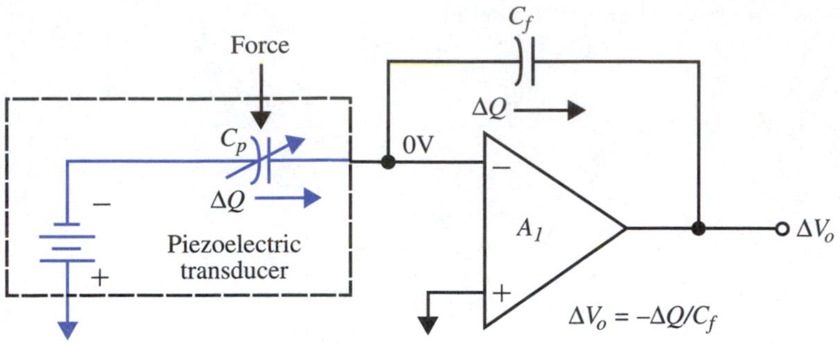

FIGURE 15–12 Piezoelectric transducer with charge amplifier

PRESSURE TRANSDUCERS

Pressure is force spread over a specified area. If we have 25lbs./in.^{2} pressure on a plate with an area of 10in.^{2}, the force on the plate is

$$25\text{lbs./in.}^2 \times 10\text{in.}^2 = 250\text{lbs}$$

Pressure can be measured by various mechanical devices such as bellows, diaphragms, manometer tubes, and Bourdon tubes. These devices effectively measure the displacement or length increase that occurs as a result of the pressure. Pressure can be specified three ways:

1) Absolute
2) Gauge
3) Differential

Absolute pressure is measured with reference to zero pressure (a vacuum). Gauge pressure is measured relative to ambient pressure (14.7lbs./in.^{2} at sea level). Gauge pressures below ambient are usually referred to as negative pressures of, or inches of, vacuum. Differential pressure is the difference in pressure between two points. Figure 15–13 shows the three pressure gauge configurations.

The mechanical change from the pressure gauge can be converted to an electrical change by coupling to a potentiometer, strain gauge, or piezoelectric transducer.

If a potentiometer (or rheostat) is used, it can be wire-wound, conductive plastic, or metal film. The wire-wound potentiometer is the most stable type, but resolution suffers as the wiper arm moves between adjacent turns. Potentiometers are sometimes used in a bridge configuration to reduce temperature effects.

The pressure can be converted to a force if the area of the pressure diaphragm is known. The force can then be measured by using a strain gauge.

Piezoelectric pressure transducers have the fastest response to pressure changes. But the electrical output might have to be passed through a low-pass filter to eliminate high-level transients.

The SenSym Company makes a series of piezoelectric pressure transducers. The LX06XXXG is a single inlet plus ambient port gauge sensor, the LX06XXXD is a dual inlet

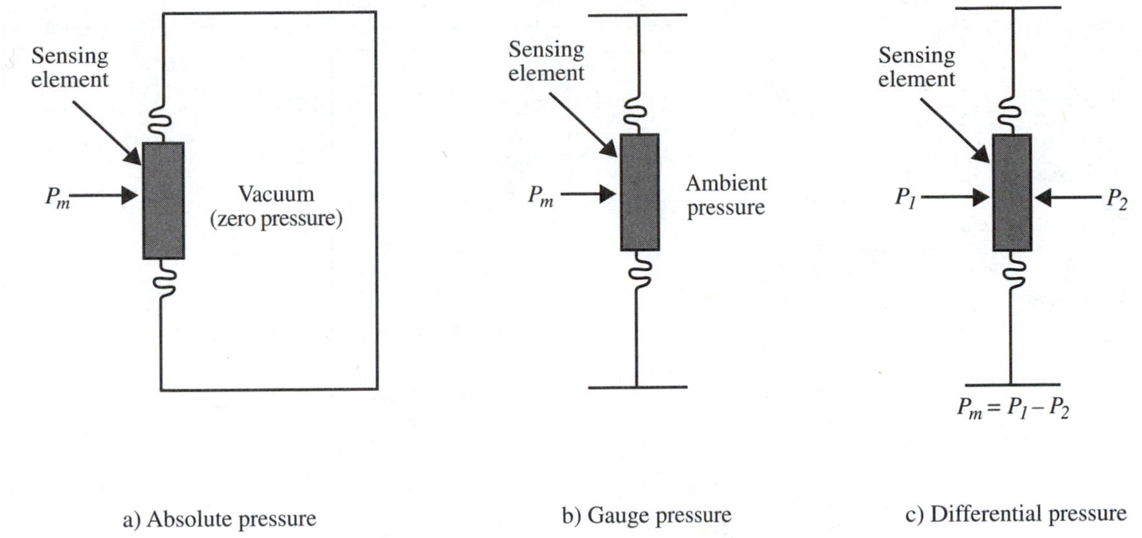

FIGURE 15–13 Different pressure measurements

differential sensor, and the LX06XXXA is a single inlet absolute sensor. These sensors are available with full-scale pressures ranging from 1psi to 100psi and repeatability of 0.1%. Sensitivities range from 1.4mV/psi to 27.7mV/psi. Package dimensions are 0.82″ × 0.75″ × 0.75″, and the device can be mounted directly on a printed circuit board.

FLOW TRANSDUCERS

Measurement of the flow of a liquid or gas passing a particular point depends on whether the volume flow (m^3/s) is required or the mass flow (kg/s). Mass flow is more difficult to measure because the mass changes with temperature—but it can be done by measuring the volume flow and correcting for temperature.

There are many ways to measure flow, and these include: the Pitot tube, which determines flow from the pressure drop between two points (pressure measurement); deflecting vanes (potentiometer, strain gauge—voltage measurement); turbines (frequency measurement); and hot-wire (resistance measurement).

FLOW MEASURED BY PRESSURE DROP

Flow measured by a difference in pressure is related to the square root of the pressure.

$$\text{Flow} = K_p\sqrt{P_1 - P_2}$$

where K_p is the flow constant determined by the physical configuration.

EXAMPLE

Determine the flow rate if the flow constant is (10 gallons)(in.)/(minute)(lb.)$^{1/2}$ and P_1 is 15lbs./in.^{2} and P_2 is 7lbs./in.^{2}.

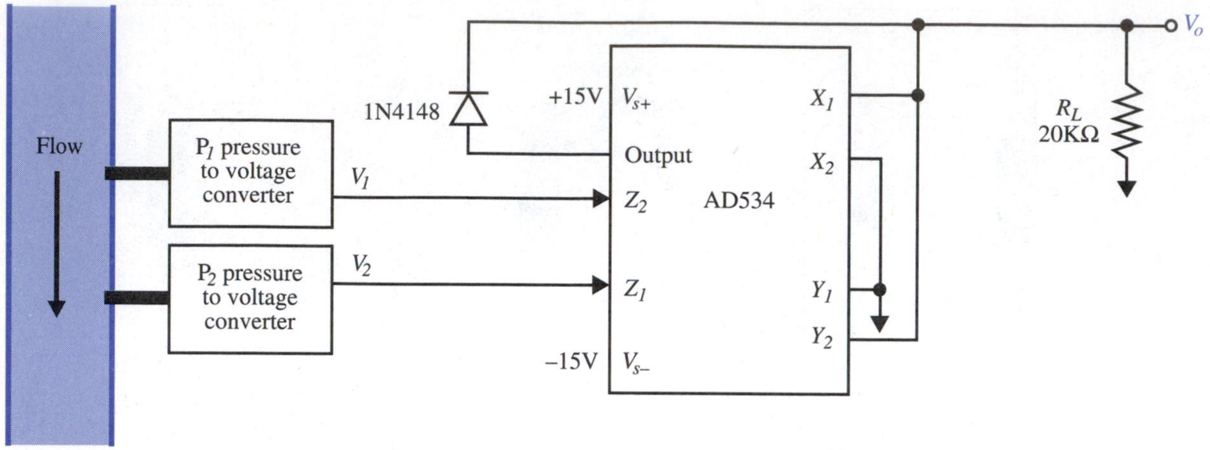

$$V_o = \sqrt{10(V_1 - V_2)} \qquad 100\text{mV} < (V_1 - V_2) < 10\text{V}$$

FIGURE 15–14 Flow measurement by pressure drop. *Courtesy Analog Devices.*

SOLUTION

Using the flow equation,

$$\text{Flow} = K_p\sqrt{P_1 - P_2}$$

$$\text{Flow} = \left[\frac{(10\text{gal})(\text{in.})}{(\text{min.})(\text{lb}^{1/2})}\right]\sqrt{\frac{15\text{lbs.}}{\text{in.}^2} - \frac{7\text{lbs.}}{\text{in.}^2}}$$

$$\text{Flow} = 28.3 \text{ gal./min.}$$

If the outputs from the pressure sensors are voltages, the flow can be determined from the square root of the difference between these voltages. The square root can be taken by a computer using a look-up table, or as an analog signal using an analog multiplier/divider. This last approach is shown in the circuit of Figure 15–14, using an Analog Devices AD534. The circuit is normalized for a 10V output with a 10V difference between the outputs of the pressure to voltage converters ($V_1 - V_2$).

HOT-WIRE FLOW MEASUREMENT

A hot wire will lose heat as liquid or gas flows by. The amount of heat lost is a function of the flow rate, temperature, and composition of the flowing medium. The cooling of the wire lowers the wire's resistance, and this can be used as a measure of the flow rate.

This flow-measuring technique can be implemented by using a platinum RTD, which will drop in resistance when cooled. However, the best way to operate the RTD is in a constant resistance mode (constant temperature). In this mode, the voltage across the RTD is increased, which increases the V^2/R heating, to compensate for the heat lost to the medium. The amount of voltage change across the bridge (ΔV_s) needed to keep the resistance constant is a measure of the flow.

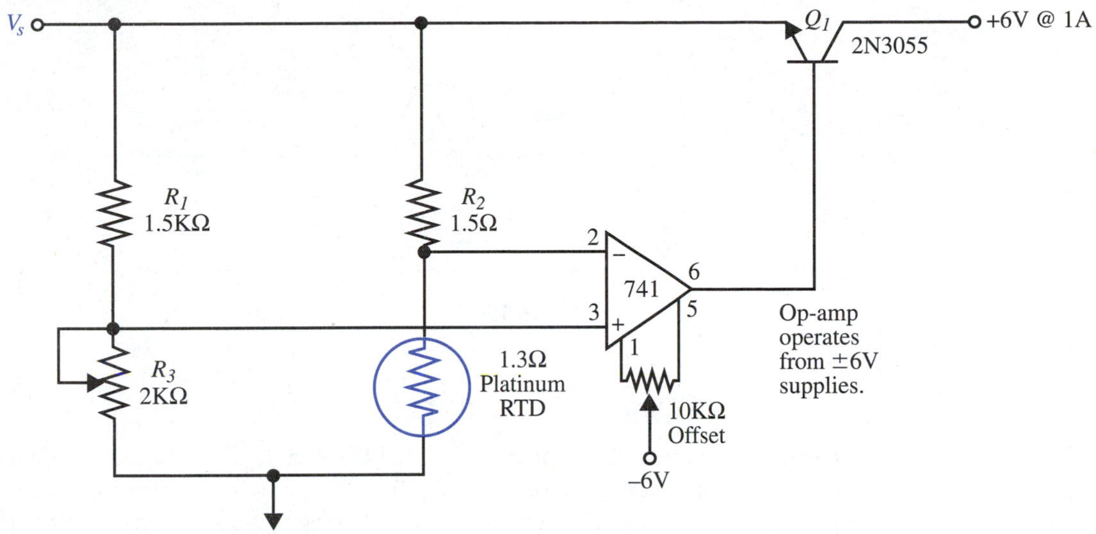

FIGURE 15–15 Hot-wire flow-measuring circuit

Figure 15–15 provides a hot-wire flow-measuring circuit. The RTD (length 0.4375in., diameter 0.002in.) is assumed to have a hot resistance of 1.3Ω and is part of a bridge configuration. If the RTD resistance changes, the op-amp maintains 0V between its two inputs by adjusting the voltage across the bridge via Q_1.

R_3 is initially set at 0Ω. The op-amp offset potentiometer is adjusted to ensure that at power turn-on the output of the op-amp goes positive and turns on Q_1. With the power turned on, R_3 is adjusted so that the RTD just glows red, and then the potentiometer is backed off a little.

As the RTD is cooled by the flow, the voltage across it drops below the voltage across R_3. This makes the voltage at the noninverting input to the op-amp more positive than the voltage at the inverting input—which causes the op-amp output to go positive. The more-positive voltage on the base of Q_1 raises the emitter voltage, and therefore the voltage across the bridge. This causes a higher voltage across the RTD, and it regains its temperature and therefore resistance.

The output voltage V_s provides an amplified version of the RTD voltage. The indication of flow, by this voltage, is nonlinear but it can be calibrated. The sensitivity is greatest at low flow rates.

FLOW MEASURED BY VANE

A vane used to measure flow deflects proportional to the flow. The vane can be connected to a potentiometer and the deflection converted to a change in voltage. Or it can be connected to a rheostat (or strain gauge) and converted to a change in resistance. Figure 15–16 shows a conceptual vane flow-measuring system.

The sensitivity of the flow measurement can be increased by using a gear box between the vane and the potentiometer, which will convert the limited angular rotation of the vane into complete rotations of the potentiometer.

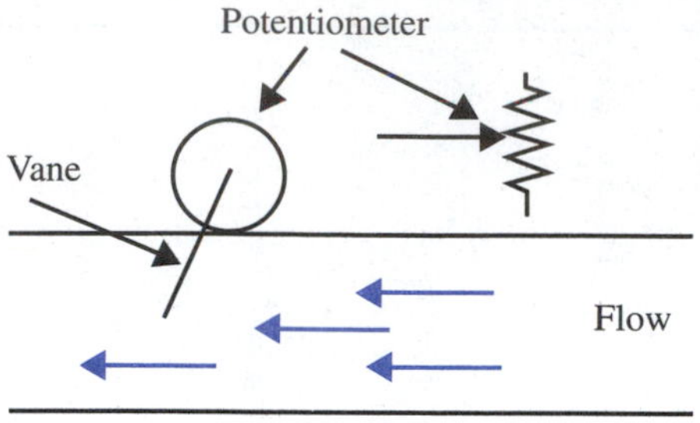

FIGURE 15–16 Vane flow measurement

FLOW MEASURED BY TURBINE

The angular rotation of a turbine can be used to sense the flow of a liquid or gas. The rotation can be converted to an electrical signal by coupling the turbine to a DC generator or by generating pulses that are counted by a tachometer. We shall look at this latter method in more detail.

Using a turbine to measure flow, how do we determine the number of rotations it is making? We can have the turbine drive a shaft, or we can use a noncontacting method. Let us assume that the turbine has a small magnet or magnets attached, and the magnetic field is detected by a magnetic sensor. Figure 15–17 shows the configuration.

The magnetic sensor could be a reed switch of a Hall-effect sensor. The reed switch consists of two strips of magnetic material that are parallel to one another and separated by a small air gap. A magnetic field directed along the axis of the strips causes the strips to be attracted to one another, and they make electrical contact.

A Hall-effect device generates a voltage in the presence of a magnetic field. Electrons flowing through a flat conductor are forced to one side of the conductor by the magnetic field. This uneven distribution of electrons results in a voltage being generated at right angles to the current flow.

Figure 15–18 shows both the reed switch and Hall-effect sensors.

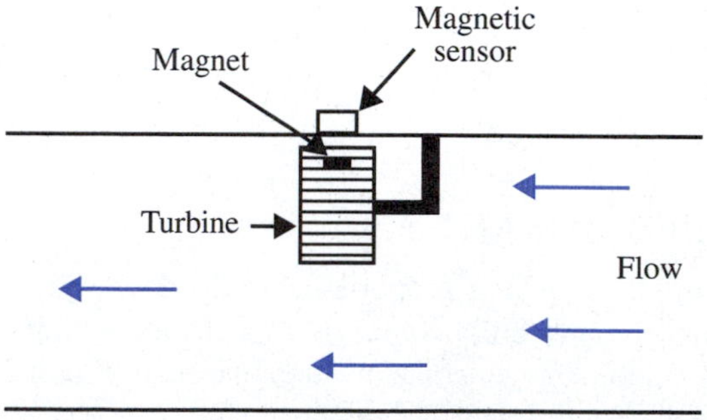

FIGURE 15–17 Magnetic sensing of turbine rotation

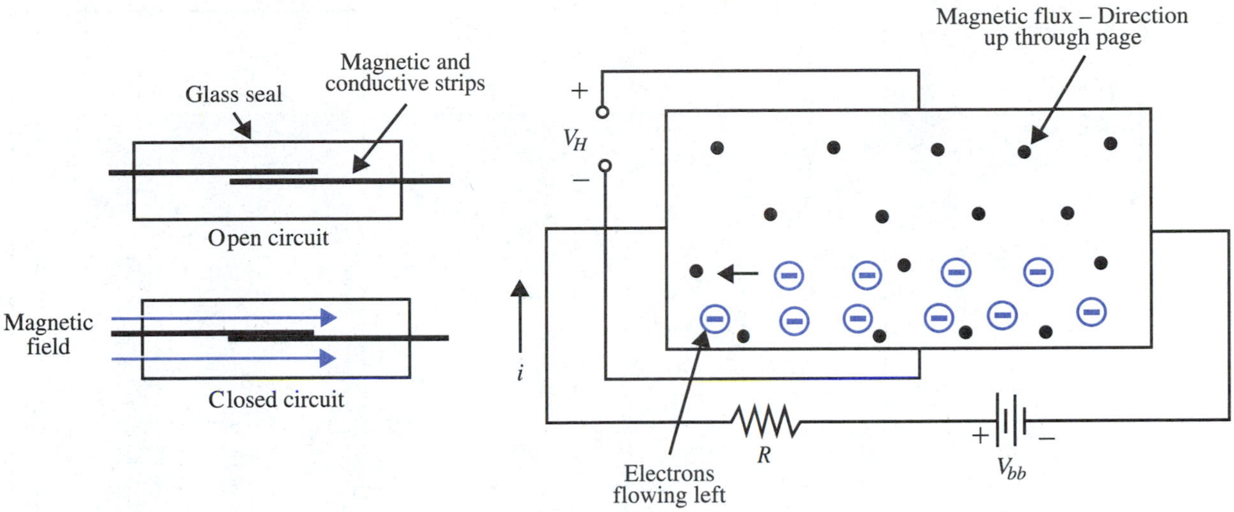

a) Reed switch b) Hall effect

FIGURE 15–18 Magnetic sensors

The polarity of the generated Hall voltage (V_H) is determined from the direction of the shift of the electrons caused by interaction of the electron magnetic field and the external magnetic field (refer to magnetic field theory).

If we sense the turbine rotation by detecting the magnetic field of an attached magnet, we need to convert the electrical pulses generated by this method into an analog signal or a digital code representing the speed of rotation. A way to perform the analog conversion is to use a frequency-to-voltage converter. The faster the turbine rotates, the greater the output voltage.

FREQUENCY-TO-VOLTAGE CONVERSION

A frequency-to-voltage converter translates the frequency of an input signal into an output voltage. This circuit can be used in speed control systems, tachometers, motor controllers, etc.

A device that can perform this function is the LM2907 frequency-to-voltage converter integrated circuit, shown in block diagram form in Figure 15–19. The input pulse to the LM2907 is generated by sensing the rotating field of the magnet attached to the turbine. The magnetic transducer can be a reed switch, or the Hall-effect sensor—both described previously.

The operation of the frequency-to-voltage converter is as follows:

- The input pulse is applied via Pin 1 to the noninverting input of a Schmitt trigger circuit (see Chapter 9), which has ±30mV hysteresis for noise immunity.

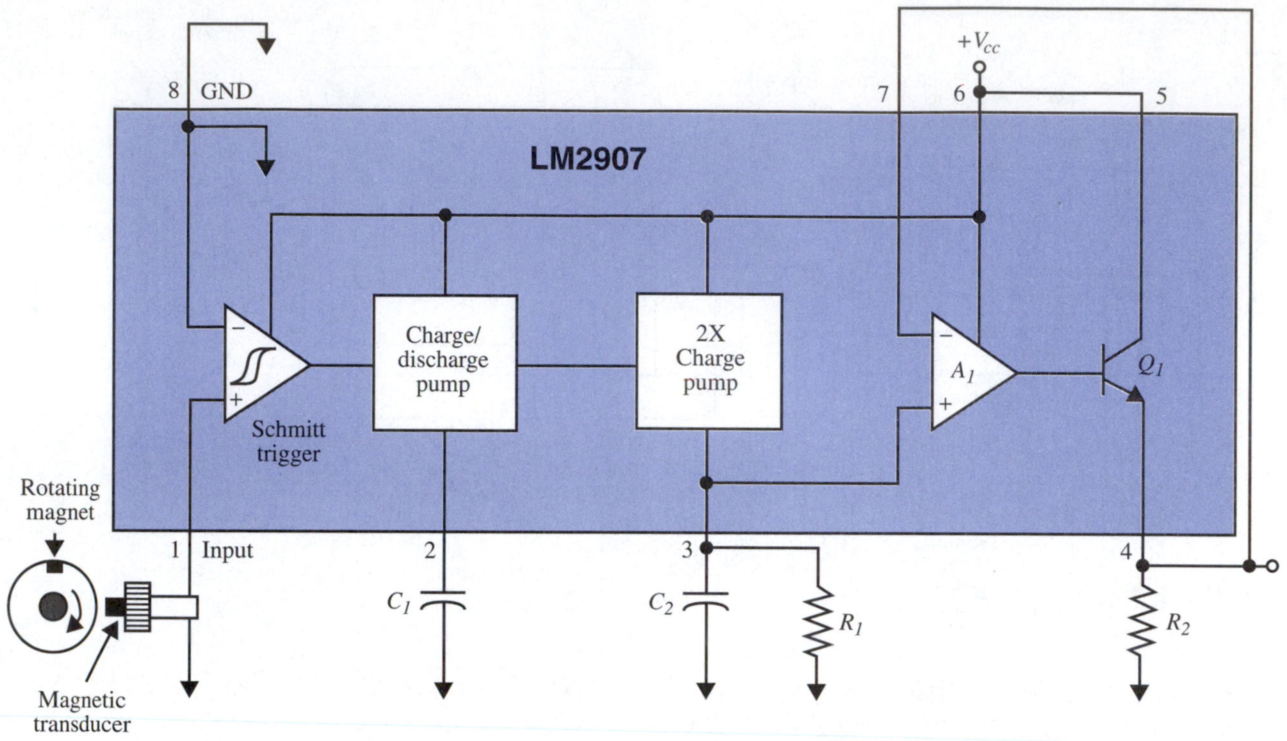

FIGURE 15–19 Frequency-to-voltage converter. *Courtesy National Semiconductor.*

- If the input signal goes negative (below ground) by more than 40mV, the output of the Schmitt trigger goes low, causing the charge/discharge pump to charge C_1 with a **constant charging current** (I_1) of approximately 180μA.
- The voltage across C_1 ramps positive from an initial level of $1/4V_{cc}$ to a final level of $3/4V_{cc}$ and holds at the higher level.
- When the input signal swings positive above ground by more than 40mV, the charge/discharge pump shifts to a discharge mode, where the capacitor is discharged at a 180μA rate.
- The voltage across the capacitor now ramps negative from $3/4V_{cc}$ down to $1/4V_{cc}$ and holds at the low level until the next negative transition of the input occurs.
- Each time C_1 is charged or discharged, C_2 is charged by the 2X pump. Between charges, some charge leaks off of C_2 via R_1. But an equilibrium is reached with the voltage (DC) on C_2 proportional to the number of input transitions and, therefore, the input frequency.
- The voltage across C_2 is applied to the follower (A_1) and booster (Q_1) combination and appears at Pin 4 across R_2. The output transistor can source or sink 50mA.

FREQUENCY-TO-VOLTAGE EQUATIONS

Let us now develop the equations for the frequency-to-voltage converter.

The charge on C_1 is

$$q_1 = C_1\left(\frac{3}{4}V_{cc} - \frac{1}{4}V_{cc}\right)$$

$$q_1 = \frac{C_1V_{cc}}{2} \qquad \textbf{(equation 15–4)}$$

Over one input cycle C_2 receives twice this charge, or

$$q_2(\text{charge}) = C_1V_{cc} \qquad \textbf{(equation 15–5)}$$

The resistor R_1 connected across C_2 causes a discharge over one input period of

$$q_2(\text{discharge}) = \frac{V_3T}{R_1} \qquad \textbf{(equation 15–6)}$$

where V_3 is the average voltage on Pin 3 and T the period of the input signal. In equilibrium,

$$q_2(\text{charge}) = q_2(\text{discharge})$$

Equating equations 15–5 and 15–6,

$$C_1V_{cc} = \frac{V_3T}{R_1}$$

Substituting f of $1/T$ and solving for the output voltage V_o (V_o equals V_3),

$$V_o = V_{cc}R_1C_1f \qquad \textbf{(equation 15–7)}$$

Remember, f is the frequency of the input signal.

What equation 15–7 indicates is that the output voltage is directly proportional to the input frequency, i.e., if the input frequency doubles, the output voltage doubles.

EXAMPLE

Determine the output voltage for an LM2907 frequency-to-voltage converter if R_1 is 10KΩ, C_1 is 10nF, V_{cc} is 15V, and the input frequency is 5KHz.

SOLUTION

Entering the given values into equation 15–7,

$$V_o = V_{cc}R_1C_1f$$

$$V_o = 15\text{V} \times 10\text{K}\Omega \times 10\text{nF} \times 5\text{KHz}$$

$$V_o = 7.5\text{V}$$

The maximum input frequency is limited by the time it takes to change the voltage across C_1 by $V_{cc}/2$, or

$$t_c = \frac{C_1 V_{cc}}{2I_{1\,(min)}} \qquad \textbf{(equation 15–8)}$$

Where $I_{1\,(min)}$ is the minimum specified current for the source/sink circuit (listed at 140μA).

A complete charge/discharge cycle for the capacitor C_1 takes $2t$, so the maximum input frequency equation becomes

$$f_{(max)} = \frac{1}{2t} = \frac{I_{1\,(min)}}{C_1 V_{cc}} \qquad \textbf{(equation 15–9)}$$

Equation 15–8 applies to symmetrical inputs, i.e., sine waves and square waves. If pulse inputs are used, the minimum pulse width is given by equation 15–8.

We can see from these relationships that the maximum input frequencies that can be converted by the LM2907 are symmetrical waveforms (equal positive and negative excursions).

REQUIREMENTS FOR THE CONVERSION CAPACITOR (C_2)

Since C_2 charges from a constant current source and discharges through R_1, a ripple voltage will appear across this capacitor. In order for us to determine the value of the ripple voltage, we must consider the current flow during the discharge time, as indicated in Figure 15–20b.

For small ripple, we can assume that

$$\Delta V = \frac{i\,\Delta t}{C_2}$$

and

$$i = V_o/R_1$$

so

$$\Delta V = \frac{V_o \Delta t}{R_1 C_2} \qquad \textbf{(equation 15–10)}$$

Δt is the time when the 2X pump is off (also the time when the charge/discharge pump is off), and it is also equal to

$$\Delta t = t_d - t_c$$

where t_d is the widest width of the input signal (half the period for a symmetrical input) and t_c the shortest time ($I_{1\,(max)}$) to charge the capacitor C_1.

Using equation 15–7 to substitute for t_c but using $I_{1\,(max)}$ because it provides the shortest time,

$$\Delta t = t_d - \frac{C_1 V_{cc}}{2I_{1(max)}}$$

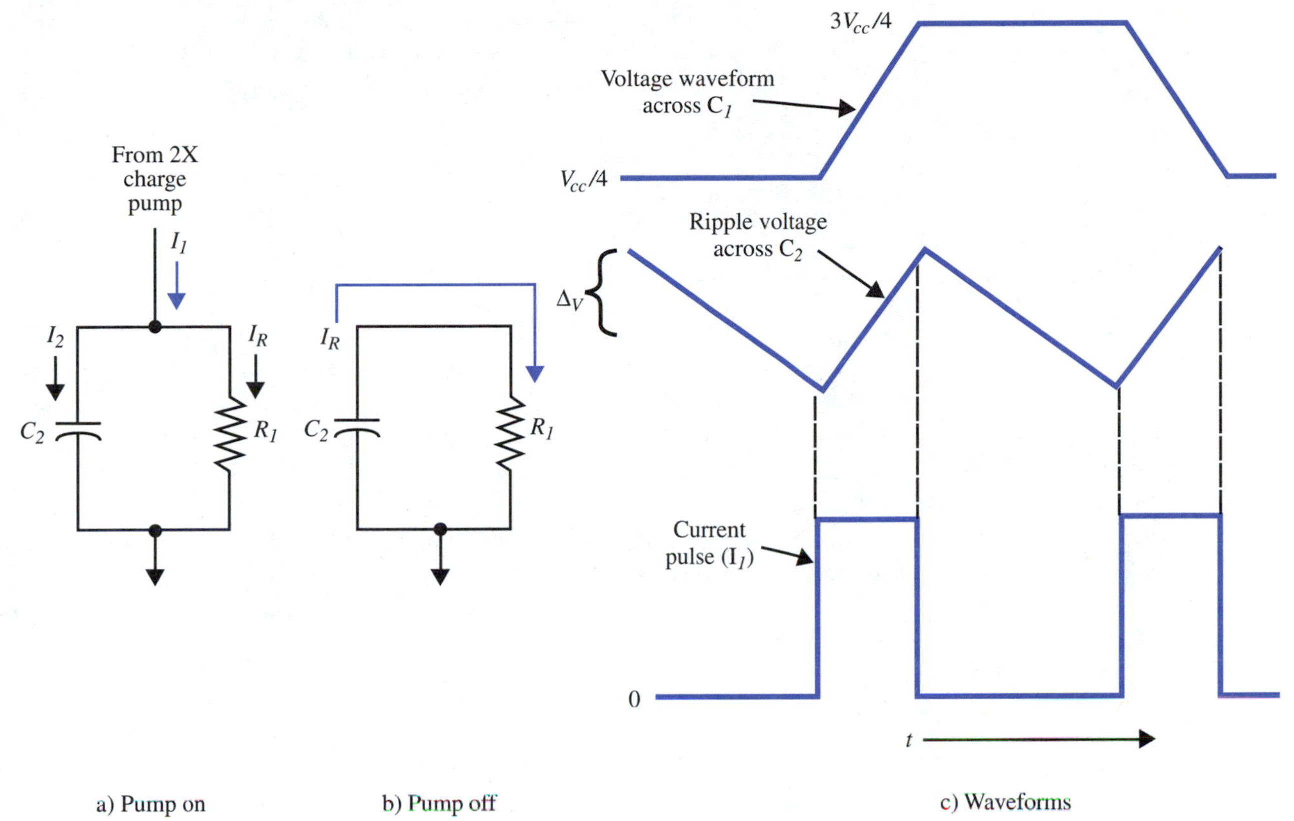

FIGURE 15–20 Current and wave forms for C_2

Substituting into equation 15–10,

$$\text{Ripple } \Delta V = \frac{V_o}{R_1 C_2}\left(t_d - \frac{C_1 V_{cc}}{2I_{1(max)}}\right) \qquad \textbf{(equation 15–11)}$$

We should use the value of t_d at the lowest frequency because ripple is greater at low frequencies.

Considering now practical values for C_1, R_1, and C_2:

- C_1 also provides internal compensation for the charge/discharge pump and is required to have a minimum value of 100pF. If we increase C_1, the output voltage increases for a given input frequency (equation 15–7). Increasing C_1 also increases the time to charge (and discharge) the capacitor by the required $V_{cc}/2$ and therefore lowers the maximum operating frequency (equation 15–9).
- R_1 must be smaller than the 10MΩ output resistance of the charge/discharge pump to maintain it as a constant current source, and so we shall assume a maximum value of 50KΩ. But don't make R_1 too low, because it lowers the output voltage at a given input frequency (equation 15–7).

- C_2 should be large to keep the output voltage ripple low. But a large value for C_2 slows the response of the circuit to fast input frequency changes. This is illustrated by the following equation for the rate of change of the output voltage:

$$\frac{\Delta V_o}{\Delta t} = \frac{I_1}{C_2}$$

EXAMPLE

Assume the turbine of Figure 15–19 has a maximum rotation rate of 6,000RPM and a minimum of 600RPM—the sense time when a pulse is present is 10 percent of the total period. We are using a +12V supply, and the output voltage across R_2 is to be +1V at the maximum RPM. The ripple voltage should be a maximum of 5 percent of the output voltage ($0.05V_o$). Find the values for C_1, C_2, R_1, and R_2 ($I_{1(max)}$ is 240μA, and $I_{1(min)}$ is 140μA).

SOLUTION

The maximum rotation rate of 6,000RPM provides a frequency of 100Hz, which has a period of 10ms. With the sense pulse present 10 percent of the time, its width (t_c) at the high RPM is 1ms. Using this value and $I_{1(min)}$ of 140μA in equation 15–8, we can solve for C_1.

$$C_1 = \frac{2t_cI_1}{V_{cc}}$$

$$C_1 = \frac{2 \times 1\text{ms} \times 140\mu\text{A}}{12\text{V}}$$

$$C_1 = 23.3\text{nF}$$

(Use 22nF.)

We can find R_1 by using equation 15–7 and the specified values for V_o, V_{cc}, and f.

$$R_1 = \frac{C_o}{V_{cc}C_1f}$$

$$R_1 = \frac{1\text{V}}{12\text{V} \times 22\text{nF} \times 100\text{Hz}}$$

$$R_1 = 37.9\text{K}\Omega$$

(Use 39KΩ.)

To find C_2, we need to use equation 15–12—but solved for C_2.

$$C_2 = \frac{V_o}{\Delta VR_1}\left(t_d - \frac{C_1V_{cc}}{2I_{1(max)}}\right)$$

The value of t_d is determined from the period of the lowest input frequency minus the pulse width at this frequency. The period at 600RPM is 100ms and the pulse width is 10 percent of this value, or 10ms.

$$t_d = 100\text{ms} - 10\text{ms}$$

$$t_d = 90\text{ms}$$

Now back to the equation for C_2.

$$C_2 = \left(\frac{1\text{V}}{(0.05 \times 1\text{V}) \times 39\text{K}\Omega}\right)\left(90\text{ms} - \frac{22\text{nF} \times 12\text{V}}{2 \times 240\mu\text{A}}\right)$$

$$C_2 = 45.9\mu\text{F}$$

(Use 50μF.)

The value for R_2 is determined by the maximum load current we want through the output transistor Q_1 at the highest input frequency. Let us pick a value of 10mA, then R_2 is 1V/10mA or 100Ω. To allow circuit calibration, it is suggested that a 91Ω resistor in series with a 20Ω potentiometer be used for R_2.

LEVEL SENSORS

What if we needed to check the level of liquid in a tank or grain in an elevator? We might want to know the actual level—or when a certain level has been reached.

The simplest form of level sensors for liquid is a float attached to a potentiometer. As the level changes, the float varies the potentiometer center tap position and therefore the output voltage from the potentiometer. Other level sensors we shall study in this section determine the level using such methods as finding the heat loss to a liquid and ultrasonic ranging.

RTD LIQUID-LEVEL SENSOR

RTD level sensors rely on the fact that heat transfer from the RTD to the liquid being measured is greater than the heat transfer to air. So, the more liquid surrounding the RTD (higher liquid level), the greater the heat transfer.

Figure 15–21 shows a metal probe and a level-measuring circuit using the National Semiconductor LM1042 Fluid Level Detector.

The probe shown in Figure 15–21a is made of nickel-iron resistance wire, which has a resistance/temperature coefficient of 0.46%/°C. In operation, a constant current is passed through the probe, and the voltage drop across the probe is immediately measured (indicates resistance). The probe is allowed to heat for a fixed time (resistance goes up), and then another probe voltage measurement is made. The difference in voltage between the two measurements indicates the temperature rise of the probe. If the probe is in air, the temperature rise will be great, because air is a good thermal insulator. But in a liquid, heat will flow from the probe, and the temperature rise will not be as great. By calibrating the temperature rise of the probe over the fixed time interval, the height of liquid covering the probe can be determined.

Looking at the circuit of Figure 15–21b in more detail, the initial voltage measurement is stored on capacitor C_1. The capacitor C_t determines the time the probe is allowed to heat before the second voltage reading is made. C_3 stores the voltage difference between the two

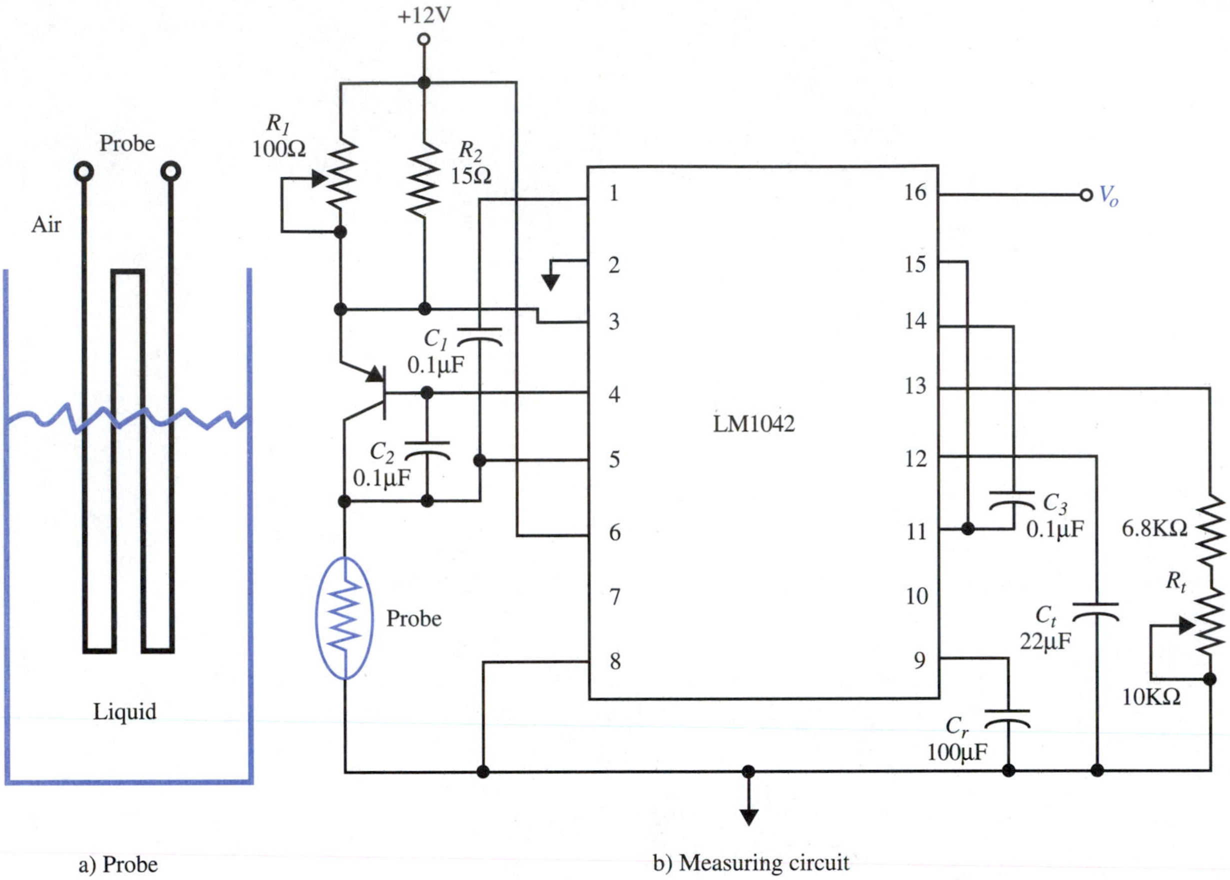

FIGURE 15–21 Liquid-level sensor. *Courtesy National Semiconductor.*

readings and provides the difference voltage amplified by 1.2 at Pin 16. The circuit takes complete measurements at a rate proportional to the value of C_r. With the 100μF value used in the schematic, the rate is about once per minute. For manual operation, Pin 9 is grounded and Pin 8 is tied high through a 1KΩ resistor. To initiate a measurement, Pin 8 is taken low.

The current through the probe is selected by the parallel combination of R_1 and R_2 divided into the 2V that is maintained across these resistors. A typical current is 200mA, and a probe voltage range of from 0.7V to 5.3V is accommodated by the LM1042, which means the probe resistance can vary from 3.5Ω to 26.5Ω with this current.

EXAMPLE

The probe used with the circuit of Figure 15–21b has an initial resistance of 10Ω, and the current source is set up for 200mA. Find the voltage at Pin 16 at the completion of the measurement cycle if the probe heats up from 35°C to 85°C (assume K_t of 0.46%/°C). Also find the sample rate if C_r is changed to 10μF.

SOLUTION

The change in the resistance of the probe for the 50°C increase in temperature will be

$$\Delta R_p = R_p \times K_t \times \Delta T$$

$$\Delta R_p = 10\Omega \times 0.0046/°C \times 50°C$$

$$\Delta R_p = 2.3\Omega$$

With the 200mA flowing through the probe, this represents a voltage of $2.3\Omega \times 0.2A$, or 0.46V stored on capacitor C_3. At the output, this translates to a voltage of $0.46V \times 1.2$, or 0.552V.

Reducing the value of C_r from 100μF to 10μF will reduce the time between samples by a factor of 10, i.e., from 60 seconds to 6 seconds, which means a 10 times higher sample rate.

ULTRASONIC LEVEL SENSOR

An ultrasonic signal is a sound wave that is above the range of human hearing, and typically consists of a narrow high-power pulse of high-frequency oscillations generated by a piezoceramic element. This element is resonant at a frequency determined by its physical size. Resonance occurs at two frequencies: one for series resonance and the other for parallel resonance. For maximum power into (and out of) the element, transmitting should be at the series resonant frequency. However, if the same element is used for receiving the ultrasonic signal, parallel resonance will allow more voltage to be developed across the element from the impinging sound. Since the two resonant frequencies are close together, a compromise is to transmit at a frequency between the two.

To determine the level of a substance in a tank, the ultrasonic pulse is transmitted to the surface, and the time to receive an echo (t_e) at the transmitter is measured. Knowing the velocity of the sound pulse (v_S), the distance between the ultrasonic transceiver and the surface can be calculated.

$$\text{Distance } (d) = v_S \times t_e/2 \qquad \textbf{(equation 15–13)}$$

The time is divided by two because the sound travels to and from the surface.

EXAMPLE

Determine the distance from the surface of grain in an elevator to the ultrasonic transmitter if the time for the return echo pulse is 20ms. Assume the velocity of sound is 1,087ft./sec.

SOLUTION

Using equation 15–13,

$$\text{Distance } (d) = v_S \times t_e/2$$

$$\text{Distance } (d) = 1087\text{ft./sec.} \times 20\text{msec.}/2$$

$$\text{Distance } (d) = 10.87\text{ft.}$$

ULTRASONIC LEVEL-SENSING CIRCUIT

An ultrasonic level sensor can be built using an LM1812 Ultrasonic Transceiver. This chip can provide 12W of power into a piezoceramic transducer, which gives a maximum sensing range of 20ft. in air and 100ft. in water. The complete transceiver schematic is provided in Figure 15–22.

Circuitry within the chip generates a series of one-microsecond, one-amp pulses, for the transmitter. These pulses are applied to the transducer at a rate determined by the oscillator controlled by L_1 and C_1.

The LM555 timer is configured as an oscillator that determines how long and how often the transmitter is keyed. The pulse width from the timer is (refer to Chapter 9)

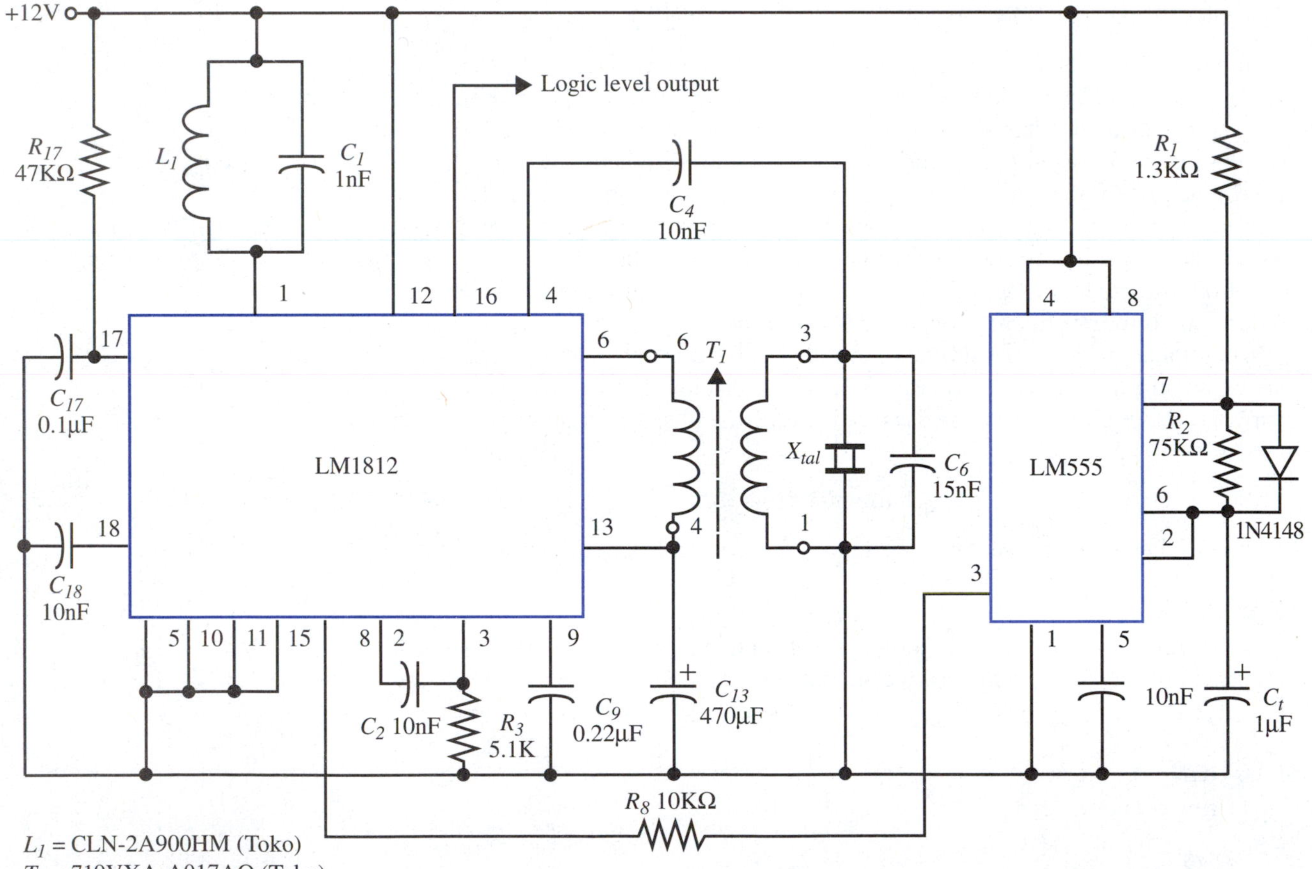

L_1 = CLN-2A900HM (Toko)
T_1 = 719VXA-A017AO (Toko)
X_{tal} = EFR-OTB40K2 (Panasonic Tel. 201-392-4651)

FIGURE 15–22 40KHz ultrasonic transceiver. *Courtesy National Semiconductor.*

$$t_p = 0.693 R_1 C_t$$

$$t_p = 0.693 \times 1.3\text{K}\Omega \times 1\mu\text{F}$$

$$t_p = 900\mu\text{s}$$

The period of the timer is

$$T = 0.693(R_1 + R_2)C_t$$

$$T = 0.693(1.3\text{K}\Omega + 75\text{K}\Omega)1\mu\text{F}$$

$$T = 52.9\text{ms}$$

This provides a keying time of 900μs at a frequency of 1/52.9ms, or 18.9Hz.

The receiver section of the chip contains circuits to inhibit reception when transmission is occurring and to restore full amplification when a return signal is expected. To ensure that the receiver and transmitter are tuned to the same frequency, the L_1, C_1 tuned circuit used in the transmitter is also used as a band-pass filter in the receiver.

The purpose of the other components shown in the schematic is as follows:

- C_2 couples the output of the first receiver amplifier stage into the second stage, and with R_3, acts as an attenuator.
- C_4 limits the current spikes into the receiver from high-level voltages applied to the transducer.
- T_1 power matches the output of the transmitter to the piezoceramic transducer.
- R_8 limits the current supplied by the 555 for transmitter keying.
- C_9 sets the receiver turn-on delay after transmit. With the 220nF value shown, the delay is about 2ms.
- C_{13} decouples the transmitter power supply from the receiver.
- C_{17} and R_{17} form a receiver noise filter circuit (integrator) with a time constant from 20% to 50% of the transmit time.
- C_{18} controls the delay time to discharge the integrator after the receiver input signal has dropped below a $1.4V_{p-p}$ threshold level.

The circuit shown in Figure 15–22 operates at a transmit frequency of 40KHz—but by changing the values of L_1, C_1, and the transducer, frequencies up to 325KHz can be used. Higher frequency means a narrower transmit pulse can be used to improve target resolution; but available power output is less at the higher frequencies, which means a shorter detection range.

DISTANCE OR PRESENCE DETECTION

Transducers for sensing distance or presence of an object use similar detection methods. These include ultrasonic, magnetic, capacitive, and optical.

The ultrasonic sensors used for distance and object detection are similar in operation to the ultrasonic level sensor we studied. Sensor units available from the Turck Company (model CP40 ultrasonic sensor) can measure distance from 0.2m to 1.0m, or detect an object—at a preset distance—over the same range. The output provided for the distance measured is either 0V–10V in 256 increments (eight-bits) or 0mA–20mA in 256 increments. The output for object sensing is a PNP transistor capable of sourcing up to 200mA with the object present. Operation is possible with a 10V–30V DC power source.

MAGNETIC DISTANCE AND PRESENCE DETECTION

Magnetic determination of object distance can be accomplished noncontacting by induction (metal detector) or contacting by a variable coupling transformer.

If metal is introduced into the changing magnetic field of a coil, eddy currents are induced in the metal. If the coil is part of the tuned circuit of an oscillator, the power loss from the eddy currents lowers the amplitude of oscillation (lower Q). The closer the metal is to the coil, the greater the effect; so the amplitude change can be related to distance. An example of this type of sensor is the Turck K90, which has a sensing range of 6cm.

A variable coupling transformer used for distance measurement is called a linear variable differential transformer (LVDT). This sensor consists of a transformer with a primary coil and two secondary coils and a movable core, as shown in Figure 15–23.

In Figure 15–23a, the magnetic flux from the primary is coupled equally to both secondaries. Since the secondaries are wired out-of-phase and added, the output voltage is zero.

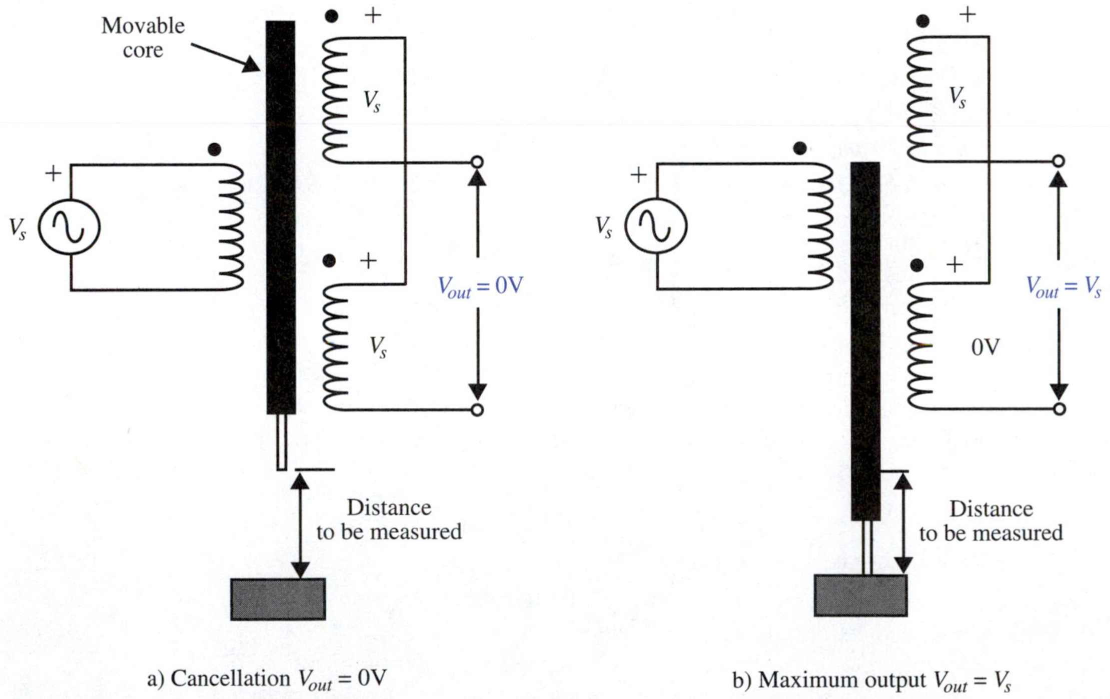

FIGURE 15–23 LVDT distance measurement

TABLE 15–3 Dielectric constants

Material	*Dielectric Constant (Range)*
Acetone	19.5
Air	1.000264
Cement Powder	4.0
Cereal	3.0–5.0
Glass	3.7–10
Mica	5.7–6.7
Paper	1.6–2.6
Porcelain	5–7
Salt	6
Sugar	3.0
Wood, dry	2–6
Wood, wet	10–30

When the core shaft is moved into contact with the object, at the distance to be measured, there is no coupling to the upper primary, as shown in Figure 15–23b. The secondary voltage is now only provided by the lower coil, which is in phase with the input, and so the output voltage is V_S. As the core is moved down, between the two extremes, a decreasing output voltage from the upper coil will be subtracted from the full voltage of the lower coil.

CAPACITIVE DISTANCE AND PRESENCE DETECTION

Capacitive sensors sense an object by using the change in **dielectric constant,** which occurs when an object is placed in the electrostatic field of a capacitor. Objects with dielectric constants greater than 1.2 can be detected with this method (dielectric constant of air is 1.0). The greater the dielectric constant of the object, the greater the detection range.

Capacitive sensing is used on nonmetals, which are difficult to detect with magnetic induction. The dielectric constants of some selected materials are given in Table 15.3.

As an example of a commercial capacitive sensor, the Turck model K40SR sensor has sensing distances ranging from 0.8cm for cardboard to 4cm for water and metal. The sensor operates by the change in capacitance causing a change in the frequency of an internal RC-oscillator.

OPTICAL DISTANCE AND PRESENCE DETECTION

Measuring distance to an object using optics can be accomplished by pulsing a laser diode (or gas laser) and measuring the time it takes for the reflected light, from the object, to reach a detector. This method is used for military range finders.

The equation for the distance can be determined knowing the time (t) and the speed of light ($c = 3 \times 10^8$m/s):

$$\text{Distance } (d) = c \times t/2$$

$$\text{Distance } (d) = 3 \times 10^8 \text{m/s} \times t/2 \qquad \textbf{(equation 15–14)}$$

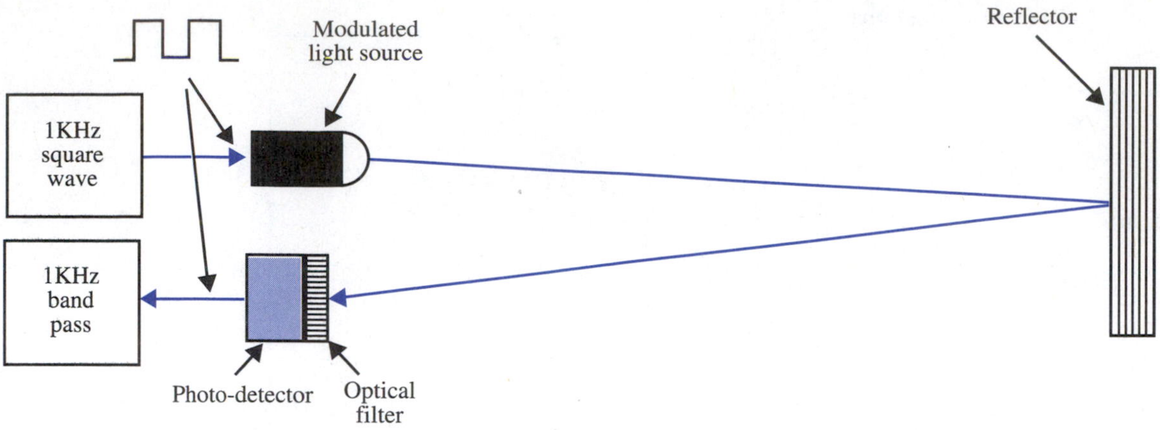

FIGURE 15–24 Optical detection system

EXAMPLE

Find the distance to an object using optical ranging if the time for the return signal to reach the detector is 20ns.

SOLUTION

Using equation 15–14,

$$\text{Distance } (d) = 3 \times 10^8\text{m/s} \times t/2$$

$$\text{Distance } (d) = 3 \times 10^8\text{m/s} \times 20\text{ns}/2$$

$$\text{Distance } (d) = 3\text{m}$$

Objects can be sensed using optics by reflecting light from the object or by having the object interrupt the path of light from the source to the detector. To overcome ambient light problems, the light source is modulated light source by a sine wave or square wave (at frequencies not harmonics of 60Hz), and the detector has a band-pass filter with a center frequency at the modulating frequency (see Figure 15–24).

If a narrow-band light source like an LED or laser diode is used, further rejection of ambient light can be accomplished by putting an optical filter over the detector.

A much used optical detection system is the "bar code reader." Figure 15–25 shows the optical system. Light from the LED is reflected from the white stripes to the detector diode and blocked by the black stripes. The code is sensed from the relative widths of the reflected stripes.

Range for detection of objects using optics is limited only by the power level of the source and sensitivity of the detector. However, dirt and moisture must be considered.

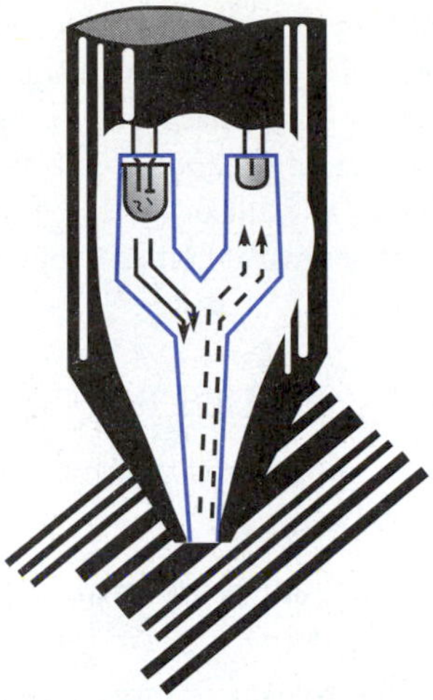

FIGURE 15–25 Bar code reader

VELOCITY MEASUREMENT

There are two types of velocity, rotational and linear. Measurement of rotational velocity can be accomplished by contacting or noncontacting methods. A contacting rotational sensor is typically a tachometer, which is a DC generator with an output voltage proportional to RPM. An example of a noncontacting sensor is a Hall-effect sensor (or reed switch), which senses iron teeth on a wheel (see Figure 15–17), or an optical sensor that senses reflections from the rotating member.

Linear velocity can be measured, noncontacting, by Doppler frequency shift of a radar signal (police radar) or by measuring the time to travel between fixed markers (Maglev and air-cushion people movers). Contacting methods of measuring linear velocity use rotational sensors to measure the rotational velocity of a known-diameter wheel and then compute the linear velocity from this measurement.

ACCELERATION MEASUREMENT

Acceleration is the rate of change of velocity ($\Delta v/\Delta t$), and for acceleration to occur, a force must be present, as indicated by the fundamental equation of dynamics, which states

$$\text{Force } (F) = m \times a$$

where m is the mass (the weight divided by the acceleration due to gravity—32ft./sec^2) and a is the acceleration. The greater the force acting on an object capable of moving, the faster it accelerates. Acceleration can be positive (velocity increasing) or negative (velocity decreasing).

A sensor for measuring acceleration is called an **accelerometer**. Most accelerometers measure the force due to acceleration. This is accomplished by noting the change in resistance of a strain gauge attached to a known mass when acceleration occurs.

Analog Devices has developed a sensor for automobile air bags that measures changes in capacitance rather than resistance as a result of the force caused by acceleration. The advantage of this approach, which is used with their ADXL50 acceleration sensor, is less sensitivity to temperature change.

TROUBLESHOOTING TRANSDUCERS

The basic approach to checking a transducer is to provide an input, and check for the required response at the transducer output.

- Temperature sensors will require some form of heat input (heat gun) or heat removal (ice).
- Strain gauge types of force sensors simply require an ohmmeter check. Other types could require the application of force, and one technique is to use (with care) a machine vise.
- Pressure sources (air or hydraulic) are usually available for checking pressure sensors.
- Flow can usually be simulated to check flow sensors. Vane types can be manually deflected and turbine types caused to rotate with an air gun. Hall-effect sensors can be checked with a small magnet.
- Level sensors can be checked by a physical measurement of the level or distance to the surface.
- Distance sensors can also be checked by physical measurement of the distance and presence sensors by testing with a known object.
- Velocity sensors can be checked by measuring the rotations, or distance traveled, in a fixed time.
- Acceleration sensors require a check against a standard for calibration. To simply check for an output, the accelerometer can be tapped.

SUMMARY

- A transducer is a device that converts other forms of energy into electrical energy.
- A thermocouple consists of two (couple) wires of different metals connected together that can be used to measure temperature.
- Ambient temperature can be canceled by using a thermocouple cold junction.
- An RTD is a metal temperature transducer that increases in resistance with temperature in a rather linear fashion.
- Platinum is a common metal used for RTDs.
- A thermistor is made from a material that reduces in resistance as the temperature increases.
- Linearizing networks are available that make the thermistor more linear at the expense of about a factor of ten reduction in sensitivity.

- A semiconductor temperature transducer produces an output voltage or current that is proportional to temperature.
- A resistance strain gauge consists of a metal strip or wire that changes in length, and therefore resistance, when a force is applied.
- To reduce temperature effects, strain gauges are typically used in a bridge configuration.
- A load cell consists of strain gauges mounted on a precisely machined spring element.
- Semiconductor strain gauges have the advantage of greater resistance change with a deformation of the semiconductor material. But the semiconductor strain gauge is less accurate, much more temperature sensitive, and is harder to compensate.
- Piezoelectric transducers are used to measure forces that are continuously changing over, perhaps, millisecond time intervals.
- Pressure is force spread over a specified area.
- Pressure can be measured by various mechanical devices such as bellows, diaphragms, manometer tubes, and Bourdon tubes.
- Pressure can be specified three ways: absolute, gauge, and differential.
- Absolute pressure measures pressure with reference to zero pressure (a vacuum).
- Gauge pressure is measured relative to ambient pressure (14.7 lbs./in.^{2} at sea level). Gauge pressures below ambient are usually referred to as negative pressures, or inches, of vacuum.
- Differential pressure is the difference in pressure between two points.
- Measurement of the flow of a liquid or gas passing a particular point depends on whether the volume flow (m^3/s) is required, or the mass flow (kg/s).
- Flow measured by a difference in pressure is related to the square root of the pressure.
- A vane used to measure flow deflects proportional to the flow.
- The angular rotation of a turbine can be used to sense the flow of a liquid or gas.
- A frequency-to-voltage converter translates the frequency of an input signal into an output voltage.
- A Hall-effect device generates a voltage in the presence of a magnetic field.
- A magnetic field directed along the axis of the strips of a reed switch causes the strips to be attracted to one another and the switch closes.
- The cooling of a hot-wire lowers the wire's resistance, and this can be used as a measure of the flow rate.
- The simplest form of level sensors for liquid is a float attached to a potentiometer.
- RTD level sensors rely on the fact that heat transfer from the RTD to the liquid being measured is greater than the heat transfer to air.
- To determine the level of a substance in a tank, an ultrasonic pulse can be transmitted to the surface, and the time to receive an echo (t_e) at the transmitter measured.

- Transducers for sensing distance or presence of an object include ultrasonic, magnetic, capacitive, and optical.
- Eddy current sensors for metal use the fact that adjacent metal can lower the *Q* of a tuned circuit.
- A variable coupling transformer used for distance measurement is called a linear variable differential transformer (LVDT).
- Capacitive sensors sense an object by using the change in dielectric constant, which occurs when an object is placed in the electrostatic field of a capacitor.
- Measuring distance to an object using optics can be accomplished by pulsing a laser diode (or gas laser) and measuring the time it takes for the reflected light, from the object, to reach a detector.
- A contacting rotational sensor is typically a tachometer, which is a DC generator with an output voltage proportional to RPM.
- An example of a noncontacting rotational sensor is a Hall-effect sensor (or reed switch) that senses iron teeth on a wheel.
- Linear velocity can be measured, noncontacting, by Doppler frequency shift of a radar signal, or by measuring the time to travel between fixed markers.
- Acceleration is the rate of change of velocity ($\Delta v/\Delta t$), and for acceleration to occur, a force must be present.
- A sensor for measuring acceleration is called an accelerometer. Most accelerometers measure the force due to acceleration.

EXERCISE PROBLEMS

1. Define a transducer.
2. Describe the operation of a thermocouple.
3. Find the measured temperature for the configuration in Figure 15–A if the voltage read at the meter is 20mV.

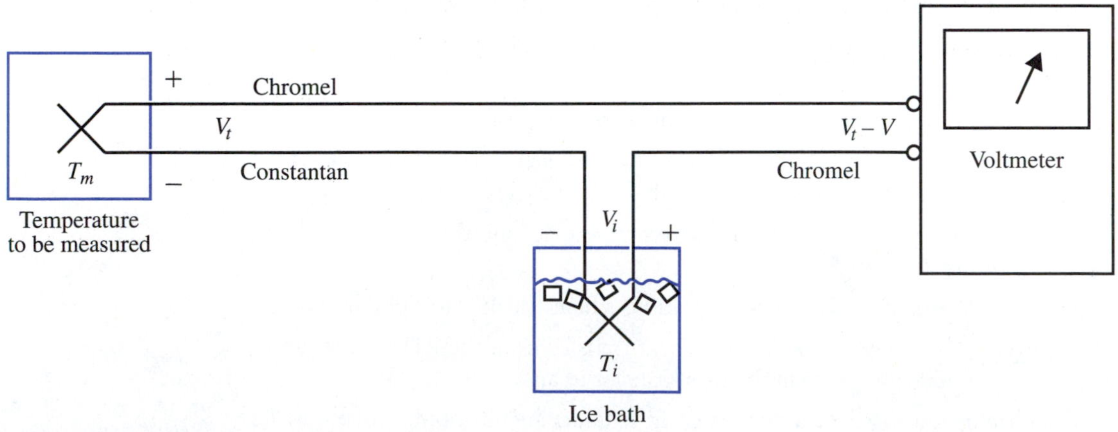

FIGURE 15–A (text figure 15–1)

4. Determine the value of R_c for the circuit in Figure 15–B if an iron-constantan thermocouple is used.

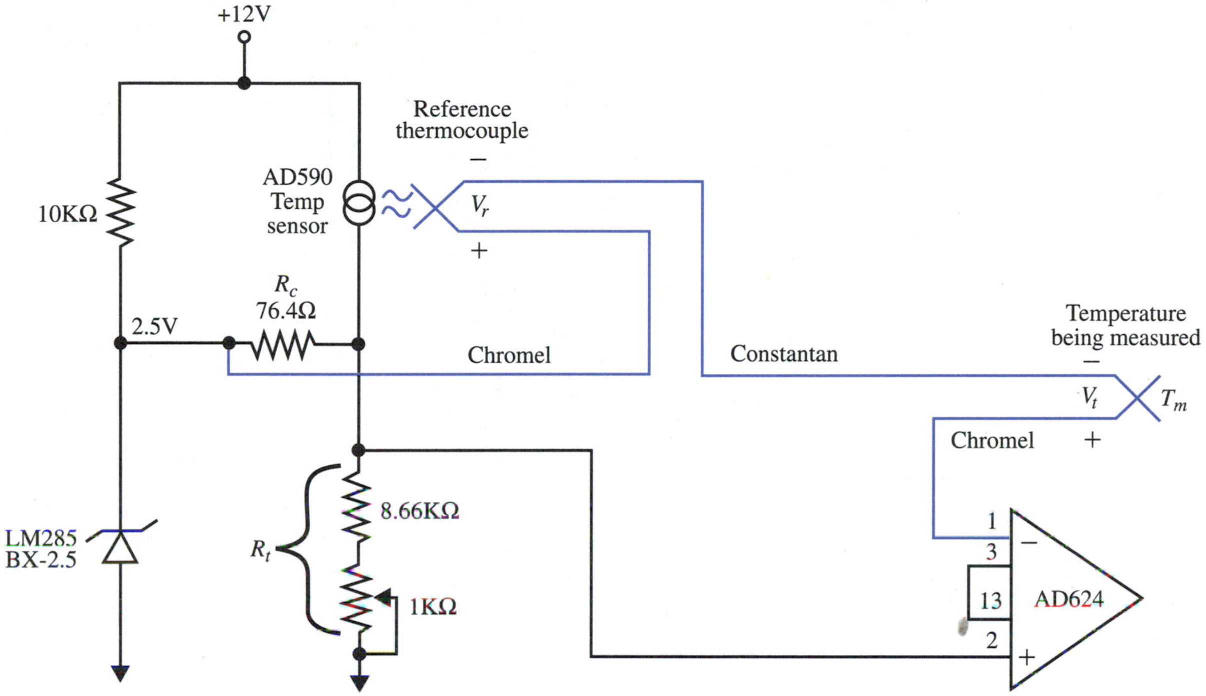

Note: $R_c = 76.4\Omega$ assumes a thermocouple temperature sensitivity K_t of 76.4μV/°C over the operating range of this circuit (15–35 degrees Celsius).

FIGURE 15–B (text figure 15–2)

5. Describe the purpose and operation of an RTD.
6. Determine the output voltage at 100°C for the circuit in Figure 15–C if a 100Ω nickel RTD is used instead of platinum.
7. Describe the purpose and operation of a thermistor.
8. If the voltage reference in Figure 15–D fails and is replaced by a reference with a voltage of 2.6V, what is the output voltage at 100°C and how can it be brought back to 1.00V?
9. Variations of which circuit components can be accommodated with the Zero Adj potentiometer in the circuit in Figure 15–E?
10. Find the output voltage for the circuit in Figure 15–F if the #2 sensor has a temperature of 85°C and the #1 sensor a temperature of 25°C.
11. Determine the output voltage from an LM34 at 150°F.
12. Determine the output voltage from an LM35 at 150°F.
13. Describe the operation of a strain gauge.

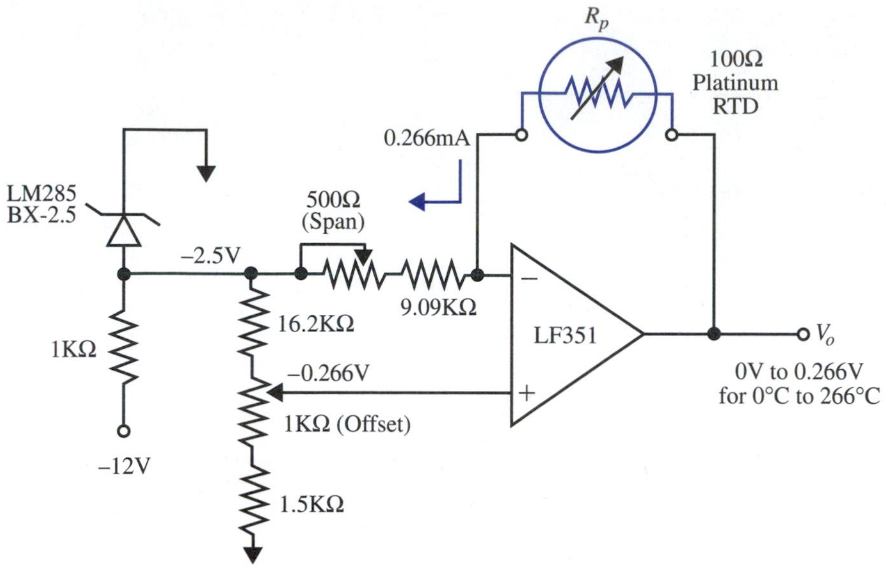

FIGURE 15–C (text figure 15–3)

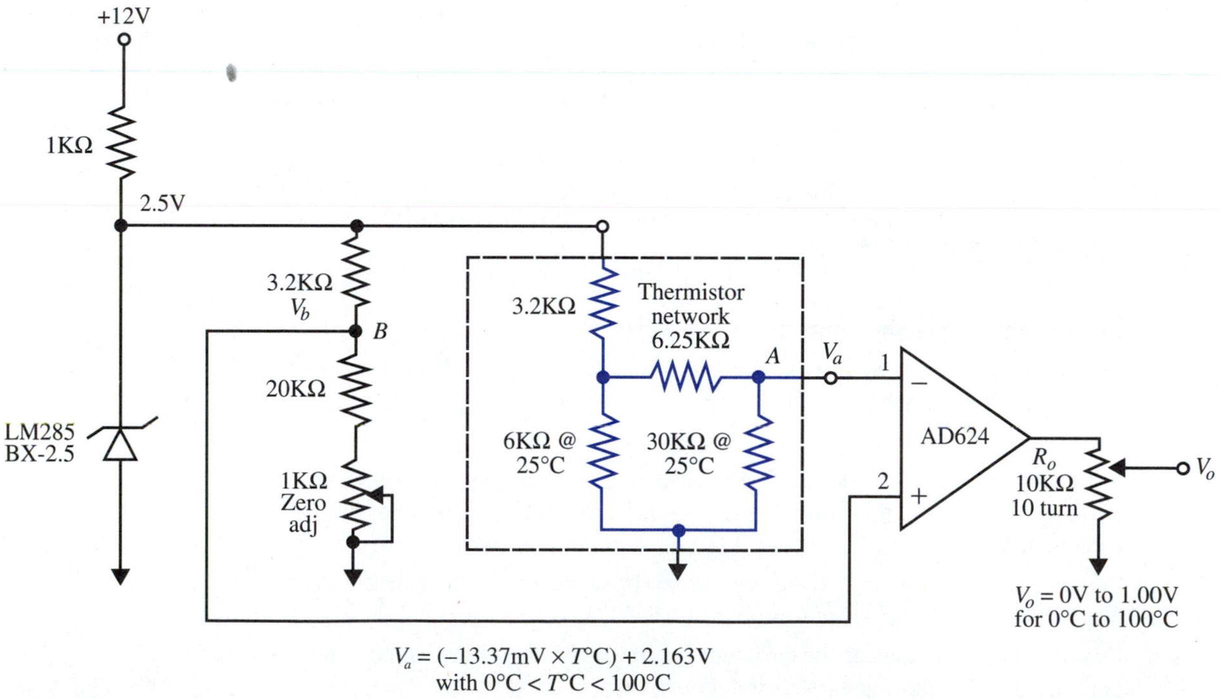

FIGURE 15–D (text figure 15–7)

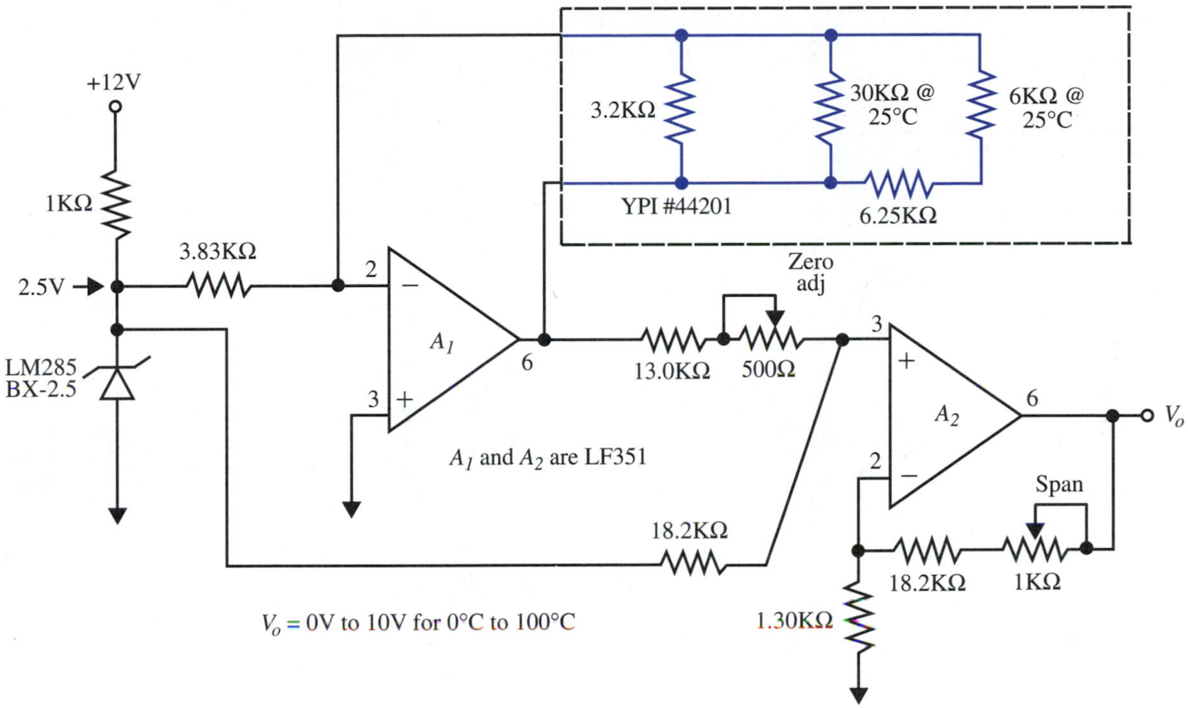

FIGURE 15–E (text figure 15–8)

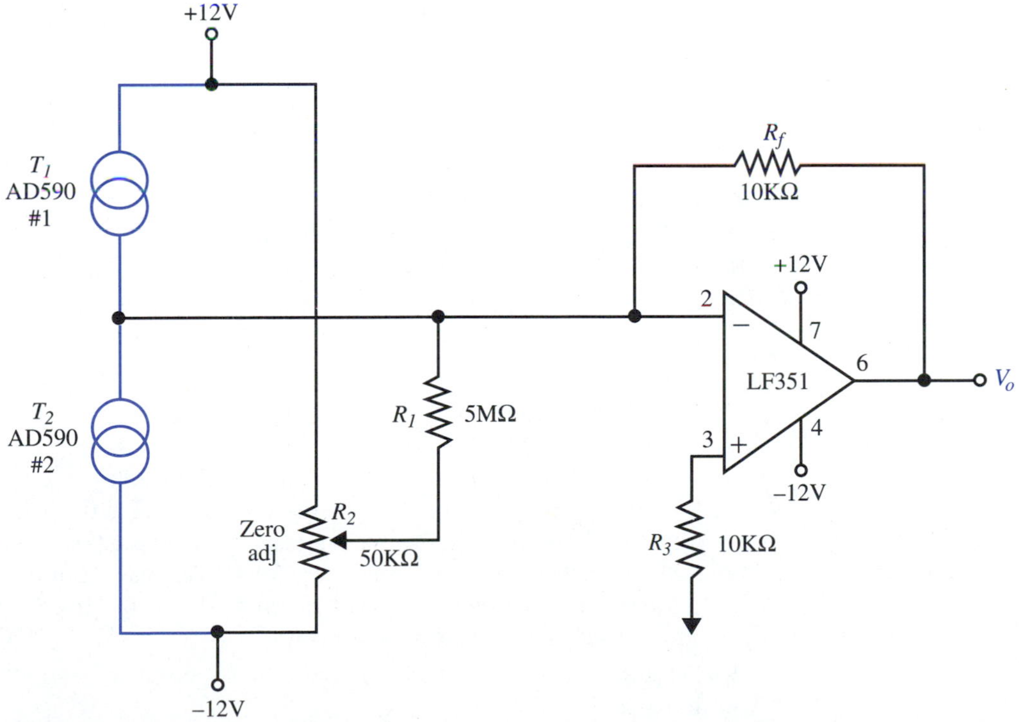

FIGURE 15–F (text figure 15–9)

14. If the load cell in Figure 15–G is changed to one with a sensitivity of 2.5mV/V, what is the required value of R_g to obtain the same output voltage with a 100lb. load?

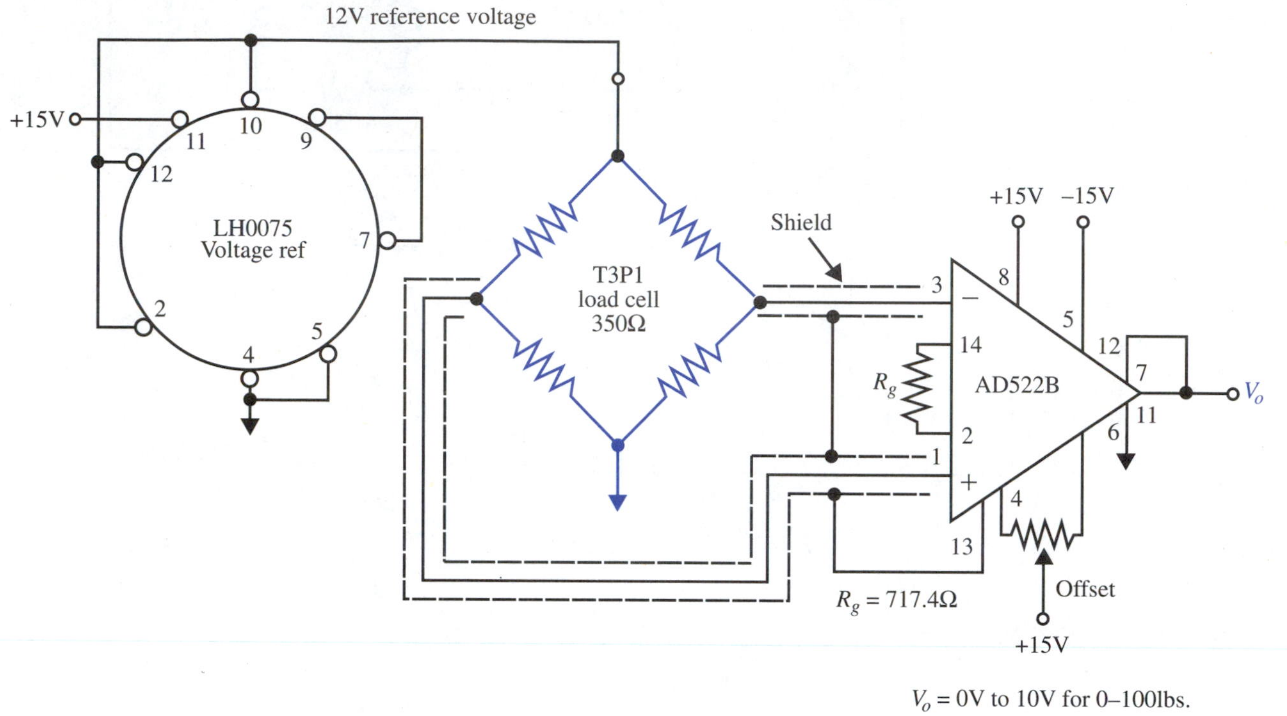

FIGURE 15–G (text figure 15–11)

15. What is a null bridge?
16. Describe absolute, gauge, and differential pressure.
17. Determine the flow rate if the flow constant is (6 gallons)(in.)/(minute)(lb.)$^{1/2}$ and P_1 is 15lbs./in.^{2} and P_2 is 4lbs./in.^{2}.
18. Describe a reed switch and a Hall-effect sensor.
19. Determine the input frequency for an LM2907 frequency-to-voltage converter if R_1 is 10KΩ, C_1 is 10nF, V_{cc} is 15V, and the output voltage is 6V.
20. Compute the size of C_2 for an LM2907 frequency-to-voltage converter if C_1 is 22nF, R_1 is 15KΩ, V_{cc} is 12V, the lowest input frequency is a 25Hz square wave, and the maximum ripple 3%.
21. Design a tachometer circuit for a six-cylinder automobile using an LM2907 that is to be triggered from the coil (assume a square wave at the highest RPM) and a 1mA meter movement. The maximum engine RPM is 5,000 and the minimum 500 (remember there are 3 sparks per RPM). The maximum ripple should be 5 percent and the maximum voltage across R_1 should be 10V. Battery voltage of 12V is available for the chip power source.
22. What is the anticipated R_3 potentiometer resistance final setting for the circuit in Figure 15–H?

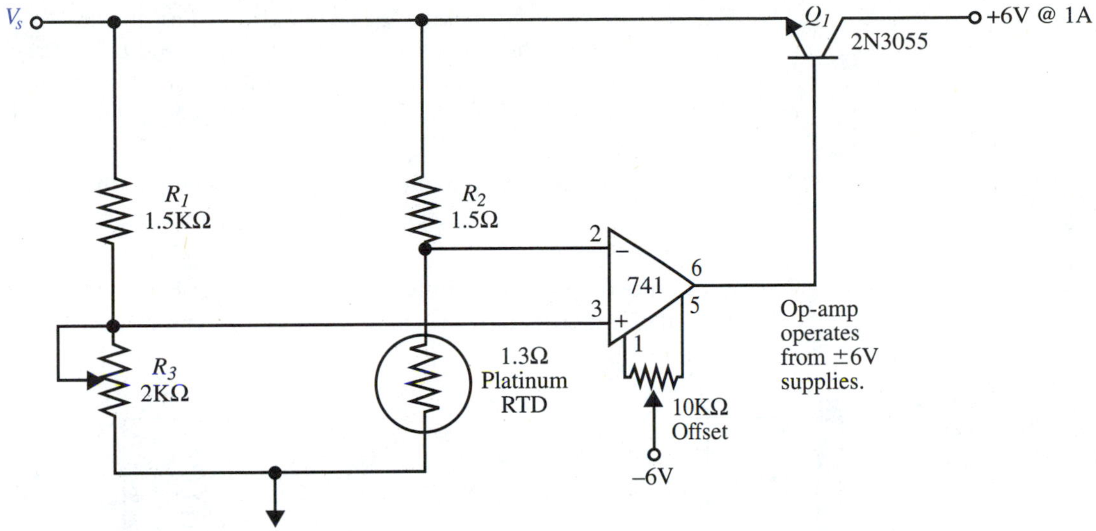

FIGURE 15–H (text figure 15–15)

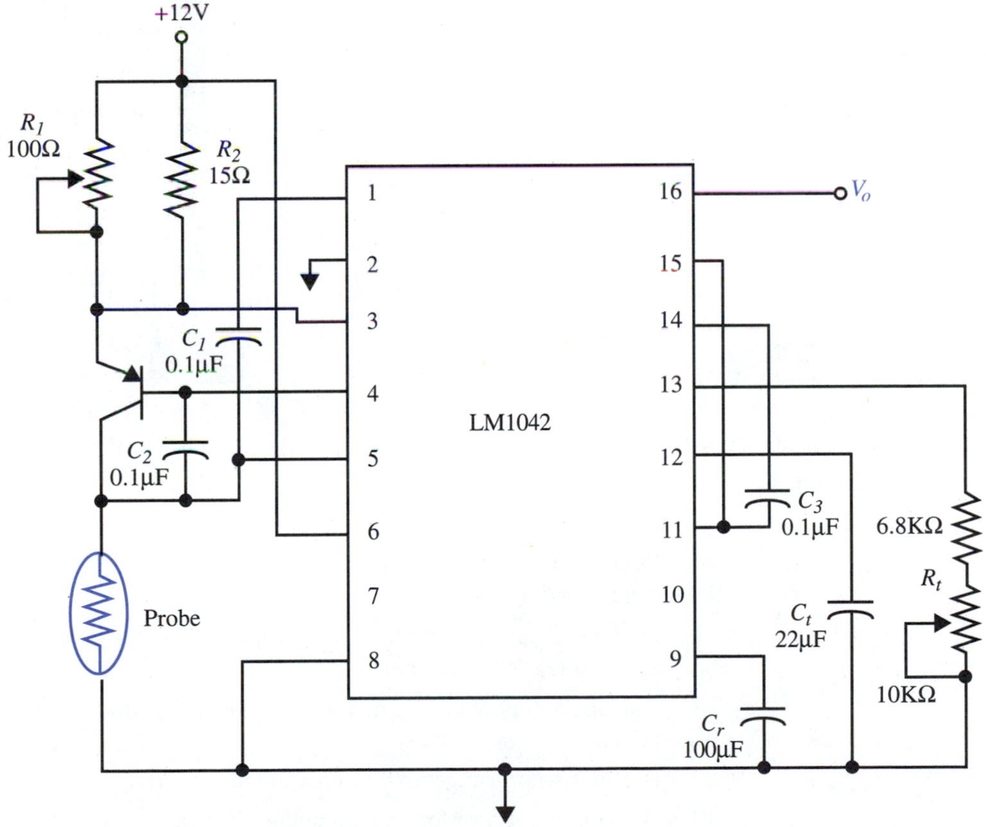

FIGURE 15–I (text figure 15–21b)

23. The probe used with the circuit in Figure 15–I has an initial resistance of 12Ω, and the current source is set up for 200mA. Find the voltage at Pin 16 at the completion of the measurement cycle if the probe heats up from 30°C to 70°C (assume K_t of 0.46%/°C). Also find the sample rate if C_r is changed to 20μF.

24. Determine the distance from the surface of oil in a tank to the ultrasonic transmitter, if the time for the return echo pulse is 35ms. Assume the velocity of sound is 1,087ft./sec.

25. What is the output voltage from the LVDT circuit in Figure 15–J if the core has traveled one-fourth of the maximum distance that can be measured?

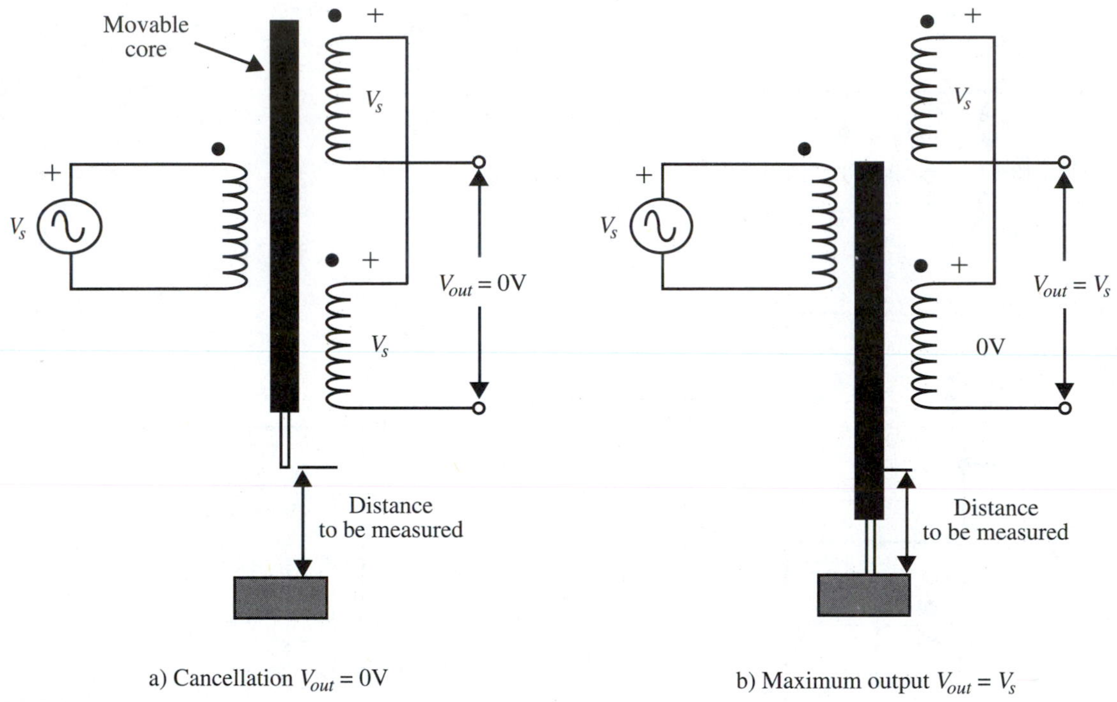

FIGURE 15–J (text figure 15–23)

26. Which material can be detected at a greater range with a capacitive sensor: cement powder or salt?

27. Find the distance to an object, using optical ranging, if the time for the return signal to reach the detector is 32ns.

28. Describe a noncontacting method of measuring rotational velocity and linear velocity.

29. How is a strain gauge used to measure acceleration?

30. Describe the sensor for an automobile air bag.

ABBREVIATIONS

a = Acceleration
ΔR_p = Change in probe resistance with temperature change
K_p = Flow pressure constant
K_t = Temperature coefficient
LVDT = Linear variable differential transformer
P = Pressure
RTD = Resistance temperature device
t_c = Shortest time to charge capacitor
t_d = Widest width of input signal (e.g., square wave half the period)
t_e = Time to receive echo
t_p = Timer pulse width
V_H = Hall-effect voltage
v_S = Velocity of sound—1,087ft./sec. at 25°C
V_t = Voltage from temperature-measuring thermocouple
V_i = Voltage from reference (ice) thermocouple

ANSWERS TO PROBLEMS

1. Device that converts other forms of energy to electrical energy
2. Dissimilar metals in contact develop a voltage that varies with temperature.
3. T_m = 262°C
4. $R_c = 53\Omega$
5. Measures temperature by increase in resistance of a metal
6. V_o = 17.82mV
7. Device indicates temperature by resistance; nonlinear resistance drop with temperature increase
8. V_o = 1.04V; adjust R_o
9. 3.83KΩ, 18.2KΩ, and the network resistance
10. V_o = 0.6V
11. V_o = 1.5V
12. V_o = 0.656V
13. Distortion of material caused by strain causes a resistance change.
14. $R_g = 602\Omega$
15. Bridge always nulled for nonlinearities
16. See text.
17. Flow rate = 19.9 gal./min.
18. See text.
19. f = 4KHz
20. C_2 = 43μF (use 50μF)
21. C_1 = 0.047μF; R_1 = 71.5KΩ (use 7.15KΩ); C_2 = 5μF
22. R_3 = 1.3KΩ
23. V_{16} = 0.528V; sample rate = 8.333/sec.
24. d = 19ft.

25. $V_o = -V_s/4$
26. Salt
27. $d = 4.8\text{m}$
28. See text.
29. Length related to force; force related to acceleration
30. Semiconductor de-acceleration sensor

CHAPTER 16

Signal Conversion

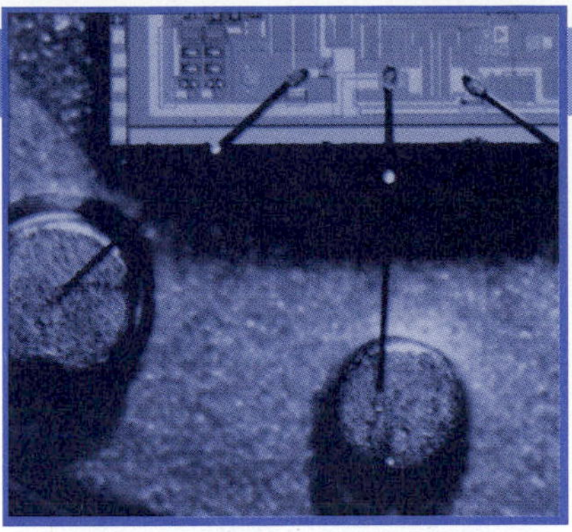

CHAPTER OBJECTIVES

After completing this chapter, you should be able to describe the following signal conversion concepts:

- Purpose of signal conversion
- Conversion accuracy
- Multiplexing of signals
- Sampling and storing of signals
- Analog-to-digital conversion (ADC)
- Successive approximation ADC
- Dual slope ADC
- Flash ADC
- Sigma-delta ADC
- Digital-to-analog conversion (DAC)
- Current mode DAC
- Sigma-delta DAC
- Troubleshooting signal conversion circuits

INTRODUCTION

Signal conversion is used to convert analog signals to digital signals and vice versa. This chapter will explain why and how this is accomplished.

First the need for signal conversion will be established, and then the factors to be considered with the conversion will be explained. Next the common methods for converting from analog to digital will be examined. Finally, the reverse process of converting from digital signals to analog signals will be covered.

WHY CONVERT ANALOG SIGNALS TO DIGITAL SIGNALS

An analog signal is one that can assume any voltage value within the limitations of power supply voltage, e.g., sine waves, speech voltage waveforms, etc. In contrast, binary digital signals can assume only two values called *1* and *0*; for transistor transistor logic (TTL) the *1* state is approximately +5V and the *0* state is 0V.

If we wish to transfer analog signals from one point to another, or store analog signals, with minimum interference from noise, we need to convert analog signals to digital. The reason for this is indicated in Figure 16–1. Notice that the analog signal is much less toler-

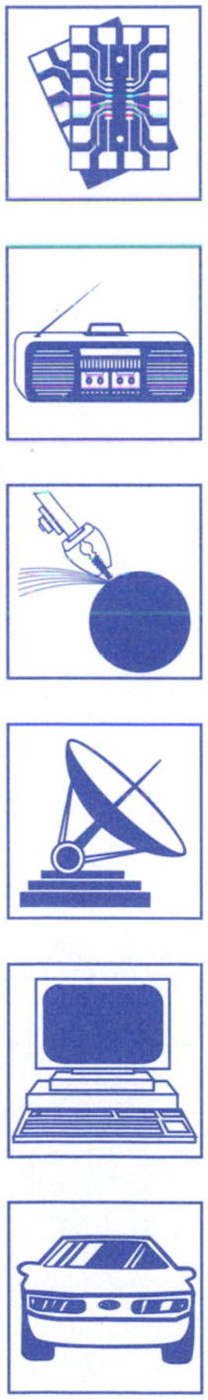

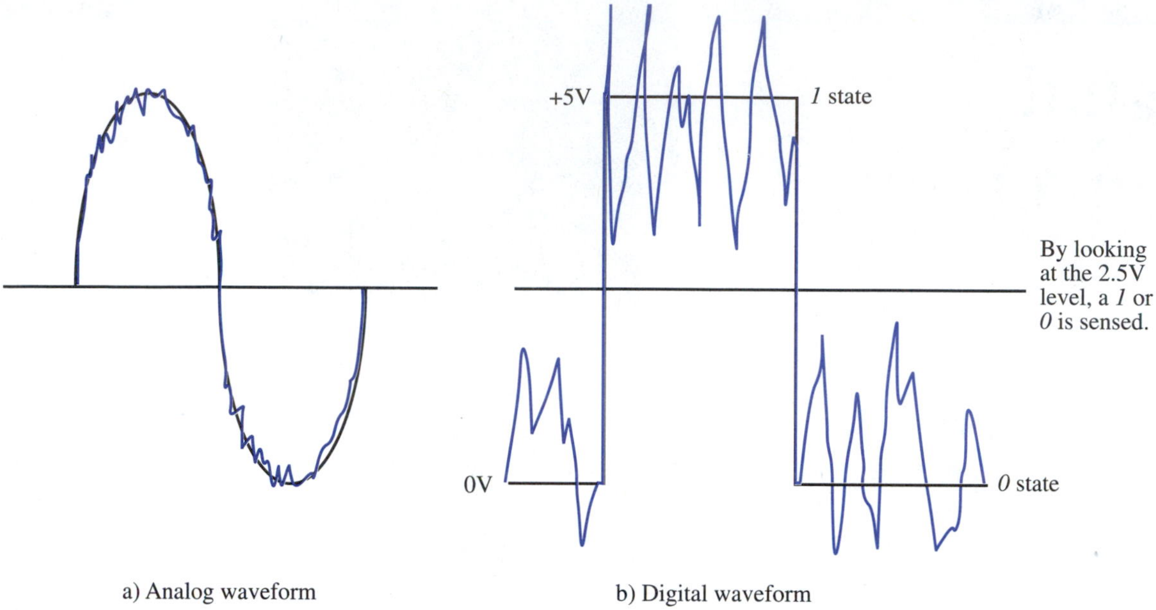

FIGURE 16–1 Noise effects on analog and digital signals

ant to noise interference, since the digital signal can tolerate noise levels up to almost half the voltage excursion (2.5V) and still detect the *1* state by looking at the 2.5V level, but details in the analog waveform are lost with small levels of noise interference.

This noise advantage is why high-speed data transmission uses digital signals, and by converting low-level audio signals to digital before recording (compact discs), they are not lost in noise.

HOW ANALOG SIGNALS ARE CONVERTED TO DIGITAL

Analog signals are converted to digital by sampling the analog voltage value at fixed intervals of time and converting each sample to a digital word (series of bits). Figure 16–2 indicates the process. In the figure, the sampling is performed at times t_0 through t_{13}, and the sampled voltage is determined from the intercept of a vertical line at these times with the curve. For example, at time t_1, the sampled voltage is 3.5V; at time t_6, 13.5V; and so on. Note that some of the shape of the analog waveform is lost in the sampling process. The jog in the waveform between t_1 and t_2 is not sampled. This demonstrates a key fact in the sampling of analog waveforms: The more often samples of the analog are made (higher sampling rate), the better the fidelity of the data.

The number of samples that can be taken is limited by how fast the sampled voltage can be converted to a digital word. In other words, there is no point in getting a new sample from the analog waveform if the last sample has not yet been converted.

We have seen that because of the time required for the sample process, we can miss information in the analog waveform. Another loss in information occurs in the digital conversion process. Let us assume that the digital word that represents the sampled voltage has

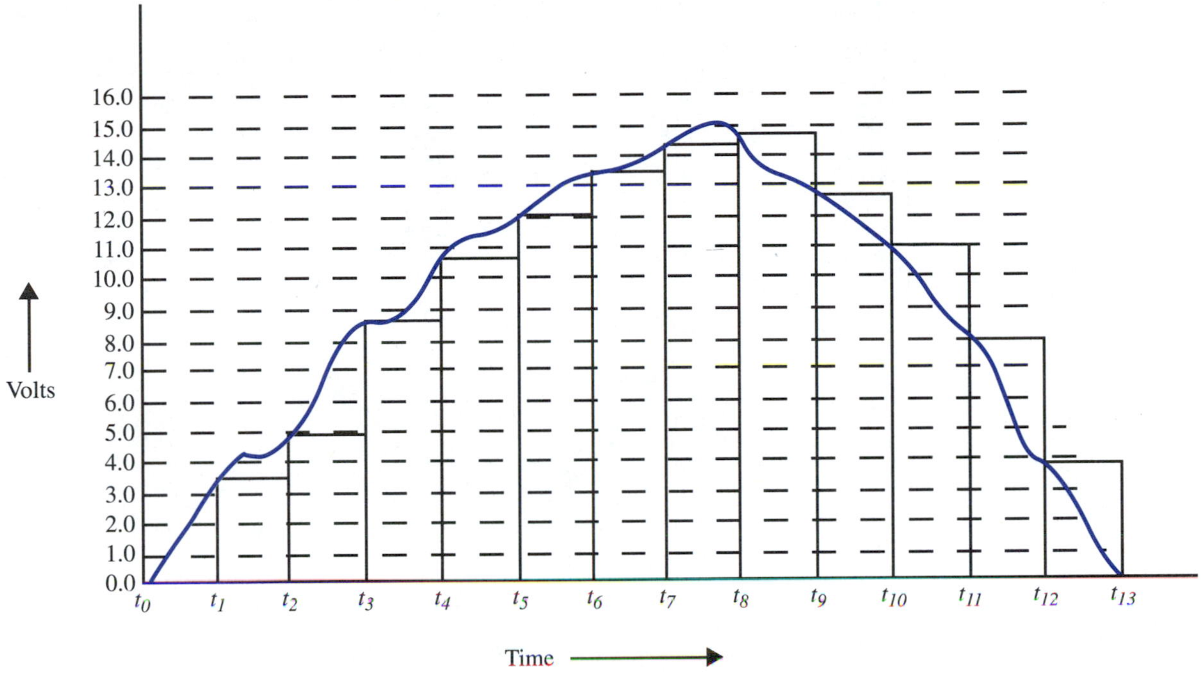

FIGURE 16–2 Sampling an analog waveform

4 bits. This means that 2^4 or 16 different voltage levels can be recorded. If we assume that each step is 1V, then the 16 possible digital converted levels range from 0V to a maximum of 15V, $(2^n - 1)V$. Now if we look back at Figure 16–2, we see that the samples taken can have any value between 0V and 15V; but this is lost in the conversion process since only 1V changes are recorded.

Let us assume that the code 0000 represents 0V to 0.5V. The code 0001 would then represent (for our example) 0.5V–1.5V. Table 16–1 shows the complete representation.

Notice that the highest code 1111 represents a full-scale voltage of 15V, but the actual maximum voltage that can be converted, with the same code, is half a step higher, or 15.5V (this half a step is called the **quantization error**).

If we use the Table 16–1 range to convert the sampled data points of Figure 16–2, we can determine the digital code for each sampled voltage, as listed on the right side of Table 16–2. Using the data from these tables, we can construct an analog representation of these digital codes, as shown in Figure 16–3.

The stepped waveform of Figure 16–3 indicates the midpoint voltage of the digital code (see Table 16–1), e.g., t_6 sample is at 13V rather than 13.4V. The curve derived from these points is shown along with the original curve. Notice that the derived curve can lead the original waveform when this curve is increasing in amplitude and always lags when the waveform is decreasing.

What Figure 16–3 shows is the distortion that can result from analog-to-digital conversion. This distortion can be minimized, however, by increasing the number of time sam-

TABLE 16–1 Voltage-to-Digital Conversion

4-Bit Digital Code	*Voltage Range*
0000	0.0V–0.5V
0001	0.5V–1.5V
0010	1.5V–2.5V
0011	2.5V–3.5V
0100	3.5V–4.5V
0101	4.5V–5.5V
0110	5.5V–6.5V
0111	6.5V–7.5V
1000	7.5V–8.5V
1001	8.5V–9.5V
1010	9.5V–10.5V
1011	10.5V–11.5V
1100	11.5V–12.5V
1101	12.5V–13.5V
1110	13.5V–14.5V
1111	14.5V–15.5V

TABLE 16–2 Codes for the Sampled Data of Figure 16–2

Sample Time	*Sampled Voltage*	*Digital Code*
t_0	0.0V	0000
t_1	3.5V	0100
t_2	5V	0101
t_3	8.7V	1001
t_4	10.6V	1011
t_5	12.0V	1100
t_6	13.4V	1101
t_7	14.2V	1110
t_8	14.7V	1111
t_9	12.7V	1101
t_{10}	11.0V	1011
t_{11}	8.0V	1000
t_{12}	3.9V	0100
t_{13}	0.0V	0000

ples taken over the waveform (shorter sample time) and by decreasing the size of the voltage steps (more steps) used in the process. More samples means a higher sample clock frequency, and smaller voltage steps require a higher number of bits for the digital word.

EXAMPLE

If we increased the number of digital bits to 8 in our previous example of Figure 16–2, what would be the size of each step (the **resolution**), and what is the maximum voltage that can be converted to digital (output code 11111111)?

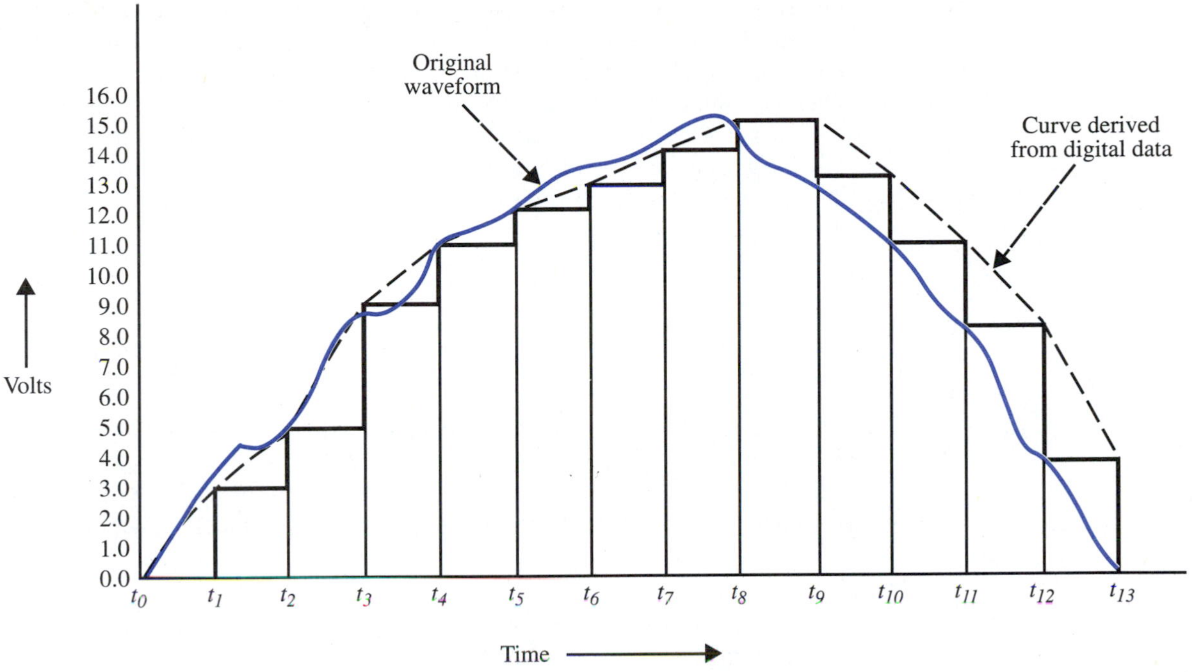

FIGURE 16–3 Analog curve derived from digital data

SOLUTION

With 2^8 we have 256 steps, which is 16 times the number of steps we used previously. Each 1V step is now divided into 16 smaller steps, and the size of the new step is

$$V_{step} = 1V/16 = 62.5mV$$

The binary code 11111111 in decimal is

$$(2^n) - 1 = 255$$

Since each step is 62.5mV, this code represents

$$62.5mV \times 255 = 15.935V$$

The full-scale voltage is this voltage plus half a voltage step, or

$$V_{max} = 15.935V + 62.5mV/2$$

$$V_{max} = 15.969V$$

MINIMUM SAMPLING FREQUENCY (ALIASING)

When determining the minimum sampling frequency for a complex waveform (music voltage waveform), we need to remember from Fourier analysis that all such waveforms are made up of harmonics (multiples) of the lowest (fundamental) frequency.

Let us assume that we are listening to an FM-stereo station. The highest frequency (f_{max}) transmitted, which could be a harmonic of a musical instrument, is 15KHz. If we now wanted to convert this music signal to digital to store on a compact disc, the question could be "What is the minimum sample frequency that can be used?" This question was answered by H. Nyquist in 1924 when he stated that the minimum number of samples to describe a sine wave is two, i.e., one sample for the positive excursion of the sine wave and one for the negative excursion. If we apply these criteria to the 15KHz harmonic, we need a minimum sampling frequency f_s of

$$f_s = 2f_{max} \text{ (Nyquist criterion)} \qquad \textbf{(equation 16–1)}$$

$$f_s = 2 \times 15\text{KHz}$$

$$f_s = 30\text{KHz}$$

A disadvantage of sampling right at the Nyquist frequency is that the waveform could be sampled at the zero crossings of the wave, which means the sampled voltages would be zero. Typically, sampling is done at a higher frequency with terms like "2X over sampling" being used—which means the waveform is being sampled at two times the Nyquist frequency.

EXAMPLE

Determine the actual sampling frequency if f_{max} is 10KHz, and 3X over sampling is being used.

SOLUTION

From equation 16–1, the Nyquist frequency is

$$f_N = 2f_{max}$$

$$f_N = 2 \times 10\text{KHz}$$

$$f_N = 20\text{KHz}$$

With 3X over sampling,

$$f_s = 3 \times 20\text{KHz}$$

$$f_s = 60\text{KHz}$$

If we sample at less than the Nyquist frequency, we get an effect called **aliasing**, which generates a new frequency that is lower than the frequency being sampled. Figure 16–4 shows this effect. In the figure, the new low frequency is generated from tracing through the sample points. The effect of sampling the FM-stereo music at a frequency lower than the Nyquist frequency is to generate a new frequency in the audio range, which creates distortion.

The frequency f_a of the alias signal can be determined from the equation

$$f_a = f_s - f_{max} \qquad \textbf{(equation 16–2)}$$

We can see from this equation that if f_s is greater than twice f_{max} (Nyquist criteria), the alias signal doesn't fall below f_{max}—which means that we don't get an interfering signal within the desired band of frequencies.

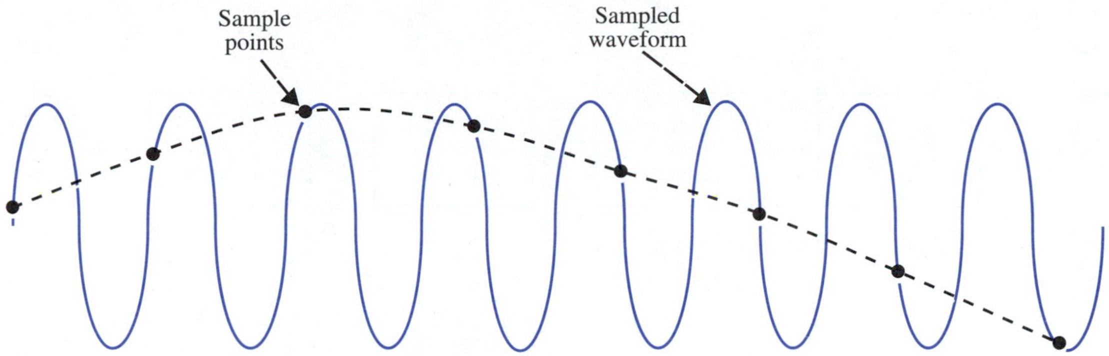

FIGURE 16–4 Aliasing caused by too-low sample frequency

Figure 16–5 shows the effect of the sampling frequency being too low, and the sampling frequency being above the Nyquist frequency ($>2f_{max}$).

The sharp cutoff of signals at the upper end of the audio band shown in Figure 16–5 is rather artificial. In practice, steep-sided low-pass filters have to be used to prevent frequencies greater than f_{max} from creating alias signals in the audio band. A higher sampling frequency alleviates this problem.

COMPLETE SIGNAL CONVERSION SYSTEM

Let us now look at the complete signal conversion system as indicated in Figure 16–6. In the figure, the analog signal is sampled by the sample and hold circuit and held at a fixed level for the time required for the analog-to-digital converter to convert the signal to a digital form. When this conversion is complete, the sample and hold (S & H) circuit takes another sample.

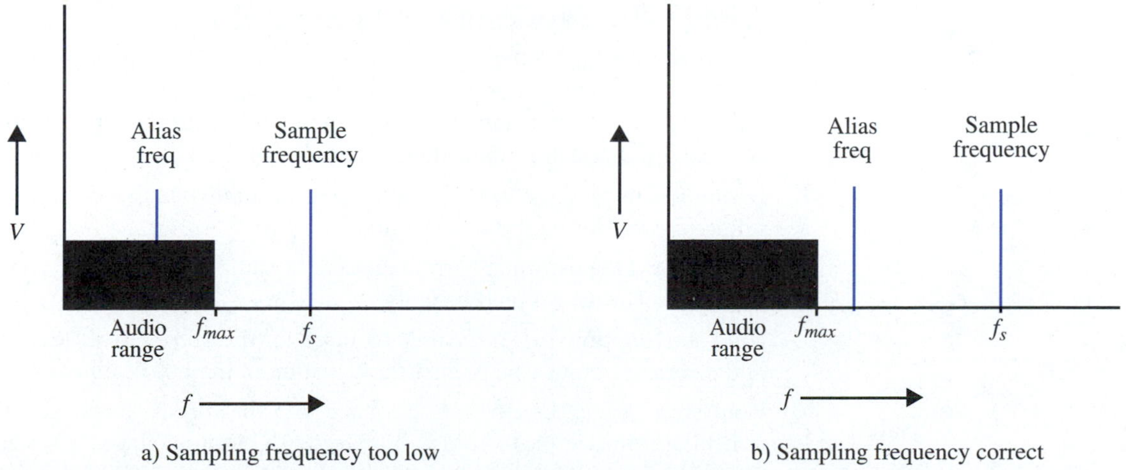

FIGURE 16–5 Sampling frequency effects on audio band

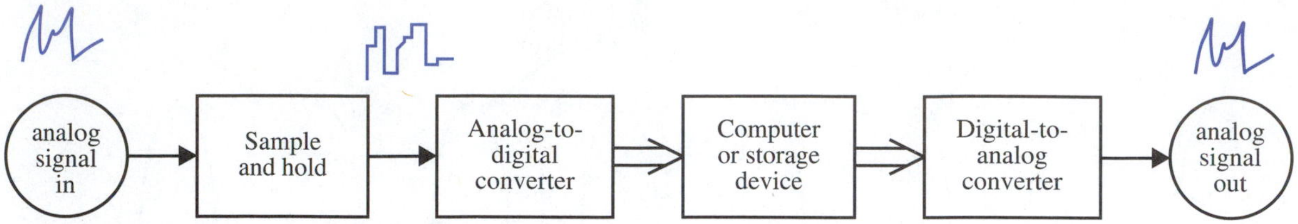

FIGURE 16–6 Complete signal conversion system

The output from the ADC is typically on parallel wires (indicated by the wide arrow), and the number of wires corresponds to the number of conversion bits of the ADC; for example, an eight-bit ADC would have eight output lines plus a return wire.

Data entering the computer would be stored in the computer memory for processing, which could, for example, be to determine whether predetermined levels have not been exceeded. In a music recording system, the digital data could be recorded on a compact disc.

After processing, or storage, the digital data are converted back to analog by a digital-to-analog converter (DAC). The output audio can be identical to the input or modified to suit particular requirements.

We shall now look in detail at the different devices that can perform the S & H, ADC, and DAC functions.

SAMPLE AND HOLD

The purpose of the sample and hold device is, again, to sample the analog input waveform and hold the sampled voltage steady while the ADC makes the conversion to digital.

There are several important terms associated with the S & H circuit that describe its operational capabilities. These terms are listed and defined here.

1) **Acquisition time** (t_a) is the time required to acquire a new analog input voltage after the sample command is given. Usually, the amount of voltage change from the previous sample is specified because the time it takes the input circuits to settle can be a function of the amount of change.
2) **Aperture time** (t_{ap}) is the time during which the input voltage must not change after the hold command is given, otherwise the held signal will be in error. It is the time from the initiation of the hold command to the time it takes to disconnect the input circuit from the hold storage device.
3) **Dynamic sampling error** (V_d) is the change in millivolts that occurs across the storage device during the aperture time.
4) **Gain error** is the percentage error between the sampled input signal and the output signal. The gain should be unity.
5) **Hold settling time** (t_{hs}) is the time required for the output to settle within 1mV of the correct output voltage after the initiation of the hold command.
6) **Hold step** (V_h) is the amount of voltage step that occurs in the output when switching from sample to hold with a steady DC input voltage. This step comes from feed-through from the logic circuitry.

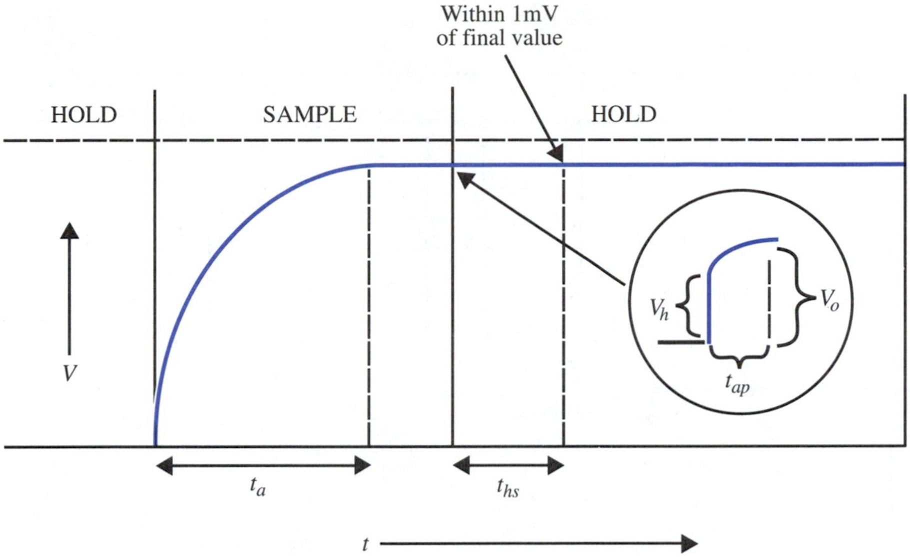

FIGURE 16–7 Sample and hold parameters

A graphical representation of these terms is given in Figure 16–7. During the sample time, the voltage change is exponential. This is because the sampled voltage is stored (held) with a capacitor. The sample time must be equal to or greater than the acquisition time.

A TYPICAL SAMPLE AND HOLD DEVICE

A device that can perform the sample and hold operation is the LF398. The internal block diagram for this device is shown in Figure 16–8.

Let us assume that the voltage reference is tied to logic ground for the discussion of the device operation. If now the logic input is set high, the U_3 comparator output goes high, closing the S/H switch, and puts the device in the sample mode.

The analog signal to be sampled is applied to the input terminal (Pin 3) and appears at the output of op-amp U_1. This op-amp has open loop gain until the output voltage (connected via R_a to inverting input of U_1) gets close to the input voltage (noninverting input of U_1) or diode D_1 or D_2 conducts (U_1 output 0.6V greater or less than the input). So initially, the output of U_1 is 0.6V different from the input and then becomes equal to the input when the output voltage catches up with the input voltage.

DETERMINING THE ACQUISITION TIME

With the sample and hold switch closed, the external hold capacitor charges to the output voltage of U_1 via the switch "on" resistance R_b. The output of U_2 initially overshoots (speeds up charging C_H) in response to the open loop output from U_1 and then follows the voltage on C_H. The time it takes the hold capacitor C_H to charge to the input voltage is related to both the value of R_b and C_H. It also is related to how close we want the capacitor voltage to

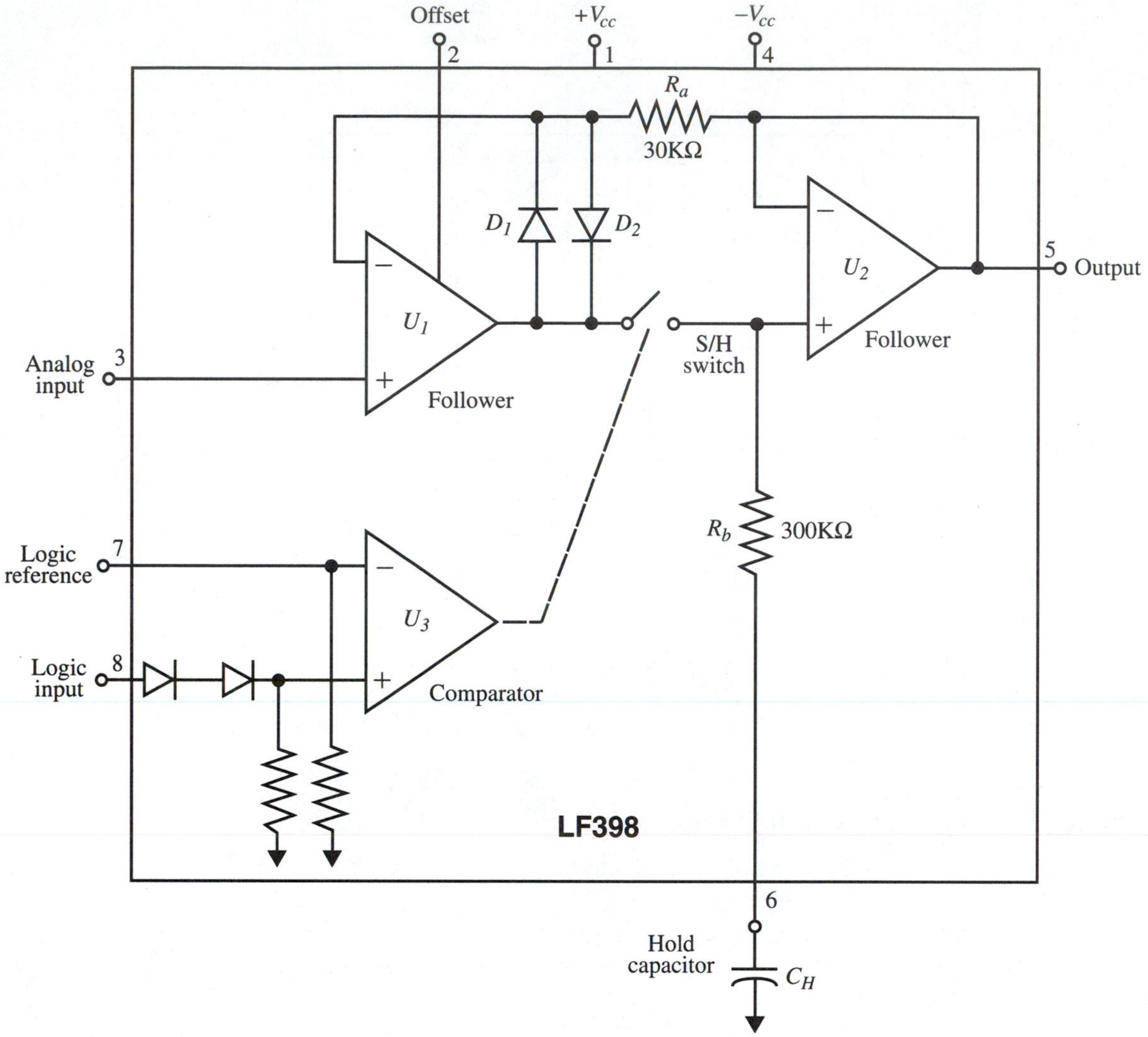

FIGURE 16–8 LF398 block diagram with external hold capacitor. *Courtesy National Semiconductor.*

be to the input voltage. If for simplicity we ignore the initial overshoot at the output of U_1, then the charging equation for the capacitor voltage V_H is

$$V_H = V_s[1 - \epsilon^{-(t_a/R_bC_H)}] \qquad \textbf{(equation 16–3)}$$

where V_s is the input voltage and t_a the acquisition time.

The smaller the term $\epsilon^{-(t_a/R_bC_H)}$, the closer the capacitor voltage V_H is to the input voltage V_s. If we want this term to go to zero, we have to wait an infinite amount of time. In a practical situation, we can assign some maximum value to this term, based on the allowable error between the input and the capacitor voltage, and compute the time required to reach this value. To do this, we solve for the acquisition time t_a.

$$t_a = -R_bC_H \ln(\text{Error}) \qquad \textbf{(equation 16–4)}$$

where

$$\text{Error} = \frac{(V_s - V_H)}{V_s}$$

EXAMPLE

Determine the acquisition time for an LF398 sample and hold circuit if the maximum allowable error is 0.1% (0.001) and the hold capacitor is 10nF.

SOLUTION

Using equation 16–4,

$$t_a = -300\Omega \times 10\text{nF} \times \ln(0.001)$$

$$t_a = 20.7\mu\text{s}$$

This is a conservative value, because the overshoot at the output of U_1 speeds the charging of C_H.

OTHER SPECIFICATIONS FOR THE LF398

The specifications for the LF398 at the indicated conditions are listed in Table 16–3. The dynamic sampling error indicates how much the voltage will change across a 10nF hold capacitor (C_H) with a positive-going slope at the input of 1mV/μs during the aperture time. The slope of the input indicates how much the input is changing during the conversion process. As expected, if the input is changing at twice the rate, the dynamic sampling error will double.

The hold settling time must be allowed to occur before the analog-to-digital conversion can be initiated.

Gain, dynamic sampling, and hold step errors must be considered, along with the acquisition error, when determining the total error assignment for the sample and hold device.

TABLE 16–3 LF398 Specifications

Parameter	*Conditions*	*Worst-Case Value*
Aperture time (t_{ap})	Negative-going input	200ns
Aperture time (t_{ap})	Positive-going input	150ns
Dynamic sampling error	1mV/μs slew rate, C_H = 10nF	−2mV
Gain error	25°C	0.01%
Hold settling time (t_{HS})	25°C	0.8μs
Hold step (V_H)	C_H = 10nF	2.5mV

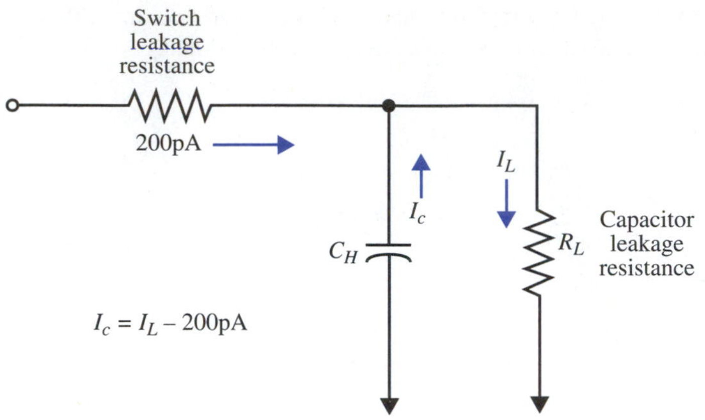

FIGURE 16–9 Hold capacitor circuit

CAPACITOR VOLTAGE CHANGE DURING HOLD TIME

During the hold time, the hold capacitor must retain its initial charge, otherwise the input to the ADC will be changing. However, all capacitors have some leakage and some exhibit a sag-back effect—where the voltage across the capacitor will drop by as much 0.2 percent after the charging source is removed. Capacitors made of polystyrene, polypropylene, and Teflon exhibit low values of sag-back and so are good for this application. The sag-back time for polypropylene capacitors is greater than 10ms. So, using this type of capacitor for a short hold-time application will minimize the effect of sag-back.

When in the hold mode, the LF398 can provide as much as 200pA of S/H-switch leakage current into the hold capacitor. The equivalent circuit for the capacitor during the hold time is shown in Figure 16–9.

If we assume that the switch leakage current is flowing out of the LF398, then the hold capacitor will be discharging if its leakage current is greater than 200pA and charging if its leakage current is less than 200pA. With the current directions as shown in Figure 16–9, the equation for the capacitor current is

$$I_c = I_L - 200\text{pA} \qquad \textbf{(equation 16–5)}$$

We can determine I_L by dividing the voltage on the capacitor by the leakage resistance.

There are conflicting requirements on C_H. For short acquisition times, the value of C_H should be small, but for long hold times, the capacitance should be high because the voltage change on the capacitor during the hold time is

$$\Delta V_H = \frac{I_c t_H}{C_H} \qquad \textbf{(equation 16–6)}$$

where t_H is the hold time.

EXAMPLE

Find the droop (ΔV_H) across the capacitor at the end of the hold time if the hold time is 1ms, the capacitor is 10nF, the capacitor leakage resistance is 1000MΩ, and the voltage being held is 10V.

SOLUTION

$$I_L = \frac{10\text{V}}{1000\text{M}\Omega}$$

$$I_L = 10\text{nA}$$

From equation 16–5,

$$I_c = I_L - 200\text{pA}$$

$$I_c = 10\text{nA} - 200\text{pA}$$

$$I_c = 9.8\text{nA}$$

Using equation 16–6,

$$\Delta V_H = \frac{I_c t_H}{C_H}$$

$$\Delta V_H = \frac{9.8\text{nA} \times 1\text{ms}}{10\text{nF}}$$

$$\Delta V_H = 0.98\text{mV}$$

Relative to the 10V on the hold capacitor, this represents an error of one part in 10,204 (10 bits). Again, this voltage change at the input to the ADC during conversion must be considered in the overall accuracy computations.

METER APPLICATION FOR A SAMPLE AND HOLD CIRCUIT

Another application where a sample and hold circuit is used is with meter circuits. It is difficult sometimes to hold a probe steady on a small test point while reading the meter. By using a switch on the voltmeter probe to activate a sample and hold circuit, the voltage is stored. Now the probe can be removed from the circuit, and the voltmeter reading will remain displayed.

Referring back to Figure 16–8, activating the switch on the probe would take the logic input high and put the S & H into the sample mode, allowing the capacitor to charge to the metered voltage. Releasing the switch would preserve the voltage on the hold capacitor.

ANALOG-TO-DIGITAL CONVERTER

The function of the sample and hold circuit was to hold the input steady while the conversion was being performed. The analog-to-digital converter (ADC) performs the actual conversion.

Earlier, we looked at the actual conversion process with the graphs in Figures 16–2, 16–3, and also Table 16–1. From the table, we saw that conversion generates a quantization error of one-half step. In other words , the same digital code will be provided at the output of the ADC for voltages as much as $\pm 1/2$ step from the center value. With our previous 1V steps, the quantization error was 0.5V.

Let us now determine the requirements for the ADC with respect to the number of digital bits needed and the required accuracy.

For an example, we shall assume that an analog input range from 0V to 5.12V is applied to an eight-bit converter. Since an eight-bit ADC has 256 steps, each step has a magnitude V_{st} of

$$V_{st} = \frac{5.12\text{V}}{256}$$

$$V_{st} = 20\text{mV}$$

The quantization error is then 20mV/2 or 10mV. So a $\pm$10mV analog input variation from the center value will still provide the same output digital code. If we express this variation as a percent error based on the full-scale input voltage, we have

$$\text{Full-scale percent error} = \frac{10\text{mV}}{5.12\text{V}} \times 100\% = 0.195\%$$

But if we express the variation as a percent error based on the minimum step input of 20mV, we have

$$\text{Minimum step percent error} = \frac{10\text{mV}}{20\text{mV}} \times 100\% = 50\%$$

As we can see from these results, quantization error is a problem with low-level input signals, and special techniques are used with analog signals that have a large dynamic range (ratio of largest amplitude signal over smallest), e.g., audio signals. The basic approach to solving this problem is to use nonlinear conversion, where the steps for low-level signals are smaller than those for larger signals. Since the conversion error is the step voltage divided by two, smaller steps mean smaller error.

In addition to the quantization error, ADCs have a specified accuracy in LSB (least significant bit). This accuracy must be added to the quantization error to determine the overall error.

$$\text{Total ADC LSB error} = \text{quantization (1/2LSB)} + \text{ADC accuracy} \qquad \textbf{(equation 16–7)}$$

EXAMPLE

The maximum input voltage to a 10-bit ADC is 10.24V. If the accuracy of the ADC is *1/4*LSB, find the volts variation of analog input that could provide the same output digital code.

SOLUTION

Using the relationship of equation 16–7,

$$\text{Total ADC bits error} = 1/2\text{LSB} + 1/4\text{LSB}$$

$$\text{Total ADC bits error} = 3/4\ \text{LSB}$$

A 10-bit ADC has 2^{10} steps or 1,024 steps.

Each step has a magnitude of

$$V_{st} = \frac{10.24\text{V}}{1024}$$

$$V_{st} = 10\text{mV}$$

This is the magnitude of the least significant bit.

For *3/4LSB*, the voltage range V_r is

$$V_r = 3/4\ \text{LSB} \times 10\text{mV/LSB}$$

$$V_r = 7.5\text{mV}$$

What this indicates is that a variation of ±7.5mV from the center value could provide the same output code. But a variation of more than 2.5mV from the worst case center value could also produce the next code! Another way of looking at this is that in order to guarantee that the correct code will appear at the output with this particular ADC, we have to limit the input variation to ±*1/4LSB*, or ±2.5mV.

TYPES OF ADCS

There are four basic types of analog-to-digital converters:

1) Successive approximation
2) Dual slope
3) Flash
4) Sigma-delta

The successive approximation ADC is used where reasonably fast conversion is required. Dual conversion is slow but offers excellent rejection of power supply noise. The flash type is used for fast changing analog signals. Sigma-delta conversion is the newest type and has high resolution.

SUCCESSIVE APPROXIMATION ADC

The basic block diagram for an eight-bit Successive Approximation Register (SAR) ADC is shown in Figure 16–10.

The overall SAR operation consists of successively comparing the analog input voltage to an internal reference voltage divider and, when a match is achieved between the input and a tap on the divider, an output digital word is generated.

Let us now look at the detailed operation of an SAR:

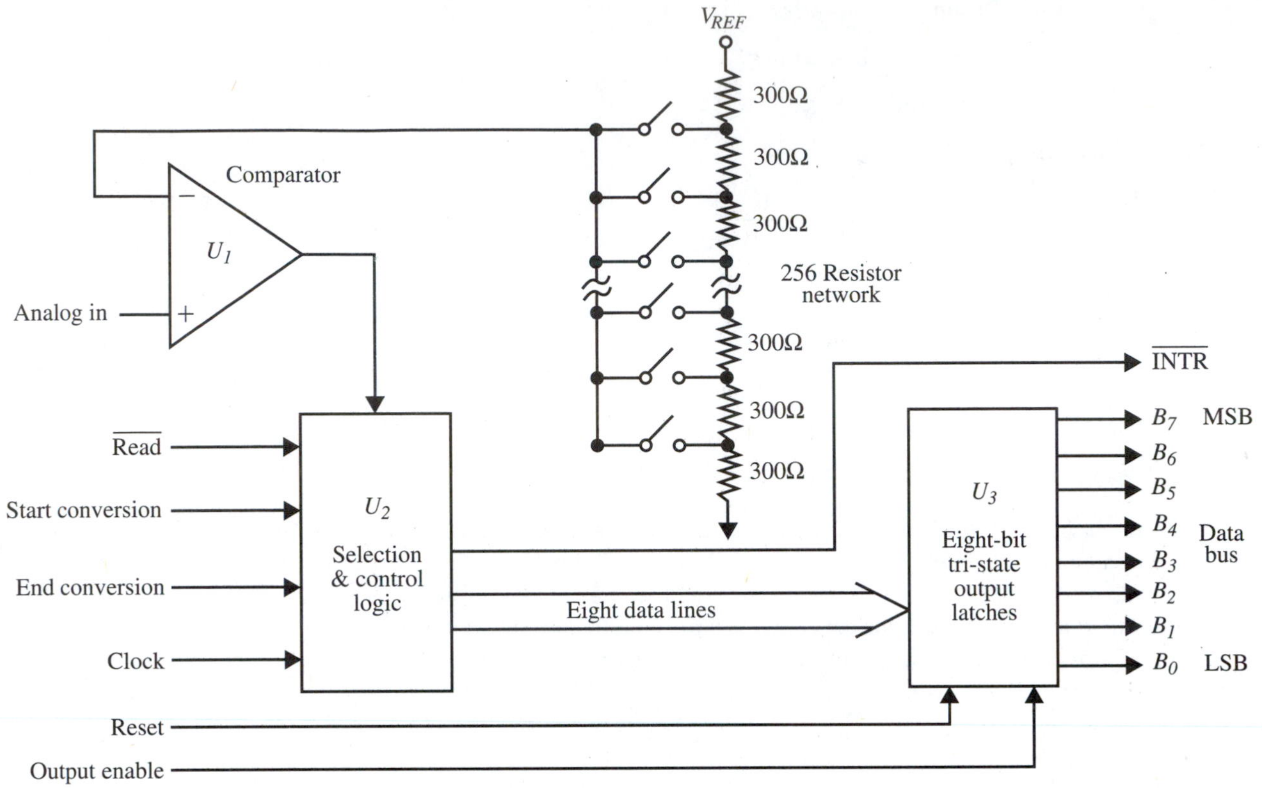

FIGURE 16–10 Successive approximation converter

- The analog signal is applied to the analog in terminal, and the conversion process is initiated by the start conversion signal along with the clock signal, with all bits in the eight-bit word used for the output digital data set to *0*.
- If we assume that V_{REF} is +5.12V, then there is a 20mV across each 300Ω resistor.
- The conversion starts with the closure of the center switch on the resistor network (only one switch closes at a time), which provides 2.56V (half of 5.12V) to the inverting input to the comparator. Also a *1* is placed in the MSB position of the eight-bit output word.
- If we assume the actual analog input is +3.5V, then the *1* will be retained in the eight-bit output word because 3.5V is greater than 2.56V. The switch midway between the middle and top of the divider will now be closed, and a *1* will now be placed in the second-most-significant position of the eight-bit output word.
- The voltage at this new position will be

$$(5.12\text{V} \times 0.75) = +3.84\text{V}$$

which is greater than the 3.5V analog input, so the output of the comparator goes low, causing the *1* to be removed from the second-most-significant position of the eight-bit output word.

TABLE 16–4 Binary-to-Decimal conversions (Eight-Bit)

Bit #	B_7	B_6	B_5	B_4	B_3	B_2	B_1	B_0
2^n	2^7	2^6	2^5	2^4	2^3	2^2	2^1	2^0
Decimal	128	64	32	16	8	4	2	1

- A switch is now closed between 0.75 and 0.5 of the resistive divider, yielding a voltage of 3.2V, and a *1* is placed in the third-most-significant position of the eight-bit word.
- Since the divider voltage is less than the analog input, the output of the comparator will be high, and the *1* bit will be retained in the third-most-significant position of the eight-bit word.
- The process continues (takes up to eight samples—one for each output data bit) until the sampled voltage is the closest match to the input; which for our example will be when the top of the 175th resistor is sampled (175 × 20mV).
- At the completion of the sampling, the eight-bit code is transferred from the selection and control logic to the eight-bit tri-state output latches. Activating the tri-state output enable line will cause these data to be available on the eight-line output data bus.

If we want to determine which digital code will appear on the output data bus with the +3.5V input used for our example above, we need to look at the binary representation of the step at which the match between the analog input and divider voltage occurred. Remember, the match was at the top of the 175th resistor, so we need the binary representation of 175. The significance of each bit is as presented in Table 16–4.

The most significant bit is $2^7 = 128$ and the least is $2^0 = 1$. We can determine the binary equivalent of 175 by going through the following process:

- The number 175 is greater than 128, therefore the 2^7 bit is a *1*. If we subtract 128 from 175, we are left with 47.
- The remainder (47) is less than 64, so the 2^6 bit is *0*.
- The 2^5 bit (32) is less than 47, so this bit is *1*. If we subtract 32 from 47, we are left with 15.
- The remainder (15) is less than 16, so the 2^4 bit is *0*.
- The 2^3 bit (8) is less than 15, so this bit is *1*. If we subtract 8 from 15, we are left with 7.
- The remainder (7) is greater than 4, so the 2^2 bit is *1*. If we subtract 4 from 7, we are left with 3.
- The remainder (3) is greater than 2, so the 2^1 bit is *1*. If we subtract 2 from 3, we are left with 1.
- Finally, 2^0 is equal to the remainder (1), so this bit is a *1*.

The complete representation in binary for 175 is

10101111

Note that the three most significant bit states agree with those determined when we went through the first three steps of the successive approximation process.

EXAMPLE

Determine the binary code if the match between the analog input and the resistive divider voltage occurred at the 202 tap on the divider.

SOLUTION

Using Table 16–4, and mathematically following the process outlined above, we obtain

$$202 - 128 = 74 \qquad 2^7 \text{ bit} = 1$$

$$74 - 64 = 10 \qquad 2^6 \text{ bit} = 1$$

$$10 - 32 = -22 \qquad 2^5 \text{ bit} = 0$$

$$10 - 16 = -6 \qquad 2^4 \text{ bit} = 0$$

$$10 - 8 = 2 \qquad 2^3 \text{ bit} = 1$$

$$2 - 4 = -2 \qquad 2^2 \text{ bit} = 0$$

$$2 - 2 = 0 \qquad 2^1 \text{ bit} = 1$$

$$0 - 1 = -1 \qquad 2^0 \text{ bit} = 0$$

The binary output is then

11001010

EXAMPLE

Determine the output binary code for an eight-bit ADC if the analog input is 2.44V and the reference divider input voltage (V_{REF}) is 10.24V.

SOLUTION

The voltage V_{st} for each of the 256 (2^8) steps is

$$V_{st} = \frac{10.24\text{V}}{256}$$

$$V_{st} = 40\text{mV}$$

To find the number of steps N, we divide the voltage per step into the analog input.

$$N = \frac{2.44\text{V}}{40\text{mV}}$$

$$N = 61$$

Converting this decimal number to binary, using the process from the previous example, we get the following binary output:

00111101

AN ACTUAL SUCCESSIVE APPROXIMATION CONVERTER

An actual successive approximation ADC converter is the ADC 0801, shown in block diagram form in Figure 16–11.

This converter is an eight-bit device and is microprocessor compatible—which means it has inputs and outputs provided to allow control-signal transfer between the ADC and the computer. The chip operates off of a nominal +5V power supply and has a total unadjusted error of ±1/4bit. The external clock can be as high as 1.46MHz and is divided down by a factor of 8 to provide the internal clock. We shall go through the operational steps with reference to Figure 16–11:

- The converter is initialized by the microprocessor taking the chip select ($\overline{CS}$) and the write ($\overline{WR}$) lines low (the bar over the abbreviation indicates active low). This sets the start flip-flop U_2, and the resulting high on the Q output resets the shift register (counter) U_4. It also sets the interrupt flip-flop U_5 and inputs a *1* to flip-flop 1 (U_3).
- The clock transfers the input *1* to the output of flip-flop 1, which is connected to the D input of the eight-bit shift register U_4 and also the AND gate U_6.
- When the clock signal goes high, and either $\overline{CS}$ or $\overline{WR}$ goes high, the U_6 gate output resets the start flip-flop U_2. This removes the reset signal from the eight-bit counter, and the *1* at the input to this counter from flip-flop 1 can now be shifted through the counter to start the conversion process.
- The counter stage outputs cause switches on the ladder network to close, and as described previously, the comparisons with the input signal are made until the best match is obtained.
- After eight clock pulses (64 external clock pulses), the *1* at the input to the counter will have reached the counter output. This provides a *1* to an input of AND gate U_7—which, along with the high on the $\overline{Q}$ output of Latch 1 (U_8), causes the AND gate output to go high, which puts a high on the transfer line. A high on this XFER line causes the digital word, corresponding to the analog input, to be transferred from the ladder network to the tri-state output latch.
- The *1* at the *D* input to Latch 1 (U_8) is shifted by the next clock pulse to the output of the latch. This causes the $\overline{Q}$ output of the latch to go low, setting the interrupt flip-flop, which in turn makes the $\overline{INTR}$ line go low. A low on the $\overline{INTR}$ line signals to the microprocessor that data are available.
- When the microprocessor is ready to take the data, it causes the $\overline{CS}$ and $\overline{RD}$ (read) lines to go low, which enables the tri-state output latch and the digital word to appear on the data bus. With $\overline{CS}$ and $\overline{RD}$ low, the interrupt flip-flop is reset and the $\overline{INTR}$ line goes high.
- If both the $\overline{CS}$ and $\overline{WR}$ lines go low while a conversion is in process, the conversion stops; and when either $\overline{CS}$ or $\overline{WR}$ goes high, a new conversion starts.

The timing diagram of Figure 16–12 shows the signals on the various ADC lines for the conversion sequence.

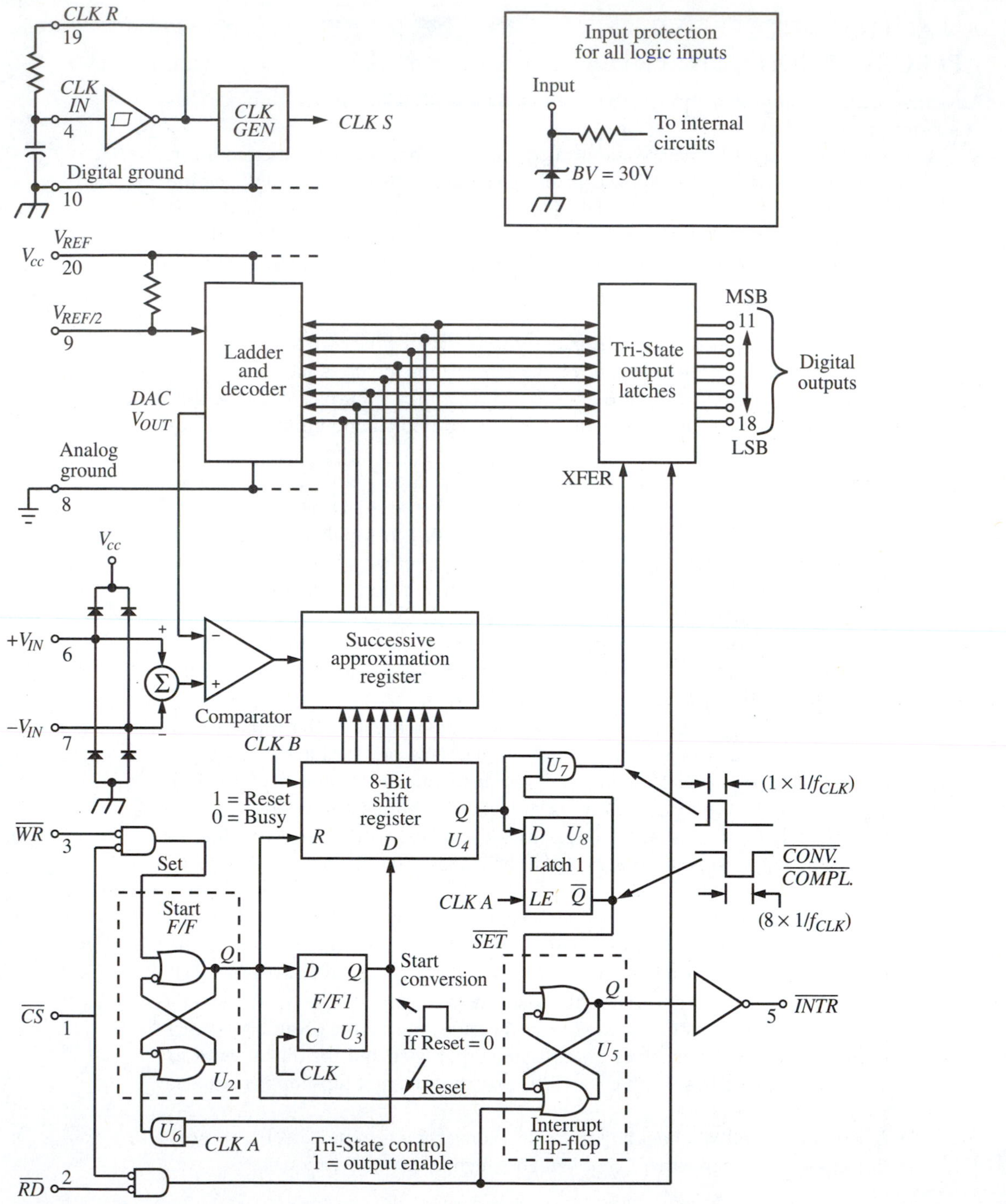

FIGURE 16–11 Block diagram of ADC 0801. *Courtesy National Semidconductor.*

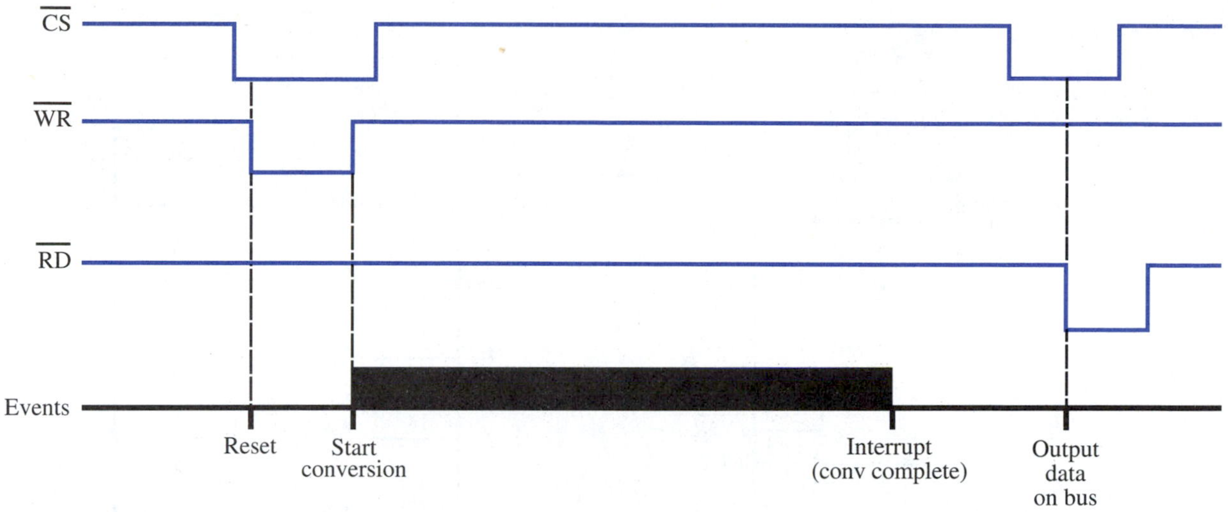

FIGURE 16–12 ADC 0801 timing diagram

FREE-RUNNING ADC

The converter can be operated in a free-running mode by tying the $\overline{CS}$ line low and connecting the $\overline{INTR}$ output to the $\overline{WR}$ terminal. Now when the $\overline{INTR}$ goes low at the end of conversion, it takes the $\overline{WR}$ low and a new conversion is initiated. A chip test circuit connected in this fashion is shown in Figure 16–13.

Referring to Figure 16–13, the analog input signal is connected to V_{IN^+}, and write line $\overline{WR}$ is momentarily grounded to start the conversion process.

The power supply voltage used is +5.12V. Input voltages on any other terminal of the chip should not exceed the power supply value or be less than 0V (requires TTL level clock), or the device will be damaged (consider spikes riding on any input signal). The differential input V_{IN^+} and V_{IN^-} can be used to reduce common mode noise, or the V_{IN^-} input can be used to subtract a voltage from the analog input, e.g., removal of an offset voltage. In the circuit shown, the V_{IN^-} terminal is grounded, which provides an unbalanced input.

Additional flexibility is provided by allowing the ladder reference voltage to be changed. Let us assume that the analog input voltage variation is from 0V to 1V. If we use a 5.120V reference, then our input varies only over about one-fifth of the available range, so our resolution is reduced by a factor of 5. But, by connecting an external reference voltage of 0.5V to the $V_{REF}/2$ terminal (Pin 9), we force the full-scale reference to be 1V and recover our resolution. If we don't connect a voltage to the $V_{REF}/2$ terminal, then one-half of the supply voltage (2.560V) will appear at this terminal.

EXAMPLE

Determine the voltage to be connected to the $V_{REF}/2$ terminal of an ADC 0801 if the maximum input voltage is 3V.

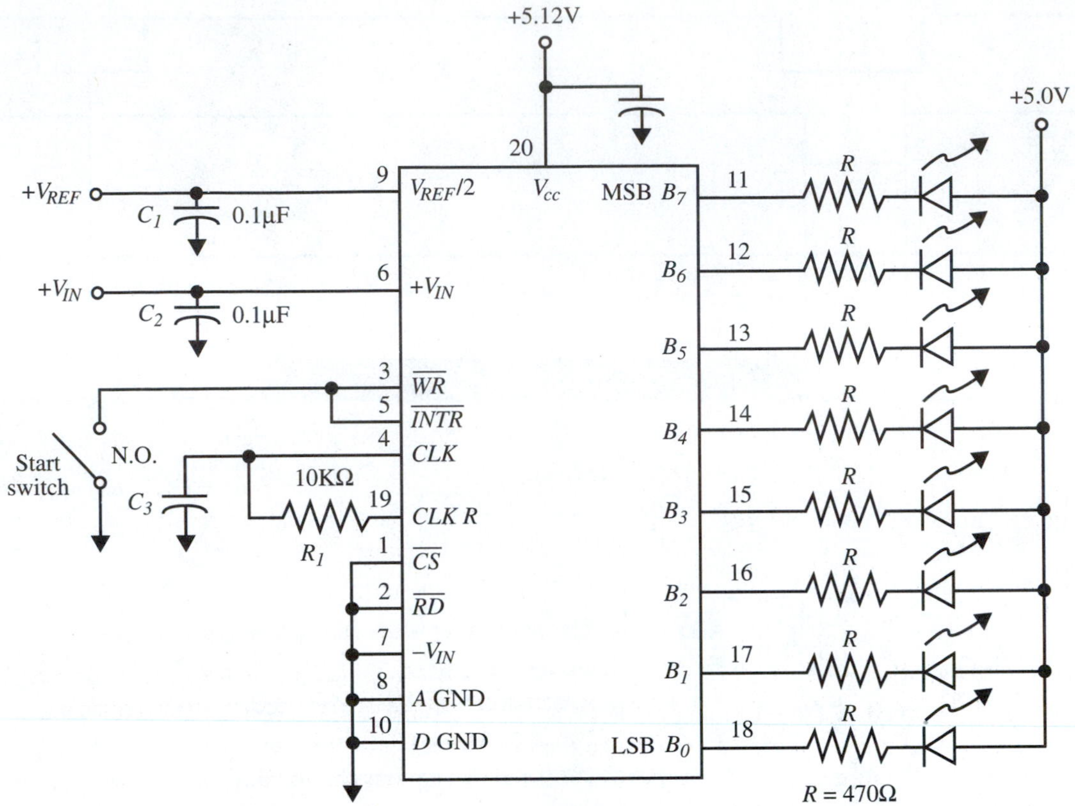

FIGURE 16–13 Free-running analog-to-digital converter

SOLUTION

To get a full-scale reading of 3V, we need to have a voltage of 3V/2 or 1.5V connected to the $V_{REF}/2$ terminal.

DETERMINING THE ADC OUTPUT STATE

In Figure 16–13, an LED is connected to each digital output. These LEDs will illuminate when the output goes low (sink connection). We can determine the output state by treating the illuminated LED as a *0*.

Another way of decoding the digital output is by dividing the eight bits into two hex characters—the four most significant bits and the four least significant bits. This hex code can be converted to the equivalent decimal number *N*. We can now determine what the analog input voltage had to be to generate this code by multiplying *N* by the volts per step (V_{st}).

EXAMPLE

If the output code is 10101111 for the circuit of Figure 16–13, find the analog input voltage that generated this code.

SOLUTION

The code 10101111 breaks down into 1010 (10 in decimal, A in hex) and 1111 (15 in decimal, F in hex), which is AF in hex.
Converting from hex to decimal

$$(10 \times 16) + (15 \times 1) = 175$$

Since the voltage per step is

$$5.12\text{V}/256 = 20\text{mV}$$

the input analog voltage must be

$$175 \times 20\text{mV} = 3.5\text{V}$$

The clock signal for the ADC 0801 can be provided externally (by connecting to the CLK terminal) or internally by connecting a timing resistor R_1 and capacitor C_3, as shown in Figure 16–12. The equation for the clock frequency is

$$f_{CLK} = \frac{1}{1.1R_tC_t} \qquad \textbf{(equation 16–8)}$$

With the components shown in Figure 16–13, the clock frequency is about 600KHz.

The time for the complete conversion process time t_c is a maximum of 73 clock pulses (64 plus up to 9 start pulses) times the period of 600KHz.

$$t_c = 73 \times 1/f_{CLK} \qquad \textbf{(equation 16–9)}$$

$$t_c = 73 \times 1/600\text{KHz}$$

$$t_c = 122\mu\text{s}$$

The higher the clock frequency, the shorter the conversion time.

OTHER SUCCESSIVE APPROXIMATION CONVERTERS

Although the ADC 0801 is a common eight-bit analog-to-digital converter, there are newer successive approximation ADCs that have faster conversion times. For example, the Maxim MAX165, which is an eight-bit analog-to-digital converter, has a conversion time of 5μs when a 4MHz clock is used. Conversion of full-scale input signals up to 50KHz is possible with this device. This converter is also microprocessor compatible and has an on-board sample and hold circuit.

The Analog Devices ADC-908 (another eight-bit analog-to-digital converter) has a conversion time of 6μs when used with a 1.35MHz clock. This device is also microprocessor compatible.

FLASH ANALOG-TO-DIGITAL CONVERTER

We discussed, previously, the number of voltage samples required for a particular waveform in order to get a good representation of the signal. Recall that the Nyquist criterion requires a minimum of two samples for a sine wave.

What if we wanted to convert a 500KHz sine wave to digital? The minimum number of samples is determined by twice the frequency for a sample rate of 1MHz. With this sample frequency, the conversion time must be no longer than 1/1MHz or 1μs. If we try using an ADC 0801 converter with a clock frequency of 1MHz, the conversion time (using equation 16–9) is 73μs. If we add to this a sample and hold circuit hold time of 20μs, we have a total conversion time of 93μs. This would allow two conversions for a 5KHz sine wave, but not for the 500KHz we wish to convert. We need, then, to find a faster converter.

Flash converters are able to perform conversions in less than one microsecond; so, using one of these devices would allow us to perform the required conversion of the 500KHz (or higher frequency) signal. The fast speed of a flash converter is achieved by performing the conversion in one clock period rather than the 73 required by the ADC 0801. We shall see how this is achieved by looking at the detailed flash converter circuit description.

FLASH CONVERTER CIRCUIT OPERATION

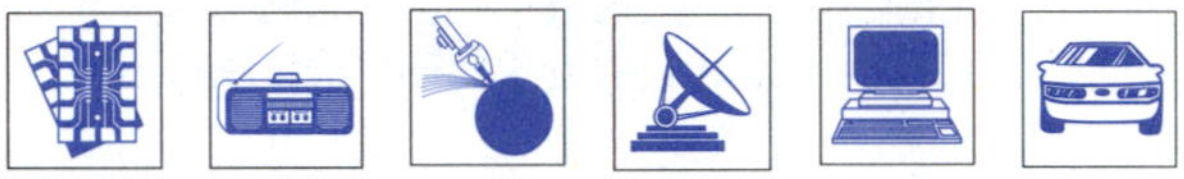

A functional block diagram of a flash converter is shown in Figure 16–14. As with the ADC 0801, a 256 resistor divider is used—but instead of sampling the divider in a step-by-step sequence, all samples are made at the same time.

A phase 1 (ϕ_1) and phase 2 (ϕ_2) clock are generated in the circuit with a 180° phase difference between the two signals (U_3 provides the phase shift). The particular state of these two clock signals generates an auto-balance or a sample phase by closing the appropriate switches (indicated by a crossed circle in the diagram).

Auto-Balance Phase

The conversion starts with the phase 1 clock high, which closes the switches connecting the divided reference voltage to the amplifier (comparator) input capacitors C_a and also shorts the output of each amplifier to its input. This is called the auto-balance phase and serves to balance the comparators at their trip points. During this phase, the input capacitors C_a charge to the voltage taken off the reference voltage divider V_{RN} (see Figure 16–15 for a single path). The second set of amplifiers (U_b) are also balanced during this phase to remove any voltage offset.

Sample Phase

To start the sample phase, the ϕ_1 clock goes low and the ϕ_2 clock goes high. So, all the switches closed by the ϕ_1 clock are opened, and the ϕ_2-controlled switches that connect the input capacitors to the analog input are closed.

At the completion of the auto-balance phase, the input capacitor can be thought of as a battery charged to the divider voltage (V_{RN}). If we consider the input voltage (V_s) also as a battery, then the voltage appearing at the high impedance input to the amplifier is the algebraic sum of these two voltages. Figure 16–16 shows the equivalent circuit.

As shown in Figure 16–16, the voltage at the input to the amplifier (V_a) is

$$V_a = V_s - V_{RN} \qquad \textbf{(equation 16–10)}$$

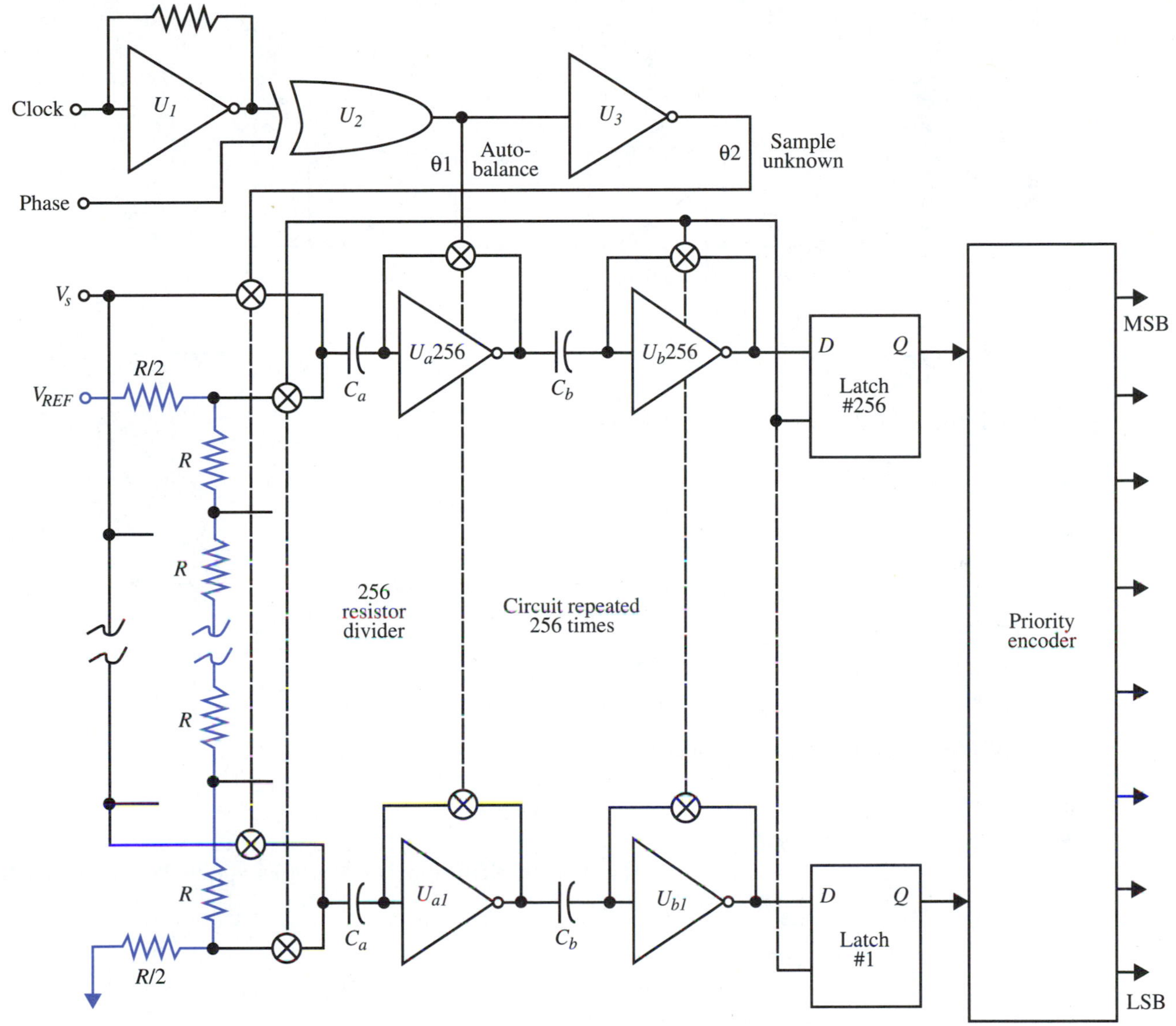

FIGURE 16–14 Flash converter block diagram

This voltage into the amplifier U_a can be positive or negative or even zero (unlikely). If it is positive, a *1* will enter the latch; if negative, a *0*.

V_{RN} can be expressed with reference to the resistive divider of Figure 16–14 as

$$V_{RN} = (V_{REF} \times N/256) - (V_{REF}/512) \qquad \textbf{(equation 16–11)}$$

The last part of the equation is the half-step voltage that is the result of the *R*/2 resistor at the bottom of the divider.

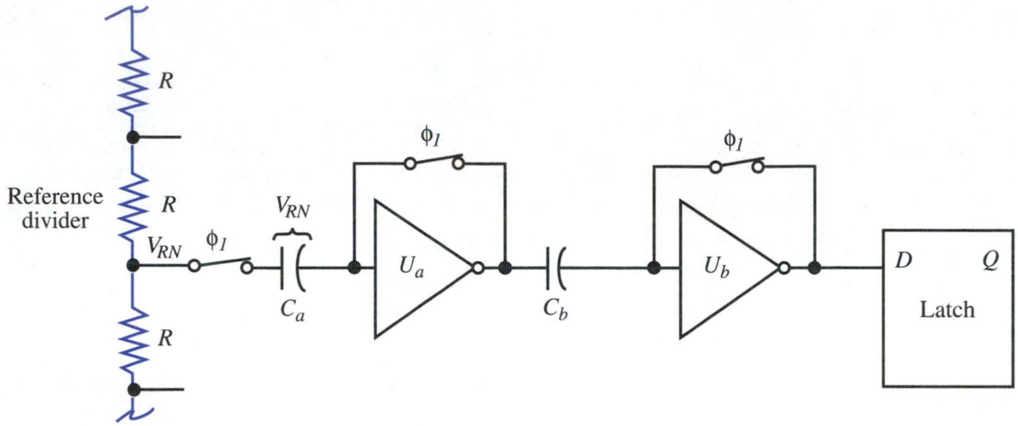

FIGURE 16–15 Auto-balance for 1 of 256 paths

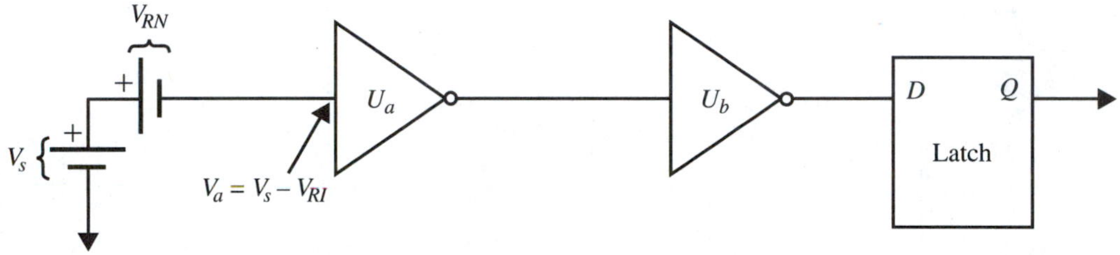

FIGURE 16–16 Sample phase input equivalent circuit

EXAMPLE

Find the input voltage V_a at the 96th amplifier (N is 96) if V_{REF} is 4V and the analog input V_s is 3V.

SOLUTION

First we must find the voltage V_{R96} at the 96th tap, using equation 16–11; then we can use equation 16–10 to find V_s.

$$V_{RN} = (V_{REF} \times N/256) - (V_{REF}/512)$$

$$V_{R96} = (4\text{V} \times 96/256) - (4\text{V}/512)$$

$$V_{R96} = 1.4922\text{V}$$

Now, using equation 16–10,

$$V_a = V_s - V_{RN}$$

$$V_a = 3\text{V} - 1.4922\text{V}$$

$$V_a = +1.5078\text{V}$$

Since this number is greater than zero, it will cause a *1* to be shifted into the 96th latch.

EXAMPLE

Find the voltage divider tap in an eight-bit flash converter at which the voltage V_a changes from positive to negative if V_{REF} is 4V and the analog input V_s is 3V.

SOLUTION

By using equation 16–10, we try to find the tap where V_a is 0.

$$V_a = V_s - V_{RN} = 0$$

Then solving for V_{RN},

$$V_{RN} = V_s$$

$$V_{RN} = 3\text{V}$$

Now solving for N in equation 16–11,

$$N = 256\left(\frac{V_{RN}}{V_{REF}} + \frac{1}{512}\right)$$

$$N = 256\left(\frac{3\text{V}}{4\text{V}} + \frac{1}{512}\right)$$

$$N = 192.5$$

Since this is not an integer, V_a will not be zero at any amplifier input. It will be negative at the 192 input and positive at the 193 input.

During the next auto-balance cycle, the 256 bits of information in the data latches are converted by the priority encoder (see Figure 16–14) into an eight-bit data word. The encoder provides a digital output code corresponding to the highest (N number) decoder input with a *1* state.

The sequence described here is with the Phase input in Figure 16–14 to the "exclusive or" (high output when both inputs are different) circuit in the high state. Holding this line low will reverse the ϕ_1 and ϕ_2 operations.

PRACTICAL FLASH CONVERTER

An actual eight-bit flash converter is the Harris CA3318. This device has a conversion time of 66.6ns when a clock frequency of 15MHz is used. This allows conversion of signals up to 7.5MHz. Power supply voltage can range up to a maximum +7.5V, and the reference voltage can range from +4V to +7.5V. For best accuracy, it is suggested that the reference voltage not exceed the power supply voltage by more than 1.5V. When using a +5V supply, total power consumption is less than 150mW.

The operation of this chip is similar to the previous description of the flash converter. A ninth digital output bit is provided, however, to signal an overflow (over-range) condition.

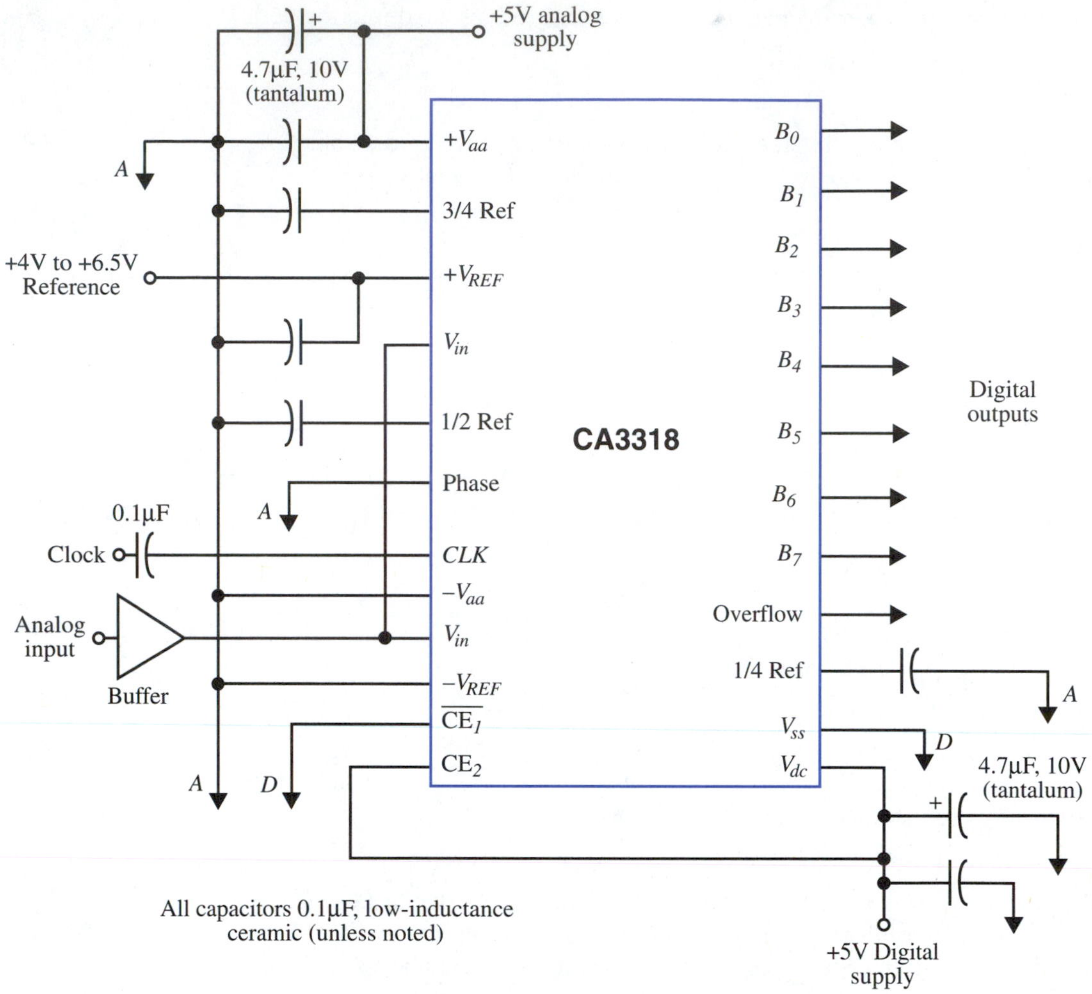

FIGURE 16–17 Typical CA3318 operational configuration. *Courtesy National Semiconductor.*

Figure 16–17 shows a typical application configuration for a CA3318. Notice in the diagram that the analog and digital power supplies and grounds are kept separate. This is to prevent noise interaction between the analog and digital circuits within the chip, which can be a problem at the high operating frequency for this converter.

When $\overline{CE_1}$ is low, or CE_2 is high, the tri-state buffers at the digital outputs are enabled, and data are put on the external bus. With CE_2 high, the overflow bit is also enabled.

HALF-FLASH CONVERTER

A problem with the flash converter is that the 256 input comparators require chip space, and they represent a rather heavy load for the signal source to drive. A modified flash con-

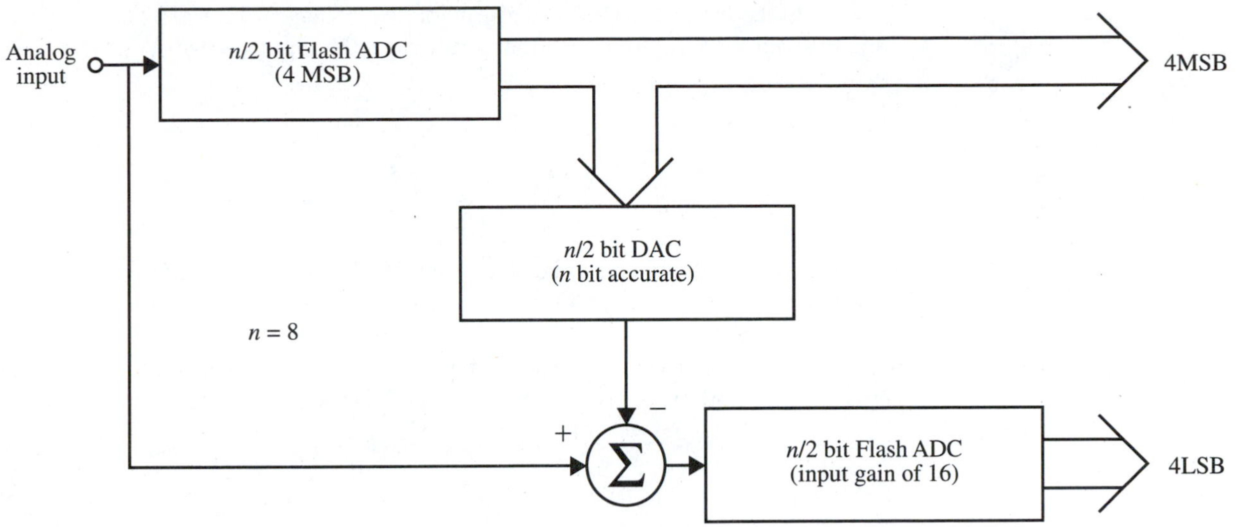

FIGURE 16–18 Basic eight-bit half-flash converter

verter that uses less comparators and is capable of greater resolution, at the cost of increased complexity and longer conversion time, is the half-flash.

A simplified block diagram of a half-flash converter is shown in Figure 16–18. The input signal is applied to the four most significant bits (MSB) flash converter and to a summing circuit. The four-digital-bit output of the flash converter is converted by the digital-to-analog converter (DAC) back to an analog voltage representing the 4 MSB. This voltage is subtracted from the input signal, and the voltage entering the four least significant bits (LSB) flash converter is an amplitude containing only the 4 LSB information of the input voltage.

I think this can be made clearer by using a numerical example. Let us assume that the reference voltage is 5.12V and the analog input signal is 3.5V and that we are using an eight-bit half-flash converter. From our previous examples, we would expect the output code with this input to be 10101111.

The four-bit MSB converter has a 16-step reference divider, and each voltage step V_s is

$$V_s = 5.12\text{V}/16$$

$$V_s = 0.32\text{V}$$

Dividing this voltage into our input voltage to find the N number,

$$N = 3.5\text{V}/0.32\text{V}$$

$$N = 10.93$$

Since we are computing the tap on the reference voltage divider, we have to stay with integers, and so N is 10 (input voltage is not high enough for the 11th tap). Converting 10 to digital, we have **1010** for the most significant bits.

Feeding this code into the DAC will give an output voltage of 10 × 0.32V, or 3.2V. Subtracting this voltage V_{DAC} from the input V_s provides the input voltage (V_{LSB}) to the 4 LSB flash converter.

$$V_{LSB} = V_s - V_{DAC}$$

$$V_{LSB} = 3.5 - 3.2\text{V}$$

$$V_{LSB} = 0.3\text{V}$$

To maximize sensitivity and to be able to use the same 5.12V reference for the 4 LSB ADC, this voltage needs to amplified by 16. The actual voltage that is converted is now 16 × 0.3V, or 4.8V.

Since the 4LSB converter has the same step voltage of 0.32V (sometimes the same converter is used for both operations), the N number for this converter is

$$N = 4.8\text{V}/0.32\text{V}$$

$$N = 15$$

Converting this N number to digital we get **1111**. Combining the most significant bits with the least significant bits, we have **10101111**, which agrees with our previous computations.

ACTUAL HALF-FLASH CONVERTERS

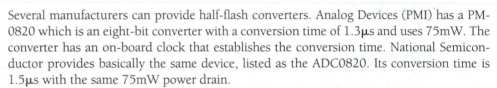

Several manufacturers can provide half-flash converters. Analog Devices (PMI) has a PM-0820 which is an eight-bit converter with a conversion time of 1.3μs and uses 75mW. The converter has an on-board clock that establishes the conversion time. National Semiconductor provides basically the same device, listed as the ADC0820. Its conversion time is 1.5μs with the same 75mW power drain.

Figure 16–19 shows a microprocessor interface circuit for the ADC0820.

ADC0820 MODE CONTROL

The ADC0820 has a mode control pin to allow flexibility in the read/write process. With the MODE pin grounded, the converter is set for the Read (RD) mode. In this configuration, a complete conversion is done by taking $\overline{\text{RD}}$ low until data appear at the output—signified by the $\overline{\text{INTR}}$ line going low. If the MODE pin is tied high, the ADC is set up for the WR-RD mode. Now a conversion starts when the $\overline{\text{WR}}$ goes low.

There are two options for reading the output data. With the interface driven option, the interrupt signal is sensed before the $\overline{\text{RD}}$ line is taken low. The other option provides a faster conversion by taking the $\overline{\text{RD}}$ line low 600ns after the rising edge of the $\overline{\text{WR}}$ signal.

DUAL SLOPE CONVERTER

As mentioned previously, the dual slope analog-to-digital converter has good noise immunity but is restricted to measuring slow varying or DC inputs. This is accomplished by allowing the analog signal to input typically for a time equal to a multiple of the period of the

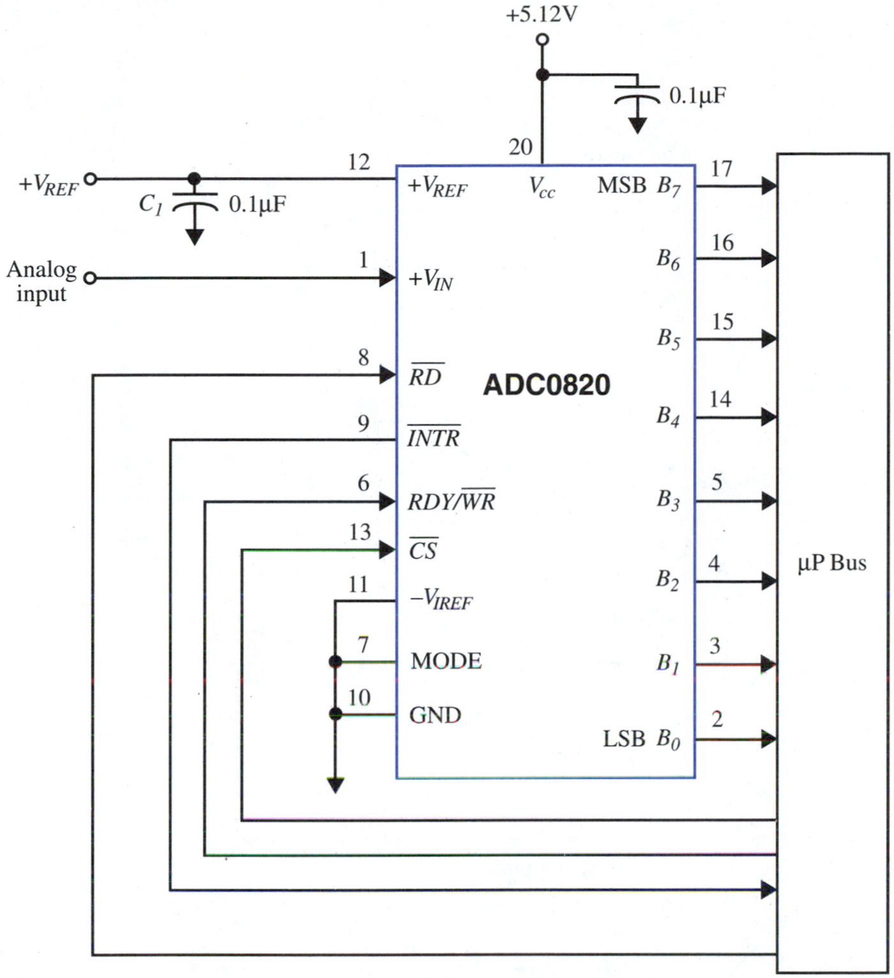

FIGURE 16–19 ADC0820 WR-RD mode configuration. *Courtesy National Semiconductor.*

AC line frequency (16.67ms). Line noise riding on the positive and negative alternations of the AC line sine wave will tend to cancel over this time interval.

A simplified block diagram for a dual slope converter is shown in Figure 16–20. If a positive analog input V_s is applied to the integrator input, the integrator output voltage will fall in accordance with the equation (see Chapter 7)

$$V_i = \frac{-V_s t}{RC}$$

At the end of the sample period $n \times 16.67ms$ (where n is a positive integer), the integrator output voltage will be

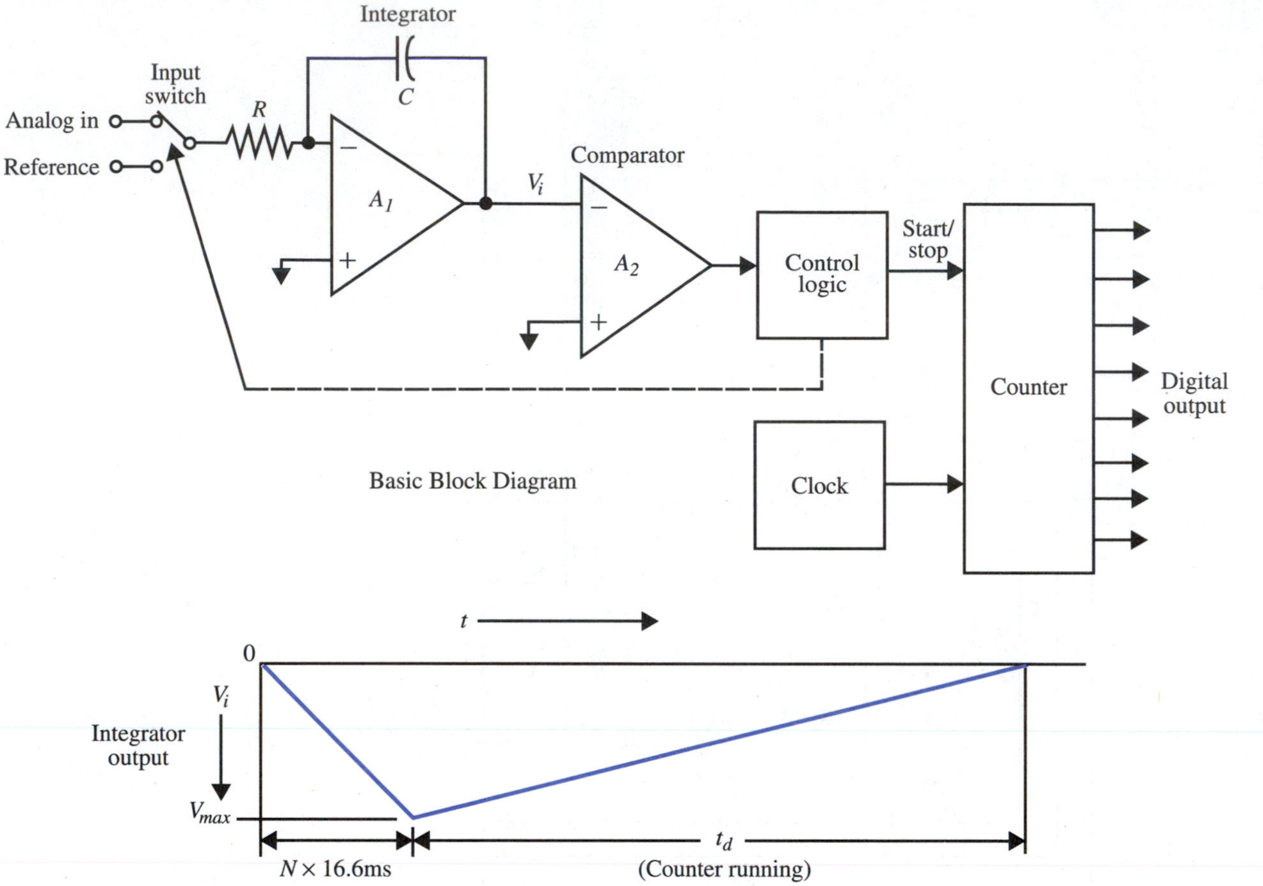

FIGURE 16–20 Dual slope ADC

$$V_{max} = \frac{-V_s(n \times 16.67\text{ms})}{RC} \quad \textbf{(equation 16–12)}$$

With the completion of this sample period, the internal counter is started; also at this time, the integrator input is switched to the reference (negative voltage if the input is positive). The negative input to the integrator causes the output to now ramp positive from the minus V_{max} value.

When the integrator output voltage ramps up to 0V (takes time t_d), the comparator output changes state, and the control logic stops the counter (final count N). The number N, stored in the counter at this time, is proportional to the time t_d the integrator takes to ramp back to 0V (actually N equals t_d/T_c, where T_c is the clock period) and N is converted to its binary equivalent to provide the ADC output.

If we double the input voltage V_s, then from equation 16–12, we see that V_{max} doubles. Since the positive-going slope is fixed, it will take twice as long to ramp back to 0V, and so t_d doubles. Considering the positive slope equation with the integrator input now $-V_{REF}$,

$$V_{max} = \frac{V_{REF} \times t_d}{RC} \qquad \textbf{(equation 16–13)}$$

If we equate equations 16–12 and 16–13, we get

$$\frac{-V_s(n \times 16.67\text{ms})}{RC} = \frac{V_{REF} \times t_d}{RC}$$

Solving for t_d,

$$t_d = \frac{V_s(n \times 16.7\text{ms})}{V_{REF}} \qquad \textbf{(equation 16–14)}$$

Finally, since $(n \times 16.7ms/V_{REF})$ is a constant, which we shall call K, and the count (N) in the counter is proportional to t_a, then we can say

$$N = K \times V_s \qquad \textbf{(equation 16–15)}$$

So if we are given the K factor for a particular dual slope converter, we can predict the count N for a particular input voltage.

The dual slope converter gets it name from the two output slopes of the integrator during a conversion. This type of converter is much used in DMMs, and in this application the count is converted to seven-segment display readouts.

EXAMPLE

Find the count N for a dual slope converter if the reference voltage is 1V, the input voltage is 0.5V, the clock is 30KHz, and there are 2 readings per second.

SOLUTION

First, we need to find t_d, using equation 16–14.

$$t_d = \frac{V_s(n \times 16.7\text{ms})}{V_{REF}}$$

$$t_d = \frac{0.5\text{V}(2 \times 16.7\text{ms})}{1\text{V}}$$

$$t_d = 16.7\text{ms}$$

Now since

$$N = t_d/T_c$$

and

$$T_c = 1/30\text{KHz}$$

$$N = 16.7\text{ms} \times 30\text{KHz}$$

$$N = 501$$

EXAMPLE OF A DUAL SLOPE CONVERTER

A typical dual slope converter is the Harris ICL7136. This device is a 3-1/2 digit C-MOS display (the half digit allows only a *1* or *0* in the most significant bit position). It includes seven-segment decoders, LSD drivers, voltage reference, and a clock. The full-scale input voltage can be as high as 2V (1.999V). A small, required supply current of 100μa at 9V means the device can be used in battery-powered meters.

In our description, we shall divide the device into the analog section and the digital section.

ANALOG DESCRIPTION OF ICL7136

Operation of the analog section is divided into three phases:

1) Auto-Zero Phase (**A/Z**)
2) Signal Integrate Phase (**INT**)
3) Reference Integrate or De-Integrate Phase (**DE**)

The block diagram for the analog section is shown in Figure 16–21. Switches on the block diagram are indicated with a crossed circle with the phase abbreviations alongside. Each set of switches is closed during the particular phase, i.e., all A/Z switches are closed during the auto-zero phase. We will now look at the operations during each phase.

Auto-Zero Phase

Several things happen during the auto-zero phase. Analog input HI is disconnected from the chip and internally shorted to the analog input LO, and the reference capacitor C_{REF} is charged to the external reference voltage. Finally, the output of the comparator is shorted to the inverting input of the integrator. This allows the capacitor C_{AZ} to charge to a voltage created by the total voltage offset of the buffer, integrator, and comparator.

Signal Integrate Phase

The signal integrate phase connects the analog input to the integrator for a time corresponding to 1,000 clock pulses. At the end of this phase, a voltage is developed on the integrating capacitor (C_{INT}) that is proportional to the input voltage. The capacitor C_{AZ} subtracts out any offset error from this voltage.

De-integrate Phase

During the de-integrate phase, the reference voltage replaces the analog voltage as an integrator input. The reference voltage stored on capacitor C_{REF} is used rather than the reference source directly to allow for ease of polarity reversal.

If the analog input is positive, then the reference input must be negative in order to discharge the integrating capacitor (and vice versa). This is controlled by the reversing configuration of the DE(+) and DE(−) switches. The sense of the comparator output tells the logic which set of switches to close. It can take a period of time from 0 to 2,000 clock pulses

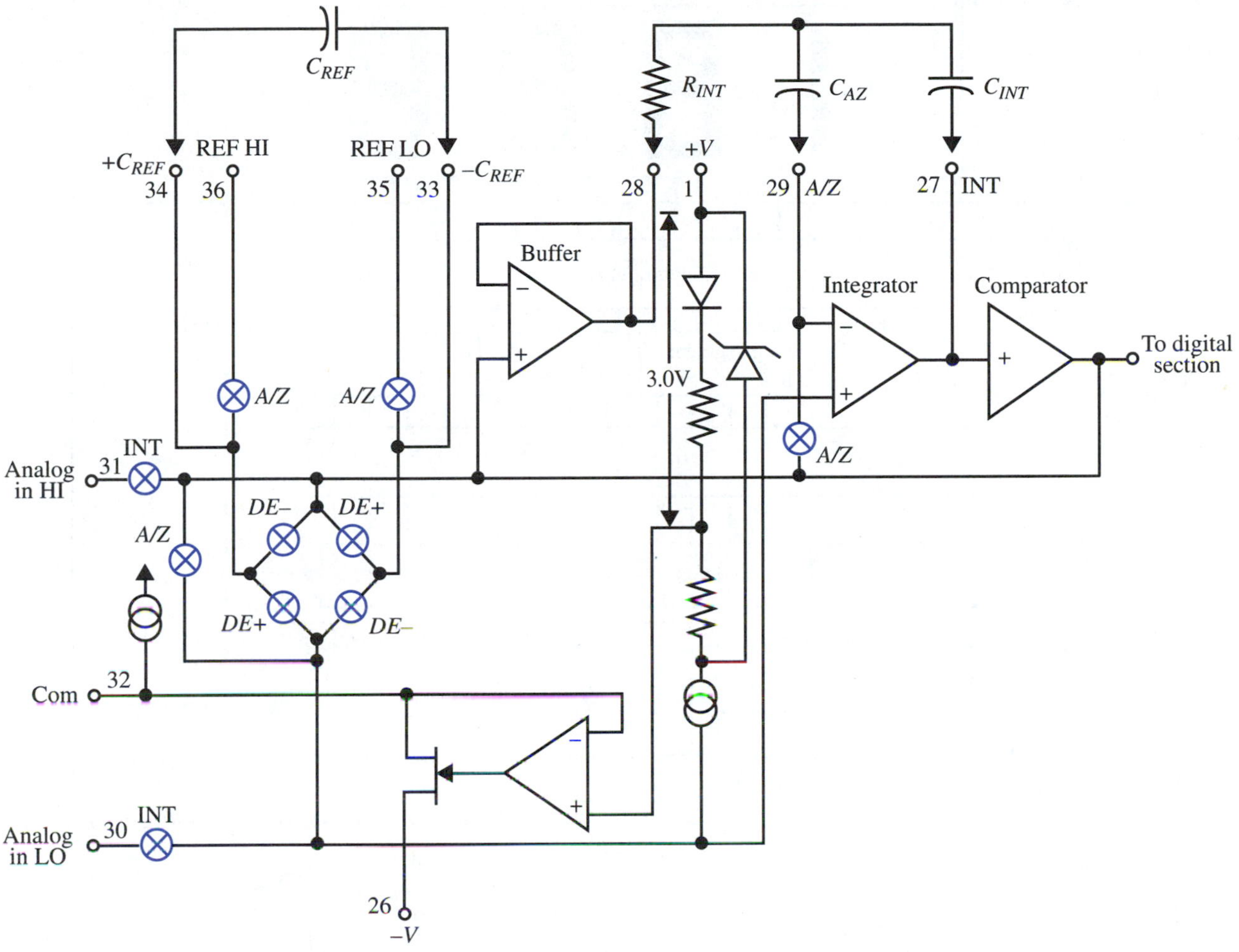

FIGURE 16–21 Analog section of ICL7136. *Courtesy Harris Semiconductor.*

for the integrating capacitor to discharge, with the time being proportional to the analog input (V_s). The actual digital reading displayed is

$$\text{Digital Count } N = 1000(V_s/V_{REF})$$

With a maximum count of 2,000 at a full-scale V_s of 2V, we can see from this equation that V_{REF} must be 1V. If we desire a full-scale value less than 2V, we simply reduce V_{REF}.

DIGITAL SECTION OF ICL7136

The digital section of the converter is shown in Figure 16–22.

Carrying on the description from the analog section, the comparator output indicates to the logic control the polarity of the analog input and when the integrating capac-

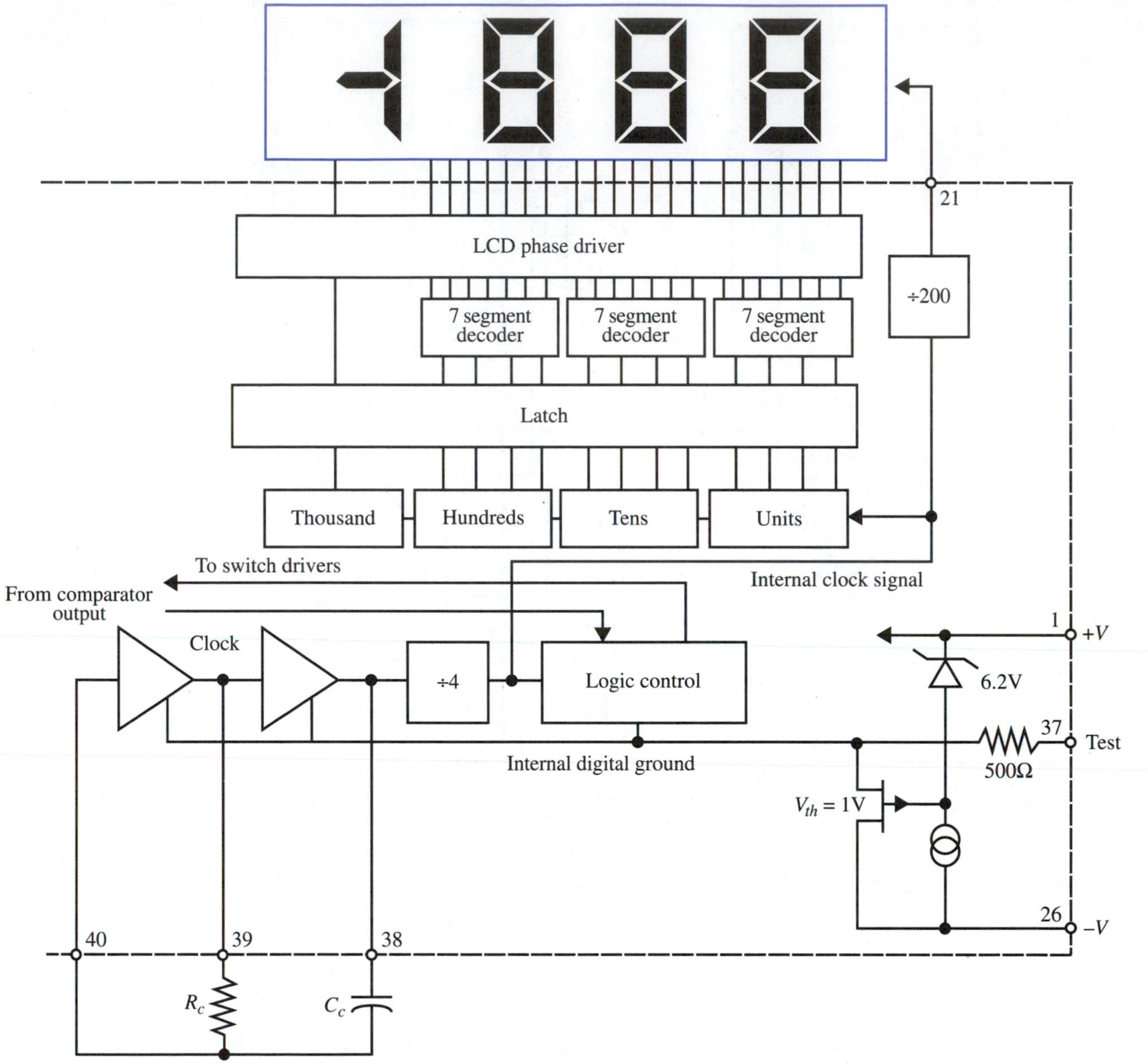

FIGURE 16–22 Digital section of ICL7136. *Courtesy Harris Semiconductor.*

itor voltage is zero (comparator output changes state), which terminates the de-integrate phase.

The clock signal is derived from an external oscillator connected to Pin 40, or a crystal between Pins 39 and 40, or an RC oscillator using Pins 38, 39, and 40, as shown in Figure 16–22. Division of this oscillator by four is performed to obtain the internal clock signal. It takes 4,000 counts of the internal clock (or 16,000 oscillator cycles) for a complete conversion cycle. The cycle consists of

1) signal integrate phase, one-fourth of the total—which is 1,000 counts
2) reference de-integrate phase—0 to 2000 counts
3) auto-zero phase, the remainder of 1,000–3,000 counts

Again, it is recommended that for maximum rejection of 60Hz line pickup, the signal integrate time period (1,000 counts) should be a multiple of 16.67ms (the period of 60Hz). Since this integrate time period is one-fourth of the total, the shortest conversion time is $4 \times 16.67ms$, or 66.67ms. With this time, we would have 1/66.67ms, or 15 conversions per second. The external oscillator runs for 16,000 clock pulses during a conversion, which means the highest oscillator frequency, corresponding to the shortest conversion time, is

$$f_{max} = \frac{16{,}000}{66.67\text{ms}}$$

$$f_{max} = 240\text{KHz}$$

The required oscillator frequency f_{osc} for n conversions per second is then

$$f_{osc} = \frac{240\text{KHz} \times n}{15} \qquad \textbf{(equation 16–16)}$$

EXAMPLE

Find the external oscillator frequency for an ICL7136 converter if it is required to have four conversions a second (n is 4).

SOLUTION

Using equation 16–16,

$$f_{osc} = \frac{240\text{KHz} \times n}{15}$$

$$f_{osc} = \frac{240\text{KHz} \times 4}{15}$$

$$f_{osc} = 64\text{KHz}$$

Referring back to Figure 16–22, the control logic allows the clock signal to be counted during the de-integrate phase. At the end of this phase, the accumulated count is stored in the latch. The stored count is converted into a seven-segment code and applied to the individual liquid crystal segments via the drivers.

To prolong the life of the liquid crystal display, an AC voltage must be developed between the segment and the crystal back plane. This is accomplished by applying to the back plane the divided down (by 200) internal clock signal, which generates a 5V square wave. If, for example, the external oscillator frequency is 48KHz (3 conversions per second), the frequency of the signal applied to the back plane is $48KHz/(4 \times 200)$, or 60Hz.

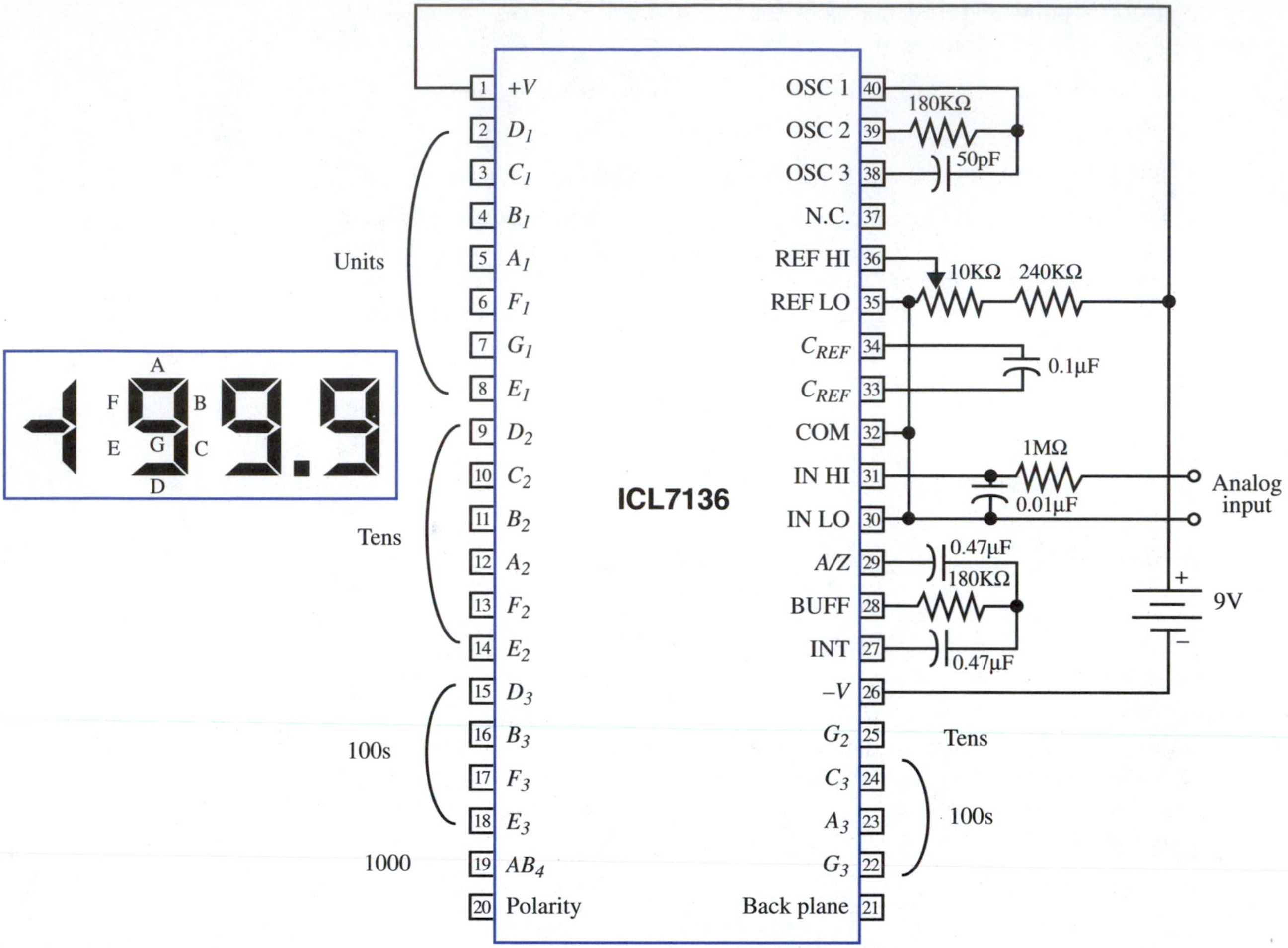

FIGURE 16–23 ICL7136 200mV meter circuit

If the LCD segments are driven at the same amplitude and frequency as the back-plane signal and are in phase with the back-plane signal, the segment will be off (white). When the signals are out of phase, the segment will be on (black).

A meter application of the ICL7136 is shown in Figure 16–23. The outputs to drive the seven-segment displays are divided into units, tens, hundreds, and thousand (half segment). The letters within each grouping refer to the individual segment—for example, A1 will activate the top segment in the right-most display.

With the reference set for 100mV, the full-scale reading is 200mV.

Looking at the components selected for this circuit:

- The 1MΩ resistor and the 0.01μF capacitor between the analog input lines serve as a noise filter.

- An integrating resistor of 180KΩ is recommended for the full-scale reading of 200mV. With 2V full scale, the resistor would have to be increased to 1.8MΩ to maintain the same charging current.
- For 3 readings per second (48KHz oscillator), the integrating capacitor (C_{INT}) is 0.047μF. If we reduced the readings to 1 per second (16KHz oscillator), C_{INT} would have to be increased by a factor of 3 to 0.15μF (inverse relationship).
- A 48KHz clock (f_{osc}) is established from the approximate equation:

$$f_{osc} \approx 0.45/R_tC_t \qquad \textbf{(equation 16–17)}$$

With a recommended C_t of 50pF, the R_t schematic value of 180KΩ is obtained.

- The auto-zero capacitor (C_{AZ}) should have a value of 0.47μF for minimum system noise.
- Finally, a range of 0.1μF–1.0μF is allowable for the reference capacitor—with the 1μF value recommended when the REF LO pin is not tied to IN LO and a 200mV full scale is used.

SIGMA-DELTA ANALOG-TO-DIGITAL CONVERSION

In our look at the analog-to-digital conversion process, we arbitrarily divided the input signal into 256 steps—which meant we were using an eight-bit converter. But in order to recognize low-level signals, we may have to make the steps smaller, and that means that more bits are required for the conversion process. Table 16–5 lists the number of bits required for different types of analog input signals. Notice in the table that the bits required is independent of the frequency bandwidth of the signal (greater bandwidth requires higher sampling rate), but proportional to the dynamic range.

The dynamic range is defined as 20 times the log of the ratio of the largest signal encountered over the smallest encountered. For example, digital audio has a dynamic range of 100db. The ratio is then

$$\text{Ratio} = \text{Inv Log (dynamic range/20)}$$

$$\text{Ratio} = \text{Inv Log (100 /20)}$$

$$\text{Ratio} = 100{,}000$$

TABLE 16–5 ADC Bit Requirements for Various Signals

Application	*Signal Bandwidth*	*Dynamic Range*	*ADC Bits Required*
Seismology	10Hz	146db	24
Digital audio	20KHz	100db	17
Speech processing	4KHz	74db	12
V.32 Modems	4KHz	74db	12
Ultrasound	15MHz	60db	10
Radar	5MHz	74db	12
Broadband receivers	5MHz	86db	15

This means the largest signal we encounter in audio can be 100,000 times greater than the smallest. To determine the bits required for this dynamic range, we assume that the smallest step in the conversion process is equal to the smallest signal. We convert from ratio to bits by using the equation

$$\text{Digital Bits Required} = \frac{\text{Log Ratio}}{\text{Log } 2} \qquad \textbf{(equation 16–18)}$$

$$\text{Digital Bits Required} = \frac{\text{Log } 100000}{\text{Log } 2}$$

$$\text{Digital Bits Required} = 16.6$$

Since we require integers, we use 17 bits.

The bit requirements listed in Table 16–5 are difficult to implement with the ADCs we have studied up to this point. To appreciate the problem, using a successive approximation converter for digital audio would require a 131,072 (2^{17}) resistor divider network.

A converter that has the capability for high digital bit conversion is the sigma-delta ADC. This is the newest type of converter and is rapidly becoming the ADC of choice for high-bit conversion. In basic operation, a sigma-delta ADC digitizes an analog signal using a one-bit ADC at a very high over-sampling rate (over-sampling is sampling at a rate higher than the Nyquist value). Using over-sampling, noise shaping, and digital filtering (decimation), the effective resolution (bits) is increased. The problem of filtering out the alias generating signals above the signal band is tremendously reduced with over-sampling because the filter slope doesn't have to be so steep (see Figure 16–5).

BASICS OF A SIGMA-DELTA ADC

The block diagram for a first-order sigma-delta is shown in Figure 16–24.

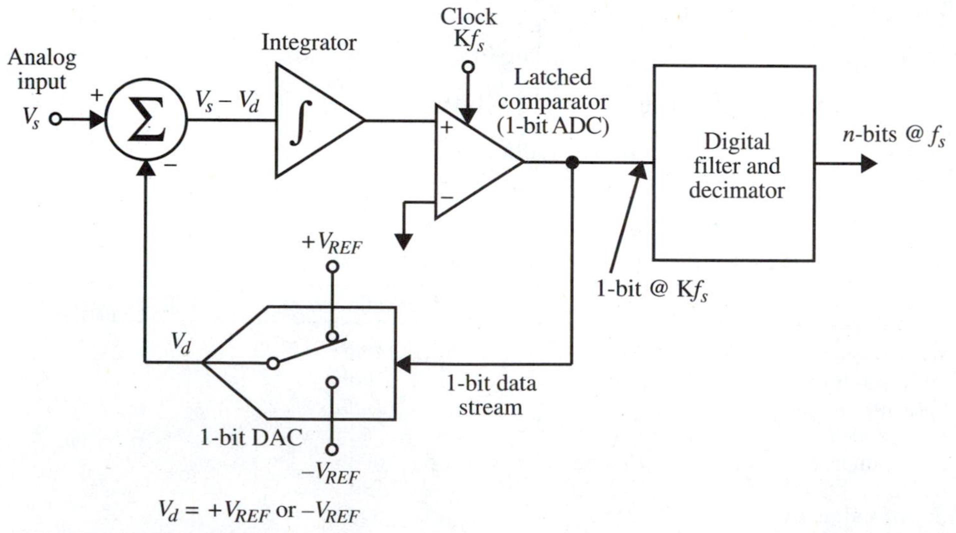

FIGURE 16–24 First-order sigma-delta ADC

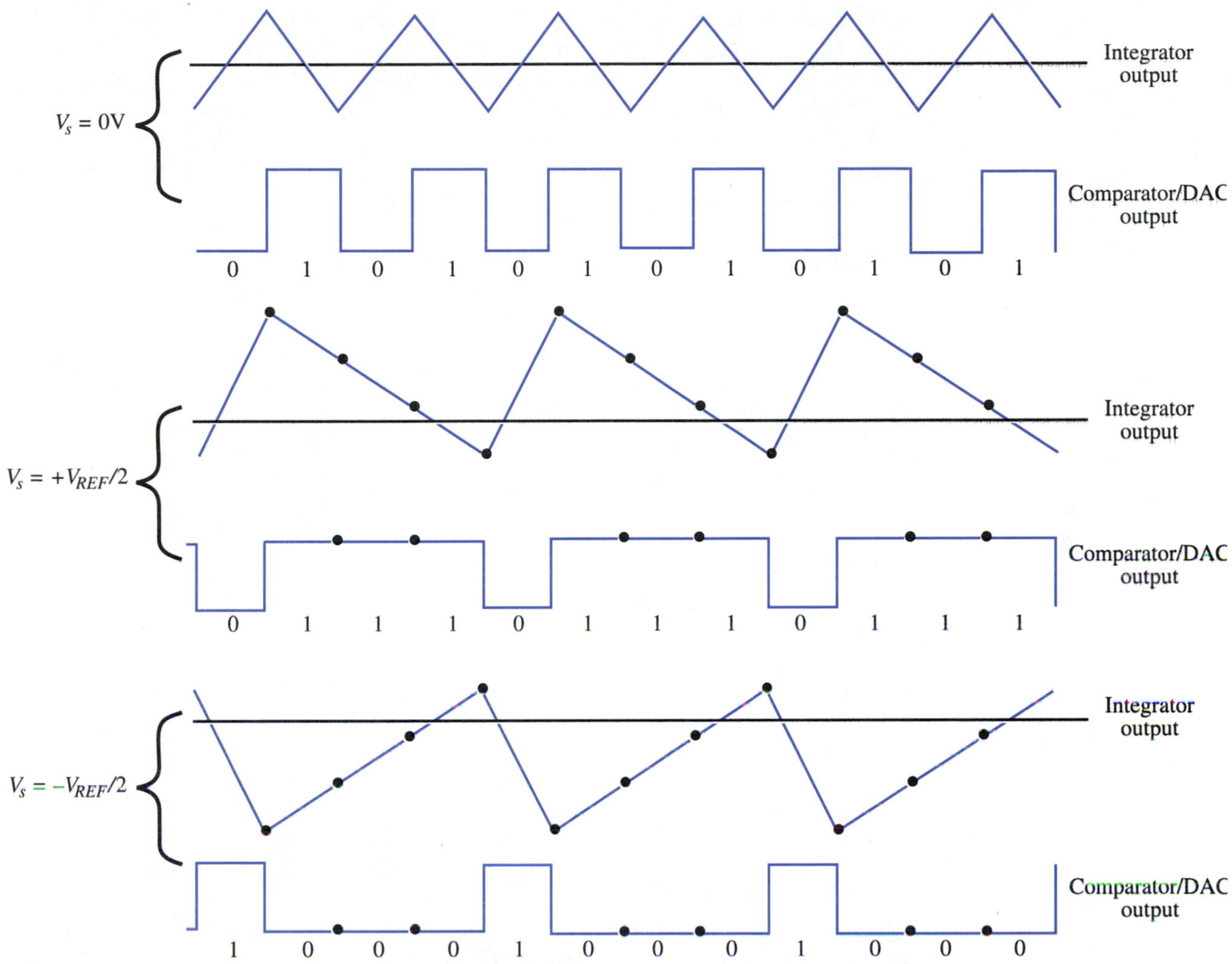

FIGURE 16–25 Sigma-delta modulator waveforms

The latched comparator compares the output of the integrator output to ground potential when activated by the clock signal. The integrator integrates the difference $V_s - V_d$ between the input analog signal V_s and the output of the DAC, which is V_d. Remember the slope of the signal out of an integrator is proportional to the amplitude of the input signal. When the comparator output signal is high, the DAC output is $+V_{REF}$, and when it is low, it is $-V_{REF}$.

The operation can best be described with reference to the waveforms at each point in the circuit. Figure 16–25 shows the waveforms with a DC input voltage of 0V, with the input equal to $+V_{REF}/2$, and finally with the input equal to $-V_{REF}/2$.

With the input V_s at 0V, the initial $-V_{REF}$ input from the DAC causes the integrator output to ramp positive until a clock-initiated comparison with 0V is made. This causes the

DAC output to change state to $+V_{REF}$, and the integrator now ramps negative. The output from the comparator with this analog input is a series of *1*'s and *0*'s. If we take this output and average four pulses, we would end up with 2/4 of the amplitude of each pulse to represent 0V input. Averaging four clocked output pulses gives an output resolution of 2 bits (2^2). Averaging eight pulses yields a resolution of 3 bits and so on. The greater the number of pulses that are averaged, the greater the effective converter resolution. But, more pulses averaged means a slower conversion time. This process of digital averaging the pulses is called **decimation.**

With the input V_s of $+V_{REF}/2$, the middle waveforms of Figure 16–25 are generated. Now when the comparator output is low, the output of the DAC is $-V_{REF}$, and subtracting this signal from the input signal provides $+3/2V_{REF}$ into the integrator. The integrator positive output slope is now 3/2 greater.

When the comparator output is high, the signal into the integrator is now $-V_{REF}/2$, and the negative slope is 1/4 as steep. Since it takes longer for the output to ramp negative to 0V, the comparator output stays high for a longer period of time—giving three *1*'s rather than two. The four-bit average of the comparator output is now 3/4.

Changing the input to $-V_{REF}/2$ inverts the waveforms and gives a bit pattern of a single *1* over the four bits. Now the average is 1/4.

If the number of *1*'s at the comparator output is counted over a long period of time, the counter output will represent the digital value of the input. This is the basis for the analog-to-digital conversion.

The example inputs given above gave predictable outputs that did not require many clock cycles (4 max) in order to determine the digital equivalent of the input. With voltages slightly different from the ones we used, the comparator output could pass through hundreds of clock pulses before the output digital pattern repeats. The smaller the step away from the values we used, the greater the number of pulses for a repeated pattern, and the higher the bits of resolution.

AN ACTUAL SIGMA-DELTA ADC

Motorola is one of several companies that supply sigma-delta ADCs. We will look at their DSP56ADC16 16-bit converter. This device provides output data at a rate of 100,000 per second when using 64 times over-sampling to provide a dynamic range of 96db (16–bits; see equation 16–18). If the data output rate is increased to 400,000 per second, the resolution drops to 12 bits. Unlike the other ADCs we have studied, the DSP56ADC16 digital output is in a serial rather than parallel form—which means the 16 bits of output data are time-shared on a single output wire, with the MSB appearing first and the LSB last.

Figure 16–26 provides the internal block diagram of the DSP56ADC16. The analog circuits block contains the integrator, comparator, and DAC, shown in detail in Figure 16–24. The one-bit sampling rate is 6.4MHz.

As described previously, decimation performs a digital average of the one-bit data from the analog section to increase the resolution. It also filters out noise that could be aliased back into the desired analog signal frequency range. The comb filter samples and digital averages 4,096 bits from the analog circuit at a rate of one-sixteenth the 6.4MHz rate, or 400,000 times per second—which means that 256 bits of the total of 4,096 change per sample. This provides a resolution of 12 bits.

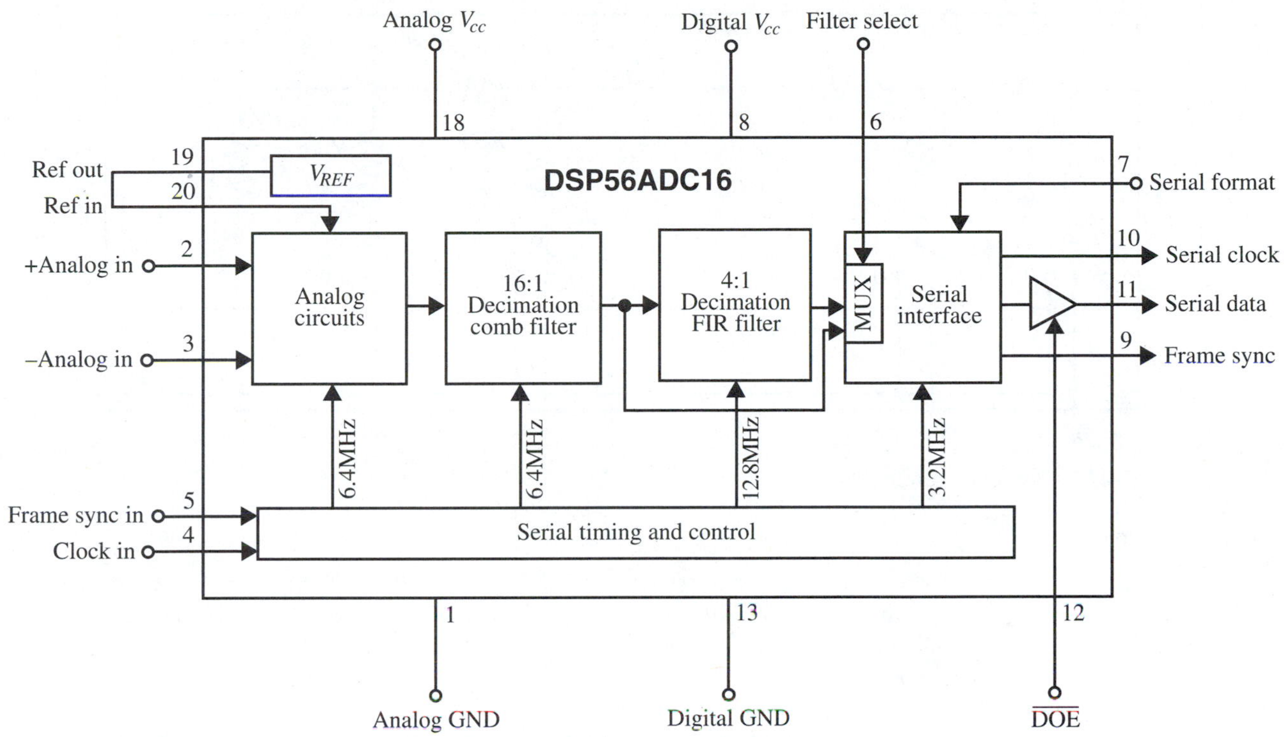

FIGURE 16–26 DSP56ADC16 block diagram. *Courtesy Motorola Semiconductor.*

The second decimation filter is a finite impulse filter (FIR), which performs a weighted average of the input data. It effectively averages 16 outputs from the comb filter at one-fourth of the comb filter sample rate, or 100,000 times per second (this is four new inputs per sample)—which provides four bits. The total resolution out of the FIR is

$$12 \text{ bits} + 4 \text{ bits} = 16 \text{ bits}$$

The multiplexer (MUX) allows selection of 12 bits at a rate of 400KBS or 16 bits at a rate of 100KBS. With filter select low, 16 bits is selected; with it high, the resolution is 12 bits.

The serial interface provides a data stream that can be transmitted easily on a single pair of wires. Three outputs are provided at this interface:

1) Serial data out—high resolution output, shown here. (The sign bit indicates the polarity of the analog signal.) This output is enabled when the data output enable line is low.

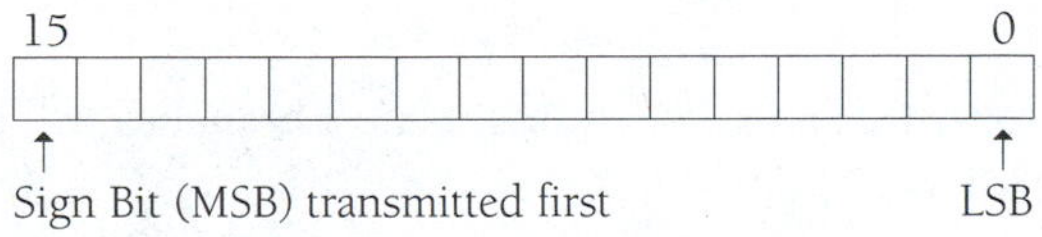

2) Frame sync out—indicates the beginning of an output word.
3) Serial clock—used for converting the serial data back to parallel at the receive end of the transmission line

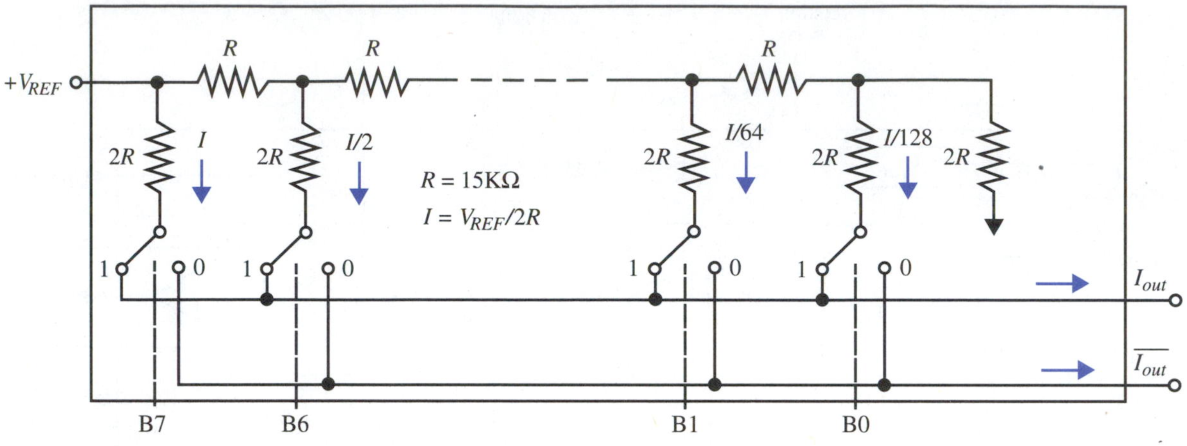

FIGURE 16–27 Internal current paths

The DSP56ADC16 requires a 12.8MHz, TTL-level clock, and a +5V power source for both the digital and analog power source to operate. It is recommended that the analog and digital power sources be kept separate to isolate digital noise from the analog section.

DIGITAL-TO-ANALOG CONVERSION

To hear the music recorded on a compact disc, we have to convert the digital bits back to an analog voltage. This requires a digital-to-analog converter (DAC). The conversion is accomplished by having the individual digital bits control switches that allow weighted currents to flow into a current summing resistor, as shown in Figure 16–27.

The eight-bit resistor network configuration provides the maximum current I ($I = V_{REF}/2R$) through the left-most resistor and decreases by a factor of two each time we move one resistor to the right. The right-most switched resistor has a current of $I/128$. Current flows through a resistor into the I_{out} terminal if the associated bit is a *1*, or into the $\overline{I_{out}}$ terminal if the bit is a *0*. Since current is flowing from a branch into I_{out} or $\overline{I_{out}}$, the current never changes through any resistor in the circuit.

EXAMPLE

Find the current flowing into the I_{out} terminal of Figure 16–27 if the reference voltage V_{REF} is +12V and the input code is 10000001 (R is 15KΩ).

SOLUTION

With this code, only the left-most and right-most branches will be providing current to the I_{out} terminal. So

$$I_{out} = I + I/128$$

But

$$I = V_{REF}/2R$$

$$I = 12V/30K\Omega$$

$$I = 0.4mA$$

And

$$I_{out} = 0.4mA + 0.4mA/128$$

$$I_{out} = 0.403125mA$$

The maximum current that can flow into the I_{out} terminal occurs when the digital input is all *1*'s. This is called the full-scale current I_{fs} and can be determined from the following equation derived by summing all the branch currents:

$$I_{fs} = \frac{V_{REF}}{R} \times \frac{255}{256} \qquad \textbf{(equation 16–19)}$$

This full-scale current is 1/256 less than the current supplied by the reference, with the lost current flowing to ground through the right-most resistor in Figure 16–27.

The current I_{out} for a given digital input is determined by first determining the decimal equivalent (N) and then using the equation

$$I_{out} = \frac{V_{REF}}{R} \times \frac{N}{256} \qquad \textbf{(equation 16–20)}$$

EXAMPLE

Determine the current I_{out} if V_{REF} is +10V, R is 15KΩ, and the input code is 10101010.

SOLUTION

The code 10101010 converted to decimal is 170 (N).

Using equation 16–20,

$$I_{out} = \frac{V_{REF}}{R} \times \frac{N}{256}$$

$$I_{out} = \frac{10V}{15K\Omega} \times \frac{170}{256}$$

$$I_{out} = 0.44271mA$$

If we desire a voltage output rather than current from the converter, we can use a circuit like the one shown in Figure 16–28.

The current $\overline{I_{out}}$ flows to ground and the current I_{out} flows not into the high impedance input of the op-amp but rather through the op-amp feedback resistor. This gives an output voltage of

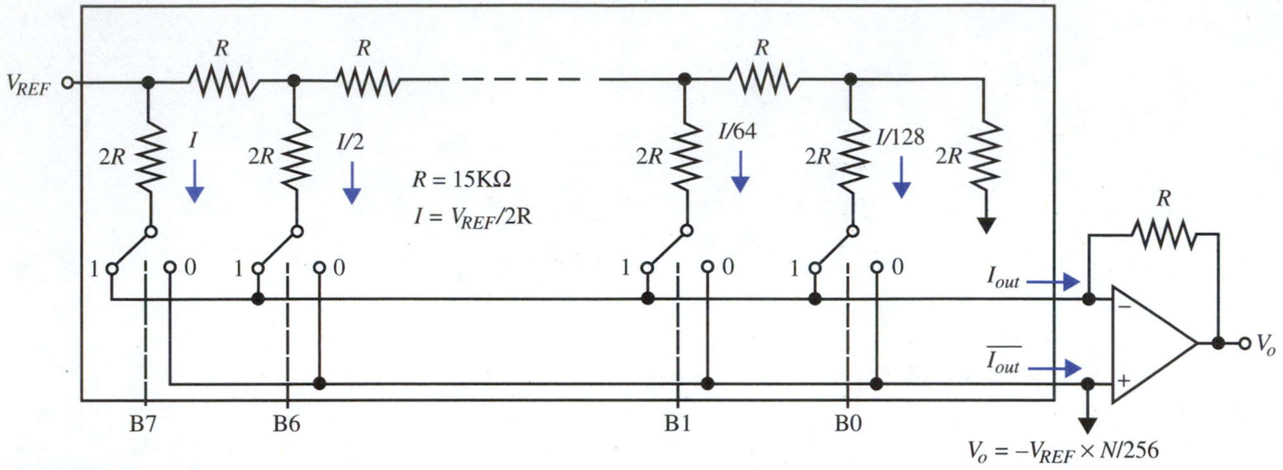

FIGURE 16–28 DAC circuit to provide a voltage output

$$V_o = -I_{out} \times R$$

But from equation 16–20,

$$I_{out} = \frac{V_{REF}}{R} \times \frac{N}{256}$$

So

$$V_o = -\frac{V_{REF}}{R} \times \frac{N}{256} \times R$$

$$V_o = -V_{REF} \times \frac{N}{256} \qquad \textbf{(equation 16–21)}$$

EXAMPLE

Find the output voltage from the circuit of Figure 16–28 if V_{REF} is -8V and the input digital code is 11001100.

SOLUTION

First, we find that the decimal equivalent of the digital input is

$$N = 204$$

Now using equation 16–21,

$$V_o = -V_{REF} \times \frac{N}{256}$$

$$V_o = -(-8\text{V}) \times \frac{204}{256}$$

$$V_o = 6.375\text{V}$$

Notice that with a negative reference voltage, we get a positive output voltage because of the inverting action of the op-amp.

AN ACTUAL DIGITAL TO ANALOG CONVERTER

A common DAC that operates in the manner previously described is the National Semiconductor DAC0830. This chip is an eight-bit converter that is designed to be microprocessor compatible.

As shown in the diagram of Figure 16–29, the device has two registers to store the incoming digital data. The first register allows the computer to download the data, and the second register holds the bits while the conversion is in process. The actual DAC portion

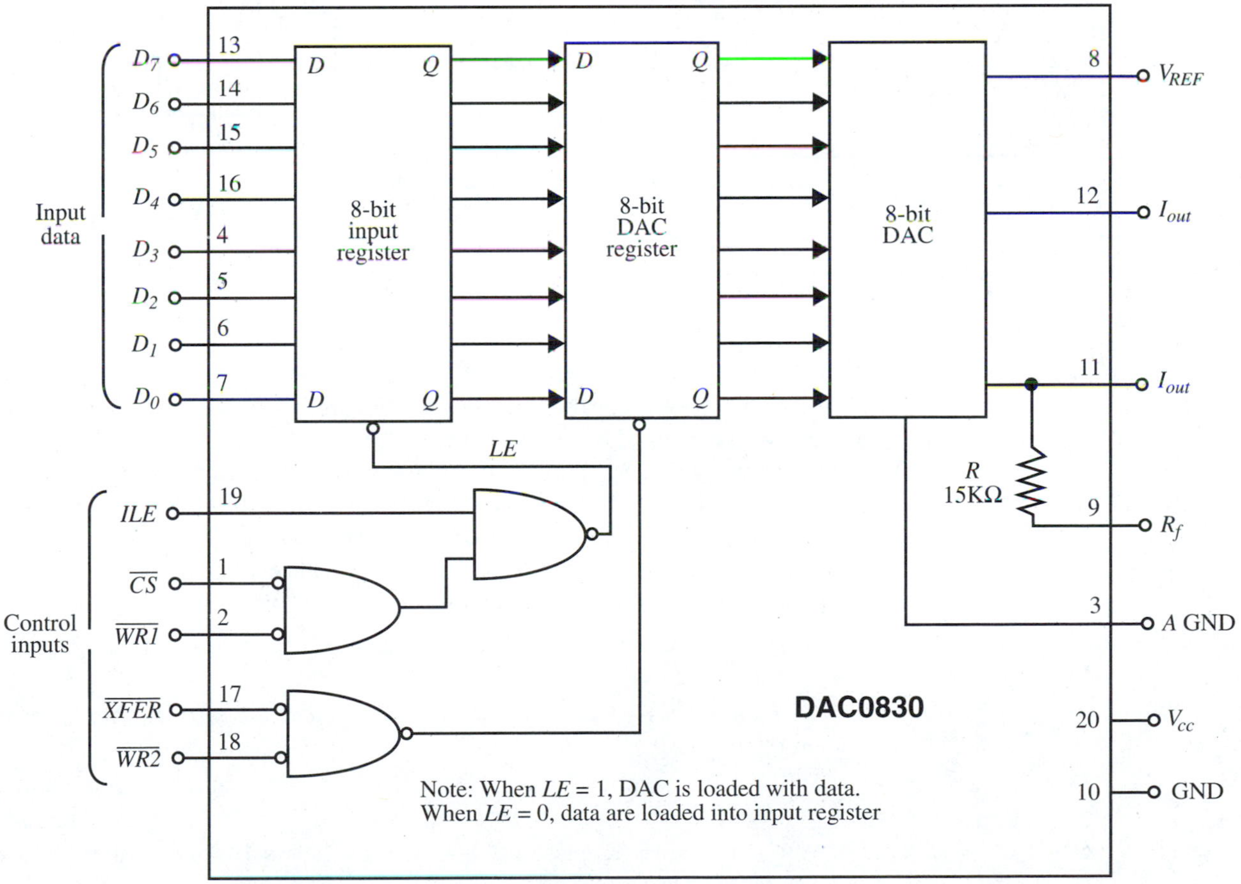

FIGURE 16–29 DAC0830 block diagram. *Courtesy National Semiconductor.*

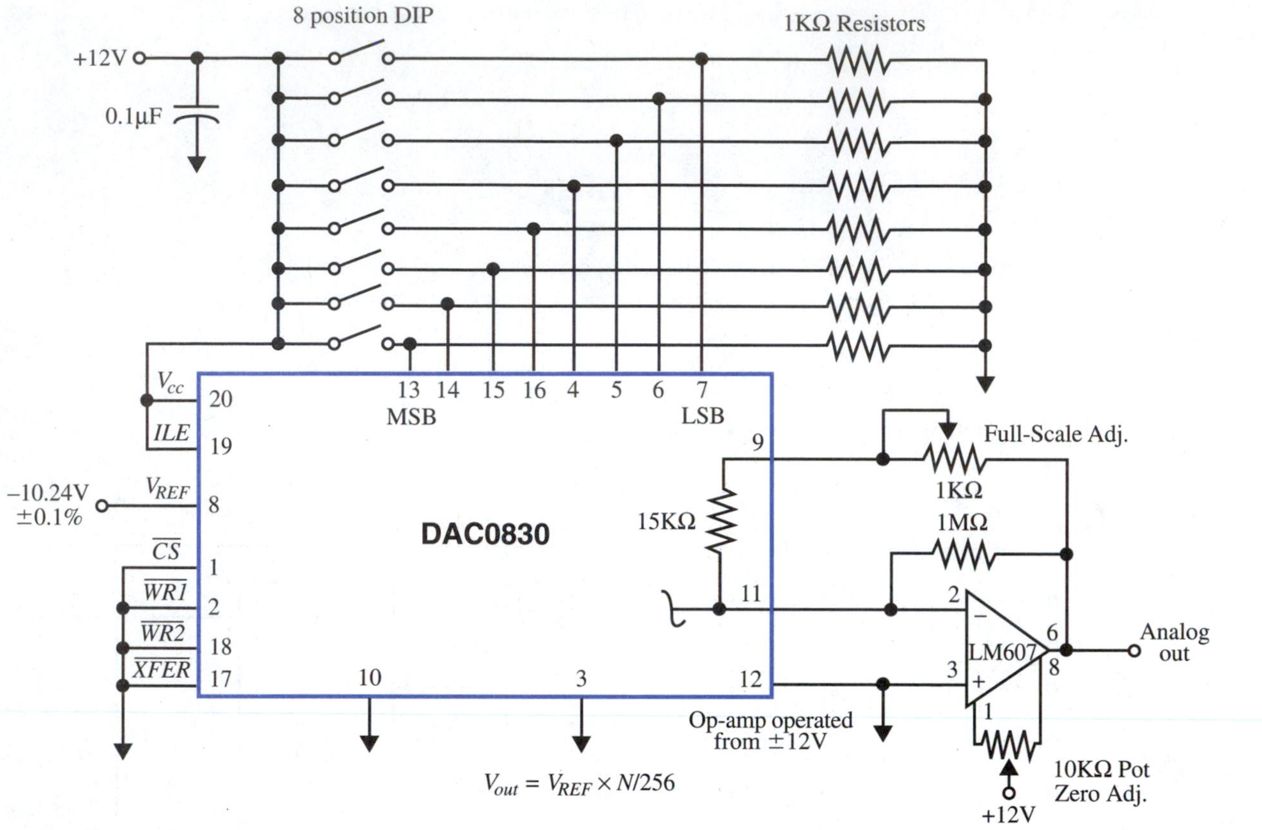

FIGURE 16–30 Test circuit for DAC0830

contains a resistor network similar to that shown in Figure 16–28—except that the op-amp feedback resistor is contained within the chip.

This DAC has a multiplier capability. By applying the multiplying voltage to the reference terminal, the output voltage can be scaled (see equation 16–21). One possible use could be volume (gain) control of a converted audio signal. If straight digital-to-analog conversion is required, then V_{REF} is tied to fixed precision voltage reference.

The function of the control lines is as follows:

- **ILE** (input latch enable)—allows input data to latch into the input register with $\overline{\text{CS}}$ and $\overline{\text{WR}_1}$ low.
- $\overline{\textbf{CS}}$ (chip select)—active low provides selection of a particular chip by microprocessor.
- $\overline{\textbf{WR}_1}$ (write into input register [1])—active low causes data to be written into input register.
- $\overline{\textbf{WR}_2}$ (write into DAC register [2])—active low causes data to be written into DAC register from input register.
- $\overline{\textbf{XFER}}$—active low enables WR_2.

These signals can originate from the microprocessor's control bus or can be generated from discrete logic decoding of address inputs (memory mapping).

Figure 16–30 shows a test circuit for operating the DAC0830 without a microprocessor interface. By grounding the $\overline{CS}$, $\overline{WR_1}$, $\overline{WR_2}$, and $\overline{XFER}$ terminals, and tying the ILE terminal high, the DAC0830 can be operated in a flow-through configuration, where input data passes directly into the DAC register.

To calibrate the DAC, the input switches are left open to provide an all *0* input, and the Zero Adj potentiometer is turned to provide 0V out. Next, all switches are closed to provide an all *1* input, and the Full-Scale Adj potentiometer is turned to provide $V_{REF} \times 255/256$.

To check control functions, a switch that allows connection between the positive supply or ground can be added to the $\overline{CS}$, $\overline{WR_1}$, $\overline{WR_2}$, $\overline{XFER}$, and ILE lines to allow either *1* or inputs. By operating the switches in a given sequence, data transfer through the registers can be observed.

TROUBLESHOOTING DATA-CONVERSION CIRCUITS

Understanding the data-conversion process is vital to intelligent troubleshooting of circuits involved in this operation. For example, if you could not determine what the output code should be for a given analog input to an ADC, how could you determine whether it is operating correctly?

Sample and Hold Circuit Troubleshooting

Sample and hold circuits are used to hold the signal constant steady while conversion is being performed. So the input to an S & H must be a time-varying waveform. The output should be a sampled (stepped) waveform, as shown in Figure 16–3.

With no output waveform, we need to verify the presence and correct amplitude of the sampling logic signal. If all signals are correct at the chip pins, then we must replace the chip.

ADC Troubleshooting

When troubleshooting ADC circuits, we again need to be aware of the correct state of the input and output signals.

Some high-speed ADCs don't require a sample and hold, or the S & H is part of the ADC chip. So it is possible to have a time-varying input to the ADC.

With a DC input to the ADC, the output digital code can be determined and then verified by looking at the logic state of the output data lines.

All ADCs require a clock to operate, so this signal must be confirmed. Also, the reference voltage value should be verified.

If the ADC is microprocessor compatible, the interrupt signal should be present each time the conversion is completed. Also, verify that the chip selection and read/write signals have the correct amplitude and occur in the right time sequence.

DAC Troubleshooting

In many applications, the DAC is used with a microprocessor. If this is the case, programming the microprocessor for a known output code to the DAC will aid in the troubleshooting process.

Again, the amplitude and timing of the chip selection and write signals should be verified.

SUMMARY

- An analog signal is a signal that can assume any voltage consistent with the power supply voltage, but a binary digital signal can only be one of two voltages.
- Analog signals are converted to digital for better noise immunity.
- Analog signals are converted to digital by sampling the analog voltage and changing it to a digital code representative of the particular voltage value.
- The greater the number of bits used in the digital code, the better the analog waveform can be represented.
- The more often the analog waveform is sampled (higher sampling rate), the better the representation of the analog signal.
- The minimum number of samples that should be taken from a sine wave is two.
- Sampling at a rate lower than two per sine wave results in aliasing.
- A typical data-conversion system consists of a sample and hold circuit, ADC converter, computer, and DAC.
- A sample and hold circuit samples an analog voltage and holds the sampled voltage steady while processing of the voltage is performed.
- The smaller the hold capacitor, the faster the input voltage can be sampled; however, a small hold capacitor means more voltage droop during the hold time.
- Accuracy of the sample and hold circuit has to be consistent with the overall data conversion accuracy requirement.
- An eight-bit flash converter uses a 256-resistor reference voltage divider to establish the voltage on 256 capacitors. These voltages are simultaneously compared with the analog input to establish the digital output.
- The successive approximation ADC converter also uses a 256-resistor network, but comparisons with the analog input take eight steps.
- Although the slowest, the dual slope converter has the best noise immunity. The device consists of an integrator, comparator, and a counter. A higher-level input voltage results in a higher-level output from the integrator in a given time (16.67 ms). The count to bring the output back to zero with a fixed reference input is the digital output corresponding to the analog input.
- The sigma-delta ADC has the highest resolution, along with a high sample rate. It uses only a one-bit ADC, but a very high sampling frequency and decimation raise the effective resolution.
- Decimation raises the effective resolution by digital averaging.
- Integrated circuit digital-to-analog converters translate a digital input signal into a current output.
- Basically, a DAC consists of a ladder network with digital control of the scaled current through the branches, the output current being the sum of the branch currents.
- The number of bits for the DAC defines the resolution; for example, a 10-bit device can provide 1,024 different analog output levels.

- Accuracy of a DAC can be no greater than the accuracy of the voltage reference used.
- The analog output can be multiplied by a signal applied to the voltage reference terminal.
- Since DACs are often used with microprocessors, compatibility with these devices is important.

EXERCISE PROBLEMS

1. Why are analog waveforms converted to digital?
2. What is the minimum number of samples that can be taken from a sine wave?
3. What is aliasing and how is it generated?
4. What determines the number of digital bits used for the converted data?
5. Determine the voltage step size (resolution) for a 10-bit ADC if the internal reference voltage is 10.24V.
6. Find the maximum voltage that can be converted to digital for a 12-bit ADC with an internal reference voltage of 5.12V.
7. Find the sampling frequency for an ADC if the maximum frequency to be converted is 20KHz and 5X over-sampling is used.
8. Define acquisition and hold time for an S & H.
9. What is the relationship between acquisition time and sample time?
10. Determine the required hold capacitor for an LF398 sample and hold if the maximum allowable error is 0.2% and the acquisition time should be no longer than 10μs.
11. If the hold capacitor is 0.1μF, the voltage held is 6V, and the hold time is 20ms, find the droop (ΔV_H) in the LF398 output voltage. Assume a capacitor leakage resistance of 500MΩ.
12. Show a circuit for a voltmeter probe using an LF398. The sample is to be taken by activation of a push-button switch. The input voltage range is from 0V to +10V.
13. A data-conversion system consists of an S/H and an ADC. What must occur before valid data are available at the ADC?
14. Show the block diagram for an ADC conversion circuit with an LF398 and an ADC0801. The termination of the hold time is to be controlled from the ADC0801. (*Hint:* Use a 555 timer for sample pulse.)
15. It is desired to convert an analog voltage to digital with a resolution of at least one part per thousand (0.1%). How many bit ADC is required?
16. The reference voltage for a 12-bit ADC is 10.24V. If the accuracy of the ADC is 1/4 LSB, find the volts variation of analog input that could provide the same output digital code.
17. Demonstrate that an eight-bit successive approximation ADC requires eight samples to find a match. Use a 10.24V reference for a numerical demonstration.

18. Determine the digital output code for an ADC0801 ADC with a +5.12V reference and an input of 3V. Find the maximum conversion time if the clock is running at 800KHz.
19. Determine the analog input voltage range to an ADC0801 if the output is 01101101 and the reference is +5.12V (include the accuracy of the ADC0801).
20. Determine the reference voltage to be applied to Pin 9 of an ADC0801 if the input signal has a maximum value of 2V.
21. Explain why a flash converter is so much faster than the successive approximation type.
22. Find the voltage divider tap in an eight-bit flash converter at which the voltage V_a changes from positive to negative if V_{REF} is 5.12V and the analog input V_s is 3V.
23. Explain the difference between a half-flash and a flash-type converter.
24. Determine the output voltage from the DAC within an eight-bit half-flash converter if the reference is 5.12V and the input voltage is 4.4V.
25. State the advantage of a dual slope ADC.
26. Find the external oscillator frequency for an ICL7136 converter if it is required to have three conversions a second (n is 3).
27. Find the count N for an ICL7136 dual slope converter if the reference voltage is 1V and the input voltage 0.25V.
28. Determine the required value of R_t if the desired clock frequency for an ICL7136 is 64KHz.
29. A digital voltmeter is required using an ICL7136 with a full-scale reading of 2V and two displays per second. Determine the oscillator frequency and all circuit components. Use a 9V battery as a power source.
30. Explain how a sigma-delta ADC achieves high resolution.
31. With reference to Figure 16–A, determine the four-bit digital code if the analog input voltage is $-V_{REF}$.

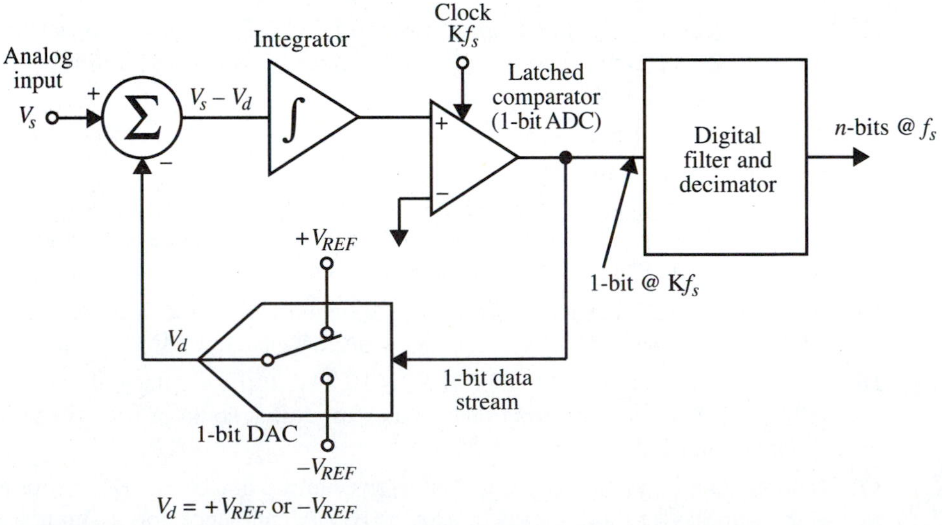

FIGURE 16–A (text figure 16–24)

32. Sketch the serial data out if the filter select line for the DSP56ADC16 is high.

33. What is the function of a DAC?

34. Find the current flowing into the I_{out} terminal of Figure 16–B if the reference voltage V_{REF} is +12V and the input code is 10100001 (R is 15KΩ).

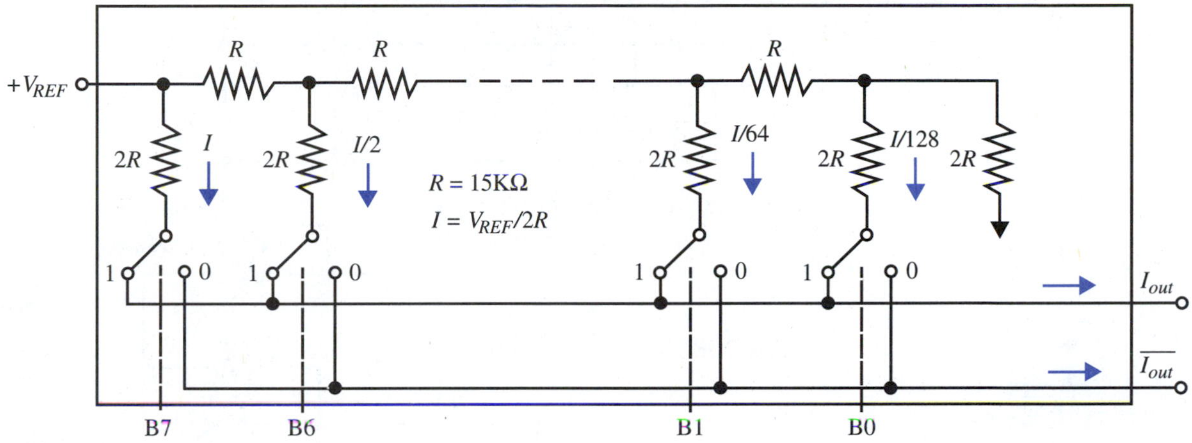

FIGURE 16–B (text figure 16–27)

35. Find the output voltage from the circuit of Figure 16–C if V_{REF} is −6V, and the input digital code is 10001100.

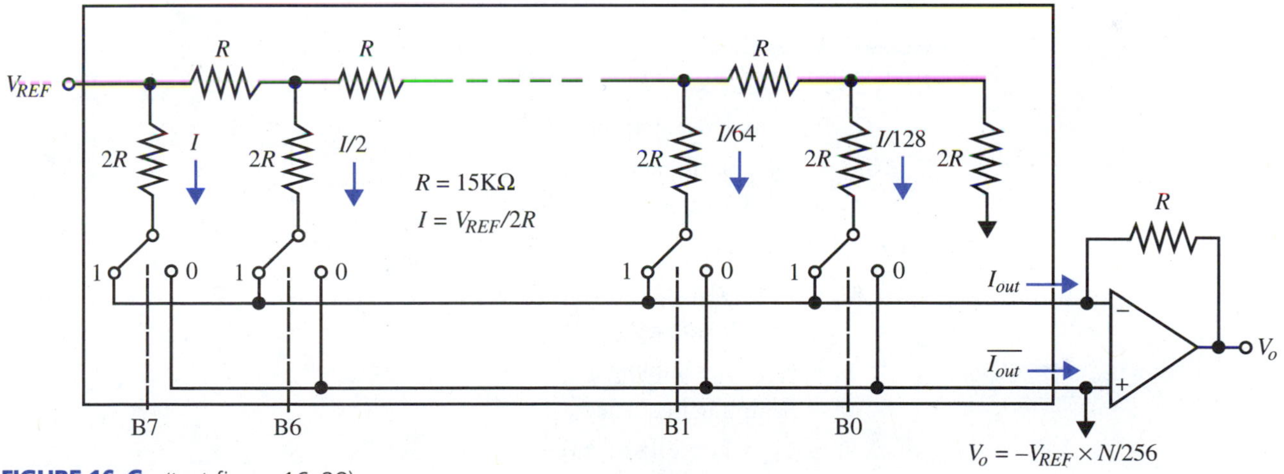

FIGURE 16–C (text figure 16–28)

36. With reference to Figure 16–D, find the range of variation in the internal 15KΩ resistor that can be accommodated with the Full Scale ADJ potentiometer.

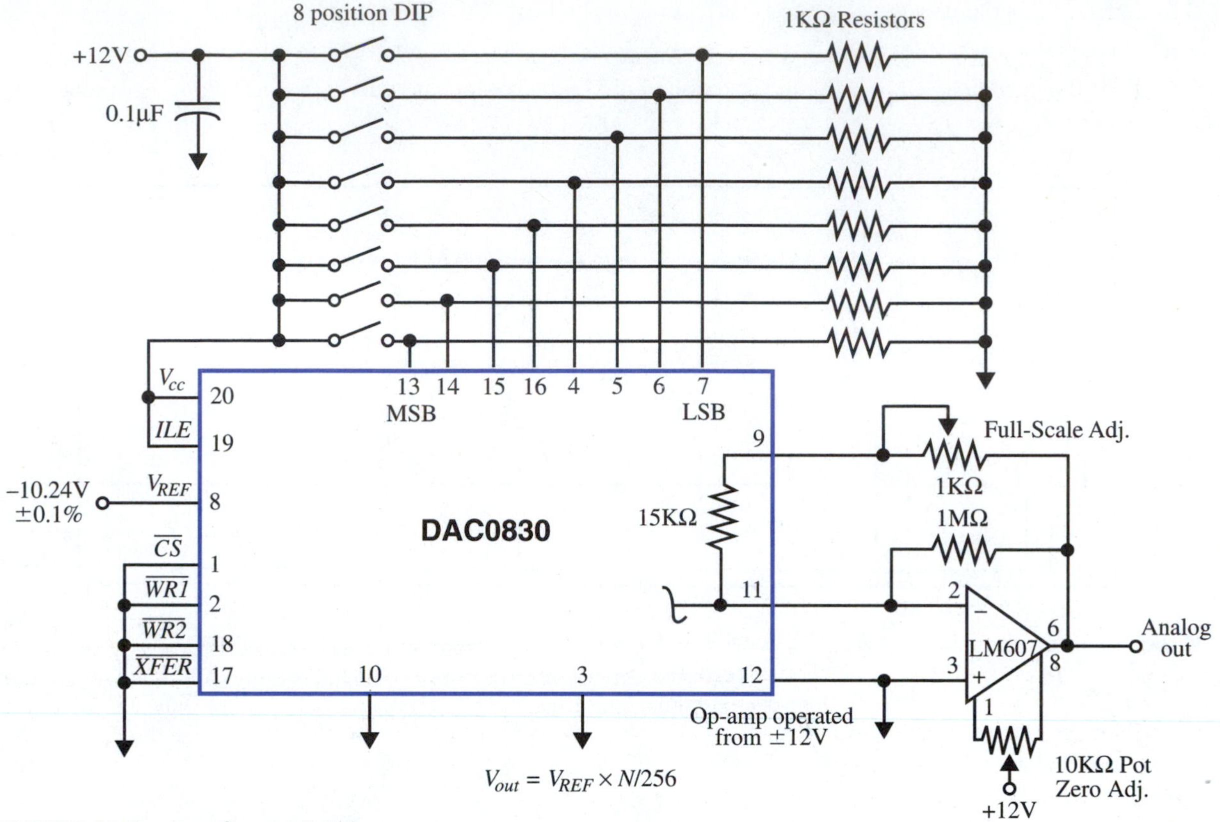

FIGURE 16–D (text figure 16–30)

ABBREVIATIONS

C_{AZ} = Auto-zero capacitor

C_{INT} = Integrating capacitor

C_H = Hold capacitor

C_t = Timing capacitor

CS = Chip select

f_a = Alias frequency

f_N = Nyquist sampling frequency

f_s = Sampling frequency

I_L = Capacitor leakage current

INTR = Interrupt

RD = Read

R_t = Timing resistor

t_a = Acquisition time

t_{ap} = Aperture time
t_c = Conversion time
t_H = Hold time
t_{hs} = Hold settling time
V_a = Input voltage to amplifier
V_d = Dynamic sampling error
V_h = Hold step
V_H = Voltage on hold capacitor
V_s = Input voltage
V_{st} = ADC step voltage
V_{RN} = Voltage at N tap on resistor network
V_{REF} = Reference voltage
WR = Write

ANSWERS TO PROBLEMS

1. Noise improvement or processing
2. Two
3. See text.
4. The required resolution
5. $V_{step} = 10\text{mV}$
6. 5.1194V
7. 200KHz
8. See text.
9. See text.
10. $C_H = 5.36\text{nF}$
11. $\Delta V_H = 2.36\text{mV}$
12. Put push-button switch between logic input and +5V source. Put 10KΩ from Pin 8 to ground.
13. All transients must be over.
14. ADC0801 interrupt signal trigs 555. Output signal from 555 is sample pulse.
15. 10-bit
16. ±1.875mV
17. Dividing reference by 2 eight times reduces 10.24 to 40mV step.
18. 10010110. Conversion time is 91.25μs
19. 2.165V–2.195V
20. 1V
21. Conversion takes just one clock cycle.
22. $N = 150$
23. See text.
24. 4.16V
25. Noise Immunity
26. $f_{osc} = 48\text{KHz}$
27. $N = 250$
28. $R_t = 141\text{K}\Omega$
29. $f_{osc} = 32\text{KHz}$; $C_{int} = 75\text{nF}$; $C_{az} = 0.47\mu\text{F}$; $C_{REF} = 0.1\mu\text{F}$; and $R_t = 281\text{K}\Omega$
30. See Text.
31. 0000
32. Twelve-bit output with MSB first
33. To convert digital data to analog
34. $I_{out} = 0.503\text{mA}$
35. $V_o = 3.28125\text{V}$
36. $R = 14.23\text{K}\Omega$ to $15.23\text{K}\Omega$

CHAPTER 17

Communication Circuits

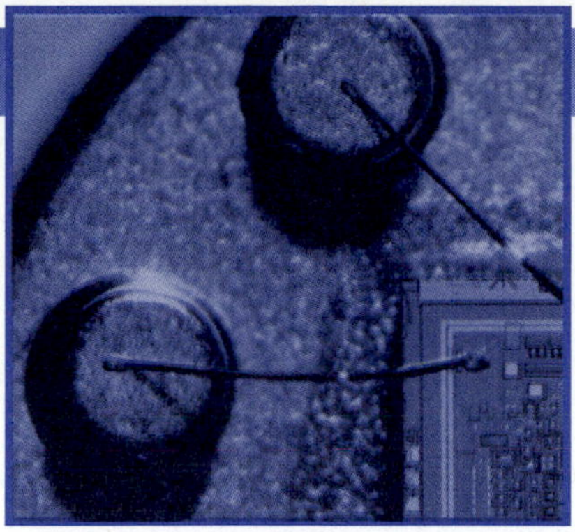

CHAPTER OBJECTIVES

After completing this chapter, you should be able to explain the following communications concepts and the operation of communication circuits:

- Definition of a communication signal
- Significance of bandwidth
- Types of noise sources
- Reason for modulation
- AM modulation
- How AM modulation is achieved
- Demodulation of AM signals
- FM modulation
- How FM modulation is achieved
- Demodulation of FM signals
- Operation of a voltage-controlled oscillator
- Operation of a phase locked loop
- Frequency synthesis
- Modulation methods for digital data
- RS233 specifications
- Advantages of fiber-optic data transmission

INTRODUCTION

Certainly, whole texts can be written on the topic of electronic communications; but the object of this chapter is to give an introduction to this broad topic and then show some integrated circuit devices that are used in communications.

First, we will be introduced to the fundamentals of electronic communications. Then a communications signal will be defined along with its bandwidth requirements, and then a definition of noise and how its effect can be reduced. The reason for modulation will be discussed, and amplitude and frequency modulation will be covered. The fast-growing topic of data communications will also be introduced. Finally, integrated circuit chips that perform communications functions will be described.

DEFINITION OF SIGNAL AND BANDWIDTH

A communications signal is that which conveys information from a specific transmitter to a specific receiver. The term *specific* is used because signals that contain information for one receiver may not for another receiver; for example, if two people are carrying on a conversation, a third person may not understand and, therefore, may not get any information from the communication.

Bandwidth of a signal is the band of frequencies required to be sent to fully describe the signal. With audio signals, the required bandwidth to cover the range of the ideal human ear is from 20Hz to 20KHz. If a person's ear has this ideal response, that person would possibly be missing sounds (information) if the full bandwidth is not transmitted.

The receiver must have sufficient bandwidth to include all the frequencies that are transmitted—otherwise, again, information will be lost.

NOISE

Any unwanted signals within the bandwidth of the wanted signal constitutes noise. If someone is talking while we are trying to listen to music, that person's voice is considered noise because it is an unwanted signal. But, the ultrasonic signal emitted by a bat is not considered noise because it is not within the bandwidth of the human ear.

Noise can be broken up into two basic types: that which contains information that is unwanted (e.g., cross-talk on a telephone line) and random noise. Random noise can be the most troublesome because it is wideband and it can be generated in so many ways. We can further break down random noise into two basic types: external and internal. Examples of external noise are atmospheric (lightning), and transient (voltage and current spikes). Internal noise, which is generated within electronic components, can be thermal, shot, flicker, or transit.

- Thermal noise is wide-bandwidth noise and is generated in anything that has electrical resistance. As the name suggests, it increases with temperature.
- Shot noise is also wideband noise and results from carriers arriving in a random manner at a collector. It is proportional to the amount of current.
- Flicker noise or $1/f$ noise occurs in semiconductor devices, and its amplitude reduces as the frequency increases. Above about 1KHz, flicker noise is insignificant.
- Transit noise results from inadvertent changes in the number of charge carriers within an amplifier. This noise is more a problem at high frequencies.

Combining these four types of internal noise results in a bathtub-shaped curve of noise voltage versus frequency.

What methods can be used to reduce the effect of noise on signal reception?

1) Keep the bandwidth only wide enough to pass the signal—no wider!
2) Obtain the strongest possible signal at the receiver.
3) Shield the circuit and/or cabling from external noise sources.
4) Shift the signal frequencies to a portion of the frequency spectrum where less noise is present.
5) Cool the receiver to reduce thermal noise when very weak signals are being received.

CARRIER MODULATION

As the name suggests, a carrier "carries" the signal information. The signal is put on the carrier by an operation called **modulation**. The carrier is a sine wave that is at a frequency much higher than the modulating signal. For example, when we tune a station on a radio, we are selecting the carrier—which is at a much higher frequency than the carrier modulating audio signal.

Why do we need a carrier? There are three prime reasons:

1) **Selectivity**—without a carrier to select, all radio stations' transmissions would be heard at the same time.
2) **Ease of transmission**—higher radio carrier frequencies use a smaller antenna.
3) **Noise reduction**—can shift to a quieter portion of the frequency spectrum.

There are two common ways of modulating a carrier: **amplitude modulation** (AM) and **frequency modulation** (FM). First we shall look at how amplitude modulation is accomplished.

AMPLITUDE MODULATION

The amplitude of the carrier is changed with amplitude modulation, as shown in Figure 17–1. The variation in the amplitude of the carrier in Figure 17–1a is called the **modulation envelope**. Tracing the peaks (negative or positive) of the carrier waveforms gives the shape of the modulation signal—which, for our figure, is a sine wave. The unmodulated carrier has a peak amplitude midway between V_{MAX} and V_{MIN} or $(V_{MAX} + V_{MIN})/2$. The greater the amplitude of the modulating signal, the greater the change in the carrier amplitude.

From Figure 17–1a, we can predict that if the amplitude of the modulating signal is too great, the carrier can actually be cut off, and the result will be clipping distortion. The maximum peak amplitude of the modulating signal, then, is equal to the peak amplitude of the carrier.

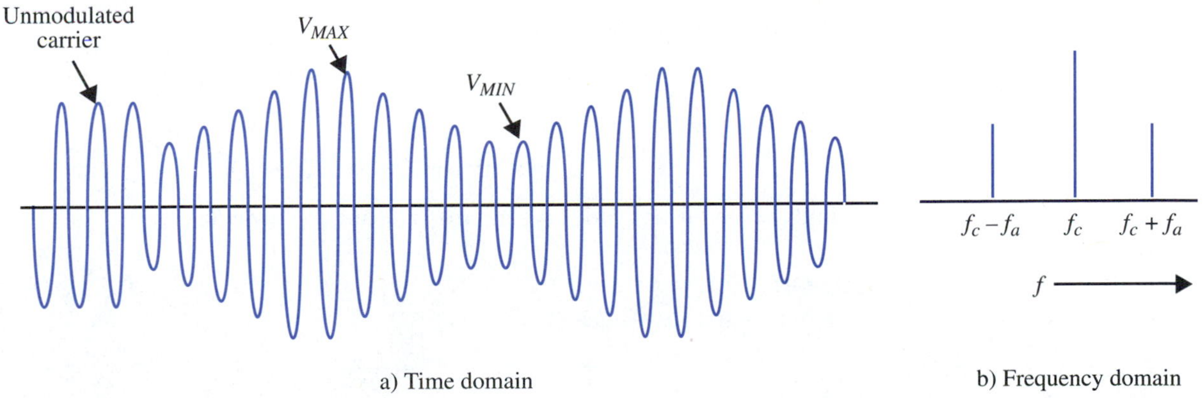

FIGURE 17–1 Amplitude modulation

The term used to indicate the degree of modulation is the amplitude modulation index (m_a).

$$m_a = \frac{V_a}{V_c} \qquad \textbf{(equation 17–1)}$$

where V_a is the amplitude of the modulating signal and V_c the amplitude of the carrier. The maximum value of m_a is unity. Modulation envelope breaks up if m_a exceeds 1.

The amplitude modulation of a carrier by a sine wave creates two new frequencies (called sidebands), as shown in Figure 17–1b. The lower sideband is the frequency of the carrier minus the modulating frequency ($f_c - f_a$); the upper sideband is the carrier plus the modulating frequency ($f_c + f_a$).

EXAMPLE

Find the upper and lower sideband frequencies if a 1MHz carrier is amplitude modulated by a 10KHz sine wave.

SOLUTION

The upper sideband f_u is

$$f_u = f_c + f_a$$

$$f_u = 1\text{MHz} + 10\text{KHz}$$

$$f_u = 1.01\text{MHz}$$

The lower sideband f_L is

$$f_L = f_c - f_a$$

$$f_L = 1\text{MHz} - 10\text{KHz}$$

$$f_L = 990\text{KHz}$$

EXAMPLE

A carrier signal is described by

$$v_c = 10\text{V} \sin (2\pi \times 760\text{KHz} \times \text{t})$$

If the modulating signal is

$$v_a = 5\text{V} \sin (2\pi \times 2\text{KHz} \times \text{t})$$

find the modulation index m_a and the sideband frequencies.

SOLUTION

From the sine wave equations,

$$V_c = 10\text{V}$$

and

$$V_a = 5\text{V}$$

The modulation index is found from equation 17–1.

$$m_a = \frac{V_a}{V_c}$$

$$m_a = \frac{5}{10}$$

$$m_a = 0.5$$

Also from the sine wave equations,

$$f_c = 760\text{KHz}$$

and

$$f_a = 2\text{KHz}$$

The sidebands are then

$$f_u = 760\text{KHz} + 2\text{KHz}$$

$$f_u = 762\text{KHz}$$

$$f_L = 760\text{KHz} - 2\text{KHz}$$

$$f_L = 758\text{KHz}$$

A pair of frequency sidebands is created for every modulating frequency. So a complex modulating waveform, having many different frequencies, will have many sidebands. The bandwidth required to transmit an amplitude-modulated carrier is twice the maximum modulating frequency, or

$$f_u - f_L = 2f_a$$

For AM radio, the maximum modulating frequency is 5KHz and so the station-allocated bandwidth is 10KHz.

The basic advantage of amplitude modulation is its simplicity. Disadvantages are susceptibility to external noise interference and inefficiency of power transmission—signal information is only contained in the sidebands but most of the transmitted power is in the carrier. This is why single-sideband transmission is so popular.

AMPLITUDE MODULATION CIRCUIT

Amplitude modulation can be accomplished by several different methods. For high-power radio transmitters, the plate voltage of the output vacuum tube is varied. Medium-power transmitters use combinations of collector and base modulation of the carrier output stage. Low-power transmitters can use low-level modulation followed by linear amplifiers that bring the modulated carrier up to the required transmitted power level.

We shall look at an integrated circuit device that can perform low-level modulation. Figure 17–2 shows the internal block diagram for the Exar XR-2206. This device is a function

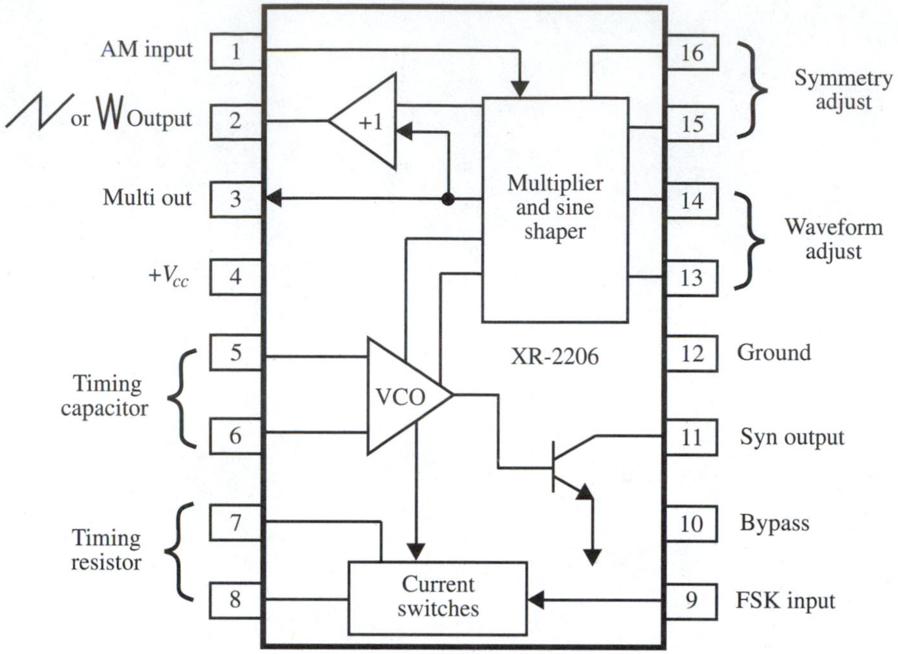

FIGURE 17–2 XR-2206 function generator. *Courtesy Exar.*

generator that can provide sine wave, square wave, triangle, ramp, and pulse waveforms as well as amplitude- and frequency-modulated output signals.

The carrier frequency is set by the selection of the timing resistor between Pins 7 and 8 and the timing capacitor between Pins 5 and 6 in accordance with the frequency equation

$$f_c = \frac{1}{R_t C_t} \text{ Hz} \qquad \textbf{(equation 17–2)}$$

Frequency range for the chip is from 0.01Hz to 1MHz.

The modulating signal is applied to Pin 1, and the modulated carrier output appears at Pin 2. Figure 17–3 provides the external circuit components for an AM modulator with a carrier frequency of 100KHz. Also included in the figure is a graph of the transfer characteristic of modulating voltage versus the carrier voltage and carrier phase.

Notice that the transfer graph shows that the output voltage drops to zero when the modulating voltage at Pin 1 is at one-half the power supply voltage (+6V). From 1V to 6V, the output voltage decreases as the modulating voltage increases; also from 6V to 11V it increases.

If, for example, we apply a 3.5V DC bias voltage to Pin 1 of the circuit of Figure 17–3, the carrier output voltage will be at one-half its maximum possible value. Now, if we superimpose an AC voltage on this DC value, the carrier output voltage will vary above and below its nominal half maximum value in response to the modulation signal. If the positive peak of the modulating signal is +2.5V, this will cause the voltage on Pin 1 to be +6V—

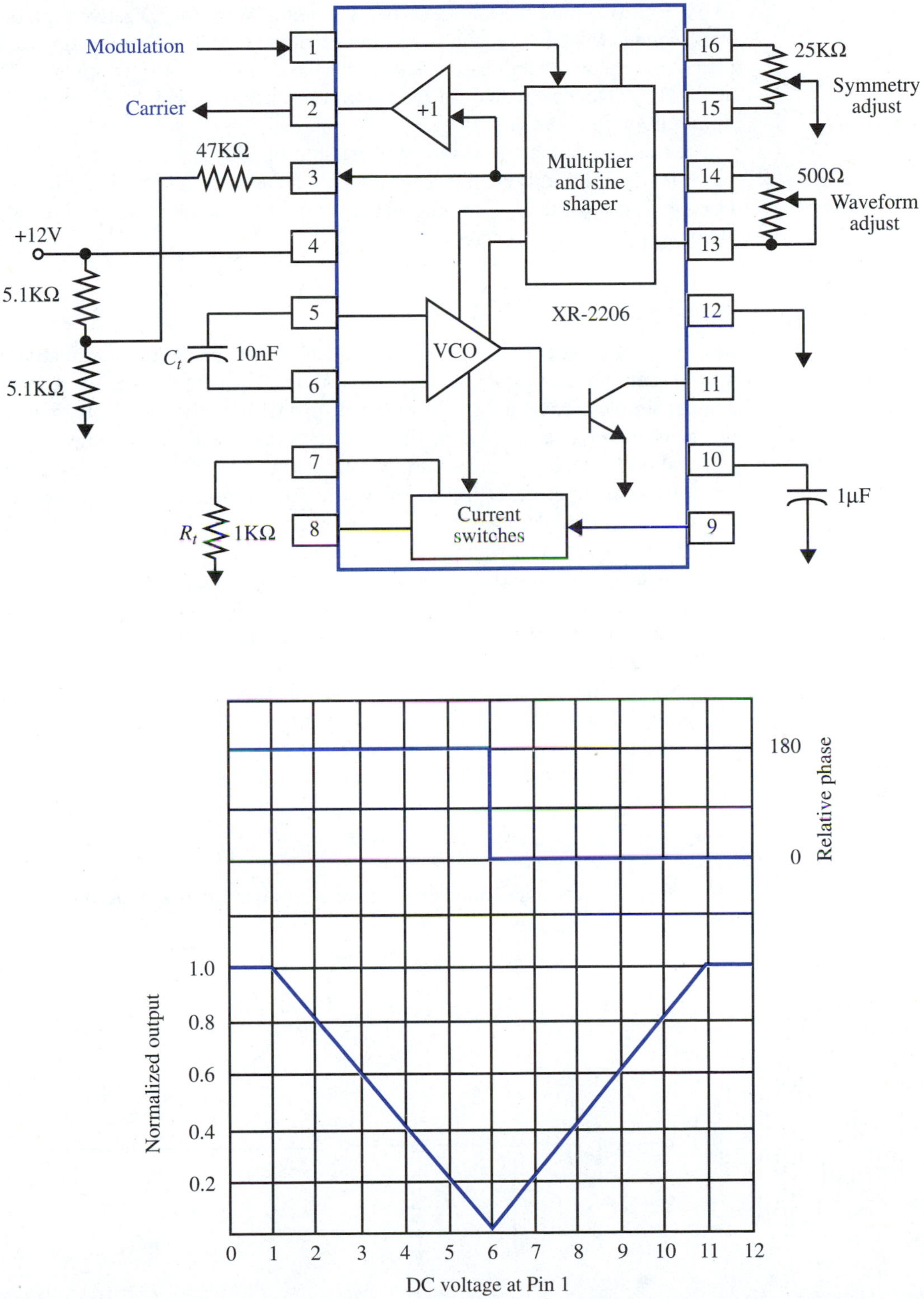

FIGURE 17–3 AM modulator circuit

(3.5V + 2.5V)—and the output carrier will go to 0V. With a negative peak of −2.5V, the Pin 1 voltage will be +1V, and the output carrier will go to maximum amplitude. We have achieved amplitude modulation! The results would be similar if we applied a DC bias voltage of 8.5V. Now we would get maximum carrier output with an AC peak of +2.5V and minimum output with a peak of −2.5V.

The transfer characteristic abrupt phase shift of 180° at 6V has no significance for AM; but it is useful for a data communications modulation method called **binary phase shift keying**. The *1*'s and *0*'s of the digital data cause the carrier phase to shift by 180°. More about this later in this chapter.

AMPLITUDE DEMODULATION

Before we can listen to the sound from an AM radio station, the modulation must be removed (demodulated) from the carrier and returned to the base-band frequencies. This is a rather simple process and can be accomplished without the use of integrated circuits. Figure 17–4 shows a basic AM demodulator circuit (also called a detector).

The diode rectifies the carrier, and the capacitor charges to the positive peaks of the carrier waveform. Since the capacitor follows the variations in the peaks, it generates a reproduction of the modulating waveform. The purpose of the resistor is to allow the capacitor to discharge when the carrier peak amplitude is dropping.

The selection of C_d and R_d is rather critical. If C_d, or R_d, is too large, the output amplitude will be reduced at high frequencies. With C_d, or R_d, too small, ripple at the carrier frequency will appear at the output. The actual selection of these components is based on just passing the highest slew rate—which will occur at the highest demodulated frequency along with the highest modulation index. The equation for determining the components is derived by equating the maximum voltage change in the demodulated sine wave during a time equal to the period of the carrier to the capacitor discharge voltage over the same time. The resulting equation is

$$R_dC_d = \frac{1}{2\pi m_a f_{a(max)}} \qquad \textbf{(equation 17–3)}$$

where $f_{a(max)}$ is the maximum expected modulation frequency and m_a the modulation index.

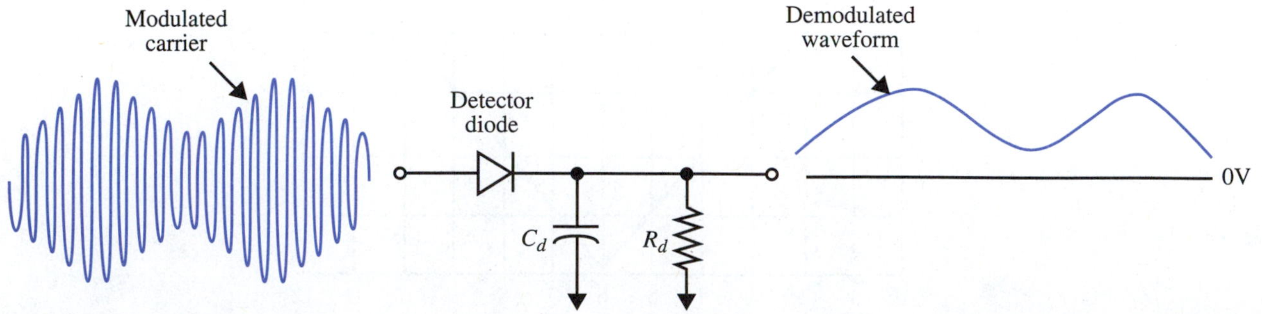

FIGURE 17–4 AM demodulator

EXAMPLE

Determine the value of R_d for an amplitude demodulator in an AM receiver if the highest audio frequency transmitted for AM is 5KHz and the maximum modulation index m_a is 0.9. Assume a value of 10nF for C_d.

SOLUTION

Solving for R_d in equation 17–3,

$$R_d = \frac{1}{2\pi m_a C_d f_{a(max)}}$$

$$R_d = \frac{1}{2\pi \times 0.9 \times 10\text{nF} \times 5\text{KHz}}$$

$$R_d = 3.54\text{K}\Omega \qquad (\text{Use } 3.6\text{K}\Omega.)$$

FREQUENCY MODULATION

Frequency modulation (FM) is a more complex modulation method and requires greater bandwidth than amplitude modulation does, but it does reduce the noise and improve transmitted power efficiency.

The frequency of the carrier is changed with frequency modulation with the amount of change (Δf_c) in carrier frequency proportional to the amplitude (v_a) of the modulating signal, or

$$\Delta f_c = K v_a \qquad \textbf{(equation 17–4)}$$

K is called the frequency deviation sensitivity (freq/volts).

Figure 17–5 shows the carrier frequency change with modulating signal amplitude. Notice that in the figure, the most-positive voltage of the modulating waveform results in the highest carrier frequency, and the most-negative, the lowest carrier frequency. The direction of frequency change could be the reverse! The point being that a change in polarity of the modulating signal causes the carrier frequency to shift in the opposite direction. How fast the carrier shifts in FM is proportional to the frequency of the modulation signal.

The FM modulation index m_f, for frequency modulation, is defined as the peak carrier frequency deviation divided by the frequency of the modulating signal.

$$m_f = \frac{\Delta f_c}{f_a} \qquad \textbf{(equation 17–5)}$$

or from equation 17–4,

$$m_f = \frac{K v_a}{f_a}$$

Unlike the AM modulation index, the FM modulation index can be less, equal to, or greater than unity.

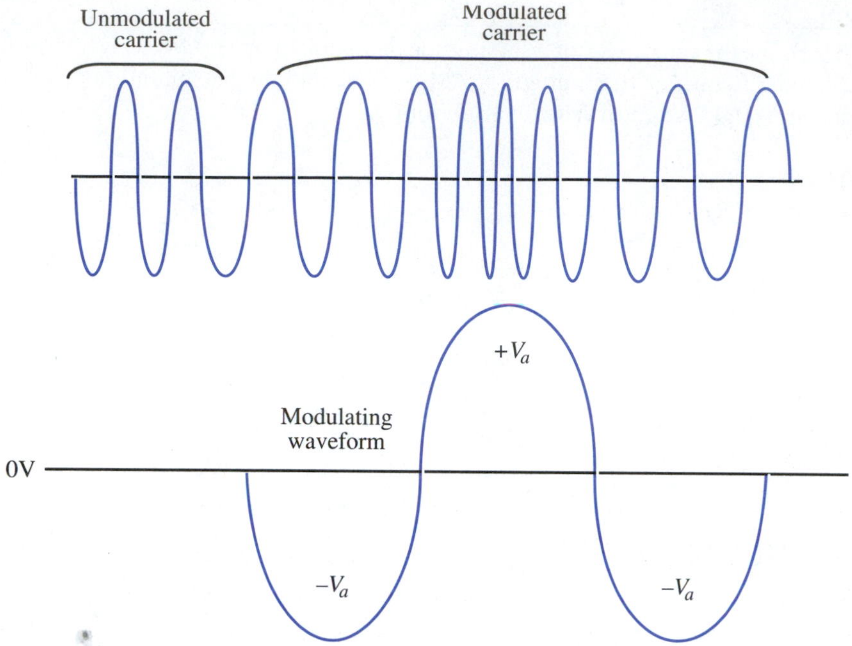

FIGURE 17–5 Frequency modulation

EXAMPLE

Find the maximum modulation index if the modulating waveform is

$$v_a = 1\text{V} \sin(2\pi \times 4\text{KHz} \times t)$$

and the deviation sensitivity K is 10KHz/V.

SOLUTION

Using equation 17–5 with the peak voltage of 1V and f_a of 4KHz,

$$m_f = \frac{Kv_a}{f_a}$$

$$m_f = \frac{10\text{KHz/V} \times 1\text{V}}{4\text{KHz}}$$

$$m_f = 2.5$$

Determining the sideband frequencies that result from modulation is not as straightforward with FM as it is with AM. Frequency modulating a carrier with a single frequency can result in an infinite number of sideband frequencies. The number of significant amplitude sidebands is related to the modulation index—but not in a simple manner. The mathematical process to determine the sidebands uses Bessel functions of the first kind. The Bessel coefficients give the amplitude of the different sidebands, but only those greater than

1 percent of the unmodulated carrier amplitude are considered. Examples of the number of significant frequency sidebands for various value of m_f are given in the following chart.

Modulation Index	Number of Significant Sidebands (N)
$m_f = 0.25$	2
$m_f = 1.0$	6
$m_f = 5.0$	16

The required bandwidth is determined by multiplying the number of sidebands (N) by the modulation frequency.

$$BW = N \times f_a \qquad \textbf{(equation 17–6)}$$

EXAMPLE

Determine the required bandwidth for an FM system modulated with a 15KHz sine wave if the modulation index is 5.

SOLUTION

With the modulation index of 5, we have 16 sidebands. Using equation 17–6,

$$BW = N \times f_a$$

$$BW = 16 \times 15\text{KHz}$$

$$BW = 240\text{KHz}$$

Amplitude modulating a carrier with a 15KHz signal requires a bandwidth of just 30KHz.

The noise advantage of FM results from the fact that the information in an FM carrier is in the zero crossing points of the waveform. Noise has an effect on the amplitude of a signal but has little effect on the zero crossing points.

Since most circuits used to remove the information from the FM carrier (demodulate) are sensitive to amplitude variations, the modulated carrier amplitude is limited (top and bottom clipped) to remove amplitude variations (noise) before demodulation is performed.

The power advantage of FM occurs because, unlike AM, power in the carrier is shifted to the sidebands during modulation.

FREQUENCY MODULATION CIRCUIT

The fundamental requirement of frequency modulation (FM) is to vary a carrier oscillator frequency with an input voltage (modulating voltage). Initially, this was accomplished by oscillators designed with a reactance tube—a device whose capacitance can be varied by changes in its input bias voltage. When the reactance tube input bias voltage (modulating voltage) would change, so would the oscillator output frequency and provide FM. Today, the reactance tube has been replaced by the varactor (voltage variable capacitor) diode for high-frequency applications, but at lower frequencies the integrated-circuit type of voltage-controlled oscillator (VCO) is used. This section will describe this type of device.

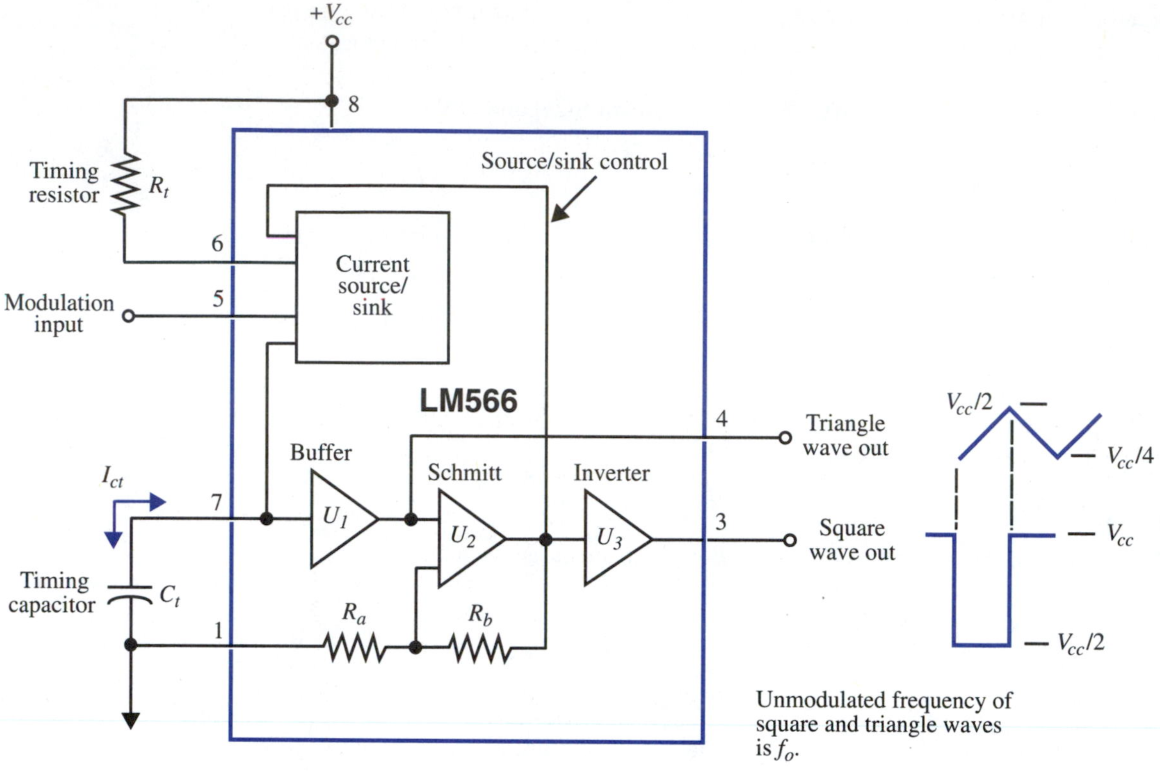

FIGURE 17–6 LM566 voltage-controlled oscillator. *Courtesy National Semiconductor.*

The XR-2206 function generator we studied with AM has the capability to provide a frequency-modulated signal. But in the interest of getting exposure to the maximum number of integrated circuit devices, we shall look at another chip that is specifically used for this operation.

One of the more common types of integrated circuit VCO devices is the LM566 (NE/SE566). This device has a maximum operating frequency of 1MHz and can provide a 10-to-1 range of frequency variation with a change in the modulating input voltage. Figure 17–6 illustrates a basic block diagram of the 566 VCO.

Referring to Figure 17–6, the current source/sink circuit provides a constant charging or discharging current to the timing capacitor C_t. The amount of current from the source/sink is controlled by the timing resistor R_t and the voltage across this resistor. Increasing the resistance of R_t decreases the capacitor charging/discharging current, which lowers the VCO frequency. Control of the charging current is also possible by changing the voltage at the modulating input. The voltage at Pin 6 is internally maintained at the same voltage as Pin 5. So if the modulating voltage at Pin 5 is increased, the voltage at Pin 6 increases, resulting in less voltage across R_t and, therefore, less charging current.

The charging voltage developed across capacitor C_t is applied to the Schmitt trigger circuit U_2 via the buffer amplifier U_1. The output voltage swing on the Schmitt trigger is approximately V_{cc} to $0.5\ V_{cc}$. Resistors R_a and R_b form a positive feedback loop from the output of U_2 to its noninverting input. With equal dividing resistors R_a and R_b, the noninverting input swing is from $0.5\ V_{cc}$ to $0.25\ V_{cc}$.

If the voltage on the timing capacitor C_t exceeds $0.5\ V_{cc}$ during charging, it will cause the Schmitt trigger output to go low ($0.5\ V_{cc}$). A low level on the output of U_2 causes the current source to change to a sink (discharging C_t). When C_t discharges to $0.25\ V_{cc}$, the output of the U_2 will swing high (V_{cc}), causing the current sink to return to a source (charging C_t). Since the source and sink currents are equal, it takes the same amount of time to charge C_t as it does to discharge this capacitor. This results in a triangular voltage waveform across C_t, which is available as a buffered output at Pin 4. A square wave appears at the output of the Schmitt trigger and is inverted by inverter U_3 for a second output at Pin 3. If the current from the source/sink is increased, the charge/discharge time for the capacitor is reduced, and the output frequency is increased (shorter period).

The nominal (unmodulated carrier) output frequency of the VCO can be changed by one of three methods:

1. Changing the value of C_t
2. Changing the value of R_t (changes current)
3. Changing the voltage at the modulating input terminal (changes current)

We can determine the actual frequency of oscillation from the time it takes to charge and discharge the capacitor. The basic equation for charging a capacitor from a constant current source is

$$i = \frac{C\,\Delta V}{\Delta t}$$

So

$$\frac{\Delta V}{\Delta t} = \frac{i}{C}$$

where ΔV is the voltage change on the capacitor during the time Δt.

The total voltage on capacitor changes from $0.25V_{cc}$ to $0.5V_{cc}$, so

$$\Delta V = 0.5V_{cc} - 0.25V_{cc}$$

$$\Delta V = 0.25V_{cc}$$

and

$$\frac{0.25V_{cc}}{\Delta t} = \frac{i}{C_t}$$

Therefore

$$\Delta t = \frac{0.25V_{cc}C_t}{i}$$

The triangular waveform on the capacitor has a period T of $2\,\Delta t$ (equal charging and discharging time). The frequency of oscillation f_o is

$$f_o = 1/T$$

$$f_o = 1/2\,\Delta t$$

Substituting for Δt,

$$f_o = \frac{i}{0.5V_{cc}C_t}$$

But i is determined from the current through R_t.

$$i = \frac{V_{cc} - V_5}{R_t}$$

where V_5 is the voltage at Pin 5.

Making the substitution for i in the frequency equation,

$$f_o = \frac{2(V_{cc} - V_5)}{C_t R_t V_{cc}} \qquad \textbf{(equation 17–7)}$$

For best operation, the resistance of R_t should be in the range of 2KΩ to 20KΩ.

In actual operation, R_t and C_t are selected for the correct carrier frequency f_o, and the modulating input voltage is varied to give the desired frequency modulation of the carrier. The range of allowable variation of the modulating signal is from $0.75V_{cc}$ to V_{cc}, which yields an output frequency variation of greater than 10 to 1.

With no input modulation signal applied, the voltage at Pin 5 is biased midway between $0.75V_{cc}$ and V_{cc}, or $0.875V_{cc}$. This allows us to simplify equation 17–7 to give the carrier frequency f_o

$$f_o = \frac{2(V_{cc} - 0.875V_{cc})}{C_t R_t V_{cc}}$$

$$f_o = \frac{1}{4C_t R_t} \qquad \textbf{(equation 17–8)}$$

EXAMPLE

Find the required value of R_t for a 566 if the desired carrier frequency is 20KHz and C_t is 10nF.

SOLUTION

Rearranging equation 17–8 to solve for R_t,

$$R_t = \frac{1}{4C_t f_o}$$

$$R_t = \frac{1}{(4 \times 10\text{nF} \times 10\text{KHz})}$$

$$R_t = 2.5\text{K}\Omega$$

If we want to determine what input modulation voltage change (ΔV_a) is required to give a given output frequency deviation (Δf_o), we need to realize that reducing the voltage at Pin

5 puts more voltage across R_t and, therefore, increases the charging current, which charges the capacitor quicker and results in an increase in frequency.

We have assumed the original frequency is f_o. Let the new frequency due to the modulating input be f_1, then the frequency will change by the amount Δf_o.

$$\Delta f_o = f_1 - f_o$$

$$\Delta f_o = \frac{2(V_{cc} - 0.875V_{cc} - \Delta V_a)}{C_t R_t V_{cc}} - \frac{2(V_{cc} - 0.875V_{cc})}{C_t R_t V_{cc}}$$

$$\Delta f_o = \frac{-2\,\Delta V_a}{C_t R_t V_{cc}}$$

Solving for ΔV_a,

$$\Delta V_a = -\frac{\Delta f_o C_t R_t V_{cc}}{2}$$

Substituting for $R_t C_t$ using equation 17–8,

$$\Delta V_a = -\frac{\Delta f_o V_{cc}}{8 f_o} \qquad \textbf{(equation 17–9)}$$

or

$$\Delta f_o = -\frac{\Delta V_a 8 f_o}{V_{cc}} \qquad \textbf{(equation 17–10)}$$

and the VCO frequency deviation sensitivity K is

$$K = \frac{\Delta f_o}{\Delta V_a}$$

$$K = -\frac{8 f_o}{V_{cc}} \qquad \textbf{(equation 17–11)}$$

EXAMPLE

Determine the required voltage change at the modulating input (Pin 5) of a 566 if it is required to increase the frequency of a 10KHz carrier by 200Hz. Supply voltage is +12V. Also find the deviation sensitivity K.

SOLUTION

Using equation 17–9,

$$\Delta V_a = -\frac{\Delta f_o V_{cc}}{8 f_o}$$

$$\Delta V_a = -\frac{200\text{Hz} \times 12\text{V}}{8 \times 10\text{KHz}}$$

$$\Delta V_a = -30\text{mV}$$

The deviation sensitivity is from equation 17–11.

$$K = -\frac{8f_o}{V_{cc}}$$

$$K = -\frac{8 \times 10\text{KHz}}{12\text{V}}$$

$$K = -6.667\text{KHz/V}$$

FREQUENCY MODULATION INPUT CIRCUIT

A typical input modulation circuit for a 566 is shown in Figure 17–7. The two resistors R_1 and R_2 are used to set up the required operating bias of $0.875V_{cc}$. The actual modulating input signal is coupled through coupling capacitor C_c.

We will now go through the process of determining the actual component values for the input circuit.

To determine the values of R_1 and R_2 we need to be aware that the input resistance at Pin 5 is typically 1MΩ. We need to make R_2 about one-hundredth of this value in order to have

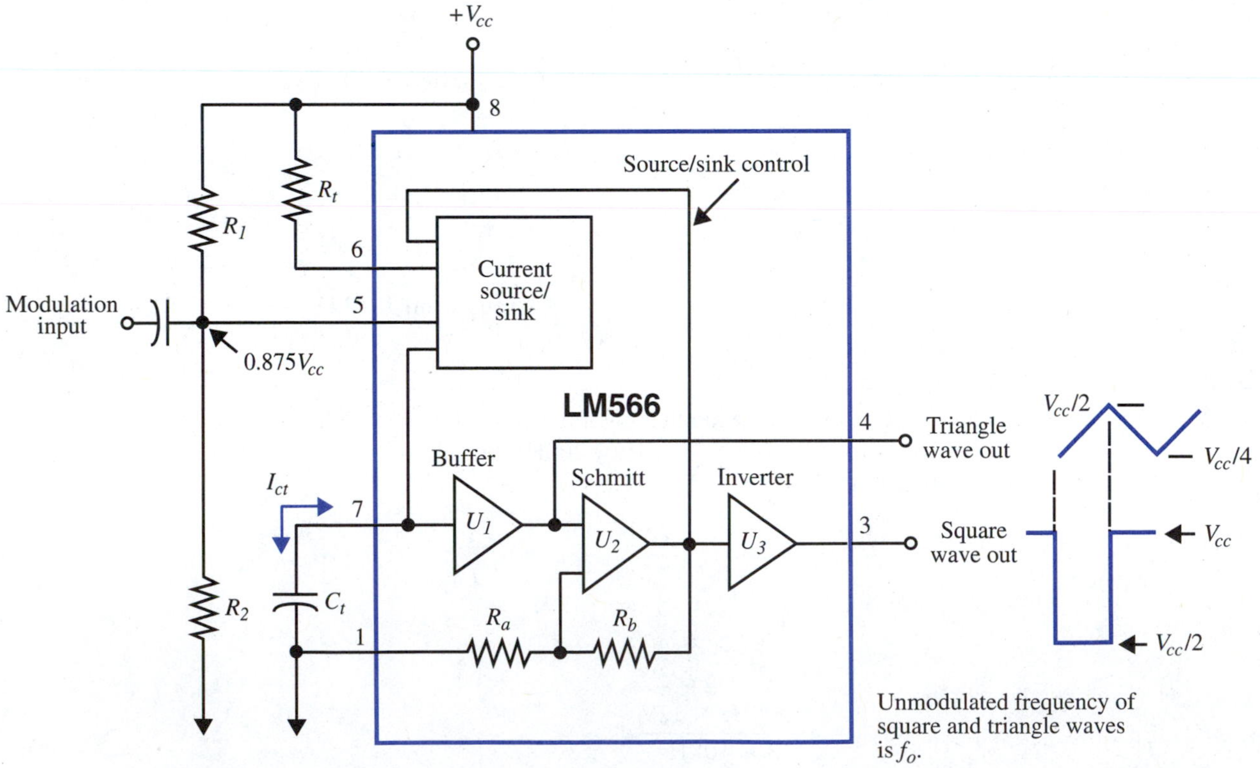

FIGURE 17–7 Modulation input circuit for the 566

a predictable DC voltage at Pin 5. Let us assume a power supply voltage of 15V and 1mA of current through the R_1,R_2 divider. The required DC voltage at Pin 5 is $15V \times 0.875$, or 13.13V. Dividing this voltage by 1mA, we get 13.13KΩ. Staying with standard resistor values, we can use a 13KΩ resistor for R_2 and a 2KΩ resistor for R_1. This still makes the DC voltage very close to the desired DC value.

The coupling capacitor can be determined by making the capacitor reactance equal to one-tenth of the parallel combination of R_1 and R_2, or

$$X_c = \frac{R_1 \| R_2}{10}$$

The capacitor equation is then

$$C_c = \frac{1}{2\pi f_a X_c}$$

$$C_c = \frac{10}{2\pi f_a (R_1 \| R_2)} \qquad \textbf{(equation 17–12)}$$

If we arbitrarily assume that the minimum modulating frequency is 100Hz, then the coupling capacitor is

$$C_c = \frac{10}{2\pi \times 100\text{Hz} \times 1.73\text{K}\Omega}$$

$$C_c = 9.2\mu\text{F}$$

(Use 10μF.)

A TYPICAL FREQUENCY MODULATOR

To further illustrate the operation of a VCO, we shall work through an expanded example of a frequency modulator. Let us assume the following values:

1) Carrier frequency is 50KHz.
2) Modulating signal frequency is 200Hz.
3) Modulating signal voltage is 20mV peak.
4) Power supply voltage is 12V.

Let us also assume the following requirements:

1) Draw the complete schematic with all component values specified.
2) Determine the maximum frequency excursions of the carrier.

Taking care of the first requirement, the values for R_1 and R_2 can stay the same since we still need $0.875V_{cc}$ at Pin 5.

The coupling capacitor size is determined from equation 17–12.

$$C_c = \frac{10}{2\pi f_a(R_1 \| R_2)}$$

$$C_c = \frac{10}{2\pi \times 200\text{Hz} \times (1.73\text{K}\Omega)}$$

$$C_c = 4.6\mu\text{F}$$

(Use 5μF.)

Values of C_t and R_t can be determined from equation 17–8.

$$f_o = \frac{1}{4C_tR_t}$$

We have to assume a value for one of the components since we have one equation and two unknowns. Let us make C_t equal to 1nF and solve for R_t.

$$R_t = \frac{1}{4C_tf_o}$$

$$R_t = \frac{1}{4 \times 1\text{nF} \times 50\text{KHz}}$$

$$R_t = 5\text{K}\Omega$$

The frequency equation for the 566 is not precise, so we need to make part of R_t variable. Suggest a 3.9KΩ fixed resistor and a 2KΩ potentiometer.

This completes the selection of external components, so now we shall determine the carrier frequency excursions.

Since we are looking for frequency change, we should use equation 17–10.

$$\Delta f_o = -\frac{\Delta V_a 8 f_o}{V_{cc}}$$

With ΔV_a of 20mV, f_o of 50KHz, and V_{cc} of 12V,

$$\Delta f_o = -\frac{20\text{mV} \times 8 \times 50\text{KHz}}{12\text{V}}$$

$$\Delta f_o = 667\text{Hz}$$

The actual carrier frequency excursions are the center frequency of 50KHz ± 667Hz or a low frequency of 49.33KHz and a high frequency of 50.67KHz.

The output can be taken from Pin 3 or Pin 4, depending on whether it is desired to have a frequency-modulated square wave or triangular wave.

The schematic, which includes the component values for this example, is shown in Figure 17–8.

DIGITAL FREQUENCY MODULATION

If the modulation input signal is digital, the carrier will assume a frequency corresponding to a *1* and another frequency corresponding to a *0*; thus, an abrupt change occurs in the

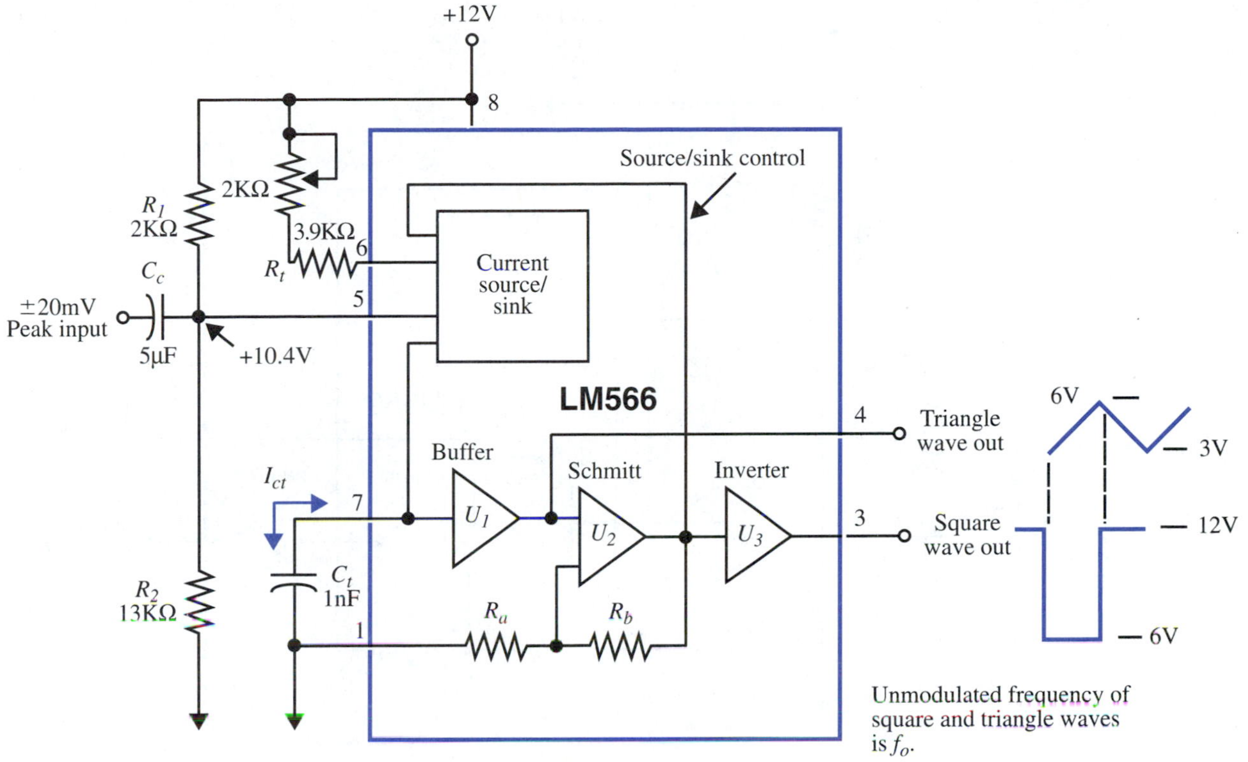

FIGURE 17–8 Frequency modulator circuit example

carrier frequency. This method of digital modulation is known as **frequency shift keying** (FSK) and is used in data communications.

Remember that when a signal passes through a capacitor, there is no DC component passed, and the waveform will center itself so there is equal signal area above and below the waveform. So, if we are coupling a digital signal and the code changes, the *1* and *0* voltage values will change; i.e., if the number of *1*'s in the code increases, the signal will shift down. This means that the actual voltage for a *1* and *0* will change, which will cause a change in the *1/0* frequency. Therefore, AC coupling should not be used for digital input signals.

Let us look at a digital input circuit that can provide the "originate" channel signals for a Bell 103 (300 baud) **modem** (modulator-demodulator). The modem uses a transmit frequency for a *1* (called a mark) of 1270Hz and the frequency for a *0* (called a space) is 1070Hz.

The circuit using the necessary DC coupling is shown in Figure 17–9. The resistors R_1 and R_2 are selected to provide a VCO frequency of 1070Hz when Q_1 is not conducting (logic *0* input).

To determine the value of these resistors, we need to know the voltage required at Pin 5 to provide this frequency. Let us assume that R_t and C_t are selected for a mid-frequency of 1170Hz and that the modulation input will cause a ±100Hz (Δf_o) change from this value.

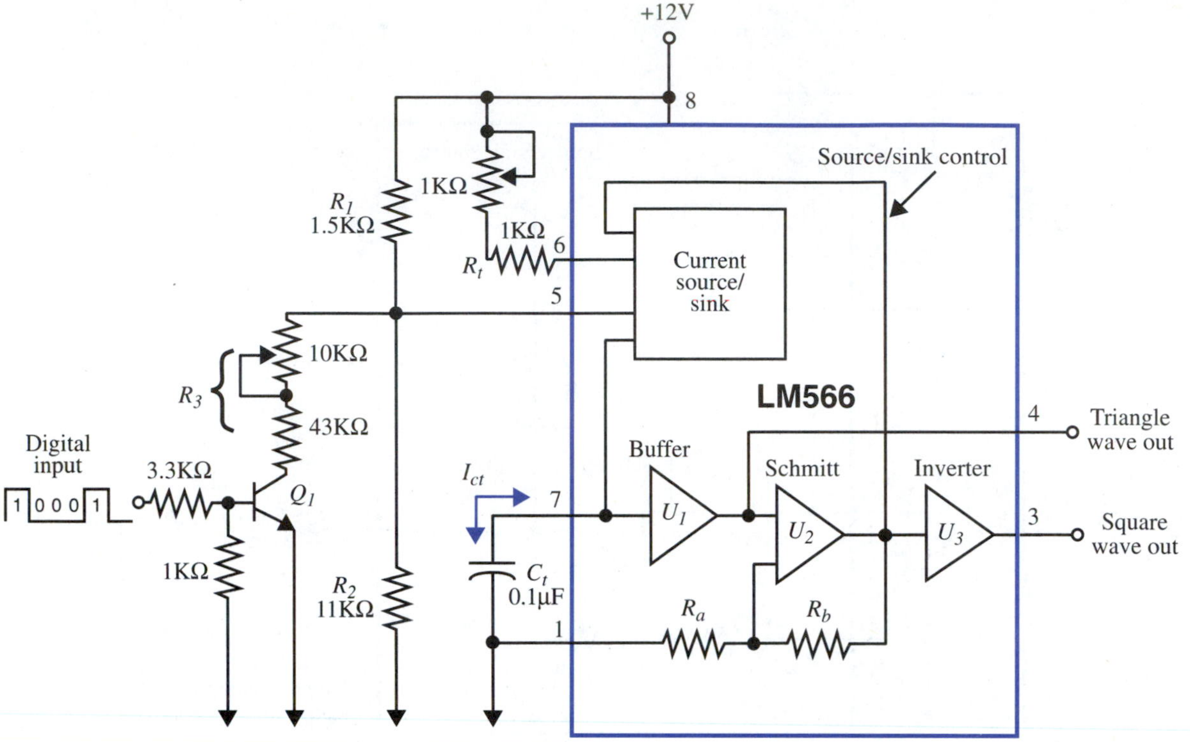

FIGURE 17–9 Modem configuration

Using equation 17–9, we can find the input voltage change ΔV_a to cause this frequency change.

$$\Delta V_a = -\frac{\Delta f_o V_{cc}}{8f_o}$$

$$\Delta V_a = -\frac{100\text{Hz} \times 12\text{V}}{8 \times 1170\text{Hz}}$$

$$\Delta V_a = 0.128\text{V}$$

The nominal DC voltage at Pin 5 is *12V × 0.875*, or 10.5V. So to provide a VCO output frequency of 1070Hz, the voltage should be *10.5V + 0.128V*, or 10.628V. If we stay with 1mA through the R_1,R_2 divider, then R_2 would be 10.628KΩ and R_1 would be 1.37KΩ. Scaling these values closer to standard resistor values gives R_2 of 11KΩ and R_1 of 1.4KΩ.

To provide the VCO output frequency of 1270Hz, the voltage at Pin 5 should drop to *10.5V − 0.128V*, or 10.372V. To obtain this voltage, we need to determine the value of resistor R_3 to put in parallel with R_2 (Q_1 conducting). With this condition, we have a series/parallel circuit with 10.372V across the parallel portion. The current through R_1 then is

$$I_1 = \frac{(12V - 10.372V)}{1.4K\Omega}$$

$$I_1 = 1.163mA$$

This same current flows through the parallel equivalent of R_2, R_3.

$$R_{eq} = 10.372V/1.163mA$$

$$R_{eq} = 8.919K\Omega$$

Solving for R_3,

$$R_3 = \frac{R_{eq}R_2}{(R_2 - R_{eq})}$$

$$R_3 = \frac{8.919K\Omega \times 11K\Omega}{(11K\Omega - 8.919K\Omega)}$$

$$R_3 = 47K\Omega$$

To accommodate tolerances, R_3 is shown in the schematic as a 43KΩ resistor in series with a 10KΩ potentiometer.

FM DEMODULATION

There are several ways to demodulate an FM carrier, but they all basically rely on a concept of providing an output amplitude, which is a function of carrier frequency. The typical S-curve detector response is shown in Figure 17–10.

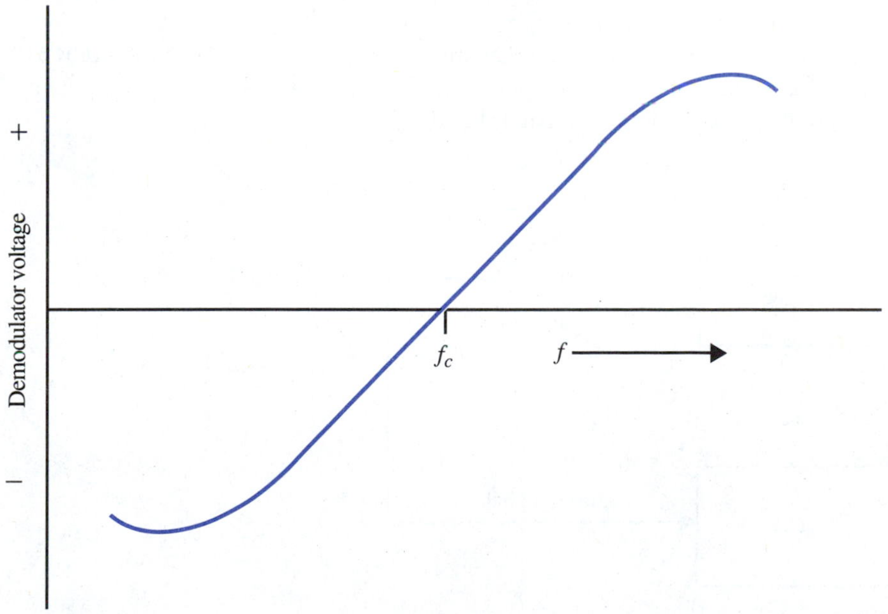

FIGURE 17–10 S-curve detector response

The curve shows that at the carrier frequency, there is no output voltage—corresponding to no frequency modulation. If the carrier frequency is modulated and shifted higher, the demodulator output voltage increases in a positive direction.

Lowering the carrier frequency increases the demodulator output voltage in the negative direction. The greater the deviation of the carrier, the greater the output voltage—which corresponds to: The greater the input voltage to the frequency modulator, the greater the frequency deviation of the carrier.

FM demodulators like the slope detector, ratio detector, and discriminator, are all adaptations of the S curve. The phased lock loop also has an S-curve response; but it has the additional advantages of ease of tuning and excellent rejection of unwanted frequencies. We shall concentrate on the integrated circuit phase locked loop as an FM demodulator and also study its operation as a frequency synthesizer.

PHASE LOCKED LOOP

A phase locked loop (PLL) is a circuit that contains a voltage controlled oscillator (VCO), a phase comparator, and an amplifier. Figure 17–11 shows the basic block diagram of a PLL.

The device is made to operate as an FM demodulator by selecting the oscillator *R,C* components so that the VCO oscillates at the same frequency as the unmodulated carrier. When the carrier is modulated, the VCO frequency is made to follow the frequency changes in the carrier. The error signal voltage required to change the frequency of the VCO to match the carrier becomes the demodulated output signal.

When the VCO is at the same frequency as the carrier, it is said to be locked in. The frequency of the VCO is controlled by the error signal that is generated by a comparison of the carrier frequency with the VCO output frequency. Comparison of the signals is accomplished by a phase comparator within the PLL. The amplifier serves to amplify the output error signal from the phase comparator.

A phase locked loop, which is the companion to the LM566 VCO, is the LM565. This chip contains a VCO identical to the 566, as well as a phase detector and amplifier. The block diagram for the LM565 is shown in Figure 17–12.

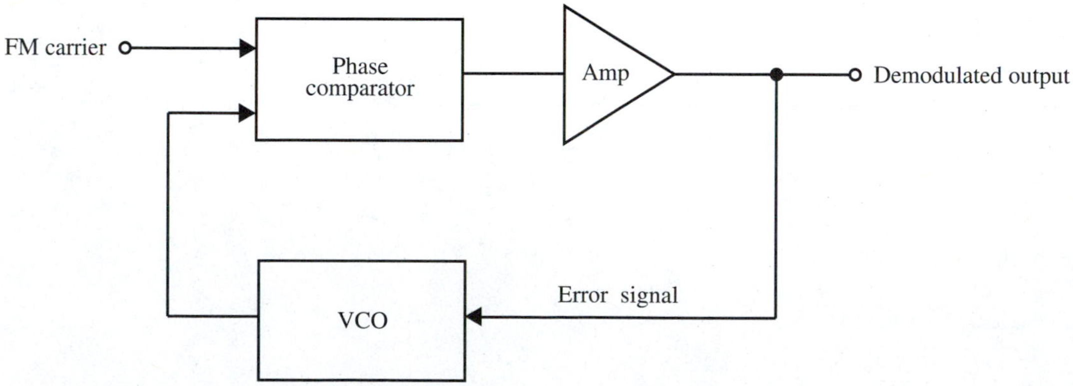

FIGURE 17–11 Basic block diagram of a PLL

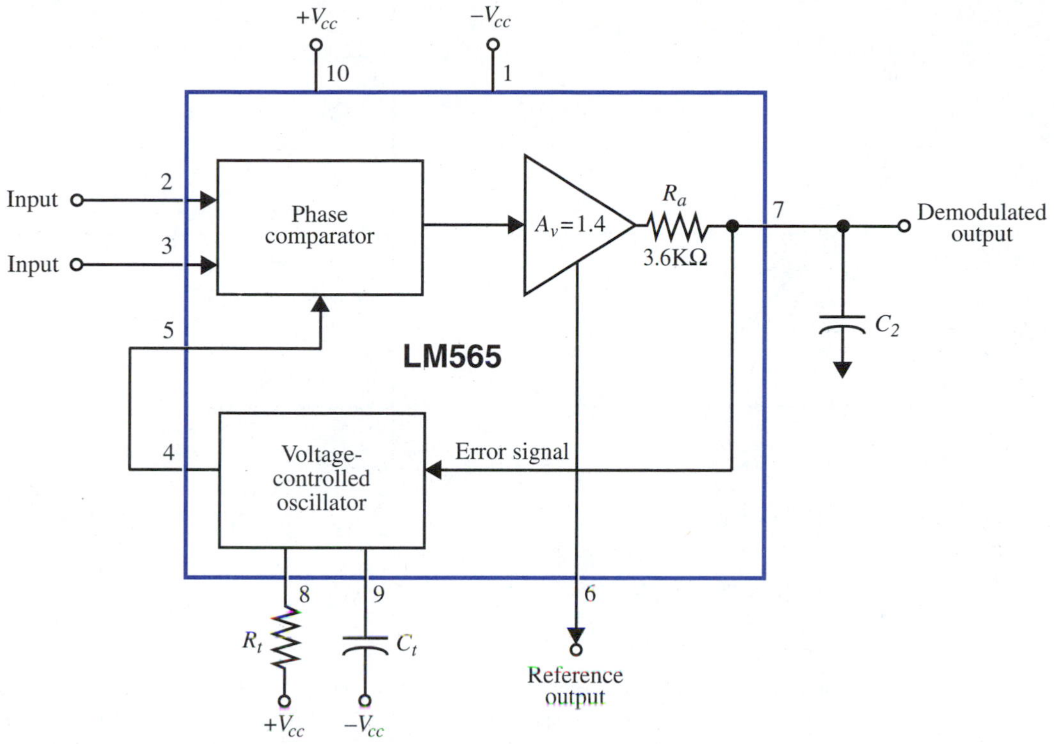

FIGURE 17–12 LM565 phase locked loop. *Courtesy National Semiconductor.*

Just as with the LM566, the VCO frequency is set by R_t and C_t, using equation 17–8

$$f_o = \frac{1}{4C_tR_t}$$

For best operation, R_t should be within the range of 2KΩ to 20KΩ, with 4 KΩ being the optimum value.

The error signal input to the VCO (Pin 7) is the same as Pin 5 on the 566, and the DC voltage at this point is 0.875 of the total power supply voltage $[+V_{cc} - (-V_{cc})]$.

Pin 6 tracks the DC voltage of Pin 7, and this pin is used if it is desired to have a differential output feeding a differential op-amp configuration, which will remove the DC component.

The dual inputs to the phase comparator (Pins 2 and 3) allow for a differential input. If a single-ended input is required, one of these pins is grounded.

PHASE LOCKED LOOP OPERATION

With reference to Figure 17–12, an input signal (carrier) is applied to the phase comparator (Pins 2 and 3) and is compared with the signal from the internal VCO. If a frequency or, therefore, a phase difference occurs between the incoming and VCO signals, an error

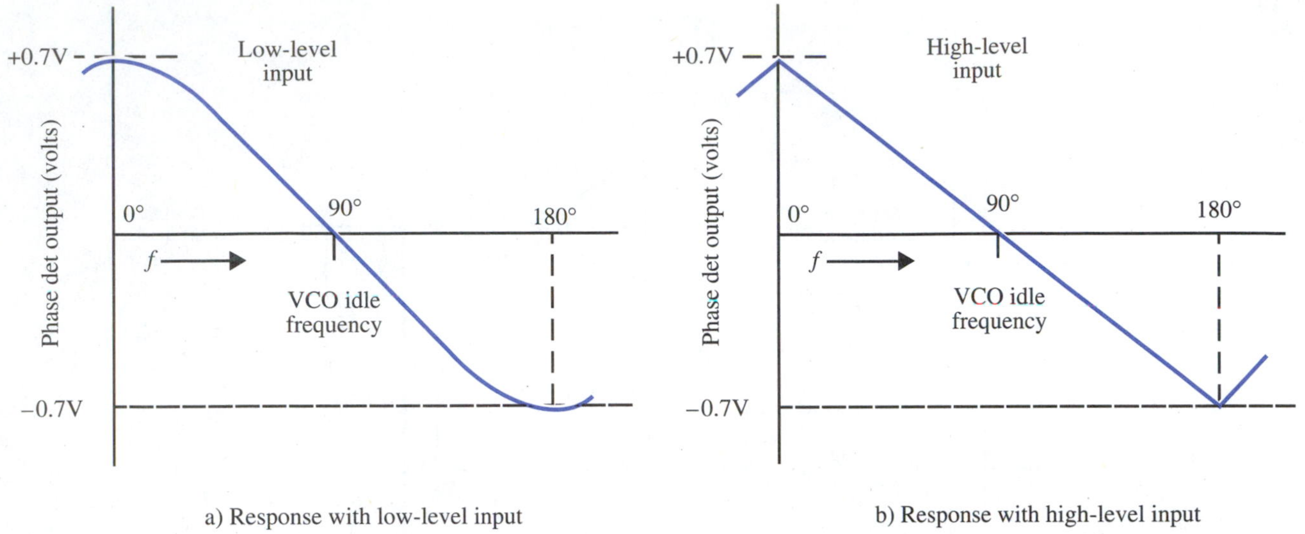

FIGURE 17–13 LM565 phase detector characteristics

signal is generated at the output of the phase comparator. This error signal is amplified and filtered by the low-pass filter, consisting of R_a (internal to the PLL) and the external capacitor C_2, and fed back to the VCO as a control (error) signal. The filtered signal is also used as the output signal when the PLL is operated as an FM demodulator.

If the input frequency and VCO output frequency differ, the feedback control signal shifts the VCO frequency until it matches the incoming frequency. When a match occurs, the VCO is in the locked condition. If the incoming signal frequency is varied, the VCO output frequency will follow, provided that the incoming signal frequency stays within a hold-in (lock-in) range of the PLL. A phase difference of 90° exists if the incoming signal is at the same frequency as the VCO idle frequency (determined by R_t and C_t).

Figure 17–13 shows the S curves for the LM565. The high-level curve results from clipping in the phase detector, which occurs when the input signal is greater than 200mV_{p-p}.

The horizontal axis is scaled for both frequency and phase shift (between VCO and incoming signal), while the vertical axis indicates the phase detector output voltage.

If the incoming signal frequency is below the VCO idle frequency, the lock-in phase ranges from 90° to 0°; with the incoming signal frequency above the VCO idle frequency, the phase is from 90° to 180°. When in lock with a low-level input, the output voltage from the phase detector follows the cosine of the phase difference between the incoming signal and the VCO frequency. With incoming signals greater than the 200mV_{p-p} level, the input sine wave will become a square wave, and the phase detector output response will become linear rather than cosine.

If the input signal and the VCO idle frequencies are the same, the output of the phase detector is zero (cos 90° is 0). When the incoming signal frequency is above the VCO idle frequency, the phase detector output ranges from 0V to −0.7V; when the incoming signal frequency is below the VCO idle frequency, the phase detector output ranges from 0V to

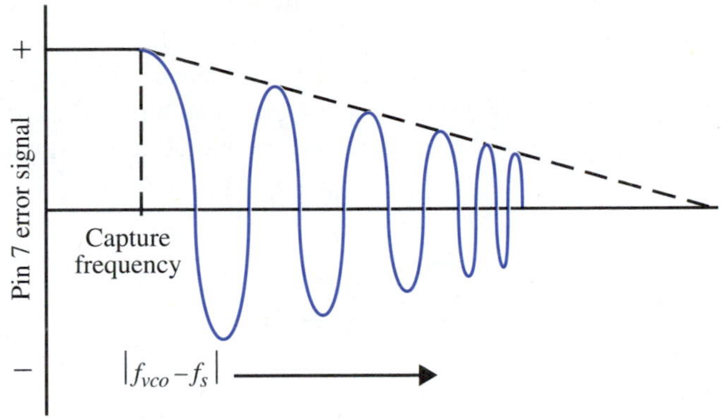

FIGURE 17–14 Capture of the 565 VCO

+0.7V. If the incoming signal frequency moves the output of the phase detector beyond its peak of 0.7V, in either direction, the output voltage reduces. The error voltage is now not great enough to move the VCO to match the incoming signal frequency, and the PLL falls out of lock.

Let us assume that the frequency of the incoming signal is far removed from the idle frequency of the VCO. The output from the phase detector will now be the difference frequency between the two signals, but it will be attenuated at Pin 7 by the low-pass filter (R_a,C_2). Now if we move the incoming signal frequency closer to the VCO frequency, the difference frequency will reduce and more voltage will appear across the low-pass filter. As we move the incoming frequency still closer, a point will be reached where the voltage at the filter is sufficient to move the VCO to match the frequency of the incoming signal. At this point, the incoming frequency is at the **capture frequency**—the VCO is captured by the incoming signal. Figure 17–14 shows graphically how the capture frequency is reached.

The VCO frequency swings back and forth with greater excursions at the slower rate as the signal frequency comes closer. When the error voltage is sufficient to cause the VCO to reach the incoming signal frequency, capture occurs. Once captured, the VCO will follow the incoming frequency over the lock-in range.

The fact that Figure 17–14 shows capture with a positive error voltage indicates that the incoming frequency was below the VCO idle frequency (review operation of the 566).

EQUATIONS FOR LM565 OPERATION

We will now develop the equations for the operation of the LM565 by first determining the equation for the lock-in range for the 565.

At the extremes of the lock-in range, the output voltage from the phase detector is ±0.7V. Now, the phase detector is followed by an amplifier with a gain of 1.4; so the maximum error voltage at Pin 7 is $\pm 0.7V \times 1.4$, or ±0.98V. To determine what frequency shift of the VCO will occur with these voltages, and therefore the lock-in range, we go back to the equation 17–10.

$$\Delta f_o = -\frac{\Delta V_a 8 f_o}{V_{cc}}$$

If we put in the error voltages of ±0.98V for ΔV_a, we obtain the equation for the deviation Δf_L of the lock-in frequencies' limits from the center VCO idle frequency (f_o).

$$\Delta f_L = \pm\frac{0.98\text{V} \times 8 f_o}{V_{cc}}$$

$$\Delta f_L = \pm\frac{7.84\text{V} \times f_o}{V_{cc}} \qquad \textbf{(equation 17–13)}$$

This equation shows that the higher the center frequency, the greater the lock-in range, but raising the supply voltage decreases the lock-in range.

EXAMPLE

Determine the lock-in range for a 565 if the R_t is 2.5KΩ, C_t is 10nF, and the supply voltages are ±6V.

SOLUTION

First, we must find the VCO idle frequency f_o, using equation 17–8:

$$f_o = \frac{1}{4C_tR_t}$$

$$f_o = \frac{1}{4 \times 10\text{nF} \times 2.5\text{K}\Omega}$$

$$f_o = 10\text{KHz}$$

Equation 17–13 requires the total supply voltage of $+6V - (-6V)$, or 12V.

$$\Delta f_L = \pm\frac{7.84\text{V} \times f_o}{V_{cc}}$$

$$\Delta f_L = \pm\frac{7.84\text{V} \times 10\text{KHz}}{12\text{V}}$$

$$\Delta f_L = \pm 6.53\text{KHz}$$

This means the actual frequency range is from 10KHz ± 6.53KHz, or 3.47KHz to 16.5KHz.

Figure 17–15 shows in the frequency domain, the lock-in frequencies relative to the VCO idle frequency for this example.

The lock-in range can also be reduced by lowering the amplifier gain. This is done by connecting a resistor between the demodulated output (Pin 7) and the reference output (Pin 6). Using resistors from 30KΩ down to 0Ω reduces the lock-in range from +60% down to +20% of f_o.

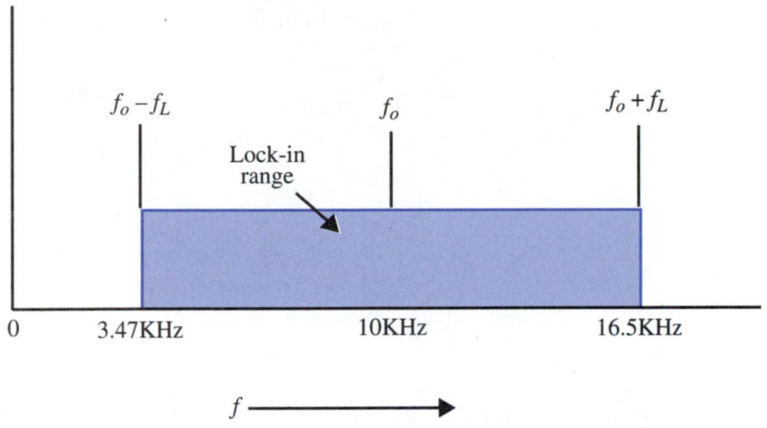

FIGURE 17–15 Lock-in range for example problem

As described before, if the incoming signal frequency is far removed (outside of the lock-in range) and is then brought closer, a point is reached where the VCO frequency is captured by the incoming signal frequency. Again, this occurs when the error voltage developed across the low-pass filter formed by R_a and C_2 is sufficient to move the VCO to the incoming frequency.

Before the VCO has been captured by the incoming signal, the error voltage is a sine wave with a frequency that is the difference between the VCO and the input signal. This sine wave becomes lower in frequency as the signal frequency gets closer to the VCO (Figure 17–14). With a lower-frequency error signal, the error voltage across the low-pass filter capacitor C_2 increases. Remembering the circuit operation just described, we will develop an equation for the capture frequencies.

On the slope portion of a low-pass filter, the output voltage (V_o) can be approximated by

$$V_o = V_e \frac{X_c}{R_a}$$

where V_e is the error voltage, and X_c the reactance of C_2.

The maximum output error voltage V_e was determined previously to be 0.98V, and R_a is given as 3.6KΩ. So

$$V_o = \frac{0.98\text{V}}{3.6\text{K}\Omega \times 2\pi f_d C_2} \qquad \textbf{(equation 17–14)}$$

with f_d the difference frequency between the VCO and the incoming frequency.

At the point of capture, let

$$f_d = f_{cap}$$

The following relationship holds:

$$\frac{f_{cap}}{f_L} = \frac{V_o}{0.98\text{V}}$$

Substituting for V_o, and solving for the capture frequency f_c,

$$f_{cap} = \pm\sqrt{\frac{f_L}{2\pi R_a C_2}} \qquad \textbf{(equation 17–15)}$$

or

$$f_{cap} = \pm\sqrt{\frac{7.84 f_o}{2\pi R_a C_2 V_{cc}}} \qquad \textbf{(equation 17–16)}$$

It should be realized that the capture frequency must always be equal to the lock-in frequency limit (without C_2) or within the lock-in frequency range.

EXAMPLE

Predict the capture frequencies for an LM565 if the VCO frequency is 50KHz, C_2 is 47nF, and the total supply voltage is 15V.

SOLUTION

Since we are provided f_o, we can use equation 17–16 to find the capture frequencies.

$$f_{cap} = \pm\sqrt{\frac{7.84 f_o}{2\pi R_a C_2 V_{cc}}}$$

$$f_{cap} = \pm\sqrt{\frac{7.84 \times 50\text{KHz}}{2\pi \times 3.6\text{K}\Omega \times 47\text{nF} \times 15\text{V}}}$$

$$f_{cap} = \pm 4.958\text{KHz}$$

The actual capture frequencies are 50KHz $\pm$ 4.958KHz, or 45.042KHz and 54.958KHz.

To prevent noise capturing of the PLL, the capture frequency should be as close as possible to the center frequency f_o of the VCO (small f_{cap}) but still accommodate drift in the input carrier frequency.

Most PLLs clip the input signal, which results in harmonics of the input frequency. So the PLL can false lock on harmonics of input signals that have frequencies f_o/n, where n is a positive integer. The solution is to pass input signals through a high-pass filter (in front of the PLL) with a cut-off frequency a little greater than $f_o/2$.

PLL DOUBLE-FREQUENCY REJECTION

The phase comparator within the PLL can be considered a multiplier for low-level inputs that multiplies the input signal ($V_c \sin 2\pi f_c t$) by the VCO signal ($V_c \sin 2\pi f_c t + \theta_o$). The product of the multiplied signals is the output voltage v_e from the phase comparator.

$$v_e = K V_c V_o (\sin 2\pi f_c t)(\sin 2\pi f_c t + \theta_o)$$

where K is the phase detector constant and θ_o the phase shift between the incoming and the VCO when in the lock-in range.

Using trigonometric manipulation and setting f_c equal to f_o (in lock), the product breaks down into the following two terms:

$$v_e = KV_cV_o\cos(-\theta_o) - KV_cV_o\cos[2\pi(2f_ot) + \theta_o] \quad \textbf{(equation 17–17)}$$

What we have here first is a DC term, the amplitude of which varies with the phase shift (θ_o) between the incoming signal frequency and the VCO frequency (this is the error or demodulated output signal of Figure 17–13); the second term is double the locked-in VCO frequency, with a peak amplitude of 0.98V. Both of these signals appear at the LM565 Pin 7 output—but the double-frequency term contains no modulation information and therefore is undesirable.

The low-pass filter consisting of R_a and C_2 can reject the PLL double-frequency signal, the amount of rejection being a function of the value of C_2 and the frequency. So now we find that C_2 affects the capture frequency (equations 17–15 and 17–16) and the PLL double-frequency rejection. But we also have to consider the effect of C_2 on the highest demodulation frequency. We want the demodulated output but not the double-frequency signal. This means we need to separate the frequencies of these two signals as much as possible. Since the maximum demodulated signal is fixed by the system requirements, we should make the carrier frequency as high as possible. For example, if we make the carrier frequency 100 times greater than the highest modulating signal and set the −3db point of the low-pass filter at twice the highest modulating frequency, the rejection of the doubled carrier frequency at Pin 7 will be about 200 times greater than the modulation signal (reactance of C_2, and therefore the output voltage, drops by $1/f$).

It is possible that the requirement for C_2 to establish a particular capture frequency is in conflict with the value required for double-frequency rejection. The solution is to add a second low-pass RC filter to the output that will provide the necessary rejection. To prevent interaction, the value of the new resistor should be at least 10 times greater than 3.6KΩ.

FREQUENCY DEMODULATOR EXAMPLE

In order to more fully illustrate PLL operation, we shall go through a more complete example of a frequency demodulator.

Let us assume that we wish to FM demodulate a 100KHz carrier modulated with a 1KHz sine wave that deviates the carrier frequency ±10 KHz.

First, we choose a 565 PLL and set up the VCO frequency f_o to 100KHz. If we pick a 1nF capacitor for C_t, then R_t—from rearrangement of equation 17–8—is

$$R_t = \frac{1}{4C_tf_o}$$

$$R_t = 2.5\text{K}\Omega$$

In order to determine the output voltage we will get with a 10KHz deviation of the carrier, we first need to determine the lock-in frequency deviation Δf_L by using equation 17–13. (Assume that the total supply voltage V_{cc} is +12V.)

$$\Delta f_L = \pm\frac{7.84\text{V} \times f_o}{V_{cc}}$$

$$\Delta f_L = \pm\frac{7.84\text{V} \times 100\text{KHz}}{12\text{V}}$$

$$\Delta f_L = \pm 65.3\text{KHz}$$

This frequency deviation provides an output voltage of ±0.98V. With a deviation of ±10KHz, we get a peak sine wave output voltage v_a of

$$v_a = \frac{0.98\text{V} \times 10\text{KHz}}{65.3\text{KHz}}$$

$$v_a = 150\text{mV}$$

This PLL output voltage is the FM demodulated output we desire and, in this case, is a 1KHz sine wave with a peak-to-peak amplitude of

$$2 \times 150\text{mV} = 300\text{mV}$$

Also appearing at the amplifier output is the unwanted double-frequency component (see equation 17–17) of peak amplitude 0.98V. The low-pass filter capacitor (C_2) must be picked to reduce this component below the wanted signal. If we pick the −3db down frequency of the filter at 2KHz (twice the modulation frequency), then C_2 becomes

$$C_2 = \frac{1}{2\pi \times 3.6\text{K}\Omega \times 2\text{KHz}}$$

$$C_2 = 22\text{nF}$$

With this capacitor value, we can now determine the rejection of the unwanted 200KHz double frequency and also the capture frequency.

The double-frequency amplitude V_D can be determined from the voltage divider action of R_a and C_2—assuming that the reactance of the capacitor is much less than the resistance of R_a.

$$V_D = 0.98\text{V}\left(\frac{X_c}{R_a}\right)$$

$$V_D = \frac{0.98\text{V}}{2\pi \times 200\text{KHz} \times 3.6\text{K}\Omega \times 22\text{nF}}$$

$$V_D = 9.85\text{mV}$$

This is a 100-to-1 reduction in the double-frequency amplitude.

The capture frequency can be determined from equation 17–15.

$$f_{cap} = \pm\sqrt{\frac{f_L}{2\pi R_a C_2}}$$

$$f_{cap} = \pm\sqrt{\frac{65\text{KHz}}{2\pi \times 3.6\text{KHz} \times 22\text{nF}}}$$

$$f_{cap} = \pm 11.4\text{KHz}$$

This means that if the signal carrier frequency is within ±11.4KHz of the PLL, VCO frequency capture will occur. The complete circuit schematic is shown in Figure 17–16.

Pin 2 is grounded rather than Pin 3 to provide a demodulated output that is in phase with the original modulating signal.

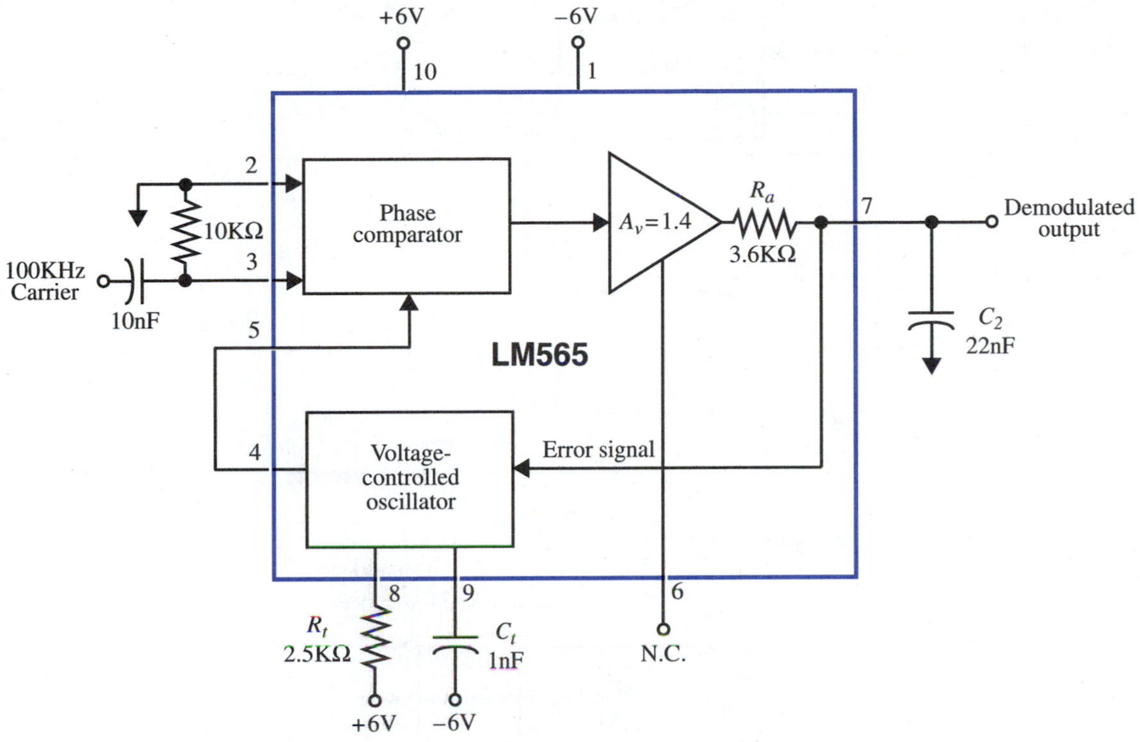

FIGURE 17–16 100KHz FM demodulator

PLL FREQUENCY SYNTHESIS

Frequency synthesis involves generating new frequencies from a fixed source—typically, a crystal oscillator. The new frequencies have the same long-term frequency stability as the crystal, but without the need for a different crystal for each frequency. A phase locked loop can be used for this application.

If we desire a frequency that is a multiple of the crystal frequency, we can use the PLL circuit shown in Figure 17–17.

Notice that the VCO output is not fed directly to the phase detector but passes through a digital divide-by-*N* counter first, which reduces the VCO output frequency by the count-down factor *N*. Since the closed loop ensures that the two frequencies into the phase detector are the same, the VCO output is *N* times the crystal frequency.

$$f_{out} = Nf_{in} \qquad \textbf{(equation 17–18)}$$

HETERODYNE SYNTHESIZER

The required output signals from the synthesizer are not always simple multiples of the crystal frequency. A heterodyne conversion system, shown in Figure 17–18, allows for non-harmonically related outputs.

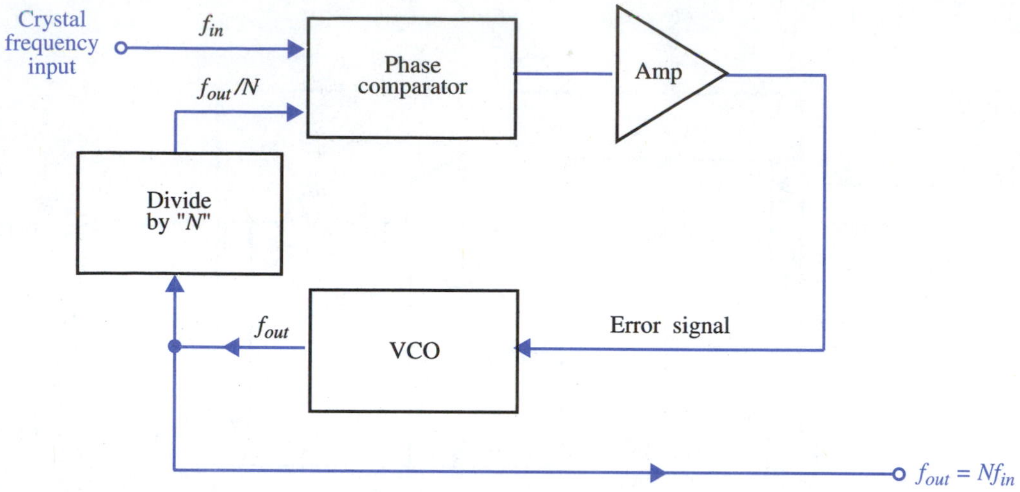

FIGURE 17–17 Frequency multiplier

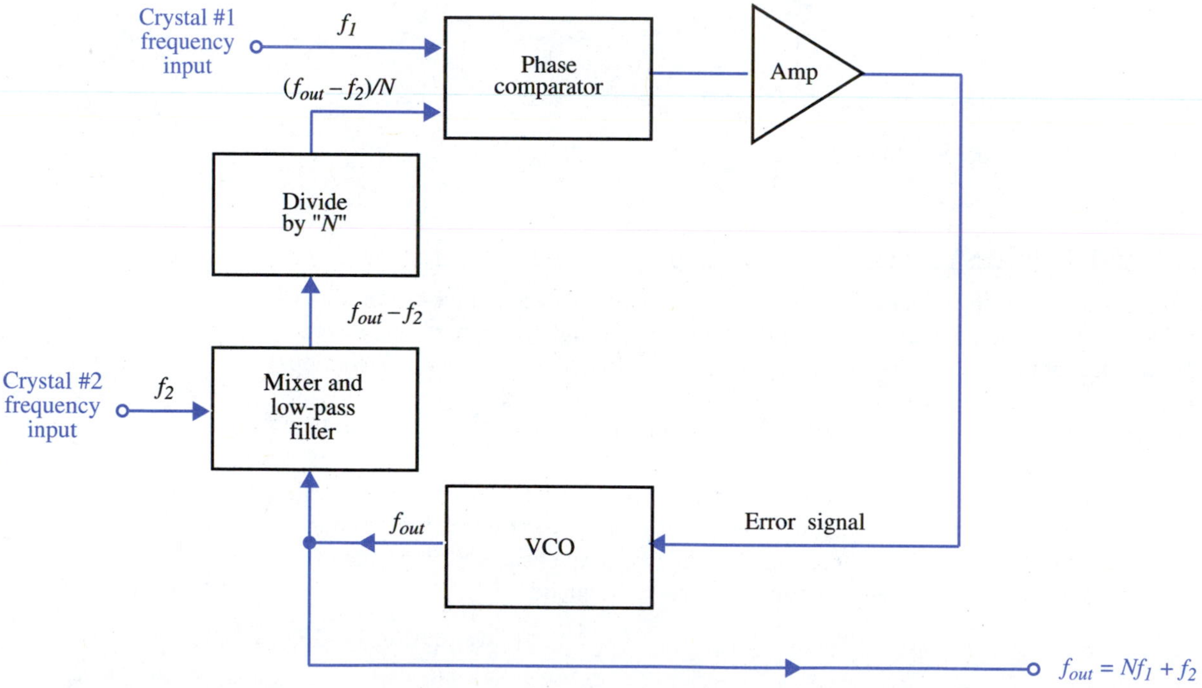

FIGURE 17–18 Heterodyne frequency conversion

Added components from the previous circuit are the second crystal frequency f_2 and the mixer. A mixer is a nonlinear device that generates the sum and difference of two input frequencies as well as harmonics. In this application, except for the difference frequency $f_{out} - f_2$ (f_{out} is greater than f_2), all output frequencies from the mixer are filtered out by the low-pass filter. Since the input frequencies to the phase detector must be equal, f_1 is

$$f_1 = \frac{f_{out} - f_2}{N}$$

Then f_{out} is

$$f_{out} = Nf_1 + f_2 \qquad \textbf{(equation 17–19)}$$

Selecting suitable choices for N, f_1, and f_2, we can generate the desired output frequency. In practice, the counter factor N is made variable so we can select several output frequencies.

EXAMPLE OF A HETERODYNE SYNTHESIZER

As an example of a heterodyne synthesizer, let us assume that we need to generate local oscillator frequencies for a 44-channel CB receiver with an intermediate amplifier frequency (*IF*) of 7.8MHz. The CB channel range is from 26.965MHz to 27.405MHz in 10KHz increments. So the required local oscillator frequency range is from 34.765MHz to 35.205MHz (adding 7.8MHz *IF* frequency to incoming frequency).

With the required 10KHz increments, f_1 should be at this frequency. Crystal-controlled oscillators are at higher frequencies than 10KHz—so let us assume we have a 1MHz crystal available for the oscillator. We can then reduce this frequency by a factor of 100 by use of decade counters.

The frequency f_2 can be determined by assuming that the lowest value of N is 1 (corresponding to the lowest output frequency of 34.765MHz). Then, rearranging equation 17–19, f_2 is

$$f_2 = f_{out} - Nf_1$$

$$f_2 = 34.765\text{MHz} - 10\text{KHz}$$

$$f_2 = 34.755\text{MHz}$$

In order to obtain the full range of possible signals, N must be varied from 1 to 44. This is accomplished by using a digital divide-by-N counter, i.e., CD4526, etc. These counters are programmed by changing the levels on four control lines, which allows selection of up to 16 combinations. Using two cascaded counters allows the 44 selections required for this example. The frequency requirements for this example are beyond the capability of the LM565. However, the Signetics NE564, which has a maximum operating frequency of 50MHz, or the NE568, with a maximum operating frequency of 150MHz, could be used for this application.

COMPLETE RADIO RECEIVER

We have looked in detail at modulation and demodulation of a carrier signal. I think it will be helpful at this point to briefly look at a complete radio receiver. The block diagram for

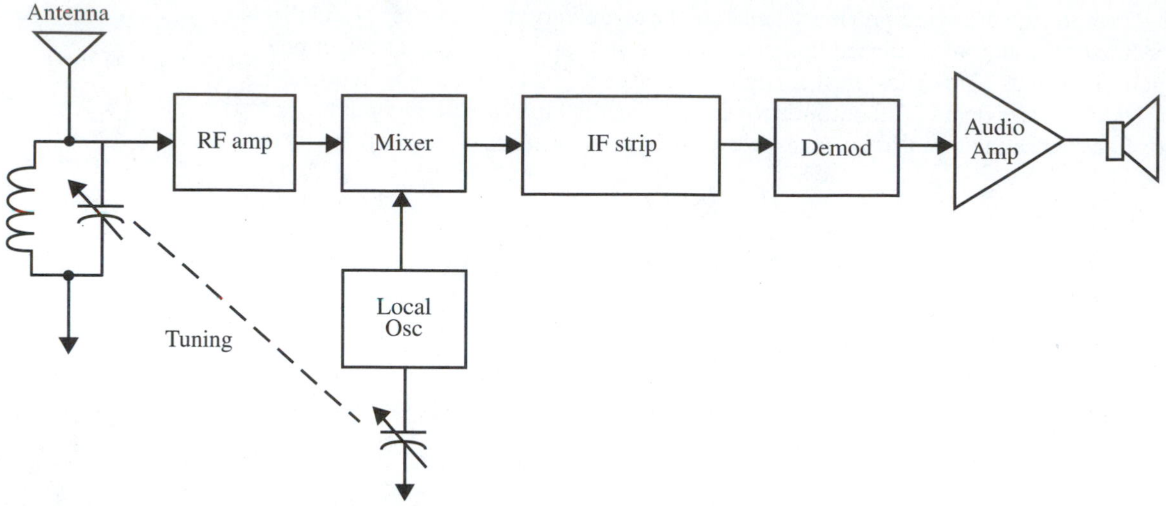

FIGURE 17–19 Radio receiver

a typical receiver is shown in Figure 17–19. The incoming radio frequency carrier induces a signal into the antenna, and the signal appears across the parallel resonant circuit (tank circuit) tuned to the incoming frequency. Amplification of the weak signal (microvolts) occurs in the radio frequency amplifier (budget radios have no RF stage).

The purpose of the mixer is to reduce the carrier to a lower frequency that is easier to amplify and is the same frequency regardless of the carrier frequency. Again, mixing is the process of taking two different frequencies through a nonlinear device that creates sum and difference frequencies and all the harmonics. In this case, the two frequencies are the incoming carrier and the local oscillator frequency.

The local oscillator is tuned to always be above the carrier by a fixed amount. For AM radios, the difference frequency between the carrier and the local oscillator is 455KHz, for FM, 10.7MHz. The mixing action, just described, is also called heterodyning, and a receiver using this conversion is called a superheterodyne.

The intermediate amplifier (*IF*) is set up to only pass the difference frequency (455KHz or 10.7MHz), and since the difference frequency is constant, no tuning of the *IF* is required. The main amplification in a superheterodyne receiver occurs in the *IF* amplifier.

The demodulator, following the *IF*, consists of a simple diode detector for an AM radio, or a slope detector—e.g., PLL—for an FM receiver.

Integrated circuit chips exist for complete radio receivers and usually only require the addition of crystals and tuning components. For example, Motorola provides MC33XX series of FM receivers, and National Semiconductor has a group of AM, AM/FM, and FM receiver chips.

DIGITAL DATA COMMUNICATIONS

One of the fastest growing areas of electronics is data communications. It covers the area of serial (one bit follows the other) data transmission via radio links, metallic cables, and fiber optics. The topic is a broad area to cover so we shall just introduce some of the basics (hopefully to stimulate your interest in this field), and then concentrate on devices associated with RS-232 and fiber-optic transmission.

When we studied the ADC, one of the needs for converting analog signals to digital was the noise advantage of digital during signal transmission. To further improve noise performance and obtain higher data transmission rates, various modulation methods are used. Two modulation schemes were briefly mentioned earlier: frequency shift keying and biphase modulation (XR2206).

To obtain high data rates on transmission links with limited bandwidth, e.g., voice-quality telephone lines, higher-level modulation is used where a transition in the carrier represents more than one bit of digital data. One technique, which allows a data rate of 7,200 bits per second on 3KHz bandwidth telephone lines, is 8PSK (phase shift keying). With this modulation, digital data are broken into three-bit groupings. Depending on the particular code for the group (2^3 is eight combinations), a sine wave carrier with one of eight possible 45° phase shifts is generated.

The number of phase transitions of the carrier (signal changes) per second is called a **baud**. Each transition contains three bits of data. So the baud rate is one-third the bit rate—or for our example, 2,400 baud. The significance of the speed in baud is that it indicates the required bandwidth for the signal. For our example, the bandwidth requirement is 2400Hz, which is within the 3KHz bandwidth limitation of telephone lines.

Figure 17–20 shows the 8PSK phase shifts versus codes—called a **constellation diagram**.

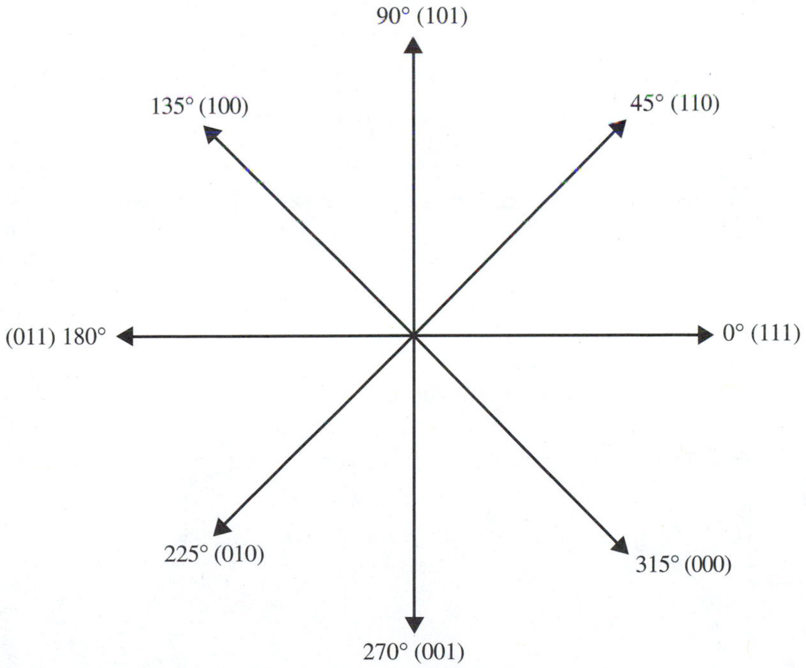

FIGURE 17–20 8PSK phase shift versus digital code

Other higher-level modulations use combinations of carrier phase shift along with amplitude modulation to go to even higher data rates. Devices that include both the modulator and demodulator on one chip are called modems (*mo*dulator-*dem*odulator).

The Exar XR-2900 is a fax/modem chip pair that can process data at bit-per-second (BPS) rates of 9,600, 7,200, 4,800, 2,400, and 300 over telephone lines. The chip set contains telephone interface circuits as well as the modem functions.

RS232 TRANSMISSION

Over short distances (up to 50ft. at 20,000 BPS), modulation of a carrier by serial digital data is not required. Rather, the level of the basic digital signal can be increased and converted to a bipolar form to minimize noise problems. This is the approach taken under the RS232 specifications, which require that the transmitted signal swing both positive and negative with a logic *1* represented by −5V to −15V and a logic *0* by +5V to +15V.

A simple circuit that converts from TTL levels to RS232 levels and then back to TTL is shown in Figure 17–21.

When the TTL input exceeds +2V (*1* level), the inverting input to the op-amp is more positive than the noninverting input, and the output swings to the required −15V level. If the TTL input is less than +0.8V (*0* level), the output of the op-amp is at +15V.

At the receiving end, a −15V level on the transmission line causes the transistor to be reverse biased (diode prevents too much reverse bias). This negative bias turns off the transistor, and the output (collector) rises to +5V. If the level on the transmission line is +15V, the transistor is biased into saturation, and the output voltage is close to 0V.

In addition to the voltage swing, the electrical specifications for RS232 require the following:

1) Terminating resistance at the receiver to be in the range of 3KΩ to 7KΩ.
2) Receiver output to be at the *1* level if transmission line becomes open.
3) Transmitter voltage, if the line load is disconnected, to be less than 25V.
4) Current with a short circuit on the transmitter not to exceed 0.5A.

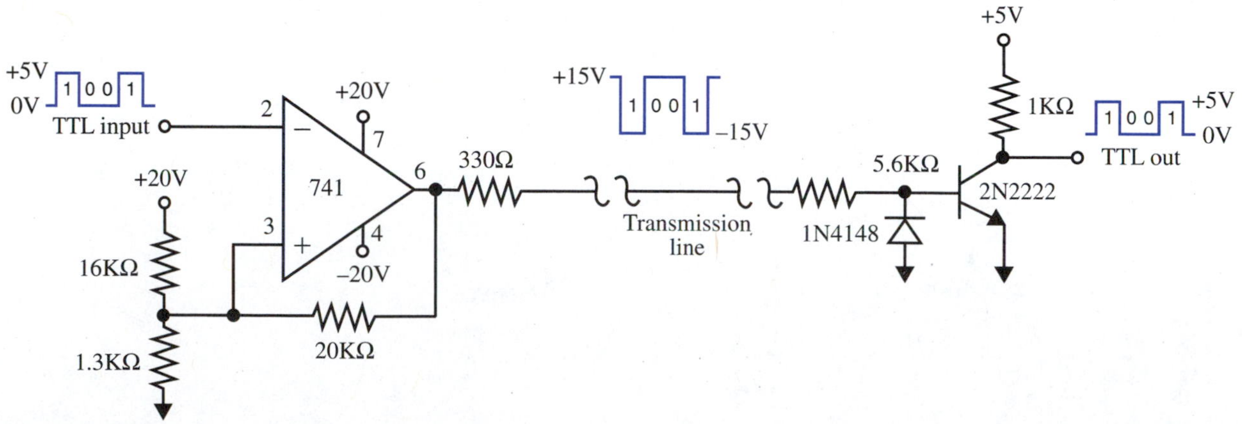

FIGURE 17–21 TTL-to-RS232 conversion

5) Transmitter output rate of change (slew rate) not to exceed 30V/ms. However, the time to pass through the −3V to +3V dead zone is not to exceed one millisecond (6mV/μs) or 4 percent of the data bit time.
6) The cable and termination (receiver) capacitance loading on the transmitter shall not exceed 2500pF.
7) The resistance seen looking back into the transmitter with the power off should be greater than 300Ω.

Comparing these specifications with our simple circuit of Figure 17–21, we can see by inspection that items 1, 2, 3, and 7 are met. Item 4 is also met because the maximum short circuit current for a 741 op-amp is 20mA.

The maximum slew rate for a 741 is about 1.5V/μs—well within the allowed maximum of 30V/μs. A minimum slew rate of 0.3V/μs for the 741 is greater than the minimum allowable value of 6mV/μs.

With the minimum 0.3V/μs slew rate for the 741, it takes 20μs to go from −3V to +3V. For this to not exceed 4 percent of the bit time requires the minimum bit width to be 20μs/0.04, or 500μs. This allows a maximum bit rate of 2KBS (kilobits per second) using the 741 op-amp.

Maximum capacitance loading of 2500pF (@ data rate 20KBS) results from the capacitance between the wires of the transmission line. To find the maximum length of the line, we divide 2500pF by the capacitance per foot. Reducing the data rate increases the allowable length (inverse relationship).

EXAMPLE

Find the maximum allowable length of transmission line used for RS232 if the capacitance per foot is 19pF and the data rate is 20KBS; and also with 2KBS.

SOLUTION

At a data rate of 20KBS,

$$L = \frac{2500\text{pF}}{19\text{pF/ft.}}$$

$$L = 131.6\text{ft.}$$

At a data rate of 2KBS,

$$L = \frac{2500\text{pF} \times 10}{19\text{pF/ft.}}$$

$$L = 1{,}316\text{ft.}$$

INTEGRATED CIRCUIT RS232 TRANSMITTERS AND RECEIVERS

Rather than use the circuit shown in Figure 17–21, we can use an integrated circuit driver (transmitter) such as the Motorola MC1488, and as a receiver, the MC1489. The connections for these two chips are shown in Figure 17–22.

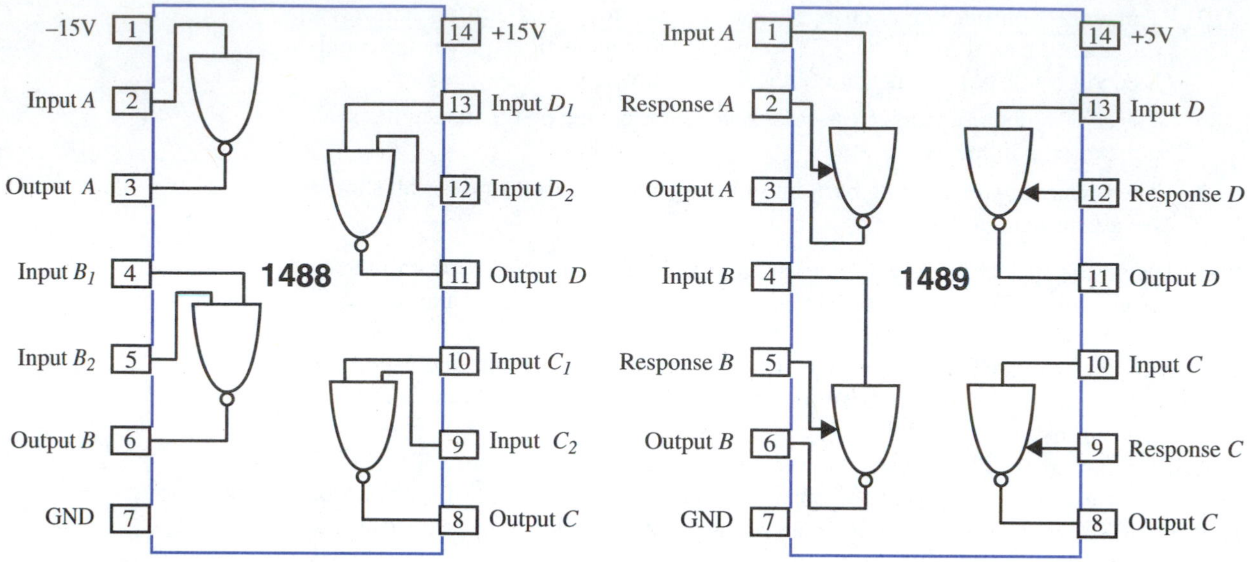

FIGURE 17–22 1488 driver and 1489 receiver

Notice that there are three dual input drivers and one single input driver contained within the 1488 package. We can disable the dual input drivers by taking the unused input low.

The 1488 meets all the electrical requirements of RS232 except for slew rate. However, by connecting a 330pF capacitor from output to ground, the maximum 30V/μs is not exceeded. This capacitor is not required if the total transmission line capacitance exceeds 330pF.

Four receivers are contained within the 1489 chip, and they meet all RS232 specifications for receivers. In addition, they have built-in hysteresis of 0.25V (1.1V for the 1489A)—which means they can reject up to 0.25V of noise riding on the signal. Additional noise rejection can be obtained by connecting a capacitor from the response pin to ground.

Normal power supply voltages for the 1488 are ±15V, and a single +5V supply for the 1489.

SINGLE POWER SUPPLY RS232 DRIVER AND RECEIVERS

Maxim has developed a series of chips that contain both the drivers and receivers on one chip, and they also operate from a single +5V power supply. For example, the MAX240 chip has five drivers and five receivers on a 44-lead plastic flat pack. The MAX241 contains four drivers and five receivers on a 28-pin DIP. These chips have a power-down feature when not in use, which can be activated to drop the supply current to less than 10μA. This is a good feature for battery-powered equipment.

Single power supply operation of the drivers is accomplished by using the same technology of the MAX680/681 (described in Chapter 12), which converts +5V to ±10V. With this power source, the typical output voltage swings for the drivers is ±9V.

Maxim also provides a +5V-powered transceiver chip (MAX252) that contains two drivers and two receivers but also provides complete electrical isolation between the logic signals and the RS232 levels. This ground isolation is ideal for interfacing in a noisy industrial environment.

RS232 CONNECTOR AND SIGNALS

As well as defining signal levels, RS232 defines the connector and the serial interface signals. The 25-pin connector is shown in Figure 17–23.

The purpose of the RS232 specifications is to define the interface between a computer serial port (DTE—data terminal equipment) and the telephone modem (DCE—data communication equipment). The interface includes the serial data lines, to and from the computer, along with handshake signals to allow an orderly exchange of data between CTE and DCE.

The RS232 signals used most often are listed in Table 17.1

Let us look at each signal line:

- Pin 2—Transmit Data. This is the serial data stream from the computer to the modem.
- Pin 3—Receive Data. This is the serial data stream from the modem to the computer.
- Pin 4—Request to Send. This line is taken to an active level when the computer wants to transmit data.
- Pin 5—Clear to Send. After receiving the RTS from the computer, the modem will make this line active if it can now receive transmit data from the computer.
- Pin 6—Data Set Ready. Continuous status signal from the modem notifying the computer that the modem is operational.
- Pin 8—Receiving Line Signal Detect. Status signal from the modem telling the computer that the modem is detecting a good carrier level on the telephone lines.
- Pin 11—Printer Busy Signal. Notifies the computer that the printer cannot receive data at this time.
- Pin 20—Data Terminal Ready. Continuous status signal from the computer notifying the modem that the computer is operational.
- Pin 22—Ring Indicator. Signal sent from the modem to the computer notifying the computer that a remote location (computer, terminal, etc.) wishes to communicate.

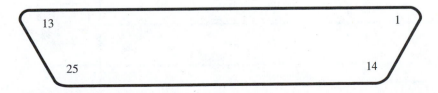

FIGURE 17–23 RS232 connector female mating side

TABLE 17–1 Common RS232 Signals

PIN	CIRCUIT	NOTES
1	Protective Ground	Connect to Earth Ground
2	Transmit Data (TD)	Data from DTE
3	Receive Data (RD)	Data from DCE
4	Request to Send (RTS)	Handshake from DTE
5	Clear to Send (CTS)	Handshake from DCE
6	Data Set Ready (DSR)	Handshake from DCE
7	Signal Ground	Reference point for signals
8	Received Line Signal Detector (RLSD)•	Handshake from DCE
11	Printer Busy Signal••	Handshake from Printer
20	Data Terminal Ready	Handshake from DTE
22	Ring Indicator	Handshake from DCE

•Also called *Data Carrier Detect* (DCD)

••Not listed in RS232 specs

RS232 TRANSMIT SEQUENCE

To further clarify the RS232 signal lines, we shall look at a typical telephone line transmit sequence. Figure 17–24 shows the timing diagram for a transmit sequence from the computer (DTE).

We shall assume that the computer wishes to send a data message, via telephone, to a remote location. This is initiated by the computer taking the RTS line active low. The modem senses this condition and rings up the remote location and also puts the unmodulated carrier on the telephone line. When the remote location goes "off hook" (answers), an unmodulated carrier is sent back to the modem. The presence of this carrier (RLSD) notifies the modem that communication has been established; so it takes the CTS line active low. Now the computer puts the digital data on the TXD line.

When the computer has completed its message, it takes the RTS line high (inactive). This signals the modem to take the CTS line high and turn off the carrier. Loss of

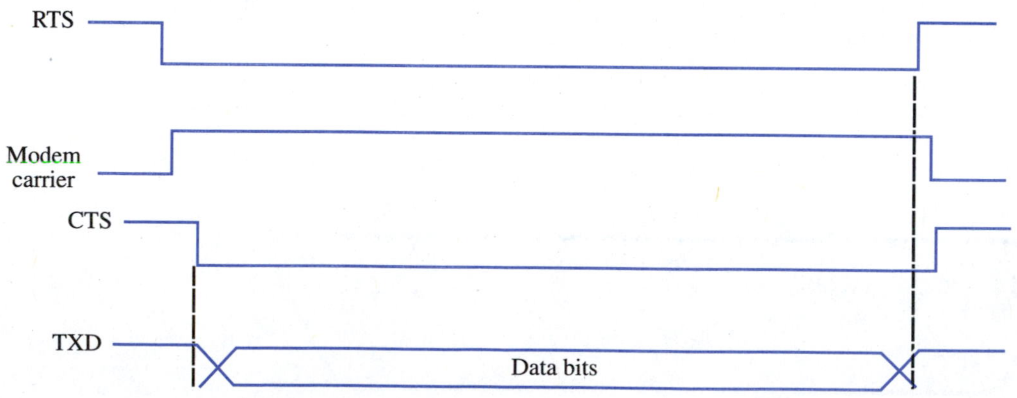

FIGURE 17–24 RS232 transmit sequence

this carrier at the remote location causes it to hang up and remove its carrier from the telephone line.

FIBER-OPTIC TRANSMISSION OF DATA

The great future needs in data communications are high-speed transmission with a minimum of noise interference. Fiber-optic transmission systems satisfy these requirements because the high-frequency (light) carrier (about 3×10^{14}Hz) allows an extremely wide bandwidth—which means a very high data rate. Also, the fiber-optic system is immune to electromagnetic interference, and optical interference can be eliminated with an opaque cover over the fiber.

Simply, a fiber-optic system consists of an amplitude-modulated light source, a fiber-optic cable, and a light detector. The light source can be an LED or a laser diode. The LED is cheaper and more reliable, but the laser diode can provide about ten times more light power. The cable can be made from plastic or glass. Plastic is cheaper, but glass transmits the light with less loss. Detectors can be phototransistors or diodes—with diodes most used because of their lower capacitance and therefore higher speed.

To become more familiar with fiber-optic communication, we shall look at a practical transmitter and receiver system, as shown in Figure 17–25.

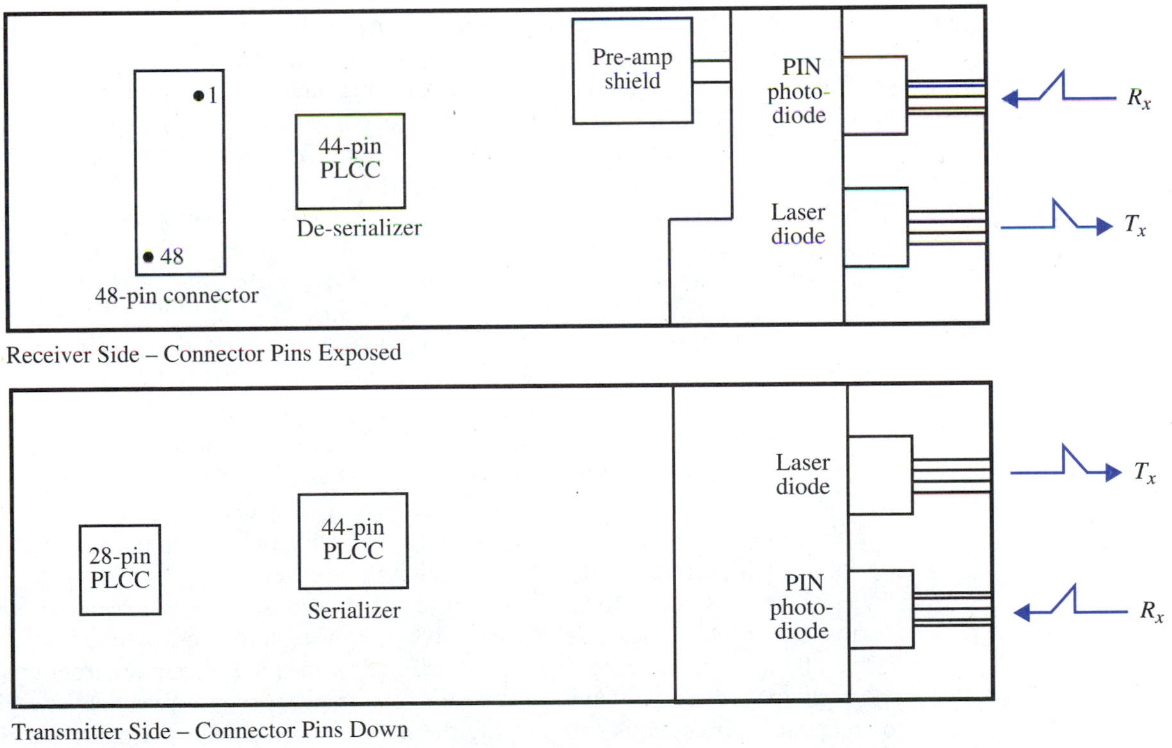

FIGURE 17–25 Fiber-optic transceiver card. *Courtesy Hewlett Packard.*

The layout shown in the figure consists of the Hewlett Packard HOLC-0266 optical link card transmitter and receiver. Data rates of 266 megabaud, for distances up to 2 kilometers, can be transmitted by systems using this card. Lower data rates will allow proportionally longer transmission distances.

The transmitter accepts 10-line TTL parallel input data at the 48-pin connector. This data is converted to serial form in the serializer and is shifted out by a 265.625 Mb/s phase locked loop frequency multiplier that is locked to an external 26.5625MHz clock signal. The serial data modulate the optical transmitter (turn it on and off).

The transmitter optics consists of a laser diode with maximum emission at an optical wavelength of 0.780μm (780nm), at a typical power level of −5dbm to 0dbm (dbm is referenced to a milliwatt). The laser is mounted in a receptacle with a lens and is coupled to an SC-type (push-on) fiber-optic connector.

A fiber-optic cable typically consists of a glass or plastic core surrounded by a covering of optical material (**cladding**) of lower index of refraction. The object of the cladding is to refract (bend) light that tries to leave the core back into the core. If light passes through the cable by refracting from the sides rather than straight down the center (single mode), it is called **multimode propagation**. This mode is used because it is easier to couple light from the diode into this type of cable, but it does have about three times more attenuation (db/km) than single mode because some light is lost in the cladding.

Two things limit the maximum distance that light can be transmitted over a fiber cable: attenuation (loss) and pulse spreading (widening). Losses occur because of defects in the fiber, which occur naturally during manufacturing, and because of sharp bends in the cable (light ray is trapped in the cladding). Pulse spreading results from light paths of individual rays traveling over a different distance during multimode transmission—the longer the cable, the greater the time difference and therefore the greater the spreading (doesn't occur with single mode because there is only one ray). The effect of pulse spreading could be that a transmitted code of 101 is received as 111.

Figure 17–26 shows the differences in light path length that can occur with multimode transmission and the resultant distortion of the received signal. The distortion is more apparent with the narrow pulses associated with high data rates, and with higher data rates seeing more use, transmission could shift more to single mode.

The 50/125μmeters (core diameter/cladding diameter) multimode fiber-optic cable recommended for this application has become an international (CCITT) standard. It is a graded (different across the diameter) refractive index cable that operates multi-mode and has an attenuation factor of 4db/km at an optical wavelength of 850nm and 2db/km at 1300nm. To minimize coupling losses, the fiber end must be polished, square, and in complete alignment (without air gaps) with the laser and photodetector.

In the receiver portion of the HOLC-0266, a PIN diode is used as a photodetector. This is a special diode created for optical detection. It is faster responding than a regular photodiode, it is more sensitive to light, and it generates less noise. The name comes from its *P* (narrow region), *I*ntrinsic (wide region) and *N* (narrow region) construction.

The PIN diode is mounted in a receptacle to allow maximum coupling from the fiber-optic cable SC connector. On the electrical side, a transimpedance preamplifier (current-to-voltage converter) is used to amplify the detected signal from the diode. To reduce electrical noise pickup, the preamplifier is covered with a metal shield. The sensitivity of the receiver is from −17dbm to 0dbm, with a bit error rate (BER) of $<10^{-12}$ over this range. A BER of 10^{-12} means one bit in error for every trillion bits transmitted!

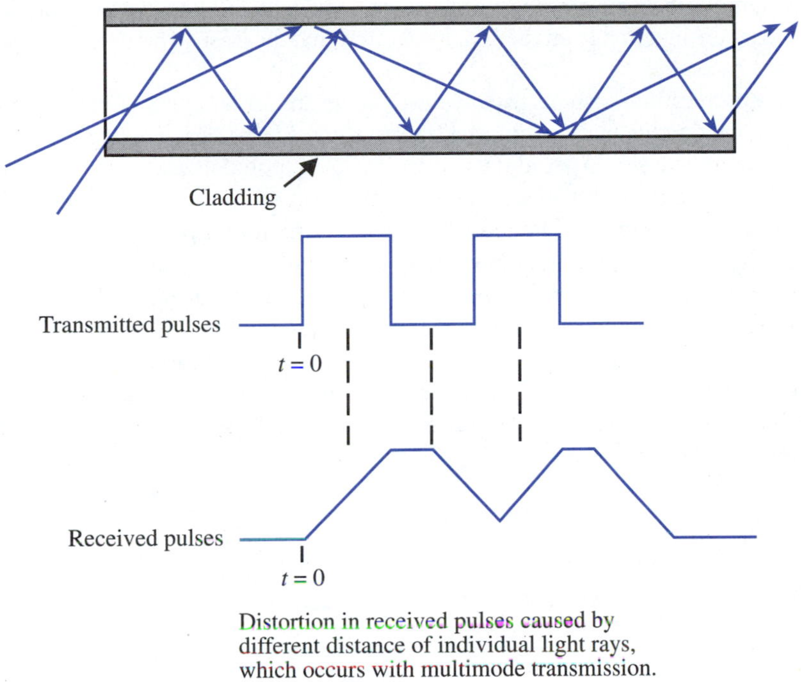

FIGURE 17–26 Distortion in fiber-optic multimode transmission

The serial data from the receiver are converted back to 10-line parallel form (TTL) in the deserializer and provided at the 48-pin connector.

An open loop fiber control system is provided that senses the received power level, and if it drops below a certain level, it is assumed that the fiber cable is disconnected. With this condition, the transmitted laser power level is reduced to a level considered safe for operating and maintenance personnel.

Both the receiver and transmitter operate from a +5V supply over a controlled temperature range of 10°C to 50°C.

TROUBLESHOOTING COMMUNICATIONS DEVICES

An understanding of the unique requirements of communications equipment is essential for the troubleshooting process. These requirements include high-frequency signals and very stable frequency sources. Also, it requires concern about bandwidth, noise, and distortion.

When dealing with high frequencies, we must be aware of the limitations of the test equipment and the effect of lead inductance and stray capacitance. Precision frequency sources require frequency counters with a greater degree of measurement resolution and accuracy.

Measurement of bandwidth, noise, and distortion can require sweep frequency generators, spectrum analyzers, standard noise sources, and distortion analyzers.

As with the other linear devices we have studied, troubleshooting communications chips requires a knowledge of the expected signal (both AC and DC values) to be seen at each pin. For example, if we are viewing a digital signal on an RS232 line, we must be aware that the serial data are inverted from the TTL levels, and that the least significant bit appears first on the oscilloscope.

Data communications systems usually have provisions for remote loop-back testing. This involves connecting the output from the transmitting side, at various points in the system, back into the receive side. Special data codes are used to initiate the loop back at selected points (one at a time), and then test codes are transmitted. Correct reception of these test codes indicates the system is operational up to the turn-around point.

Troubleshooting fiber-optic systems is rather straightforward because of the simplicity of the optics—which consists of a modulated light source, an optical path, and a detector. Isolation of a problem with the optics requires verification of light source output, continuity of optical coupling over the fiber-optic link, and detector response to an optical input.

If repairs have to be made on the fiber-optic cable, the cable manufacturer's recommendations need to be followed to ensure that the excessive losses are not introduced by incorrect termination techniques.

SUMMARY

- A communication signal is one that contains information.
- Any unwanted signal is considered noise.
- Noise is of two basic types: unwanted information and random.
- Random noise can be broken down into external noise and internal noise.
- Internal noise can be thermal, shot, flicker, or transit.
- Keeping the bandwidth as narrow as possible reduces noise effects.
- The amplitude of a higher-frequency carrier is varied with amplitude modulation.
- The ratio of the peak modulating signal amplitude over the carrier is the AM modulation index.
- AM creates sidebands on each side of the carrier separated by the frequency of the modulating signal.
- The XR-2206 can operate as an amplitude modulator.
- The frequency of a higher-frequency carrier is varied with frequency modulation.
- The ratio of the frequency deviation over the modulating frequency is called the FM modulation index.
- A voltage-controlled oscillator enables us to convert voltage variations into frequency variations.
- VCOs can be used to provide frequency-modulated output signals and single-frequency tones.
- A common VCO is the 566, which has a maximum operating frequency of 1 MHz.
- A sine wave and a square wave output are provided by the 566.
- When digital signals are used to modulate a VCO, the output shifts between two discrete frequencies. This is called frequency shift keying (FSK).

- Because the average value of a digital data stream changes with time, DC coupling must be used when the VCO is modulated with this type of signal.
- To establish the center frequency of the VCO, the modulation input to the 566 must be DC biased to $7/8V_{CC}$.
- The KHz/V factor for a VCO indicates how much frequency deviation of the output occurs with a given input voltage change.
- A phase locked loop is used to demodulate FM signals or to generate new frequencies.
- Once it is captured by the incoming signal, a VCO in a phase locked loop has the same frequency as the signal.
- The phase error between the incoming signal and the VCO creates the voltage necessary to shift the VCO frequency so that it is identical to the incoming frequency.
- If the incoming signal is injected at the same frequency as the VCO, the phase shift for the 565 will be 90°, and the error voltage will be zero.
- The lock-in range is the range of frequencies over which the VCO will follow the incoming signal.
- At the limits of the lock-in range, the phase shift is either 0° or 180°.
- The capture frequency is that frequency at which there is sufficient error voltage to move the VCO frequency to the same frequency as the incoming signal.
- Capture frequency is determined by the low-pass filter capacitor C_2 and the lock-in range and can never be outside of the lock-in range.
- When used for FM demodulation, the output signal is the error voltage created by the VCO tracking the input carrier.
- By putting a frequency divider (N) between the VCO and the phase detector, we can scale up the frequency of the VCO by the factor N. This is used in frequency synthesis.
- Receivers with a mixer and local oscillator are called superheterodyne.
- The main amplification in a receiver occurs in the fixed tuned *IF*.
- The IF is tuned to the difference frequency between the incoming carrier and the local oscillator frequency.
- The fastest growing area of communications is data communications.
- To overcome bandwidth limitations of telephone lines, higher-level modulation is used.
- Baud represents a signaling change and is not the same as bit rate.
- Baud is always less than or equal to the bit rate but always equal to bandwidth.
- RS232 is a specification that defines serial data transmission levels and handshaking signals.
- A logic *1* is converted to −5V to −15V, and logic *0* is converted to +5V to +15V on RS232 transmission lines.

- Fiber-optic transmission of data has the advantage of very large bandwidth and noise immunity.
- The core of a fiber-optic cable can be glass or plastic.
- Light is confined in a fiber-optic cable by refraction in the cladding.

EXERCISE PROBLEMS

1. Define a communication signal.
2. What is the significance of signal bandwidth?
3. Describe the two basic types of noise.
4. List the steps to minimize the effects of noise.
5. Why is a carrier used?
6. Sketch an amplitude modulation waveform.
7. Determine the AM modulation index if the RMS value of the carrier is 10V and the peak of the modulating signal is 5V.
8. Provide the sideband frequencies if a 3MHz carrier is AM modulated by a 50KHz sine wave.
9. A carrier signal is described by the equation

$$v_c = 25\text{V} \sin(2\pi \times 1270\text{KHz} \times t)$$

If the amplitude-modulating signal is

$$v_a = 8\text{V} \sin(2\pi \times 3\text{KHz} \times t)$$

find the modulation index m_a and the sideband frequencies.
10. Determine the value of R_t for an XR-2206 if the desired carrier frequency is 40KHz and C_t is 10nF.
11. With reference to the graph in Figure 17–A, determine the peak output carrier voltage with a DC voltage at Pin 1 of 4V if the peak carrier output voltage is 2V with 1V applied to this pin.
12. Determine the value of R_d for an amplitude demodulator in an AM receiver, if the highest audio frequency transmitted for AM is 3KHz and the maximum modulation index is 0.8. Assume a value of 10nF for C_d.
13. Find the deviation sensitivity K for an FM modulator if the carrier shifts by 10KHz when a 1.4V RMS signal is applied to the modulator.
14. Find the maximum FM modulation index m_f if the modulating waveform is

$$v_a = 2\text{V} \sin(2\pi \times 5\text{KHz} \times t)$$

and the deviation sensitivity K is 12.5KHz/V.
15. Determine the required bandwidth for the conditions of problem 14.
16. Why is amplitude clipping used in an FM receiver?
17. What would happen to the output frequency of the 566 VCO if the internal resistor R_a was changed to a value of twice R_b (refer to Figure 17–B)?

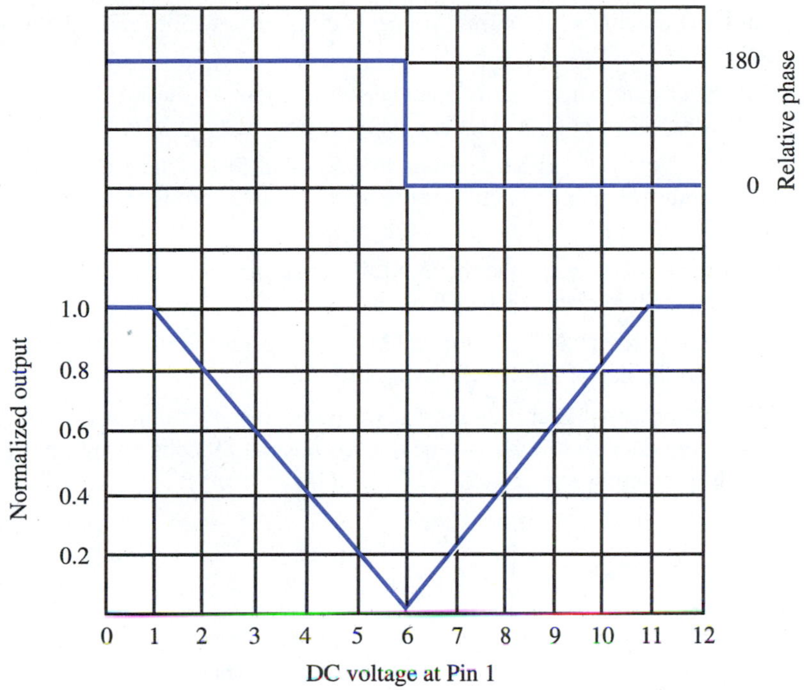

FIGURE 17–A (text figure 17–3b)

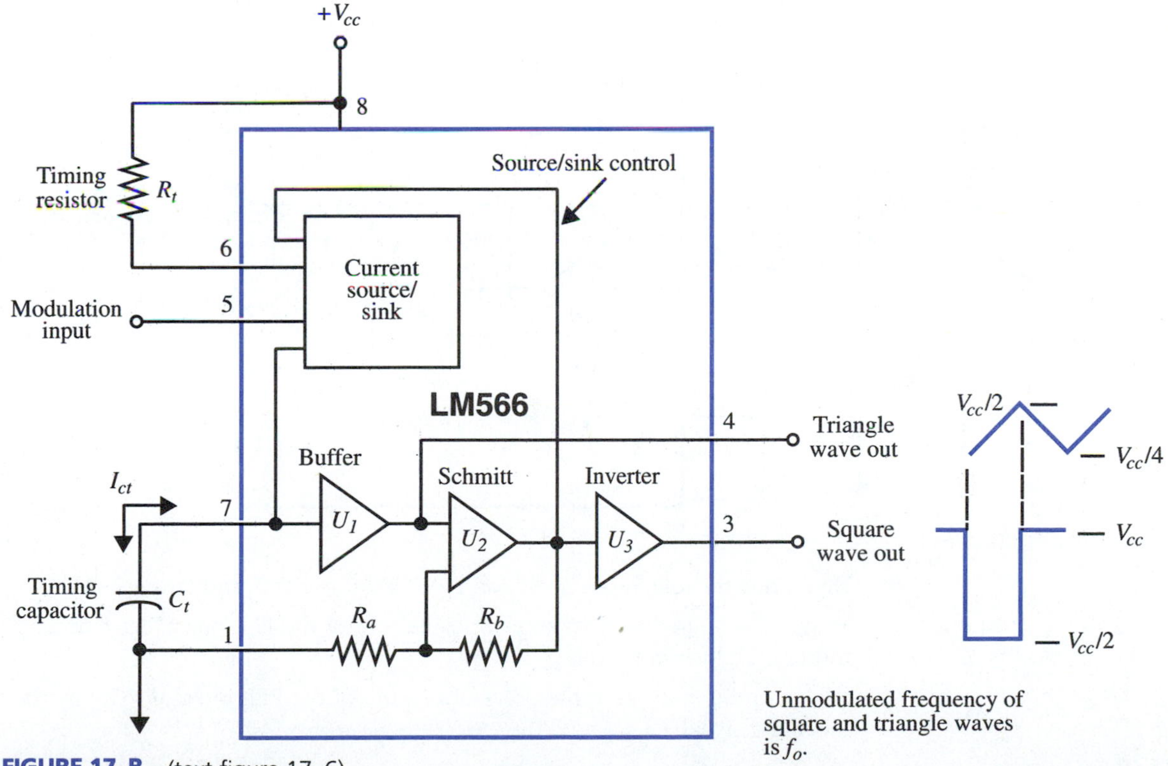

FIGURE 17–B (text figure 17–6)

18. Find the square wave output voltage swing from a 566 VCO if the supply is 12V.
19. Show a circuit that will allow manual switching of an output tone from 5KHz to 10KHz. Use a 566 VCO with a +12V supply and C_t of 10nF.
20. Design a 566 VCO with a nominal frequency of 5KHz and show the AC coupled input modulation circuit for the 50Hz modulation signal. Specify all parts and use V_{CC} of +12V. Assume a value of 10nF for C_t.
21. For problem 20, determine the input modulation voltage swing to give an output frequency increase of 100Hz.
22. Determine the KHz/V factor for an LM566 at a center frequency of 5KHz and a center frequency of 10KHz with a V_{CC} of 12V.
23. With reference to Figure 17–C, what would be the new value of R_3 to obtain the same frequency shift if we assume that Q_1 has a 0.2V collector-to-emitter voltage drop when in saturation?

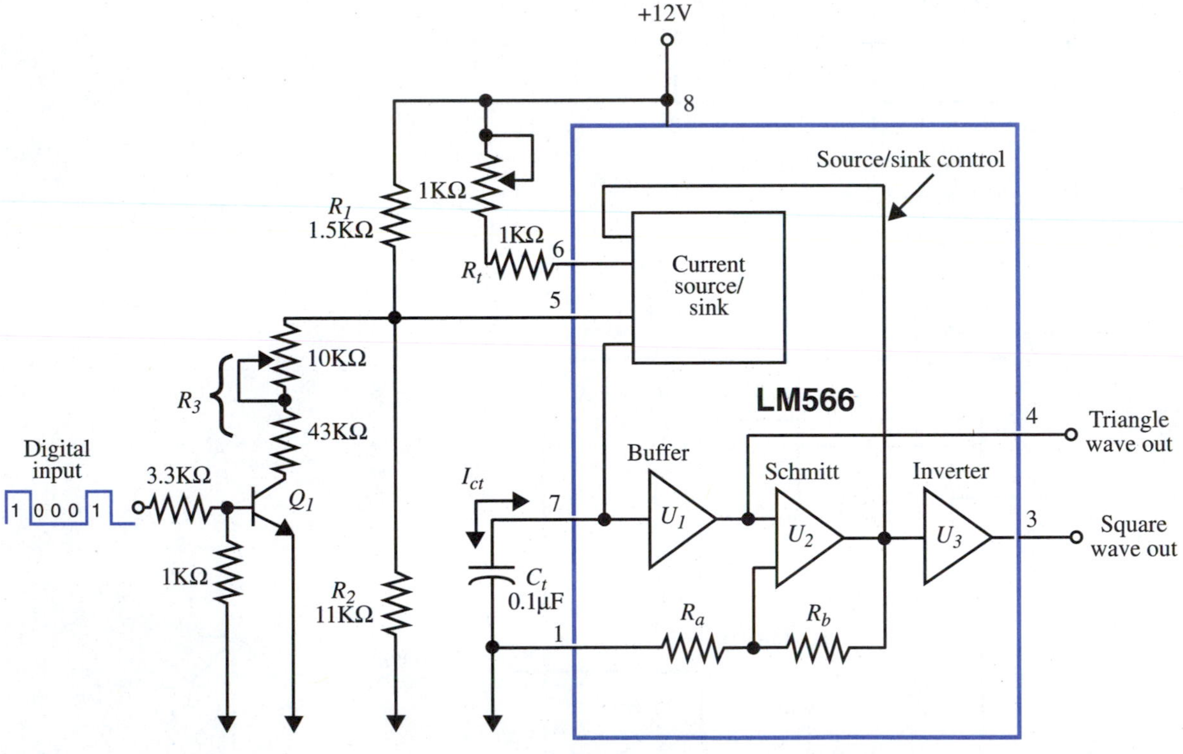

FIGURE 17–C (text figure 17–9)

24. Determine the idle frequency for an LM565 if C_t is 10nF and R_t is 4.3KΩ.
25. Why is it required to have an error signal out of the PLL phase detector over most of the lock-in range?
26. What condition exists at the point of capture of the PLL internal VCO by the incoming signal?

27. Determine the lock-in range for a 565 if R_t is 4KΩ, C_t is 10nF, and the supply voltages are ±6V.
28. Find the capture frequencies for an LM565 if the VCO frequency is 20KHz, C_2 is 10nF, and the total supply voltage is 15V.
29. Determine the value of C_2 required and the value of the double-frequency component voltage if the locked-in frequency is 150KHz and the highest modulating frequency is 5KHz.
30. With reference to Figure 17–D, determine the frequencies f_1 and f_2 for a heterodyne synthesizer if it is required to generate frequencies from 10MHz to 12MHz in 50KHz increments.

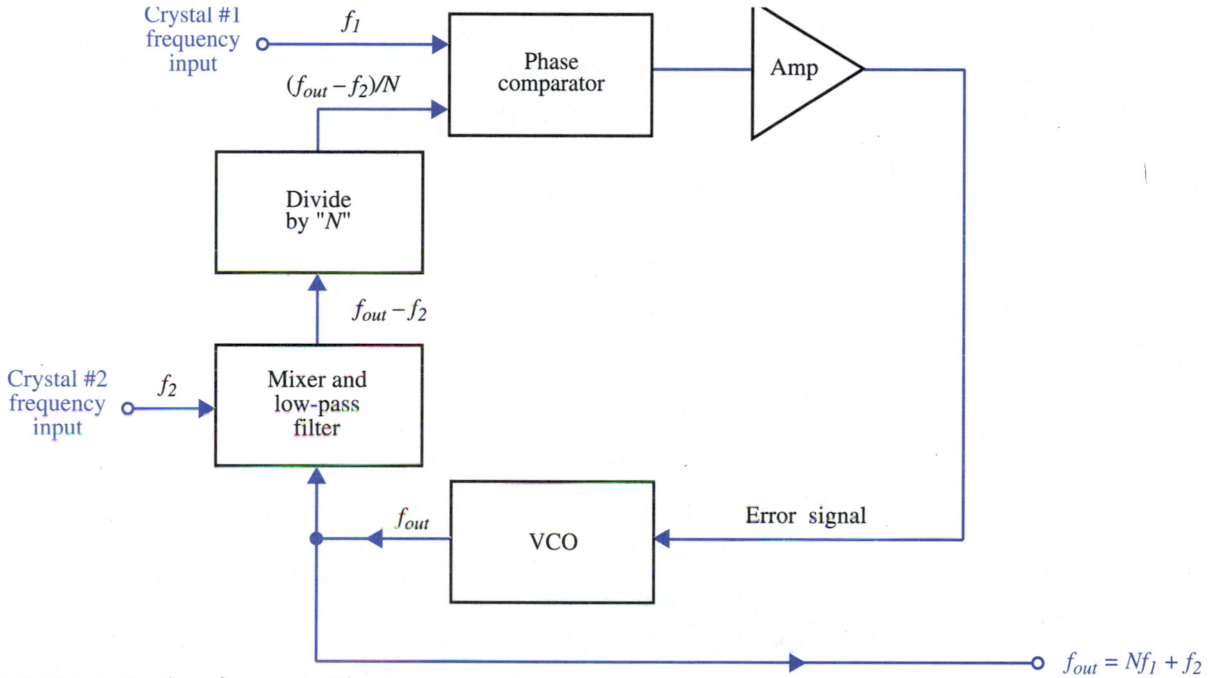

FIGURE 17–D (text figure 17–18)

31. Compute the local oscillator frequency for an FM radio tuned to 105.1MHz.
32. What is a superheterodyne receiver?
33. In data communications, what does higher-level modulation mean?
34. What is a constellation diagram?
35. State the signal levels for *1* and *0* data bits on an RS232 transmission line.
36. Find the maximum allowable length of transmission line used for RS232 if the capacitance per foot is 17pF and the data rate is 20KBS, and also with 2KBS.
37. How much capacitance should be connected to the output of a 1488 RS232 driver if the total line capacitance is 150pF?
38. How many pins are there on an RS232 connector?

39. Describe the following RS232 control lines: CTS, RTS, DSR, and DCD.
40. State the advantages of fiber-optic transmission.
41. What is the purpose of the cladding in a fiber-optic cable?
42. Determine the attenuation for one mile of 50/125μm fiber-optic cable at a wavelength of 1300nm.

ABBREVIATIONS

ΔV = Voltage change
Δf_o = Change in the idle frequency
Δf_L = Limit of lock-in range (±) of PLL
f_a = Frequency of modulating signal
f_c = Frequency of carrier signal
f_{cap} = Capture frequency (±) of PLL
f_o = Center or idle frequency
f_u = Upper sideband frequency
f_L = Lower sideband frequency
K = Deviation sensitivity (KHz/V)
m_a = Amplitude modulation Index
m_f = Frequency modulation Index
V_a = Amplitude of modulating signal
V_c = Amplitude of carrier
V_D = Double-frequency amplitude
VCO = Voltage-controlled oscillator

ANSWERS TO PROBLEMS

1. See text.
2. See text.
3. External and random
4. See text.
5. Selectivity and ease of transmission
6. See text.
7. $m_a = 0.35$
8. $f_l = 2.95\text{MHz}; f_U = 3.05\text{MHz}$
9. $m_a = 0.32; f_l = 1267\text{KHz}; f_U = 1273\text{KHz}$
10. $R_t = 2.5\text{K}\Omega$
11. $V_c = 0.8\text{V}$
12. $R_d = 6.63\text{K}\Omega$
13. $K = 5.05\text{KHz/V}$
14. $m_f = 5$
15. BW = 90KHz
16. To remove noise
17. Frequency reduces.
18. $V_o = 6\text{V}_{p-p}$
19. Use R_t of 5KΩ for 5KHz; switch second 5KΩ in parallel for 10KHz.

20. See Figure 17–8. $R_t = 5\text{K}\Omega$; $C_c = 20\mu\text{F}$
21. $\Delta V_a = -30\text{mV}$
22. $K = 3.33\text{KHz/V}$; $K = 6.67\text{KHz/V}$
23. $R_3 = 46.2\text{K}\Omega$
24. $f_o = 5.8\text{KHz}$
25. To shift VCO to match incoming frequency
26. Error voltage is sufficient to shift VCO to carrier frequency.
27. Lock-in range 2.17KHz to 10.33KHz
28. Capture frequencies: 13.2KHz and 26.8KHz
29. $C_2 = 8.84\text{nF}$ (Use 10nF.)
30. $f_1 = 50\text{KHz}$; $f_2 = 9.95\text{MHz}$
31. $f_{LO} = 115.8\text{MHz}$
32. See text.
33. See text.
34. See Figure 17–20.
35. *1* level: −5V to −15V; *0* level: +5V to +15V
36. $L_{20KBS} = 147\text{ft.}$; $L_{2KBS} = 1470\text{ft.}$
37. $C = 180\text{pF}$
38. Twenty-five pins
39. See text.
40. Less noise
41. Refracts light back into core
42. 3.22db

CHAPTER 18

Ground Loops and Noise

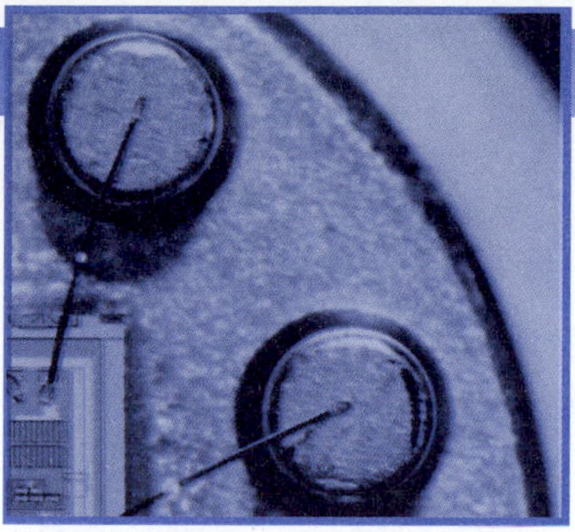

CHAPTER OBJECTIVES

After completing this chapter, you should be able to explain the following concepts about grounding techniques and noise coupling:

- How voltage drops in conductors can be computed
- How voltage drops in conductors can be reduced
- Problems with CAD printed circuit board layout
- Use of sense leads to reduce voltage drop
- Reducing voltage drops in ground connections
- Measurement of voltage drops
- PC board leakage currents
- Computing inductance of straight conductors
- Sources of circuit noise
- How to reduce conductive noise
- Reduction of capacitive noise
- How to reduce inductive noise
- Reduction of radiative noise

INTRODUCTION

Over the years, the accuracy of electronic circuits has increased tremendously. Tolerances of components have been tightened so that 1% resistors and capacitors are now quite common. The accuracy of devices such as analog-to-digital converters has been extended, and circuits with resolutions of 20 bits (one part in a million) are now available. When dealing with these types of accuracies, we have to be extra conscious of voltage drops, ground loops, and noise that can degrade circuit performance.

In this chapter, we will first consider inadvertent voltage drops in circuits; then we will see the effect of ground loops and how they can be avoided; and finally, how the effect of noise can be minimized.

Much of the material in this chapter has been derived from material provided by Analog Devices and is based on this company's extensive experience in precision analog circuitry.

THE PROBLEMS WITH VOLTAGE DROPS

The voltage drops that we shall examine in this section are the result of current flowing through leads that form the connections between circuit components.

The resistance per inch of a conductor can be determined from the following equation:

$$R/\text{inch} = \frac{\rho}{\text{Area}} \qquad \textbf{(equation 18–1)}$$

Where ρ is the resistivity of the conductor.

If we consider a high-density printed circuit board with a printed circuit track of 0.001″-thick copper, 0.01″ wide, and with the resistivity ρ of copper 0.6788μΩ in., then the resistance per inch is

$$R/\text{inch} = \frac{0.6788\mu\Omega \times \text{in.}}{0.01\text{in.} \times 0.001\text{in.}}$$

$$R/\text{inch} = 67.88\text{m}\Omega/\text{in.}$$

This resistance per inch must be considered in the layout of a printed circuit board that uses precision components.

As an example, assume that we are using a PC board with this size of circuit track to connect the output from an op-amp into a 16-bit analog-to-digital converter. Also assume that the length of the track between the output of the op-amp and the input to the ADC is 2″ (Figure 18–1 shows the layout). The input resistance to the ADC is specified to be 10KΩ.

Since we have 2″ of track, the PC track resistance R_T is

$$R_T = 2\text{in.} \times 67.88\text{m}\Omega/\text{in.}$$

$$R_T = 0.13576\Omega$$

Now if the op-amp provides the full-scale voltage to the ADC of 5.1200V, the voltage per step of the ADC is

$$V_{st} = \frac{5.1200\text{V}}{2^{16}}$$

$$V_{st} = 78.125\mu\text{V}$$

Using Ohm's law to compute the voltage drop V_d across the two inches of track.

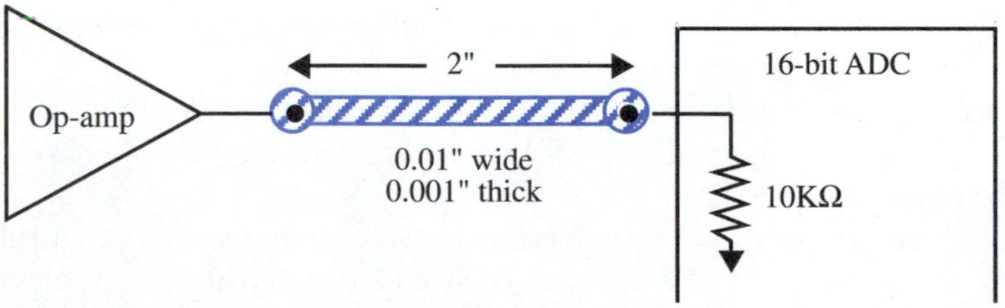

FIGURE 18–1 Printed circuit board voltage drop

$$V_d = \frac{5.1200\text{V} \times 0.13576\Omega}{10000.13576\Omega}$$

$$V_d = 69.5\mu\text{V}$$

Comparing this value with the voltage per step, we see it is almost equal—in other words the resistance of our two inches of PC track creates a 1LSB error! The solution to this problem is to shorten the distance between the op-amp and the ADC and/or widen the track. Either of these changes would lower the resistance and therefore the voltage drop.

A PROBLEM WITH CAD

Computer programs for the design of printed circuit boards normally treat all points in the same metal track as being a **node** and assume that they are at the same potential. This is quite a reasonable assumption for logic circuitry where noise immunities are greater than a volt, but it can be a problem in precision analog applications. Consider the effect of a small change in the circuit of Figure 18–1 as shown in Figure 18–2—where points A and B must be joined. A computer-aided design (CAD) printed circuit board program is quite likely to join the points with a link and then route the previous track around it. This has increased the track distance between the op-amp and the ADC and has really aggravated our voltage drop problem. Raising the input resistance levels of critical circuits will reduce the effect of track resistance voltage drop, but this will make the circuit more sensitive to noise pick-up.

NEED FOR SEPARATE VOLTAGE SENSE LEADS

One method of reducing the effect of conductor resistance (and the effects of output resistance) is to use separate leads for sensing the voltage at the point where the voltage is being applied. If in our previous example a separate voltage sensing lead was taken from the input of the ADC, and no current flowed through this lead, then the voltage drop in the printed circuit track would be ignored. Figure 18–3 shows the configuration.

From our knowledge of op-amps, we know that the left side of the 100KΩ resistor is at 2.5600V, and the right side is at 5.1200V. Since the R_f resistor is ten times greater than the

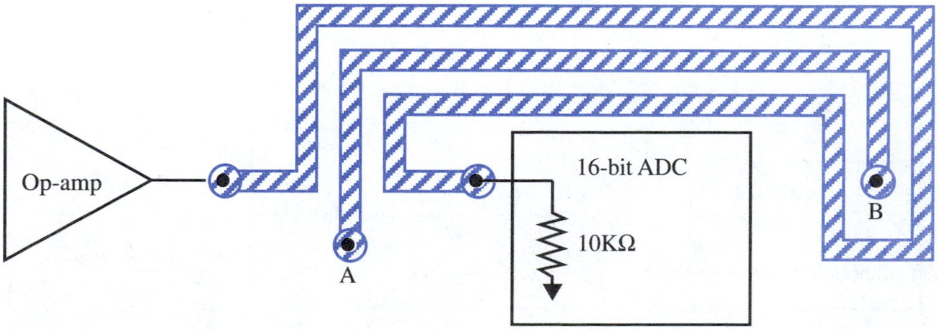

FIGURE 18–2 Problem with CAD layout

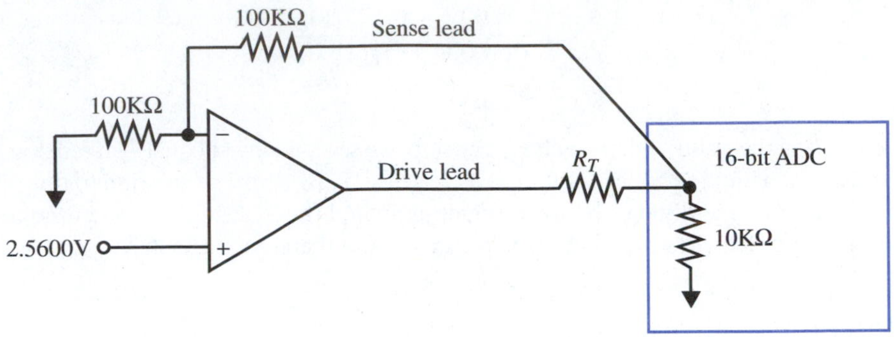

FIGURE 18–3 Use of voltage sensing lead

ADC, 10KΩ input resistor, and the voltage across it is one-half, the current through this R_f resistor is one-twentieth of the current through the ADC input resistor. This makes the voltage drop per inch in the sense lead track one-twentieth of that in the drive lead. Since the sense lead establishes the voltage at the ADC, the error voltage now is one-twentieth of the LSB voltage with the same length of track.

EFFECT OF GROUND CONNECTIONS

In the analysis of the voltage drops across the circuit conductors, we ignored the effect of current flowing through the ground leads. This needs to be considered, because we tend to think of ground as having 0Ω resistance and being at 0V potential.

If in our previous example the bottom end of the ADC input resistor was connected to the op-amp ground reference point by two inches of track, then another LSB voltage would be lost in this ground lead. To avoid this problem, we should single point ground the op-amp at the bottom of the ADC input resistor, as shown in Figure 18–4.

Since the same current flows through R_1 as R_f (no current into op-amp terminal), the current through the op-amp ground is the same small value as through the sense lead, so the

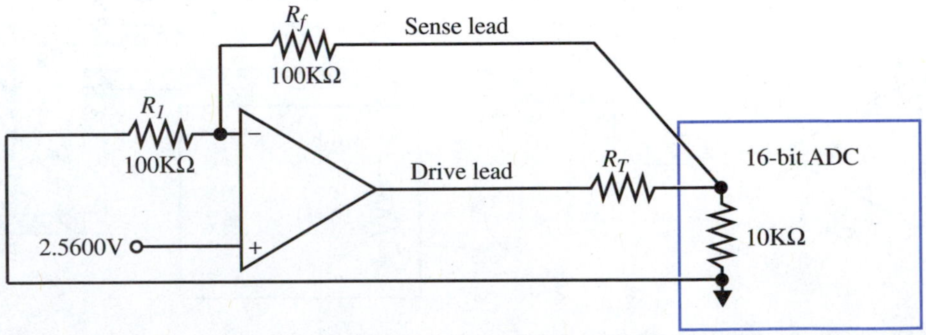

FIGURE 18–4 Reducing voltage drop in ground lead

voltage drop per inch of track is the same. But, again, it is one-twentieth of the volt drop that would have appeared in the ADC, 10KΩ terminating resistor ground track.

Another point to consider with the circuit of Figure 18–4 is that if we have equal lengths of track for the sense and ground leads, then the same value of resistance is added to both R_f and R_1. This means the op-amp gain, for our example, remains the same, and the voltage at the ADC is 5.1200V. In other words, the voltage drops in the sense and ground leads have been canceled, and there is no error due to track voltage drop.

USE OF A GROUND PLANE

A ground plane is created by having a large area of copper for the return path of current. Large area implies low resistance and therefore negligible voltage drops.

The resistivity of copper again is 0.6788μΩ-in. and the resistance of 0.001″-thick copper is

$$R = \frac{(0.6788\mu\Omega\text{-in.})L}{(0.001'')W} \qquad \textbf{(equation 18–2)}$$

where L is the length and W the width of the copper sheet.

If we make the length equal to the width (a square), the resistance becomes 0.6788mΩ/square regardless of the size of the square.

To obtain a good ground plane, a typical approach is to make one side of the printed circuit board all copper. If the board is 4 inches wide by 4 inches long, the resistance from one end to the other is 0.6788mΩ. If the board is 4 inches wide by 8 inches long, the resistance from one end to the other is 2 × *0.6788mΩ,* or 1.3576mΩ.

One thing to consider with ground planes is that, even with low resistance, significant volt drops can occur if the current is high. Let us look at an example, as shown in Figure 18–5, of a circuit board with a 15A power stage mounted at the top of the card to enhance heat flow.

In Figure 18–5, the effective ground plane size is 3 inches by 5 inches. With the width of 3 inches, the resistance of the ground plane per inch of length is, from equation 18–2,

$$R = \frac{(0.6788\mu\Omega\text{-in.})L}{(0.001'')W}$$

$$R = \frac{(0.6788\mu\Omega\text{-in.})1''}{(0.001'')3''}$$

$$R = 0.22627\text{m}\Omega/\text{in.}$$

The 15A current flowing through this resistance will create a voltage drop per inch of

$$V_d/\text{in.} = 0.22627\text{m}\Omega/\text{in.} \times 15\text{A}$$

$$V_d/\text{in.} = 3.394\text{mV/in.}$$

This voltage drop will cause differences in ground point voltages for various parts of the precision analog circuitry—which will result in accuracy problems. The solution, as shown in Figure 18–5b, is to isolate the ground path for the power circuitry from the precision cir-

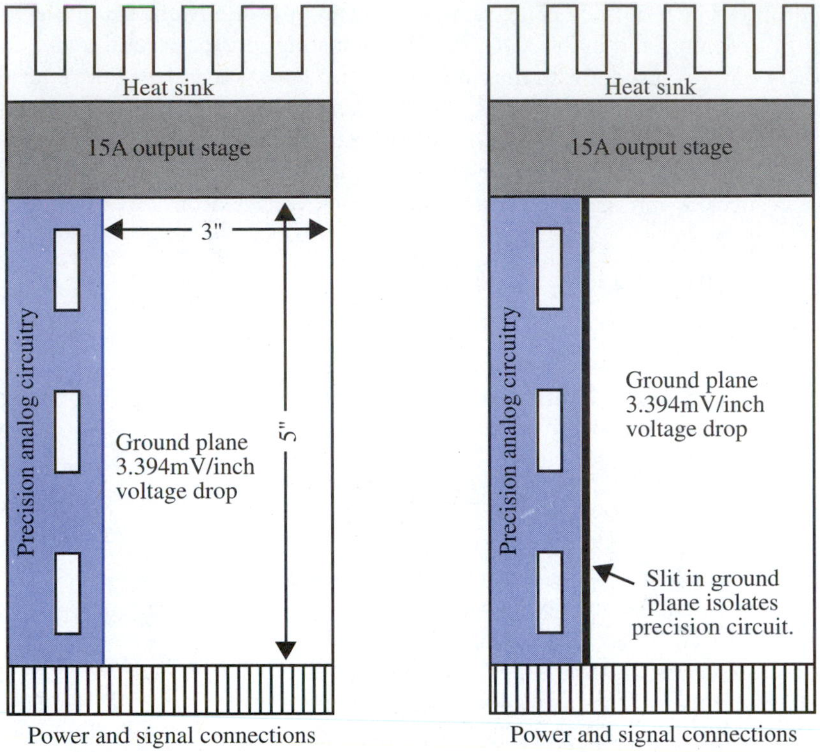

a) Improper ground plane

b) Correct ground plane

FIGURE 18–5 Solution to ground plane problem

cuitry by cutting a slit in the copper. This provides a separate ground plane for each circuit (3 inches wide for the power stage and 1 inch wide for the precision circuit), and the high current no longer flows through the precision circuit ground.

GROUND DIFFERENCES BETWEEN PRINTED CIRCUIT CARDS

When a signal is transmitted from one circuit card to another, differences in ground potentials on the two cards can cause signal errors. One solution is to use a differential amplifier to drive the signal from the first card into the differential input of an amplifier on the second card. Figure 18–6 shows the circuit configuration.

Without the differential configuration, the noise voltage difference between the two grounds (V_n) would be added to the V_s signal and appear on the second card. With the differential arrangement, the noise is added equally to both differential inputs and is therefore canceled by the differential amplifier A_2 (Refer to Chapter 7.)

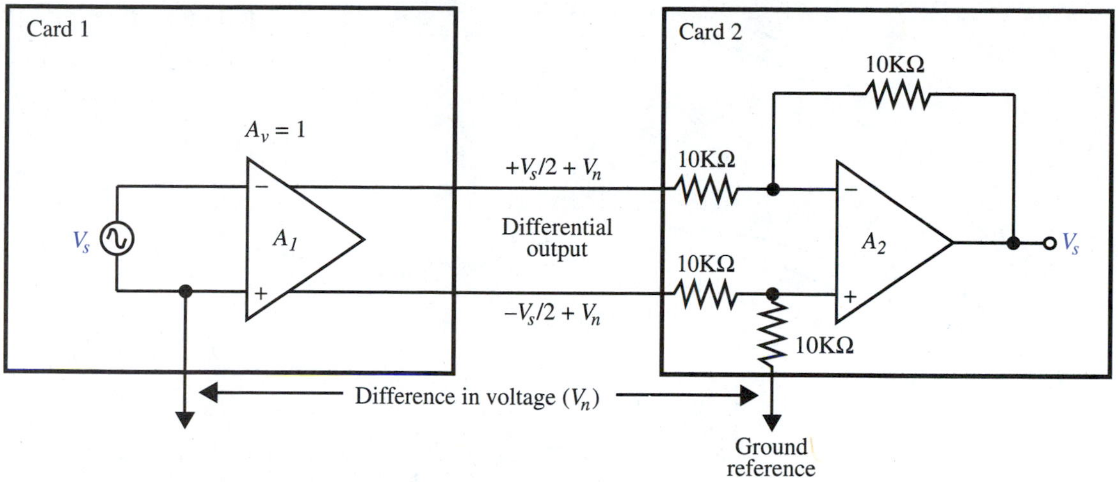

FIGURE 18–6 Overcoming ground voltage differences

MEASURING SMALL VOLTAGE DROPS

A circuit suggested by Analog Devices for measuring small voltage drops in ground or signal conductors is shown in Figure 18–7.

The AD624 instrumentation amplifier has low noise ($0.2\mu V_{p-p}$ to 10Hz), good common mode rejection (130db), and a programmable gain up to 1000. It may be used to measure microvolt signals at bandwidths up to 25KHz.

By connecting the probes on different parts of a printed circuit track or at different ground points, we can measure very small voltage drops. The high input resistance of the

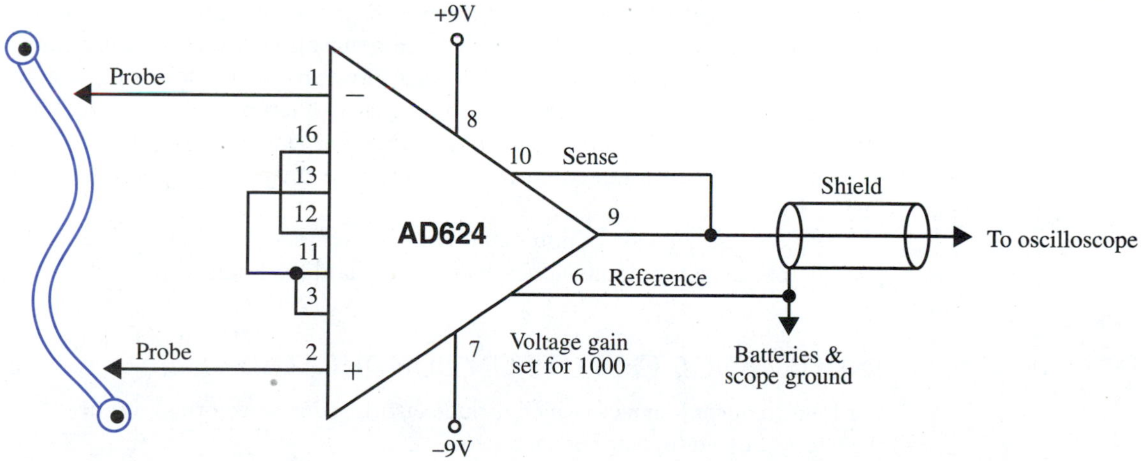

FIGURE 18–7 Circuit for voltage drop measurement. *Courtesy Analog Devices.*

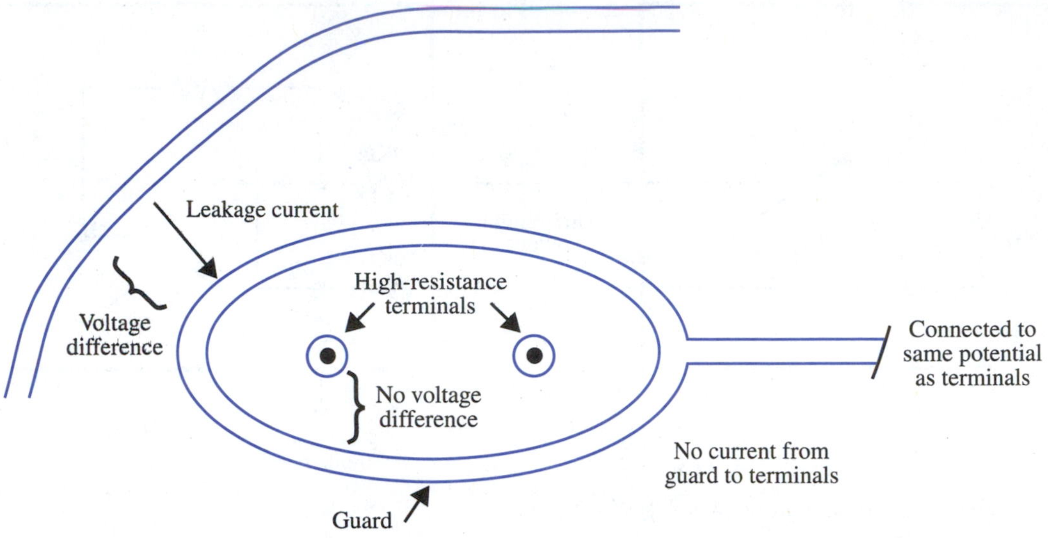

FIGURE 18–8 Use of a guard ring to reduce leakage currents

AD624 precludes circuit loading from the probes. Twisting and shielding the probe leads will reduce higher-frequency noise pickup.

The circuit can be mounted in a metal case and be self-contained by using two 9V batteries. The center point of the batteries should be tied to the ground reference.

PC BOARD LEAKAGE RESISTANCE

Another source of unwanted voltage results from leakage currents across the surface of a printed circuit board. These leakage currents, even in the picoamp range, can generate voltages if they flow through the megohms or giga-ohms resistance of some circuits.

Since current requires a voltage difference, we can put a guard ring around critical terminals on the board to eliminate this voltage difference. We can accomplish this by making the voltage of this guard ring at the same potential as the terminals we wish to protect, and the unwanted current is reduced. A guard ring application is shown in Figure 18–8.

In critical cases of leakage into input terminals, it is suggested that the terminals be virgin teflon insulated standoffs rather than the standard PC terminals.

INDUCTANCE EFFECTS ON CONDUCTORS

For high-frequency signals (which include fast rise time waveforms), we need to consider the effect of the inductance of conductors.

The inductance of a round wire can be approximated by the following formula:

$$L(\mu\text{H}) = 5.08 \times 10^{-3}[\ln(4y/d) - 0.75]y \qquad \textbf{(equation 18–3)}$$

where y is the length of the wire, and d is the diameter—both in inches.

The inductance for a printed circuit board track can be approximated by the following formula:

$$L(\mu H) = 5.08 \times 10^{-3}\left[\ln\left(\frac{2y}{w+h}\right) + 0.5 + 0.2235\left(\frac{w+h}{y}\right)\right]y \qquad \textbf{(equation 18–4)}$$

where y is the length of the track, w the width, and h the thickness—all in inches.

EXAMPLE

Find the inductive reactance for one inch of 22-gauge wire (diameter 0.0253 inch) at a frequency of 60MHz.

SOLUTION

First, we need to find the inductance of the wire, using equation 18–3.

$$L(\mu H) = 5.08 \times 10^{-3}[\ln(4y/d) - 0.75]y$$

$$L(\mu H) = 5.08 \times 10^{-3}[\ln(4 \times 1/0.0253) - 0.75]1$$

$$L(\mu H) = 0.0219\mu H$$

Now to find the reactance at 60MHz,

$$X_L = 2\pi f L$$

$$X_L = 2\pi \times 60\text{MHz} \times 0.0219\mu H$$

$$X_L = 8.26\Omega$$

To bring this into perspective, a signal with a rise time of 3.5ns includes frequencies up to 100MHz (f_c is $0.35/t_r$).

EXAMPLE

Find the inductive reactance for one inch of 0.001″-thick, and 0.01″-wide, printed circuit track at a frequency of 60MHz.

SOLUTION

First, we need to find the inductance of the wire, using equation 18–4.

$$L(\mu H) = 5.08 \times 10^{-3}\left[\ln\left(\frac{2y}{w+h}\right) + 0.5 + 0.2235\left(\frac{w+h}{y}\right)\right]y$$

$$L(\mu H) = 5.08 \times 10^{-3}\left[\ln\left(\frac{2 \times 1}{0.01 + 0.001}\right) + 0.5 + 0.2235\left(\frac{0.01 + 0.001}{1}\right)\right]1$$

$$L(\mu H) = 0.029\mu H$$

Now to find the reactance at 60MHz,

$$X_L = 2\pi fL$$
$$X_L = 2\pi \times 60\text{MHz} \times 0.029\mu\text{H}$$
$$X_L = 10.9\Omega$$

The results of these examples indicate that a ballpark figure for the reactance of one inch of #22 wire or 0.01″-wide printed circuit track at 60MHz is 10Ω. This value can be remembered and scaled for other frequencies, e.g., 1Ω at 6MHz.

We should now be aware that the inductance of wiring can create inductive reactance values that affect circuit operation. A quick calculation of the reactance can determine whether it is going to be a problem in a given application.

Inductance in leads can also affect the operation of a bypass capacitor. For example, the combination of one inch of lead length and internal inductance, causes a 0.01μF ceramic capacitor to series resonate at 13MHz. What this means is that at frequencies above 13MHz, the capacitor acts as an inductor and is no longer able to bypass signals.

EXAMPLE

Find the series self-resonant frequency for an ideal (no internal inductance) 0.1F bypass capacitor with a total of one inch of #22 leads.

SOLUTION

The lead inductance from the earlier example is 0.0219μH. Solving for the series resonant frequency,

$$f = \frac{1}{2\pi}\sqrt{\frac{1}{LC}}$$
$$f = \frac{1}{2\pi}\sqrt{\frac{1}{0.0219\mu\text{H} \times 0.1\mu\text{F}}}$$
$$f = 3.4\text{MHz}$$

Again, above this frequency the capacitor doesn't act as a bypass.

CIRCUIT NOISE PROBLEMS

In this section we shall study how to reduce noise introduction into a circuit by

- Conductive coupling
- Capacitive coupling
- Inductive coupling
- Radiative coupling

Earlier, in Chapter 17, we defined noise as any unwanted signal and, in electronic equipment, the high-frequency signals associated with switching devices tend to couple into the analog circuitry and become a source of noise.

Conductive Coupling

Current (conventional) flows from the most-positive voltage to the most-negative voltage point in a circuit. It will take all possible paths—but sometimes that path can be through a conductor, which can create interference with other circuits. The interference occurs because the current flows through circuit conductor resistance (and/or reactance) that is common to both circuits and develops a voltage drop.

Figure 18–9 illustrates the problem of a shared ground between a digital circuit and an analog circuit.

The common impedance in the ground circuit consists of the wiring resistance and its inductance. This inductance explains the differentiated waveform at the junction point of the common impedance.

The noise voltage V_n, developed across the common impedance by the current from the digital circuit, becomes an input voltage to the noninverting input of the op-amp A_1. At the output of the op-amp is both the amplified input signal and the amplified unwanted noise.

The solution to this noise problem is to use a separate ground return path to prevent any of the digital circuit current from flowing through an analog ground. Figure 18–10 provides an approach for separation of analog and digital grounds for circuit cards.

Notice in Figure 18–10 that the analog ground and digital grounds are connected at only one point. There is no current though the jumper that makes this connection because the power supplies are isolated and there is no ground loop—so no digital current flows through an analog ground or vice versa.

The analog grounds are tied together at a single point, and current from this point returns to the analog power supply. The jumpers from the analog board to the ADC board should be kept as short as possible to reduce ground voltage drops.

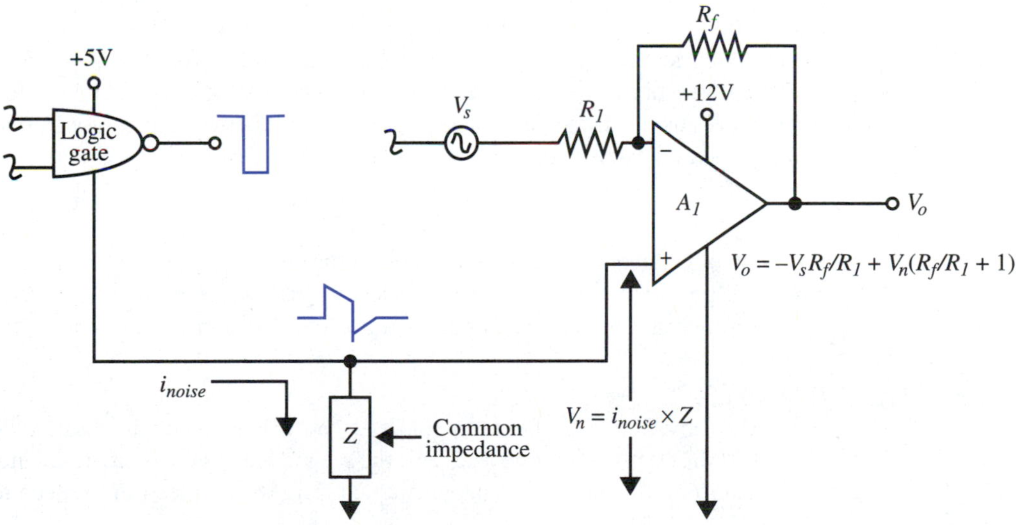

FIGURE 18–9 Common digital and analog ground

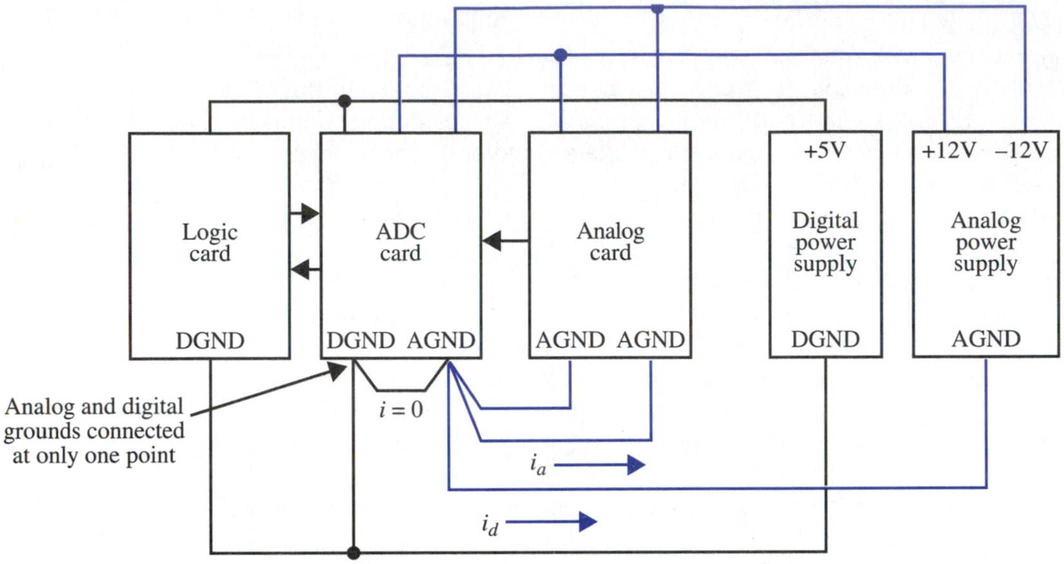

FIGURE 18–10 Correct grounding technique

Because the ADC circuit board contains circuitry that is most sensitive to ground voltage drops, the connections of all analog grounds, and the connection between digital and analog ground, are made at this board.

Capacitive Noise Coupling

Capacitive noise coupling is the result of an electric field coupling unwanted signals from one circuit to another. All capacitive coupling can be represented by a voltage noise source, the coupling capacitor, and the circuit impedance, as shown in Figure 18–11.

The amount of noise coupled into the circuit (consisting of R and C) is a function of the amount of coupling capacitance C_n. More capacitance between the noise source and the circuit means more noise voltage transfer. Looking at the capacitance equation

$$C_n = \mu A/d$$

the value of C_n increases with a reduction in distance d between the noise source and the circuit. Capacitance is least when the dielectric μ between the noise source and circuit is air, but it increases when the coupling surface area A is increased.

EXAMPLE

The capacitance C_n between the leads on a DIP device is about 0.5pF. If a 2V, 100MHz signal is on one lead, and the second lead has a 10KΩ resistance to ground, determine the amount of signal coupled from the first lead to the second lead.

SOLUTION

First, we need to find the reactance between leads at 100MHz.

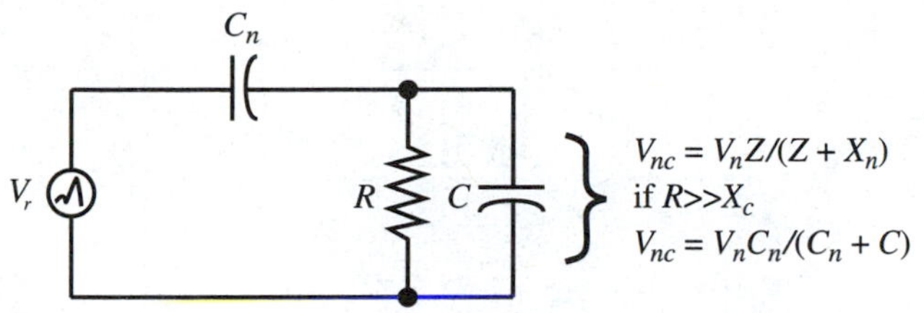

FIGURE 18–11 Capacitive coupling equivalent circuit

$$X_c = \frac{1}{2\pi f C_n}$$

$$X_c = \frac{1}{2\pi \times 100\text{MHz} \times 0.5\text{pF}}$$

$$X_c = 3.18\text{K}\Omega$$

The voltage V_c coupled across the 10KΩ will be

$$V_c = \frac{2\text{V} \times R}{\sqrt{(R)^2 + (X_c)^2)}}$$

$$V_c = \frac{2\text{V} \times 10\text{K}\Omega}{\sqrt{(10\text{K}\Omega)^2 + (3.18\text{K}\Omega)^2}}$$

$$V_c = 1.91\text{V}$$

This is not much less than the voltage at the first lead.

Ways to reduce the noise voltage coupling also include filtering the noise, and shielding. Use of a metallic capacitive shield (Faraday shield) is illustrated in Figure 18–12.

What the shield does is to break the noise coupling capacitor into two separate capacitors. C_{n1} has the full noise voltage developed across its plates, and the noise current is diverted harmlessly to the noise source common. C_{n2} has no noise voltage, and so no noise current flows into the signal circuit.

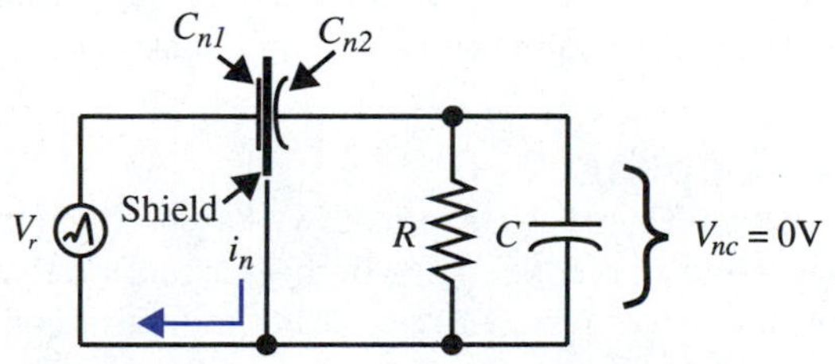

FIGURE 18–12 Shielding capacitance coupling

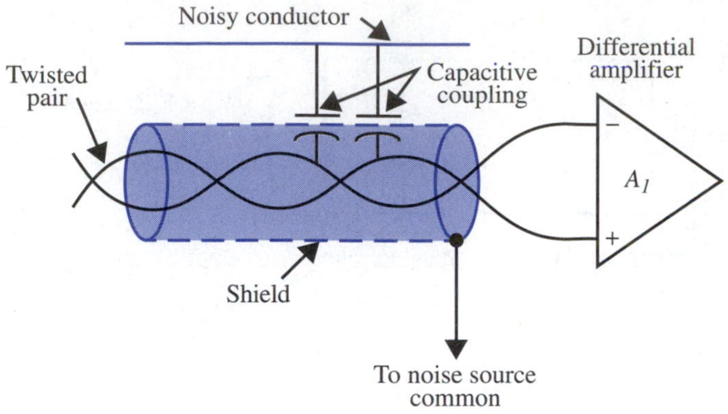

FIGURE 18–13 Shielding twisted pair

To ensure that the noise currents follow a predictable path, capacitive shields must always be connected to the noise source common, and then only connected at one point.

An example of capacitive coupling is cross-talk. If the signal cable is a twisted pair, then the coupling capacitance from the noise source will be the same for each wire. If the twisted pair is feeding the input to a balanced (differential) amplifier, then the noise is common mode and is rejected by the amplifier. Further improvement can be achieved by using a shield around the twisted pair, as shown in Figure 18–13. Again, the shield around the twisted pair should only be connected at one point to the noise source common.

Transformer Capacitive Coupling Transformers can have capacitive coupling from the primary winding to the secondary winding. This is because of distributed capacitance, which results from the primary and secondary windings being in close proximity. So a noise voltage that is coupled to the transformer primary can appear at the secondary winding.

A way to prevent this transformer capacitive coupling is to have the transformer constructed with a copper sheet separating the primary and secondary windings (Faraday shield). The copper sheet allows the magnetic flux to pass from primary to secondary—so transformer action still occurs, but the shield prevents the noise voltage from appearing at the secondary. A transformer with and without a Faraday shield is shown in Figure 18–14.

In Figure 18–14a, the noise current (i_n) flows through the primary to secondary capacitance, and a part of this current (i_{n1}) flows through the load resistor, creating a noise voltage at the output.

In contrast, we see in Figure 18–14b that the noise current flows through the primary-to-shield capacitance and back to the noise source without passing through the load resistor.

Inductive Noise Coupling

Whenever we have a changing current in a circuit, we have a changing magnetic field that can induce noise voltages into adjacent circuits. In this section, we shall look at the ways to reduce this effect. But first, let us review the basics of inductive coupling. Figure 18–15 shows the factors affecting coupling.

From Figure 18–15, we see that the current in the interfering loop creates a magnetic flux that couples into the signal circuit. This flux level is directly related to the amount of noise

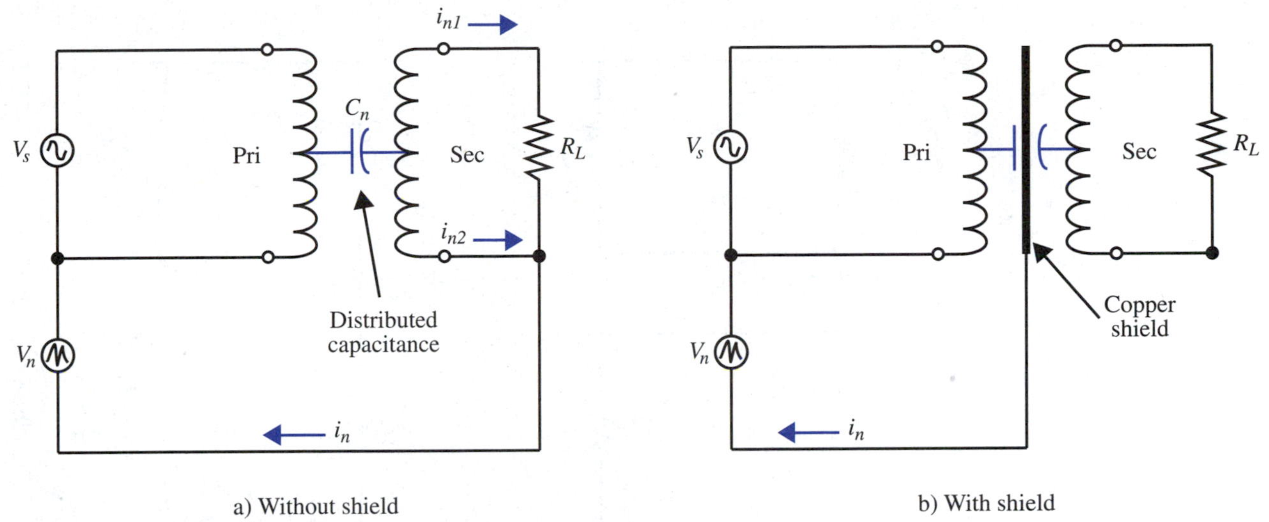

FIGURE 18–14 Transformer capacitance shielding

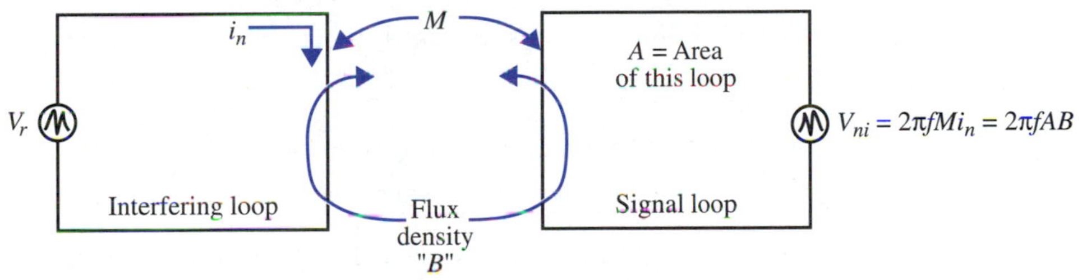

FIGURE 18–15 Inductive noise coupling

current. The portion of the total flux that is coupled into the signal circuit is defined by the mutual coupling coefficient M. With M of 1, all of the noise flux cuts through the signal loop. With lower values of M, less flux is coupled. The flux per unit area (flux density) is called B.

The amount of noise voltage that is induced in the signal loop is related to the coupled flux—but it is also related to how fast the flux is changing. So, the higher the frequency of the noise source, the higher the induced noise voltage. Putting these relationships into an equation form for the induced noise voltage V_{ni},

$$V_{ni} = 2\pi fMi_n = 2\pi fAB \qquad \textbf{(equation 18–5)}$$

Looking at the two equations, we see that we can reduce the induced noise voltage reduction by

1) Reducing the frequencies in the noise source
2) Reducing the mutual coupling M

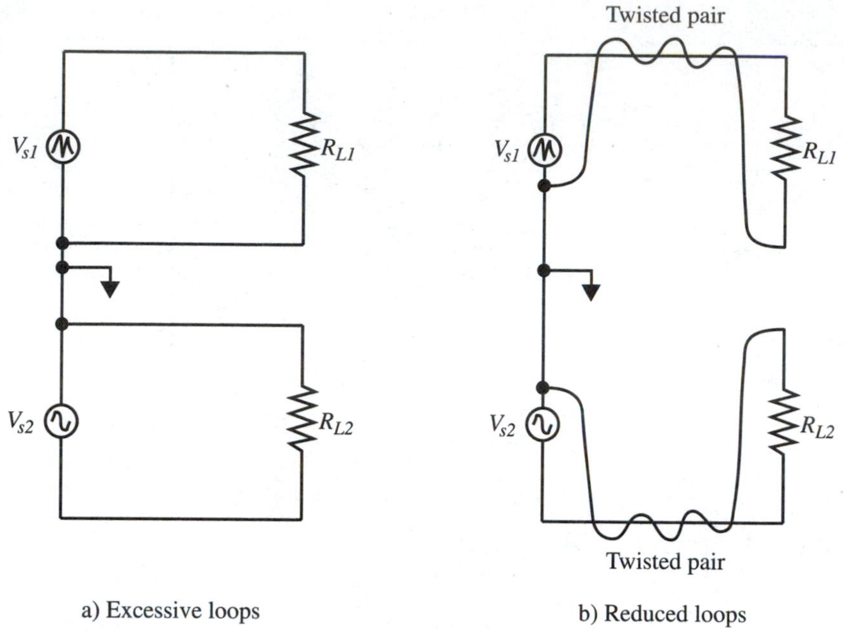

FIGURE 18–16 Reducing noise coupling

3) Reducing the noise current i_n (determines flux)
4) Reducing the area A of the signal loop

Frequencies in the noise source can be reduced if they are the result of rise times of signals that are faster than is actually required for correct circuit operation. An example of this is the addition of capacitance to the output of an RS232 driver; to increase the rise time and thereby reduce the noise frequencies (discussed in Chapter 17).

Reducing the mutual coupling requires greater separation of the two circuits or making the effective circuit loops perpendicular to one another.

To determine whether the noise source current can be reduced requires a knowledge of the circuit operation of the source. Remember, the circuit causing the noise is also performing a useful function. However, circuit operations can be implemented more than one way, and it is possible a lower-current circuit approach is possible.

Reducing the loop area is usually fairly simple to implement and is very effective. Figure 18–16 shows first an arrangement with excessive loop area and then the preferred small loops with twisted conductors. By making the loop small, the pick-up is reduced, and moving the wires in the loop closer together cancels out the magnetic field because of equal and opposite current in the wires. Twisting the wires reduces the voltage induced in the wires. Relative to just running the two wires side by side, two twists per foot provides a reduction of voltage pick-up of about 5:1, three twists per foot 15:1, and twelve twists per foot 140:1.

Problems with Ribbon Cable A ribbon cable consists of parallel signal conductors which form loops with the return lead as shown in Figure 18–17.

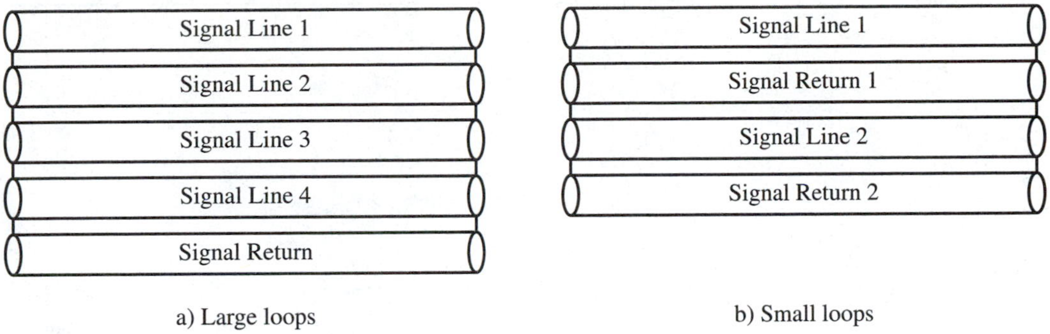

FIGURE 18–17 Ribbon cable current loop

Notice in Figure 18–17a that signal line 1 and the return line form a large loop area that encloses the other loops (maximizes intercircuit coupling). An improvement is shown in Figure 18–17b: Each signal line has its own return line, which minimizes the loop area and reduces coupling.

Using twisted pairs (signal and return), or coax cable, is an improvement over ribbon cable. Both configurations confine the magnetic field to the pair of wires.

Grounds on Printed Circuit Boards If one side of a printed circuit board is used as a ground plane, current will flow in the copper foil in a way to minimize the area of the current loop. Figure 18–18 illustrates this effect. The dotted line shows the main current path in the ground plane. Notice that it automatically stays just below the conductor on the circuit side of the board to minimize the encountered inductance.

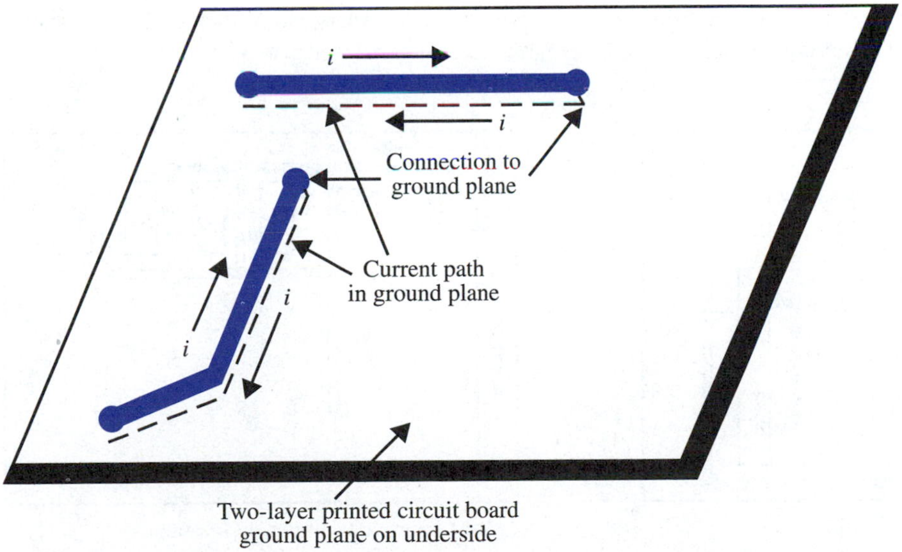

FIGURE 18–18 Current flow in ground plane

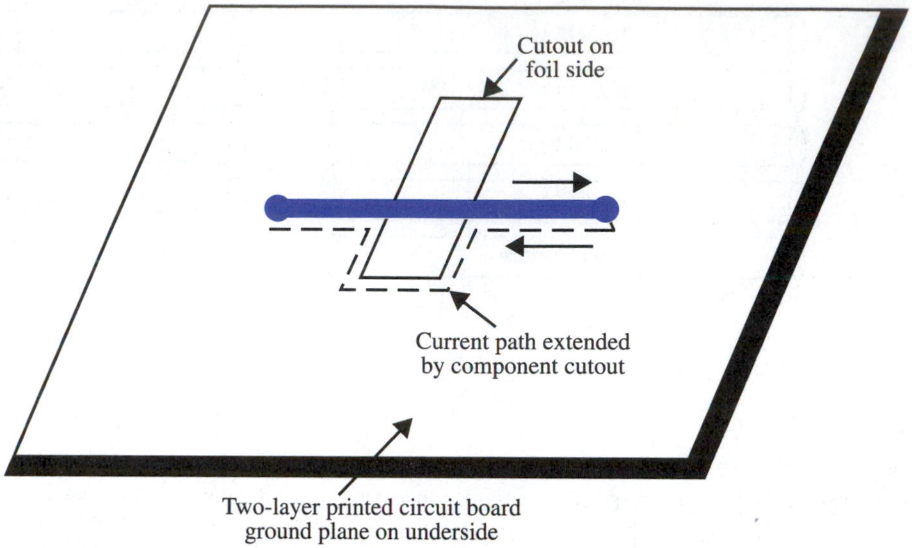

FIGURE 18–19 Problem of ground plane cutouts

If a cutout is made in the copper foil ground plane to mount a component, it could interfere with a ground plane current path. Figure 18–19 illustrates the problem. The cutout in the ground foil has increased the area of the current loop and can cause possible noise coupling.

When a printed circuit board layout is made, the component placement and connector assignment should be made with a consideration of ground loops. Figure 18–20a is an example of both poor component location and too much spacing between the signal connector terminal and the return terminal, which creates a large loop. Figure 18–20b shows an improved layout and connector assignment that generates a small current loop.

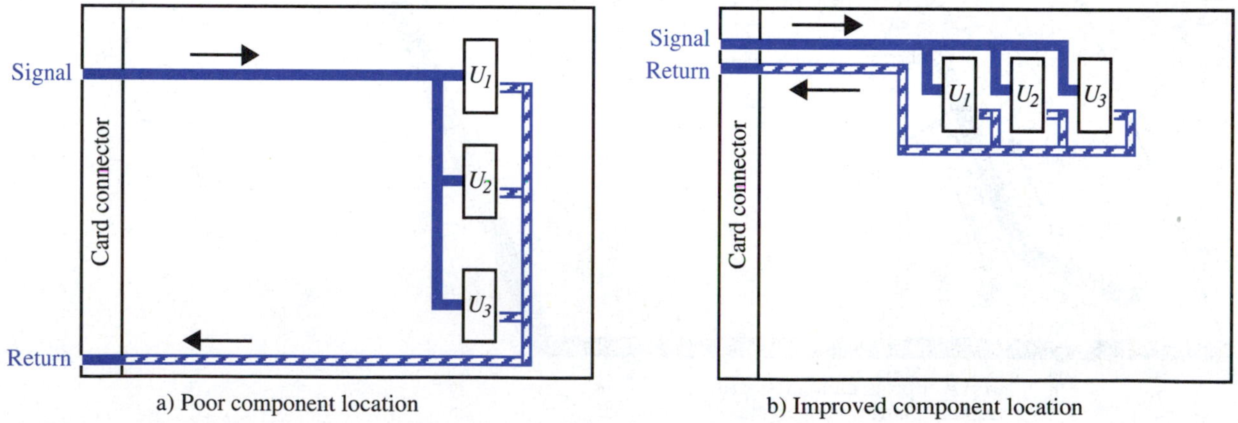

FIGURE 18–20 Printed circuit board layout

TABLE 18–1 Relative Magnetic Permeability

Material	*Relative Permeability to Air*
Aluminum	1
Beryllium	1
Copper	1
Nickel	50
Cast iron	60
Powdered iron	100
Machine steel	300
Ferrite (typical)	1,000
Permalloy 45	2,500
Transformer iron	3,000
Silicon iron	3,500
Iron (pure)	4,000
Mumetal	20,000
Sendust	30,000
Supermalloy	100,000

Magnetic Shielding Another way of decreasing magnetic coupling is through shielding. The capacitive shield discussed previously intercepts the electric field and reroutes the noise currents back to their source. A magnetic shield can be used to redirect the magnetic flux by providing a path of higher permeability (magnetic conductance).

Permeability is a measure of how easily magnetic flux can be established in a material. For nonmagnetic materials—such as copper, aluminum, wood, or glass—the permeability is, practically speaking, the same as that for air. For magnetic materials, the permeability is a hundred or a thousand times that of air. In other words, these materials conduct magnetic flux easier than air. The relative (to air) permeability of some selected materials is listed in Table 18–1.

The magnetic flux will take the path that has highest permeability, just like current flowing through the lowest resistance. This characteristic can be used to construct a magnetic shield. Figure 18–21 shows the magnetic field (flux) pattern for a circuit shielded by a low-permeability aluminum case and also by a steel case.

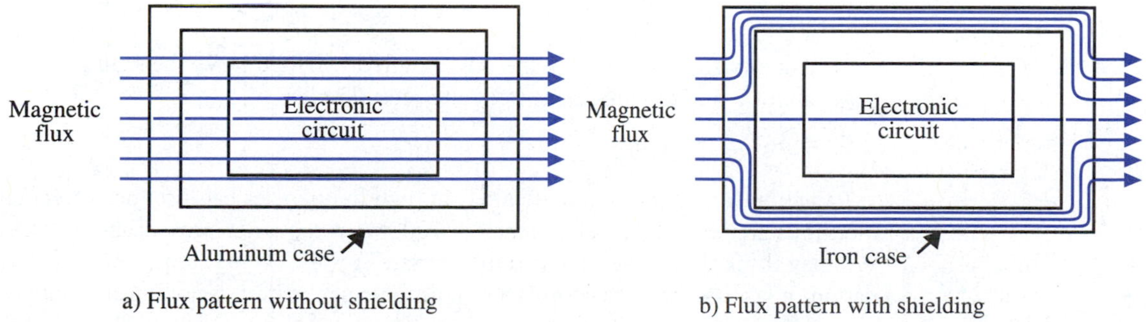

FIGURE 18–21 Effect of magnetic shielding

The aluminum case is a good capacitive coupling shield, but the magnetic flux lines pass right through the case and induce a noise voltage into the electronic circuitry. With the steel case, almost all the flux lines follow the low-permeability path around the electronic circuit, rather than through it, and the induced noise pickup is very small.

The typical shielded wire and coax cable have copper or aluminum shields, and as you can see from the relative permeability value of Table 18–1, they provide no magnetic shielding.

The most-effective magnetic shields are those made out of highest-permeability materials, but these materials have a cost penalty and are not always easy to form in the desired shape.

Radiative Coupling

The term **radiative coupling** means transmission or reception of electromagnetic radiation. Electromagnetic radiation occurs when the dimensions of a conductor are close to a wavelength of a frequency present in the circuit and it becomes an antenna. When this condition exists, energy leaves the conductor and propagates through space. Close to the circuit, a capacitive and magnetic coupling field exists; but greater than a wavelength away, these fields become negligible and radiation is predominant. The relationship between wavelength and frequency is given by the equation

$$\text{Wavelength } (\lambda) = \frac{3 \times 10^8 \text{ meters/sec}}{f} \qquad \textbf{(equation 18–6)}$$

EXAMPLE

Find the wavelength λ at 100MHz.

SOLUTION

Using equation 18–6,

$$\lambda = \frac{3 \times 10^8 \text{meters/sec}}{f}$$

$$\lambda = \frac{3 \times 10^8 \text{meters/sec}}{100\text{MHz}}$$

$$\lambda = 3 \text{ meters } (9.84\text{ft.})$$

Antennas are formed with the wavelength down to λ/4—so the right configuration of about 2.5 ft. of wire at this frequency could radiate.

Radiation from circuits is more likely at high frequencies because the correct circuit dimensions are possible (at a frequency of 1000MHz, a quarter wave is about three inches). We may think that most circuits don't operate at these higher frequencies—but a combination of fast rise time waveforms and harmonic generation in nonlinear circuits can generate these higher frequencies.

Ways to prevent radiation:

- Keep circuit conductor loop dimensions small.
- Use ceramic (short-lead) bypass capacitors on wires leaving and entering circuits.
- Contain the circuit within a steel box (bypass wires leaving box).
- Use ferrite beads on wires that have a radiation-inducing length.

The ferrite beads attenuate by acting as a lossy element for the high-frequency signals that are capable of radiating.

If a steel box is used, it typically requires holes for ventilation. To maintain shielding, the individual holes (mesh) should be smaller than a quarter wavelength of the highest frequency that will be encountered.

Radio Transmitters as a Source of Interference Since radio transmitters are required to radiate, they can become a source of interference to other electronic equipment. The tremendous amount of transmitters in use today require (FCC rules) that equipment be designed to operate in such an environment. The amount of protection required can be a function of the radiated frequency (will it affect circuit operation?) and power level. For example, a computer in an automobile could encounter an on-board CB radio or a cellular phone. So the computer should be designed with the appropriate shielding for the known power levels and frequencies of these signals.

Since an antenna can radiate and receive a frequency with the same efficiency, the techniques used to reduce susceptibility to radiation are basically the same as those used to prevent radiation. In other words, if a circuit can radiate at a particular frequency it can equally well receive interfering radiation at that frequency.

UNWANTED CIRCUIT OSCILLATIONS

There is an old adage that states, "Amplifiers oscillate and oscillators amplify." The difference between an amplifier and an oscillator is that an oscillator is an amplifier with positive feedback. So a typical problem is that an amplifier has inadvertent positive feedback that causes the amplifier to become an oscillator. If the amount of output signal fed back to the input of an amplifier is sufficient to sustain the output of the amplifier, oscillation will occur.

Oscillation Caused by Incorrect Wiring

Wiring can generate feedback by capacitance between amplifier output leads and input leads. At high frequencies, picofarads of capacitive coupling is all that is required for oscillation. Capacitance can be reduced by methods such as shielding and by not running input and output leads close to one another—particularly parallel; better to cross at right angles.

Another wiring problem that can create oscillation is caused by ground loop current from the output developing a feedback voltage to the input. The solution is, as mentioned previously, to keep the input and output grounds separate, with a single chassis ground point.

Oscillation Caused by Power Supply Impedance

We typically think of a DC power supply impedance as having 0Ω impedance. But the power supply current has to flow through a conductor in order to reach the circuit. This conductor has both resistance and inductive reactance, which can cause an AC voltage drop to develop. Since most of the current drawn by the circuit is from the output stage, the voltage drop is a portion of the circuit output voltage. If the same conductor supplies current to both the input and output of the circuit, then feedback and possible oscillations will occur.

Solutions to the power supply impedance problem are to run separate power supply conductors to the input and output circuits and/or connect a bypass capacitor (0.1 μF typical) from between where the power supply conductor feeds the output stage and the output stage ground. This capacitor bypass technique effectively AC bypasses the conductor and the power supply.

TROUBLESHOOTING VOLTAGE DROP AND NOISE PROBLEMS

Because we are dealing with very low voltages, troubleshooting circuits for possible voltage drop and/or noise problems requires extreme care in the measurement of these parameters.

It is assumed that measurements will be made with a sensitive oscilloscope, and the differential amplifier of Figure 18–7 might be required. Considerations in making the measurements are as follows:

1) Be sure the bandwidth of the measuring equipment is sufficient to sense the signal being measured. (High-frequency oscillations could be missed.)
2) Keep the leads to the measurement probe as short as possible because false noise signals could be induced in the leads. Check by shorting the leads and looking for noise.
3) Connect the ground lead on the probe to a common, right where the voltage is being measured—in the case of an amplifier, right on the input terminals.

An open filter capacitor should be suspected if oscillations occur in a piece of equipment. This can be the bypass capacitor on a printed circuit board or the capacitor in the power supply. Low-frequency oscillation (motorboating) occurs in radios and audio amplifiers if a power supply capacitor becomes open-circuited.

SUMMARY

- Knowledge of the resistance per inch is important in the layout of printed circuit boards.
- CAD programs for PC board layout do not consider the resistance of board conductors.
- Separate sense leads for an amplifier can overcome lead voltage drop.
- Voltage drops in ground leads must be considered.
- A ground plane will reduce ground lead voltage drop.
- The effect of high currents in a ground plane must be considered.
- Differences in ground potentials on circuit cards can be overcome by using a differential amplifier.

- Board leakage currents can be a problem with high-impedance circuits.
- Inductance of leads must be considered with high-frequency (MHz) circuits.
- Types of noise conduction are
  1) Conductive coupling
  2) Capacitive coupling
  3) Inductive coupling
  4) Radiative coupling
- Conductive coupling occurs when circuits share the same ground lead.
- Capacitive coupling is the result of stray capacitance.
- Inductive coupling results from stray magnetic fields.
- Radiative coupling is most likely to occur when circuit dimensions are close to the wavelength of the radiated signal.
- When looking for voltage drop and coupling problems, great care must be taken with the measurements.

EXERCISE PROBLEMS

1. Why is there more concern today about conductor voltage drops and noise pickup?
2. Determine the resistance of a 0.001″ copper track if the width is 0.02″ and the length 4″.
3. If the track in problem 2 is between the output of an amplifier (no sense leads) and the 5KΩ input resistance to a 16-bit ADC with a 10.24000V reference, determine the LSB error.
4. State why voltage sense leads are used at the output of an amplifier.
5. Why should the ground reference connection for an amplifier be made at the load resistor?
6. Why is a ground plane used?
7. What is the resistance from side to side of a 0.002″-thick square of copper foil?
8. Why is a differential amplifier sometimes used to couple an analog signal from one printed circuit card to another?
9. Why is a guard ring used around high-impedance terminals on a PC card?
10. Determine the inductance for a one-foot length of #26 wire (diameter is 0.0159″).
11. Determine the reactance for a 4″ length of 0.001″-thick, 0.02″-wide, printed circuit track at a frequency of 10MHz.
12. Compute the resonant frequency for an ideal 0.47μF capacitor with 3/8″-long #22 leads.
13. State how capacitive coupling occurs.
14. Suggest a method for reducing capacitive coupling.

15. Determine the voltage coupled to a DIP lead with 20KΩ resistance to ground from an adjacent pin that is routing a 1V, 40MHz signal.
16. List the factors that affect inductive coupling.
17. Suggest a method for reducing inductive coupling.
18. Describe the optimal assignment of signal wires and returns in a ribbon cable.
19. What problem can occur if a section of a PC card ground plane is removed?
20. How does magnetic shielding reduce inductive coupling?
21. What is radiative coupling?
22. How can radiative coupling be reduced?

ABBREVIATIONS

A = Area of circuit loop
B = Magnetic flux density
CAD = Computer-aided design
h = Height or thickness of conductor
M = Mutual magnetic coupling between circuits
R_T = Track resistance
y = Length of conductor
V_c = Coupled voltage
V_d = Voltage drop
V_{st} = Voltage per step
w = Width of conductor
ρ = Resistivity
λ = Wavelength of electromagnetic radiation = 3×10^8m/sec/f

ANSWERS TO PROBLEMS

1. Circuits have greater precision.
2. $R = 0.13576\Omega$
3. $V_d = 0.278$mV; almost 2LSB
4. To overcome lead drop
5. To reduce ground drop
6. To reduce ground drop
7. $R = 0.3394$mΩ
8. To eliminate ground difference
9. To prevent leakage currents
10. $L = 0.443\mu$H
11. $L = 0.131\mu$H
12. $f = 1.813$MHZ
13. See text.
14. Faraday shield
15. $V_{coup} = 0.929$V
16. i, area, M.
17. See text.
18. Signal and return adjacent
19. Current loop increases
20. Provides an easier path for flux
21. See text.
22. Shielding and circuit loop size

CHAPTER 19

Thermal Considerations

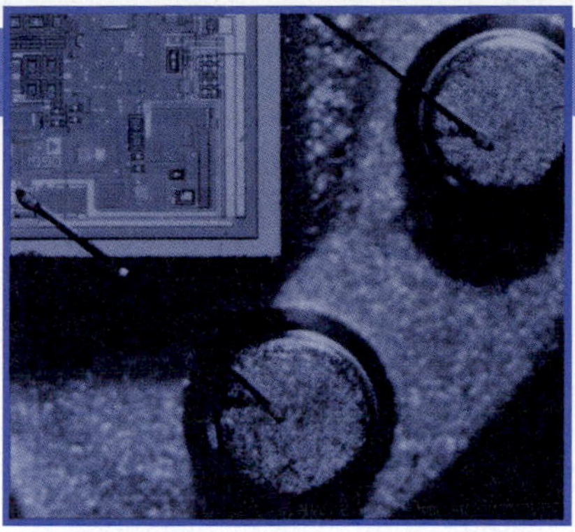

CHAPTER OBJECTIVES

After completing this chapter, you should be able to explain the following concepts:

- The effects of heat
- What controls temperature rise
- Component temperature considerations
- Resistor temperature effects
- Inductor temperature effects
- Integrated circuit temperature effects
- Heat flow and thermal Ohm's law
- Thermal equivalent circuit
- Determining thermal resistance
- Effect of thermal resistance on device power rating
- Changing thermal resistance

INTRODUCTION

In earlier chapters, we have at times briefly touched on the effect of heat on circuit operation. This chapter will go into much greater depth on the nature of heat, how temperature affects component operation, the damaging effects of heat, and how temperatures can be controlled to maintain correct circuit performance—even with severe ambient conditions.

Examples of the effect of heat on specific practical circuits will be given to further the understanding of this topic. The application of commercially available heat sinks for temperature control will also be covered.

NATURE OF HEAT

Heat is a form of energy that is generated one of three ways: from mechanical energy by friction, chemical energy by conversion, or electrical energy by resistive heating. The application of heat energy to a material causes an agitation of molecules within the material. The amount of heat energy in an object is indicated by the temperature—but it is also a function of the size of the object and its composition.

Let us consider, as an analogy, a tank that is being filled with water. The height of the water in the tank corresponds to the temperature, and the amount of water in the tank the heat energy. The larger the tank, the more water it can hold. Similarly, the larger the piece of material, the more heat energy it can hold. If we made a hole in the bottom of the tank, water would flow out and the tank level would drop. If we allow heat energy to flow out of an item, the temperature will drop. The rate of water flow from a tank is related to the level of water in the tank, which creates a pressure. The higher the level, the greater the pressure at the bottom of the tank, and the faster the water flows out. If a hot object is placed in contact with a cooler object, the rate of heat flow is related to the temperature difference. The greater the temperature difference, the faster the heat flow out of the hot object.

When filling a water tank with a hole in the bottom, a point could be reached where, because of the increased pressure at the bottom of the tank, as the water level rises, the flow out of the tank equals the flow into the tank. At this point the water level will stay constant. A similar effect will occur with an object being heated. As the temperature rises, it increases the heat flow out of the object. A point is reached where the heat flow out equals the heat flow in, and the temperature no longer increases but stays constant. This is **thermal equilibrium**.

EFFECT OF HEAT ON ELECTRICAL COMPONENTS

Adding heat to an electrical component can occur one of two ways: heat can transfer from heated objects or air; and/or self-heating of the component can occur from internal power dissipation. The addition of heat to a component can have two negative effects: parameters of the component can change; or with too much heat, the component can be permanently damaged.

Changes in component parameters occur with the temperature (heat level) of the component. This must be factored into the design of electronic circuits. This must include both the temperature changes due to external heat transfer (ambient temperature changes) as well as the often neglected self-heating induced change. All electronic components decay or change with time. These effects are accelerated at the higher temperatures.

We will now look at the effect of the temperature on specific electronic components, starting with the resistor.

RESISTOR TEMPERATURE EFFECTS

Most of us know that the value of a resistor changes with temperature. The amount of change is determined from the specified temperature coefficient. The coefficient (K_t) is usually given as parts per million (ppm) per degree Celsius. To determine the actual percent change ΔR in the value of a resistor with a given temperature change ΔT, we use the equation

$$\Delta R = K_t \Delta T(100\%) \qquad \textbf{(equation 19–1)}$$

Table 19–1 lists the temperature coefficients for various types of resistors. The minus sign for carbon indicates that the resistance goes down as the temperature goes up.

EXAMPLE

Find the change in the value of a resistor with a temperature change of 50°C if the resistor is carbon composition; wire-wound (use 20ppm/°C).

TABLE 19–1 Temperature Coefficients for Resistors

Resistor Type	*Temperature Coefficient (K_t) ppm/°C*
Carbon composition	−1500
Carbon film	to −1000
Wire-wound	1 to 30
Metal film	<1 to 200
Metal foil	to <1
Thick film	>100
Thick Film on Substrates	<100

SOLUTION

The value of K_t for the carbon composition resistor is from Table 19–1, −1500ppm/°C. Using equation 19–1,

$$\Delta R = K_t \Delta T(100\%)$$

$$\Delta R = \left(\frac{-1500}{10^6 \times °C}\right) \times 50°C \times 100\%$$

$$\Delta R = -7.5\%$$

The value of K_t for the wire-wound resistor is given as 20ppm/°C. Again using equation 19–1,

$$\Delta R = K_t \Delta T(100\%)$$

$$\Delta R = \left(\frac{20}{10^6 \times °C}\right) \times 50°C \times 100\%$$

$$\Delta R = 0.1\%$$

This example indicates how much more stable a wire-wound resistor is with temperature change.

If we are using two resistors of the same type in a voltage divider application, we would think that as the temperature changes, both resistors would change by the same percentage, and that the voltage at the junction of the two resistors would stay the same. But resistors of the same type could have slightly different temperature coefficients, and so there could be a voltage change.

EXAMPLE

Two carbon-film resistors are used in a voltage divider application with one resistor having a temperature coefficient of −800ppm/°C and the other −850ppm/°C. Find the percent change in the divided voltage if the temperature changes by 60°C.

SOLUTION

Change will occur because of the difference in the temperature coefficient of 50ppm/°C. Using this value in equation 19–1,

$$\Delta R = K_t \Delta T(100\%)$$

$$\Delta R = \left(\frac{50}{10^6 \times °C}\right) \times 60°C \times 100\%$$

$$\Delta R = 0.3\%$$

Since the voltage will reflect the change in resistance, the voltage will also change by 0.3%.

In critical applications, the resistors in the divider should be located so that they see the same ambient temperature, and they should also be mounted away from other heat-producing devices.

RESISTOR SELF-HEATING

Another problem that can occur with a voltage divider is when self-heating in one of the resistors exceeds that in the other resistor. This causes the temperature of one resistor to be greater than the other; even if the resistors have equal temperature coefficients, the divided voltage will change.

We determine the amount of self-heating from the temperature derating curve for the particular resistor. Figure 19–1 shows the curve for a type RN55D, 60D, or 65D metal-film resistor. This curve shows how the allowable dissipation in a resistor drops as the temperature increases; for example, at 120°C, the allowable power dissipation is one-half that at 70°C.

Let us assume that we are using an RN60D resistor. The power rating for this type is 0.25W at 70°C. As we can see from Figure 19–1, the power rating drops to 0W at 170°C. From these two points, we can determine the °C/W (K_d) constant for this resistor.

$$K_d(°C/W) = \frac{170°C - 70°C}{0.25W}$$

$$K_d(°C/W) = 400°C/W$$

With this constant, we can determine the temperature rise ΔT for a given power dissipation P_d within the resistor by using the relationship

$$\Delta T = K_d P_d \qquad \textbf{(equation 19–2)}$$

EXAMPLE

Find the change ΔR in the value of an RN60D, 1KΩ metal-film resistor after 10V is applied. Assume K_t is 100ppm/°C.

SOLUTION

First, we need to find the power dissipation P_d in the resistor.

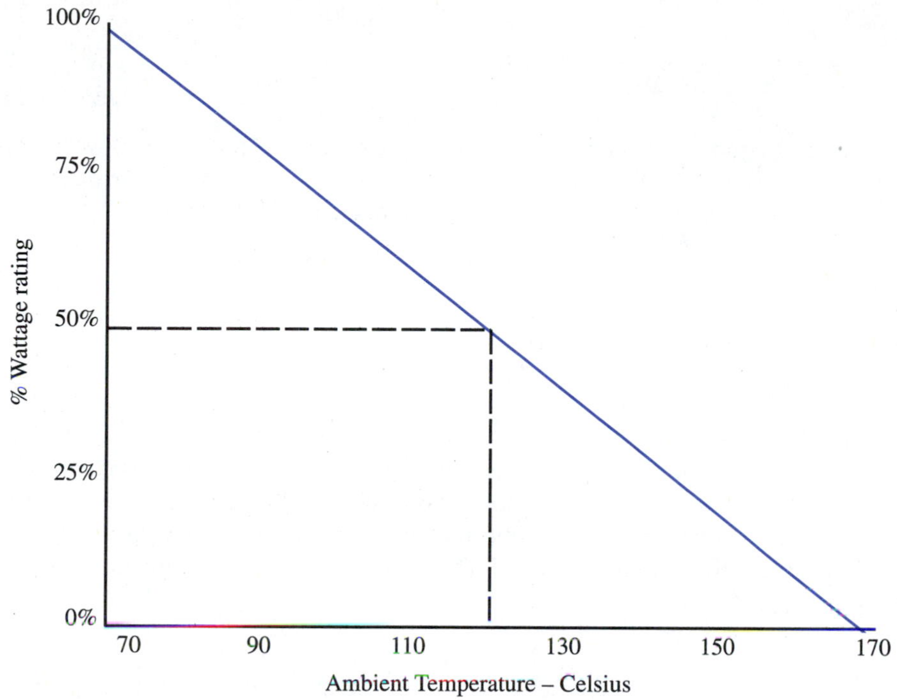

FIGURE 19–1 Power derating curve RN"D"-type resistors

$$P_d = V^2/R$$
$$P_d = (10)^2/1\text{K}\Omega$$
$$P_d = 0.1\text{W}$$

Now using equation 19–2 to find the temperature rise in the resistor with this power dissipation,

$$\Delta T = K_d P_d$$
$$\Delta T = 400°\text{C/W} \times 0.1\text{W}$$
$$\Delta T = 40°\text{C}$$

To find the change in resistance with this temperature rise, we use equation 19–1.

$$\Delta R = K_t \Delta T(100\%)$$
$$\Delta R = \left(\frac{100}{10^6 \times °\text{C}}\right) \times 40°\text{C} \times 100\%$$
$$\Delta R = 0.4\%$$

This percentage represents an actual change ΔR in resistance of

$$\Delta R = 1\text{K}\Omega \times 0.004$$

$$\Delta R = 4\Omega$$

Considering again the significance of the derating curve of Figure 19–1, if we are dissipating 0.25W in an RN60D resistor at an ambient temperature of 70°C, the temperature of the resistor will be 170°C. But what will be the temperature of the resistor if the ambient temperature is different from 70°C?

EXAMPLE

Determine the temperature of an RN60D resistor at an ambient temperature of 25°C if the resistor dissipation is 0.25W.

SOLUTION

We need to recognize that the increase in temperature is independent of the ambient temperature. So, if a power dissipation of 0.25W in an RN60D resistor causes a 100°C increase in temperature with an ambient temperature of 70°C, then the same increase will occur with an ambient of 25°C. The temperature we are looking for will then be *25°C + 100°C*, or 125°C.

If self-heating causes too much change in the resistance value, then a larger wattage and/or a more temperature-stable resistor should be used. The effect of self-heating is not instantaneous—it takes time for the resistor to heat up (thermal time constant).

HEAT EFFECTS ON DISCRETE SEMICONDUCTOR DEVICES

When we studied the diode, we learned that temperature can change the forward drop of a diode with a change of −2.2mV/°C. The reverse leakage current also changes with temperature, and roughly doubles for every 6°C increase in temperature.

Temperature has the same effect on bipolar transistors with the change in forward drop across the base-emitter junction changing the bias point. Leakage current change of the reverse biased collector-base junction is usually not a problem—except for high-power output transistors where thermal runaway can occur. (Leakage current causes an increase in power dissipation, which further increases the leakage current.)

HEAT EFFECTS ON INTEGRATED CIRCUIT DEVICES

Since integrated circuits consist of diodes and transistors, they are also subject to temperature effects. However, internal temperature compensation schemes are often used with ICs to make them less susceptible to thermal changes.

The high density of integrated circuits can result in quite a bit of heat generation in a small area—particularly with the power integrated circuits such as regulators. The concern is that temperatures can reach the point where the semiconductor is damaged. To avoid this problem, a heat sink is built into the device to take heat away from the hot spot within the device. This hot spot is usually the collector-base junction of the power stage.

The heat sink typically has a mounting hole so that it can be mounted to the chassis or to a larger heat sink.

We shall delve further into heat sinks by determining the required size and type for various applications. But first we shall look at the basics of heat transfer.

THERMAL OHM'S LAW

Since at this point we are very familiar with the electrical Ohm's law, it is helpful to use equivalent relationships when we study heat transfer. We shall call these relationships thermal Ohm's law, and they are listed in Table 19–2 along with the electrical equivalents.

The thermal Ohm's law equations can now be listed by equivalency from the electrical Ohm's law.

$$\text{Heat/sec. } (P) = \frac{\text{Temperature } (T)}{\text{Thermal resistance } (\theta)} \qquad \textbf{(equation 19–3)}$$

$$\text{Temperature } (T) = \text{Heat/sec. } (P) \times \text{Thermal resistance } (\theta) \qquad \textbf{(equation 19–4)}$$

$$\text{Thermal resistance } (\theta) = \frac{\text{Temperature } (T)}{\text{Heat/sec. } (P)} \qquad \textbf{(equation 19–5)}$$

Heat/sec. is represented by P (power) because in electrical circuits heat comes from electrical watts. The unit for temperature is degrees Celsius, and the unit for thermal resistance is degrees Celsius per watt.

These relationships can be shown in an equivalent circuit as indicated in Figure 19–2.

The constant current symbol is used for the heat source to indicate that the amount of heat generated is independent of the thermal resistance or temperature. It is only a function of the power dissipation (watts) in the circuit.

The temperature drop is the temperature difference across the thermal resistance and is a function of how much heat is flowing through the thermal resistance and the value of the thermal resistance.

The thermal resistance is the resistance to heat flow. Metals allow heat to pass easily, so they have low thermal resistance. Fiberglass is noted for its insulation or resistance to heat flow, so it has high thermal resistance. Dry air is also a very good insulator of heat. Some materials, like mica, have low thermal resistance but very high electrical resistance. These materials are useful when we want to allow heat to flow out of an electronic component that is not at ground potential.

TABLE 19–2 Thermal/Electrical Equivalents

Thermal	*Electrical Equivalent*
Heat/sec. (P)	Current (I)
Temperature (T)	Voltage (V)
Thermal resistance (θ)	Electrical resistance (R)

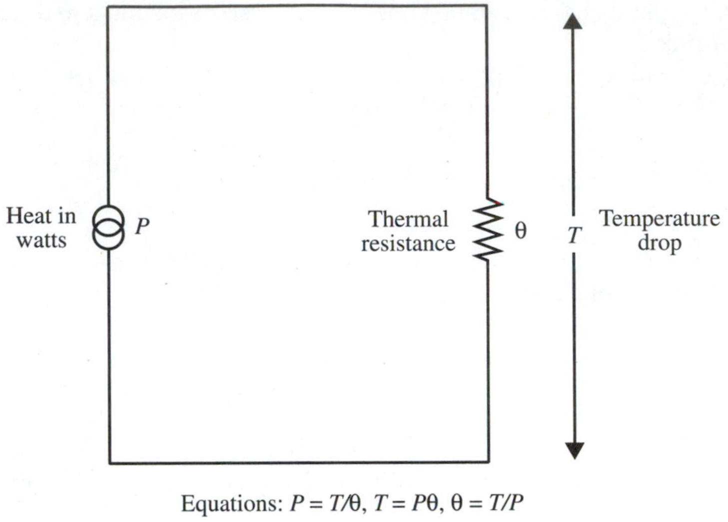

FIGURE 19–2 Thermal equivalent circuit

USING THERMAL OHM'S LAW

To show how thermal Ohm's law is used, we shall first look at a simple example of resistor heating. The resistor we shall use is a CMC power type, chassis mount, as shown in Figure 19–3. We shall go through an example to determine the temperature on the body of the resistor under specified conditions.

The fins on the resistor do provide some heat transfer to air. But let us assume that all the heat leaving the resistor passes into the chassis. Let us also assume the following values:

θ_{rc} = 0.5°C/W (thermal resistance from the resistor body to the chassis)
R = 10Ω
V_r = 10V (voltage across resistor)
T_c = 50°C (chassis temperature)

First, we must find the heat introduced into the resistor by the power dissipation.

$$P = \frac{(V_r)^2}{R}$$

$$P = \frac{(10\text{V})^2}{10\Omega}$$

$$P = 10\text{W}$$

Now, the temperature at the body of the resistor will be the chassis temperature T_c plus T_{rc}, the temperature drop across the resistor-chassis thermal resistance.

Figure 19–4 shows the equivalent circuit.

The equation to determine the resistor body temperature is

$$T_r = T_{rc} + T_c$$

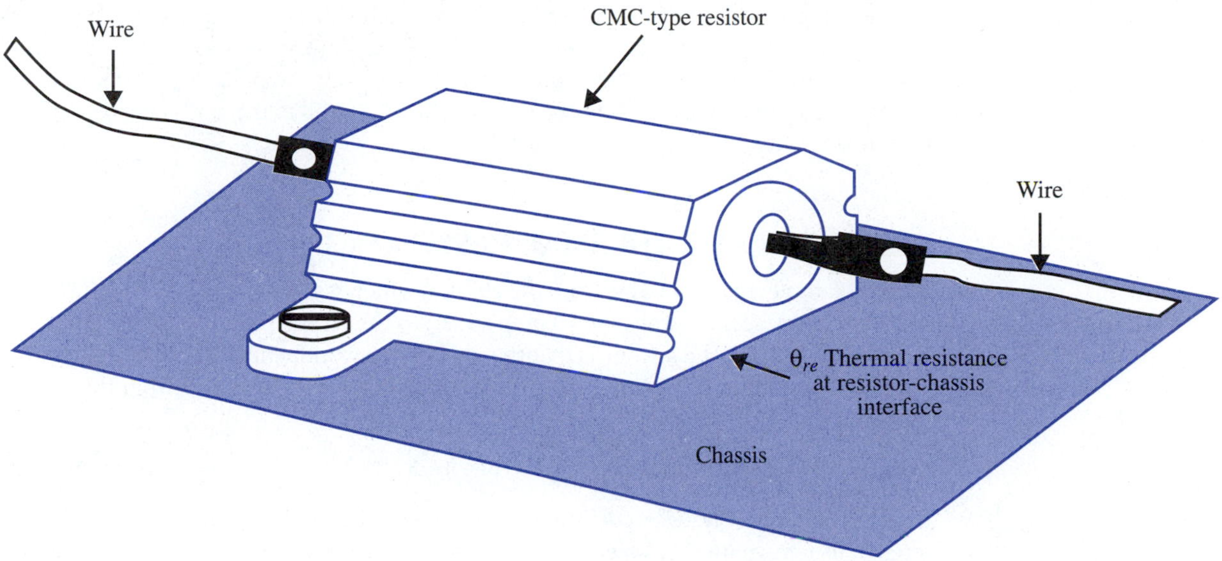

FIGURE 19–3 Power resistor heating

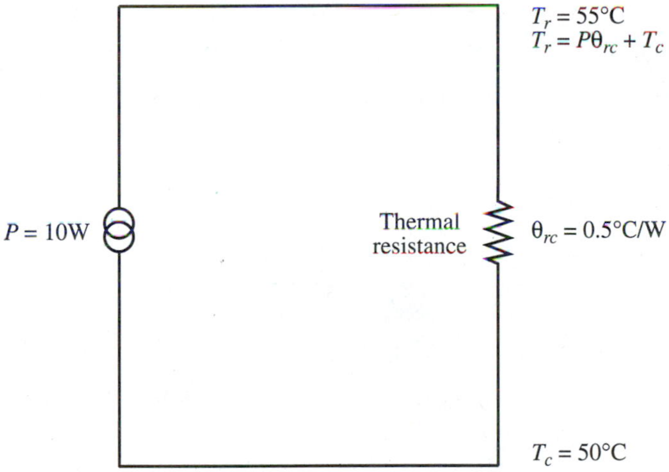

FIGURE 19–4 Equivalent circuit for resistor heating

but

$$T_{rc} = P \times \theta_{rc} \qquad \textbf{(equation 19–6)}$$

$$T_r = (P \times \theta_{rc}) + T_c$$

$$T_r = (10\text{W} \times 0.5°\text{C/W}) + 50°\text{C}$$

$$T_r = 55°\text{C}$$

or a 5°C rise over the chassis temperature.

The thermal resistance of the junction between the chassis is a function of the surface conditions. Flat polished metal surfaces will give the lowest resistance and therefore the lowest temperature rise. Another option, if it is not possible to polish the surfaces, is to use a thermal conducting grease between the two surfaces to fill in the voids on a rough surface (eliminates trapped air, which is an insulator).

CHASSIS-TO-AIR THERMAL RESISTANCE

In our example, the heat flowed into the chassis, which was assumed to have a temperature of 50°C. But if heat was constantly being added to the chassis, why didn't the temperature of the chassis rise? The answer is that the chassis was conducting heat to the surrounding air. The amount of watts transferred to the air (P_t) can be determined from the following equation:

$$P_t = hA\,\Delta T \qquad \textbf{(equation 19–7)}$$

where h is the heat transfer coefficient, A the area of the surface (chassis), and ΔT the difference in temperature between the chassis surface and the ambient air. The h factor was determined from empirical data and is equal to

$$h = \frac{0.009\text{W}}{°\text{Cin.}^2}$$

In equilibrium, the watts dissipated in the resistor must be the same as that leaving the chassis. So now the only unknown in equation 19–7 is the area of the chassis. Rewriting equation 19–7 to solve for this area,

$$A = \frac{P_t}{h\,\Delta T}$$

$$A = \frac{10\text{W}}{(0.009\text{W}/°\text{Cin.}^2) \times 5°\text{C}}$$

$$A = 222\text{in.}^2$$

or roughly a 10 inch by 20 inch chassis. If we had used a smaller chassis than this, the chassis would have run hotter, and the temperature of the resistor would have raised by the same amount.

We can think of the thermal interface between the chassis and the air as being a thermal resistance θ_{ca}.

$$\theta_{ca} = \frac{T_c - T_a}{P}$$

where T_c is the chassis temperature and T_a the ambient air temperature.

$$\theta_{ca} = \frac{50°\text{C} - 25°\text{C}}{10\text{W}}$$

$$\theta_{ca} = 2.5°\text{C/W}$$

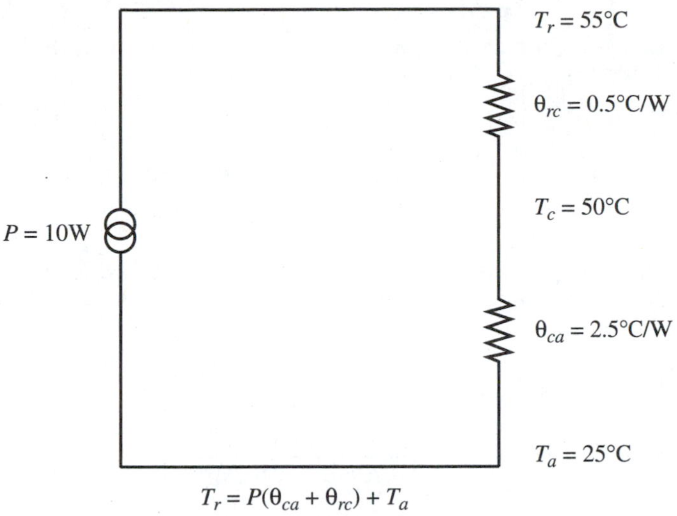

FIGURE 19–5 More complete thermal circuit

A more complete equivalent circuit, which includes this chassis-to-air thermal resistance, is shown in Figure 19–5. Notice in this figure that the bottom of the resistor network is at the ambient temperature of 25°C rather than the chassis temperature of 50°C in Figure 19–4.

TEMPERATURE DROP ACROSS A CHASSIS

When we said that the chassis was at a temperature of 50°C, we assumed that the temperature was uniform over the whole surface. Actually, the chassis has some thermal resistance, which causes a temperature drop as heat flows from the location where the resistor is mounted, to the outside of the chassis. But heat is leaving the surface as it flows to the outside of the chassis and so the heat flow becomes zero at the edge of the chassis. To determine the temperature drop to the outside, we shall assume, then, an average heat flow from the heat source to the outside of one-half.

We can determine the temperature drop ΔT_c from the following conductivity equation:

$$\Delta T_c = \frac{P \times L}{AK_c} \qquad \text{(equation 19–8)}$$

where P is the power in watts, L the length that the heat travels (in.), A the cross-sectional area (in.^{2}), and K_c the thermal conductivity constant (watts × °C/in.^{3}).

EXAMPLE

Find the temperature at the long ends of the 5 inch by 10 inch chassis if the temperature in the center, where the resistor (dissipation 10W) is mounted, is 50°C and the chassis is 16-gauge aluminum.

SOLUTION

Aluminum has a conductivity constant K_c of 5.2W°C/in.^{3}, and the thickness of 16 gauge is 0.057″. The cross-sectional area A is then

$$A = h \times w$$

$$A = 0.057'' \times 5''$$

$$A = 0.285\text{in.}^2$$

Now putting these known values into equation 19–8 and solving for the temperature drop at the end, which is 5 inches from the center,

$$\Delta T_c = \frac{P \times L}{AK_c}$$

$$\Delta T_c = \frac{2.5\text{W} \times 5''}{0.285\text{in.}^2 \times 5.2\text{W°C/in.}^3}$$

$$\Delta T_c = 8.4°\text{C}$$

Notice that 2.5W was used rather than 10W for two reasons—half the source heat flowed each way from the center; and our assumption that the average flow to the outside is one-half the initial flow from the source.

The temperature at the ends with the 8.4°C drop is *50°C − 8.4°C*, or 41.6°C.

It would appear that the value used for ΔT in the chassis-to-air transfer equation (19–7) should take into account the temperature drop across the chassis, and that an average temperature (45.8°C for the example here) should be used for the chassis temperature. However, we can't find the temperature drop across the chassis until we know the size of the chassis. The exact determination of the chassis size, then, requires successive approximations—but equation 19–7 is conservative enough not to require this in most applications.

COOLING A POWER SUPPLY REGULATOR

As an example of heat transfer with an active electronic device, we shall study the cooling requirements for an LM317K power supply regulator. (Refer to Chapter 11.) Heat sinking affects the current capability of a regulator because the more heat we can remove from the device, the higher the allowable internal power dissipation.

Since the LM317 regulator is internally thermally protected, the available current from the device is a function of the

- Device internal current limiting
- Voltage across the regulator (difference between input and output)
- Heat sinking

For our thermal example, we shall assume the following power supply requirements:

- Output voltage range—variable from 2V to 20V
- Output current capability—0.8A
- Maximum input voltage to regulator—28V (at 10% high line)
- Maximum ambient temperature—40°C
- Regulator to be mounted on a 16-gauge aluminum chassis

With the LM317, the internal current limiting is set at 2.2A with up to 12V across the regulator and decreases to 0.4A with 40V differential between the input voltage and output voltage. (See Figure 19–6.)

Setting the power supply for the minimum output voltage of 2V, the voltage across the regulator with our input of 28V is

$$V_R = (28V - 2V)$$

$$V_R = 26V$$

Looking at the current available with this voltage drop across the regulator, we read from Figure 19–6 a maximum output current value of about 0.9A—which is fine, because it is greater than our required output current of 0.8A. With the minimum output voltage, we have the maximum voltage across the regulator and therefore the maximum internal power dissipation.

$$P_R = I_{max} \times V_R$$

$$P_R = 0.8A \times 26V$$

$$P_R = 20.8W$$

Let us now look at the heat sink requirements. The LM317 will go into thermal shutdown when the collector-base junction temperature of the internal power transistor exceeds 170°C. But for more reliable device operation, we will restrict the junction temperature to 150°C (thermal shutdown can still occur with a short circuit on the regulator output).

The thermal resistance of the junction to case (θ_{jc}) specified for the LM317, TO-3 (K) package is 3°C/W (TO-220 package is 5°C/W). If we use a mica insulator coated with thermally-loaded silicon grease between the regulator case and the chassis (heat sink), the thermal resistance (θ_{cs}) of this mounting is (from Table 19–3) 0.33°C/W.

The last thing we need to determine is the required thermal resistance of the chassis to air. We can do this with reference to the thermal equivalent circuit of Figure 19–7.

We can find the value of the sink-to-air thermal resistance by finding the total thermal resistance and then subtracting the junction-to-case, and the case-to-sink, resistances from this value. The total thermal resistance θ_t is

$$\theta_t = \frac{T_j - T_a}{P}$$

$$\theta_t = \frac{150°C - 40°C}{20.8W}$$

$$\theta_t = 5.29°C/W$$

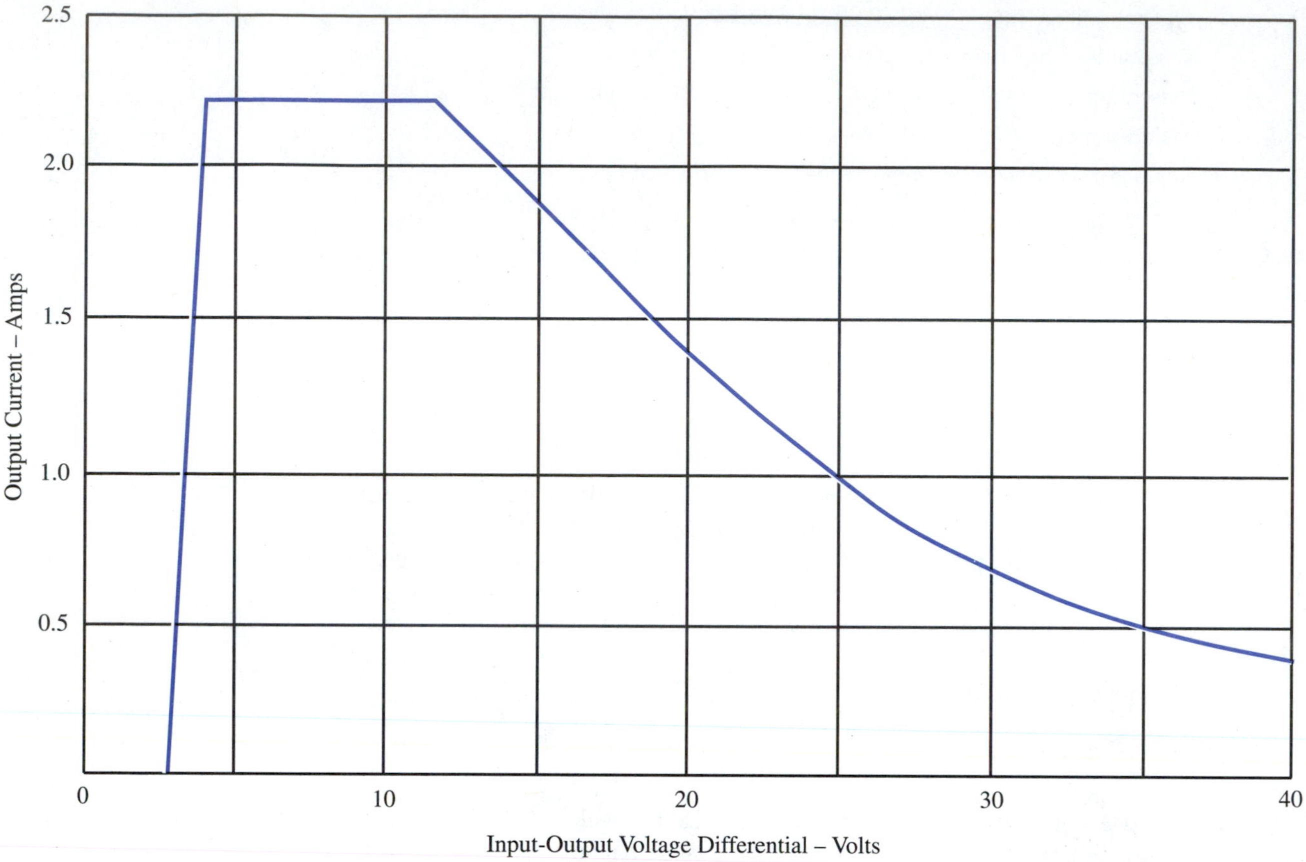

FIGURE 19–6 LM 317 max output current versus voltage drop (T and K package). *Courtesy National Semiconductor.*

TABLE 19–3 Mounting Thermal Resistance (θ_{cs})

Case Type	*Direct Contact—clean metal*	*With "Loaded" Silicon Grease**	*With "Loaded" Silicon Grease & 0.002≤ Mica Insulator*
TO-3 (K)	0.5 °C/W	0.14 °C/W	0.33 °C/W
TO-220 (T)	1.1 °C/W	0.95 °C/W	1.5 °C/W

*Thermalloy "Thermalcote" or Aavid "Ther-O-Link 100000" Mounting bolts torqued to 6"-lbs

Then the required sink (chassis)-to-air thermal resistance is

$$\theta_{sa} = \theta_t - \theta_{jc} - \theta_{cs}$$

$$\theta_{sa} = 5.29°\text{C} - 3.0°\text{C/W} - 0.33°\text{C/W}$$

$$\theta_{sa} = 1.96°\text{C/W}$$

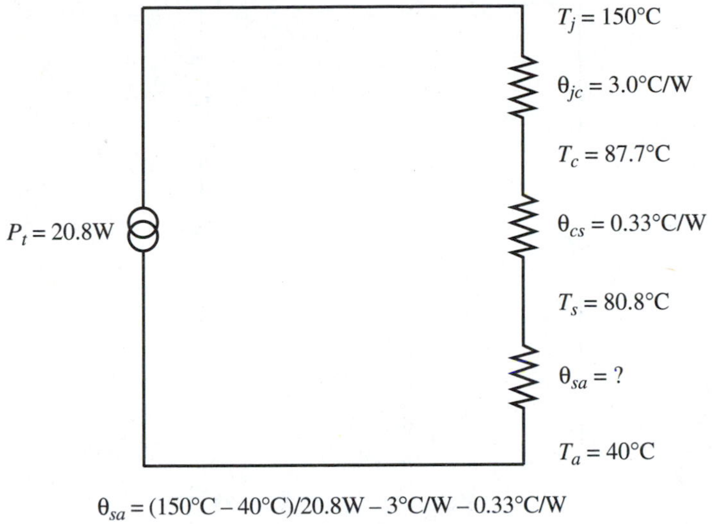

FIGURE 19–7 Regulator thermal equivalent circuit

From this value we can determine the temperature of the chassis T_c by multiplying the thermal resistance by the power and adding the result to the ambient air temperature.

$$T_c = (\theta_{sa} \times P_t) + T_a$$

$$T_c = (1.96°\text{C/W} \times 20.8\text{W}) + 40°\text{C}$$

$$T_c = 80.8°\text{C}$$

We specified that the regulator should be mounted on a 16-gauge aluminum chassis. We need to find the size of the chassis to provide the required thermal resistance of 1.96°C/W.

Going back to equation 19–7 and solving for the required areas,

$$A = \frac{P_t}{h\Delta T}$$

ΔT is the temperature difference between chassis and air.

But we can see from Figure 19–7 that $\Delta T/P_t$ is equal to θ_{sa}, so

$$A = \frac{1}{\theta_{sa}h} \quad \textbf{(equation 19–9)}$$

$$A = \frac{1}{1.96°\text{C/W} \times 0.009\text{W/°Cin.}^2}$$

$$A = 56.7\text{in.}^2$$

The analysis we have conducted on regulator heat sinking applies to any type of semiconductor. All we need to know about the device is the maximum junction temperature, the internal power dissipation, and the thermal resistance from the junction to the case.

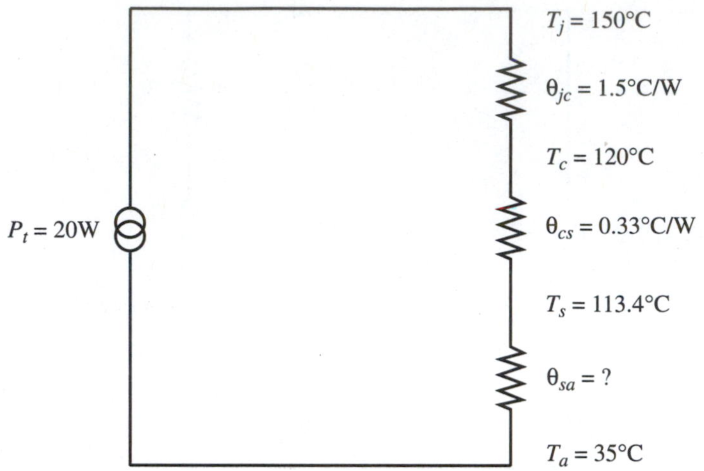

$\theta_{sa} = (150°C - 35°C)/20W - 1.5°C/W - 0.33°C/W$

FIGURE 19–8 2N3055 power transistor thermal circuit

EXAMPLE

A 2N3055 (TO-3 [K] package) power transistor is operating in a circuit with 2A flowing through it, and 10V from collector to emitter. Find the size of heat sink required, and the temperature at each point, if the ambient temperature is 35°C, the maximum junction temperature is 150°C, and the transistor is mounted to the heat sink with a mica insulator and silicon grease. The junction-to-case thermal resistance is specified to be 1.5°C/W.

SOLUTION

The first thing we need to do is to generate the thermal schematic as shown in Figure 19–8 with the computed power of 2A × 10V or 20W, and the given thermal resistances.

The case temperature T_c was computed by taking the temperature drop across the junction-to-case thermal resistance θ_{jc} and subtracting it from the junction temperature.

$$T_c = 150°C - (1.5°C/W)(20W)$$

$$T_c = 120°C$$

Similarly, the heat sink temperature T_s was computed by taking the temperature drop across the case-to-sink thermal resistance θ_{cs} (taken from Table 19–3) and subtracting it from the case temperature.

$$T_s = 120°C - (0.33°C/W)(20W)$$

$$T_s = 113.4°C$$

We can now determine the required heat sink-to-air thermal resistance θ_{sa} by simply dividing the temperature drop across this resistance by the power dissipation.

$$\theta_{sa} = \frac{\Delta T_{sa}}{P_t}$$

$$\theta_{sa} = \frac{113.4°C - 35°C}{20W}$$

$$\theta_{sa} = 3.92°C/W$$

Now, using equation 19–9, we can determine the required size of the heat sink.

$$A = \frac{1}{\theta_{sa} h}$$

$$A = \frac{1}{3.92°C/W \times 0.009W/°Cin.^2}$$

$$A = 28.3in.^2$$

It should be pointed out that all the thermally connected parts of the equipment case or chassis can be considered as part of the heat sink. So it is not always necessary to provide a separate heat sink.

Commercial heat sinks are available with finned structures and mounting provisions for different devices' packages, which range from small transistors and DIPs to power configurations (TO-3, TO-220, etc.). For example, Thermalloy Company provides various heat sinks for the TO-3 package with thermal resistances ranging from 0.77°C/W (size 5.5″ × 4.75″ × 2.63″) to 11.5°C/W (size 1.88″ × 1.4″ × 0.5″).

The main advantage of the commercial heat sinks is that they are smaller for a given thermal resistance than the sheet metal heat sinks. This is because they provide a larger surface area, using complex configurations. Thermal resistance can be further reduced by blowing air over the heat sink—with about a five-to-one reduction in thermal resistance with air flows past the sink of 1000ft./min. versus still air.

TROUBLESHOOTING THERMAL PROBLEMS

When troubleshooting electronic equipment, it should be remembered that heat is a prime cause of malfunction or failure of the equipment. This is not always apparent, but if the malfunctioning equipment starts to operate normally when it is allowed to cool (or when it warms), suspect thermal problems.

Reasons for heat problems are as follows:

1) Equipment is not designed for the temperature range that is being encountered.
2) Parts are not installed on heat sinks correctly.
3) Air filters or components are dirty.
4) Ventilation holes are blocked.

5) There is insufficient clearance between equipment and other items.
6) Cooling fan (if used) is inoperative.

The first reason listed—design deficiency—is a common cause of thermal failure, because design engineers typically do not have much training in thermal design. This results in equipment where temperature rise due to localized heating is not considered. A circuit may work fine in a lab situation, but when it is buried in a piece of equipment, the environment changes. This problem can be avoided if the complete equipment package is temperature cycled during the design phase.

Solutions for design deficiencies could include use of heat sinks (or larger heat sinks), relocation of parts away from hot spots, added ventilation, and the use of a fan.

When parts are installed on heat sinks, thermal grease should be used, and both the mating surface of the part and the heat sink should be clean (no paint), flat, and free of burrs.

The remaining reasons for thermal problems can be cured by maintenance and/or equipment relocation.

SUMMARY

- Heat can enter an object by conduction or can be generated within the object by internal power dissipation.
- Temperature indicates the level of heat in an item.
- Heat can cause temporary change or permanent change to an electronic component.
- The temperature effect on resistance depends on the type of resistor.
- Carbon resistors change the most with temperature, metal foil the least.
- Temperature change in a component is usually listed in parts per million (ppm) per °C.
- Matched ppm/°C is important for resistors in a precision voltage divider.
- The power rating curve shows how the resistor power dissipation drops with temperature.
- Devices heat up with a thermal time constant.
- A temperature effect on silicon semiconductor diodes and transistors is a decrease in forward voltage drop of 2.2mV/°C.
- Silicon diodes and transistors experience a doubling of leakage current as temperature increases by 6°C.
- A heat sink protects by conducting heat away from a semiconductor device.
- Thermal Ohm's law allows us to apply electrical concepts to solve thermal problems.
- Heat/sec. (watts) is equivalent to electrical current, temperature to voltage, and thermal resistance to electrical resistance.
- Heat/sec. is represented by a constant current symbol in thermal equivalent circuits because the heat generated is only a function of power dissipation in the component.

- Metals are good conductors of heat, but dry air is a good heat insulator.
- The thermal resistance between conducting solids is a function of how close they contact without air gaps.
- Thermal conducting grease can fill in air voids and reduce thermal resistance.
- Components run cooler with lower total thermal resistance.
- Cooling a component allows for greater internal power dissipation.

EXERCISE PROBLEMS

1. What does temperature indicate?
2. What are the two ways that heat can enter an object?
3. Which type resistor has the greatest change value with temperature?
4. Compute the value of a carbon resistor at 60°C if its value at 25°C is 1KΩ.
5. Compute the value of a wire-wound resistor at 60°C if its value at 25°C is 1KΩ (use K_t of 10ppm/°C).
6. Two metal-film resistors are used in a voltage divider application with one resistor having a temperature coefficient of 100ppm/°C and the other 105ppm/°C. Find the percent change in the divided voltage if the temperature changes by 60°C.
7. Determine the change ΔR in the value of an RN60D, 475Ω metal-film resistor after 10V is applied. Assume that K_t is 200ppm/°C.
8. Find the temperature of an RN60D resistor at an ambient temperature of 35°C if the resistor dissipation is 0.25W.
9. What effect does heat have on a silicon transistor?
10. Determine the body temperature of a 40Ω CNC power resistor under the following conditions:
    - 20V across resistor
    - 0.7°C/W thermal resistance from resistor body to chassis
    - Chassis temperature of 60°C
11. Determine the body temperature of a 20Ω CNC power resistor under the following conditions:
    - 16V across resistor
    - 0.4°C/W thermal resistance from resistor body to chassis
    - Chassis temperature of 40°C
12. A power component generating 2W is mounted at one end of a 2 inch by 8 inch, #16 gauge, aluminum chassis, and raises the temperature at that end to 60°C. Find the temperature at the other end.
13. A power component generating 5W is mounted in the center of a 4 inch by 6 inch, #16 gauge, aluminum chassis. The temperature at one end, 3 inches from the center, is 45°C. Find the temperature at the chassis center.

14. An LM317K regulator is required to provide 5V–20V at a current of 1A. Find the required thermal resistance θ_{sa} (sink to air) and the temperature at the case and heat sink, with the following conditions:
    - Input voltage to regulator is 30V max.
    - LM317K is mounted to heat sink using thermal grease (no insulator).
    - Ambient air temperature is 30°C.
    - Regulator junction temperature should not exceed 150°C.
15. Determine the minimum area of the heat sink required for problem 14 using #16 aluminum.
16. A 2N3055 (TO-3 package) power transistor is operating in a circuit with 2A flowing through it, and 15V from collector to emitter. Find the minimum #16 aluminum chassis size required, and the temperature at the transistor case, if the ambient temperature is 25°C, the maximum junction temperature is 150°C, and the transistor mounting surface is coated with thermal grease. The junction-to-case thermal resistance is specified to be 1.5°C/W.
17. What are possible solutions to improving equipment that has thermal design deficiencies?
18. What precautions should be taken when installing a component on a heat sink?

ABBREVIATIONS

K_c = Conductivity coefficient

K_d = Temperature per watt of resistor dissipation

K_t = Temperature coefficient

P_d = Power dissipation

P_r = Power dissipated in regulator

P_t = Heat power transferred to air

θ_{rc} = Thermal resistance resistor to chassis

θ_{ca} = Thermal resistance case to air

θ_{cs} = Thermal resistance case to heat sink

θ_{jc} = Thermal resistance semiconductor junction to case

θ_{sa} = Thermal resistance sink to ambient air

θ_t = Total thermal resistance

T_r = Temperature of resistor

T_j = Temperature of semiconductor junction

T_c = Temperature of case

T_s = Temperature of sink

T_a = Temperature of ambient air

V_r = Voltage drop across regulator

ANSWERS TO PROBLEMS

1. Amount of heat
2. Conducted and internally generated
3. Carbon
4. $R_{60} = 947.5\Omega$
5. $R_{60} = 1000.35\Omega$
6. % change = 0.03%
7. $\Delta R = 7.93\Omega$
8. $T_R = 135°C$
9. Reduces V_{be}; increases c-b leakage
10. $T_R = 67°C$
11. $T_R = 45.12°C$
12. $T_{end} = 46.5°C$
13. $T_{cen} = 51.3°C$
14. $\theta_{sa} = 1.66°C/W$; $T_s = 71.5°C$; $T_c = 75°C$
15. $A = 66.9 in.^2$
16. $\theta_{sa} = 2.53°C/W$; Area = 43.9 $in.^2$, $T_c = 105°C$
17. See Troubleshooting section.
18. Clean, smooth surface

CHAPTER 20

Circuit Components

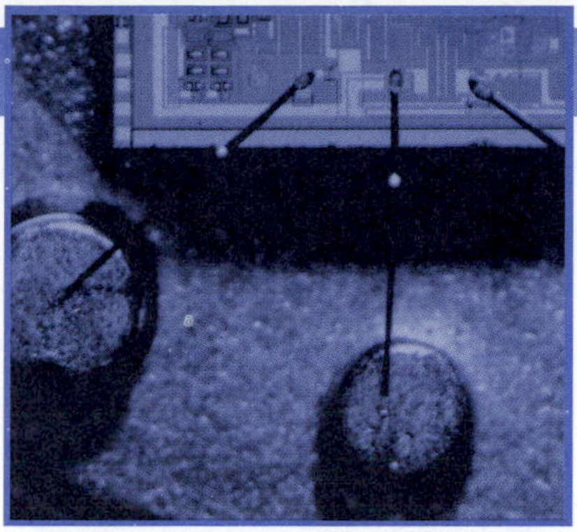

CHAPTER OBJECTIVES

After completing this chapter, you should be able to describe the following concepts:

- Why a specification for a component is necessary
- Need for specification control drawings
- Meaning of standard conditions
- Significance of typical, minimum, and maximum values
- How to interpret graph data
- Importance of maximum ratings
- Significance of part number designations
- Considerations with part interchangeability
- Process of part substitution
- How to take data
- Need to compare data with something
- Significance of percent deviation
- Why an oscilloscope should be used
- Need for an experimental conclusion
- Types of resistors and obtainable resistance values
- Types of capacitors and obtainable capacitor values

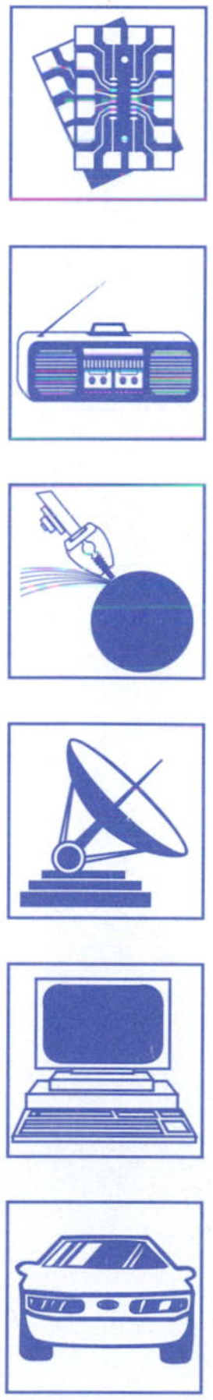

INTRODUCTION

This chapter will first look at manufacturer's specifications, which define both the capabilities and limitations of electronic devices. Next, we shall learn how to interpret a data sheet and the considerations for interchangeability. Then, we shall look at suggested steps to follow when conducting device experiments and taking data. Finally, we shall become familiar with the standard values used for resistors and capacitors.

DEVICE SPECIFICATIONS

Manufacturers' specifications define the electrical and mechanical features of a particular linear integrated circuit. These specifications serve as a basis for applying the devices in a particular circuit design.

If the manufacturer doesn't specify a particular characteristic of a component, there is no guarantee that it will be the same from device to device. This means that equipment may no longer operate if a failed IC is replaced. Equipment designers get around this problem by writing their own specifications that define the critical values on a specification control drawing. In this chapter, we will look at a typical data sheet, become aware of the significance of maximum ratings and tolerances, and also look at the different types of available packages.

TYPICAL DATA SHEET

A data sheet contains the key device parameters under given standard conditions. The conditions can be operation at room temperature (25°C) or at elevated or lower temperatures. Power supply voltages can cause changes in the specifications, so these have to be known along with the specific circuit load conditions. The operating frequency can also have an effect on the device characteristics.

We can become familiar with specifications by going through an actual data sheet example. The data sheet for the LM2878 dual 5W power amplifier will be used, Table 20–1.

The manufacturer suggests that this amplifier can be used for a stereo phonograph, an output stage in an AM-FM receiver, a power amplifier or comparator, or as a servo amplifier.

We shall go down through each item on the data sheet—but first, we should note that the specifications are given under the following standard conditions:

1) $V_{CC} = 22$ V
2) T_{AMB} (ambient temperature) = 25°C
3) R_L (load resistor) = 8Ω
4) A_v (voltage gain) = 50

If the conditions are different from these values, the specifications could change.

The first item listed in Table 20–1 is the total power supply current with no output power. This is also called the standby current and is used to determine how much drain there is on the power supply (or battery) when the amplifier is not in use. If we are concerned with battery life, we must assume that the amplifier is drawing the maximum standby current of 50mA. Using the typical value of 10mA could result in the battery running down before we predicted it would.

Operating supply voltage of 6V–32V determines the possible range for the power source—but remember that the rest of the values listed in the data sheet are for a supply voltage of 22V.

The output power for each channel has a typical value of 5.5W, but a minimum value of 5W. Again we must assume the worst case and use the 5W value. We must also consider the 10% THD (total harmonic distortion). If this is too high, we must operate at a lower output power level.

The distortion versus power data shows that at low power levels (less than 2W), the distortion actually increases as the output power level is decreased. This is because distortion at the crossover point of the internal push-pull transistors at turn-on has a fixed value, so

TABLE 20–1 LM2878 Data Sheet *Courtesy National Semiconductor*

Absolute Maximum Ratings	
Supply Voltage	35V
Input Voltage (Note 1)	+0.7V
Operating Temperature (Note 2)	0°C to 70°C
Storage Temperature	−65°C to 150°C
Junction Temperature	150°C
Lead Temp (Soldering 10 seconds)	300°C

Electrical Characteristics (Conditions):

V_{cc} = 22V, T_a = 25°C, R_L = 8Ω, A_v = 50 (34 dB) unless otherwise specified (Note 2).

Parameter	*Condition*	*Min*	*Typ*	*Max*	*Units*
Total Supply Current	P_o = 0W		10	50	mA
Operating Supply		6		32	V
Output Power/Channel	f = 1KHz, THD = 10%	5	5.5		W
Distortion	P_o = 50mV, f = 1KHz, R_L = 8Ω		0.20		%
	P_o = 0.5W, C = 1KHz, R_L = 8Ω		0.15		%
	P_o = 2W, f = 1KHz, R_L = 8Ω		0.14		%
Output Swing	R_L = 8Ω		V_{cc} − 6V		V_{p-p}
Channel Separation	C_{BYPASS} = 50μF, C_{in} = 0.1μF C = 1KHz, output referred. V_{ripple} = $4V_{rms}$	−50	−70		db
PSRR Power Supply Rejection Ratio	C_{BYPASS} = 50μF, C_{in} = 0.1 μF, f = 120Hz, output referred. V_{ripple} = $1V_{rms}$	−50	−60		db
PSRR Negative Supply	Measured at DC, input referred		−60		db
Common-Mode Range	±15V Supplies, Pin 1 tied to Pin 11		±13.5		V
Input Offset Voltage			10		mV
Noise	Equivalent input noise, R_s = 0, C_{in} = 0.1μF.				
	BW = 20 − 20KHz		2.5		μV
	CCIR-ARM		3.0		μV
	Output Noise Wide band, A_v = 200		0.8		mV
Open Loop Gain	R_s = 51Ω, f = 1KHz, R_L = 8Ω.		70		db
Input Bias Current			100		nA
Input Impedance	Open Loop		4		MΩ
DC Output Voltage	V_{cc} = 22V	10	11	12	V
Slew Rate			2		V/μs
Power Bandwidth	3db Bandwidth at 2.5W		65		KHz
Current Limit			1.5		A

Note 1: ± 0.7V applies to audio applications; for extended range see Application Hints.

Note 2: For operation at temperatures greater than 25°C, the LM2878 must be derated based on a maximum 150°C junction temperature using a thermal resistance which depends upon device mounting techniques.

that when the output voltage is reduced, the distortion becomes a greater percentage of the output.

A typical output swing of the supply voltage −6V is listed using a load resistor of 8Ω. With the supply voltage of 22V, this provides an output peak-to-peak voltage swing of 16V. Converting this to RMS by dividing by $2\sqrt{2}$ gives 5.66V. If we square this value and divide by the load resistor, we can find the power out as

$$P_o = \frac{(5.66V)^2}{8\Omega}$$

$$P_o = 4W$$

This is less than the minimum value of 5W specified, so we run into our first inconsistency in manufacturers' specifications. The other problem we have is that the output swing listed was "typical," but what is the worst case? This could mean a power output of less than 4W at a power supply of 22V.

With two amplifiers on the same chip, there is a possibility that the signal being amplified by one amplifier could cross-couple into the other. The specified stereo channel separation is with a 4V RMS, 1KHz signal at the output of one amplifier, and no signal fed into the other amplifier. A minimum separation of 50db (316) means the coupled voltage at the output of the second amplifier could be as high as 4V/316 or 12.6 mV RMS.

Power supply rejection was defined earlier when we covered op-amps. A minimum value of 50db (316) for the LM2878 can be compared with the 77db (7079) for a 741 op-amp. If a negative supply is used, the typical rejection is the same as with a positive supply (60db). However, no minimum value is listed for this situation, so it is impossible to determine the worst-case situation.

The value of 10mV given for the offset voltage is a typical value, and maximum offset is not provided. No offset pins are provided on the amplifier, so if offset is a problem in a DC amplifier configuration, an opposing DC voltage should be injected into the input as shown in Figure 20–1.

Connecting the 5KΩ potentiometer between the positive and negative supplies allows for cancellation of an offset of either polarity.

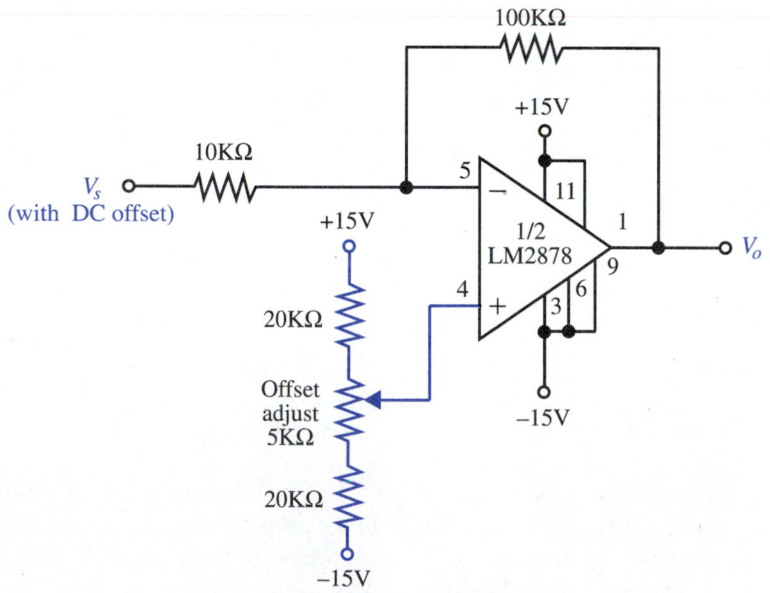

FIGURE 20–1 Offset adjust circuit for LM2878

The noise specification is first given with reference to the input because output noise depends on the circuit gain. Using a source resistor (R_S) of zero eliminates noise contribution from this resistor, so we will only see amplifier noise. In the Chapter 8 section on op-amp noise, we saw that increasing the bandwidth increased the noise (by a square root factor). The 2.5μV input noise value listed covers the audio bandwidth of 20Hz to 20KHz. Output noise of typically 800μV is given for an amplifier gain of 200. This is listed as wideband noise, but the bandwidth is not listed. Using the given 2.5μV figure for a bandwidth of 20KHz, a bandwidth of 50KHz is computed using an input referred noise of 800μV/200 or 4μV (remembering again the square root factor on bandwidth). CCIR-ARM is a weighted method of noise measurement developed by Ray Dolby and others, with the response peaking up in the 6KHz region.

Open loop gain of 70db corresponds to a voltage ratio of 3,162. This is low compared to the typical op-amp gain of 100db (100,000). It will be remembered that lower open loop gain means less-accurate closed loop gain, more distortion, and also a reduction in the gain bandwidth product.

The input bias current of 100nA is close to the 80nA figure for a 741 op-amp, while the input resistance of 4MΩ is twice that of the 741.

An internal equal resistance voltage divider connected to the noninverting input sets up the DC output voltage. With a 22V supply, the divider provides a nominal 11V at the output, which gives the maximum output voltage swing. The possible ±1V variation means a corresponding reduction in this output voltage swing.

The slew rate of 2V/μs listed is four times better than the 0.5V/μs value for a 741 op-amp, and this allows a greater output voltage swing at the higher frequencies.

Power bandwidth is the frequency range over which the output is within 3db (half power) of the maximum value of 5W. The 65KHz typical value listed easily exceeds the audio range.

Finally, the internal current limit of 1.5A protects both the load and amplifier from excessive dissipation.

What we should learn from going through this typical spec sheet is that not all parameters are well defined, and if a minimum or maximum is not provided, then we have no guarantee what the value will be. A way around this problem is to generate a specification control drawing that defines the range of variation of all critical parameters. If the IC manufacturer agrees to these specs, they become the basis for incoming inspection's acceptance of the devices.

READING A GRAPH

In addition to the specification sheet, manufacturers provide curves of the various parameters as a function of voltage, temperature, frequency, etc. Figure 20–2 shows graphs for the LM2878.

The first graph shown in the figure relates device dissipation to ambient temperatures, but also includes the effect of various heat sinks. It should be clear that internal device dissipation is not the same as output power. The positioning of the heat sink (horizontal or vertical) is not given.

Looking at the two curves for power supply rejection, we can see that rejection is quite poor at low frequencies (below 1KHz) and with low values of power supply bypass capacitors.

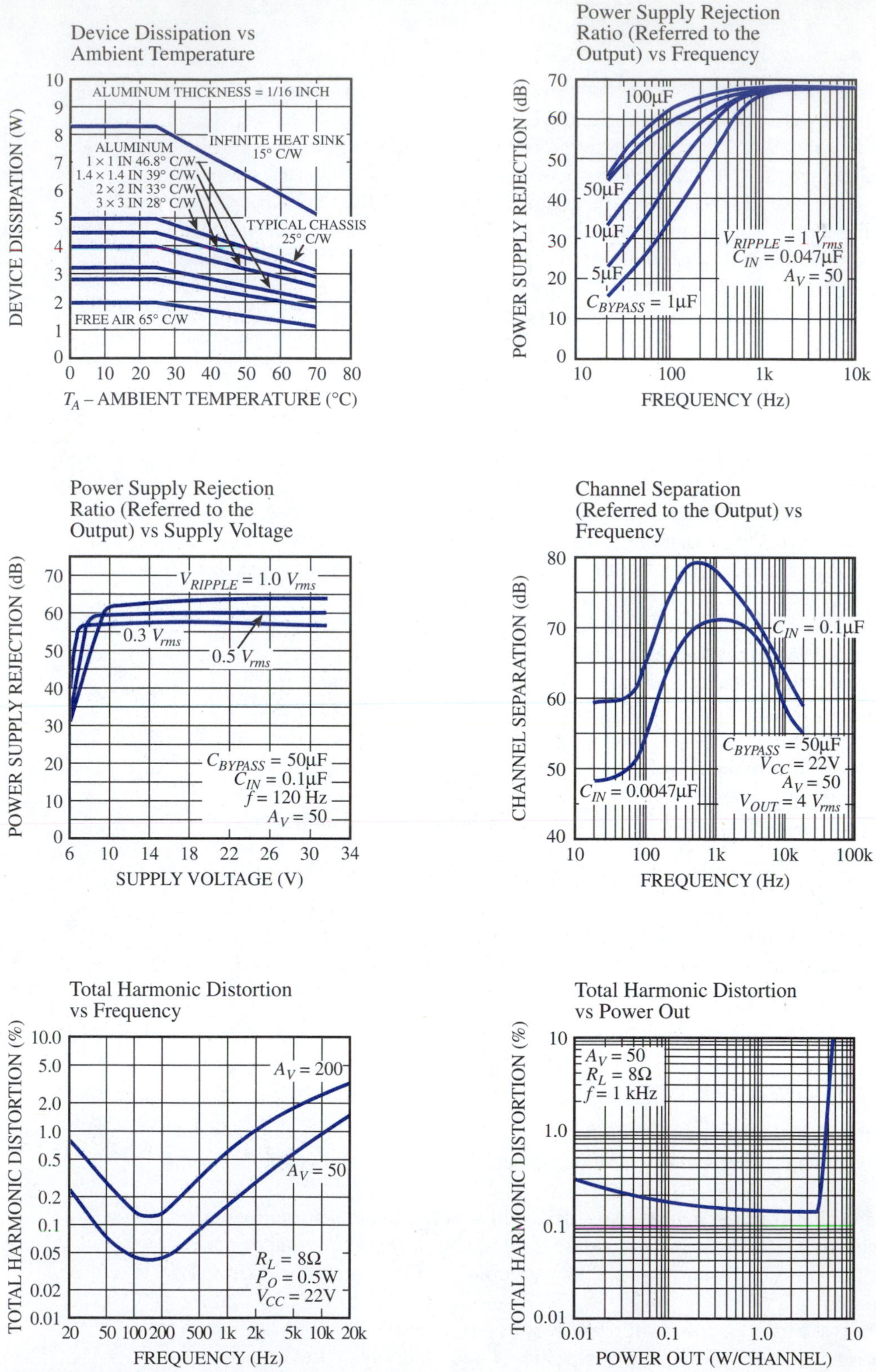
Device Dissipation vs
Ambient Temperature
ALUMINUM THICKNESS = 1/16 INCH
INFINITE HEAT SINK
15° C/W
ALUMINUM
1 × 1 IN 46.8° C/W
1.4 × 1.4 IN 39° C/W
2 × 2 IN 33° C/W
3 × 3 IN 28° C/W
TYPICAL CHASSIS
25° C/W
FREE AIR 65° C/W
DEVICE DISSIPATION (W)
10
9
8
7
6
5
4
3
2
1
0
0 10 20 30 40 50 60 70 80
T_A – AMBIENT TEMPERATURE (°C)
Power Supply Rejection
Ratio (Referred to the
Output) vs Frequency
100µF
50µF
10µF
5µF
C_{BYPASS} = 1µF
V_{RIPPLE} = 1 V_{rms}
C_{IN} = 0.047µF
A_V = 50
POWER SUPPLY REJECTION (dB)
70
60
50
40
30
20
10
0
10 100 1k 10k
FREQUENCY (Hz)
Power Supply Rejection
Ratio (Referred to the
Output) vs Supply Voltage
V_{RIPPLE} = 1.0 V_{rms}
0.3 V_{rms}
0.5 V_{rms}
C_{BYPASS} = 50µF
C_{IN} = 0.1µF
f = 120 Hz
A_V = 50
POWER SUPPLY REJECTION (dB)
70
60
50
40
30
20
10
0
6 10 14 18 22 26 30 34
SUPPLY VOLTAGE (V)
Channel Separation
(Referred to the Output) vs
Frequency
C_{IN} = 0.1µF
C_{IN} = 0.0047µF
C_{BYPASS} = 50µF
V_{CC} = 22V
A_V = 50
V_{OUT} = 4 V_{rms}
CHANNEL SEPARATION (dB)
80
70
60
50
40
10 100 1k 10k 100k
FREQUENCY (Hz)
Total Harmonic Distortion
vs Frequency
A_V = 200
A_V = 50
R_L = 8Ω
P_O = 0.5W
V_{CC} = 22V
TOTAL HARMONIC DISTORTION (%)
10.0
5.0
2.0
1.0
0.5
0.2
0.1
0.05
0.02
0.01
20 50 100 200 500 1k 2k 5k 10k 20k
FREQUENCY (Hz)
Total Harmonic Distortion
vs Power Out
A_V = 50
R_L = 8Ω
f = 1 kHz
TOTAL HARMONIC DISTORTION (%)
10
1.0
0.1
0.01
0.01 0.1 1.0 10
POWER OUT (W/CHANNEL)

FIGURE 20–2 LM2878 graphs. *Courtesy National Semiconductor.*

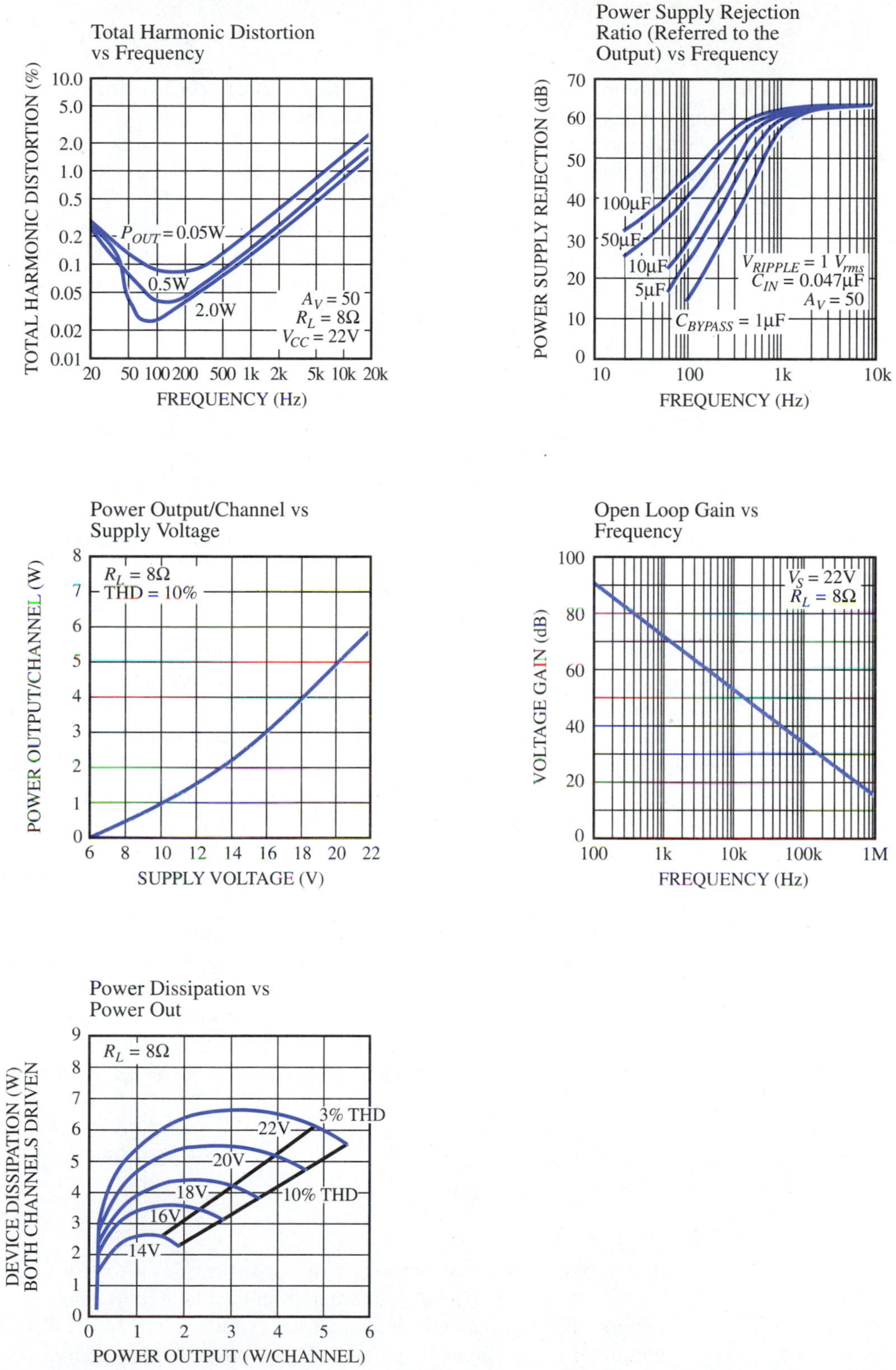

FIGURE 20–2 *(continued)*

In addition, reducing the input coupling capacitor decreases the rejection. Above a 10V supply voltage, the rejection is quite constant.

Channel separation is best with an operating frequency of about 600Hz. It is interesting that from the $C_{in} = 0.1\mu F$ curve, the separation reads 77db at 1KHz, whereas the typical value listed on the data sheet is 70db.

Harmonic distortion below about 1% is considered undetectable by the human ear. This level is exceeded, as shown on the next three graphs, with input frequencies greater than 5KHz with gains greater than 50, or greater than 2KHz with a gain of 200. Power levels greater than 4W cause a large increase in harmonic distortion.

As expected, power output is a function of a supply voltage. A power output of 6W is indicated for a power supply of 22V. The specification sheet indicates a typical value of 5.5W for the same voltage (different people generated the graphs).

The open loop gain curve shows that the amplifier has a stable 20db per decade (6db per octave) slope out to a frequency of 1MHz.

Finally, the "internal" power dissipation versus "output" power is graphed. Knowing the required power output, this curve can be used in conjunction with the first graph to determine the required heat sink. Notice that distortion increases as the amplifier internal dissipation increases (device gets hotter).

MAXIMUM RATINGS

Actually, the first specifications we should consider when using a device are the absolute maximum ratings, because if we exceed any of these values, the device is no longer able to meet other specifications. A safe approach is to not exceed 75 percent of the maximum specified value.

Design engineers don't always take notice of the maximum ratings. It was brought to the attention of the author that an LM723 voltage regulator used in a piece of medical equipment had a very high failure rate. Examination of the circuit revealed that a continuous 50V was being applied to the input to the regulator, which has an absolute maximum rating of 40V!

For the LM2878, the maximum power supply voltage is 35V. If this is not to be exceeded, we must consider the power supply regulation, voltage spikes on the power supply, and possible power supply failure modes.

A maximum input voltage of +0.7V is allowed without circuit damage. This means care must be taken to avoid high-level inputs. Or, in other words, the gain should be high enough (greater than 15) so that the maximum output power can be achieved without exceeding 0.7V.

The maximum operating temperature range is given as 0°C–70°C. This is based on 5.5W internal dissipation at 70°C with an infinite heat sink. Practically, we are limited by the heat sink available.

EXAMPLE

Let us assume that we use a heat sink that provides a total thermal resistance of 28°C/W (3 × 3 × 1/16-in. aluminum from the first graph), and we require a power output of 3W with a maximum 3 percent distortion into an 8Ω load. Find the maximum operating ambient temperature if the power supply is 18V.

SOLUTION

We find from the Power Dissipation versus Power Output graph (bottom right), that with a supply voltage of 18V, the internal dissipation is 4.3W. Using this value with the Power Dissipation versus Ambient Temperature graph (top left) for a 28°C/W heat sink, we find that the maximum operating temperature is about 30°C (86°F). To operate at a higher ambient temperature requires a larger heat sink (lower thermal resistance).

A storage temperature range of from −65°C to 150°C is specified for a device that is not being operated, while the maximum junction temperature of 150°C controls the amount of allowable dissipation with a given heat sink. Using numbers from our previous example, we have an allowable junction temperature rise of (150 − 30)°C, or 120°C. Using the equations developed in Chapter 19, and dividing this temperature rise by the thermal resistance of 28°C/W, we get a maximum allowable internal power dissipation (P_m) of

$$P_m = \frac{120°\text{C}}{28°\text{C/W}}$$

$$P_m = 4.29\text{ W}$$

which is close to the 4.3W read from the graph.

The last item under maximum ratings is component lead temperature, which is listed as 300°C for a soldering time of 10 seconds. This is based on the internal thermal time constant keeping the junction from rising instantaneously to 150°C. But if the 10 seconds is exceeded, the junction will rise above 150°C, and the chip will be damaged. Good soldering practice of a hot, clean tinned iron, along with clean connections for low thermal resistance, will prevent damage during chip installation or removal.

SIGNIFICANCE OF PART NUMBER DESIGNATIONS

Letter codes are added to the basic part number to identify the required package and differences in the specifications. For example, the LM317 National Semiconductor regulator comes in the following choice of four packages:

LM317K Metal Can, TO-3 (20W)

LM317H Metal Can, TO-39 (2W)

LM317T Plastic, TO-220 (20W)

LM317MP Plastic, TO-202 (2W)

These device designators indicate the package configuration that controls the allowable power dissipation and also the thermal resistance.

Additional designators indicate the specific device specifications. The LM317 operates over a commercial junction temperature range of 0°C to +70°C, while the LM317A industrial version operates from −40°C to +125°C. In addition the "A" version has tighter specifications. Within the same regulator family with the LM317 is the LM117 and LM117A. These are military versions of the regulator and can operate over a junction temperature range of −55°C to +150°C.

Similarly the 741 op-amp comes in four versions. 741A and 741E have better electrical characteristics than the 741 and 741C, and the characteristics of the 741 are somewhat better than those of the 741C. The 741 and 741A can operate over a temperature range of −55°C to +150°C, while the 741C and 741E operate from 0°C to +70°C.

Packages available for the 741 are the metal TO-5 (H-designator), the ceramic DIP (J-designator), the plastic DIP (N-designator), and the surface mount (M-designator).

Suppose we want a 741 op-amp in a plastic DIP package that has a minimum voltage offset and can operate from 0°C to 70°C. From the characteristics of the 741 data sheets, we find that we need a 741E, and the plastic DIP requirement makes it a 741EN.

COMPONENT INTERCHANGEABILITY

When a piece of equipment stops operating because of an IC failure, we have to find a replacement part. If the failed part is a 741 op-amp, we can't simply purchase a 741, but we must specify the full part number in order to guarantee electrical and mechanical compatibility. For example, replacing a 741AH with a 741CN will create two problems: First, the 741CN has looser specs; second, the N is a DIP, while the H is a round metal can. But if we had a failed 741CN, we could replace it with a better and mechanically equivalent 741AN or 741EN type if we are willing to pay the extra cost.

Sometimes it is desirable to replace a failed component with a different type. This would be the case if the failure was determined to be caused by exceeding the device maximum rating. Suppose we found that an LM317 voltage regulator failed because the input voltage was 5V greater than the specified 40V (input-output differential). It would make no sense to replace the failed regulator with the same type. Rather, the LM317HV should be considered because it has a maximum input voltage of 60V.

If we were trying to fix a radio transmitter on a sinking ship, we couldn't wait until the right part came along. In this extreme example, we would find the closest part that would work and put it in the circuit.

Ideally, we select a component equal to or better than the failed part. When the intent is to replace the substituted part when the correct part is available, a visible red tagging should be placed on the equipment as a reminder.

We need to know how to read device specifications in order to make an intelligent part selection or substitution. It is hoped that the material in this section has accomplished this objective.

TAKING DATA—THE LAB EXPERIMENT

The purpose of a lab experiment is to reinforce the concepts described in the text and to develop the skills necessary for equipment repair. Also, an experiment should teach how devices are applied and how their characteristics can be verified by measurement. "Does the device do what it is supposed to do?" is a question that should be answered by the lab experiment.

Experiments suggest a particular application or verification of specifications. But, there is no point in taking data if it can't be compared against something—either a computed value or specified value.

A certain discipline in the approach to conducting the experiment should be followed in order to prevent damage to components or equipment and to achieve valid results. A suggested generalized procedure is given here.

1) Always consult the manufacturer's data sheet to determine the absolute maximum rating for the devices. Try to stay at least 25 percent (multiply by 0.75) below these values in the experiment.
2) If not provided, generate a complete schematic of the required test circuit, listing all component values, considering item 1 above. If the experiment is conducted in a formal lab situation, the schematic should be available prior to the start of the lab.
3) Wire the circuit from the schematic, checking off (or yellowing out) on the schematic each component and wire as it is installed. I would suggest using the following color code for hookup wires (based on military specs) to avoid confusion when the invariable troubleshooting takes place:
    a) Red—most-positive power supply voltage
    b) Orange—less-positive power supply voltage (if used)
    c) White—signal
    d) Black—power supply return
    e) Brown—signal return
    f) Green—less-negative supply voltage (if used)
    g) Blue—most-negative supply voltage
4) Before connecting power to the circuit, verify the power supply voltages—after connecting to the circuit, check to see that they haven't changed! If a signal generator is used, consider whether a coupling capacitor is required to prevent the internal resistance of the generator (typically low) from changing the DC voltage at the circuit input. Also determine whether the generator will be excessively loaded by the input circuit, causing the signal voltage to change.
5) Think through and determine which waveshape and DC voltage you expect to see at key points in the circuit (this requires an understanding of the circuit operation). It is helpful to mark them on the schematic.
6) Always use an oscilloscope to check circuit operation, because a voltmeter will not indicate unwanted oscillation or distortion (remember, most AC meters are calibrated with pure sine waves). If a signal generator is used, verify that the same generator frequency exists at expected points in the circuit.
7) Take a complete set of circuit measurements and then repeat the measurements. This tends to catch the always-present errors in measurement.
8) Compare measured values with computed or specified values; always compute percent error or deviation. An error as small as 1mV could be a 100-percent error. If an error is larger than expected, repeat the measurement.
9) Get in the habit of writing a conclusion that considers and explains deviations of the measured from specified values. Also in the conclusion, list the

significance and practical applications for the circuit, its advantages, and its disadvantages.

10) Critique the experiment; consider how you could have avoided the problems encountered with the experiment. Suggest how the experiment could be improved to meet the objectives of text support and practical application.

COMPONENT VALUES

The components we are concerned with in this section are resistors and capacitors. Both of these components come in various types to meet the requirements of different applications. These types range from tiny semiconductor versions to large power components. They are also available in configurations where the resistance or capacitance can be varied—variable capacitors, however, are restricted to picofarad values.

RESISTOR VALUES

Resistors are the most common circuit element used and consist of four basic types:

1) Resistor, carbon composition
2) Resistor, carbon film
3) Resistor, metal wire-wound
4) Resistor, metal film

Carbon composition resistors are the least-expensive type and are available in 1/4, 1/2, 1, or 2 watt ratings. Tolerances of carbon resistors are 5, 10, and 20 percent (10 percent most used).

Standard carbon resistor values are listed here for 5 percent tolerance. The 10 percent values are limited to those with an asterisk.

Resistor Values for Carbon Resistors

Ohms							
1.0*	5.1	27*	130	680*	3,600	18,000*	91,000
1.1	5.6*	30	150*	750	3,900*	20,000	100,000*
1.2*	6.2	33*	160	820*	4,300	22,000*	110,000
1.3	6.8*	36	180*	910	4,700*	24,000	120,000*
1.5*	7.5	39*	200	1,000*	5,100	27,000*	130,000
1.6	8.2*	43	220*	1,100	5,600*	30,000	150,000*
1.8*	9.1	47*	240	1,200*	6,200	33,000*	160,000
2.0	10*	51	270*	1,300	6,800*	36,000	180,000*
2.2*	11	56*	300	1,500*	7,500	39,000*	200,000
2.4	12*	62	330*	1,600	8,200*	43,000	220,000*
2.7*	13	68*	360	1,800*	9,100	47,000*	
3.0	15*	75	390*	2,000	10,000*	51,000	
3.3*	16	82*	430	2,200*	11,000	56,000*	
3.6	18*	91	470*	2,400	12,000*	62,000	
3.9*	20	100*	510	2,700*	13,000	68,000*	
4.3	22*	110	569*	3,000	15,000*	75,000	
4.7*	24	120*	620	3,300*	16,000	82,000*	

Megohms							
0.24	0.45	0.75	1.3	2.4	4.3	7.5	13.0
0.27*	0.47*	0.82*	1.5*	2.7*	4.7*	8.2*	15.0*
0 30	0.51	0.91	1.6	3.0	5.1	9.1	16.0
0 33*	0.56*	1.0*	1.8*	3.3*	5.6*	10.0*	18.0*
0.36	0.62	1.1	2.0	3.6	6.2	11.0	20.0
0.38*	0.68*	1.2*	2.2*	3.9*	6.8*	12.0*	22.0*

Carbon film resistors come in 5 percent tolerance values (see chart above) and are smaller and more stable than the composition type.

Wire-wound metal resistors are more stable than the carbon types and are available in higher wattage ratings, tighter tolerance, and lower resistance values.

Tolerances are typically 1, 5, or 10 percent. Standard resistance values for 5 percent and 10 percent (asterisk) resistors are tabulated here.

Resistance Values for Wire-Wound Resistors

0.10*	0.27~	0.68*	1.8*	4.7*	12*	−3*	82*	220*	560*	1.5K*
0.11	0.30	0.75	2.0	5.1	13	36	91	240	620	1.6K
0.12*	0.33*	0.82	2.2*	5.6*	15*	39*	100*	270*	680*	1.8K*
0.13	0.36	0.91	2.4	6.2	16	43	110	300	750	2.0K
0.15*	0.39~	1.0*	2.7*	6.8*	18*	47*	120*	330*	820*	
0.16	0.43	1.1	3.0	7.5	20	51	130	360	910	
0.18*	0.47*	1.2*	3.3*	8.2*	22*	56*	150*	390*	1K*	
0.20	0.51	1.3	3.6	9.1*	24	62	160	430	1.1K	
0.22*	0.56	1.5*	3.9*	10*	27*	68*	180*	470*	1.2K*	
0.24	0.42	1.6	4.3	11	30	75	200	510	1.3K	

Metal film resistors have a 1 percent tolerance and are very temperature stable compared to the carbon types. They also generate much less electrical noise than carbon and should therefore be used in low-noise amplifiers. Standard values are tabulated here (multiply by powers of ten):

Resistance Values for Metal Film Resistors

1.00	1.21	1.47	1.78	2.15	2.61	3.16	.283	4.64	5.67	6.81	8.25
1.02	1.24	1.50	1.82	2.21	2.67	3.24	3.92	4.75	5.76	6.98	8.45
1.05	1.27	1.54	1.87	2.26	2.74	3.32	4.02	4.87	5.90	7.15	8.66
1.07	1.30	1.58	1.91	2.32	2.80	3.40	4.12	4.99	6.04	7.32	8.87
1.10	1.33	1.62	1.96	2.37	2.87	3.48	4.22	5.11	6.19	7.50	9.09
1.13	1.37	1.65	2.00	2.43	2.94	3.57	4.32	5.23	6.34	7.68	9.31
1.15	1.40	1.69	2.05	2.49	3.01	3.65	4.42	5.36	6.49	7.87	9.53
1.18	1.43	1.74	2.10	2.55	3.09	3.74	4.53	5.49	6.65	8.06	9.76

Variable resistors (potentiometers) are available in a smaller range of values. Standard values are tabulated here (multiply by powers of ten):

Potentiometer Values

1, 1.5, 2.0, 2.5, 3.5, 5, 7.5

CAPACITOR VALUES

Capacitors are much more varied than resistors and range from the high-capacitance electrolytic to low-capacitance ceramic types. Electrolytic capacitors have wide tolerances, such as +75% and −10% (tantalum types can be ±10%). They are used when a large capacitor is required (power supply filtering) but the value is not critical. Most electrolytic capacitors are polarized, which requires consideration of the polarity when they are placed in a circuit. The smaller-value capacitors (mica, ceramic, polystyrene, film, etc.) have tighter tolerances, which range from 1% to 20%, and can provide capacitance values from 1 or 2 picofarads to about 5 microfarads.

The actual capacitance value is not marked on some of the smaller capacitors. A numerical code is used, similar to the system of precision resistors but with a picofarad baseline. For example, a capacitor marked 103 would have a 10,000 picofarad value (or 0.01 microfarad). Standard values for most capacitors (nonelectrolytic) are one of the following numbers multiplied by powers of ten:

Capacitor Values

1.0, 1.5, 2.2, 3.3, 4.7, 6.8, 8.2

For example, a 22nF capacitor is 2.2 multiplied by 10^{-8}.
Electrolytic capacitors have a greater and less predictable set of values.

Tolerance of some capacitors is indicated by a letter following the number. The letter code is:

$$K = \pm 10\%$$
$$J = \pm 5\%$$
$$G = \pm 2\%$$

Other capacitors give a numerical tolerance, while others use a color code like resistors.

SUMMARY

- Specifications set limits on a device so we can predict its performance.
- Without specifications, we would have no guarantee that a replacement part would work.
- A specification control drawing defines equipment-specific requirements for a component.
- Standard conditions allow duplication/verification of specified values.
- The worst-case value on a specification sheet should be assumed.
- Inconsistencies can exist in manufacturer's data sheets.
- Typical values are the most probable value, but minimum and maximum value should also be provided.

- Part number codes identify the part, its particular specifications, and the mechanical package.
- Parts can be substituted, provided they have more stringent specifications than the part being replaced provided physical mounting is the same.
- If a substituted part is to be replaced when the correct part is available, a visible red tag should be attached to the equipment.
- An experiment should teach how devices are applied and how their characteristics can be verified by measurement.
- "Does the device do what it is supposed to do?" is a question that should be answered by the lab experiment.
- There is no point in taking data if they can't be compared against something.
- Before connecting power to a circuit, verify the power supply voltages—after connecting to the circuit, check to see that they haven't changed!
- Think through and determine which waveshape and DC voltage you expect to see at key points in the circuit under test.
- Always use an oscilloscope to check circuit operation, because a voltmeter will not indicate unwanted oscillation or distortion.
- Always take a complete set of circuit measurements and then repeat the measurements.
- Get in the habit of writing a conclusion that considers and explains deviations of the measured from specified values. Also in the conclusion, list the significance and practical applications for the circuit, its advantages and disadvantages.
- Critique the experiment; consider how you could have avoided the problems encountered with the experiment.
- Resistors are the most-used common circuit element.

EXERCISE PROBLEMS

1. State the purpose of a component specification.
2. What is a specification control drawing and why is it used?
3. What is the problem if only typical values are provided?
4. Why are standard conditions specified?
5. Using a graph of Figure 20–2, determine the maximum operating temperature for an LM2878 mounted on a 2″ × 2″ × 1/16″ aluminum plate if the internal dissipation is 3W.
6. Find the power supply rejection ratio for an LM2878 set for a gain of 50 and using a 10μF bypass capacitor, a 0.1μF input capacitor, and amplifying a signal with a frequency of 100Hz.
7. Find the channel separation for an LM2878 set for a gain of 50 and using a 50μF bypass capacitor, a 0.1μF input capacitor, and amplifying a frequency of 100Hz.
8. Determine the total harmonic distortion (THD) at 1KHz for an LM2878 set for a gain of 50 with a power output of 2W into 8Ω and a supply voltage of 22V.

9. Determine the total harmonic distortion (THD) at 1KHz, with a power output of 3W into an 8Ω load, if the LM2878 is set for a gain of 50.
10. Find the maximum output power for an LM2878 with a power supply voltage of 12V; with 20V.
11. Let us assume that we use a heat sink that provides a total thermal resistance of 25°C/W, and we require a power output of 4W with a maximum 10% distortion into an 8Ω load. Find the maximum operating ambient temperature.
12. Describe the package configuration for the following designators: *N*, *K*, and *T*.
13. When testing a device, how close should the levels be relative to the maximum ratings?
14. When wiring a circuit, how can you keep track of which components and wiring are installed?
15. Why should you use different color wiring in a circuit?
16. Why should you recheck the voltage level of a generator or power supply after it is connected to the circuit?
17. What is the significance of listing percent deviation?
18. Why should you critique an experiment?

ANSWERS TO PROBLEMS

1. Defines and controls the component's characteristics
2. A specification sheet generated by the user to control the component's characteristics
3. No limits are provided.
4. To enable test results to be duplicated
5. 52°C
6. 45db
7. 58db
8. 0.13%
9. 0.13%
10. 1.4W; 4.8W
11. 45°C
12. See text.
13. 75% of max
14. Yellow out on schematic
15. To ease troubleshooting
16. Loading could change voltage.
17. More meaningful
18. Increases your lab efficiency

APPENDIX A

ELEMENTARY CALCULUS

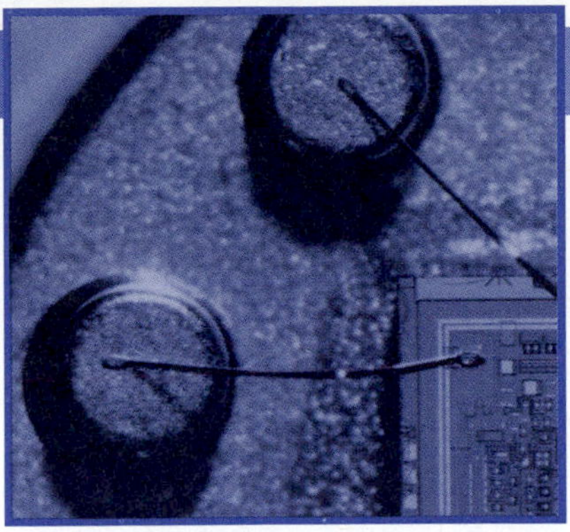

THE INTEGRAL

If we wish to find the area of a rectangle, we simply multiply length by width. But if we have a more complex area, it is helpful to use integral calculus. We can illustrate the use of integrals by first determining the area of a rectangle shown in Figure A–1.

Note that we can determine the total area by breaking the large area into known small areas $(dx)(Y_1)$ and adding them together. We can indicate this mathematically as

$$Y_1X_1 = \sum_{Y_1}^{n} dx$$

where n is the total number of rectangles of width dx along the x-axis, and Σ^n indicates the summing process of n rectangles of area Y_1dx.

If we let dx approach zero $(dx \rightarrow 0)$ we approach an infinite number of samples over the width X_1. In the limit as $dx \rightarrow 0$ the sum (Σ) becomes an integral and is represented by the symbol $\int$. So the area

$$Y_1X_1 = \int_0^{X_1} Y_1 dx$$

where 0 and X_1 are the limits over which the integral is evaluated. In order to find the area Y_1X_1, we need to evaluate the integral

$$\int_0^{X_1} Y_1 dx.$$

From math tables, the integral

$$\int Y_1 dx = yx.$$

The integral shown without limits is called an improper integral. If we put in the limits, the total area is

$$\int_0^{X_1} Y_1 dx = (Y_1)(x)\Big|_0^{X_1} = Y_1X_1 - Y_1 0 = Y_1X_1$$

Note that the upper limit is inserted for x and then the lower limit is also inserted for x and the indicated subtraction performed. The advantage of using integrals is not apparent from this example since we could have achieved the same result just by multiplying length

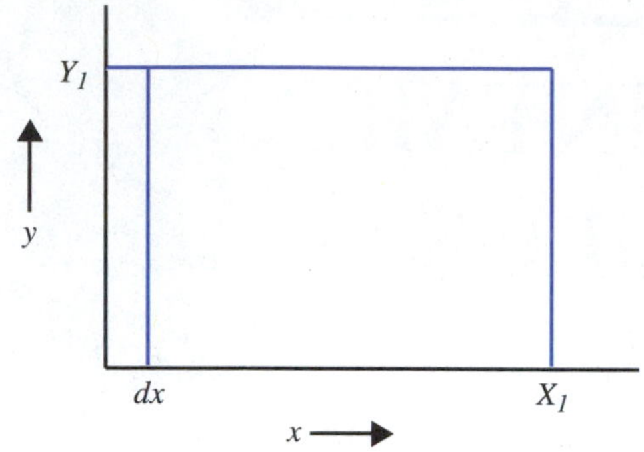

FIGURE A–1.

by width. Let us determine the area under a half sine curve, shown in Figure A–2. We can't simply multiply y by Θ in this case because y is variable.

Using integrals with $y = 1$, the area is

$$A = \int_0^{\pi} \sin \Theta d\Theta$$

From integral tables,

$$A = \int_0^{\pi} \sin \Theta d\Theta = -\cos \Big|_0^{\pi} = -\cos \pi - (-\cos 0) = 2$$

So the area under the sine curve is equal to 2. If we wish to find the average value of y, we divide by the width π.

$$y \text{ average} = 2/\pi = 0.637$$

This value is the average value of a sine wave alternation used in AC theory.

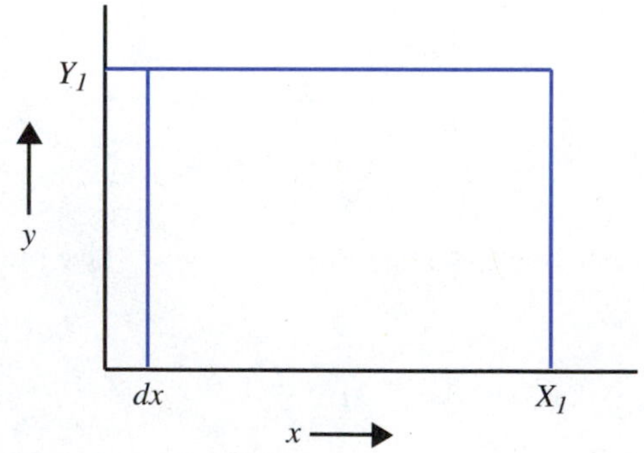

FIGURE A–2.

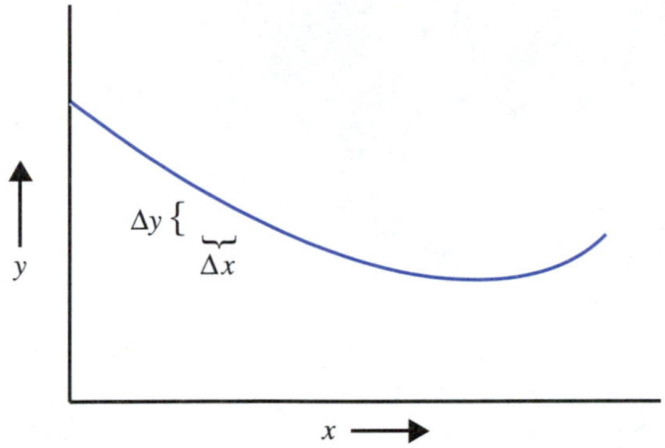

FIGURE A–3.

In summary, the integral is the area under the curve and can be determined for all continuous functions of y.

THE DERIVATIVE

If we are interested in the slope at any point of the curve in Figure A–3, we can take a small increment (**dx**) of x and find the resultant change (Δy) in y. The slope is defined as $\Delta y/\Delta x$, and this is also the derivative of the curve.

If we can determine the derivative at any point of the curve, we can then determine the slope. Referring back to the sine curve of Figure A–2, let us determine the slope at $\Theta = 0$

From derivative tables, derivative $d(\sin \Theta) = \cos \Theta$, with $\Theta = 0°$, $\cos 0° = 1$. So at 0° the slope of the sine curve is 1 or $\Delta x = \Delta y$.

At $\Theta = \pi/2$ the slope is 0 and at $\Theta = \pi$ the slope is -1. As we expect, the sine wave slope is positive, reduces to zero, and goes to increasing negative values.

Applications of the derivative are to find the maximum slope of a function and to find the maximum and minimum values of a function—slope is zero at maximum and minimum points.

COMMON INTEGRALS

$$\int dx = x \quad \text{(A–1)}$$

$$\int c\,dx = cx \quad \text{(A–2)}$$

$$\int x\,dx = \frac{x^2}{2} \quad \text{(A–3)}$$

$$\int kx^n dx = \frac{kx^{n+1}}{n+1} \quad \text{(A–4)}$$

$$\int \sin kx\,dx = -\frac{\cos kx}{k} \quad \text{(A–5)}$$

$$\int \cos kx dx = \frac{\sin kx}{k} \tag{A–6}$$

$$\int \epsilon^{kx}\, dx = \frac{\epsilon^{kx}}{k} \tag{A–7}$$

COMMON DIFFERENTIALS

$$dx(c) = 0 \tag{A–8}$$

$$dx(x) = 1 \tag{A–9}$$

$$dx(kx) = k \tag{A–10}$$

$$dx(kx^n) = knx^{n-1} \tag{A–11}$$

$$dx(\sin kx) = k \cos kx \tag{A–12}$$

$$dx(\cos kx) = -k \sin kx \tag{A–13}$$

$$dx(\epsilon^{kx}) = k\epsilon^{kx} \tag{A–14}$$

Index

A

D

E

F

G

H

I

J

K

L

M

N

T

U

V